International Symposium on Physical Sciences in Space 2011

Journal of Physics: Conference Series Volume 327

Bonn, Germany
11 – 15 July 2011

ISBN: 978-1-61839-350-0
ISSN: 1742-6588

Printed from e-media with permission by:

Curran Associates, Inc.
57 Morehouse Lane
Red Hook, NY 12571

Some format issues inherent in the e-media version may also appear in this print version.

Printed by Curran Associates, Inc. (2011)

For permission requests, please contact the Institute of Physics
at the address below.

Institute of Physics
Dirac House, Temple Back
Bristol BS1 6BE UK

Phone: 44 1 17 929 7481
Fax: 44 1 17 920 0979

techtracking@iop.org

Additional copies of this publication are available from:

Curran Associates, Inc.
57 Morehouse Lane
Red Hook, NY 12571 USA
Phone: 845-758-0400
Fax: 845-758-2634
Email: curran@proceedings.com
Web: www.proceedings.com

TABLE OF CONTENTS

Melting Kinetics In Microgravity 1
M. Glicksman

In-Situ And Real-Time Investigation Of The Columnar-Equiaxed Transition In The Transparent Alloy System Neopentylglycol-Camphor Onboard The Sounding Rocket TEXUS-47 16
L. Sturz, G. Zimmermann

Investigation Of Columnar-To-Equiaxed Transition In Solidification Processing Of AlSi Alloys In Microgravity – The CETSOL Project 25
G. Zimmermann, L. Sturz, B. Billia, N. Mangelinck-Noel, H. Thi, C. Gandin, D. Browne, W. Mirihanage

Phase-Field Modeling Of The Columnar-To-Equiaxed Transition In Neopentylglycol-Camphor Alloy Solidification 37
A. Viardin, L. Sturz, G. Zimmermann, M. Apel

Metastable Solidification In Undercooled Liquid Droplets Of Fe-Ni Based Soft-Magnetic Alloys Under Terrestrial And Microgravity Conditions 45
W. Loser, T. Woodcock, O. Shuleshova, R. Hermann, H. Lindenkreuz, B. Gehrmann, S. Schneider, T. Volkmann

Investigation Of Growth Of Single Crystal SRR99 Superalloy Under Microgravity Using 50-Meter-High Drop Tube 51
S. Feng, X. Luo

Containerless Solidification and Characterization of Industrial Alloys (NEQUISOL) 57
A. Ilbagi, H. Heneine, J. Chen, D. Herlach, R. Lengsdorf, C. Gandin, D. Tourret, A. Garcia-Escorial

Preparation Of ZrNiSn Half-Heusler Compounds With Crystalline Alignment By Unidirectional Solidification In Short-Duration Microgravity And Their Thermoelectric Properties 76
H. Nagai, R. Muroi, T. Okutnal

Phase Selection in the Mushy-zone: LODESTARS and ELFSTONE Projects 83
D. Matson, R. Hyers, T. Volkmann, H. Fecht

Effect Of Cooling Rate On Solidification Of Al-Ni Alloys 90
A. Ilbagi, P. Khatibi, H. Henein, R. Lengsdorf, D. Herlach

Versatile Compact X-Ray Radiography Module For Materials Science Under Microgravity Conditions 102
F. Kargl, M. Balter, C. Stenzel, T. Gruhl, N. Daneke, A. Meyer

Investigation Of Gravity Effects On Solidification Of Binary Alloys With In Situ X-Ray Radiography On Earth And In Microgravity Environment 110
H. Nguyen-Thi, A. Bogno, G. Reinhart, B. Billia, R. Mathiesen, G. Zimmermann, Y. Houltz, K. Loth, D. Voss, A. Verga, F. Pascale

Fluid-Flow Effects on Phase Selection and Nucleation in Undercooled Liquid Metals 121
R. Hyers, D. Matson, K. Kelton, D. Holland-Mortiz, T. Volkmann

In-Situ Characterization Of Droplets During Free Fall In The Drop Tube-Impulse System 128
P. Khatibi, A. Ilbagi, D. Beinker, H. Henein

Materials Science Investigations using Electromagnetic Levitation 139
A. Seidel, W. Soellner, C. Stenzel

Diffusion Limited Silicon Dissolution into Germanium Melt 150
N. Armour, S. Dost

Homogeneous SiGe Crystal Growth In Microgravity By The Travelling Liquidus-Zone Method 155
K. Kinoshita, Y. Arai, Y. Inatomi, H. Miyata, R. Tanaka, T. Sone, J. Yoshikawa, T. Kihara, H. Shibayama, Y. Kubota, T. Shimaoka, Y. Warashina, K. Sakata, M. Takayanagi, S. Yoda

Morphological Transition in Crystallization of Si from Undercooled Melt 161
K. Watanabe, K. Nagayama, K. Kuribayashi

Metastable Phase Formation from Undercooled Melt of Oxide Material 170
K. Kuribayashi, M. Kumar

Temperature Dependence of Surface Tension of Molten Iron under Reducing Gas Atmosphere 180
S. Ozawa, S. Takahashi, H. Fukuyama, M. Watanabe

The Effect Of Rotation On Resonant Frequency Of Interfacial Oscillation Of A Droplet Using Electrostatic Levitator 186
R. Tanaka, S. Matsumoto, A. Kaneko, Y. Abe

The Viscosity Of Eutectic Pd-Si Alloys 196
L. Egry

Non-Equilibrium Fluctuations On Earth And In Micro-Gravity. The GRADFLEX Experiment 203
A. Vailati, R. Cerbino, S. Mazzoni, M. Giglio, C. Takacs, D. Cannell

Foam Stability in Microgravity ... 216
N. Vandewalle, H. Caps, G. Delon, A. Saint-Jaimes, E. Rio, L. Saulnier, M. Adler, A. Biance, O. Pitois, S. Addad, R. Hohler, D. Weaire, S. Hutzler, D. Langevin

Foam Generation and Sample Composition Optimization for the FOAM-C Experiment of the ISS ... 224
R. Carpy, G. Picker, B. Amann, H. Ranebo, S. Vincent-Bonnieu, O. Minster, J. Winter, J. Dettmann, L. Castiglione, R. Hohler, D. Langevin

Droplet Collisions After Liquid Jet Breakup In Microgravity Conditions ... 230
F. Sunol, R. Gonzalez-Cinca

Numerical Modelling of Liquid Droplet Dynamics in Microgravity ... 237
S. Easter, V. Bojarevics, K. Pericleous

Numerical Study Of Bubble Dynamics With The Boundary Element Method ... 250
N. Mendez, R. Gonzalez-Cinca

Space Experiment On The Instability Of Marangoni Convection In Large Liquid Bridge - MEIS-4: Effect Of Prandtl Number ... 259
T. Yano, K. Nishino, H. Kawamura, I. Ueno, S. Matsumoto, M. Ohnishi, M. Sakurai

Thermocapillary And Shear Driven Flows In Gas/Liquid System In Annular Duct ... 266
G. Yu, A. Nepomnyashchy, V. Shevtsova

IVIDIL: On-Board G-Jitters And Diffusion Controlled Phenomena ... 276
V. Shevtsova, T. Lyubimova, Z. Saghir, D. Melnikov, Y. Gaponenko, V. Sechenyh, J. Legros, A. Mialdun

Ground-Based Activities In Preparation Of SELENE ISS Experiment On Self-Rewetting Fluids ... 286
R. Savino, Y. Abe, D. Castagnolo, G. Celata, O. Kabov, M. Kawaji, M. Sato, K. Tanaka, J. Thome, S. Vaerenbergh

Long Range Boundary Effect Of 2D Intermediate Number Density Vibro-Fluidized Granular Media In Micro-Gravity ... 299
C. Yanpei, P. Evesque, M. Hou, C. Lecoutre, F. Palencia, Y. Garrabos

Directed Clustering In Driven Compartmentalized Granular Gas Systems In Zero Gravity ... 309
Y. Li, M. Hou, P. Evesque

Dynamical Regimes of a Granular Gas in Microgravity : a Molecular Dynamics Study ... 316
E. Opsomer, F. Ludewig, N. Vandewalle

Agglomeration of Ni-nanoparticles in the Gas Phase under Gravity and Microgravity Conditions ... 327
S. Losch, G. Iles, B. Schmitz, B. Gunther

Structures And Dynamics Of Fine Particles In Fine Particle Plasmas Under Microgravity And Friction Between Two-Dimensional Layers Of Charges ... 336
H. Totsuji, C. Totsuji

Effects of Soot Formation on Shape of a Nonpremixed Laminar Flame Established in a Shear Boundary Layer in Microgravity ... 350
H. Wang, J. Merino, P. Dagaut

Boiling In Variable Gravity Under The Action Of An Electric Field: Results Of Parabolic Flight Experiments ... 365
P. Marco, R. Raj, J. Kim

Experiment On Nucleate Pool Boiling In Microgravity By Using Transparent Heating Surface – Analysis Of Surface Heat Transfer Coefficients ... 378
C. Kubota, O. Kawanami, Y. Asada, Y. Wada, T. Nagayasu, Y. Shinmoto, H. Ohta, O. Kabov, P. Queeckers, S. Chikov, J. Straub

Mass Transfer Of Organic Substances In Supercritical Carbon Dioxide ... 386
M. Hu, R. Benning, O. Ertunc, J. Neukam, T. Bielke, A. Delgado, V. Nercissian, A. Berger

Microgravity Experiments On ISS In Order To Examine A New Atomization Theory Discovered Through Normalgravity And Microgravity Environments ... 402
J. Osaka, S. Suzuki, Y. Suzuki, A. Umemura

Particle Temperature Measurement Using Pair Distribution Function In Complex Plasmas ... 416
S. Adachi, H. Totsuji, K. Takahashi, Y. Hayashi, M. Takayanagi

Photophoretic Forces On Chondrules In Drop Tower Experiments ... 427
A. Hesse, J. Teiser. G. Wurm

Electrodeposition Experiments In Microgravity Conditions ... 435
K. Nishikawas, Y. Fukunaka, E. Chassaing, M. Rosso

Microgravity Experiments In The Field Of Physical Chemistry In Japan ... 443
M. Natsuisaka, K. Tsujii, M. Shimomura, H. Yabu, Y. Hirai, T. Mashiko, S. Deguchi, S. Mukai, Y. Inoue, Y. Nishiyama, M. Sawada, K. Okumura, K. Sakamoto

Light Induced Erosion of Dusty Planetesimals and Mars: µg Experiments ... 450
C. Beule, T. Kelling, G. Wurm, J. Teiser, T. Jankowski

Electrostatic Precipitation Of Dust In The Martian Atmosphere: Implications For The Utilization Of Resources During Future Manned Exploration Missions ... 458
C. Calle, S. Thompson, N. Cox, M. Johanson, B. Williams, M. Hogue, J. Clements

Atomic Clock Ensemble in Space 472
L. Cacciapuoti, C. Salomon
The Space Atom Interferometer Project: Status And Prospects 485
F. Sorrentino, K. Bongs, P. Bouyer, L. Cacciapuot, M. Angelis, H. Dittus, W. Ertmer, J. Hartwig, M. Hauth, S. Herrmann, K. Huang, M. Inguscio, E. Kajari, T. Konemann, C. Lammerzahl, A. Landragin, G. Modugno, F. Santos, A. Peters, M. Prevedelli, E. Rasel, W. Schleich, M. Schmidt, A. Senger, K. Sengstock, G. Stern, G. Tino, T. Valenzuela, R. Walser, P. Windpassinger
ATLAS-M and Batt-M: Development of Flight Hardware for MAPHEUS Sounding Rocket 498
G. Blochberger, J. Drescher, C. Neumann, P. Penkert, A. Griesche, F. Kargi, A. Meyer
Compact High-Temperature Shear-Cell Furnace for In-Situ Diffusion Measurements 504
C. Neumann, E. Sondermann, F. Kargi, A. Meyer
MSL Compatible Isothermal Furnace Insert For High Temperature Shear-Cell Diffusion Experiments 508
D. Heuskin, F. Kargl, A. Griesche, C. Stenzel, D. Mitschke, D. Brauer, A. Meyer
Entering "A NEW REALM" Of KIBO Payload Operations - Continuous Efforts For Microgravity Experiment Environment And Lessons Learned From Real Time Experiment Operations In KIBO - 516
K. Sakagami, M. Goto, S. Matsumoto, H. Ohkuma
Proposal of Experimental Setup on Boiling Two-phase Flow on-orbit Experiments Onboard Japanese Experiment Module "KIBO" 529
S. Baba, T. Sakai, K. Sawada, C. Kubota, Y. Wada, Y. Shinmoto, H. Ohta, H. Asano, O. Kawanami, K. Suzuki, R. Imai, H. Kawasaki, K. Fujii, M. Takayanagi, S. Yoda
Cyclonic Two-Phase Flow Separator Experimentation and Simulation for Use in a Microgravity Environment 541
N. Hoyt, M. Kang, A. Kharraz, J. Kadambi, Y. Kamotani
EML - An Electromagnetic Levitator for the International Space Station 555
A. Seidel, W. Soellner, C. Stenzel
Magnetic Ejection Of Submilimeter-Sized Diamagnetic Grains Observed In A Chamber-Type Drop Shaft 569
K. Hisayoshi, C. Uyeda, K. Kuwada, M. Mamiya, H. Nagai
Author Index

PREFACE

ISPS is the major international scientific forum for researchers in physics utilizing the space environment, in particular microgravity. It is intended to inspire and encourage cross-cutting discussions between different scientific communities working in the same environment. Contributions discussing results of experiments carried out on drop towers, parabolic aircraft flights, sounding rockets, unmanned recoverable capsules and, last but not least, the International Space Station ISS, are the backbone of this conference series, complemented by preparatory ground-based work, both experimentally and theoretically.

The first International Symposium on Physical Sciences in Space (ISPS) sponsored by the International Microgravity Strategic Planning Group (IMSPG) took place in 2000 in Sorrento, Italy. IMSPG seeks to coordinate the planning of space for research in physical sciences by space agencies worldwide. AEB (Brazil), ASI (Italy), CNES (France), CSA (Canada), DLR (Germany), ESA (Europe), JAXA (Japan), NASA (USA), NSAU (Ukraine) and RSA (Russia) are members, and CNSA (China) and ISRO (India) are also invited to join IMSPG meetings.

ISPS-4 was the fourth symposium in that series, following ISPS-2 organized by CSA in 2004 in Toronto, Canada, and ISPS-3 organized in 2007 by JAXA in Nara, Japan. ISPS-4 was jointly organized by ESA and DLR on behalf of the IMSPG and was held in Bonn from 11–15 July 2011.

230 participants from 17 different countries attended ISPS-4. Recent microgravity experiments were presented, analysed, and set in context to results from Earth bound experiments in 16 plenary and 68 topical talks. Lively discussions continued during two dedicated poster sessions and at the exhibition booths of space industry and research centers with new flight hardware on display.

The oral presentations at ISPS4 were selected exclusively on the basis of scientific merit, as evidenced through the submitted abstracts. The selection was performed by the International Scientific Program Committee. It is our pleasure to thank the members of this committee for their excellent support in setting up a high-quality, well-balanced program.

We also thank our sponsors, the German Aerospace Center and the European Space Agency, the Bundesministerium für Wirtschaft und Technologie (German Federal Ministry of Economics and Technology), the ZARM Center of Applied Space Technology and Microgravity as well as our industrial sponsors EADS-Astrium and Kayser-Threde, for their generous contributions.

Our special thanks go to the authors and reviewers of the papers in these proceedings. Together we were able to realize up-to-date, peer reviewed conference proceedings, containing new and original data. Thanks to their efforts and that of the IOP Publishing staff, we succeeded in publishing the proceedings within six months of the conference.

We are confident that this collection of papers will provide a useful reference for all workers in the field.

Andreas Meyer and Ivan Egry
Chairmen ISPS-4
Institute of Materials Physics in Space German Aerospace Center, Cologne

Professor Dr Andreas Meyer, Chairman ISPS 4, Professor Dr Johann-Dietrich Wörner, CEO DLR, and Dr Martin Zell, Head of ESA ISS Utilisation (left to right), at the opening ceremony of ISPS-4.

Journal of Physics: Conference Series **327** (2011) 012001 doi:10.1088/1742-6596/327/1/012001

Melting kinetics in microgravity

M E Glicksman

Allen S. Henry Chair and University Professor of Engineering, Department of Mechanical & Aerospace Engineering, Florida Institute of Technology, Melbourne, FL 32901 6975 USA

Present address: Professor and Florida 21st Century Scholar, Materials Science & Engineering Department, University of Florida, Gainesville, FL, 32611-6400, USA

E-mail: mglicksman@fit.edu

E-mail: mglic@mse.ufl.edu

Abstract. We analyzed the large data archive obtained from microgravity experiments called the Isothermal Dendritic Growth Experiment (IDGE). The final orbital mission of the IDGE allowed streaming of real-time video of crystal growth and melting in low-Earth orbit of high-purity pivalic acid anhydride (PVA). Melting kinetics was observed in detail for the first time under microgravity conditions. Normally, terrestrial gravity precludes observation and analysis of conduction-limited melting by stimulating buoyancy flows and inducing collapse of the solid-liquid structure. The initial dendritic mushy zone melts under steady heating into isolated crystallite clusters that remain stationary within the melt, and eventually shrink toward their complete extinction. Near the end of conduction-limited melting, a few needle-like crystallites remain, exhibiting various axial ratios, C/A, of their major ellipsoidal axis, C, to their minor axis, A. Analysis of their convection-free melting, based on quasi-static thermal conduction, yields rates in agreement with the microgravity video and IDGE thermal data. Theory and experiment compare well, when the melting Stefan numbers, based on thermal data telemetered from the space-borne thermostat are used. Isolated PVA fragments shrink in scale at an accelerating rate toward total extinction. The 'point effect' of thermal conduction operates during most of the melting process when C/A remains almost constant. Prior to extinction, when the radius of a typical crystallite's minor axis approaches circa 100 micrometers, capillary effects become important. Then the crystallite's C/A ratio abruptly falls toward unity, as the needle-like crystals contract into spheres by accelerated melting near their poles. These experiments, and the accompanying field-theoretic analysis, introduce some new considerations, and raise questions concerning the roles of capillarity, heat conduction, interface shape, and solid-liquid phase change kinetics.

1. Introduction

The recent final launch, and return to Earth, of space shuttle *Endeavor* to the International Space Station (ISS) leaves just one remaining active member of the Space Transportation System (STS) fleet, *Atlantis*, now scheduled for its final launch in mid-2011. That will end the 30 year era of orbital space flights by US Space Shuttles. It seems fitting, therefore, to report some additional, but important, details on data obtained through NASA's microgravity sciences program, derived in that period when NASA was striving to achieve up to 25 Space Transportation System (STS) missions per year.

The author in those halcyon days was Principal Investigator of the Isothermal Dendritic

Growth Experiment (IDGE), which was launched three times aboard shuttle *Columbia*.[1] The third orbital mission of the IDGE (on USMP-4) allowed, for the first time, K-band video sequences to be transmitted in real-time at 30 fps of more than 100 sequences of dendritic crystal growth of high purity (4-9s) pivalic anhydride (PVA), each followed by melting. Details concerning the IDGE series of microgravity experiments, and the science derived therefrom, may be found in [3]-[7].

PVA is an organic crystal that melts to form a transparent melt phase of slightly lower density than its FCC crystalline phase. When terrestrial melting experiments were attempted under normal gravity, the melting dendritic crystals detached from the mushy zone and sank through the surrounding melt, quickly settling at the bottom of the sample chamber. The relative motion, or sedimentation, of PVA crystallites sinking through their high-Prandtl number melt precluded quantitative analysis of the heat transfer, as there was significant hydrodynamic interactions between the heat-absorbing crystals and the liquid phase. In microgravity, by contrast, the acceleration of gravity is greatly reduced to circa $g \leq 10^{-7}g_0$, where $g_0 = 9.8$ m-s^{-2} is its terrestrial value, and detached crystal fragments remain motionless within the melt throughout the entire process of melting. Advective melt motions at the solid-liquid interface arising from the small (3%) mass density difference between the solid and melt phases have a negligible affect on heat transfer to or from the solid-liquid interface [8], as does thermal expansion of the PVA melt responding to the release of latent heat [9].

2. Initial freezing and melting

Total melting time (typically 10-30 min.) of the partially solidified dendritic mass depends on the initial melt supercooling, which determines the relative fractions of the solid and liquid phases that comprises the mushy zone structure immediately after its formation. Adiabatic freezing of a supercooling melt constrains the volume fraction of solid to equal the freezing Stefan number, St^-, or dimensionless supercooling, namely,

$$St^- = \frac{(T_m - T_0)C_p}{\Delta H_f}, \tag{1}$$

where T_0 is the initial (supercooled) melt temperature, $T_m = 35.97 \pm 0.005$ C is the equilibrium melting point of PVA, and $\Delta H_f/C_p = 11.0$ K is the ratio of the latent heat of fusion to the specific heat of the melt. This ratio defines the solid-to-liquid transformation's 'characteristic' temperature.

Figure 1 provides four micrographs from a video sequence of melting imposed on a dendritic mushy zone comprised of about 9% solid. The melt supercooling, $\Delta T = T_m - T_0$, applied prior to initiation of adiabatic dendritic solidification was $\Delta T = 0.97K$, so from Eq. (1) one finds $St^- = 0.088$, i.e. a starting solid fraction of about 9%. The subsequent melting sequence for this cycle then occurred over a duration of 30 min., which was determined by the melt superheating applied to the thermostat to terminate the prior experimental growth cycle by destroying the mushy zone. The value of the superheating for every cycle eventually reached 1.8 K above PVA's equilibrium melting point, T_m, ensuring that every vestige of solid was melted prior to the start of the next freeze-melt cycle. It is also convenient to report the actual superheating, $T_\infty - T_m$,

[1] *Columbia*, the heaviest member of the STS fleet, was incapable of reaching the near-polar orbit of the ISS. Consequently, *Columbia* was dedicated, in large measure, to launches into low-inclination orbits for the purposes of conducting a wide variety of microgravity experiments in the physical and biological sciences. Before its tragic accident when flying as STS-107, with loss of all crew upon re-entry into the atmosphere on February 1, 2003, *Columbia* provided power, cooling, communications, S- and K-band digital data streams, all in an excellent microgravity environment to support teleoperated scientific experiments on the United States Microgravity Payload (USMP), and astronaut-mediated experiments in SpaceLab and the Shuttle Mid-deck [1] [2].

present during crystallite melting as the dimensionless quantity St^+, which is the melting Stefan number, defined here as,

$$St^+ = \frac{(T_\infty - T_m)C_p}{\Delta H_f}, \tag{2}$$

where T_∞ is the temperature impressed by the external thermostat on the melt at the walls of the sample chamber, which are distant from the melting crystals.

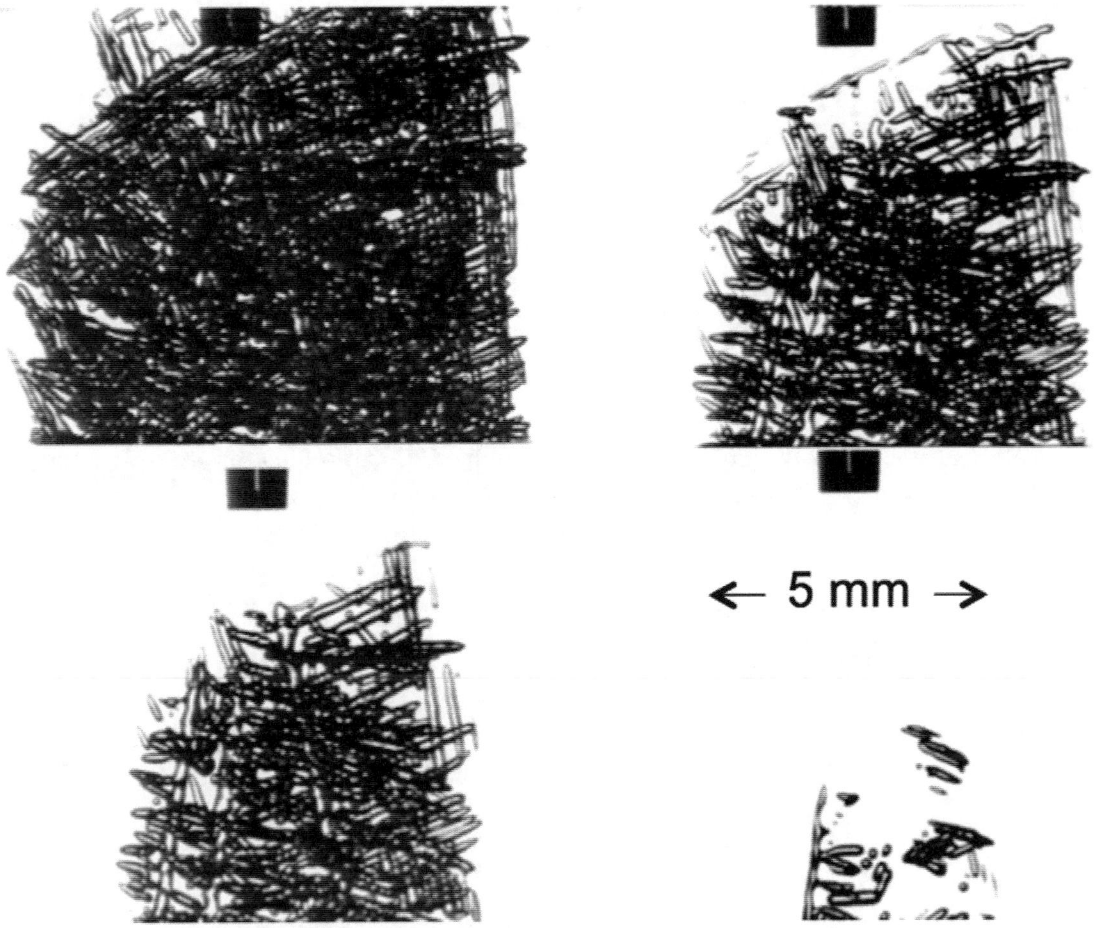

Figure 1. Progressive melting in microgravity of a dense dendritic mushy zone of high-purity PVA. **Upper left:** Dense mushy zone immediately after adiabatic crystallization contains approximately 9% PVA crystallites by volume. **Upper right:** Mushy zone coarsens slowly and detaches from the tip of the 1 mm hollow fused quartz injector (centered at the top of each micrograph) through which dendrites emerged into the growth chamber. **Lower left:** Mushy zone remains stationary and loses crystallites by progressive detachments and melting. **Lower right:** Remaining crystalline fragments prior to their total extinction by melting

The melting of a dendritic mushy zone in microgravity involves several distinct stages as suggested by the structures observed in Figure 1:

(i) *Peripheral melting* initiates at the boundary of the mushy zone as shown in the upper left panel of Figure 1. The interior regions do not melt because they are well shielded from the applied superheating by the dense spatial distribution of interconnected dendrites, as localized peripheral melting converts all the sensible heat reaching it to latent heat. Marsh has discussed the thermal shielding properties of dense mushy zones [10].

(ii) *Coarsening* of the zone interior occurs as neighboring dendritic branches grow and shrink simultaneously. The average thickness of the dendritic elements comprising the mushy zone increases over time in accord with Ostwald ripening kinetics, but the mean volume fractions of solid and melt within the zone remain constant [11]-[15].

(iii) *Detachment* occurs as dendritic side arms pinch off and the mushy zone recedes by melting away from the surrounding hotter melt. The thinning of the density of solid crystallites allows more sensible heat to penetrate the zone, which, in turn, further reduces the volume fraction of solid. This process accelerates until only a small cluster of isolated dendritic fragments remain, as shown in the lower left and right panels of Figure 1. The kinetics of dendritic detachment and melting interactions among neighboring crystallites has been discussed by Lupulescu et al. [16].

(iv) *Capillary-mediated melting* occurs when crystallites melt to a length $C \leq 2$ mm. For a typical ellipsoidal crystallite with $10 \leq C/A \leq 20$, the minor axis length, or equatorial diameter, A, would fall in the range 100-200 micrometers. These size scales are in the appropriate range where solid-liquid surface tension effects become noticeable. When a typical dendritic fragment enters this size regime, its C/A ratio falls toward unity, and the crystallite shrinks into a small sphere that rapidly melts to extinction.

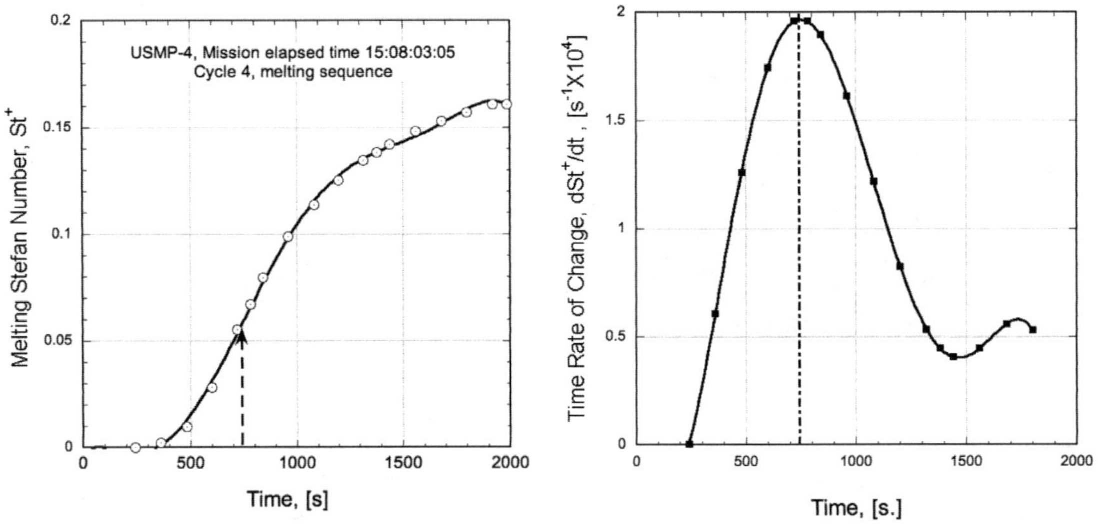

Figure 2. (*a*) The Stefan numbers (○) during melting plotted against melting time were obtained from temperature data furnished by telemetry from within the IDGE sample chamber. Data are shown for the first 33 minutes of IDGE cycle 04. Vertical dashed arrow indicates time at which the maximum time rate of change, dSt^+/dt, occurs. (*b*) Rate of change of the melting Stefan number (■) versus time, obtained by differentiating the curve $St^+(t)$. The applicable melting Stefan number at the peak of its rate of change corresponds to the value $St^+ = 0.055 \pm 0.004$ for the measured kinetics during this cycle. This value remains nearly constant over the 40 s interval around 740 s. These kinetics are displayed in Figure 4

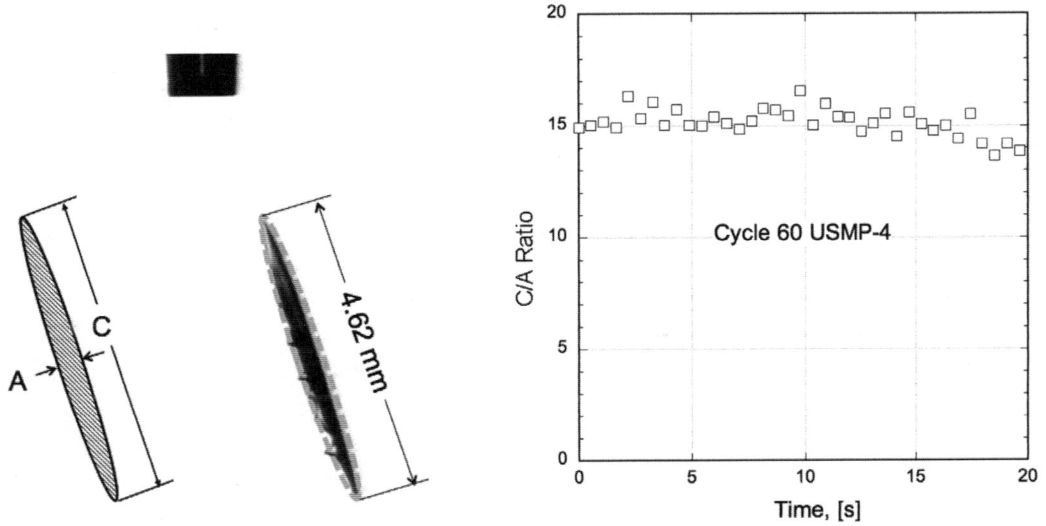

Figure 3. (a) Video frame of an isolated (stationary) PVA crystallite undergoing melting in microgravity. Software draws a border onto the digitized image (dashed curve surrounding the crystallite) that locates the solid-liquid interface and measures the major and minor axes, yielding the ellipsoid's C/A ratio at the time corresponding to the video frame. (b) C/A ratio versus time for a crystallite melting in cycle 60. Data (□) confirm an example of quasi-static melting at nearly constant $C/A = 15$ over a 20 s interval

2.1. Experimental melting kinetics

The IDGE automated thermostat was designed to provide a constant freezing Stefan number, but not a constant melting Stefan number. An example of the thermal data telemetered from within the sample chamber is shown in Figure 2(a). Here one sees the first 33 min. of mushy zone melting ending IDGE cycle 04. During the first 400 s much of the dense dendritic mushy zone gradually melted, absorbing as latent heat all the heat conducted into the chamber from the surrounding thermostat. Consequently, little temperature rise (or melting Stefan number increase) is experienced initially within the melting solid-liquid mixture. The melting Stefan number increases rapidly after about 500 s as the density of unmelted crystals steadily decreases, lowering the mixture's ability to absorb the added sensible heat as latent heat. The vertical arrow placed at 740 s indicates the time at which the last few crystallites melted. Beyond this time, and up to about 1900 s, the temperature of the now fully melted chamber rises toward its maximum of 1.8 K above the melting point of PVA. The melt temperature uniformly approaches the thermostat temperature, reducing the thermal gradients between the chamber and the thermostat. This slows, and eventually halts, the rise of the melting Stefan number at its final value of $St^+ = 0.16$. Figure 2(b) shows the corresponding time derivative of the melting Stefan number, dSt^+/dt, versus melting time, which peaks around $t = 740$ s. These data confirm the precise time interval over which the last few isolated crystallites underwent *isolated* melting, and establish quite precisely the applicable melting Stefan number responsible for melting the remaining crystallites.

2.2. Isolated crystallites

Near the end of each melting sequence, a few dendritic fragments remain that undergo melting as isolated crystallites. Figure 3 (a) shows a typical isolated dendritic fragment melting. For those crystallites oriented so its video image appears in true length, software determines a bordering ellipse defining the solid-liquid interface (dashed curve surrounding the particle) that yields the major and minor axes, and the C/A ratio at that moment during the melting sequence. If a crystallite was well-isolated from all others, and its initial shape lacked strong irregularities or branches, it would continue to shrink by melting at a nearly constant C/A ratio. Figure 3 (*b*) shows C/A ratio data (□) versus time during such an event for an isolated ellipsoidal crystallite melting near the end of melting cycle 60. The remaining crystallites at this stage of melting a mushy zone in microgravity typically exhibited C/A ratios in the range 10-20.

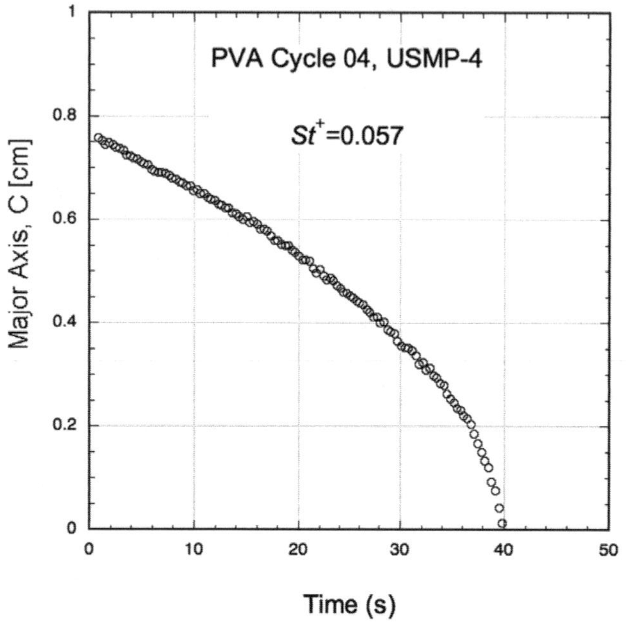

Figure 4. Final 40 s of melting a well-isolated ellipsoidal crystallite starting near $t = 720$ s, cycle 04. Data points (○) are the crystallite's major axis length, C, versus time, as this crystallite's C/A ratio slowly increased from 10-15, with a time-averaged value of $\langle C/A \rangle = 12 \pm 1$. The melting Stefan number ($St^+ = 0.057 \pm 0.002$) for the applied superheating of the melt during this interval was obtained from temperature data furnished via telemetry from the IDGE thermostat. See again Figures 2(*a*) and (*b*)

Isolated ellipsoidal crystallites that were longer than about 2 mm maintained a nearly constant, or slowly changing, C/A ratio during their initial melting. Such long crystallites exhibited smooth 'trajectories' relating how their size, or volume, diminished over time. An example is shown in Figure 4, where the length of a crystallite's C-axis is plotted versus melting time. This ellipsoidal crystallite was almost 0.8 cm in length when it began melting ($t = 0$). It shrank to 0.2 cm at $t = 36$ s, when its axial ratio began a rapid decrease from about 15 toward 1. Note that this melting trajectory also displays a slight 'kink' at 36 s. The last video frame and data point (○) recorded on that trajectory, prior to its rapid extinction as a sphere, occurred at 40 s where $C = A = 0.01$ cm.

3. Theoretical interpretations of melting in microgravity

Two interesting melting phenomena observed by performing IDGE microgravity experiments require modeling and theory for their interpretation. They include: 1) establishing the basis for self-similar melting, i.e., melting at constant C/A of ellipsoidal crystallites with $C \geq 2$ mm, and 2) explaining the sudden onset of spheroidization as the $C/A \to 1$ when the length of ellipsoidal crystallites became smaller than about 2 mm.

The author and former collaborators analyzed basic mesoscopic heat transfer theory pertaining to the melting of isolated ellipsoidal crystallites suspended in a melt undergoing quasi-static melting [17]. Quasi-static melting of a crystal is based on calculating the thermal current, based on Laplace's equation, entering or leaving a *static* solid-liquid interface. Kinematic equations are then formulated to deal with the crystal's actual loss of volume by melting. The rate of change of the particle's volume is assumed to occur slowly enough that the surrounding Laplace transport field has sufficient time to adjust to the changing particle shape. Here, quasi-static melting is limited to melting under the condition that $St^+ \ll 1$, and that the solid-liquid interface remains isothermal at $T = T_m$. The maximum value of the melting Stefan number in the experiments was $St^+ \leq 0.16$. As discussed in Section 2.1, melting of isolated ellipsoidal crystallites was observed when the melting Stefan number was much smaller, namely, $St^+ \approx 0.06$, which adequately satisfies the thermal requirements for quasi-static melting. Only a summary of our detailed potential theoretic analysis [17] will be used here in discussing self-similar melting.

The same authors suggested in a later publication [16] that capillarity was probably responsible for the terminal features of melting in microgravity, where the nearly constant C/A ratio during self-similar melting suddenly plummets toward unity, resulting in the extinction by melting of a spherical crystal. No direct proof, however, was offered at that time showing that the extremely weak thermal influences of capillarity were, in fact, responsible for altering the shapes of melting crystallites from needle-like ellipsoids to spheres.

3.1. Quasi-static theory

The axis-symmetric temperature field, $T(\vec{r})$, surrounding an ellipsoidal crystallite, may be non-dimensionalized as a Laplacian potential, $\vartheta(\vec{r})$, where the axis threading the crystallite's north and south poles defines the distance vector $\vec{r} = 0$. The heat flow intensity to the crystal is determined by the magnitude of the potential gradient at the solid-liquid interface, which may be determined from the potential's definition

$$\vartheta(\vec{r}) \equiv \frac{T(\vec{r}) - T_m}{T_\infty - T_m}. \tag{3}$$

It is useful to express the potential defined in Eq. (3) in orthogonal, curvilinear prolate spheroidal coordinates, $\vartheta(\xi, \eta, \varphi)$. Consequently, the solid-liquid interface of an ellipsoidal crystallite is considered at present to be the coordinate surface labeled $\xi = \Xi$, over which the interface potential remains uniform, namely, $\vartheta(\Xi, \eta, \varphi) = 0$. This key assumption ignores the effect of interface curvature on the local equilibrium melting point, T_m. The influence of capillarity, which appears towards the end of each melting trajectory, will be considered later in Section 5. The prolate spheroidal coordinate system is discussed in detail by Morse and Feshbach [18], for which we chose to scale the dimensionless coordinates ξ and η by the interfocal length of the ellipsoidal crystallite, a, defined in [18] as $a = C/\Xi = A/\sqrt{\Xi^2 - 1}$. Prolate spheroidal coordinates are displayed as the $\varphi = 0$ projection shown in Figure 5. The cross-hatched elliptical 'crystallite' placed within Figure 5 has a C/A ratio of almost 10, which is at the lower end of the axial ratios of PVA crystallites studied here; its solid-liquid interface is specified by the coordinate surface $\Xi = 1.008$.

In brief, heat flow from the hotter melt to the solid-liquid interface is responsible for melting the ellipsoidal crystal, the centroid of which is located at the origin of these coordinates, $\eta = 0$, $\xi = 1$. Heat transfer occurs by three-dimensional conduction along the hyperboloidal surfaces $\eta =$const. The isotherms of the resultant quasi-static field in the melt form concentric prolate ellipsoids, $\xi =$ const. $\geq \Xi$. The solid-liquid interface, $\xi = \Xi$, itself remains uniformly at the equilibrium melting point given by the potential surface $\vartheta(\Xi, \eta, \varphi) = 0$. The enclosed solid region is consequently isothermal.

The Laplace potential defined in Eq. (3) is based on the total temperature difference through the melt. This potential may be recast using the melting Stefan number, already defined in

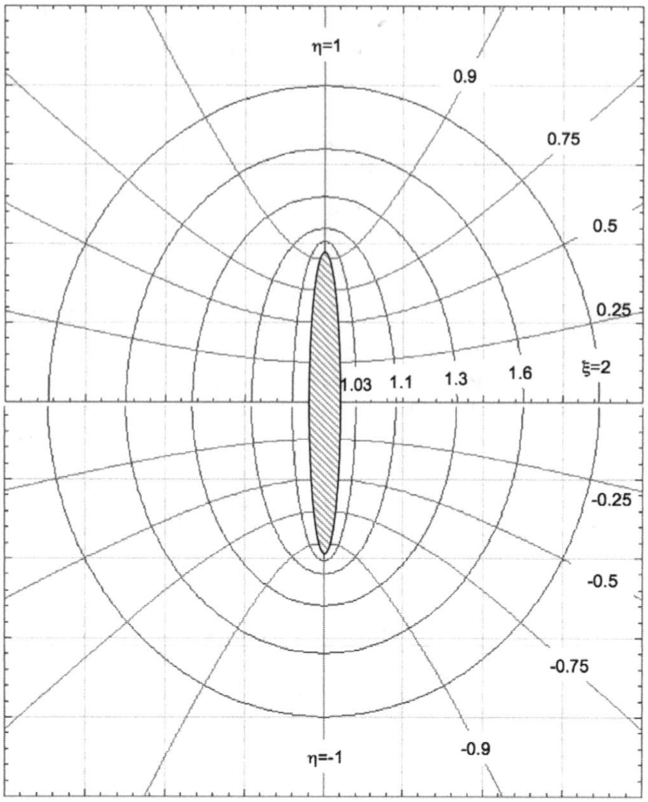

Figure 5. Prolate spheroidal coordinates, (ξ, η, φ), projected on the plane $\varphi = 0$, as described by Morse and Feshbach [18]. A solid-liquid interface is inserted that projects as the profile $\Xi = 1.008$ surrounding the cross-hatched 'crystallite', shown here with an aspect ratio of $C/A = \Xi/\sqrt{(\Xi^2-1)} \approx 10$. Isotherms in the melt are denoted as the ellipsoidal surfaces $\xi \geq \Xi =$ const., which vary in their potential between $\vartheta = 0$ on $\xi = \Xi$, and $\vartheta = 1$ on $\xi = \infty$. The heat responsible for melting the crystallite moves along the hyperboloidal sheets $\pm\eta =$ const.

Eq. (2), as the measure of melting intensity. Thus, the thermal potential in prolate spheroidal coordinates transforms to

$$\vartheta(\xi, \eta, \varphi) = \frac{1}{St^+} \frac{(T(\xi, \eta, \varphi) - T_m)}{\Delta H_f / C_p}, \;\; (\xi \geq \Xi) \tag{4}$$

everywhere throughout the melt, and within the solid to

$$\vartheta(\xi, \eta, \varphi) = 0, \;\; (\xi \leq \Xi). \tag{5}$$

Morse and Feshbach show that the Laplace potential surrounding a prolate spheroid with the fixed interfacial potential $\vartheta(\Xi, \eta, \varphi) = 0$ embedded in a quiescent melt with the far-field potential $\vartheta(\infty, \eta, \varphi) = 1$, is

$$\vartheta(\xi, \eta, \varphi) = 1 - \frac{\log\left(\frac{\xi+1}{\xi-1}\right)}{\log\left(\frac{\Xi+1}{\Xi-1}\right)}. \tag{6}$$

The gradient field in the melt surrounding the solid-liquid interface associated with this thermal potential is found by taking the partial derivative of Eq. (6) with respect to ξ, and then dividing the result by the ξ-coordinate's metric, $h_\xi = (a/2)\sqrt{(\xi^2 - \eta^2)/(\xi^2 - 1)}$. These operations yield a dimensionless, axis symmetric gradient field, $\vec{\nabla}\vartheta(\xi, \eta, \varphi)$, which when evaluated at the solid-liquid interface, $\xi = \Xi$, gives for its normal component,

$$\vec{\nabla}\vartheta(\Xi, \eta, \varphi) \cdot \vec{n} = \frac{4}{a}\left(\sqrt{(\Xi^2-1)(\Xi^2-\eta^2)}\log\frac{(\Xi+1)}{(\Xi-1)}\right)^{-1}, \tag{7}$$

where $\vec{n}$ is a unit normal vector on the solid-liquid interface. The relationship of the normal component of the interfacial gradient, Eq. (7), to the thermal gradient in the melt surrounding an ellipsoidal crystallite may be found by differentiating Eq. (4) with respect to ξ, dividing through by the curvilinear coordinate metric, h_ξ, to form the Jacobian transform, and setting $\xi = \Xi$.

$$\nabla\vartheta(\Xi,\eta,\varphi) = \frac{1}{St^+}\frac{C_p}{\Delta H_f}\left(\frac{1}{h_\xi}\frac{\partial T(\xi,\eta,\varphi)}{\partial \xi}\right)_{\xi=\Xi}. \qquad (8)$$

Rescaling the normal component of the Laplace potential gradient, Eq. (8), through the characteristic temperature of the melting transformation, $\Delta H_f/C_\ell$, multiplied by the system's melting Stefan number, and substituting the ξ-coordinate metric, h_Ξ, that acts normal to the solid-liquid interface, yields the desired form for the normal component of *dimensional* interfacial thermal gradient, $\vec{\nabla}T_\Xi \cdot \vec{n}$, namely,

$$\left(St^+\frac{\Delta H_f}{C_\ell^p}\right)\nabla\vartheta(\Xi,\eta,\varphi) = \frac{2}{a}\sqrt{\frac{\Xi^2-1}{\Xi^2-\eta^2}}\cdot\left(\frac{\partial T(\xi,\eta,\varphi)}{\partial\xi}\right)_{\xi=\Xi} = \vec{\nabla}T_\Xi\cdot\vec{n}. \qquad (9)$$

3.2. Enthalpy current

The enthalpy flux, $\vec{J}$ entering the solid-liquid interface may be found using Fourier's law of heat conduction,

$$\vec{J} = -k_\ell\vec{\nabla}T_\Xi, \qquad (10)$$

where k_ℓ is the thermal conductivity of the melt. The thermal gradient normal to the solid-melt interface, Eq. (9), can be substituted into Fourier's law to yield the melting flux,

$$\vec{J} = \left[-\left(k_\ell\frac{4}{a}\frac{\Delta H_f}{C_\ell^p}St^+\right)\left(\sqrt{(\Xi^2-1)(\Xi^2-\eta^2)}\log\frac{\Xi+1}{\Xi-1}\right)^{-1}\right]\cdot\vec{n}. \qquad (11)$$

The sensible heat current, $\dot{Q}_{tot}$, melting the prolate spheroid is the integral of the flux, $\vec{J}$, over its solid-liquid interface, namely,

$$\dot{Q}_{tot} = \iint_{interface}\vec{J}\cdot\vec{n}\,dS, \qquad (12)$$

where the differential element of interface area on the prolate spheroid, $\xi = \Xi$, is

$$dS = r(\Xi,\eta)d\sigma(\Xi,\eta)d\varphi = r(\Xi,\eta)h_\eta d\eta d\varphi. \qquad (13)$$

Here $r(\Xi,\eta) = (a/2)\sqrt{(\Xi^2-1)(1-\eta^2)}$ denotes the radial distance from the major C-axis to an arbitrary location, η,φ, on the interface, and $d\sigma = h_\eta d\eta$ is the differential of the curvilinear arc length. The η-coordinate metric needed to scale arc length, σ, along an ellipsoidal interface is defined by Morse and Feshbach as $h_\eta \equiv a/2\sqrt{(\Xi^2-\eta^2)/(1-\eta^2)}$. Using these transformations with Eq. (13) yields the differential element of interfacial area on a prolate spheroidal crystal as,

$$dS = \frac{a^2}{4}\sqrt{(\Xi^2-1)(\Xi^2-\eta^2)}d\eta d\varphi. \qquad (14)$$

Substituting Eq. (11), for the normal flux, and Eq. (14), for the differential interfacial area, into the formula for $\dot{Q}_{tot}$, Eq. (12), yields the definite integral,

$$\dot{Q}_{tot} = -\frac{ak_\ell\Delta H_f St^+}{C_\ell^p}\left(\log\frac{(\Xi+1)}{(\Xi-1)}\right)^{-1}\int_{-1}^{1}d\eta\int_0^{2\pi}d\varphi. \qquad (15)$$

The integration limits on η $(-1 \leq \eta \leq 1)$ and on φ $(0 \leq \varphi \leq 2\pi)$ correspond to integrating the heat fluxes from pole-to-pole over the prolate spheroid. Carrying out the integrations indicated in Eq. (15) introduces the expected spherical image, 4π, and shows that the instantaneous heat current entering a prolate spheroid with an interfocal length a is

$$\dot{Q}_{tot} = -\frac{4\pi a}{\log\frac{(\Xi+1)}{(\Xi-1)}}\frac{k_\ell \Delta H_f}{C_\ell^p} St^+. \tag{16}$$

3.3. Kinematic energy balance

The volume of a spheroidal crystal is given by the standard formula [19],

$$V(t) = \frac{\pi}{6}A^2(t)C(t) = \frac{\pi}{6}\left(\frac{C}{A}\right)^{-2} C^3(t), \tag{17}$$

Sensible heat, $\dot{Q}_{tot}$, provides the latent heat for melting. The time-dependence of the major axis, $C(t)$, is explicitly noted in the RHS of Eq. (17), whereas the spheroid's aspect ratio, C/A, is considered to be constant. The validity of this important approximation, $C/A =$ const., will be demonstrated later. The time-rate of change of the major-axis length as melting proceeds is related kinematically to the volume derivative at constant C/A as

$$\dot{V}(t) = \frac{\pi}{2}\left(\frac{C}{A}\right)^{-2} C^2(t)\,\dot{C}(t), \tag{18}$$

where the overdot indicates time differentiation. The rate of enthalpy change associated with the volume change is

$$\dot{H}_{tot} = \frac{\pi}{2}\frac{\Delta H_f}{\Omega}\left(\frac{A}{C}\right)^2 C^2(t)\,\dot{C}(t). \tag{19}$$

where Ω is the molar volume of the solid phase. This enthalpy rate expression ignores the imperceptible advection occurring in the melt near the interface caused by the small ($\approx 3\%$) density differences of the phases. Coriell and McFadden have shown that the effect of this advection on the transport of energy to or from a moving solid-liquid interface is negligible for real materials [8]. Work by Karma and Plapp [20] using phase-field computations also confirm the results found by Coriell and McFadden.

The kinematic energy balance under quasi-static conditions is therefore $\dot{H}_{tot} = \dot{Q}_{tot}$. Equating Eq. (19) with Eq. (16) gives

$$\frac{\pi}{2}\frac{\Delta H_f}{\Omega}\left(\frac{A}{C}\right)^2 C^2(t)\,\dot{C}(t) = -\frac{4\pi a}{\log\frac{(\Xi+1)}{(\Xi-1)}}\frac{k_\ell \Delta H_f}{C_\ell^p} St^+. \tag{20}$$

3.4. Melting kinetics in microgravity

Equation (20) provides the differential equation describing quasi-static evolution of a stationary prolate spheroidal crystal of constant axial ratio surrounded by an infinite superheated melt at a Stefan number, St^+. Equation (20) may now be solved for the kinetic response, $C(t)$, which was measured during the IDGE microgravity melting experiments on PVA by employing the following steps:

i. cancel common terms;
ii. recognize that $k_\ell \Omega / C_\ell^p \equiv \alpha$, where α is the thermal diffusivity of the melt;
iii. apply the transformation for the interfocal length, $a = C/\Xi$;

Journal of Physics: Conference Series **327** (2011) 012001 doi:10.1088/1742-6596/327/1/012001

iv. separate the dummy variables, C' and t';

v. integrate between the start of melting, $t' = 0$, when $C' = C_0$, and some arbitrary later time, $t' = t$, when $C = C(t)$.

After applying each of the steps enumerated above, one obtains

$$\int_{C_0}^{C(t)} C'dC' = -\frac{8}{\Xi \log \frac{(\Xi+1)}{(\Xi-1)}} \left(\frac{C}{A}\right)^2 \alpha St^+ \int_0^t dt'. \tag{21}$$

The integrations in Eq. (21) yield, after several steps of algebra,

$$C(t) = \sqrt{C_0^2 - \frac{16\alpha St^+}{\Xi \log \frac{(\Xi+1)}{(\Xi-1)}} \left(\frac{C}{A}\right)^2 t}. \tag{22}$$

Equation (22) may be non-dimensionalized as

$$\frac{C(t)}{C_0} = \sqrt{1 - K \cdot St^+ \cdot Fo}, \tag{23}$$

where a melting rate constant, K, is defined as

$$K \equiv \frac{16}{\Xi \log \frac{\Xi+1}{\Xi-1}} \left(\frac{C}{A}\right)^2, \tag{24}$$

and the Fourier number (dimensionless melting time) is defined as

$$Fo \equiv \frac{\alpha}{C_0^2} t. \tag{25}$$

4. Quasi-static melting results

The key theoretical results for predicting the kinetics of quasi-static melting of isolated spheroidal crystallites in microgravity are captured in Eqs. (22)-(25). As an example, Figure 6 shows the crystallite major axis length, $C(t)$, scaled by the starting length, C_0, Eq. 23, versus Fourier number, or dimensionless melting time. The value for the rate constant ($K = 360.6$) applied in Eq. (23) is for the same C/A ratio as the PVA crystallite measured in cycle 04, the melting trajectory for which appears in Figure 4, as $C(t)$ versus real time, t. The value of the coefficient that converts real time, t, to Fourier number ($\alpha/C_0^2 = 1.22$ ms) is based both on the thermal diffusivity for molten PVA at its melting point ($\alpha = 7.0 \times 10^{-4}$ cm^2/s) and on the crystallite's starting length indicated in Figure 4 ($C_0 = 0.758$ cm). The dashed trajectory appearing in Figure 6 is calculated for the experimental Stefan number, $St^+ = 0.057$ measured from the IDGE telemetry data during melting cycle 04. See again Figures 2(a) and 2(b). One notes general agreement between the theoretical prediction and the observed melting trajectory, considering that there are not any adjustable parameters in the theory. More specifically, the initial linear slope of this melting trajectory, when the crystallite was relatively large, is predicted accurately. The reason that the initial melting kinetics are predicted so well by quasi-static potential theory is that the heat flux to the poles ($\eta = \pm 1$) is *exactly* C/A times larger than the heat flux to the crystallite's equator ($\eta = 0$). The influence of the crystallite's geometry on shaping and concentrating the external potential near the poles is called the 'point effect' of diffusion, or, in this case, of heat conduction. Thus the crystal initially melts self-similarly, maintaining a constant axial ratio, because the external heat fluxes to the poles and the equator remain in the fixed ratio C/A.

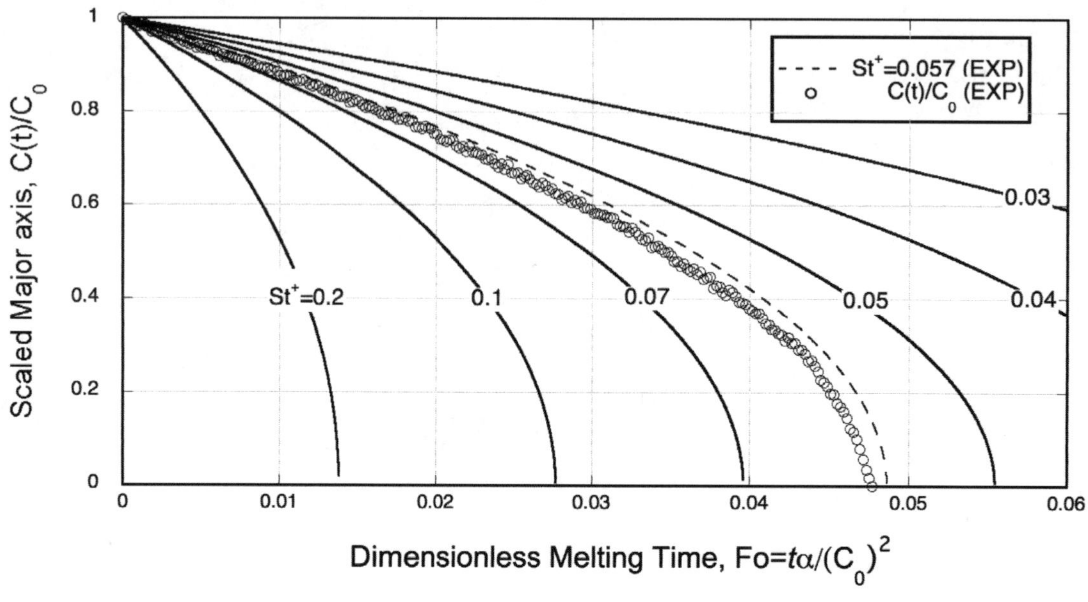

Figure 6. Melting trajectories for various melting intensities, St^+, for spheroidal PVA crystallites with $C/A = 12$. Curves are plotted as scaled length, $C(t)/C_0$, versus dimensionless time or Fourier number. $C(t)/C_0$ experimental data are for the crystallite trajectory, $C(t)$, in Figure 4.

The data in Figure 6 also show that the initial close fit gradually deteriorates as the crystallite shrinks further, with the observed melting rate gradually occurring faster than that given by quasi-static theory. The slight 'kink' occurring over the last 30% of the observed melting trajectory, as the crystallite's length decreased below about 2 mm, causes additional departure from the theoretical predictions.

5. Capillarity

5.1. Background

Capillarity—the influence of solid-liquid interfacial energy—occurs during melting in sufficiently small crystallites. Observations from IDGE experiments in microgravity indicate that when the equatorial radius of an ellipsoidal PVA crystallite falls below about 100 micrometers, so that the crystallite is approximately 2 mm in length, capillarity phenomena become significant and begin to influence the melting kinetics. See again Section 2.2. Quasi-static melting theory, outlined in Section 3.1, ignores capillarity through its approximation that the solid-liquid interface remains at a constant potential. The temperature shift, or capillary correction to the equilibrium melting temperature, δT, is given by the Gibbs-Thomson-Herring equation [24]-[26].

$$\delta T(\Xi,\eta) = T_e(\mathcal{H}) - T_m = -\frac{\gamma_{s\ell}(\theta) + \gamma_{\theta,\theta}}{\Delta S_f/\Omega}\mathcal{H}(\Xi,\eta), \qquad (26)$$

where $\mathcal{H}$ is the mean curvature [2] at an arbitrary point (Ξ,η) on the solid-liquid interface. $\gamma_{s\ell}(\theta)$ denotes the interfacial free energy per unit area at some location (Ξ,η), where the energy at that point varies anisotropically with the local normal angle θ. The term $\gamma_{\vartheta,\vartheta}$ is the second angular

[2] We adopt the convention that the mean curvature, $\mathcal{H}$, of a convex solid-liquid interface, e.g., where melt surrounds a spheroidal crystal, is positive.

derivative of the energy, also known as the torque or 'interfacial stiffness'. The solid-liquid free energy of pure PVA was measured by several investigators [21]-[23]. The mean interfacial energy of PVA is $\langle\gamma_{s\ell}\rangle = 2.8 \pm 0.3$ mJ/m^2 [21, 23], with a four-fold anisotropy of 3-5% about the $\langle 100\rangle$ zone axis [21, 22]. Equation (26) is easily converted to the capillary-dependent interface potential

$$\vartheta(\Xi,\eta) = \frac{\Omega T_m C_p}{\Delta H_f{}^2}\left(\frac{\gamma_{s\ell}(\theta) + \gamma_{\theta,\theta}}{St^+}\right)\mathcal{H}(\Xi,\eta). \quad (27)$$

5.2. Capillary influence on heat flux

The Stefan numbers applied by the IDGE thermostat during each melting cycle as the last, isolated crystallites melted, typically were in the range of $0.03 \leq St^+ \leq 0.06$. Although the equilibrium temperature shift along the curved interface does directly affect the point-to-point value of the *local* Stefan melting number, this direct influence is extremely small. Even near the poles, where the curvature is greatest, the effect of capillarity on the value of St^+ is only about 1 part in 10^4 [16]. Thus, the external sensible heat flux arriving from the melt is only trivially influenced by capillarity.

There is, however, an *indirect* effect of capillarity on the total heat flux reaching the poles. This indirect effect occurs because the non-isothermal temperature distribution along the curved solid-liquid interface causes thermal gradients to develop between the equatorial region, where the curvature and temperature shift are smaller than near the poles, where the curvature and temperature shift are much greater. In fact the curvature at the poles is $(C/A)^2$ times greater than the curvature at the equator. These temperature differences develop rapidly close to the poles, so even small temperature differences acting over small distances can stimulate significant heat fluxes *within* the crystallite. When the internal heat flux is added to the external heat flux near the poles it accentuates the melting rate at the poles.

S. Salon performed finite difference computations to evaluate the nature of the internal heat fluxes due to capillarity. (Private communication, S. Salon, Rensselaer Polytechnic Institute, Troy, NY, USA, 2005) Salon computed the internal and external fluxes reaching the poles in prolate spheroids of varying axial ratios and major axis lengths for a range of melting Stefan numbers. He numerically computed the Laplace field to high precision within the solid and melt phases, i.e., both inside and outside the solid-liquid interface. The interfacial energy parameters and the thermal diffusivity for PVA were used to check the influence of capillarity.

Figure 7(a) shows computational results for the influence of capillarity on the increase in the total flux reaching the poles of a prolate spheroid versus C/A, due to heat flow internal to the crystal. Results are displayed for two axial lengths, $C = 2$ cm and $C = 0.5$ cm. One notes that the internal heat flux increases rapidly with increasing C/A and decreasing crystallite size. Figure 7(b) exhibits a useful regularity in judging the strength of the capillary-induced internal flux responses. If the percent flux increase at the poles is multiplied by the crystallite's major axis length, C, the data sets for different size crystals collapse to a unique value for each C/A. Thus, using the curve in Figure 7(b) for the crystallite tracked in Figure 4, where $C/A \approx 12$, one finds that the percent increase in polar flux when $C = 0.2$ cm is appreciable, viz., about 15%. The capillary-mediated fluxes at the poles of this crystal would continue to increase rapidly with the inverse of the size of the shrinking crystallite. The increase in internal heat flux is then mitigated, and ultimately reduced, by the decrease in the C/A ratio as spheroidization from melting progresses.

6. Conclusions

(i) The IDGE experiments, launched aboard *Columbia* in the mid-to-late 1990s, collectively delivered a rich archive of solidification and melting data in microgravity. Those data

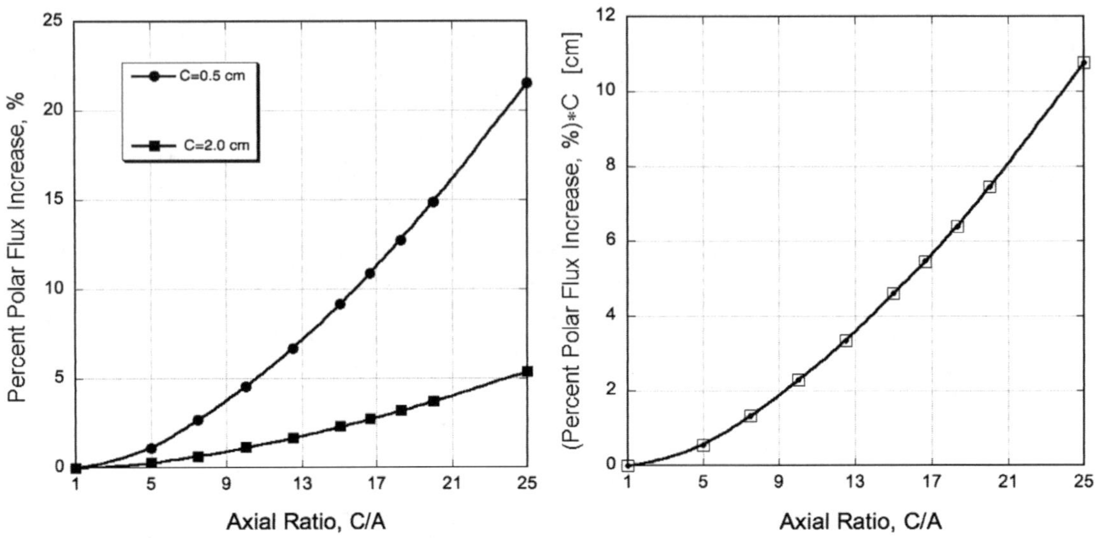

Figure 7. (*a*) Internal heat flux, resulting from capillarity, versus axial ratio, for prolate spheroids with major axes $C = 2.0$ cm (■), and $C = 0.5$ cm (●). (*b*) Percent flux increase at the poles times the major axis length, C, versus axial ratio. Both data sets from (*a*) combine here to yield identical values

accumulated during the final IDGE mission allows detailed analysis of melting kinetics for pivalic anhydride (PVA) crystallites.

(ii) The dendritic mass, or PVA mushy zone, melted initially from its periphery, and as progressive melting caused detachments and fragmentation, gradually reduced the dense mushy zone to a few isolated, needle-shaped crystallites, about 1-2 cm in length. These crystallites melted maintaining a nearly constant C/A ratio, as their length was reduced to about 2 mm.

(iii) Upon reaching a pole-to-pole length of circa 2 mm, the C/A ratio drops from values in the range 10-20, toward unity. Finite difference computations that include capillarity show that when prolate spheroids with C/A ratios in the range of 10-20 shrink to about 2 mm, the internal heat flux conducted to the poles increases. Thus, the poles receive an increase in total heat flux that accelerates their melting rate, causing rapid spheroidization. This process is substantially completed just before extinction occurs.

Acknowledgments

The author acknowledges the assistance of Dr. A. Lupulescu, G.E. Global Research Center, Niskayuna, NY, who extracted the PVA melting data from IDGE USMP-4 K-band video telemetry. Special thanks also go to Professor M.B. Koss, Physics Department, College of the Holy Cross, Worcester, MA, and to Professor Dr. Markus Rettenmayr, Faculty of Physics and Astronomy, Department of Metallic Materials, Friedrich Schiller University Jena, Germany, for reviewing this manuscript and making valuable suggestions for its improvement.

References

[1] Curreri P A 1988 Low-gravity effects during solidification *ASM Handbook Series, Casting,* vol **15** p 147
[2] Glicksman M E 2008 Solidification research in microgravity *ASM Handbook Series, Casting,* vol **15** p 398
[3] Hahn R C Koss M B Tirmizi S H, Selleck M E Velosa A Glicksman M E and Winsa E A 1991 *Adv. Space Res.* **11** p 53
[4] Koss M B Glicksman M E Hahn R C Rojas A Karthikyan M and Winsa E A 1993 *Adv. Space Res.* **13** p 209
[5] Glicksman M E Koss M B and Winsa E A 1994 *Phys. Rev. Lett.* **73** p 573
[6] Koss M B Glicksman M E and Winsa E A 1995 *J. Metals* **47** p 49
[7] Glicksman M E, Koss M B, Bushnell L T, LaCombe J C, and Winsa E A 1996 *Mat. Sci. Forum* **215-216** p 179
[8] McFadden G B and Coriell S R 1986 *J. Cryst. Growth* **74** p 507
[9] LaCombe J C Oudemool J L Koss M B Bushnell L T and Glicksman M E 1997 *J. Cryst. Growth* **173** p 167
[10] Marsh S P and Glicksman M E 1996 *Acta. Mat.* **44** p 3761
[11] Smith R Glicksman M E Marsh S P, and Kuklinski R 1992 *Metall. Trans. A* **23A** p 659
[12] Marsh S P and Glicksman M E 1996 *Metall. & Mat. Trans. A* **27A** p 557
[13] Marsh S P and Glicksman M E 1997 *Mat. Sci. and Engr.* **A238** p 140
[14] Mandyam H Glicksman M E Helsing J and Marsh S P 1998 *Phys. Rev. E* **58** p 2119
[15] Wang K G Glicksman M E and Marsh S P 2001 *J. Cryst. Growth* **230** p 318
[16] Lupulescu A Glicksman M E and Koss M B 2005 *J. Cryst. Growth* **276** p 549
[17] Glicksman M E Lupulescu A and Koss M B 2003 *J. Thermophysics & Heat Transfer* **17** p 69
[18] Morse P M and Feshbach H 1953 *Methods of Theoretical Physics* vol 2 (New York: McGraw-Hill) p 1284
[19] Eschbach O W 1961 *Handbook of Engineering Fundamentals* 2nd Edition (New York: Wiley) chapter 2, p 60
[20] Karma A and Plapp M, 2002 *Phase-field modeling of coupled growth in binary and ternary alloys* AIAA Reprint No. 2002-0600
[21] Singh N B and Glicksman M E 1989 *J. Cryst. Growth* **98** p 573
[22] Muschol M Liu D and Cummins H Z 1992 *Phys. Rev. A* **46** p 1038
[23] Bayender B Marasli N Cadirli E Sisman H and Gündüz M 1998 *J. Cryst. Growth* **194** p 119
[24] Herring C 1951 *Physics of powder metallurgy* ed Kingston W E, (New York:McGraw-Hill) p 143
[25] Johnson C A 1965 *Surface Science* **3** p 429
[26] Davis S H 2001 *Theory of solidification* (Cambridge: Cambridge University Press)

In-situ and real-time investigation of the columnar-equiaxed transition in the transparent alloy system neopentylglycol-camphor onboard the sounding rocket TEXUS-47

L Sturz, G Zimmermann

ACCESS e.V., Intzestrasse 5, 52072 Aachen, Germany

E-Mail: L.Sturz@access-technology.de

Abstract. The low-gravity experiment TRACE (TRansparent Alloys in Columnar Equiaxed solidification) has been performed onboard the sounding rocket TEXUS-47 to enable the investigation of dendritic growth and the dendrites' columnar to equiaxed transition during solidification. Low-gravity conditions provide solidification under diffusive heat and mass transfer conditions and without sedimentation or buoyancy of equiaxed dendrites or nucleation seeds to simplify the boundary conditions for dendritic microstructure simulation. In addition the transparent organic alloy system Neopentylglycol (NPG) - (D)Camphor (DC) was used to allow for a real-time and in-situ observation of the microstructure evolution with standard optics. For the flight experiment all relevant experimental parameters like thermal gradient, solidification velocity and undercooling within the bulk liquid and at the columnar dendritic tips have been determined by image analysis or from thermocouple recordings within the solidifying alloy. This allows a very detailed comparison with results of existing models for dendritic growth and for columnar-to-equiaxed transition. Here we present a summary of the experimental findings in comparison with results of some of the theoretical models.

1. Introduction

Dendritic microstructures are common to a large variety of solidification processes in technical applications. Dendrites may be subject to coarsening or phase-transformations towards their final microstructure at ambient, but the mechanical properties are often linked to the primary grain formation from the liquid phase. The grain structure of the primary alloy phase results from a competition of growth and nucleation, depending on the solidification conditions and properties of the alloy. The columnar-to equiaxed transition (CET) in grain grown of as-solidified alloys is identified as a change in microstructural features, like grain size and form and can lead to mechanical inhomogenities. As a simplified rule-of-thumb columnar dendritic growth is obtained at high thermal gradients and low cooling rates at the interface between solid dendrites and liquid melt and equiaxed dendritic growth vice versa. The columnar grains are oriented and elongated, while equiaxed grains have more isotropic properties and are formed upon nucleation and growth. As a consequence, equiaxed grains have generally smaller grain sizes with more spherical shapes and random crystallographic orientations. The CET is in general not favoured for a technical product, since microstructural heterogeneities can lead to mechanical ones. Physical mechanisms and conditions for the occurrence of CET are summarized in [1].

This article focuses on experimental results and a comparison to models of the microgravity experiment TRACE ((TRansparent Alloys in Columnar Equiaxed solidification). In the experiment the low-gravity environment of the sounding-rocket TEXUS-47 has been used to provide convection- and sedimentation/buoyancy-free solidification processing. A transparent organic alloy Neopentylglycol-Camphor has been used to observe in-situ and real-time the columnar dendritic growth of NPG-dendrites and their CET. This allows the determination of parameters like solidification velocity and dendrite tip undercooling or nucleation undercooling for equiaxed growth. We compare the experimental results with existing models for dendritic growth [2-5] and CET [6], the later with different approximations.

2. Experimental set-up

The experimental set-up, procedure, boundary-conditions and first results have been summarized in [7]. Here, we recall the main features and results. The materials NPG and DC have been purified, alloyed as NPG-37.5wt.-%DC and filled into the transparent experimental cell. The experimental volume in the cell is 20 mm width x 16 mm height x 1 mm thickness. The experiment is controlled by the temperatures at the bottom and the top of the experimental cell, defining the melting and solidification direction within the cell. Observation is carried out perpendicular to the height and width-axes. Macro-images of the size of the cell are acquired, as well as detail images, with two different CCD-cameras, field-of-views and optical resolutions. The columnar growth was started on ground before lift-off of the sounding rocket with a low solidification speed, corresponding to a cooling-rate of -0.2 $Kmin^{-1}$. Within the low-gravity period of 410 s onboard the sounding rocket TEXUS-47 a gravity level lower than ±4 mg in all directions was obtained. The low-gravity period is within +60 s and +470 s after lift-off, the initiation of CET was started by setting the cooling-rate to 2.0 $Kmin^{-1}$ simultaneously to the beginning of the low-gravity period at t=+60 s. Temperatures were measured within the bulk material by five thermocouples of diameter 0.25 mm, that were inserted through one side of the cell. From the macro-images and the thermocouple recordings the position, velocity and temperature of the dendritic interface were measured in columnar growth. After nucleation of equiaxed dendrites their density and volume fraction were determined as a function of time. The temperature gradient determined from the thermocouple recordings.

3. Experimental results

3.1 CET parameters

For comparison with CET-models the critical parameters at the CET have been determined from the experiment [7] and are summarized in table 1.

Table 1. Critical parameters at CET in NPG-37.5wt.-%DC within "TRACE" onboard TEXUS-47.

	Value	Comment
Time	435 s	After lift-off
Front undercooling	11.5±1 K	Averaged along front [7].
Front velocity	13.5±0.5 μms^{-1}	
Equiaxed volume fraction	0.36	
Thermal gradient	16.5±1 Kcm^{-1}	Along solidification direction [7]
Equiaxed grain density	0.5 mm^{-3}	

In addition to these critical parameters the distribution of nucleation temperatures are determined here, as well as the relationship between undercooling and solidification velocity.

3.2 Nucleation temperatures

In figure 1a we see the macro-image obtained at the CET at t=435 s after lift-off, in which equiaxed dendrites are present ahead and engulfed in the columnar dendritic front. The equiaxed dendrites are of different sizes and at different axial positions, their volume fraction is 0.36 [7]. We have determined the position of the equiaxed dendrites from the macro-images at smallest visible size. With an optical resolution in the focal plane of about 30 μm, that size is about 50 μm. With the assumption of a constant axial temperature gradient in between two thermocouples, the temperature at the position of these small equiaxed dendrites was determined from the thermocouple recordings above and below that position at the time of occurrence and by linear interpolation. In fact, we assume that their temperature is the same as the nucleation temperature and that release of latent heat or curvature effects during growth towards the small size do not effect the determination of the temperature. Figure 1b shows the frequency distribution of the nucleation undercooling for all 26 equiaxed dendrites within the experiment obtained by the procedure described above.

The nucleation undercooling is calculated relative to the experimentally determined liquidus-temperature of T_L=64±1°C [7]. A distribution between 8.5 and 12.0 K is found and not a single value for the nucleation temperature. The average and standard deviation of the distribution are given as 10.3±0.9 K. During the experiment a tendency towards higher nucleation temperatures was observed.

Figure 1a. Macro-image for determination of nucleation temperatures.

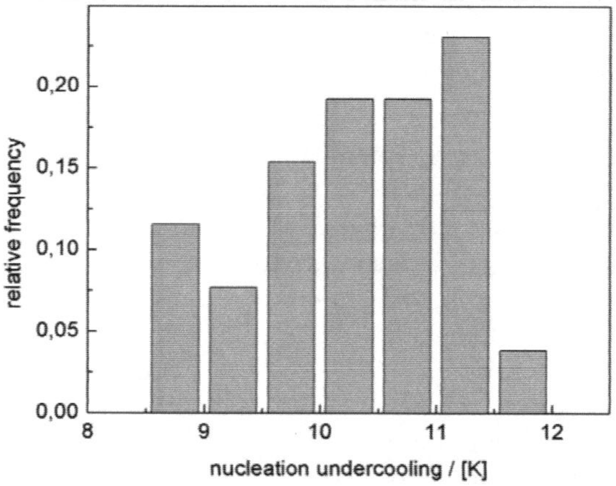

Figure 1b. Frequency distribution of the nucleation temperatures.

The evolution of equiaxed grain temperature from nucleation towards engulfment in the columnar front is shown in figure 2 in comparison to the columnar front temperature. The undercooling of the equiaxed dendrites was determined at their centers. The columnar front undercooling is increased from its steady-state value of about 7 K at -0.2 $Kmin^{-1}$ to about 11.7 K at the end of the microgravity period upon changing the cooling-rate. The "hump" from t= -100 s to +100 s corresponds to a technical mal operation and not to a physical effect. The equiaxed dendrites nucleate ahead of the columnar front at lower temperatures. Four equiaxed dendrites have been analyzed (see figure 3) from detail-images and show nucleation temperatures at the lower limit of the nucleation temperature frequency distribution (8.3-9.0 K). Small differences in the temperature determination exist between the analysis from the macro-image (figure 1) and the detail-images (figure 3) due to the higher resolution in the latter. This leads to earlier recognition of equiaxed dendrites in the detail images at lower times (293 s compared to 315 s) and lower nucleation temperatures (8.3 K compared to 8.5 K).

As the equiaxed grains increase in size, their centers move towards the growing columnar front. Thus, the undercooling of the equiaxed dendrites approach the columnar front undercooling. In some cases the undercooling of the equiaxed dendrites exceeds that of the columnar front. This is related to the slightly different temperature determination procedures.

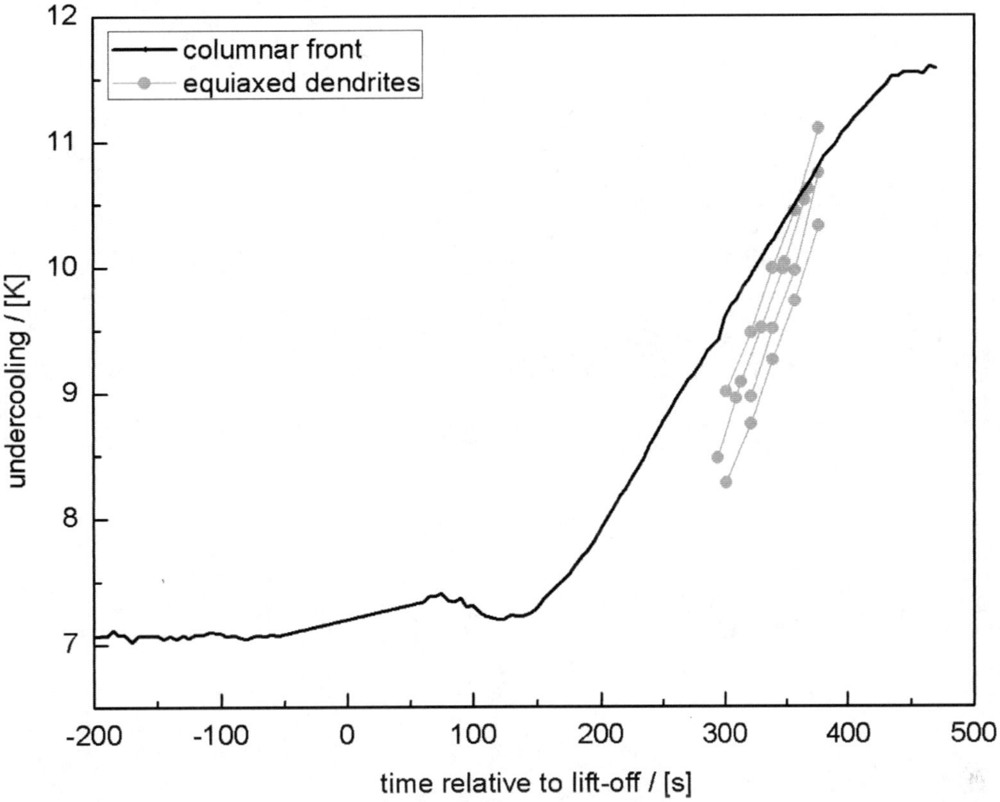

Figure 2: Evolution of the undercooling of the columnar front and of some equiaxed dendrites.

3.3 Undercooling-velocity relationships

For a comparison to existing theories on dendritic growth for alloy solidification the relationship between undercooling and solidification velocity was determined from the experiment. This was done both for columnar and equiaxed dendritic growth. In columnar dendritic growth the parameters were determined from the macro-images at the applied cooling rates of -0.2 $Kmin^{-1}$ until the first observed equiaxed grains at -2.0 $Kmin^{-1}$ (t=315 s).

The growth of equiaxed dendrites was followed in the detailed images with higher optical resolution of about 3 μm and at a field-of-view of 2.1 mm in width and 1.7 mm in height. Figure 3 shows a sequence of images with individual equiaxed dendrites growing in time. The undercooling was calculated by determining the positions of the centre of each equiaxed dendrite and interpolating their temperatures from the surrounding thermocouples. The result is shown as function of time in figure 2. The growth rate was calculated from the change of the equiaxed dendrite size in time. The size was determined by fitting an ellipse around the circumference of each equiaxed dendrite and by calculating the average length of the ellipse axes.

Figure 4 shows the result for the columnar front and the four equiaxed dendrites shown in figure 3, as individual data points. In addition x-error bars are added to the latter to capture the systematic error by determining the undercooling of the dendrites' centre and not their tips.

Journal of Physics: Conference Series **327** (2011) 012002 doi:10.1088/1742-6596/327/1/012002

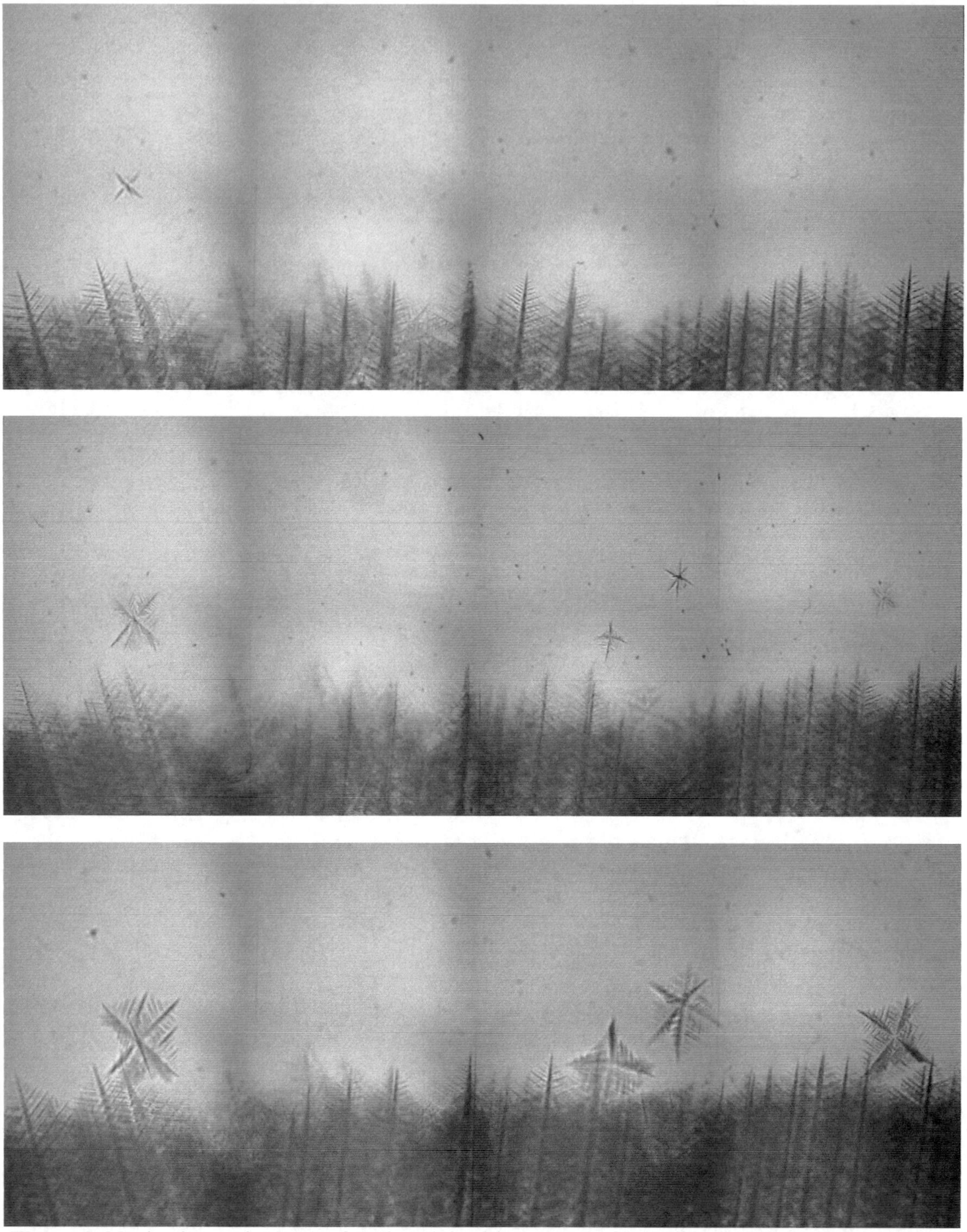

Figure 3. Nucleation and growth of equiaxed dendrites ahead of the columnar dendritic front. Each image of size 6.6 x 2.9 mm² consists of 8 individual images. The time to take the images by shifting the camera position is roughly 14 s. The images correspond to (top) t=293.0 s–307.3 s; (centre) t=331.0 s–345.2 s and (bottom) t=368.0 s–382.2 s.

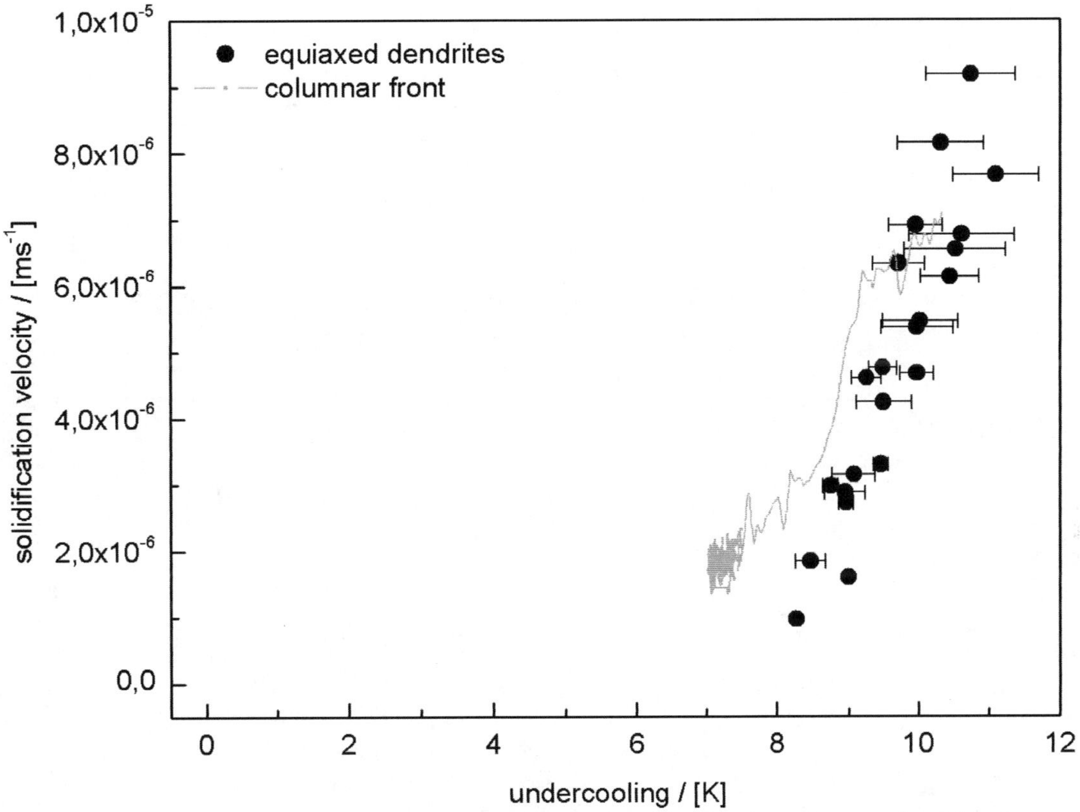

Figure 4. Undercooling-solidification velocity relationships for the columnar dendritic front and the equiaxed dendrites from figure 3.

4. Theoretical models

4.1 Undercooling-solidification velocity relationships

The models from Burden-Hunt (BH) [2], Lu-Hunt (LH) [3] and LGK/KGT [4,5] have been plotted in figure 5a in addition to the experimental data. As can be seen, none of them is able to describe the experimental data correctly. Within the LGK/KGT-theory a stability parameter σ^* is used, which is defined as $\sigma^*=1/4\pi^2=0.0253$. In literature deviations from these value for other transparent model alloys are reported, for example $\sigma^*=0.0192\text{-}0.313$ for succinonitrile-acetone [8] or $\sigma^*=0.081\pm0.02$ for NBr_4-water [9]. Therefore we decided to perform a fit of the LGK/KGT-theory to the experimental data with σ^* as fit parameter. The result is shown in figure 5b with the stability constant obtained as $\sigma^*=0.007$ for the NPG-37.5wt.-%DC alloy system. In addition a power-law ($V=A*\Delta T^P$) is fitted to the experimental data, giving an even better fit result. Nevertheless the power-law lacks of theoretical basis. For the power-law fit-values of $A=2.1*10^{-10}$ ms^{-1} and $P=4.43$ are found.

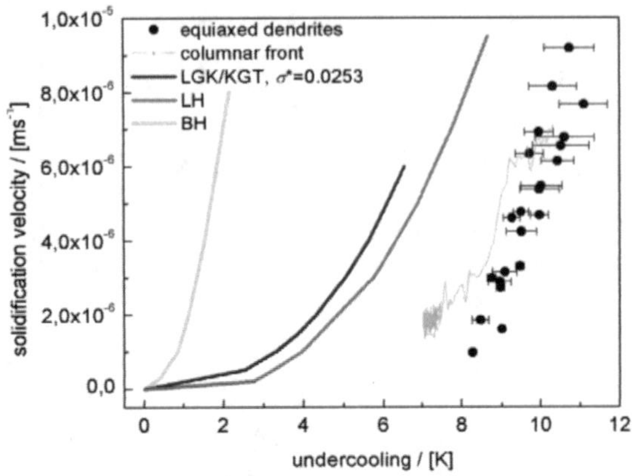

Figure 5a. Comparison of experimental data (figure 4) with theoretical models.

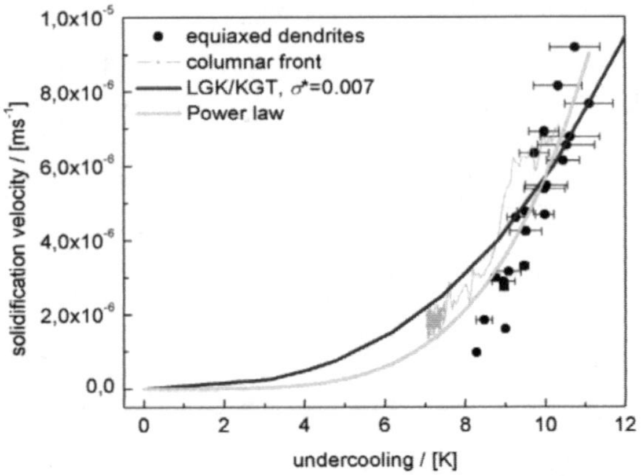

Figure 5b. Comparison of experimental data (figure 4) with theoretical models and power-law fit.

4.2 CET-model

Hunt's model [6] is based on the assumptions of steady-state growth conditions, constant alloy properties, a linearized phase diagram, diffusive transport of mass in the liquid and a constant thermal gradient without release of latent heat during solidification. The same kinetic law (undercooling-solidification velocity relationship, sec. 4.1) is employed for both columnar and equiaxed dendritic growth, while direct interaction between the dendrites is neglected. The existence of several dendrites is accounted for in an extended volume formulation. Heterogeneous nucleation at a given nucleation density with a single value for the critical nucleation undercooling is assumed. The CET-criterion is based on the equiaxed volume fraction ϕ_E, which is calculated in the model. For fractions $\phi_E>0.49$ equiaxed growth is assumed to dominate columnar growth.

The original model [6] was developed further by several authors to account for heat transport and the release of latent heat [6, appendix], LGK/KGT-kinetic law [10] and non-standard stability constants σ^* [11], non-equilibrium effects in thermodynamics [12] and a Gaussian distribution of nucleation undercooling [10, 13]. The general assumption of $\phi_E=0.49$, known as the "mechanical blocking"-criterion was discussed critically [14], while other authors take into account solutal [15] or thermal [16] interactions of the dendrites.

Here we calculate the predictions based on the original model [6], taking into account the known thermophysical properties [7] of the alloy and LGK/KGT kinetic-law [4,5] with the original as wells as the best-fit stability constant in the latter. Furthermore we investigate the effect of nucleation density N_0 and critical growth velocity V and compare the predictions for a single value of the single nucleation undercooling (at the average value of $\Delta T_N=10.3$ K) with the assumption of a Gaussian distribution fitted to the experimental distribution (figure 1b). The results for ϕ_E are summarized in table 2, together with the experimental result from TRACE.

Table 2. Results of the model predictions following [6], in comparison to experimental results.

	Kinetic law	σ^*	N_0 / [m^{-3}]	ΔT_N / [K]	V / [μms^{-1}]	ϕ_E	Comment
(1)	LGK/KGT	0.0253	5*10^8	10.3	13.5	0 (columnar)	standard
(2)	LGK/KGT	0.007	5*10^8	10.3	13.5	0.997 (equiaxed)	Best-fit σ^*

(3)	LGK/KGT	0.007	$500*10^8$	10.3	13.5	1 (equiaxed)	like (2), but $100*N_0$
(4)	LGK/KGT	0.007	$5*10^8$	10.3	11.0	0.935 (equiaxed)	like (2), changed V
(5)	Power-law	-	$5*10^8$	10.3	13.5	0.664 (equiaxed)	standard
(6)	Power-law	-	$5*10^8$	10.3	11.0	0.390 (equiaxed)	like (5), changed V
(7)	Power-law	-	$5*10^8$	10.3, ΔT_σ=3.5	13.5	0.764 (equiaxed)	like(5), Gaussian
EXP	Figure 4		N_E=$5*10^8$	10.3	13.5	0.36	CET

As can be seen from table 2, the Hunt model in combination with LGK/KGT kinetic law results in prediction of grain growth, which is either almost completely columnar ($\phi_E \cong 0$) or equiaxed ($\phi_E \cong 1$). This holds for the standard stability constant σ^* =0.0253 (1) and the calculation with the best-fit value of σ^*=0.007 (2). Assuming a reduced nucleation efficiency of 1% (100 times higher seed density) (3) or a lower solidification velocity at CET (from t=420 s) (4) do not improve the predictions.

With the power-law fit the difference between experimental result (ϕ_E=0.36) and the standard calculation is slightly decreased, while using a gaussian distribution for the nucleation temperatures does not improve. The good agreement for parameters given in (6) with the experimental result is misleading, since a critical velocity of 11.0 μms^{-1} corresponds to t=420 s, at which the experimental equiaxed volume fraction was determined to be ϕ_E=0.17. This is again significantly smaller than the model result ϕ_E=0.39.

5. Conclusions

Results of the low-gravity experiment TRACE performed onboard the sounding rocket TEXUS-47 are summarized here, in addition to previously reported data [7].

The nucleation temperature distribution was determined, showing values between 8.5 and 12.0 K with an average and standard-deviation of 10.3±0.9 K. The distribution is related to heterogeneous nucleation on substrates of different size or type.

The undercooling-solidification velocity relationship is similar for the columnar front and equiaxed dendrites growing ahead of the front. Comparison with dendrite growth models from Burden-Hunt [2], Lu-Hunt [3] and the standard LGK/KGT-model [4,5] are not capable to predict the measured relationship correctly. Possible explanations can be the simplified geometries of dendrites in the models.

A fit of the stability constant σ^* within the LGK/KGT-model results in σ^*=0.007, which is more than three times less than the standard value. This explains why high undercoolings are found at relatively low solidification velocities.

A power-law fit gives the best agreement to experimental results, but lacks of physical background. Nevertheless, it offers the advantage that alloy properties do not have to be known for predictions of CET in the model of Hunt. In contrast, the kinetic laws entering Hunt's model do depend on alloy properties for LH, BH, LGK/KGT [2-5].

Hunt's model is used to predict the equiaxed volume fractions at the columnar front. This is the first time that all input parameters for modeling are known and a direct comparison with the experiment is possible. Using the LGK/KGT-kinetics within Hunt's model results in the predictions of either complete columnar or equiaxed dendritic growth for all variations carried out. The same holds for the power-law-kinetics, which improve the predictions slightly. We conclude that none of the kinetic laws in combination with Hunt's model for mechanical blocking is able to predict the experimental CET-results quantitatively in terms of the equiaxed volume fraction. Interestingly, the conditions applied experimentally to obtain columnar growth (-0.2 $Kmin^{-1}$ corresponding to a steady-state velocity of 2.02 μms^{-1}) and to force the CET in a transient process (-2.0 $Kmin^{-1}$ corresponding to a steady-state of velocity 20.2 μms^{-1}) result in correct predictions in Hunt's model.

Consequently, proposals to tackle the disagreement at the CET are: (i) the dendrite shapes should be resolved on the microscopic scale (for example with phase-field-simulations), (ii) solutal interaction should be taken into account and (iii) the CET should be calculated in the transient state.

Acknowledgments

This work was financially supported by the German Space Agency DLR under grant no. 50 WM 0843, which is gratefully acknowledged. The authors appreciate the Team of Astrium Space Transportation GmbH in Trauen/Bremen for the hardware development and support during the preparatory and final experiments. SSC, DLR-Moraba and Kayser-Threde are acknowledged for mission support at Esrange/Sweden.

References

[1] Spittle J A 2006 *Int. Mat. Rev.* **51** 247.
[2] Burden M H, Hunt J D 1974 *J. Crys. Growth* **22** 109.
[3] Hunt J D, Lu S-Z 1996 *Metall. Mater. Trans. A* **27A** 611.
[4] Lipton J, Glicksman M E, Kurz W, 1984 *Mater. Sci. Eng.* **65** 57.
[5] Kurz W, Giovanola B, Trivedi R 1986 *Acta Met.* **34** 823.
[6] Hunt J D 1984 *Mat. Sci. Eng.* **65** 75.
[7] Sturz L, Zimmermann G 2011 subm. to *Proc. 20th ESA Symposium on European Rocket and Balloon Programmes and Related Research*, Hyeres, France, 23-27 May 2011
[8] Chopra M A, Glicksman M E, Singh N B 1988 *Met. Trans. A* **19A** 3087.
[9] Dogherty A, Gollub J P 1988 *Phys Rev. A* **38** 3043.
[10] Cockcroft S L, Rappaz M, Mitchell A, Fernihough J, Schmalz A J 1994 *Materials for Advanced Power Engineering, Part II* Coutsouradis D (Kluwer Academic Publishers, Netherlands) 1145.
[11] Rebow M, Browne D J 2007 *Scripta Mat.* **56** 481.
[12] Gäumann M, Trivedi R, Kurz W 1997 *Mat. Sci. Eng. A* **226-228** 763.
[13] Martorano M A, Biscuola V B 2009 *Acta Mat.* **57** 607.
[14] Biscuola V B, Martorano M A 2008 *Met. Mat. Trans A* **39A** 2885.
[15] Martorano M A, Beckermann C, Gandin C A 2003 *Met. Mat. Trans A* **34A** 1657.
[16] McFadden S, Browne D J 2006 *Scripta Mat.* **55** 847.

Investigation of columnar-to-equiaxed transition in solidification processing of AlSi alloys in microgravity – The CETSOL project

G Zimmermann[1], L Sturz[1], B Billia[2,3], N Mangelinck-Noël[2,3], H Nguyen Thi[2,3], Ch-A Gandin[4], D J Browne[5] and W U Mirihanage[5]

[1] ACCESS e.V., Intzestrasse 5, 52072 Aachen, Germany
[2] Aix Marseille Université, 13397 Marseille Cedex 20, France
[3] CNRS, IM2NP, Campus Saint-Jérôme, Case 142, 13397 Marseille Cedex 20, France
[4] ARMINES MINES ParisTech CEMEF, 06904 Sophia Antipolis, France
[5] University College Dublin, Belfield, Dublin 4, Ireland

E-mail: g.zimmermann@access-technology.de

Abstract. Grain structures observed in most casting processes of metallic alloys are the result of a competition between the growth of several arrays of dendrites that develop under constrained and unconstrained conditions. Often this leads to a transition from columnar to equiaxed grain growth during solidification (CET). A microgravity environment results in suppression of buoyancy-driven melt flow and so enables growth of equiaxed grains free of sedimentation and buoyancy effects. This contribution presents first results obtained in experiments on-board the International Space Station (ISS), which were performed in the frame of the ESA-MAP programme CETSOL. Hypoeutectic aluminium-silicon alloys with and without grain refiners were processed successfully in a low gradient furnace (MSL-LGF). First analysis shows that in the non grain refined samples columnar dendritic growth exists, whereas CET is observed in the grain refined samples. From analysis of the thermal data and the grain structure the critical parameters for the temperature gradient and the cooling rate describing CET are determined. These data are used for initial numerical simulations to predict the position of the columnar-to-equiaxed transition and will form a unique database for calibration and further development of numerical CET-modeling.

1. Introduction

Grain structures observed in most casting processes of metallic alloys are the result of a competition between the growth of several arrays of dendrites that develop under constrained and unconstrained conditions. Figure 1 schematically shows the solidification of a metallic melt in a mould. Typically for a high cooling rate at the beginning directional growth of columnar dendritic grains from the side walls is initiated. During solidification the temperature difference between the melt and the mould is reduced and the cooling rate decreases. This often leads to nucleation and growth of equiaxed grains in the undercooled melt. This effect is described as a columnar-to-equiaxed transition (CET) and was investigated intensively in the last decades [1-12].

On ground a far-reaching effect of the buoyancy-driven flow is observed. This leads to relative transport of the equiaxed crystals and flow-induced macrosegregation. It results in difficulties in conducting precise investigations on the origin of the formation of the equiaxed crystals and their interaction with the development of the columnar grain structure.

A microgravity environment allows for suppression of buoyancy-driven melt flow and so for growth of equiaxed grains free of sedimentation and buoyancy effects. Therefore, experiments in microgravity provide unique data for testing fundamental theories of grain structure formation [13-14]. To carry out such experiments and to model the process of columnar to equiaxed transition is the topic of the research project Columnar-to-Equiaxed Transition in SOLidification Processing (CETSOL) in the framework of the Microgravity Application Promotion (MAP) programme of the European Space Agency (ESA). At present the CETSOL team consists of four European scientific partners assisted by five partners from industry.

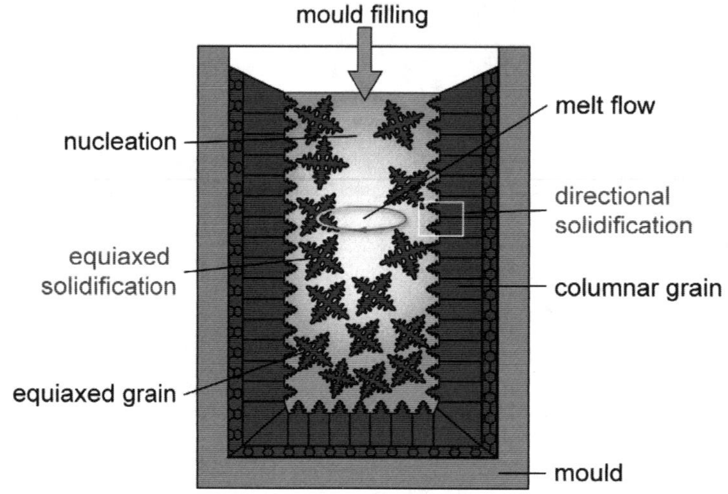

Figure 1. Schematic drawing of columnar and equiaxed grain growth during solidification

2. Set-up of the microgravity experiments

To investigate the columnar-to-equiaxed transition under diffusive conditions for heat and mass transport, experiments in microgravity were performed and will be performed. In 2006 the microgravity experiment MACE (Metallic Alloys for Columnar-Equiaxed Transition) was carried out in the experimental module TEM 01-2M during MAXUS-7 sounding rocket mission. Within this campaign three Al-7wt%Si samples were processed successfully. As a result the apparent effect of TiB_2 grain refiner particles in the alloy was to lower the critical undercooling for heterogeneous nucleation of primary equiaxed Al-dendrites in the bulk; CET was thus present at higher gradients and lower front velocities. Also, a significant delay of CET is observed on ground compared to the experiments performed in reduced gravity environment [15-19].

In continuation of these investigations experiments in microgravity were defined and successfully performed in the Materials Science Laboratory (MSL) with the Low Gradient Furnace (LGF) module onboard the International Space Station (ISS). For a Batch#1 the CETSOL team selected 6 experiments to investigate CET behaviour in microgravity. The AlSi7 sample material with and without grain refiner particles was provided by Hydro Aluminium Deutschland GmbH. Rod-like samples with diameter 7.8 mm and length 245 mm were integrated into flight cartridges by industry (EADS Astrium / Soterem). Figure 2 shows a sketch of the sample cartridge assembly (SCA) located in the furnace (LGF). The metallic Al-7wt%Si alloy is enclosed in a protective alumina tubular crucible with Shapal plugs and boron nitride spacers at both ends. The conical shape of the upper Shapal plug acts as passive volume compensation device during heating-up and melting of the alloy. At the outer surface of the alumina tube 4 longitudinal grooves were manufactured. In each groove 3 thermocouples at different axial positions were placed. This allows measuring the temperature profile along the sample axis within the crucible walls. The whole set-up is inserted leak-tight in a tantalum tube and filled with 200 mbar helium. For sample processing the cartridge is inserted into the low gradient furnace (LGF). The main parts of the LGF are the relatively cold zone which consists of 3 heaters H1 to H3, the hot zone which consists of 4 heaters, and an adiabatic zone in between (see figure 2). By controlling the temperatures of the cold and the hot zone a temperature gradient along the sample axis develops. Melting or solidification of the metallic alloy is realized by a movement of the furnace insert along the axis of the fixed sample with a defined speed. The situation in figure 2 shows the furnace at its very left position (zero furnace position), which corresponds to the maximum melting position of the sample at the beginning of the solidification phase.

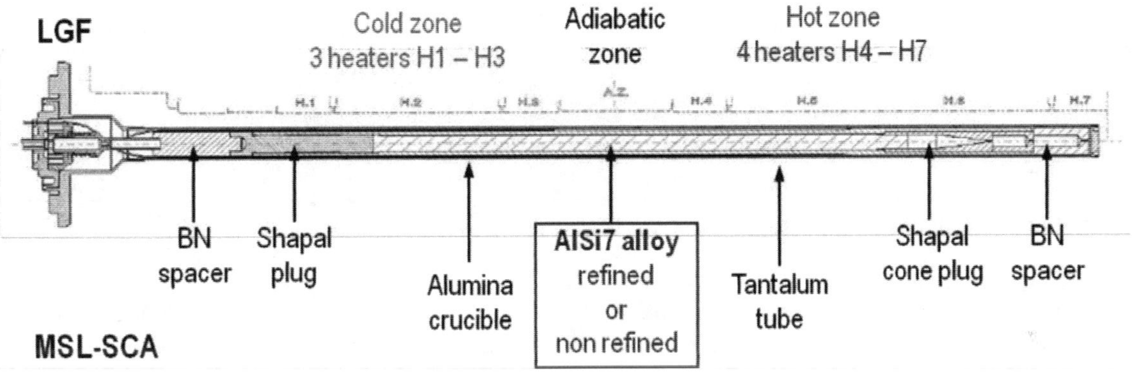

Figure 2. Sketch of a sample cartridge (MSL-SCA). Its position inside the low gradient furnace (LGF) corresponds to the beginning of the solidification phase.

Table 1 summarizes the main parameters of the flight experiments. In samples FM#3 and FM#4 non-refined AlSi7 alloys were used, while all other samples contain 0.5wt% of TiB_2 for grain refinement. At furnace position 80mm the temperatures of all heaters were increased to 500°C, followed by a further increase to 540°C for the 3 heaters of the cold zone and to 660°C for the 4 heaters of the hot zone. With this a temperature profile was established along the sample axis resulting in a partly melting of the Al-7wt%Si alloy in the hot zone region. Moving the furnace at a speed of 80µm/s to the end position (as shown in figure 2), directional melting of the sample was initiated. With an additional increase of the temperatures of all heaters via telecommand by 30K the solid-liquid region of the sample was centred in the adiabatic zone of the furnace. From the temperature measurements along the sample axis a temperature gradient of about 0.9K/mm at the liquidus temperature of 618°C was determined. This procedure was identical for all flight experiments to allow meaningful comparison of the experimental results.

After reaching thermal equilibrium the solidification process was started following different thermal homogenization times t_H. A short homogenization time of about 10 minutes just stabilizes the temperature profile. During a waiting period of 5 hours a diffusive coarsening process in the mushy zone region is expected, which may result in an axial segregation of silicon due to a temperature gradient zone melting (TGZM) effect [20].

The first solidification phase was identical for all experiments (see table 1). A furnace movement with $v1$=10µm/s for $z1$=20mm is expected to generate a columnar dendritic growth. A transition to equiaxed growth should be triggered in the second 50mm solidification phase of furnace movement. Therefore the temperatures of the heaters in the hot zone were decreased at dT/dt=4K/min. Additionally, in the experiments FM#1 to FM#3 the furnace speed is increased to $v2$=200µm/s. Finally a fast movement of the furnace with $v3$=3000µm/s to its end position and a shut down of the heaters stops the experiment.

Table 1. Process parameters of the Batch#1 flight experiments.

Sample No.	Alloy	Homoge nization time	Solidification phase 1		Solidification phase 2			Fast movement
CETSOL1		t_H (min)	$v1$ (µm/s)	$z1$ (mm)	$v2$ (µm/s)	$z2$ (mm)	dT/dt (K/min)	$v3$ (µm/s)
FM#1	AlSi7+g.r.	10	10	20	200	50	-4	3000
FM#2	AlSi7+g.r.	300	10	20	200	50	-4	3000
FM#3	AlSi7	300	10	20	200	50	-4	3000
FM#4	AlSi7	300	10	20	10	20	-4	3000
FM#5	AlSi7+g.r.	10	10	20	10	20	-4	3000
FM#6	AlSi7+g.r.	300	10	20	10	20	-4	3000

The flight experiments have been performed from November 2009 until April 2010 on the ISS. The processed cartridges have been downloaded with Space Shuttles and de-integrated by industry, followed by scientific evaluation of the CETSOL team. First results are presented in the next chapter.

3. CETSOL1 Batch#1 microgravity experiments

The performance of the flight samples consists of several steps which were agreed by the CETSOL team partners to achieve as much scientific output as possible. The main items can be summarized as follows:

3.1. Process parameters

The critical phases of each microgravity experiment, i.e. the homogenization and solidification phases, were performed during sleeping periods of the astronauts on the ISS. Thereby a microgravity level without larger disturbances was expected. The evaluation of the data of the gravity sensors close to the MSL rack confirms that a gravity level below $\pm 0.5 \times 10^{-3} g_0$ was achieved during all experiments, where g_0=9.8m/s² is the standard terrestrial value.

The housekeeping data – mainly heater and sample temperatures as well as the furnace movement – were controlled on-line during the experiments. This monitoring at the Microgravity User Support Centre (MUSC/DLR) in Cologne (Germany) also allowed the scientist to adjust the heater temperatures to shift the solid-liquid interface position in the metallic sample into the adiabatic zone of the MSL-LGF furnace. As an example figure 3 shows the temperatures of the 12 thermocouples which are placed along the sample with distances between two neighbouring thermocouples of 10mm. The position of the first thermocouple SCAtemp_tc_01 is 72mm from the non-molten end of the metallic sample. The diagram axis on the right side in figure 3 indicates the furnace position s. The graphs correspond to the solidification phases for experiment FM#2, starting with a furnace movement of $v1$=10µm/s at t=27380s, an increase to $v2$=200µm/s at t=29395s and finally a fast movement at *t*=29654s. The evaluation of the data demonstrates that all microgravity experiments were performed nominally.

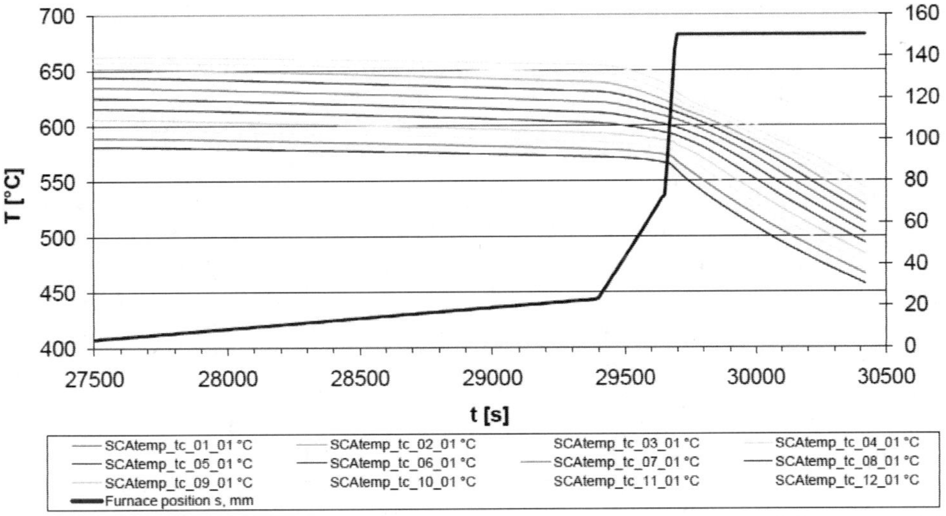

Figure 3. Temperatures of the 12 thermocouples along the sample and furnace position s (axis to the right) during the different solidification phases for experiment FM#2.

3.2. Non-destructive sample analysis

The next stage of evaluation is a non-destructive analysis of the processed samples using 3D computer tomography. As a result no macroscopic shrinkage holes were detected. This indicates that the volume compensation mechanism worked reliably. In the samples with short homogenization times of about 10min some pores with sizes up to 500μm were identified, mainly in the mushy zone region. This fact has to be taken into account for further analysis.

Figure 4 shows an image of the processed FM#2 flight sample after extraction from the alumina crucible. The maximum melting position at *z*=68mm can be identified, which separates the non-molten region and the mushy zone region. The adjacent solidification phases are also shown qualitatively.

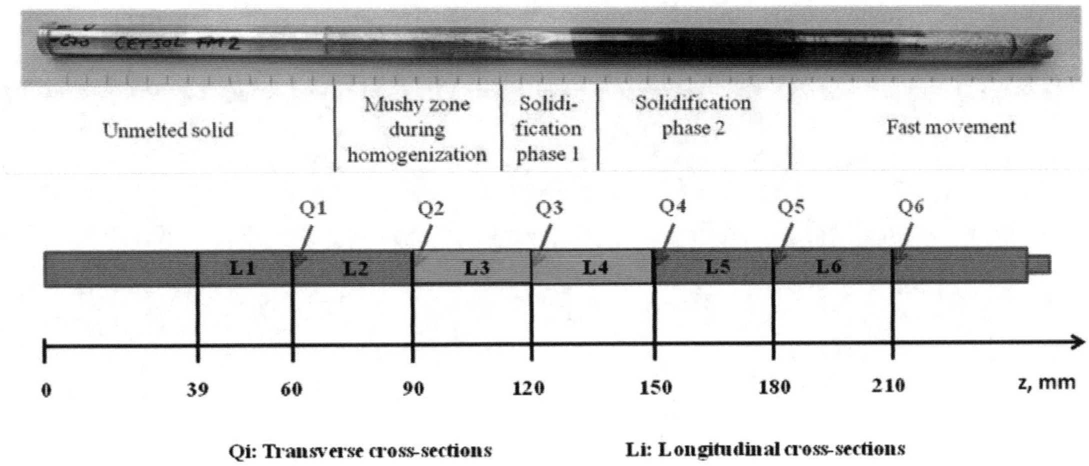

Figure 4. Processed FM#2 flight sample with related experiment phases (top) and identification of cross-sections for the metallographic preparation (below).

Journal of Physics: Conference Series **327** (2011) 012003 doi:10.1088/1742-6596/327/1/012003

3.3. Microstructure and grain structure evaluation

For analyzing the microstructure and the grain structure the samples have to be sectioned. First, the samples were cut into pieces of length 30mm to analyze the transversal cross-sections Qi (see figure 4). Second, each of these pieces was sawed along the axis to get two halves for analysis of the longitudinal cross-sections Li. To determine the microstructure the samples were polished and etched and observed with a microscope. To identify the grain structure the cross-sections were electrolytically etched and analysed in a polarized light microscope. Then, different colours represent different crystallographic orientations of the dendritic grains [21].

Figure 5 shows the longitudinal cross-sections L3 of FM#2 and FM#3 flight samples (for notation see figure 4). The lengths of the cross-sections are 30mm, the position values are referred to the non-molten end of each sample with a difference of 7mm between the FM#2 and FM#3 flight samples. Such a difference can be explained by small temperature differences (about 6K) within the two samples in the very flat gradient. The left part of each cross-section shows the developed mushy zone region. As a result of the long thermal homogenization of 5 hours in a temperature gradient (see table 1) coarsening of the primary Al-rich dendrites exists. Between refined (FM#2) and non-refined (FM#3) alloy no significant difference of the mushy zone structure is detected. After the thermal homogenization the directional solidification with $v1$=10µm/s was started. The development of a dendritic microstructure is shown in the right part of the cross-sections (figure 5). Starting from several nuclei in both samples, the non-refined alloy (FM#3) shows a more pronounced columnar structure than the refined alloy (FM#2). Concerning the grain size and grain density no substantial difference is found.

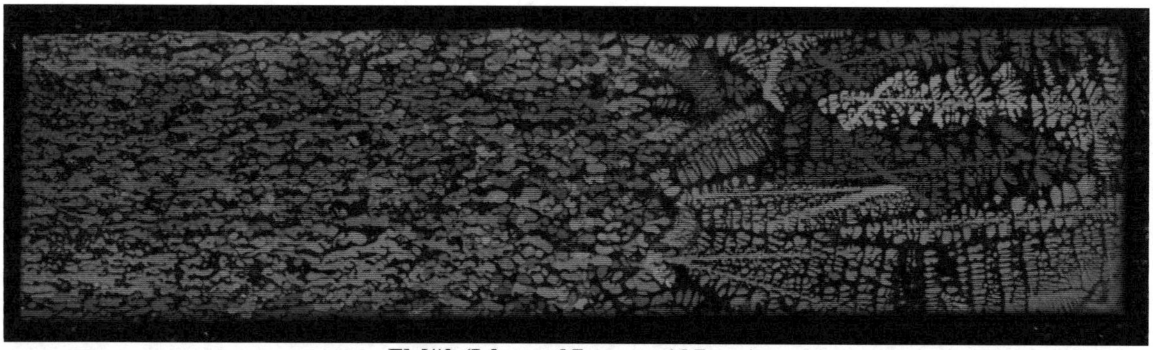

FM#3 (L3: z = 97mm to 127mm)

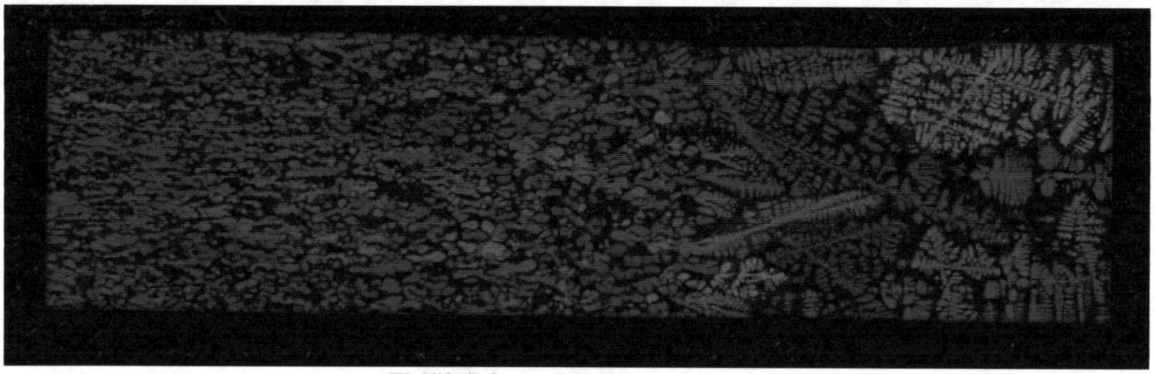

FM#2 (L3: z = 90mm to 120mm)

Figure 5. Microstructure and grain structure in longitudinal cross-sections L3 of FM#2 and FM#3 flight samples showing the mushy zone region (left) and the onset of columnar dendritic growth.

Figure 6 shows the cross-sections L4 of samples FM#2 and FM#3. The increase of the furnace velocity from $v1$=10μm/s to $v2$=200μm/s results in the development of a finer columnar dendritic microstructure in the non-refined alloy (FM#3) whereas the grain refined alloy (FM#2) shows a transition to equiaxed grain growth.

The differences in the microstructures and grain structures are also evident in the transversal cross-sections Q4 given in figure 7. Sample FM#3 shows a regular dendritic microstructure. The uniform color of the electrolytically etched image indicates the columnar growth of nearly one single grain. In the grain refined sample FM#2 however many grains with different orientations were identified.

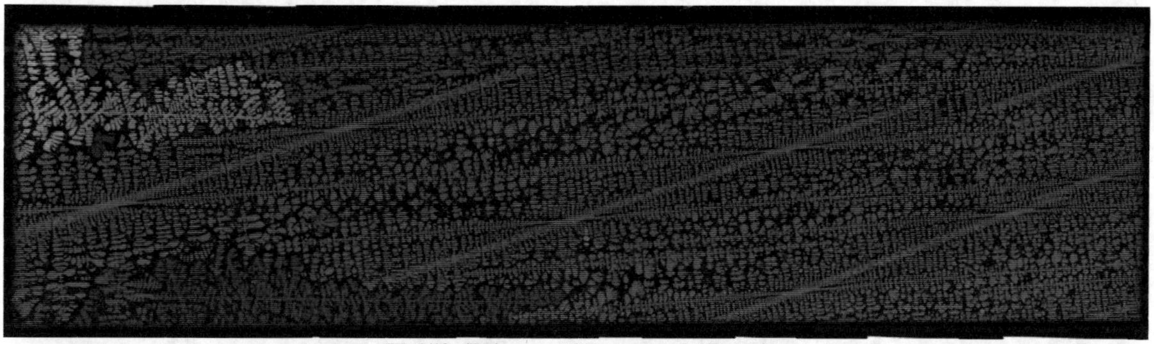

FM#3 (L4: z = 127mm to 157mm)

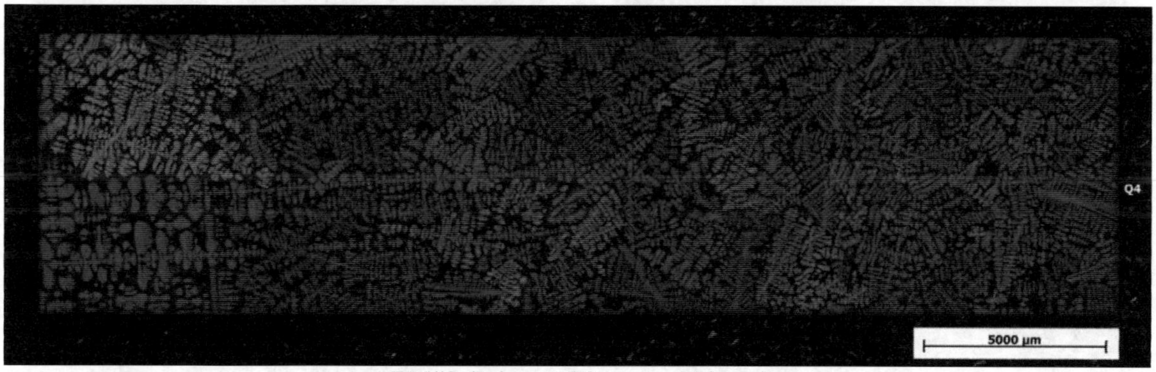

FM#2 (L4: z =120mm to 150mm)

Figure 6. Microstructure and grain structure in longitudinal cross-sections L4 of FM#2 and FM#3 flight samples showing columnar growth (FM#3) and transition to equiaxed grain growth (FM#2)

3.4. Columnar-to-equiaxed transition

For determination of the columnar-to-equiaxed transition in Batch#1 FM#2 flight sample the grain structure was evaluated quantitatively. From the longitudinal cross-sections L3 to L5 the grain sizes were measured using digital image analysis. The average grain size along the sample axis is given in figure 8, averaging was performed over all grains being totally or partly within a sheet of ±2mm around the actual position. The maximum value corresponds to columnar growth of just two large grains in the first solidification phase. Equiaxed growth is characterized by a significant decrease of the average grain size. Thus, based on this criterion and this type of evaluation, the position of CET is determined at z=126±2mm.

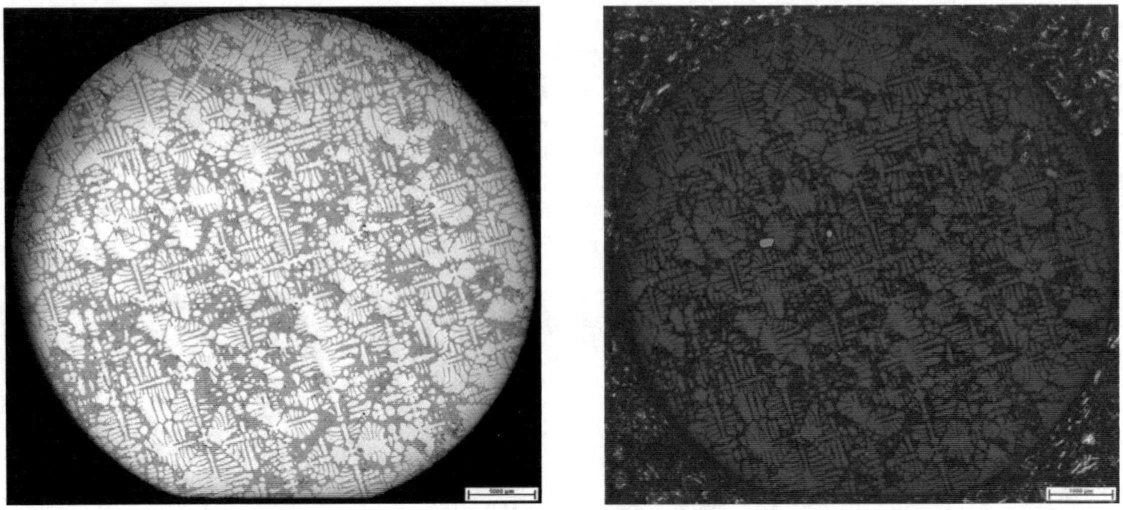

FM#3 (Q4: z =157mm)

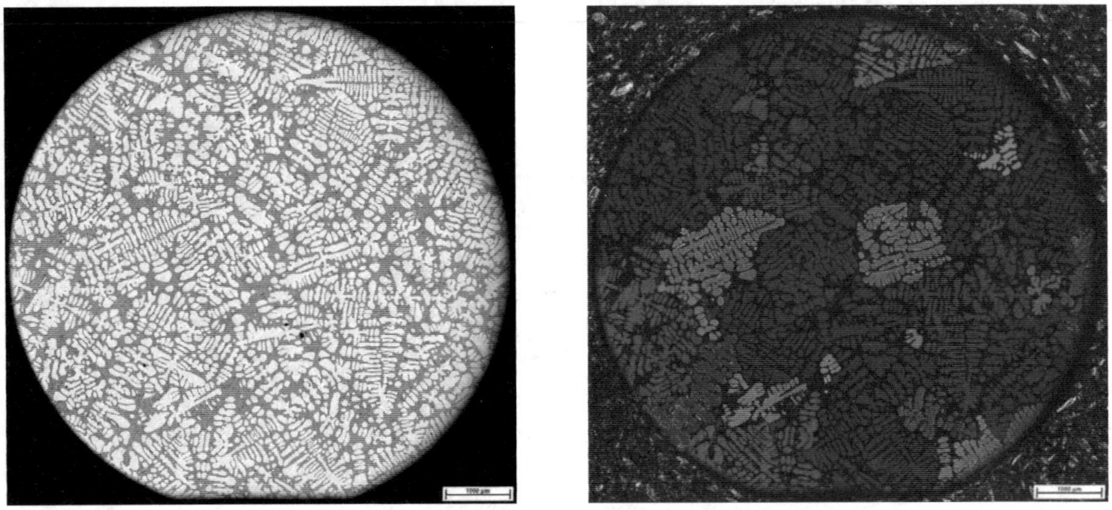

FM#2 (Q4: z =150mm)

Figure 7. Transversal cross-sections Q4 of the Batch#1 flight samples FM#3 (top) and FM#2 showing the dendritic microstructure (left) and the corresponding grain structure (right).

The relevant parameters – critical temperature gradient and critical velocity of the isotherm at CET – have to be determined from the temperature measurements along the sample axis.

To estimate the critical isotherm velocity the local cooling rate is determined. Figure 9 shows the temperature and calculated cooling rate of thermocouple SCAtemp_tc_07. This thermocouple is located at position z=132mm in the alumina crucible. This position is about 6mm ahead of the measured position of CET in the sample and is assumed to be representative for the thermal behaviour in the undercooled region ahead of the columnar front at CET. At about t=29500s the temperature of the thermocouple reaches the liquidus temperature of the alloy (T_L=618°C). Assuming a front undercooling of about 6K this temperature is reached at about t=29600s. The corresponding cooling rate in this time period is rather constant at 0.065K/s.

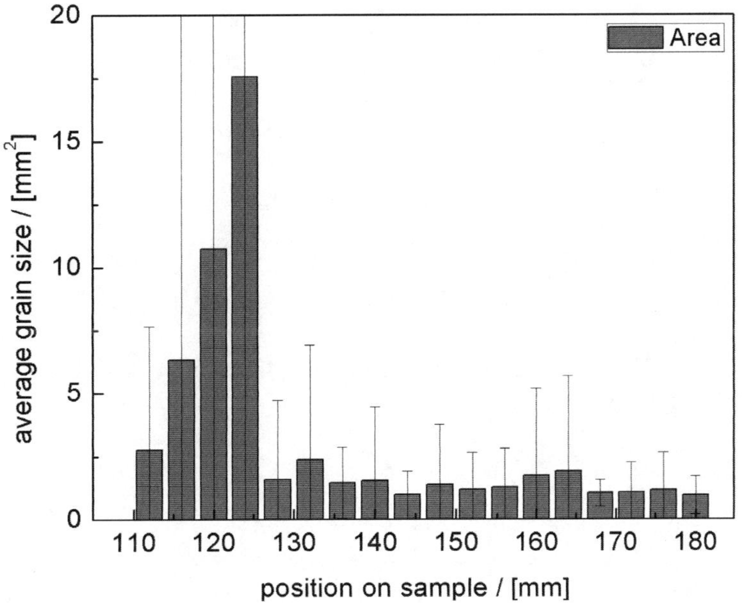

Figure 8: Grain size obtained from longitudinal cross-sections L3 to L5 of Batch#1 flight sample FM#2

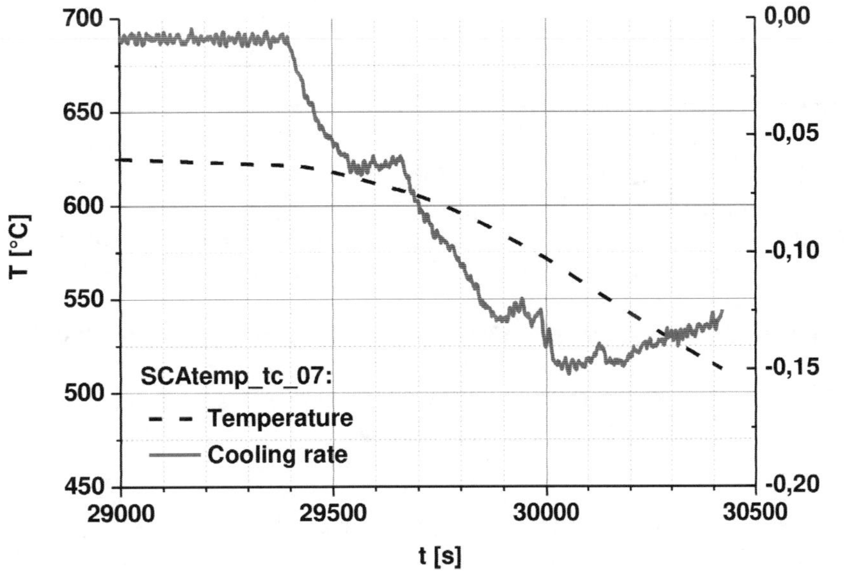

Figure 9: Temperature profile and cooling rate at thermocouple SCAtemp_tc_07 during the solidification phases

Figure 10 shows the temperature gradient at the liquidus-temperature during the solidification phases. The gradient is constant at G=0.95K/mm in the first solidification phase with $v1$=10µm/s.

During the second solidification phase with a furnace movement of $v2$=200µm/s a slight decrease of the temperature gradient is measured. Assuming CET at t=29600s the critical temperature gradient is about G_c=0.75K/mm. As a result the critical isotherm velocity is calculated from the critical values for the gradient and the cooling rate as about v_c=87µm/s.

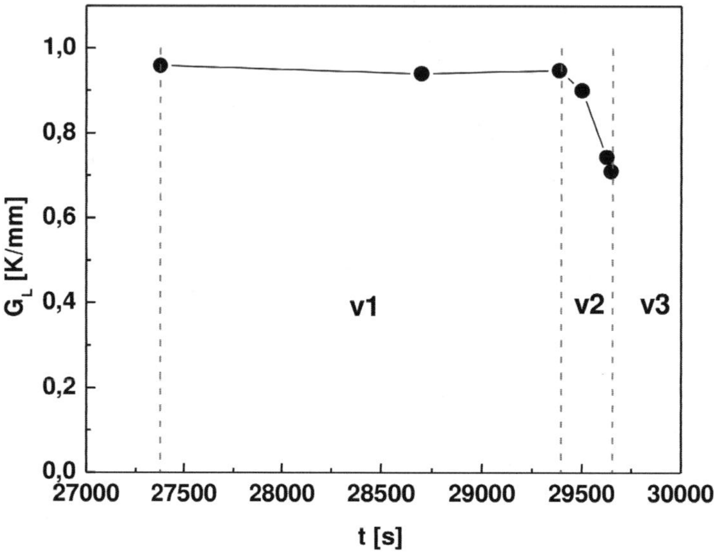

Figure 10: Temperature gradient at the solid-liquid interface for sample FM#2 during the solidification phases ($v1$=10µm/s, $v2$=200µm/s)

3.5 Numerical prediction of CET position

The modelling approach [12] that combines Front Tracking (FT) of columnar growth [5] and an equiaxed volume averaging method [22] was employed to simulate the FM#2 experiment. Among the key advantages related to this approach are: the ability to explicitly track the undercooled columnar front; the capacity to model equiaxed nucleation from grain refiners and subsequent unconstrained growth; high computational efficiency for modeling equiaxed solidification in this range of equiaxed grain sizes; and the macroscopic nature of CET.

The combined model needs to solve the macroscopic energy equation with appropriate latent heat source terms which come from the simulation of columnar and equiaxed growth. The columnar growth initiation from initial solid/mush is considered. The model tracks the evolution of the columnar dendritic front over the relevant period. In this simulation, the columnar front is assumed to commence its growth from the mushy zone (at the liquidus isotherm of the Al-7wt%Si alloy) that exists when cooling (solidification) begins. Therefore, the initial position of the computational markers that represent the columnar front was set at the computed temperature profile's liquidus isotherm. In the undercooled liquid, ahead of the columnar front, the possibility of equiaxed nucleation from grain refiners is considered. A log-normal distribution for grain refiner size [23, 24] is used to compute the number of potential equiaxed nucleation sites at a given time using the computed site temperature. More details of the equiaxed nucleation model employed are available elsewhere [22].

The simulation domain is a 132mm long section from the experimental sample and starts at position z=61.5mm (zero position corresponds to solid sample end). To define the thermal boundary conditions for the simulations, thermal data recorded during the experiment on the ISS was used. As mentioned in the experimental section, thermal data is collected from 12 thermocouple positions at 10 second intervals. Intermediate spatial and temporal input data values used to feed the model simulation were linearly interpolated. A pre-experiment chemical analysis showed that the FM#2 sample contained 30 ppm boron and 225 ppm Ti. We assumed that (i) all boron atoms appeared in TiB_2 grain refiner particles and (ii) a nearly 1% efficiency of such TiB_2 grain refiners. The effective TiB_2 density in the sample was computed as $1.25/mm^3$ using the previous literature [23, 24]. Considering that a slice with thickness equal to half of the average equiaxed dendrite diameter contributes to the 2D view, the input grain refiner density for the 2D model is obtained via multiplying the 3D grain refiner density value by a factor equal to $0.75mm^{-1}$ (the as-cast average equiaxed dendrite diameter is approximately 1.5mm).

Alloy related thermo-physical properties, for both solid and liquid phases, and the dendrite kinetics growth law (tip velocity-undercooling relationship) were taken from ref. [4]. Predicted as-cast columnar and equiaxed volume fractions together with CET position from the model simulation are presented in figure 11. The zoomed-in section of figure 11 can be directly compared with the FM#2 - L4 section of figure 6.

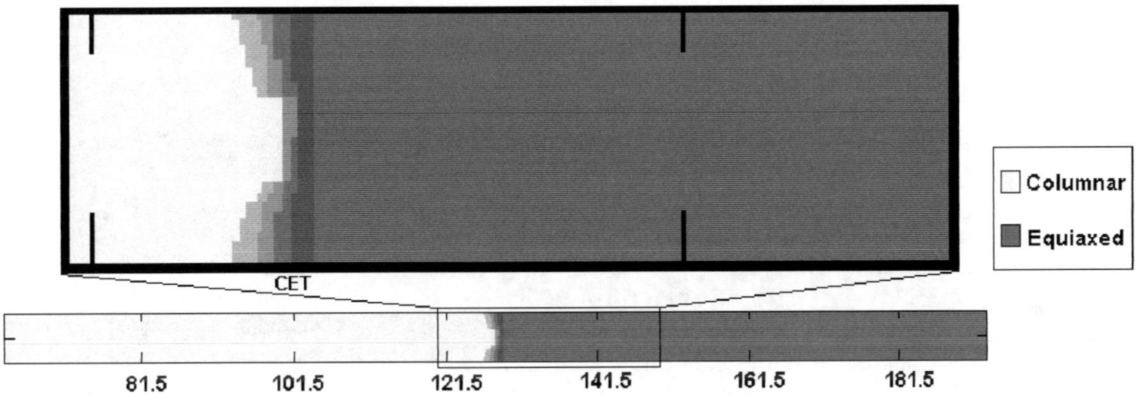

Figure 11. Predicted final as-cast columnar and equiaxed volume fractions in simulation domain. The section between z=120mm to z=150mm is magnified.

As mentioned in chapter 3.4, CET for FM#2 experiment was determined at z=126±2 mm. In the model simulation, CET is predicted at a distance of z=127.5mm. Hence a strong agreement between model simulation and experiment was found. The columnar length is approximately equal to the distance that the furnace moved at the lower velocity ($v1$) and relatively high thermal gradient (see figure 10) suggesting that CET is related to the velocity jump. However, under a similar velocity jump, FM#3 has not shown a CET. Therefore, the influence of grain refiners needs to be evaluated carefully via with further modeling.

4. Summary

This paper reports on first results obtained during metallic alloy solidification experiments onboard the International Space Station in the Materials Science Laboratory using the Low Gradient Furnace module. Within Batch#1 a set of six experiments with Al-7wt%Si alloy were performed successfully to investigate columnar-to-equiaxed (CET) solidification behaviour in microgravity. Mainly the impact of grain refiner particles in the alloy and different cooling rates on CET were investigated. First evaluations show that in the non grain refined samples columnar dendritic growth persists throughout, whereas CET is observed in the grain refined samples. Here, the microgravity environment allows for pure diffusive conditions for heat and mass transfer in the melt and therefore for investigation of CET without buoyancy convection and sedimentation of equiaxed grains in the melt. The determination of critical parameters for CET from analysis of thermal data and the grain structure will be used for calibration and further development of numerical model predicting CET.

At present a second set of microgravity experiments is in preparation. Within the Batch#2a a set of seven experiments with Al-7wt%Si alloy will be processed in the MSL using the Solidification and Quenching Furnace module (SQF). This furnace allows for higher temperature gradients and cooling rates and therefore resulting in an extension of the data basis concerning CET.

Acknowledgments

This work was conducted within the ESA MAP programme 'CETSOL', the ESA PRODEX programme, and also funded by the German Space Agency DLR and the French Space Agency CNES.

References

[1] Burden M H and Hunt J D 1974 *J. Cryst. Growth* **22** 109
[2] Kurz W, Giovanola B and Trivedi R 1986 *Acta Metall.* **34** 823
[3] Hunt J D 1990 *Acta Metall. Mater.* **38** 411
[4] Gandin Ch-A 2000 *Acta Materialia* **48** 2483
[5] Browne D J and Hunt J D 2004 *Numerical Heat Transfer, Part B: Fundamentals* **45** 395
[6] Sturz L and Zimmermann G 2006 *Materials Science Forum* **508** 419
[7] Martorano M A, Beckermann C and Gandin Ch-A 2003 *Metall. mater. trans.* A**34** 1657
[8] Banaszek J, McFadden S, Browne D J, Sturz L and Zimmermann G 2007 *Metallurgical Materials Transact.* **A38** 1476
[9] McFadden S, Browne D J and Gandin Ch-A 2009 *Metallurgical Materials Transact.* **A40** 662
[10] Noeppel A, Budenkova O, Zimmermann G, Sturz L, Mangelinck-Noël N, Jung H, Nguyen-Thi H, Billia B, Gandin Ch-A and Fautrelle Y 2009 *Int. Journal of Cast Metals Research* **22** 34.
[11] Gandin Ch-A, Blaizot J, Mosbah S, Bellet M, Zimmermann G, Sturz L, Browne D J, McFadden S, Jung H, Billia B, Mangelinck N, Nguyen-Thi H, Fautrelle Y and Wang X 2010 *Materials Science Forum* **649** 189
[12] Mirihanage W U, McFadden S and Browne D J 2010 *Materials Science Forum* **649** 355
[13] Dupouy M D and Camel D 2001 *J. Phys. IV France* **11** 119
[14] Gandin Ch-A, Billia B, Zimmermann G, Browne D J and Dupouy M-D 2006 *Materials Science Forum* **508** 393
[15] Sturz L, Zimmermann G, Jung H, Mangelinck-Noël N, Nguyen-Thi H and Billia B 2007 *ESA SP* **647** 367
[16] Sturz L and Zimmermann G 2007 *Int. J. Cast Metals Research* **20** 122
[17] Jung H, Mangelinck-Noel N, Nguyen-Thi H, Billia B, Sturz L and Zimmermann G 2007 *Proc. of the 5th Decennial International Conference on Solidification Processing* **SP07** 186
[18] McFadden S, Browne D J, Sturz L and Zimmermann G 2010 *Materials Science Forum* **649** 361
[19] McFadden S, Sturz L, Jung H, Mangelinck-Noël N, Nguyen-Thi H, Zimmermann G, Billia B, Browne D J, Voss D and Jarvis D 2008 *Journal of the Japan Society of Microgravity Application* **25** 489
[20] Nguyen-Thi H, Drevet B, Debierre J M, Camel D, Dabo Y and Billia B 2003 *J. Crystal Growth* **253** 539
[21] Schaberger E, Grote F and Schievenbusch A 2000 *Prakt. Metallographie* **37** 419
[22] Mirihanage W U and Browne D J 2009 *Computational Materials Science* **46** 777
[23] Greer A L, Bunn A M, Tronche A, Evans P V and Bristow D J 2000 *Acta Materialia* **48** 2823
[24] Quested T E and Greer A L 2005 *Acta Materialia* **53** 4643

Journal of Physics: Conference Series **327** (2011) 012004 doi:10.1088/1742-6596/327/1/012004

Phase-field modeling of the columnar-to-equiaxed transition in neopentylglycol-camphor alloy solidification

Alexandre Viardin, Laszlo Sturz, Gerhard Zimmermann and Markus Apel

ACCESS e.V., Materials and Processes, Intzestr. 5, D-52072 Aachen, Germany

E-mail: a.viardin@access-technology.de

Abstract. We have performed phase field simulations in order to explore the effect of equiaxed grain nucleation undercooling on the columnar to equiaxed transition "CET" in the NPG-DC alloy system. Our phase field model is based on the multiphase-field method. The simulation parameters are adapted to a microgravity experiment performed under purely diffusive growth conditions. The experimental investigation is made during the sounding rocket campaign TEXUS-47.

1. Introduction

Multiscale and multiphase modeling of solidification is a subject of active research in materials science. Understanding the effects of macroscopic process conditions on the microstructure evolution is important to assure e.g. the mechanical properties of the casting. The columnar to equiaxed transition (CET) in dendritic solidification describes a discontinuous change in microstructure and is a common morphological transition in technical castings. The microstructure is called columnar if the dendritic growth tends to one direction, whereas equiaxed grains are oriented in all directions. The CET depends on the thermal conditions defined on the macroscopic scale, the local undercooling near the solid-liquid interface, which is determined by the dendritic morphology on the microscale, the thermophysical properties of the alloy and the nucleation mechanism of equiaxed dendrites. The Neopentylglycol-DCamphor (NPG-DC) alloy is a transparent organic alloy which allows an in-situ observation of the solidification microstructure at temperatures closed to ambient. This alloy is used as analogue system with metallic alloys, which are opaque, and only enable a post-mortem analysis of the structure. We have performed simulations on the CET in the NPG-DC alloy system using a multi-phase field model [1, 2, 3] in 2D and 3D, in the limit of diffusive mass transport. Additional effects of thermosolutal convection and buoyancy or sedimentation can be neglected under low-gravity conditions.

2. Phase Field Model

In the present work, the multi-phase field formulation proposed by Steinbach et al [1] and further developed towards the coupling to thermodynamic databases [3] is employed to describe the morphological evolution during the solid-liquid phase transformation. The phase field method relies on a phase field parameter (ϕ) which varies continuously from liquid "l" (ϕ=0) to solid "s" (ϕ=1) along the solid-liquid interface of width η. The evolution of the phase field parameter in time is given by :

$$\dot{\phi}_s = M^a_{sl}(\vec{n})\left[\sigma^a_{sl}(\vec{n})K_{sl} + w\Delta G_{sl}\right] \tag{1}$$

with :

$$K_{sl} = \phi_s\nabla^2\phi_l - \phi_l\nabla^2\phi_l + \frac{\pi^2}{\eta^2}(\phi_s - \phi_l) \tag{2}$$

$$w = \frac{\pi}{\eta}\sqrt{\phi_s\phi_l} \tag{3}$$

$$\Delta G_{sl} = -f_s(\vec{c}_s) + f_l(\vec{c}_l) + \vec{\mu}(\vec{c}_s - \vec{c}_l) \tag{4}$$

where $M^a_{sl}(\vec{n})$ is the anisotropic interfacial mobility between liquid and solid as a function of the interface orientation (described by the normal $\vec{n}$) and $\sigma^a_{sl}(\vec{n})$ is the effective anisotropic surface energy. The anisotropic functions for the both parameters are written in expansion in terms of normalized spherical harmonics y_l selected under consideration of the specific crystal lattice periodicity [4]:

$$M^a_{sl}(\vec{n}) = M_{sl}\left(1 + \sum_l \epsilon^M_{sl}y_l(\vec{n})\right) \quad \text{and} \quad \sigma^a_{sl}(\vec{n}) = \sigma_{sl}\left(\sum_l \epsilon^\sigma_{sl}y_l(\vec{n})\right)$$

where M_{sl} and σ_{sl} are the isotropic phase field mobility and surface energy, respectively. The thermodynamic driving force ΔG_{sl} depends on the respective bulk free energies $f_s(\vec{c}_s)$ and $f_l(\vec{c}_l)$, determined using a thermodynamic database, $\vec{\mu}$ is the generalized chemical potential introduced to conserve local concentration due to the constraint of quasi-equilibrium. This constraint postulates that all the reduced chemical potentials are the same for the coexisting phases i.e $\vec{\mu} = \vec{\mu}_s(\vec{c}_s(\vec{x})) = \vec{\mu}_l(\vec{c}_l(\vec{x}))$.

The evolution of the concentration field is given by:

$$\dot{\vec{c}} = \nabla(\phi_s D_s \nabla\vec{c}_s) + \nabla(\phi_l D_l \nabla\vec{c}_l), \tag{5}$$

where D_s and D_l are the diffusion coefficients in the solid and liquid phase, respectively. We have converted the liquid concentration (c_l) into a constitutional undercooling in the melt (ΔT_c) using the relation $\Delta T_c = T_l(c_l) - T$ where $T_l(c_l)$ is the liquidus temperature and T is the local temperature defined by :

$$T = T_B - c_r \cdot t + G \cdot z \tag{6}$$

where T_B is the temperature at the bottom of the domain, c_r is the cooling rate, G is the thermal gradient.

3. Microgravity experiment

The experiment TRACE was performed in a transparent sheet-like cell with 1 mm thickness. Details of the experimental set-up and procedure are given in [5, 6]. Solidification was controlled by applying temperatures and cooling rates at the bottom and the top of the solidification volume. While the thermal gradient was almost constant, the cooling-rate was increased by a factor of 10 from -0.2 K/min for columnar growth to -2.0 K/min for the transition to equiaxed dendritic growth. Table 1 summarizes the experimental, thermodynamical and thermophysical parameters, accompanied by some numerical parameters needed for the phase-field simulations. Figure 1 shows an image taken from the experiment at steady-state columnar growth, and figure 2 during the CET, which occurs 375 s after increasing the cooling-rate to -2.0 K/min. Between the two images, the columnar front was grown 1.5-2.0 mm.

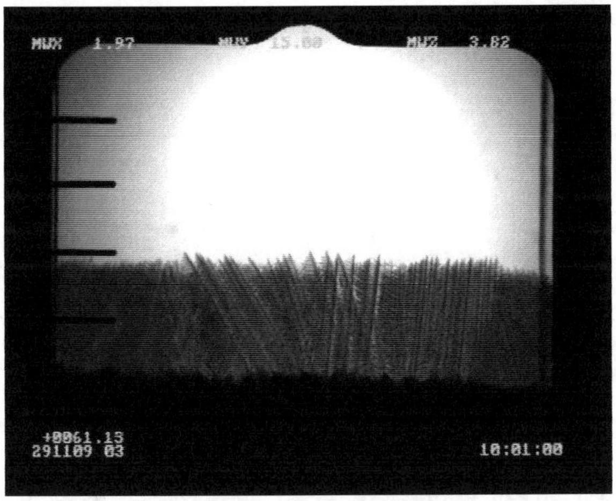

Figure 1. Microstructure observed during steady state columnar growth in the TRACE experiment. The width of the cell is 20 mm

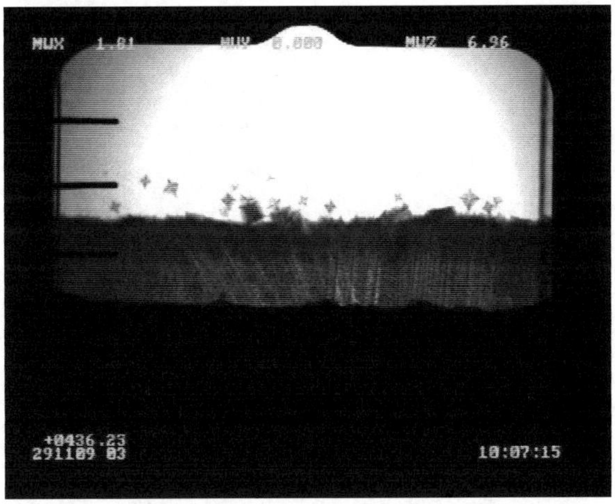

Figure 2. Microstructure observed during CET in the TRACE experiment.

4. Numerical simulation

The physical and numerical parameters that we used in our simulations are listed in table 1. The MICRESS code was used to solve numerically the phase field (Eqn 1 to 4) and diffusion equation (Eqn 5). Two-dimensional and three dimensional phase field simulations were performed. Concerning the two-dimensional calculation domain, a rectangular box of 450 μm height was used. The calculations start with a homogeneous melt at a given temperature gradient (see table 1) where three nuclei of solid primary phase are disposed at some undercooling and equal spacings on the bottom side of the calculation domain. We performed a study of the influence of dendrite spacing by varying the width of the calculation domain between 450 μm and 1350 μm. The grid spacing for all simulations is 0.75 μm. For the three-dimensional calculations, the simulation domain is a box of $150 \times 150 \times 450\,\mu\mathrm{m}^3$. Only one nuclei is disposed in the corner of the calculation domain. In all simulations, the temperature gradient is fixed and constant in one direction. Only the cooling rate is changing from -0.2 K/min to -2.0 K/min to promote CET. A moving calculation domain was used in which the distance from the dendrite tips to the top boundary of the moving calculation domain is fixed. In a first stage of the simulation, a columnar dendritic structure is allowed to evolve in order to reach a steady state growth. The cooling rate is fixed to -0.2 K/min. In a second stage, the cooling rate is increased to -2.0 K/min and nucleation is allowed, using a critical nucleation undercooling and nucleation density for randomly oriented seeds in the melt. We suppose that the formation of equiaxed dendrites is controlled by heterogeneous nucleation during the microgravity experiment. The critical nucleation of undercooling is varied during the different simulations. The nucleation density is supposed to be equal to the equiaxed grain density (N_e, Tab. 1).

Table 1. Physical and numerical parameters used in our simulations

Concentration (c^0_{DC})	37.5 wt.%
Equiaxed grain density [5] (N_e)	$5 \cdot 10^8 m^{-3}$
Temperature gradient [5] (G)	1650 K/m
Interface thickness (η)	3 μm
Diffusion coefficient in liquid [5] (D_l)	$6.1 \cdot 10^{-11}$ m^2/s
Diffusion coefficient in solid (D_s)	0 m^2/s
Surface energy (σ_{sl})	$6 \cdot 10^{-3}$ J/m^2
Phase field mobility (M_{sl})	$1 \cdot 10^{-12}$ m^4/Js
Liquidus slope [5] (m_l)	-1.65 K/wt%
Partition coefficient (k)	$7.75 \cdot 10^{-2}$
Static anisotropy (ϵ^{σ}_{sl})	1.6 %
Kinetic anisotropy (ϵ^{M}_{sl})	15%

5. Results and discussion

5.1. Steady state

Steady state columnar growth in 2D is obtained after 15000 s. In figure 3, we have represented the ΔT_c-field at t=15000s (section 2). We can distinguish three columnar dendrites with secondary branches corresponding to the three initials nuclei growing along the temperature gradient. In figure 4 undercooling profiles are shown along the dendrite tip (solid line 1) and in between the dendrites (dashed line 2). In the insert of figure 4, the solid and dashed horizontal lines correspond to the maximal undercoolings for the profiles along the dendrite tip and between the dendrites respectively. In steady state growth, the maximum of ΔT_c is constant and equal to approximately 18.1 K. As observed by Badillo and Beckermann [9] and Dong and Lee [10], the maximum undercooling is situated on the symmetry line between columnar branches. Several 2D simulations were performed using different experimentally observed dendrites spacings of 150, 300, 450 μm corresponding to 450, 900, 1350 μm calculation domain width. We have observed differences smaller than 1% concerning the value of the final maximum constitutional undercooling in the bulk. In all cases the results shows that the maximum constitutional undercooling is located between the dendrites.

To be able to describe the growth of several dendrites in 3D with a spacing of 300 μm we used symmetry conditions on the calculations domain sides to simulate the presence of the other dendrites. On figure 5, we have plotted the phase field contour (ϕ=0.5) of a dendrite at t=8000 s, for a crystallographic orientation of 45°. For the orientation of 0°, the secondary arms are growing along the domain sides. Like the 2D simulations, we observed a maximal solutal undercooling between the dendrites equal to 7.8 K. Orientation (0° or 45°) plays also a minor role on the value of undercooling. We have noticed that the time to reach constant value of the constitutional undercooling was lower in 3D (around 8000 s) than in 2D (15000 s).

We performed a comparison with the predictions of the KGT-model [7, 8] concerning the value of the constitutional undercooling in 2D and 3D. In the KGT-model, the undercooling is calculated at the dendrite tip, but the differences to the maximum bulk value are small here.

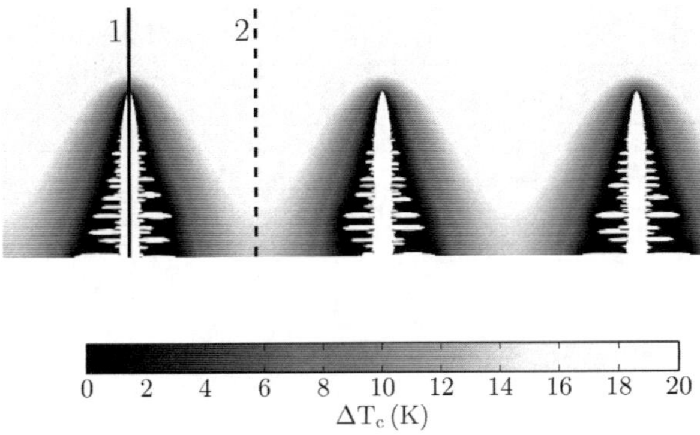

Figure 3. ΔT_c field at t=15000 s for a dendrite spacings of 450 μm

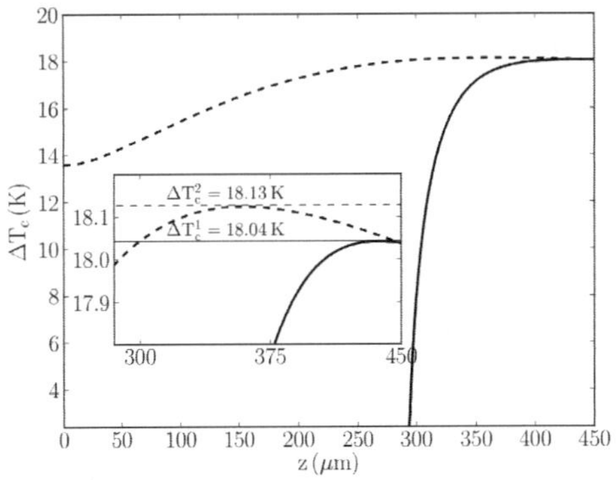

Figure 4. ΔT_c profiles along the axes 1 and 2, in the insert : zoom on the profile

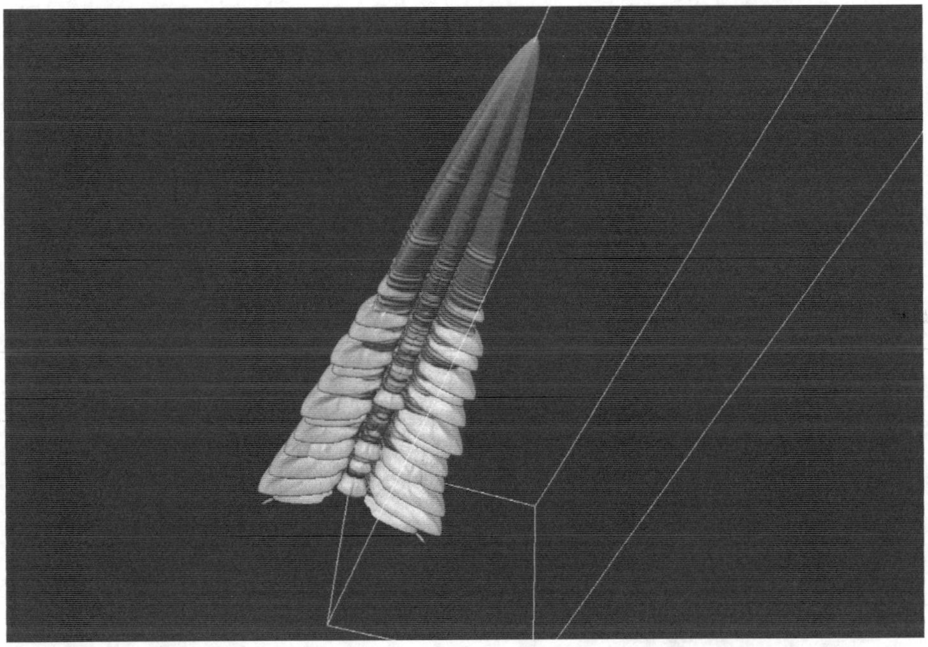

Figure 5. Dendritic structure during steady state growth at t=8000 s

Furthermore a stability constant σ^* appears, which is 0.0253 ($= 1/(4\pi^2)$) in the standard model. It was found [6] that a value of 0.007 should be used to obtain best fit to experiments. Using these two different values of σ^*, we have summarized the values for the undercooling obtained from the KGT model in 2D (ΔT_c^{2D}), in 3D (ΔT_c^{3D}) and phase field in table 2. We have seen that the phase field simulations result in values of constitutional undercooling of 7.8 K in 3D and 18.1 K in 2D giving a ratio (Q) of 2.30. The geometry that we obtained by phase field can not be described correctly by the KGT model, which assumes a parabolic shape. Nevertheless, we can remark that the 2D result of phase field is an average of the KGT values. For 3D simulation, the phase field results are above the highest KGT prediction, but close to the value (σ^*=0.007) used to fit the experimental results [6].

Table 2. Constitutional undercooling using the KGT model [7, 8] and phase field results

	ΔT_c^{2D} (K)	ΔT_c^{3D} (K)	Q=$\Delta T_c^{2D}/\Delta T_c^{3D}$
KGT $\sigma^* = 0.0253$	15.5	4.5	3.4
KGT $\sigma^* = 0.007$	21.3	7.1	3.0
Phase field	18.1	7.8	2.3

5.2. CET

We introduced seeds randomly positioned and oriented in liquid zone of the simulation domain. The initial microstructure was taken from the previous steady state simulation. We have used the same nucleation density N_e (Tab. 1) for all simulations, only the nucleation undercooling (ΔT_n) varies between 8.5 K and 20 K for 3D and 17 K and 27 K for 2D simulations. For 2D simulations, if the nucleation undercooling is lower than 27 K, we observed equiaxed grains at the front of columnar dendrites (figure 6). Above 27 K, nucleation is not possible and the microstructure remains columnar (figure 7). For 3D simulations below 20 K, CET occurs (figure 8). Above 20 K, the dendritic structure remains columnar (figure 9). It is also important to notice that 3D simulation allow to observe the change of microstructure of the columnar zone. We can clearly distinguish that ternary arms are developing orthogonal to secondary arms side because of the change of the cooling rate. The experimental measurements [6] show that CET occurs with the nucleation undercooling of nuclei situated between 8.5 K and 12 K.

Before CET occurs and the new nucleated grains blocked the columnar dendrites, a transient

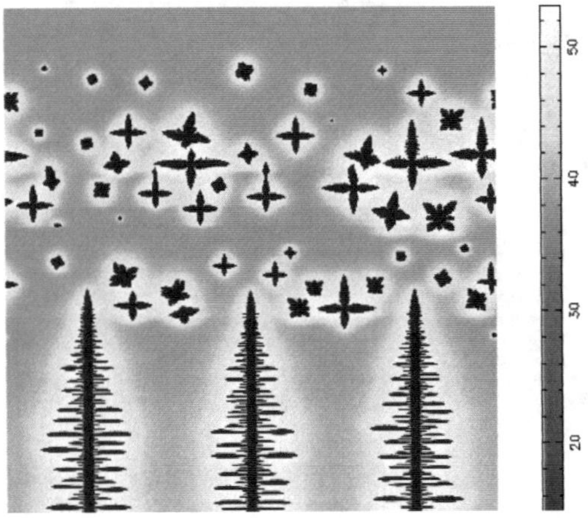

Figure 6. Concentration map (camphor %wt) during CET (ΔT_n=25 K). Width of the domain : 1350 μm

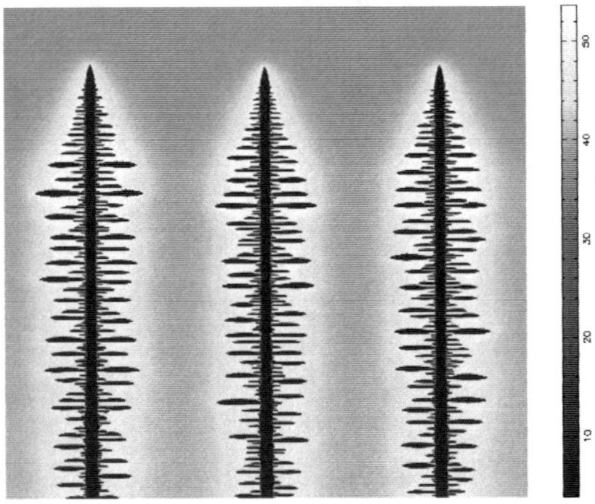

Figure 7. Concentration map (camphor %wt) when no CET occurs (ΔT_n=27 K). Width of the domain : 1350 μm

state is experimentally observed where the columnar dendrites continue to grow on a distance between 1.5 and 2.0 mm. On figure 10, we have plotted the length of this transition zone versus the different nucleation undercooling that we used for the 3D and 2D simulation. The experimental measurement are represented by the gray square. We see that for low value of nucleation undercooling the 3D-model is close to the experimental results. Nevertheless the fact that the 3D phase field results are not localised into the grey domain can be attributed to heat redistribution during the cooling rate change. In experiments, the redistribution of heat

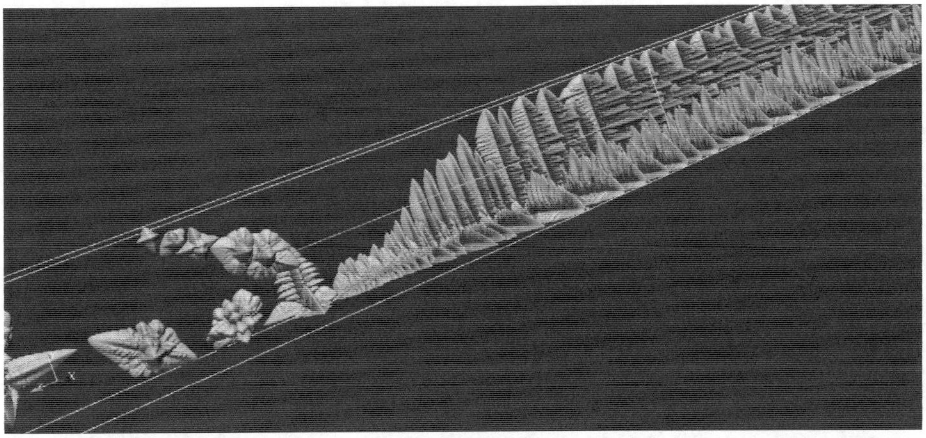

Figure 8. Phase field contour at ϕ=0.5 during CET (ΔT_n=10 K)

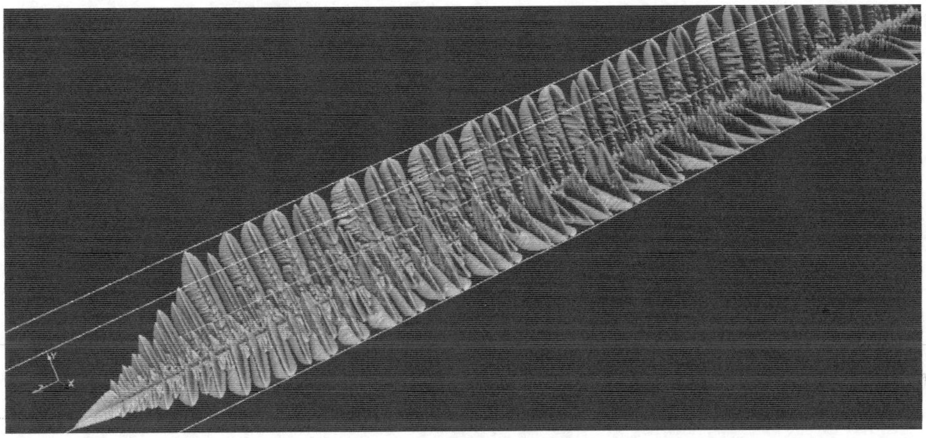

Figure 9. Phase field contour at ϕ=0.5 when no CET occurs (ΔT_n=20 K)

is continuously, not immediately as in our simulation. During this change of cooling rate the dendrite grows further during experiments than in simulations. For the 2D results, we see that the value of the nucleation undercooling are far from experiments.

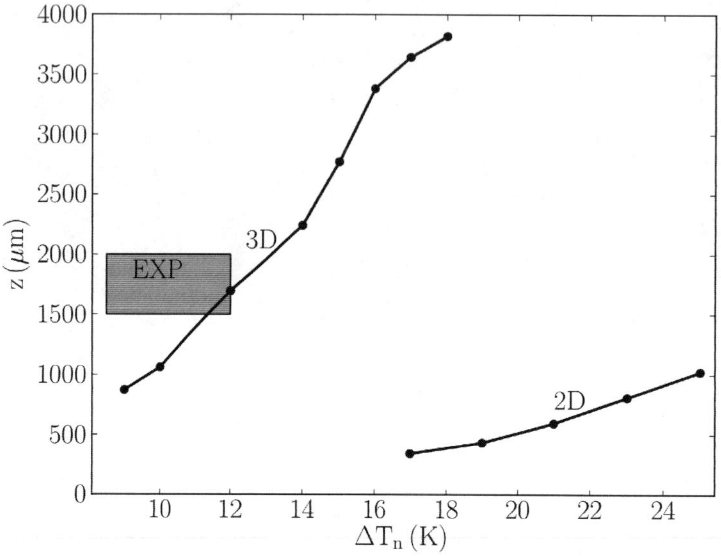

Figure 10. Length of the transition zone versus nucleation undercooling

6. Conclusion

We have shown in this work that phase field model can be very efficient in modeling the columnar to equiaxed transition. The first 3D phase field simulations of CET have been presented. The model describe correctly the CET using the experimental value of nucleation undercoolings and predicts with good accuracy the length of the columnar zone before CET. But the most important is that 2D simulations are not describing experiments as good as 3D simulations. The value of nucleation undercoolings, and the transition length for example, are very far from experiments.

Acknowledgements

This work is financially supported by the German Aerospace Center (DLR) under grant No.50WM 0843, which is gratefully acknowledged.

References

[1] Steinbach I Pezzolla F Nestler B SeesselBerg M Prieler R Schmitz G J and Rezende J L L 1996 *Phys. D* **94** 135
[2] Bottger B Eiken J and Steinbach I 2006 *Acta Mat.* **54** 2697
[3] Eiken J Bottger B and Steinbach I 2006 *Phys. Rev. E* **73** 115701
[4] Eiken J 2009 *A Phase-Field Model for Technical Alloy Solidification* Phd Thesis RWTH Aachen
[5] Sturz L and Zimmermann G 2011 *submitted to Proc. 20th ESA Symp. on European Rocket and Balloon Programmes and Related Research*
[6] Sturz L and Zimmermann G 2011 *submitted to IOP Proc. of 4th Int. Symp. on Physical Sciences in Space*
[7] Lipton J Glicksman M E and Kurz W 1984 *Mat. Sci. and Eng.* **65**
[8] Kurz W Giovanola B and Trivedi R 1986 *Acta Mat.* **34** 823
[9] Badillo A and Beckermann C 2006 *Acta Mat.* **54** 2015
[10] Dong H B and Lee P D 2005 *Acta Mat.* **53** 659

Journal of Physics: Conference Series **327** (2011) 012005 doi:10.1088/1742-6596/327/1/012005

Metastable solidification in undercooled liquid droplets of Fe-Ni based soft-magnetic alloys under terrestrial and microgravity conditions

W Löser[1], T G Woodcock[1], O Shuleshova[1], R Hermann[1], H-G Lindenkreuz[1], B Gehrmann[2], S Schneider[3], and T Volkmann[3]

[1] IFW Dresden, P.O. Box 270116, D-01171 Dresden, Germany
[2] ThyssenKrupp VDM GmbH, Development & Technical Marketing, Plettenbergerstr. 2, D-58791 Werdohl, Germany
[3] Institut für Materialphysik im Weltraum, Deutsches Zentrum für Luft- und Raumfahrt (DLR) D-51170 Köln, Germany

Email: w.loeser@ifw-dresden.de

Abstract. Microstructure evolution and metastable phase formation in undercooled melts of Fe-Ni based soft magnetic alloys were studied by the electromagnetic levitation technique in ground-based experiments and under microgravity during parabolic flight campaigns. Discontinuous changes in the microstructure were observed in samples of a commercial soft magnetic alloy quenched at distinct melt undercooling ΔT. The results were successfully interpreted within the framework of the dendrite breakup model. No metastable phase formation was detected in commercial soft magnetic alloys which possess higher Ni contents. Under microgravity conditions the lifetime of the metastable bcc phase formed for $Fe_{90}Ni_{10}$ beyond a critical melt undercooling increases. The surface tension of the levitating droplets under microgravity was determined from high speed video recordings of the oscillations of the droplets. The value obtained, $\sigma = 1.90 \pm 0.01\ Nm^{-1}$, agrees well with values from the literature.

1. Introduction

Commercial, soft magnetic, Fe-Ni based alloys are normally Ni-rich compositions which also contain various trace elements [1, 2]. The solidification behaviour of these materials at sizeable melt undercooling has rarely been studied. Better understanding of microstructure evolution can contribute to the optimisation of production routes of soft magnetic materials. Dramatic changes in the microstructure of samples processed by electromagnetic levitation (EML) and quenched at given melt undercooling (ΔT) have been reported [3]. Similar discontinuous changes in the grain size at critical ΔT have also been observed in the binary alloys Cu-Ni [4], Ni-Zr [5]. Generally, at low ΔT, a grain refined microstructure is observed, at intermediate ΔT, a coarse grained structure is present and at higher ΔT, a return to the refined grains is seen. The critical values of ΔT which define the boundaries between these transitions are known as ΔT^*_1 and ΔT^*_2. A recent theory, developed by Karma *et al.* [4, 6] is that partial remelting and breakup of the dendrites during recalescence is responsible for the grain refinement. The results from the commercial Fe-Ni alloys were successfully interpreted in terms of

Journal of Physics: Conference Series **327** (2011) 012005 doi:10.1088/1742-6596/327/1/012005

this dendrite breakup model by considering the multicomponent compositions as ternary Fe-Ni-C systems [3].

The effect of convective fluid flow on microstructural transitions resulting in grain refinement via instabilities of growing dendrites has been recently theoretically analyzed [7]. Microgravity experiments enable the study of microstructural transitions in undercooled melts under conditions where this gravity-driven convection is reduced by one order of magnitude. Changes of the critical undercoolings of the microstructural transitions, ΔT^*_1 and ΔT^*_2, are expected during these microgravity experiments [8]. Microgravity can be achieved for a short time (~ 20 s) during parabolic flights. In this time scale, melt undercoolings up to $\Delta T \sim 100$ K are accessible and the results can be used to justify the theoretical predictions concerning the first microstructural transition (*i.e.* up to and including ΔT^*_1). Longer duration of microgravity, available in the TEMPUS electromagnetic levitation facility on the International Space Station (ISS), is required in order to access the whole range of melt undercooling (> 200 K) necessary to explore these microstructure phenomena fully.

In the Fe-Ni binary system, the primary solidification switches from b.c.c. δ-Fe to the f.c.c. γ-(Fe,Ni) phase at a Ni fraction > 5 at.%. Beyond this concentration (up to 17 at.% Ni) the δ-Fe can also solidify as metastable phase if a critical melt undercooling ΔT_c is exceeded [9]. This is concluded from the occurrence of double-recalescence events, which reflect the primary solidification of the metastable phase and its subsequent decomposition into the equilibrium phase after a limited lifetime. These experiments provide an unique opportunity to measure the solidus of metastable phases. Such results from EML experiments with Fe-Co alloys have been used to provide input for computer optimisation of the Fe-Co phase diagram [10].

EML experiments in microgravity conditions, with Fe-Co for example, have indicated important differences in the behaviour of the metastable phase compared to terrestrial experiments. The ΔT_c is reduced and the arrest time before the transformation to the equilibrium phase is prolonged because of the reduced fluid flow in the levitated droplet under microgravity [11, 7].

The aim of the present work was to compare the solidification processes in undercooled melts of binary Fe-Ni alloys and commercial Fe-Ni based soft magnetic materials in EML experiments under terrestrial and microgravity conditions. The focus is on possible metastable phase formation and the variation of microstructure as function of undercooling prior to solidification.

2. Experimental methods

Ingots of $Fe_{90}Ni_{10}$ were prepared from Fe and Ni elements with 4N purity in a Hukin-type cold-crucible induction melting facility under Ar atmosphere. A commercial Fe-Ni based soft magnetic material (P36) was provided by ThyssenKrupp VDM GmbH. The nominal composition of the alloy is given elsewhere [2, 3]. The materials were sectioned into pieces of approximately 1 g mass in order to give an optimum droplet size for EML of ~ 6 mm. In ground-based EML experiments the apparatus detailed by Löser et al. [12] was used. For details of the experimental parameters used for the commercial material, refer to [3].

Undercooling experiments under microgravity conditions were carried out during parabolic flight campaigns with the TEMPUS EML facility. Temperature-time profiles were recorded from a high-speed photo diode. A high speed video camera was installed in the facility and videos showing the behaviour of the droplets during the levitation experiments were recorded at 10 kHz. The videos allowed the morphology of the solidification front to be observed in detail and the surface tension of the liquid droplets to be measured. The undercooling phase was used for the surface tension measurements. Measurements were recorded over 200 K with a mean temperature of 1583 K. An edge detection algorithm was used to extract the dimensions of the droplet along two different axes as a function of time. Oscillations of the droplet dimensions due to its liquid nature occurred and the frequency of these is proportional to the surface tension. The measurements were therefore transformed into the Fourier domain and the sum or difference of the resulting peaks was extracted for the surface tension calculation. This procedure was carried out using the software TeVi 3.41 of S.E.A. Datentechnik GmbH and DLR-MUSC, Linder Höhe, 51147 Köln, Germany.

Journal of Physics: Conference Series **327** (2011) 012005 doi:10.1088/1742-6596/327/1/012005

3. Growth kinetics and grain coarsening of commercial Fe-Ni based soft-magnetic alloys

Terrestrial EML experiments were used to achieve melt undercoolings of $\Delta T > 320$ K in a commercially available Fe-Ni based soft magnetic alloy (P36). Temperature profiles, recorded from the photo diode, showing the recalescence during solidification of this alloy at various values of ΔT are shown in figure 1. The temperature profiles become steeper with increasing ΔT, indicating higher solidification speed. A simple linear approximation was used to estimate the solidification front speed from the temperature profiles. The values shown in figure 2 are a lower limit because the surface of the spherical droplet is projected onto the diode and thus the signal measured does not take into account the curvature of the droplet. The estimated front speed increases monotonically with ΔT and at the highest values of ΔT reaches 30 ms^{-1}, meaning that solidification is complete within ~ 0.1 milliseconds.

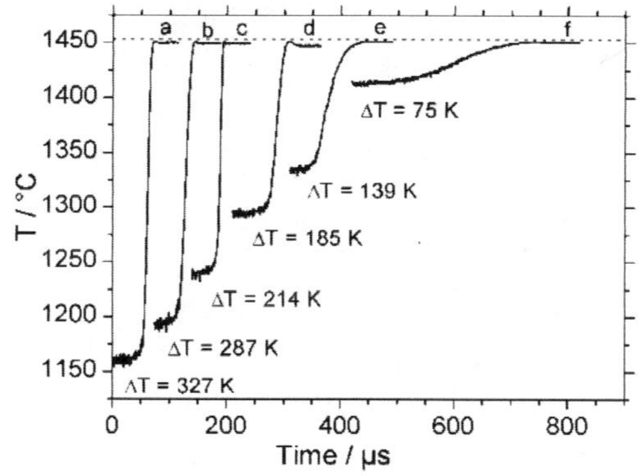

Figure 1. Recalescence curves for the Fe-Ni based alloy P36 (34 at% Ni). The solidus temperature is shown as a dotted line.

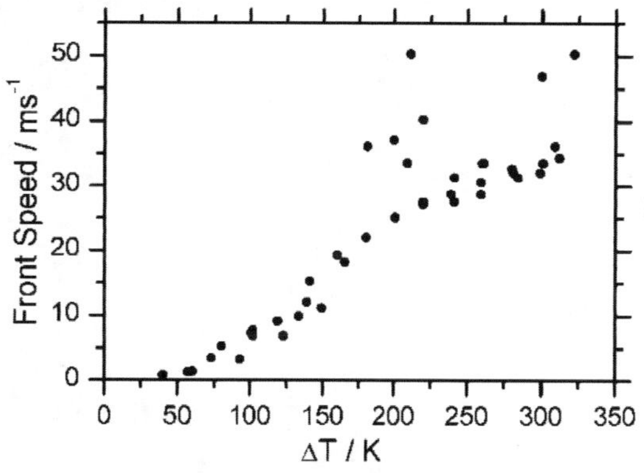

Figure 2. Front speed vs. melt undercooling ΔT evaluated from the recalescence curves of the alloy P36.

Undercooling of $\Delta T = 327$ K represents highly non-equilibrium solidification conditions but in spite of this, there is no double-recalescence behaviour, indicating that no metastable phase forms. The reason for this is the high Ni content (34 at.%) of this alloy. Solidification of the f.c.c. γ-(Fe,Ni) phase occurs because nucleation of the metastable b.c.c. δ-Fe phase is thermodynamically unfavourable, even at such extreme melt undercoolings.

The figure 3 shows optical micrographs of the commercial Fe-Ni based alloy quenched at various values of ΔT. The range of ΔT is larger than has been shown in previous microstructural investigations [3]. Microstructural transitions are apparent which can be interpreted using the Karma model [6]. At $\Delta T = 0$ K, a microstructure of refined equiaxed grains superimposed on a substructure of spherical elements was observed. Here $\Delta T < \Delta T_1^*$ and therefore $t_{bu} < t_{pl}$ (t_{bu} = dendrite breakup time; t_{pl} = plateau time during which the sample temperature is above the solidus due to recalescence). Thus remelting and breakup of the primary crystals occurred leading to the refined microstructure. At ΔT = 81 and 135 K, the microstructure consists of very large colonies of columnar dendrites. No equiaxed grains were observed. In this range, $\Delta T_1^* < \Delta T < \Delta T_2^*$ and $t_{bu} > t_{pl}$ therefore no breakup occurred. At $\Delta T = 180$ K, $\Delta T > \Delta T_2^*$ and $t_{bu} < t_{pl}$ the microstructure was again composed of refined equiaxed grains superimposed on a substructure of spherical elements. Increasing the melt undercooling to $\Delta T = 282$ K had no significant effect on the microstructure, as the inequality $t_{bu} < t_{pl}$ was still valid. It is interesting to note that the solidification front speed increased from ~ 20 ms^{-1} at $\Delta T = 180$ K to ~ 30 ms^{-1} at ΔT = 282 K (*cf.* figure 3).

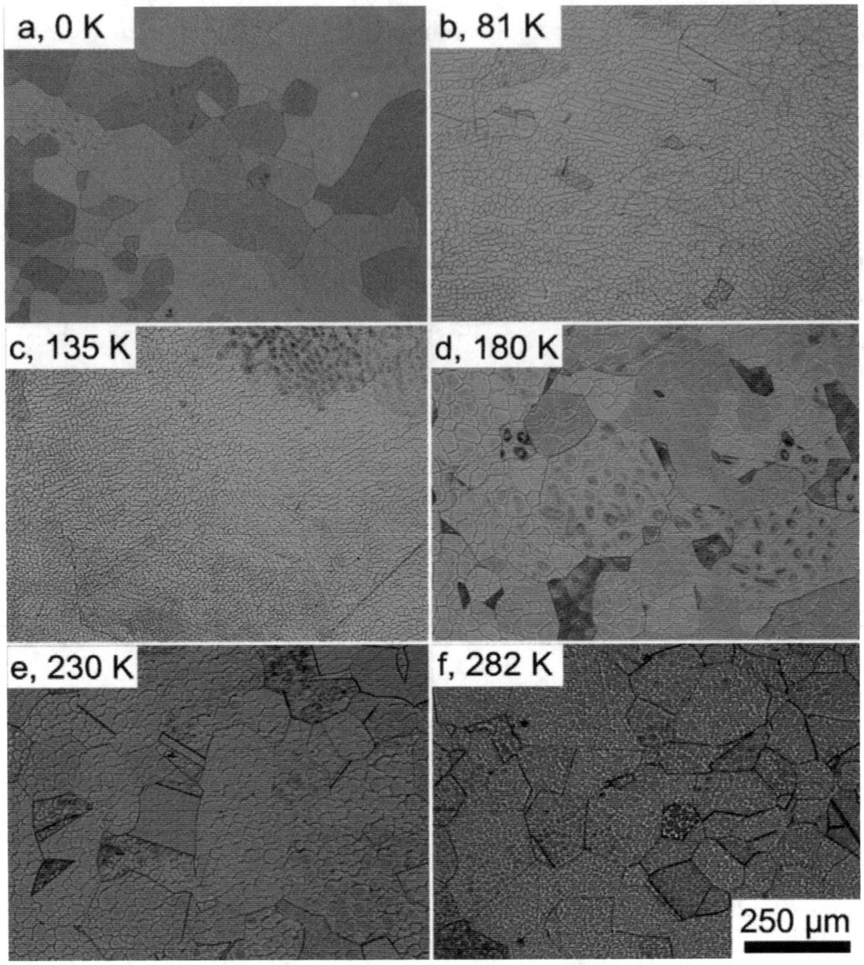

Figure 3. Optical micrographs showing the microstructure of droplets of commercial Fe-Ni soft magnetic alloy P36 processed by EML and quenched on a copper substrate at melt undercoolings in the range ΔT = 0 to 282 K.

4. Non-equilibrium solidification in $Fe_{90}Ni_{10}$ undercooled melts in microgravity

Some initial results showing the principal effects of microgravity on metastable phase formation of the $Fe_{90}Ni_{10}$ binary alloy in EML experiments are visualized in figure 4. In this example, the critical undercooling for the δ-Fe metastable phase ($\Delta T_c \approx 70$ K) appears to be comparable for both terrestrial and microgravity conditions, which is inferred from the transition from single to double recalescence.

There is some indication that the lifetime of the metastable phase increases under microgravity conditions in parabolic flights (longer plateau of the recalescence curve), as has been previously reported for other alloy systems [11, 13]. This was attributed to the reduced fluid flow in the droplet under microgravity [7]. Despite the increased lifetime the metastable phase cannot be retained to room temperature in the Fe-Ni system and there is no independent proof of its presence, as in the Fe-Ni-Cr ternary alloy [14].

The addition of a high speed video camera to the TEMPUS EML facility has allowed visualisation of the solidification front under microgravity conditions. A series of images from the high speed camera shows the growth of the solid (bright) into the undercooled melt (darker) for ΔT = 29 K during a parabolic flight (figure 5). In the images, the liquid droplet is partially obscured by the sample cup.

Solidification is complete 2.5 ms after triggering and an approximate front velocity of 0.6 ms^{-1} may be inferred from this. The striking feature of these images is that the solidification front is not smooth but contains highly angular sections, the shape of which is likely to be related to the crystal symmetry.

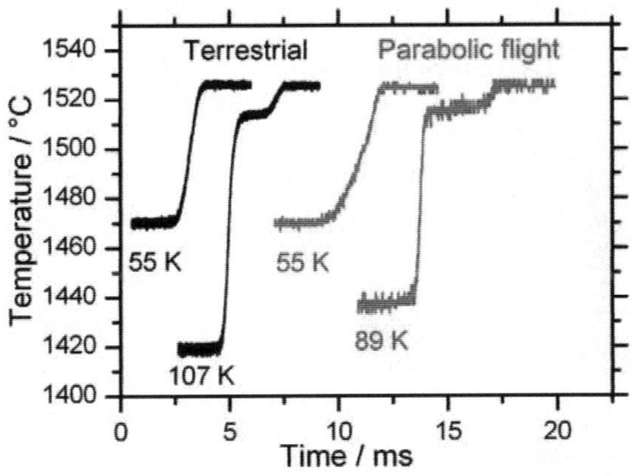

Figure 4. Temperature-time plots of a high-speed photo diode from terrestrial (left) and parabolic flight (right) levitation experiments with $Fe_{90}Ni_{10}$ droplets showing double-recalescence characteristics with an intermediate metastable phase beyond a critical melt undercooling.

Full exploitation of the spatially resolved data from the camera requires more exact modelling of the solidification front geometry within the droplet. The angular solidification front may also be one contributing factor to the sizeable scatter between the recalescence curves of different experimental trials recorded by the photodiode, especially in the derived plateau times.

A further important application of the high speed camera was to measure the surface tension of the levitating droplet. During several parabolas, the surface oscillations of the droplets during the undercooling phase were recorded. The surface tension was extracted from these oscillations by the method described above. An average value of $\sigma = 1.90 \pm 0.01$ Nm^{-1} at a mean temperature 1583°C was obtained. The surface tension of pure Ni is $\sigma = 1.73$ Nm^{-1} and that of pure Fe is $\sigma = 1.90$ Nm^{-1} at 1583°C. The value obtained for the $Fe_{90}Ni_{10}$ alloy fits well to a linear interpolation between the values for the pure elements and is in good agreement with literature data [14].

5. Summary and conclusions

The microstructure of an Fe-Ni based commercial soft magnetic alloy has been studied over a much larger range of undercooling than has been previously shown [3]. Transitions in the microstructure could be interpreted using the dendrite breakup model of Karma [6]. No further microstructural transitions were found at $\Delta T > \Delta T^*_2$. The role of convective fluid flow in producing these microstructural transitions may be studied by conducting experiments under microgravity conditions where the convection is reduced by an order of magnitude. Initial results from parabolic flight experiments showed the formation of a metastable phase during solidification of $Fe_{90}Ni_{10}$ at $\Delta T > \Delta T_c$. Some indication of differences in the lifetime of the metastable phase between terrestrial and microgravity experiments could be inferred. This may be related to the reduced fluid flow under microgravity. High speed video recordings during the parabolic flights showed that the solidification front is not planar but often highly angular. Accurate determination of the solidification front speed requires modelling of the front morphology. The surface tension of $Fe_{90}Ni_{10}$ alloy droplets was determined during the parabolic flights. The value of $\sigma = 1.90 \pm 0.01$ Nm^{-1} obtained showed good agreement with literature values.

Acknowledgement
This work is financially supported by ESA under AO99-101 ESTEC Contract No. 14980/01/NL/SH.

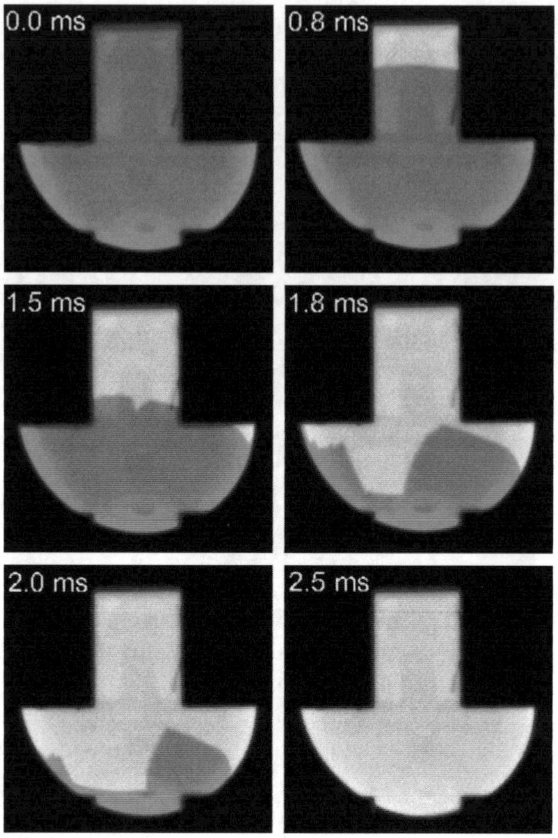

Figure 5. Picture series from the high speed camera of parabolic flight experiments for $Fe_{90}Ni_{10}$. From the moving front of the solid (bright) into the 29 K undercooled melt (dark) the solidification velocity was evaluated.

References

[1] Gehrmann B 2005 *J Magn. Magn. Mater.* **290–291** 1419
[2] Material Data Sheets No. 7001, No. 9002, and No. 9004, ThyssenKrupp VDM GmbH
[3] Woodcock TG, Shuleshova O, Gehrmann B and Löser W 2008 *Metall. & Mater. Trans.* A **39A** 2906
[4] Schwarz M, Karma A, Eckler K and Herlach DM 1994 *Phys. Rev. Lett.* **73** 1380
[5] Galenko PK, Phanikumar G, Funke O et al. 2007 *Mater. Sci. Eng.* A **449–451** 649
[6] Karma A 1998 *Int. J. Non-Equil.Process.* **11** 201
[7] Hanlon AB, Matson DM and Hyers RW 2006 *Phil. Mag. Lett.* **86** 165
[8] Herlach DM, Galenko PK 2007 *Mater. Sci. Engin. A* **449–451** 34
[9] Eckler K, Gärtner F, Assadi H, Norman AF, Greer AL and Herlach DM 1997 *Mater. Sci. Eng.* A **226** 410
[10] Woodcock TG, Hermann R and Löser W 2007 *CALPHAD* **31** 256
[11] Hermann R, Löser W, Lindenkreuz HG, Diefenbach A, Zahnow W, Dreier W, Volkmann T, Herlach DM 2004 *Mater. Sci. Eng.* A **375-377** 507
[12] Löser W, Hermann R, Leonhardt M, Stephan D and Bormann R 1997 *Mater. Sci Eng.* A **224** 53
[13] Löser W, Hermann R, Woodcock TG, Fransaer J, Krivilov M, Gránásy L, Pusztai T, Toth GI, Herlach DM, Holland-Moritz D, Kolbe M and Volkmann T 2008 *J. of The Jap. Soc. of Microgravity Appl.* **25** 495
[14] Volkmann T, Löser W and Herlach DM, 1997 *Metall. & Mater. Trans.* A **28A** 461
[15] Brillo J and Egry I 2005 *J. Mater. Sci.* **40** 2213

Investigation of growth of single crystal SRR99 superalloy under microgravity using 50-meter-high drop tube

SB Feng, XH Luo[1]

Division of Materials for Special Environments, Institute of Metal Research, Chinese Academy of Sciences, Shenyang 110016, China

E-mail: xhluo@imr.ac.cn

Abstract. The SRR99 Ni-base single crystal superalloy samples were partially remelted and then solidified during free falling in a 50-meter-high drop tube. For comparison, samples solidified statically, i.e., in unit gravity, were also conducted under equivalent condition. Under the both conditions there were areas where columnar dendrites epitaxially grew from the seed crystals. The epitaxial growth of columnar dendrites in the samples solidified during free fall was subject to the influence of microgravity. It was found that the primary dendrite arm spacing of the sample solidified under microgravity was apparently larger than that of the sample solidified under gravity. The secondary arm lengths were also longer in the microgravity sample. Besides, the distributions of heavy elements such as W and Ta along the core of a primary dendrite arm were found a little bit different between the samples solidified under the two conditions. Based on the above experimental results, the effects of gravity on solidification microstructure and solute distribution were analyzed.

1. Introduction

Nickel-base single crystal (SX) superalloys have been widely applied in advanced gas turbines and jet engines for their superior strength and fracture resistance at elevated temperatures [1, 2]. In order to obtain a combination of good mechanical strength and superior high temperature capabilities, the SX superalloys have been heavily alloyed with elements of low densities (Al, Ti) as well as elements of high densities (Ta, W, Re). In particular, the addition of these refractory elements has contributed greatly to the improvement in high-temperature performance of SX superalloys [3-5].

As the rising levels of refractory alloying additions, grain-defect formation during directional solidification of SX superalloys has become an increasingly important problem. Although the formation of these defects is resultant of a number of factors, convective instability induced by gravity during directional solidification is regarded as the precursor event [6]. Consequently, understanding the effects of convection is of great importance on the solidification microstructures of SX superalloys. However, due to the complicated nature of convection and the sophisticated interaction between fluid flow and microstructure evolution, the influence of gravity-induced convection has not been recognized thoroughly.

Owing to the absence of gravity and consequently the related phenomena, the utilization of microgravity environment is of great importance to study the effect of gravity on solidification [7, 8].

[1] To whom any correspondence should be addressed.

Among all the microgravity facilities, long drop tubes are powerful instruments in earthbound laboratories [9-11]. Therefore, in the present paper, the effect of gravity on the microstructures of a SX superalloy was investigated by using a long drop tube.

2. Experimental

The SRR99 single crystal superalloy used for the solidification experiment was prepared by the Bridgman technique. The nominal composition of this alloy is listed in table 1. The microgravity experiments were carried out in a 50 m long drop tube. Cylindrical samples of 6 mm in diameter and 20 mm in height were enveloped in high purity corundum crucibles. When doing experiment, the drop tube was evacuated to 6×10^{-3} Pa. Then, the top part of the sample was partially heated by a spiral induction heater installed at the top of the drop tube. The temperature of the top surface of the sample was monitored using an infrared pyrometer. The schematic diagram of the heating apparatus in the drop tube is shown in figure 1. When the upper part of the sample was remelted and the lower part was still solid, the power of the spiral induction coil was turned off. At the same time, the sample together with the crucible was released and dropped down, subject to a microgravity condition of about 10^{-6} g_0 (g_0 is normal gravity) for about 3.24 s. At last, the sample was quenched into silicone oil when landing at the bottom of the drop tube. In the present work, the heating time was about 13 s. Such kind of the sample was referred to as μg sample hereinafter. For comparison, the same remelting and solidification experiments were carried out, except for the 50 m free falling. The sample obtained in this case was referred to as 1g sample hereinafter. The re-solidified parts of all the samples were cut both longitudinally and transversely. The microstructures on the sections were observed under optical microscopy after mounting, grinding, polishing and etching the samples. Primary dendrite arm spacings were measured on the transverse sections at almost the same distance from the initial remelting interface. Besides, the lengths of the secondary dendrite arms were also measured on the cross sections, since on the cross sections the four secondary dendrite arms growing from one primary dendrite arm can be observed on the same plane. Electron probe microanalysis (EPMA) was applied to measure the element distributions in the primary dendrite arms on the longitudinal sections of the samples. For both types of samples, the primary dendrite arms measured were chosen to be almost at the same place on longitudinal sections. Samples were lightly etched and point analyzes were performed about every 200 μm along the dendrite core from the remelting interface to its tip. During X-ray sampling, a spot size of 1 μm was used with an accelerating voltage of 15 kV and a beam current of 100 nA. All the elements were analyzed at each point utilizing pure elemental or mineral standards. Then, the raw data were processed by a standard ZAF correction program to give normalized weight percentages of each element.

Table 1. Nominal composition of SRR99 superalloy

Element	Al	Ti	Cr	Ta	W	Co	Ni
Weight (%)	5.47	2.14	8.39	2.92	9.47	5.01	Bal

3. Results and Discussion

The remelted length is about 6 mm for both μg sample and 1g sample. Figure 2 illustrates the thermal histories of the two types of the samples. It shows that both samples were heated rapidly at the initial 6 s, which could ensure the samples to be remelted partially. Then, the temperatures increased very slowly for about 7 s, which should correspond to the melting processes of the samples. During the entire period of induction heating for 13 s the temperature characteristics of both μg sample and 1g sample were almost the same, indicating that the initial solidification conditions for both samples are almost the same. For 1g sample, as soon as the power was turned off, the temperature of the top surface of the sample kept almost constant for less than 4 s. Then, the temperature abruptly increased about 47 ℃, and soon after it decreased monotonously. Since the solid part of the sample could act as a seed crystal, the liquid part of the sample began to solidify immediately once the heating process stopped. In other words, the samples started to solidify at the moment of the 13th s. During the solidification, although the temperature at the solid-liquid interface was cooling down, because the top end of the sample was the final site to solidify, it showed there a plateau on the temperature profile of 1g sample due to the balance between the heat loss and the release of latent heat. The rapid increase of temperature after the plateau may be due to the recalescence phenomenon at the last stage of solidification. Therefore, it is reasonable to deduce from the temperature profile that 1g sample should completely solidify in 4 s. It should be pointed out that the temperature of μg sample could not be monitored after it was released to fall down in the drop tube. However, owing to the similar cooling conditions for both types of the samples, μg sample should experience a nearly identical thermal history and solidify almost totally during the 3.24 s of free fall.

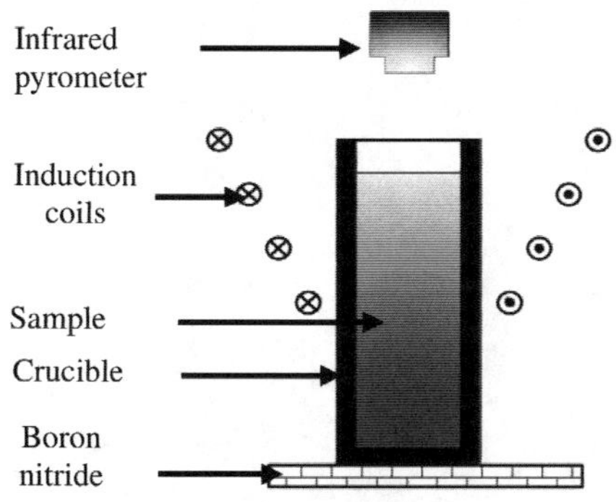

Figure 1 Apparatus for electromagnetic heating part.

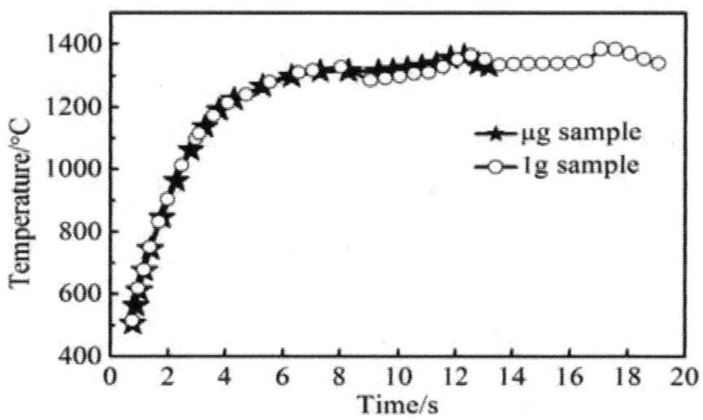

Figure 2 Temperature profiles of SRR99 samples solidified under microgravity and normal gravity.

The optical micrographs of figure 3 show the microstructures on longitudinal sections of μg sample and 1g sample, respectively. The coarse dendrites at the lower part are the microstructures of the seed crystals. At the remelting interface, straight columnar dendrites grew epitaxially from the original microstructures. Besides, at the surfaces of the cylindrical samples, some wild dendrites nucleated from the crucible surfaces and grew toward the inside of the samples, interfering and finally blocking the epitaxial growth of the dendrites from the seeds. Comparatively, the maximum length of the epitaxially grown columnar dendrites in 1g sample is about 100 μm longer than that in μg sample. It seems that the macrostructures of both samples show little difference. Since the columnar dendrites in the vicinity of the initial growth interface grew under an approximate directional solidification condition, and particularly, this part of microstructure formed undoubtedly under microgravity for μg sample, in the following the detailed features of these columnar dendrites are investigated.

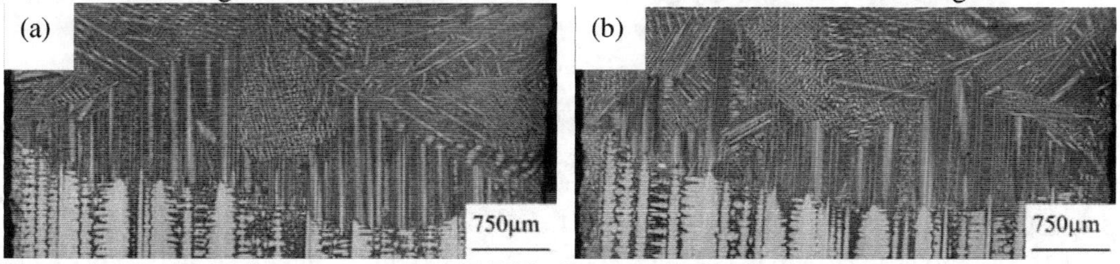

Figure 3 Typical microstructures of SRR99 alloy on longitudinal sections solidified under microgravity (a) and normal gravity (b).

The transverse microstructures of μg sample and 1g sample are shown in figure 4. It is noticeable that the dendritic trunks in μg sample are coarser than those in 1g sample. For instance, secondary dendrite arms are longer in μg sample. In addition, it is clearly seen that extensive tertiary dendrites grew from secondary branches in μg sample, whereas there are only a few tertiary dendrites observed in 1g sample, as the white arrows in figure 4 indicate. Figure 5 shows the distribution of secondary

Journal of Physics: Conference Series **327** (2011) 012006 doi:10.1088/1742-6596/327/1/012006

dendrite arm lengths. For both types of the samples, the minimum secondary dendrite arm lengths are the same. However, the maximum length in μg sample is 20 μm longer than 1g sample. Moreover, the distribution of secondary dendrite arm lengths is more scattered in μg sample. The discrepancy between the maximum and the minimum frequencies is about 20%. On the contrary, in 1g sample the distribution is more centralized, mainly concentrating in the range from 20 μm to 80 μm. The difference between the maximum and the minimum frequencies is about 37%. Another interesting phenomenon is that a great number of fine primary dendrites exist in 1g sample, while few such dendrites can be found in μg sample, as the black arrow in figure 4 (b) indicates. Figure 6 demonstrates the distribution of primary dendrite arm spacing corresponding to figure 4, the dashed lines are Gaussian fits through the data points. Although the range of primary arm spacing varying from 40 μm to 140 μm is unaltered for both types of samples, the distribution of primary arm spacing is more symmetrical in μg sample. Besides, at the peak of the Gaussian curve, the primary arm spacing is 71.5 and 58.7 μm for μg sample and 1g sample, respectively. The average primary arm spacing of μg sample and 1g sample is about 85 and 71 μm, respectively.

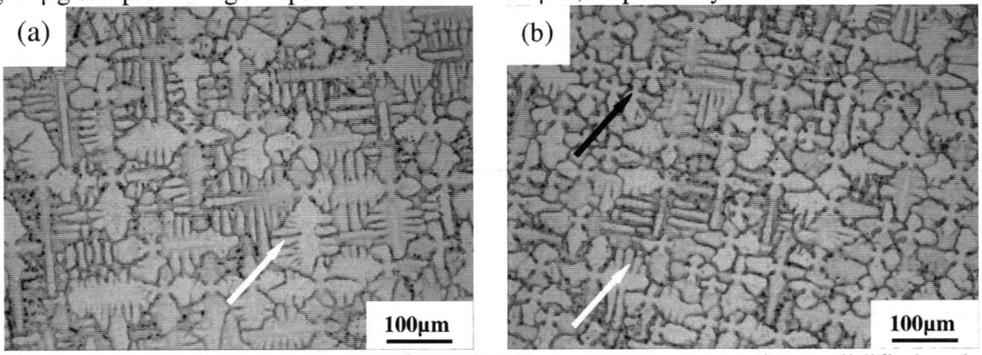

Figure 4 Typical microstructures of SRR99 alloy on transverse sections solidified under microgravity (a) and normal gravity (b).

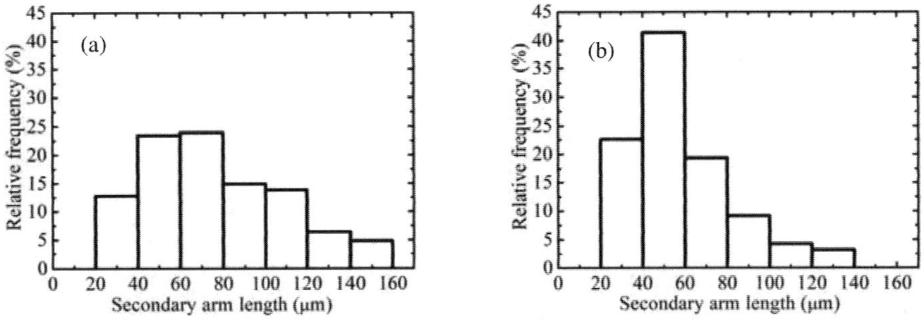

Figure 5 Distribution of secondary arm lengths on cross section of SRR99 alloys solidified under microgravity (a) and normal gravity (b).

Figure 7 illustrates the solute concentrations measured along the core of a typical epitaxially grown dendrite. It is clearly that, except for W and Ta, there are very little variations in solute compositions along the dendrite core with the distance up to 1.4 mm from the remelting interface for both types of samples. As alloy solidification proceeds, the composition of W in μg sample increases gradually from about 9% to 10%. In 1g sample however, the composition of W fluctuates at the first stage of solidification and then keeps almost constant with the increase of solidification distance. For the element Ta, it is nearly unchanged within the first half length of the columnar dendrite in μg sample, and then it oscillates with the increase of distance. While in 1g sample, the concentration of Ta decreases with the increase of the distance from the remelting interface.

The primary dendrite arm spacing in μg sample is much larger than that in 1g sample, suggesting that gravity has a significant effect on the growth of dendritic microstructures. Under microgravity conditions, the gravity driven convection becomes weak, which will decrease the mass and heat transport in the liquid phase [12]. Therefore, the growth rate of columnar dendrites will be decelerated. This is in accordance with the shorter columnar dendrites observed on the longitudinal section for μg sample. According to the model of dendrite growth, lower growth rate leads to larger primary dendrite arms spacing [13]. Resultantly, due to the decrease of growth rate, the primary dendrite spacing in μg sample solidified under microgravity increases. The increase of primary dendrite spacing provides larger spaces for the growth of secondary dendrite arms, which thus grow longer along the direction perpendicular to the primary dendrite trunks. Inversely, due to the enhanced fluid flow and consequently rapid growth rate, the primary dendrite spacing decreases in 1g sample. The decrease of primary dendrite arm spacing is characterized by the growth of tertiary dendrite arms, which are promoted by the gravity driven convection and become new primary dendrite arms. Due to the limitation of existing primary dendrites in mushy zone, the tertiary dendrite arms turn out to become fine primary dendrite arms.

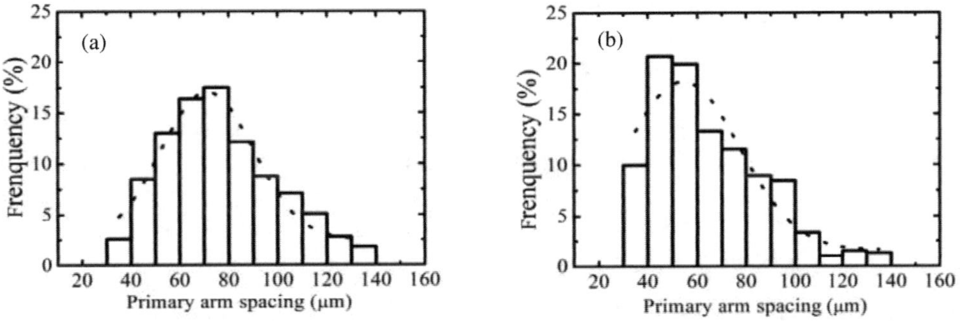

Figure 6 Distribution of primary arm lengths on transverse section of SRR99 alloys solidified under microgravity (a) and normal gravity (b).

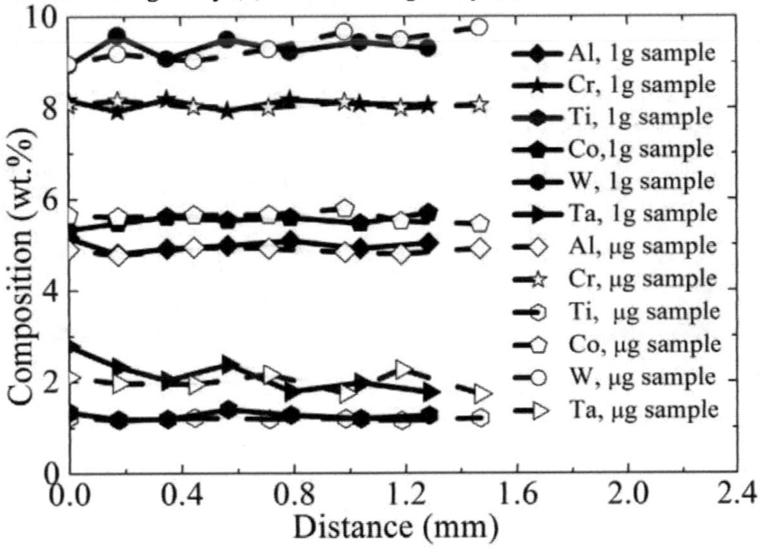

Figure 7 Solute element concentrations plotted *vs.* distance from the remelting interface along the core of a primary dendrite arm for SRR99 alloy solidified under microgravity and normal gravity.

The solute distributions along the core of a primary dendrite arm show little difference for the elements of Al, Cr, Co and Ti. However, for the heavy elements such as W and Ta, the concentrations present different features as the growth of primary dendrites proceeds. According to the study by

D'Souza et.al [1], the intrinsic measurement uncertainties of the elemental concentrations are beyond an order of magnitude smaller than the scatters. Therefore, the variations in elemental compositions of W and Ta in figure 7 should be caused by the factor of gravity, i.e., gravity may affect the distribution of higher density elements during alloy solidification, even though it shows little effects on the relative lighter elements. Since the solute concentrations along the core of the primary dendrite depend on the solute boundaries at the dendrite tip, gravity may affect the solute distribution of W and Ta by means of changing the solute boundaries. The different behaviors of W and Ta elements in both types of samples should attribute to the different partition coefficients, which are larger than 1 for W and smaller than 1 for Ta. In order to confirm these assumptions, further investigations should be done.

4. Conclusions

Solidification of SRR99 Ni-base single crystal alloy under microgravity and normal gravity conditions was carried out by using a 50 m long drop tube. Samples solidified under microgravity are characterized by coarser dendritic trunks, a little bit longer secondary dendrite arms and scattered distribution of secondary dendrite arm lengths, while samples solidified in normal gravity show shorter secondary dendrite arms and more concentrated arm length distribution. There are a great number of fine dendrites in gravity samples, whereas microgravity samples have mainly coarse dendrites. The average primary arm spacing is about 85 μm for microgravity sample, while it is about 71 μm for gravity sample. Gravity has insignificant influence on the solute distribution for the elements of Al, Cr, Co and Ti in the present work. However, for the heavy elements such as W and Ta, the solute distributions during solidification seem to be sensitive to gravity.

References

[1] D'Souza N, Beanland R, Hayward C and Dong H B 2011 *Acta Mater.* **59** 1003
[2] Volek A and Zhou Y Z 2007 *Scripta Mater.* **56** 537
[3] Seidman D N and Amouyal Y 2011 *Acta Mater.* **59** 3321
[4] Heckl A, Rettig R and Singer R F 2010 *Metall. Mater. Trans. A* **41A** 202
[5] Yang X L, Ness D, Lee P D and D'Souza N 2005 *Mater. Sci. Eng.A* **413** 571
[6] Pollock T M and Murphy W H 1996 *Metall. Mater. Trans.A* **27** 1081
[7] Jarvis D J and Voss D 2005 *Mater. Sci. Eng.A* **413** 583
[8] Cahoon J R, Chaturvedi M C and Tandon K N 1998 *Metall. Mater. Trans. A* **29** 1101
[9] Luo X H and Chen L 2008 *Sci. China Ser. E* **51** 1370
[10] Li G, Gao Y P, Chi Z H, Liu J, Zhang T and Liu R P 2008 *Phil. Mag. Lett.* **88** 543
[11] Chen L and Luo X H 2007 *Acta Metall. Sin.* **43** 769
[12] Hong Y, Tandon K N and Cahoon J R 1997 *Metall. Mater. Trans. A* **28** 1245
[13] Kurz W and Fisher D J 1981 *Acta Metall.* **29** 11

Containerless Solidification and Characterization of Industrial Alloys (NEQUISOL)

A Ilbagi[1], H Henein[1,*], J Chen[1], D M Herlach [2,3], R Lengsdorf[3], Ch-A Gandin[4], D Tourret[2,4], A Garcia-Escorial[5]

*Email: hhenein@ualberta.ca

[1]University of Alberta, Edmonton, Canada
[2]Institut für Materialphysik im Weltraum, Deutsches Zentrum für Luft- und Raumfahrt (DLR) Köln, Germany
[3]Institut für Festkörperphysik, Ruhr-Universität Bochum, Germany
[4]Ecole des Mines de Paris, CEMEF UMR CNRS 7635, Sophia Antipolis, France
[5]Dept. Metalurgia Fisica, CENIM-CSIC, Madrid, Spain

Abstract. Containerless solidification using electromagnetic levitator (EML), gas atomization and an instrumented drop tube, known as impulse atomization is investigated for Al-Fe and Al-Ni alloys. The effects of primary phase and eutectic undercooling on the microstructure of Al-Fe alloys are investigated using the impulse drop tube and parabolic flight. The TEM characterization on the eutectic microstructure of impulse-atomized Al-Fe powders with two compositions showed that the metastable Al_mFe formed in these alloys. Also, the growth undercooling that the dendritic front experiences during the solidification of the droplet resulted in variation of dendrite growth direction from <100> to <111>. For Al-4 at%Fe, it was found that in reduced-gravity and in the impulse-atomized droplets the primary intermetallic forms with a flower-like morphology, whereas in the terrestrial EML sample it has a needle like morphology. For Al-Ni, the effect of primary phase undercooling on dendrite growth velocity under terrestrial and reduced-gravity condition is discussed. It is shown that under terrestrial conditions, in the Ni-rich alloys with increasing undercooling the growth velocity increases, whereas in the Al-rich alloys the growth velocity decreases. However, the Al-rich alloy that was studied in reduced-gravity showed similar behavior to that of Ni-rich alloys. Furthermore, the effect of cooling rate on the phase fractions and metastable phase formation of impulse-atomized Al-Ni alloys is compared with EML. A microsegregation model for the solidification of Al-Ni alloys is applied to impulse atomized powders. The model accounts for the occurrence of several phase transformations, including one or several peritectic reactions and one eutectic reaction.

1. Introduction

The properties of materials used in day-to-day activities are governed by their microstructures that form during solidification. Rapid Solidification Processes (RSP) can result in final products with microstructural and constitutional features, which are very different from those obtained under near equilibrium conditions. The unique microstructural features arising from RSP may include formation of microcrystalline structures and decreased grain size or dendrite spacing. The constitutional features may include reduced microsegregation, increased solute solubility and metastable phase formation [1]. All these features have generated strong interest in RSP in recent years.

Understanding the microstructural evolution during solidification is the key to manufacturing products with desired properties. The microstructure evolution during RSP depends on the rate of motion of the solid-liquid interface, which in turn depends on the level of undercooling prior to solidification and the mode of heat extraction. Containerless solidification refers to a class of solidification processes in which large undercoolings are achieved by the complete avoidance of heterogeneous nucleation on container walls. During solidification the heat of crystallization at the solid-liquid interface is released that is conducted into the undercooled melt, i.e. the undercooled liquid acts as a heat sink itself. This gives rise to a negative temperature gradient in front of the solid/liquid interface, which destabilizes a planar solidification front and eventually leads to growth of thermal dendrites. In case of alloys, the concentration gradient enhances the destabilizing effect of the thermal gradient and causes the formation of chemical dendrites.

Presently, computer assisted modeling and simulation of casting processes is used to optimize industrial production routes. In order to develop physically relevant models of solidification, experimental validation is critical. This work focuses on the containerless solidification of industrially important materials such as Al- and Ni- based alloys in terrestrial and reduced gravity conditions [2].

Electromagnetic Levitator (EML), drop tube impulse system, and gas atomization are the containerless processes that have been utilized to investigate the effect of undercooling on eutectic and peritectic reactions. Also, the effect of fluid flow in the dendrite growth dynamics and in the microstructure evolution of solidification will be studied by using EML for comparative investigations in terrestrial and reduced gravity conditions.

EML allows for direct observation of undercooling by contactless pyrometric temperature measurements and measurements of the dendrite growth velocity by using a high speed video camera to observe directly the rapid propagation of the solidification front [3]. It can achieve very high undercoolings in relatively large droplets (~6mm) that solidify in a containerless environment. In this technique the electromagnetic levitation is applied on Earth and in reduced gravity to investigate the fluid flow effects on the dendrite growth dynamics and the microstructure evolution during solidification by post mortem analysis of the as solidified drops. However, EML is limited in the variation of cooling rate that is an important processing parameter to design metallic materials. Complementary experiments of containerless processing of small droplets in drop tube and atomization techniques allow for a change of cooling rate over several orders of magnitude since the cooling rate scales with the droplet size.

Melt disintegration into powder by use of either a mechanical driver (Impulse Atomization (IA)) or high pressure gas (gas atomization) is used to produce powders often of metastable microstructures in masses of several kg. These techniques are very suitable for investigating the formation of various microstructures as a function of droplet size and cooling rate. Using these techniques cooling rates as high as 10^6 K/s is achievable [4].

2. Experiments

2.1. Impulse Atomization

This technique has been developed at the Advanced Materials and Processing Laboratory (AMPL) at the University of Alberta, Canada. Basically, it takes advantage of the fact that a liquid jet emanating from a capillary can be destabilized when disturbed by a particular shock wave with a given frequency. The schematic view of the apparatus is shown in Figure 1. The impulse unit includes a plunger made of refractory material attached to a pulsator which impulses the molten metal through a series of capillaries on the bottom of the crucible. The impulses generate discontinuous streams of molten metal, which in turn break up into droplets. The droplets fall in a static gas atmosphere and solidify as spherical particles [5].

Rapidly solidified particles of two different alloy systems (Al-0.3, 0.93 and 4 at% Fe and Al-20 and 31.5 at%Ni) were produced using this technique. Each alloy of Al-Fe was atomized after achieving 200 K superheat above its respective liquidus temperature. For Al-Ni alloys the superheat was 100 K above the respective liquidus temperature. The falling particles lost heat during falling to the stagnant helium or nitrogen gas and were collected at the bottom of 4-meter-high drop tube. The collected particles were sieved into different size ranges according to Metal Powder Industries Federation Standard 05 [6].

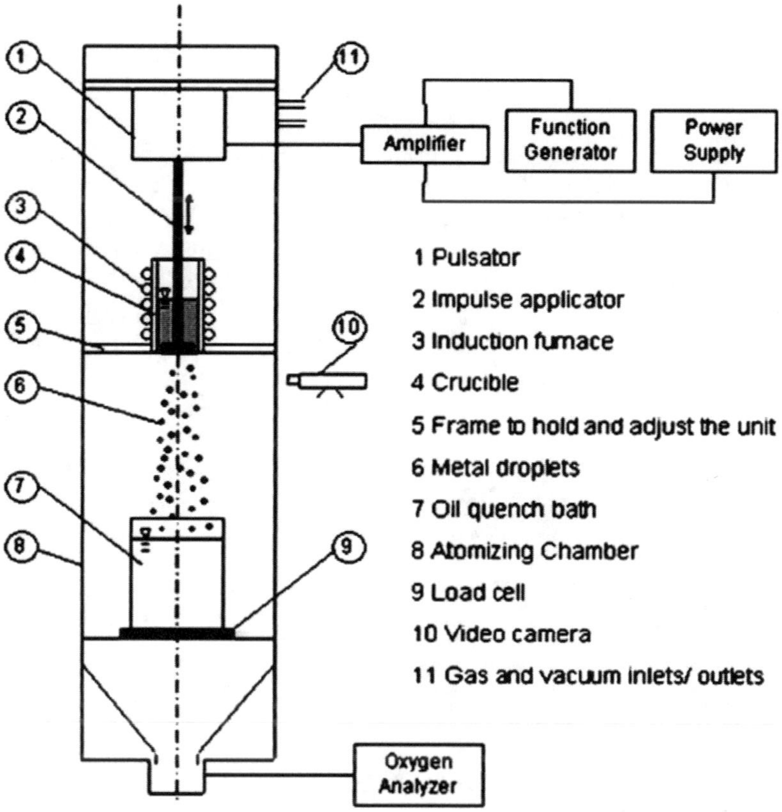

Figure 1. Schematic diagram of the Impulse Atomization tube, also known as drop tube impulse apparatus.

The microstructures of the produced particles were studied using Transmission Electron Microscope (TEM), Scanning Electron Microscope (SEM), Electron Backscatter Diffraction (EBSD), Neutron Diffraction (ND) and X-ray Diffraction (XRD). Rietveld analysis was also used to find the phase fractions in Al-Ni alloys.

The microstructure of impulse-atomized particles of Al-4 at%Fe was compared with the microstructure of an electromagnetically levitated sample solidified in terrestrial and reduced gravity condition. For the experiments under reduced gravity condition, they were conducted by using the TEMPUS facility integrated in an Airbus A300 aircraft for parabolic flight campaign (PFC) at Bordeaux, France.

2.2. Gas atomization

Inert gas atomization is an industrial containerless process where a liquid stream of a molten alloy is disintegrated by high velocity gas. This process yields powder particles in the range 10- 100 μm. The size distribution of the atomized powder particles depends on the processing parameters as gas pressure, metal/gas flow rate, melt superheat, diameter and geometry of the nozzle; on the heat capacity and thermal conductivity of the gas used; and on the composition, surface tension, density and viscosity of the material to be atomized.

In this work $Al_{68.5}Ni_{31.5}$ were successfully gas atomized in CENIM- CSIC, and the solidification path of the gas atomized powder particles was studied by means of XRD and SEM.

2.3. Electromagnetic levitation

To date EML [1, 2], schematized in Figure 2, is the most suitable technique to levitate stable metallic samples and to achieve direct measurements in deeply undercooled states. An alternating current (AC) flowing through a copper coil produces an electro-magnetic (EM) field. A metallic sample placed within the EM field is levitated by the Lorentz force, FL, which compensates the gravitational force Fg. At the same time, the sample is heated by the Eddy current, Ic. In order to limit the contamination of the melt and the evaporation of light elements the set-up is placed within an ultra-high vacuum chamber, which is evacuated to a pressure of 10^{-8} mbar before backfilling with high purity inert gas (helium 6.0). The droplet is melted and stabilized at several hundred Kelvin above its melting temperature for 10–20 s, which also removes possible residual oxides. Then the current in the coil is rapidly decreased to the minimum necessary for levitation, while cooling gas (He) jets are opened. The current in the coil and the gas jets then remain the same up to complete solidification of the droplet, while the operating pressure is maintained using a vacuum pump.

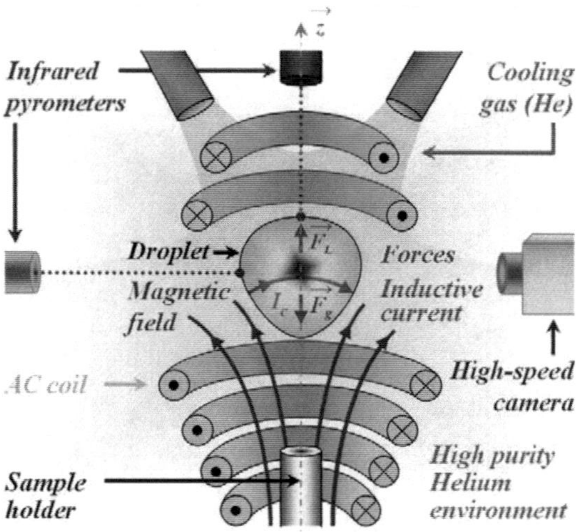

Figure 2. Schematics of the apparatus for electro-magnetic levitation (EML) [7].

During the experiment the temperature of the droplet is measured using a digital infrared pyrometer (Impac IGA10-LO). Solidification involves multiple phase transformations and the identity of the phases encountered by the pyrometer changes. Since the equilibrium temperatures of reactions can be determined from thermodynamic calculations, the emissivity of a sample can be calibrated from its melting curve. Upon heating, the melting of a solid is accompanied by a change in slope of the temperature–time profile due to the heat of melting needed for the transformation from solid to liquid and the change in heat capacities of the existing phases. In addition, levitation experiments enable also measurements of the growth velocity during rapid crystallization of undercooled drops. High-speed camera technique is utilized to measure the velocity V with which the solidification front propagates through the melt. Figure 3 shows a sequence of recorded pictures at various time intervals for an $Al_{50}Ni_{50}$ sample undercooled by 259 K. The high-speed camera is applied both in electromagnetic levitation experiments on Earth and in reduced gravity using the TEMPUS (Tiegelfreies Elektro-Magnetisches Prozessieren Unter Schwerelosigkeit) facility for containerless processing during parabolic flight campaigns. The dendrite growth velocities of congruently melting $Al_{50}Ni_{50}$ were measured both in 1 g and in reduced gravity.

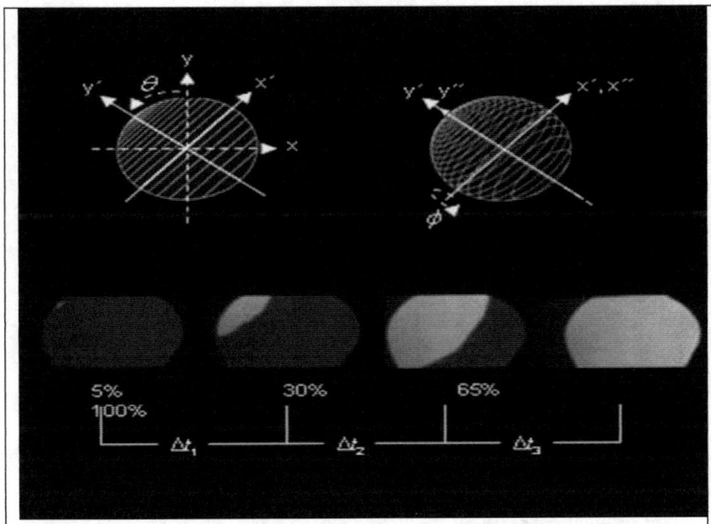

Figure 3. Sequence of high-speed video data taken during the recalescence of a $Ni_{50}Al_{50}$ specimen at an undercooling level of $\Delta T = 259$ K. The present frequency of 30 kHz provides a resolution of 256-128 pixels per frame. The growth velocity can be estimated by dividing each of the real solidification lengths – as obtained from the designated percentages– by the respective solidification time.

3. Metallurgical characterization of samples

3.1. Al-0.3 at%Fe and Al-0.93 at%Fe

The equilibrium solid solubility and diffusivity of iron in the aluminum matrix is very low. The maximum equilibrium solubility of iron in the Al matrix is 0.02 at% at 928 K, the equilibrium eutectic temperature. The eutectic reaction yielding α-Al + Al_3Fe occurs at an iron content of 0.88 at%. The Al_3Fe phase, θ, is later suggested to be $Al_{13}Fe_4$ [8]. This phase is the only equilibrium phase when Fe content is less than 24.4 at% in the binary Al-Fe system and it can be deleterious to the mechanical properties of aluminum alloys.

SEM image from a 355 μm-diameter particle of Al-0.3 at%Fe, produced in nitrogen using the IA technique is shown in Figure 4(a). The nucleation site is highlighted by a circle in this figure. Note the primary trunks radiating from the nucleation site. Figure 4(b) shows the typical eutectic structure at the boundaries of the primary cellular/dendritic α-Al. Figure 4(c) represents the SEM image of an Al-0.93 at%Fe droplet. The microstructure seems similar to Al-0.3 at%Fe, with thicker dendritic/cellular eutectic boundaries because of the higher Fe content in the composition (Figure 4(d)).

Figure 5(a) shows the TEM microstructure of the eutectic area of Al-0.3 at%Fe. Figure 5(b) is a magnified TEM image of a part of the eutectic structure seen in Figure 5(a). Selected Area Electron Diffraction (SAED) and its simulation are shown in Figure 5(d). The diffraction pattern shown in Figure 5(d) is similar to that for Al_mFe (m = 4.0 to 4.4) reported in the literature [10]. Total of 21 SAED simulations were conducted to verify this finding. Comparing the experimentally obtained diffraction patterns and the simulated patterns showed that the intensities of the simulated spots agree well with the intensities (brightness) of the experimentally obtained SAED (see arrows in both obtained [Figure 5(c)] and simulated pattern [Figure 5(d)]). Hence, the Al-Fe intermetallic in the eutectic structure of Al-0.3 at%Fe is Al_mFe. Similar result was achieved for the intermetallic observed in Al-0.93 at%Fe.

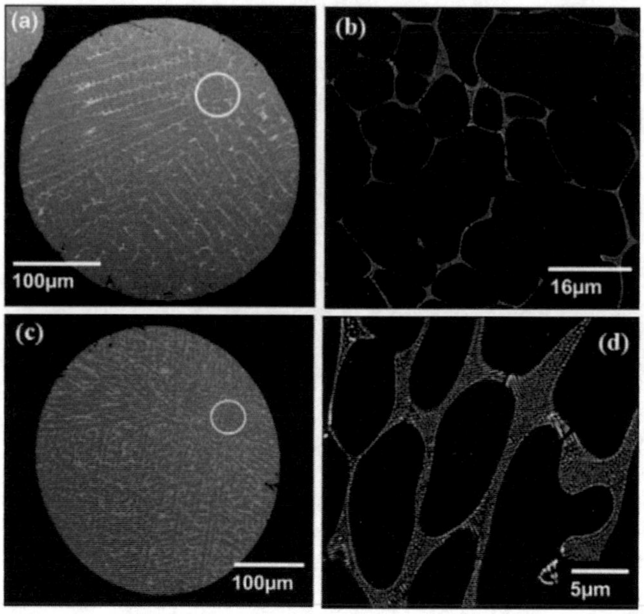

Figure 4. (a) The macrostructure of Al-0.3 at%Fe powder revealed by SEM in backscattering electron mode. Circle shows the area that contains the nucleation site. (b) illustrates the structure of the dendrites/cells at higher magnification, (c) the macrostructure of the Al-0.93 at%Fe droplet similar to that in (a), (d) the magnified image of the eutectic structure seen in (c) [9].

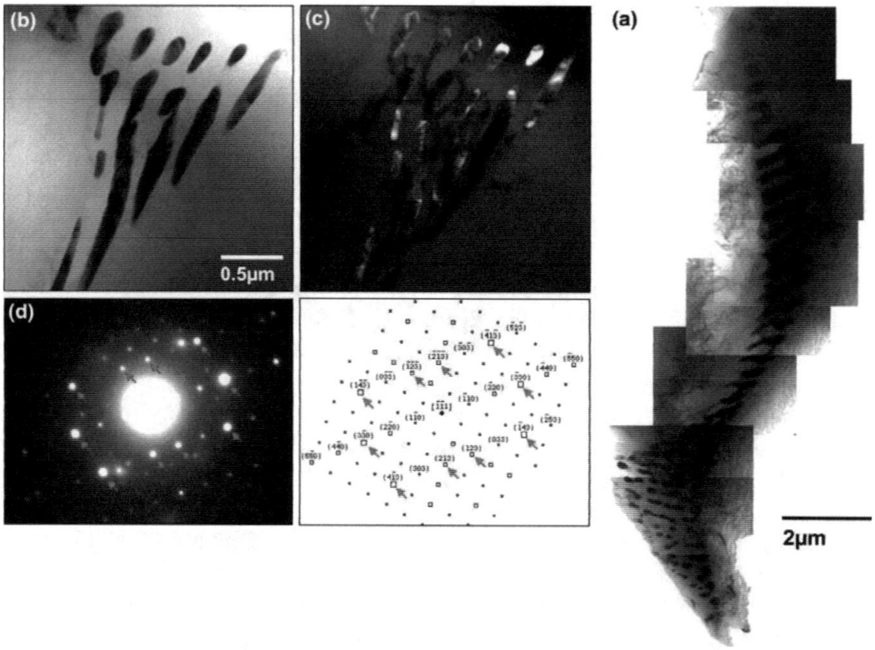

Figure 5. (a) A TEM image of the eutectic structure (b), part of the eutectic structure similar to the one shown in (a). (c) the corresponding dark-field image by choosing reflection (–330). (d) the diffraction pattern and its simulation. The arrows in the simulation correspond to the arrowed spots in the obtained patter shown on the left [9].

One might be surprised by the microstructural similarity between Al-0.3 at%Fe and Al-0.93 at%Fe. According to the equilibrium phase diagram the former is hypoeutectic and the latter is hypereutectic. However, the actual microstructure suggests that the undercooling of the droplet extends the liquidus line of α-Al into the solidus region in the phase diagram, resulting in the solidification of the hypereutectic composition of Al-0.93 at% Fe as a hypoeutectic composition in the metastable phase diagram. Therefore, the phase-formation sequence in Al-0.93 at%Fe is similar to that of Al-0.3 at%Fe (i.e., α-Al is the first solid phase solidified from the melt and eutectic Al_mFe/α-Al is the last structure formed at the end of solidification).

Using image analysis techniques on SEM micrographs the volume fraction of eutectic was measured as a function of Fe content. This measurement was used to estimate the degree of eutectic undercooling. It is found that the undercooling is about 10°C and 17°C for the 0.3 and the 0.93 at% Fe alloys, respectively [11].

Powders with microstructural features similar to those shown in Figures 4 were analyzed using high-resolution EBSD. Three regions were chosen for this analysis. The nucleation site (shown by circle in Figure 4), the trunks emanating from the nucleation site, and the dendrites/cells between the trunks spread over the surface of the droplet were analyzed using EBSD to explore the misorientation between these regions. Results for Al-0.3 at%Fe are shown in Figure 6. Different colors in this image correspond to a different misorientation as a result of different Euler angles. In Figure 6(a), all Euler angles of the nucleation center were chosen as a reference for zero misorientation. It is shown that misorientation for most of the region, especially the three trunks, is within 1 deg. Regions exhibiting colors other than blue correspond to the regions with higher misorientation. Obviously, the regions with a large misorientation are side dendrites/cells with misorientation as high as 10 deg. Also, from the EBSD analysis it was found that a cubic texture exists in both powders with a composition of Al-0.3 at%Fe or with Al-0.93 at%Fe. The dendritic growth direction within the droplets is close to <111>, which is different from the most common <100> in the solidification of cubic alloys. The change in the growth direction has been reported in the literature [12-14] to occur as a result of growth undercooling that the dendritic front experiences during the solidification. This finding confirms the nucleation undercooling discussed by Henein et al. [11].

3.2. Al-4 at%Fe

In this part the microstructure of impulse-atomized particles of Al-4 at%Fe is compared with that of same alloys solidified in EML in terrestrial and reduced gravity condition.

Figure 7 (A) and (B) show the microstructure of Al-4 at%Fe sample solidified under terrestrial and reduced gravity condition (referred as PFC), respectively. The microstructure under different condition is apparently different. The main structure of terrestrial Al-4 at%Fe displays a needle-like morphology of $Al_{13}Fe_4$ while prominent structure of PFC exhibits flower-like $Al_{13}Fe_4$. Figure 7 (C) shows the microstructure of an impulse-atomized droplet of Al-4 at%Fe with the diameter of 925 μm atomized in helium. It is obvious the microstructure is significantly refined compared with those produced using EML both in terrestrial and microgravity condition. Also significant is the morphology of the primary phase, which has the same flower-like morphology with fewer arms. It seems that the liquid stirring due to electromagnetic forces was responsible for the formation of the needle-like primary $Al_{13}Fe_4$. Further investigation on the microstructural evolution and metastable phase formation in this alloy is underway.

3.3 Al-Ni

3.3.1. Impulse Atomization: Comparison between the solidification of $Al_{80}Ni_{20}$and $Al_{68.5}Ni_{31.5}$ alloys is interesting as the former experiences only one peritectic reaction at 854 °C, while the latter goes through two peritectic reactions at 1133 and 854 °C.

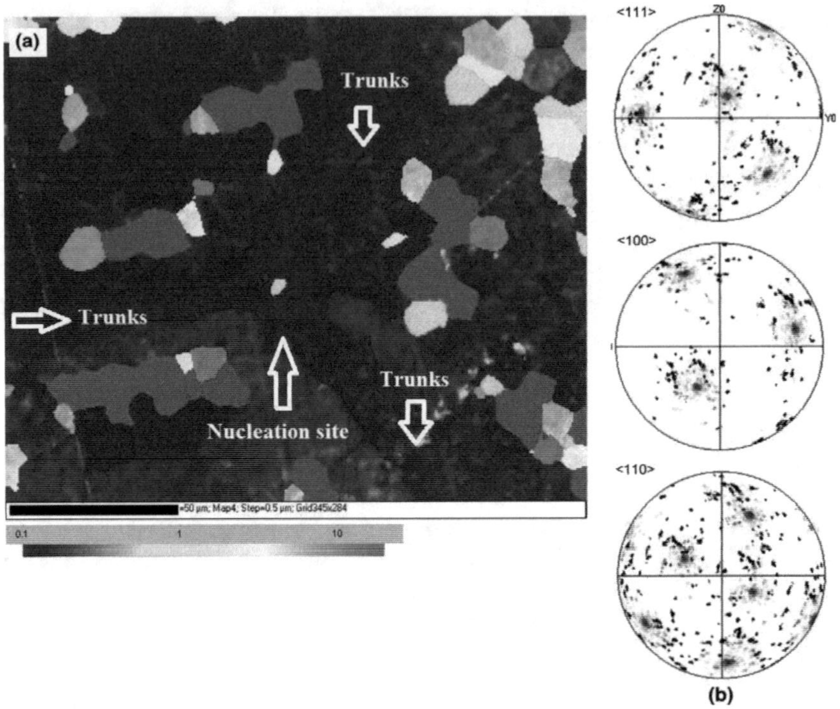

Figure 6. (a) EBSD reconstructed image of the particle shown in Figure 4(a). The orientation (Euler angles) of the nucleation center was chosen as zero misorientation, and then the color in the other area corresponds to the misorientation angle from the nucleation center. The color scheme attached at the bottom of the image shows the misorientation angle (b) illustrates <111>, <100>, and <110> pole figures obtained from the same region. Colors in the pole figures correspond to the region with the same color in (a). In the pole figures, Y_0 and Z_0 are coincident with the primary sample axes [9].

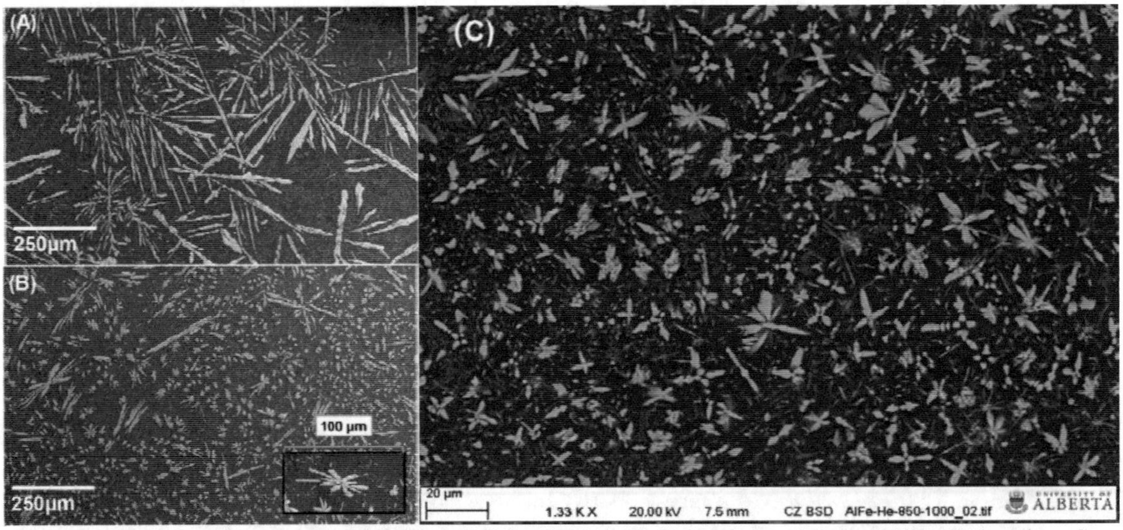

Figure 7. Microstructure of Al-4 at%Fe solidified in (A) EML in terrestrial condition, (B) reduced gravity during parabolic flight, (C) impulse-atomized in helium with diameter of 925 μm.

Powders of $Al_{68.5}Ni_{31.5}$ and $Al_{80}Ni_{20}$ were produced using the impulse atomization technique. Microstructural analysis showed that finer microstructure was achieved in the smaller particle sizes and, also, when helium was used instead of nitrogen.

In particles of $Al_{80}Ni_{20}$ with diameter of 165 μm the formation of primary Al_3Ni_2 was suppressed and Al_3Ni formed directly from the melt. The microstructure of two different sizes of this alloy is shown in Figure 8. It is clear that in Figure 8 (b) no sign of peritectic microstructure can be observed.

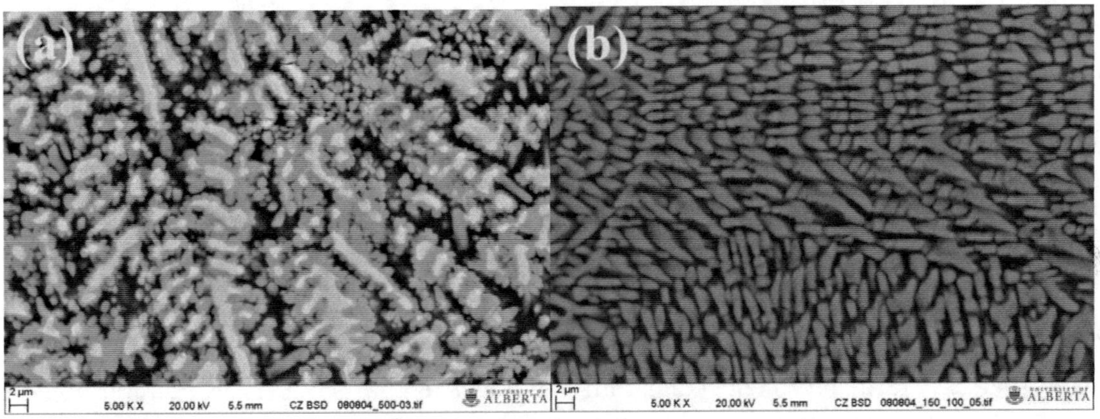

Figure 8. $Al_{80}Ni_{20}$ impulse-atomized in helium (a) D=550 μm (b) D=165 μm

Neutron diffraction investigation of small particles of $Al_{80}Ni_{20}$ also showed that particles with diameter smaller than 275 μm contain some unidentified peaks (e.g. at ~3.5Å^{-1}). Figure 9 shows the neutron diffractogram of three different particle sizes of this alloy. Clearly, the smaller the particle size the more pronounce the unidentified peak. These peaks may correspond to the formation of a quasicrystalline phase known as D-phase [15]. Shuleshova et al. [16], using electromagnetic levitation, showed that this metastable phase forms at undercoolings of 150-200 °C below the peritectic temperature 854 °C.

Secondary dendrite arm spacing was used to estimate the solidification cooling rate. The XRD and ND were successfully used to estimate the phase fractions of the solidified powders. For each alloy, the ratio of Al_3Ni to Al_3Ni_2 was calculated from the Rietveld refinement to investigate the effect of cooling rate on the phase fractions after solidification (see Figure 10). In addition, Scheil–Gulliver equation was applied to investigate the extent to which it can predict the phase fractions in these alloys. In $Al_{68.5}Ni_{31.5}$, it was found that by increasing cooling rate, the ratio of Al_3Ni to Al_3Ni_2 approaches that of Scheil–Gulliver's prediction. This was attributed to the suppression of the peritectic transformation at high cooling rates. Opposite behavior was observed in $Al_{80}Ni_{20}$, where the ratio decreased and approached Scheil–Gulliver's prediction with decreasing cooling rate. This observation was related to the direct formation of Al_3Ni from the liquid and the suppression of Al_3Ni_2 formation [17-18]. These trends have been verified with more detailed analysis by ND over a range of particle sizes.

Effect of cooling rate on the lattice parameter of Al_3Ni_2 was also investigated. It is known that the Al_3Ni_2 structure is similar to that of AlNi structure with 33% of the nickel sites vacant at the stoichiometric composition. The vacancies are ordered in a way that every third sheet of nickel atoms perpendicular to one of the threefold axes of the cube is absent. The ordering transforms the crystal structure from cubic to rhombohedral by unequal changes of the parameters *c* and *a*. The cube diagonal in AlNi becomes the *c* axis in Al_3Ni_2 ($c = a'\sqrt{3}$, where a' is the cube edge) and the cube face diagonal becomes *a* axis of Al_3Ni_2 unit cell ($a = a'\sqrt{2}$). Therefore, it is expected that the axial ratio to be *c/a*=1.225. However, due to the collapse along the *c* axis, as a result of the missing planes, it drops to 1.2132 at stoichiometric Al_3Ni_2. The *c/a* ratio falls continuously across the phase with increasing aluminium content of Al_3Ni_2 [19].

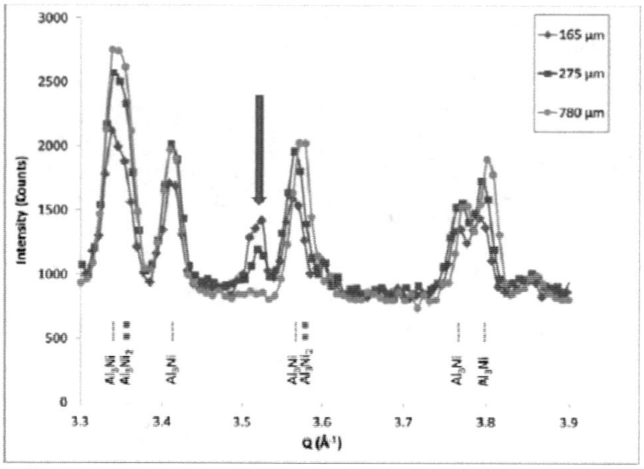

Figure 9. Part of neutron diffractogram of three different particle sizes of impulse-atomized $Al_{80}Ni_{20}$, showing peak position of different phases. Note the unindexed peak at ~3.5$Å^{-1}$.

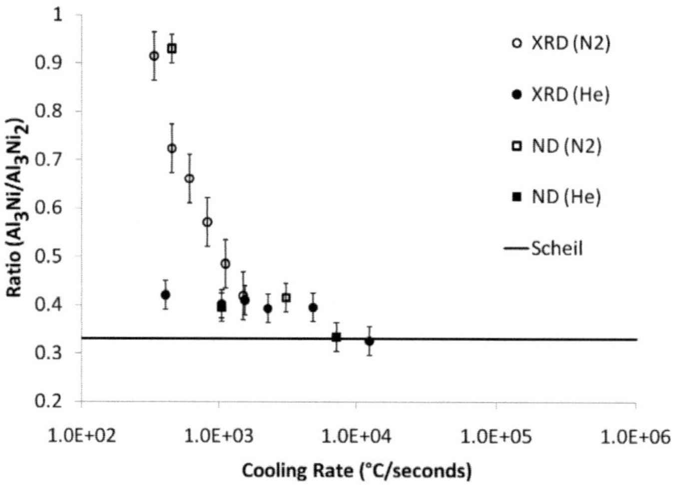

Figure 10. ND and XRD results for the ratio of Al_3Ni to Al_3Ni_2 as a function of cooling rate for two sets of Al-31.5 at%Ni samples. One cooled in helium and the other cooled in nitrogen [18]

Rietveld method was applied to the ND patterns of Al-Ni alloys to find the lattice parameters of Al_3Ni_2 and to investigate the effect of cooling rate on the *c/a* ratio in this phase. The results showed that in $Al_{80}Ni_{20}$, the *c/a* ratio is significantly smaller than the stoichiometric *c/a* ratio of Al_3Ni_2, and it decreased with increasing cooling rate. On the other hand, for $Al_{68.5}Ni_{31.5}$, the *c/a* ratio is much closer to the stoichiometric *c/a* ratio and it increases with increasing the cooling rate.

According to Al-Ni phase diagram and the shape of the Al_3Ni_2 region it is clear that the fraction of nickel within the Al_3Ni_2 that forms from AlNi is higher than that in the Al_3Ni_2 which forms directly from the liquid. This results in an increase of the *c/a* ratio as there are more nickel atoms in the lattice of Al_3Ni_2. Also, it is possible that the observed differences are due to different formation mechanisms of Al_3Ni_2 in these alloys. In $Al_{80}Ni_{20}$, the Al_3Ni_2 phase nucleates and grows directly from the liquid while in $Al_{68.5}Ni_{31.5}$ it nucleates from the peritectic reaction of AlNi with the liquid and grows both from the peritectic transformation and from the liquid.

3.3.2. EML: The equiatomic intermetallic compound $Al_{50}Ni_{50}$ was subject of recent investigations with respect to the experimental proof of the influence of forced convection on the kinetics of dendrite growth in the regime of low undercooling by comparative investigations on Earth and in reduced gravity [20]. In the present work, we report on measurements of dendrite growth velocities as a function of undercooling on a series of Al–Ni samples of various compositions. The Ni-rich alloys show the usual behavior such that the growth velocity increases with increasing undercooling. In contrast, the Al-rich alloys exhibit a very unusual growth behavior: their growth velocity decreases with increasing undercooling. Such anomalous solidification kinetics has not been reported so far in literature. High undercoolings $\Delta T = T_L - T$, where T_L is the liquidus temperature of the respective alloy, were measured ranging up to $\Delta T \leq 325$ K. Electromagnetic levitation for containerless undercooling in combination with a high-speed video camera was applied to measure the rapid dendrite growth velocity as a function of undercooling for Al–Ni alloys in terrestrial experiments. The TEMPUS facility for containerless processing in reduced gravity [21] was used during the sounding rocket mission TEXUS 44 to measure the dendrite growth velocity of Raney-type $Al_{68.5}Ni_{31.5}$ alloy. One hundred and eighty seconds of reduced-gravity time was available for this particular experiment. Three undercooling and solidification cycles were obtained. The sample was heated by more than 200 K above the liquidus temperature of this alloy, T_L = 1768 K. The sample undercooled during the subsequent solidification cycles to ΔT_1 = 185 K and ΔT_2 = 228 K, respectively. In this way two data points were obtained of dendrite growth velocity measured in reduced gravity. The TEMPUS facility is equipped with a video camera that can capture 200 pictures per second. During recalescence, three pictures were taken, which allow the determination of the growth velocity with an accuracy of ±(15–25)%.

The circles in Figure 11 show results of measurements of dendrite growth velocity as a function of undercooling performed under terrestrial conditions for the series of Al–Ni alloys. In all samples the growth velocity is measured in the undercooling range of 100 K < ΔT < 350 K. The $Al_{55}Ni_{45}$ alloy shows the usual behavior. The dendrite growth velocity increases with increasing undercooling. This behavior is understood by the fact that the driving force for crystallization scales with the undercooling obtained prior to solidification. If 5 at.% Al is added to $Al_{55}Ni_{45}$, the growth velocity of $Al_{60}Ni_{40}$ obviously passes through a minimum at an undercooling of $\Delta T \approx 175$ K. With further increasing the Al content in $Al_{65}Ni_{35}$ alloy, the minimum is shifted to larger undercoolings. In Raney-type alloy $Al_{68.5}Ni_{31.5}$ and for $Al_{70}Ni_{30}$ a monotonous decrease in the growth velocity with increasing undercooling is measured. The alloy with the largest amount of Al, $Al_{75}Ni_{25}$, shows a maximum at $\Delta T \approx 160$ K followed by a continuous decrease of $V(\Delta T)$ up to the largest undercooling ΔT = 315 K. In situ observation of phase selection in undercooled melts of $Al_{68.5}Ni_{31.5}$ alloy by energy dispersive X-ray measurements at the European Synchrotron Radiation Facility Grenoble (France) on levitation-processed drops unambiguously demonstrates that over the entire undercooling range ΔT < 320 K the intermetallic AlNi phase of B2 structure is primarily formed [16]. It subsequently transforms by a peritectic reaction L + AlNi $\rightarrow$ Al_3Ni_2. For $Al_{75}Ni_{25}$ alloy, Al_3Ni_2 primarily solidifies from the melt as evidenced by the X-ray diffraction experiments. In situ diffraction investigations indicate that this phase is primarily formed even at highest levels of undercooling. Despite the change in the primary phase formation from AlNi to Al_3Ni_2 at concentrations between 68.5 and 75 at.% Ni, all alloys in this compositional regime show a decrease in growth velocity with increasing undercooling. So far, there is no model that describes such an anomalous growth dynamics-undercooling relation.

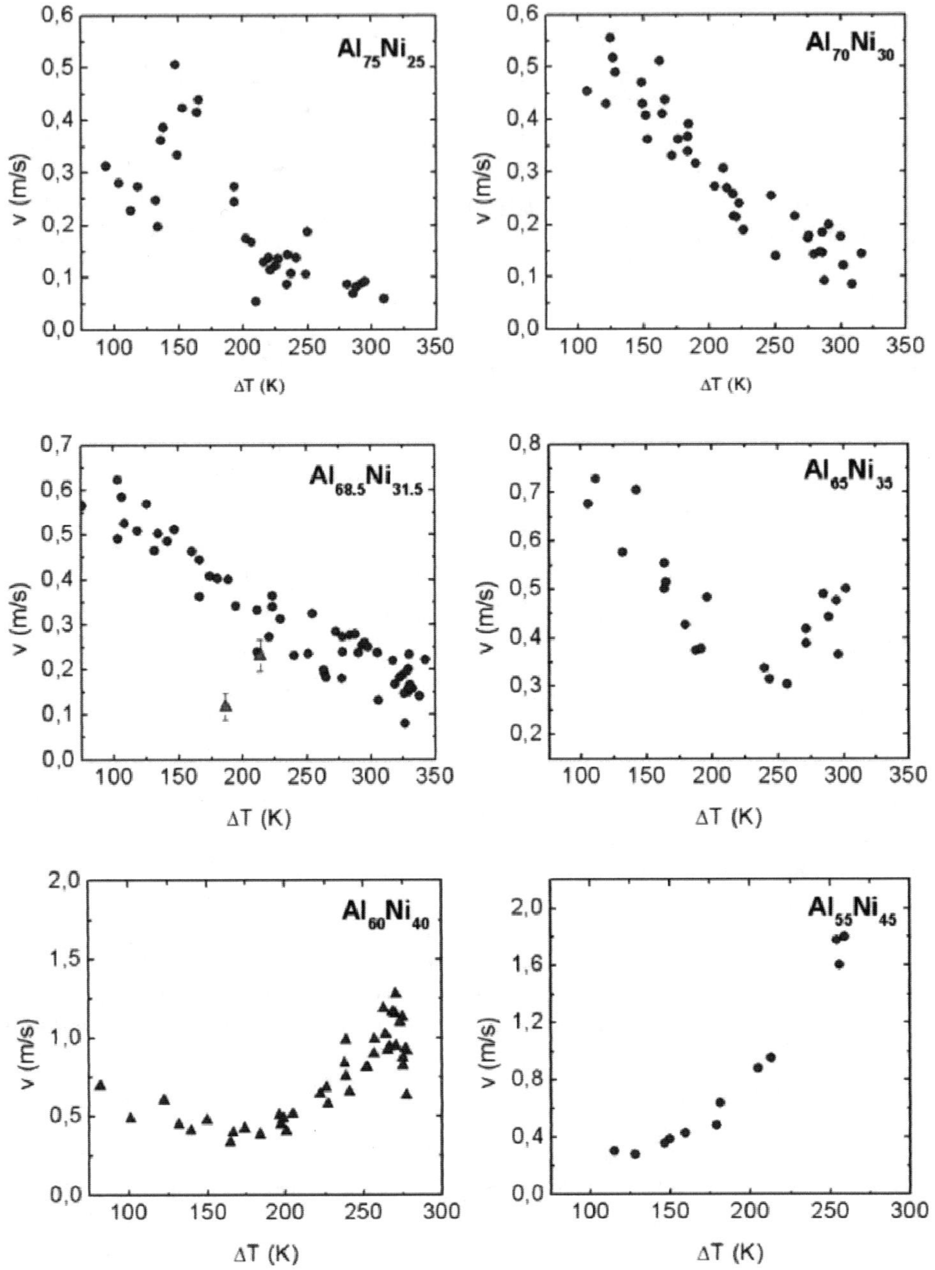

Figure 11. Dendrite growth velocity as a function of undercooling measured under terrestrial conditions (circles) by using high speed video technique on electromagnetically levitated samples of $Al_{100-x}Ni_x$ alloys of concentrations x = 45, 40, 35, 31.5, 30, 25. An anomalous dendrite growth behavior is found for Al-rich Al-Ni alloys such that the growth velocity is decreasing with increasing undercooling. The filled triangles show results obtained under reduced gravity for Raney type $Al_{68.5}Ni_{31.5}$ alloy during TEXUS 44 flight mission.

In previous investigations of the dendrite growth velocity vs. undercooling relation of equiatomic $Al_{50}Ni_{50}$, a pronounced effect of forced convection was found [20]. The growth velocity is observed to be systematically higher in terrestrial experiments than in experiments in reduced gravity. This behavior can be described by an extended sharp interface theory of dendrite growth by introducing Péclet numbers for heat and mass transport due to forced convection in the electromagnetically levitated melt. More details of the extended sharp interface theory are given elsewhere [22]. In the undercooling range in which the growth velocity is smaller than or comparable with the estimated fluid flow velocities in electromagnetically levitated liquid drops, a significant difference was observed in the two complementary sets of experiments. According to estimations by Hyers et al. the fluid flow velocity in liquid drops electromagnetically levitated on Earth is estimated by hydrodynamic calculations to be of the order of 0.5 m s^{-1} [23]. If one compares this value for fluid flow velocity with the range of dendrite growth velocities measured for the differently concentrated Al–Ni samples, it is found that the $Al_{55}Ni_{45}$ sample shows growth velocities that are comparable with or higher than the fluid flow velocity. In case of $Al_{60}Ni_{40}$ alloy the minimum in the $V(\Delta T)$ relation occurs at a velocity that is comparable with the fluid flow velocity. In all other Al-rich Al–Ni alloys the measured growth velocities are smaller than the expected fluid flow velocity over the entire range of undercooling accessible by electromagnetic levitation. In order to investigate the influence of forced convection on the dendrite growth behavior we have varied the power of the alternating electromagnetic field within the maximum range that allows levitation of the samples in terrestrial levitation experiments (variation by about 50%). Within the uncertainty of the measurements no influence of the levitation field on the growth velocities vs. undercooling relation was detected. Significantly smaller fluid flow velocities are accessible by processing under reduced-gravity conditions. The Raney-type alloy was studied with respect to dendrite growth dynamics during the sounding rocket mission TEXUS 44 of the European Space Agency (ESA) and the German Aerospace Center (DLR). TEXUS sounding rocket missions provide an excellent quality of reduced gravity of the order of 10^{-4}g. The filled triangles in Figure 11 show the growth velocity V as a function of undercooling of Raney-type $Al_{68.5}Ni_{31.5}$ alloy as retrieved from the TEXUS experiment. Two significant and important differences of the experiments under normal gravity and in microgravity are observed. First, the growth velocity measured in microgravity is substantially smaller than the data taken under 1g conditions. Second, the temperature dependence of the $V(\Delta T)$ relation differs. In reduced gravity, it is apparent that the growth velocity increases with increasing undercooling as usual for a great variety of alloys. The decrease in the velocity with undercooling as measured on the ground for a variety of Al-rich Al– Ni alloys is anomalous and unique. By taking into account the results of the TEXUS 44 flight, this anomalous behavior of dendrite growth dynamics of Al-rich Al–Ni alloys is obviously associated with a gravity dependent mechanism in crystal growth in undercooled melts. However, a detailed understanding of the experimental results obtained on Earth and in reduced gravity is still lacking.

*3.3.3. Gas Atomization:*Direct observations of gas atomized $Al_{68.5}Ni_{31.5}$ powders by SEM showed that they were nearly spherical in shape. Figure 12 shows the surface of the particles of three different sizes, less than 25 μm, 25-50 μm and 50-100 μm. The surface of the particles with diameters below 25 μm was smooth, whereas larger particles presented rougher surfaces, associated with the occurrence of a non-planar solidification front and subsequent solidification.

The X- ray diffraction patterns of the $Al_{68.5}Ni_{31.5}$ powder particles showed that the signal of the Al_3Ni_2 phase decreased when the powder particle size increased, meanwhile the signal of the Al_3Ni phase increased when the powder particle size increased. These results are in agreement with those achieved from impulse atomization of $Al_{68.5}Ni_{31.5}$ shown in Figure 10.

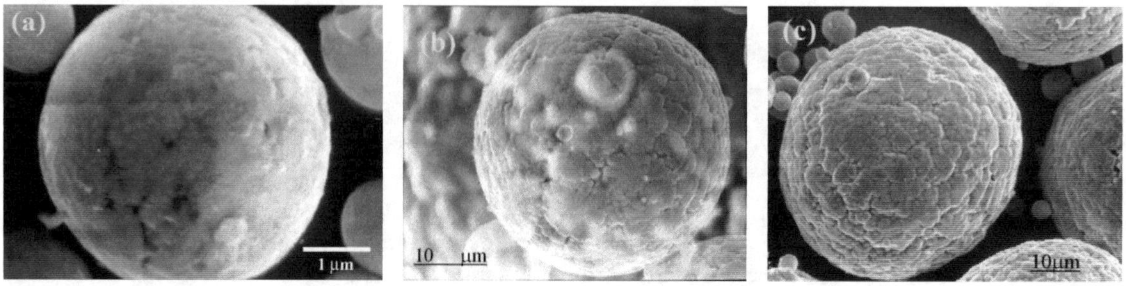

Figure 12. Particles of gas-atomized $Al_{68.5}Ni_{31.5}$ alloy with the following particle sizes: a) <25 μm, b) 25-50 μm and c) 50-100 μm.

4. Modeling

For a given alloy, thermodynamic equilibrium predicts a unique fraction of phases at room temperature, whatever the size of the system. This is obviously not the case in the experimental results presented for atomization. Not only the distribution of phases varies with the diameter of the particles, but it is also found to be a function of the atomization gas. Explanations involve competition between diffusion in phases and growth of dendritic, peritectic and eutectic microstructures, combined at various heat extraction rates. A model developed for the prediction of microsegregation taking place in a uniform temperature system [24] was later coupled with thermodynamic equilibrium calculations and applied to electromagnetic levitation [7] and gas atomization [25]. It is used for the understanding of the several physical effects involved that permit explanations of the measured fractions of phases.

Because the model is presented into details in references [7] and [24], it is only summarized hereafter. The geometry of the system is assumed spherical in shape, with radius corresponding to the particle size. A uniform temperature is considered, as well as constant densities for all phases, resulting in a diffusive regime for heat and mass flows. Nucleation of the dendritic, peritectic and eutectic microstructures takes place at the equilibrium liquidus temperature computed with the liquid composition located at the center of the system. Growth proceeds by radial propagation towards the system periphery. The growth kinetics is computed using available models in the literature [26, 27]. The average total mass and solute mass balances are written in each phase as well as at each phase interface. A global heat balance is yet used thanks to the uniform temperature approximation. Heat extraction takes place through the outer surface of the system. The heat transfer coefficient is calculated with a modified Whitaker correlation developed for impulse atomization [28]. Enthalpies of all phases and interface compositions are computed with the software ThermoCalc using the CALPHAD method [29, 30] and the public binary database PBIN (www.thermocalc.se).

Figure 13 presents the result of simulations performed for atomization of an $Al_{80}Ni_{20}$ alloy corresponding to the particle sizes given in Figure 13. The data used for the simulations are given in Table 1. Each column of Figure 13 stands for one simulation, namely (left) D = 550 μm and (right) 165 μm. The three graphs on each column share the same x-axis and show the time evolution of: (a) the temperature of the droplet, with the equilibrium temperatures for the formation of a new solid phase (horizontal lines with filled symbols) and the predicted nucleation temperatures (horizontal lines with empty symbols); (b) the average volume fraction of the phases and (c) the radius of each zone, i.e. dendritic Al_3Ni_2, peritectic Al_3Ni and eutectic (Al+ Al_3Ni), normalized by the total radius of the particle, R. The normalized radius of a zone illustrates the extent of the expansion of a microstructure within the droplet. For instance, when the radius $R^{Al_3Ni_2}/R$ reaches 1, the mushy zone – or grain envelope of the primary dendritic structure – is fully developed and no bulk liquid remains.

We notice first the difference in cooling rate and the resulting solidification time being inversely proportional to the particle size. Secondly, the predicted nucleation temperature for the peritectic phase is systematically higher than the equilibrium temperature defined by the phase diagram, while the nucleation of the eutectic occurs close to its equilibrium temperature. We detail hereafter the solidification scenarios for the two simulations, in order of complexity.

4.1. Regime of distinct successive growth
The most common solidification behavior is presented for the larger particle in Figure 13. This solidification scenario is close to those described in earlier publications [7, 24]. The cooling curve exhibits three recalescences, corresponding to the successive growths of the primary dendritic Al_3Ni_2, the peritectic Al_3Ni and the eutectic (Al+ Al_3Ni) microstructures. Direct solidification of the primary dendritic Al_3Ni_2 occurs between t = 20 and 50 milliseconds. The graph (c) with the radii of zones shows that the peritectic zone starts to grow at around t = 40 ms, but its expansion is quickly stopped by the increase in temperature due to the recalescence associated with the primary dendritic Al_3Ni_2 phase shown in graph (a). After the primary recalescence, the temperature decreases again and, at around t = 80 ms, the peritectic zone grows significantly and leads to a peritectic recalescence and a clear increase of the average volume fraction of the Al_3Ni phase as shown in (b). The decrease in primary Al_3Ni_2 fraction is first due to the remelting of the primary solid induced by the peritectic recalescence and then to the peritectic transformation $Al_3Ni_2 \rightarrow Al_3Ni$, which lasts up to complete solidification. After 100 ms, the eutectic structure grows and fills in the remaining liquid. Finally, the solidification is complete when the solid eutectic zone is fully developed, i.e. when R^{Al}/R reaches one.

4.2. Regime of shortcut of the primary growth
On the other hand, the small particle follows a different solidification path as illustrated in Figure 13. The composition gradients involved are of the same order of magnitude, and then the diffusion and growth kinetics are comparable, while the cooling rate is almost one order of magnitude higher. Hence, the growth-induced primary undercooling is so high that the temperature is already below the peritectic nucleation temperature before any primary solid has significantly started to grow. On the radii graph (c), the peritectic zone starts expanding very shortly after the primary mushy zone, but with a higher growth velocity. Hence, the peritectic zone quickly overtakes the primary zone ($R^{Al_3Ni} \geq R^{Al_3Ni_2}$), the growth of the latter then being stopped as the primary dendrites are completely surrounded by the peritectic phase. Additionally, the peritectic recalescence takes place up to temperatures higher than the peritectic equilibrium temperature $T_P^{Al_3Ni}$. After the peritectic recalescence, cooling continues and the solidification ends with eutectic growth. The final microstructure only contains peritectic Al_3Ni and eutectic (Al+ Al_3Ni) as shown in (b).

As can be concluded from a first comparison between the simulated solidification sequence presented in Figure 13 and discussed above, the model predicts well the qualitative phase fractions observed in Figure 8.

5. Summary and conclusion

Containerless solidification behavior of Al-Ni and Al-Fe alloys was investigated under terrestrial and reduced-gravity conditions. Impulse and gas atomization was utilized to investigate the effect of cooling rate on phase fractions and morphology of the evolved phases. In addition, electromagnetic levitation was used to measure the dendrite growth velocity as a function of undercooling for a series of Al-Fe and Al–Ni samples of various compositions. The following results were achieved:

1. The TEM characterization on the eutectic microstructure of impulse-atomized Al-Fe powders with two compositions showed that the metastable AlmFe formed in these alloys. The presence of metastable AlmFe indicates that eutectic nucleation undercooling occurs in the solidification. Also, the fact that the hypereutectic Al-0.93 at%Fe solidifies as a hypoeutectic composition indicates that the nucleation of the α-Al is energetically favored over nucleation of any aluminum iron intermetallics, which is due to primary phase undercooling. Also, the growth undercooling that the dendritic front experiences in the solidification of the droplet resulted in variation of dendrite growth direction from <100> to <111>.

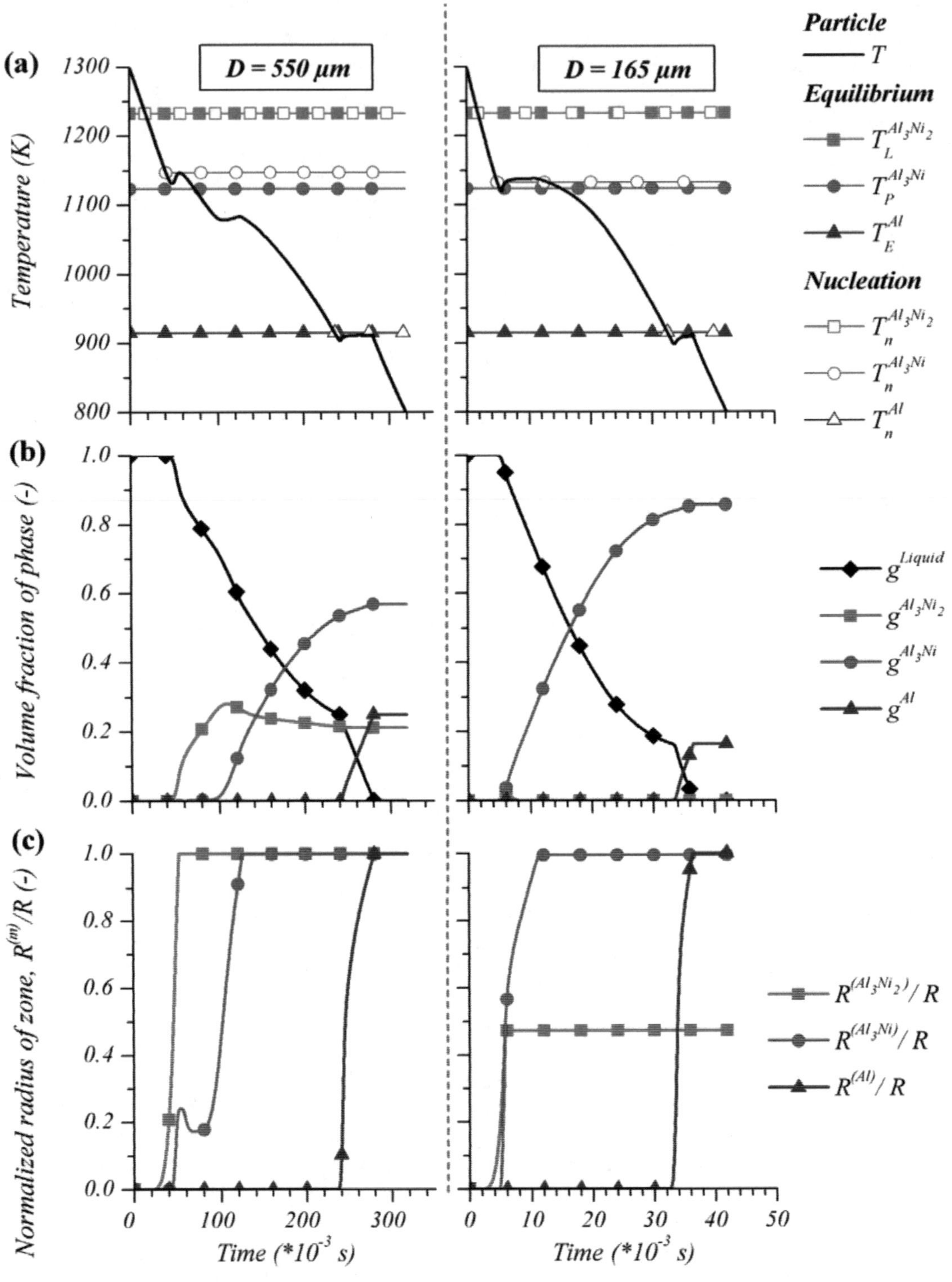

Figure 13. Simulation of the solidification of $Al_{80}Ni_{20}$ impulse-atomized in helium with particle diameter (left) D=550 μm and (right) D=165 μm, with (a) the temperature; (b) the average volume fraction of phases and (c) the normalized radii of the zones, i.e. the position of the growing dendritic, peritectic and eutectic envelopes of the microstructures.

Table 1. Simulation data for Figure 13.

Parameter	Symbol	Value	Unit	Source
ThermoCalc database	-	PBIN	-	
Nominal compositions	x_0	80	at.% Al	
	w_0	64	wt% Al	
Particle diameters	D	(Figure 13a) 550	μm	
		(Figure 13b) 165		
Solid structure diffusion coefficient	D^s	5×10^{-11}	$m^2\ s^{-1}$	[7]
Liquid phase diffusion coefficient	D^l	10^{-9}	$m^2\ s^{-1}$	[7]
Secondary dendrite arm spacing	λ_2	(Figure 13a) 10	μm	
		(Figure 13b) 1		
Solid structure nucleation undercooling	ΔT_n^s	0	K	
Gibbs-Thomson coefficient of solid/liquid interfaces	$\Gamma^{Al_3Ni_2/l}$	$2\ 10^{-8}$	m K	[7]
	$\Gamma^{Al_3Ni/l}$	3.5×10^{-8}	m K	[31]
	$\Gamma^{AlNi/l}$	1.86×10^{-7}	m K	[31]
Eutectic structure contact angle	$\theta^{Al_3Ni/l}$	19.22	°	[31]
	$\theta^{Al/l}$	11.29	°	[31]
External temperature	T_{ext}	293	K	
Droplet initial velocity	v_0	0	$m\ s^{-1}$	
Atomization gas	-	Helium	-	
Density of the particle	ρ	$(w_0/\rho_{Al} + (1-w_0)/\rho_{Ni})^{-1}$	$kg\ m^{-3}$	
Density of Aluminum	ρ_{Al}	2700	$kg\ m^{-3}$	[32]
Density of Nickel	ρ_{Ni}	8909	$kg\ m^{-3}$	[32]
Density of Helium	ρ_f	$37.303\ T^{-0.9559}$	$kg\ m^{-3}$	[28]
Viscosity of Helium	μ_f	$(4.3679\ T^{0.6702})10^{-7}$	Pa s	[28]
Thermal conductivity of Helium	k_f	$0.002778\ T^{0.7025}$	$W\ m^{-2}\ K^{-1}$	[28]
Prandtl number	Pr	0.713	-	[28]
Coefficients in Whitaker's relation	B	2.778×10^{-3}	-	[28]
	m	0.7025	-	[28]

2. For Al-4 at%Fe, it was found that in reduced-gravity and in the impulse-atomized droplets the primary intermetallic forms with a flower-like morphology whereas in the terrestrial EML sample it has a needle like morphology. This was attributed to differences in liquid convection.

3. Quasicrystalline D-phase was observed in impulse-atomized particles of Al80Ni20 with diameter smaller than 275 μm. Also it was observed that with increasing cooling the ratio of Al3Ni to

Al3Ni2 decreases. Opposite behavior was observed In Al68.5Ni31.5. The effect of cooling rate on the lattice parameter of Al3Ni2 was also studied. In Al80Ni20, the c/a ratio is significantly smaller than the stoichiometric c/a ratio and it decreases with increasing cooling rate. On the other hand, for Al68.5Ni31.5, the c/a ratio is much closer to the stoichiometric value and it increases with increasing the cooling rate.

4. Electromagnetic levitation for containerless undercooling was used to measure the dendrite growth velocity as a function of undercooling for Al–Ni alloys in terrestrial condition. The Ni-rich alloys showed the usual behavior such that the growth velocity increases with increasing undercooling while the Al-rich alloys exhibit a very unusual growth behavior. The growth velocity decreased with increasing undercooling. This phenomenon was attributed to a gravity dependent mechanism in crystal growth in undercooled melts as in reduced gravity, the growth velocity increased with increasing undercooling as usual for a great variety of alloys.

5. A microsegregation model for the solidification of binary alloys was used that accounts for diffusion in all phases, as well as nucleation undercooling and growth kinetics of the solidifying microstructures. It successively considers the occurrence of several phase transformations, including one or several peritectic reactions and one eutectic reaction. This model represents a practical tool to study the solidification of alloys that experience multiple phase transformations. It is detailed enough to take into account several phenomena occurring during the solidification (undercooling, growth kinetics, different diffusion length scales, etc.) while leading to quite efficient calculations. It is thus believed to provide a practical solution for application to metallic alloys.

6. The trends observed in phase fractions for Al-Ni alloys in gas atomized powders showed a similar trend to those observed with impulse powders.

6. Acknowledgment

The current work is part of the NEQUISOL project supported by ESA within contract number 15236/02/NL/SH and CSA within contract number 9F007-08-0154 and SSEP Grant 2008.

7. References

[1] Jones H 1983 *Rapid Solidification of Metals and Alloys* (The Institute of Metallurgists, London) p 35
[2] Herlach D M 1991 *Annu. Rev. Mater. Sci.* **21** 23
[3] Herlach D M, Galenko P and Holland-Moritz D 2007 *Metastable Solids from Undercooled Melts* (Pergamon Materials Series)
[4] Zheng B, Lin Y, Zhou Y and Lavernia E J 2009 *Metall. Mater. Trans. B* 40(5) 768
[5] Olsen K, Sterzik G, Henein H 1995 *Third Int. Symp. on Recycling of Metals and Engineered Materials* (The Minerals, Metals, Materials Society, Warrendale, PA) 67
[6] Standard test methods for metal powders and powder metallurgy products 1993 Metal Powder Industries Federation (Princeton, NJ)
[7] Tourret D, Gandin Ch-A, Volkmann T and Herlach D M 2011 *Acta Mater.* **59** 4665
[8] Zhou Y, Juarez-Islas J A, Alvarez-Fregoso O, Yoon W Y, Lavernia E J 1999 *J. Mater. Res.* **14** 729
[9] Chen J, Dahlborg U, Bao C M, Calvo-Dahlborg M and Henein H 2011 *Metall. Mater. Trans. B* Article in Press, 1
[10] Berg B S, Hansen V, Midgley P A and Gjonnes J 1998 *Ultramicroscopy* **74** 147
[11] Henein H, Buchoud V, Schmidt R-R, Watt C, Malakhov D, Gandin C-A, Lesoult G and Uhlenwinkel V 2010 *Can. Metall. Q.* **49** (3) 275
[12] Gudge K A and Jackson K A 2001 *J. Cryst. Growth* **225** 264
[13] Dragnevski K I, Cochrane R F and Mullis A M 2004 *Metall. Mater. Trans. A.* **35** 3211
[14] Battersby S E, Cochrane R F and Mullis A M 1999 *J. Mater. Sci.* **34** 2049
[15] Pohla C and Ryder P L 1997 *Acta Mater.* **45** 2155

[16] Shuleshova O, Holland-Moritz D, Loser W, Reinhart G, Iles G N and Buchner B 2009 *Europhysics Letters (EPL)*, **86**(3) 19
[17] Ilbagi A, Henein H and Phillion A B 2010 *J. Mater. Sci.*, **46** 6235
[18] Ilbagi A, Delshad Khatibi P, Swainson I P, Reinhart G and Henein H 2011 *Can. Metall. Q.*, **50**(3) 295
[19] Taylor A and Doyle N J 1972 *J. Appl. Crystallogr.* **5** 201
[20] Reutzel S, Hartmann H, Galenko P K, Schneider S and Herlach D M 2007 *Appl. Phys. Lett.* **91** 041913
[21] Piller J, Knauf R, Preu P, Herlach D M, Loho G 1986 in: *Proc. 6th Eur. Symp. on Materials Sciences under Microgravity*, (ESA SP-256, Bordeaux) p 437
[22] Herlach D M, Galenko P K 2007 *Mater. Sci. Eng. A* **449–451** 34
[23] Hyers R W, Matson D M, Kelton K F, Rogers J 2004 *Ann. NY Acad. Sci.* **1027** 474
[24] Tourret D, Gandin Ch-A 2009 *Acta Mater.* **57** 2066
[25] Tourret D, Reinhart G, Gandin Ch-A, Iles G N, Dahlborg U, Calvo-Dahlborg M, Bao C M 2011 submitted.
[26] Kurz W, Giovanola B, Trivedi R 1986 *Acta Metall. Mater.* **34** 823
[27] Jackson K A, Hunt J D 1966 *T. Metall. Soc. AIME* **236** 1129
[28] Wiskel J B, Henein H, Maire E 2002 *Can. Metall. Q.* **41** 97
[29] Hillert M 1998 *Phase Equilibrium, Phase Diagrams and Phase Transformations* (New York: Cambridge University Press)
[30] Saunders N, Miodownik A P 1998 *CALPHAD Calculation of Phase Diagrams: A Comprehensive Guide* (New York: Pergamon)
[31] Marasli N, Hunt J D 1996 *Acta Mater.* **44** 1085
[32] Smithells C J 1983 *Smithells metals reference book* (5th ed. London: Butterworths)

Journal of Physics: Conference Series **327** (2011) 012008 doi:10.1088/1742-6596/327/1/012008

Preparation of ZrNiSn half-Heusler compounds with crystalline alignment by unidirectional solidification in short-duration microgravity and their thermoelectric properties

H Nagai[1], R Muroi[2] and T Okutnai[2]
[1]National Institute of Advanced Industrial Science and Technology (AIST), Central 5, 1-1-1 Higashi, Tsukuba, 305-8565, Japan
[2]Graduate School of Environment and Information Sciences, Yokohama National University (YNU), 79-7 Tokiwadai, Hodogaya-ku, Yokohama 240-8501, Japan

E-mail: hideaki-nagai@aist.go.jp

Abstract. A ZrNiSn half-Heusler compound was prepared by unidirectional solidification in short-duration microgravity. When a ZrNiSn half-Heusler pellet with Zr:Ni:Sn=1:1:1(in mole) was solidified in microgravity (m-g) by two different cooling systems, the sample solidified in m-g had a different crystallographic alignment of ZrNiSn phase along the cooling direction. However, both sample solidified in normal gravity (1-g) had no crystalline alignment. The solidified sample in m-g had higher electrical conductivity along the cooling direction than perpendicular to the cooling direction, although these samples did not exhibit Seebeck coefficient anisotropy. This electrical conductivity anisotropy was found to be affected by crystalline alignment of the sample. The thermal conductivity was also affected by the structural difference of the sample.

1. Introduction

Ternary intermetallic compounds of general formula MNiSn (M=Ti, Zr, Hf), referred to as the half-Heusler compounds, have semiconducting properties with a band structure and have been of interest due to their potential as thermoelectric materials [1, 2]. The half-Heusler compounds usually have been prepared by arc-melting and sintering the crushed arc-melting sample because it is difficult to prepare a sample with uniform composition and ordering of the crystal structure due to phase separation. Therefore the obtained samples had no crystallographic orientation, and the research focus has been to improve the thermoelectric properties of these samples by changing composition and annealing condition.

Previously, we succeeded in directly preparing the cubic crystal system samples with <111> crystalline alignment, such as $TbFe_2$ and $SmFe_2$, by unidirectional solidification in short-duration microgravity [3-5]. The samples with crystallographic alignment had higher magnetostrictive properties than those with random orientation. ZrNiSn half-Heusler compounds are cubic crystal systems, and it is possible to synthesize samples with crystallographic alignment by using this technique. In this study, we investigated the synthesis of ZrNiSn half-Heusler compounds with crystallographic alignment by unidirectional solidification in short-duration microgravity and measured the thermoelectric properties of the samples.

2. Experimental

Starting samples of ZrNiSn were prepared by arc-melting from the constituent elements (Zr:Ni:Sn=1:1:1[mole]). Zr and Ni ingots were supplied by Rare Metallic Co., Ltd. (Zr 99.9% and Ni 99.99% purity), and Sn ingots (99.999% purity) were supplied by Kishida Chemical Co., Ltd. Because it is difficult to melt the arc-melted samples completely due to the mixture of the components with high melting points (Zr-Sn alloys), the arc-melted samples were ground, and the pellets made from this powder were annealed at 1273 K in H_2 atmosphere to eliminate of those components by solid-state reaction. These pellets were set on 2 different unidirectional solidification systems. In system (1), the sample directly contacts the copper chill block; in system (2), the sample indirectly contacts the copper plate through a quartz glass tube (figure 1). The pellet was put in the quartz glass tube; the tube was then evacuated and filled with Ar gas up to a pressure of 0.1 MPa. The sample was heated in an infrared furnace at around 1673 K until it was completely melted. The sample was then unidirectionally solidified by contact with the copper block or plate. Microgravity experiments were performed using a 10m drop tower at AIST and a 2m drop tower at YNU (figure 2). The resultant samples were 8 to 10 mm in diameter and about 2mm thick.

The crystalline structures of the obtained samples were analyzed by X-ray diffraction (XRD; MAC Science, MXP3 or RIGAKU, RINT2500), and the microstructure and element distribution of the samples were analyzed by optical microscopy (OM; OLYMPUS, BX51M) and scanning electron microscopy with energy dispersive X-ray spectrometer(SEM; JEOL, JSM-5400 or JSM-6390LV). The thermoelectric properties of the samples were evaluated by electrical conductivity, Seebeck coefficient, and thermal conductivity. The electrical conductivity and Seebeck coefficient of the sample were measured by an electrical conductivity and Seebeck coefficient measurement system (Ozawa Science, RZ2001i); the thermal conductivities of the samples were measured by the laser-flash method (ULVAC—RIKO, TC-7000).

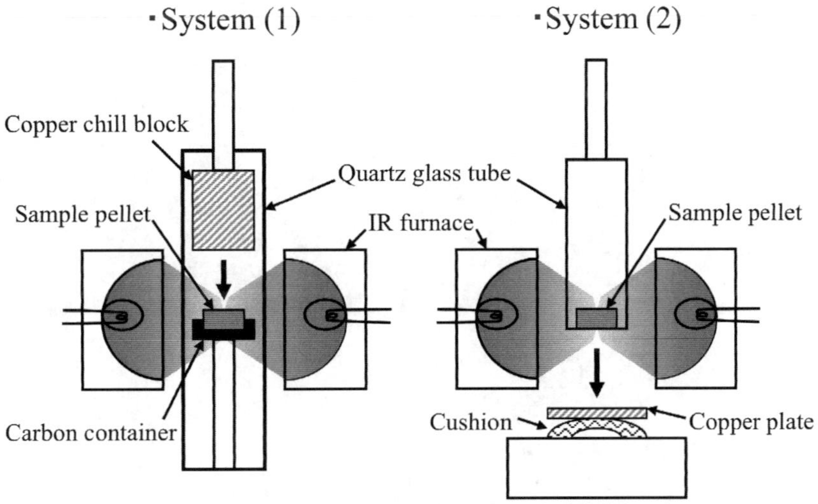

Figure 1. Schematic diagram of unidirectional solidification systems

Journal of Physics: Conference Series **327** (2011) 012008 doi:10.1088/1742-6596/327/1/012008

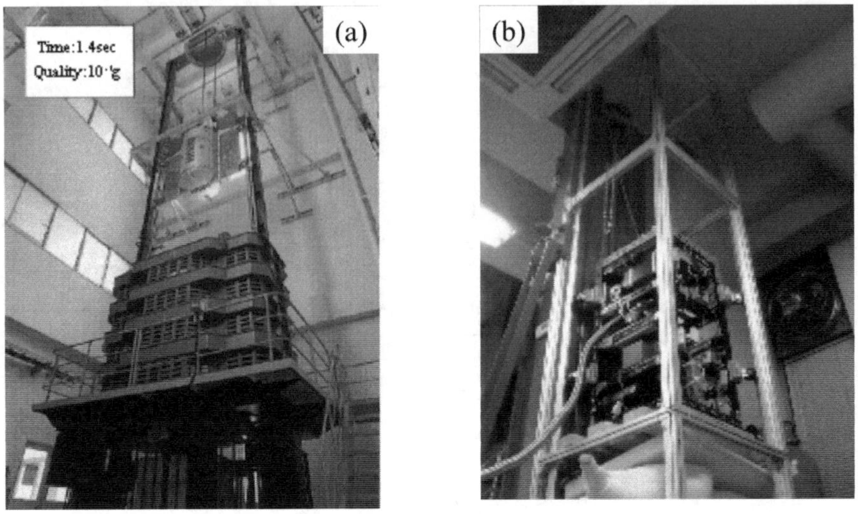

Microgravity capability: 1.43s, 10^{-3}g Microgravity capability: 0.46s, 10^{-2}g

Figure 2. AIST and YNU drop towers
(a): AIST drop tower, (b): YNU drop tower

3. Results and Discussion

Figure 3 depicts the element distribution of the arc-melted sample and typical samples unidirectionally solidified in 1-g (1-g sample) and m-g (m-g sample). The arc-melted samples had large Ni-poor regions, which XRD determined were mixtures of ZrNiSn and Zr-Sn alloys. The 1-g samples had more uniform composition distributions, and the m-g samples had almost uniform composition distributions.

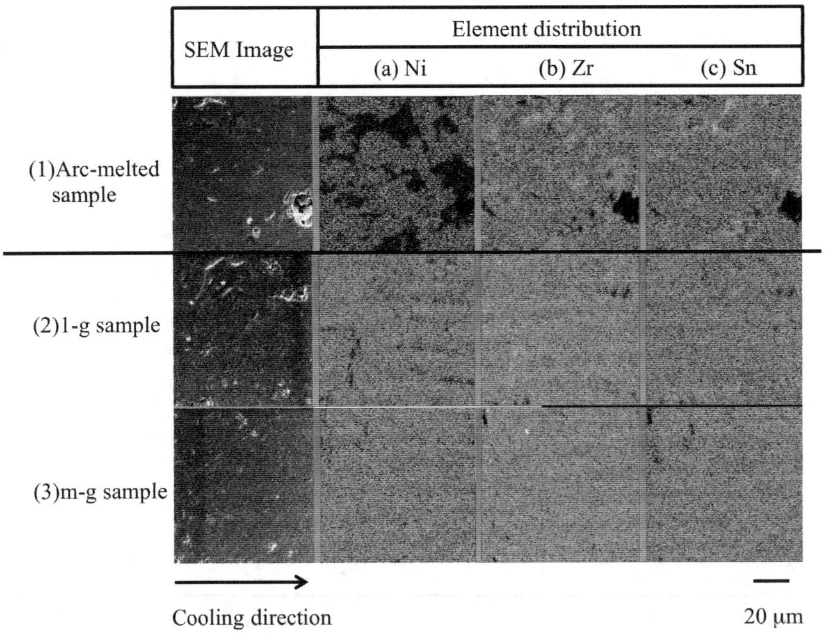

Figure 3. Element distribution of ZrNiSn sample

Figure 4 presents typical contact surface and cross section photos of the 1-g sample and the m-g sample. Both samples contained many small bubbles. Analysis of the m-g sample revealed sheet-like structure grains aligned in the cooling direction. In contrast, the 1-g sample had random columnar and granular grain structures.

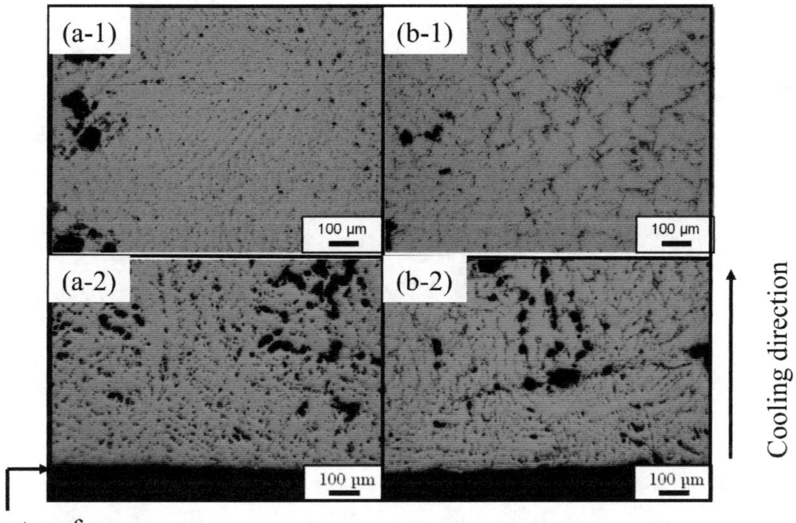

Figure 4. Microstructure of contact surface and cross section of the 1-g sample and m-g sample
(a-1): Contact surface of m-g sample, (a-2): Cross section of m-g sample
(b-1): Contact surface of 1-g sample, (b-2): Cross section of 1-g sample

Figure 5-(a) plots the XRD results of the contact surface of the 1-g sample and m-g sample acquired by using system (1). The 1-g sample had a pattern similar to that of the arc-melted sample and had no crystalline alignment along the cooling direction. The m-g sample had very strong (111) and (222) peaks. This result indicates that the m-g sample prepared using system (1) had a <111> crystallographic alignment of ZrNiSn phase along the cooling direction. Because Microgravity conditions produce a motionless and forceless melt and because the densest crystal plane in cubic system is [111], we obtained a sample with [111] alignment by using system (1). Neither sample had a crystalline alignment perpendicular to the cooling direction.

Figure 5-(b) plots the XRD results of the contact surface of the 1-g sample and m-g sample acquired by using system (2). The m-g sample had very strong (200) and (400) peaks, but the 1-g sample had no crystalline alignment along the cooling direction. This demonstrates that the m-g sample produced using system (2) had <100> crystallographic alignment along the cooling direction. When the sample was solidified using system (2), a ZrO_2 layer was formed on the contact surface of the sample with quartz glass. The obtained ZrO_2 layers had a monoclinic crystal structure, and that of the 1-g sample exhibited random alignment. However, that of the m-g sample exhibited strong (011), (022) and (033) peaks, suggesting that the ZrO_2 layer on the m-g sample had <011> crystallographic alignment. Because monoclinic ZrO_2 is transformed to high-temperature phase (tetragonal or cubic ZrO_2) when the sample melt solidifies, the aligned ZrO_2 layer of the high-temperature phase might be induced to produce a m-g sample with <100> crystallographic alignment along the cooling direction.

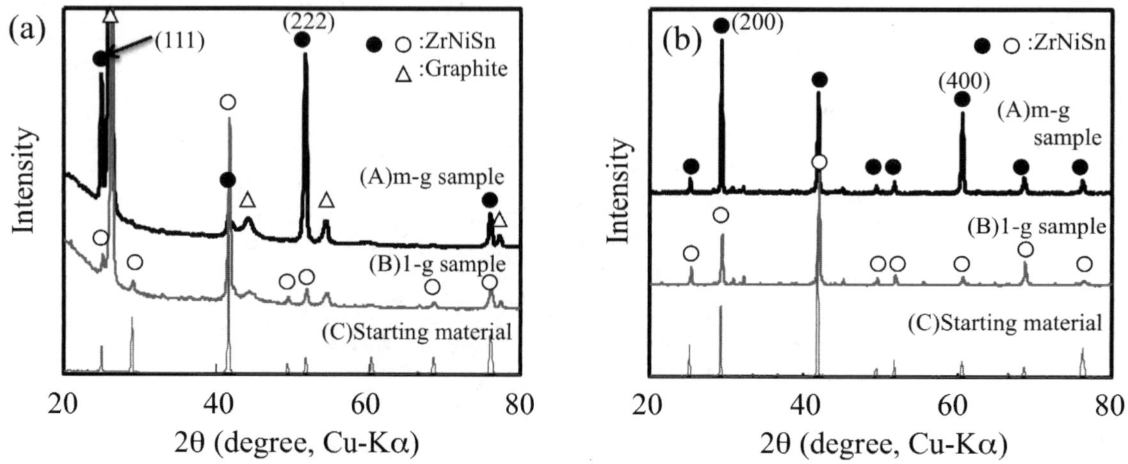

Figure 5. XRD result of the contact surface of the 1-g and m-g sample produced by different systems (a): Sample solidified by system (1), (b): Sample solidified by system (2)

Figures 6 and 7 plot the electrical conductivity (σ) and Seebeck coefficient (S) of the 1-g and m-g sample; Parameters σ and S of the m-g sample were measured along the cooling direction and perpendicular to the cooling direction, and those of the 1-g sample were only measured along the cooling direction because the 1-g sample had no crystalline alignment. Both m-g samples prepared by different systems had higher electrical conductivities along the cooling direction than perpendicular to the cooling direction, and those of the m-g sample along cooling direction were higher than those of the 1-g sample. The electrical conductivity anisotropy of the m-g sample obtained using system (2) exceeded that of the m-g sample obtained using system (1). These results suggested that this electrical conductivity anisotropy was affected by the crystalline alignment of the sample. In contrast, the sample's Seebeck coefficient did not exhibit any clear anisotropy, although there were differences between the 1-g and the m-g sample and differences of the cooling direction.

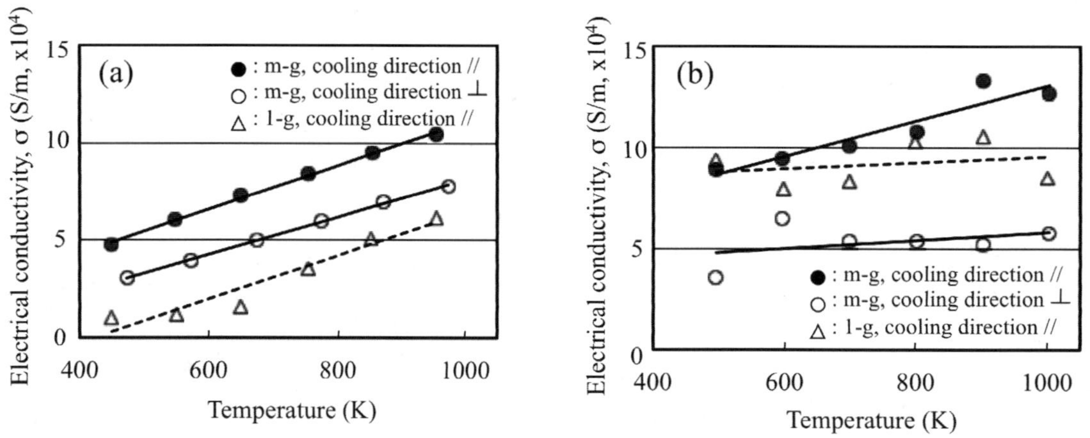

Figure 6. Electrical conductivity of the 1-g and m-g sample solidified by different systems (a): Sample solidified by system (1), (b): Sample solidified by system (2)

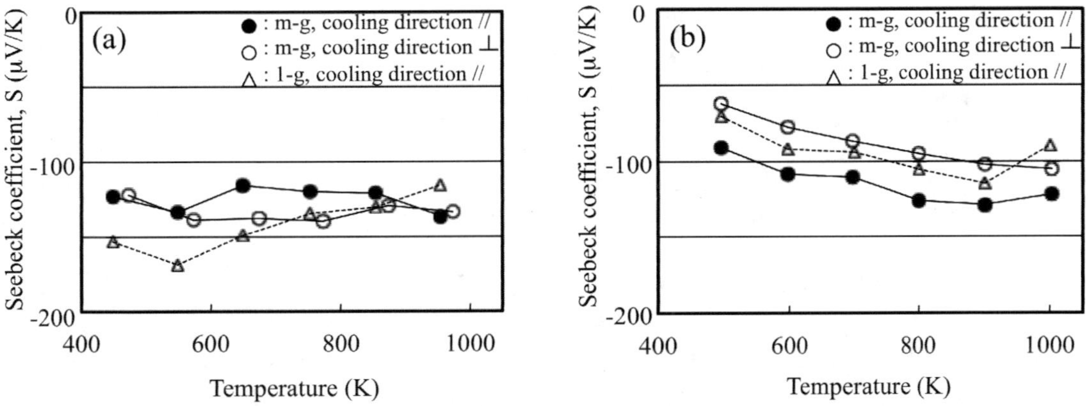

Figure 7. Seebeck coefficient of the 1-g and m-g sample solidified by different systems
(a): Sample solidified by system (1), (b): Sample solidified by system (2)

Figure 8 plots the temperature dependence of the thermal conductivities and phonon thermal conductivities of the 1-g and m-g samples along the cooling direction solidified using system (2). The thermal conductivity of the material, λ, is the sum of the electron and phonon contributions:

$$\lambda = \lambda_{electron} + \lambda_{phonon},$$

where $\lambda_{electron}$ is the electron thermal conductivity and λ_{phonon} is the phonon thermal conductivity. The electron contribution to the thermal conductivity can be estimated by the Wiedemann-Franz law:

$$\lambda_{electron} = L \cdot \sigma \cdot T,$$

where L is Lorenznumber, which is $2.445\text{x}10^{-8}$ $W \cdot \Omega \cdot K^{-2}$ for a degenerate free-electron gas system, and T is the absolute temperature. The thermal conductivity of the 1-g sample was less than those reported by Shen et al.[6] and Kawaharada et al.[7] because this sample contained many small bubbles, as indicated in figure 4. The m-g sample also contained many small bubbles, and the thermal conductivity was less than that of the 1-g sample. The m-g sample had a lower phonon contribution to the thermal conductivity than that of the 1-g sample. This lower phonon contribution of the m-g sample was considered due to phonon scattering by combination with bubbles and Ni site vacancies oriented by crystalline alignment.

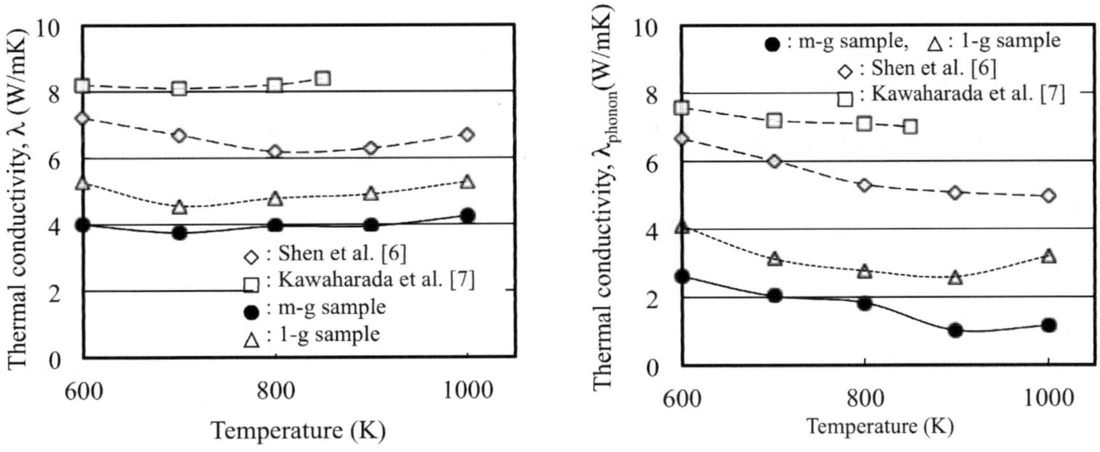

Figure 8. Thermal conductivity and phonon thermal conductivity of the 1-g and m-g sample solidified by system (2)

Table 1 lists the dimensionless figure of merit ZT ($= S^2 \cdot \sigma \cdot T/\lambda$) of the 1-g and m-g samples prepared using system (2). The ZT values of the m-g sample were about twice those of the 1-g sample because of increased electrical conductivity and reduced thermal conductivity. The maximum ZT of the m-g sample was 0.35 at 900 K. The ZT values of the 1-g sample were lower than those reported by Shen et al.[6] and Kawaharada et al.[7] because of lower Power factor ($=S^2 \cdot \sigma$) of the 1-g sample. The ZT values of the m-g sample at 900 and 1000 K were higher than the reference ones, although those of the m-g sample below 800 K were similar to those of the references.

Table 1. Dimensionless figure of merit of the 1-g and m-g samples solidified by system (2)

		ZT			
	Sample	1-g sample	m-g sample	Ref.[6]	Ref.[7]
Temperature (K)	600	0.04	0.09	0.09	0.12
	700	0.07	0.14	0.15	0.18
	800	0.13	0.23	0.19	0.23
	850				0.24
	900	0.18	0.35	0.21	
	1000	0.12	0.32	0.21	

4. Conclusion

A ZrNiSn half-Heusler sample with crystallographic alignment was prepared by unidirectional solidification in microgravity. The sample with crystallographic alignment exhibited the electrical conductivity anisotropy, and the thermal conductivity was affected by the structural difference of the sample. The ZT of the sample with crystallographic alignment exceeded those of other samples, and the maximum ZT was 0.35 at 900 K.

References

[1]Aliev F G, Kozyrkov V V, Moshchalkov V V, Scolozdra R V and Durczewski K 1990 *Z. Phys. B* **80** 353
[2]Uher C, Yang J and Hu S 1999 *Phys. Rev. B* **59** 8615
[3]Okutani T, Nakata Y and Nagai H 2004 *Ann. N.Y. Acad. Sci.* **1027** 158
[4]Okutani T, Nakata Y and Nagai H 2005 *Microgravity Sci. and Technol.* **16** 84
[5]Okutani T, Nagai H, Mamiya M, Shibuya M and Castillo M 2006 *Ann. N.Y. Acad. Sci.* **1077** 146
[6]Shen L, Goto T, Hirai T, Yang J, Meisner G P and Uher C 2001 *Appl. Phys. Lett.* **79** 4165
[7]Kawaharada Y, Uneda H, Muta H, Kurosaki K and Yamanaka S 2004 *J. Alloys Comp.* **364** 59

Journal of Physics: Conference Series **327** (2011) 012009 doi:10.1088/1742-6596/327/1/012009

Phase Selection in the Mushy-zone: *LODESTARS and ELFSTONE Projects*

D M Matson[1], R W Hyers[2], T Volkmann[3], H-J Fecht[4]

[1] Mechanical Engineering Department, Tufts University, Medford MA 02155 USA

[2] Mechanical and Industrial Engineering Department, University of Massachusetts, Amherst MA 01003 USA

[3] Institut für Materialphysik im Weltraum, Deutsches Zentrum für Luft- und Raumfahrt (DLR), D-51170 Köln, DE

[4] Institut für Mikro- und Nanomaterialien, Ulm Universität, D-89081 Ulm, DE

E-mail: douglas.matson@tufts.edu

Abstract. In a collaboration sponsored by ESA and NASA, international partners have developed a work plan to successfully address key issues relating to understanding the role of convection on alloy phase selection for commercially important structural alloys using the MSL-EML facility aboard the International Space Station. The approach is two-pronged. First, ground and space-based experiments will develop a baseline database to anchor subsequent modelling predictions. Tasks include sample preparation and verification, ground-based transformation evaluation, space-based experiments, and thermophysical property evaluation to support modelling activities. Second, modelling and theoretical analysis tasks will lead to a new understanding of the role of convection in phase selection for this class of materials. These models will allow prediction and control of microstructural evolution during solidification processing. Tasks include modelling of macroconvection induced by the EM levitation field, modelling of microconvection within the dendrite array, nucleation modelling, and modelling of the transformation kinetics specific to each alloy system. This paper outlines how two NASA-sponsored projects relate to the goals of the international collaboration.

1. Introduction

Recent advances in computerized simulation of metal casting processes allows investigators to visualize fluid flow, melt convection, channeling, and transient heat flux distribution to provide accurate predictions for process control, evolution of microstructures on a local level, and product quality. Of particular importance is the fidelity of measurements of the thermophysical properties of the melt. Evaluation of key properties near the melting temperature is challenging due to the high fluidity and reactivity of molten metal alloys. Use of containerless levitation processing techniques allows for precise measurements while minimizing the potential for melt contamination for samples in both superheated and undercooled condition [1].

Confidence in model predictions requires a better understanding of material properties. Progress in process control requires advancement in both. Feedback between experiments and simulation drives technology improvement with an emphasis on understanding the development of defect structures and evolution of localized microstructural domains. Thus, two essential components must simultaneously be addressed:

- Improve the quality of thermophysical property measurements to anchor model prediction with controlled experimental verification.
- Develop physical models to describe how fundamental processes such as phase selection and evolution of solidification structures influence local product properties.

This paper outlines efforts by the international community to organize programmatic structures to support the establishment of collaborative investigation of these two components and illustrates how this approach has been put into practice by two teams of investigators. This effort involves a broad range of activity including theoretical evaluation, multi-scale modeling, and experimental verification.

2. International collaborations

The European Space Agency (ESA) solicited proposals in September 2009 to encourage formation of International Topical Teams to address critical scientific needs to be addressed using facilities onboard the International Space Station (ISS). The goal was to reach out to the broader scientific community to consolidate research activities and foster collaborative global teamwork. AO 2009 PHYS-BIOSR identified several key topics including questions specific to materials science investigations. Two questions are highlighted in that they relate to use of the multiuser Material Science Laboratory Electromagnetic Levitation (MSL-EML) facility located in the Columbus module

- What are the thermophysical properties of high temperature melts?
- What is the influence of convection on the formation of different microstructures in alloys?

This framework resulted in the establishment of a number of topical teams requiring space-based containerless levitation processing experiments to facilitate achieving their specific scientific goals.

The International Topical Team on Thermophysical Properties (Thermolab ISS) was established under the leadership of Hans Fecht to specifically address the first of these two questions. Team members concentrate on ground and space-based investigations to develop a baseline database to anchor model predictions. Tasks include sample preparation and verification, ground-based transformation evaluation, space-based experiments, and thermophysical property evaluation to support modelling activities.

The International Topical Team on Solidification (SOL-EML) was established under the leadership of Dieter Herlach to address the second of these two questions. Team members concentrate on modelling and theoretical analysis tasks which lead to a new understanding of the role of convection in phase selection. These models will allow prediction and control of microstructural evolution during solidification processing. Tasks include modelling of macroconvection induced by the EM levitation field, modelling of microconvection within the dendrite array, nucleation modelling, and modelling of the transformation kinetics specific to a broad range of alloy systems.

The synergy attained through marriage of these two efforts satisfies the two essential components required for accurate process control modelling. Two NASA projects support these activities. The LODESTARS project (Levitation Observation of Dendrite Evolution in Steel Ternary Alloy Rapid Solidification) looks at phase selection in commercially important structural steel casting alloys with an emphasis on developing a thermophysical property database to improve and control microstructural evolution; the project involves collaboration with the Thermolab-ISS topical team. The ELFSTONE project (Electromagnetic Levitation Flight Support for Transient Observation of Nucleation Events) was established as part of the ESA PARSEC project (Peritectic Alloy Rapid Solidification with Electromagnetic Convection) [2] under the leadership of Thomas Volkmann and Douglas Matson to look at the role of convection in phase selection; these projects involve collaboration with the SOL-EML topical team.

3. Mushy-zone transformations

The mushy-zone is a two-phase region where solid and liquid exist in pseudo-equilibrium. When a molten metal is undercooled, the fraction solid that forms can be predicted by the Stefan Equation

$$C_p \, \Delta T = f_s \, \Delta H \tag{1}$$

where C_p is the specific heat, ΔT the undercooling, f_s the fraction solid and ΔH the latent heat of fusion. For short periods of time, limited heat is rejected to the environment and the fraction solid remains constant, although other processes can occur within the mushy-zone which profoundly influences the final microstructure. For some alloy classes, a metastable solid may form prior to conversion to the stable solid phase; often arising from surface energy considerations. The metastable mushy-zone solid then acts as a heterogeneous nucleation site for the stable phase [3]. Competitive nucleation and growth determine the phase composition of the final material [4].

Two classes of liquid/solid phase transformation show remarkably similar behaviour. Ternary FeCrNi steel casting alloys are eutectic systems where a cooling liquid forms two solids from the melt. During undercooling of hypoeutectic alloys two types of mushy-zone can form – the metastable body-centered cubic (bcc) ferrite or the stable face-centered cubic (fcc) austenite. [5-7] As seen in the metastable phase diagram shown in Figure 1, these alloys show significant solute partitioning and solute rejection is extremely important during solidification. When the metastable phase forms the mushy-zone, nickel is rejected while chromium is enriched and the liquid is solute lean. When the stable phase forms later, the opposite happens. Thus, clusters forming in the mushy zone do so in an environment which is solute lean.

Soft magnetic FeCo alloys are peritectic systems where a cooling liquid forms a solid and a liquid from the melt. The solid that forms may be either metastable bcc or stable fcc as was the case for the steel alloys but the influence of solute partitioning is opposite that observed previously. As seen in Figure 2, as the mushy-zone forms solute is rejected to the liquid and the growing clusters do so in an environment that is solute rich. This behaviour is seen not only in FeCo alloys [8] but also in FeNi [9], which show limited partitioning, and TiAlNb(Ta) peritectic alloys[10], which show significant partitioning.

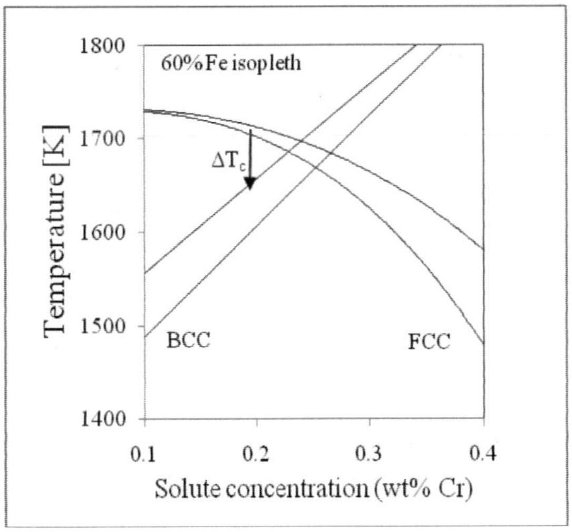

Figure 1. Metastable phase diagram for the eutectic FeCrNi ternary steel system along the 60 wt% isopleths. Hypoeutectic compositions show double recalescence.

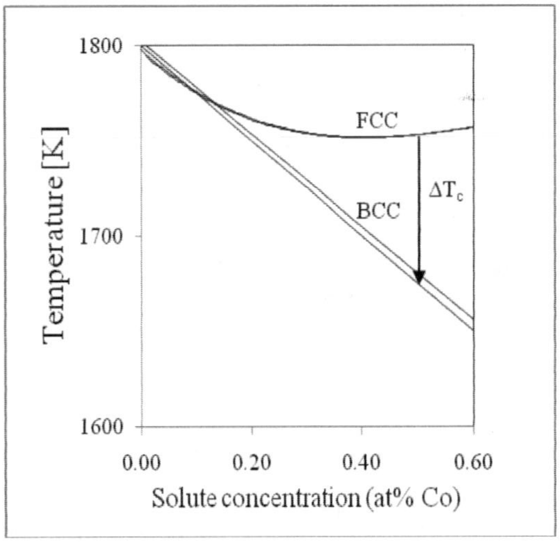

Figure 2. Metastable phase diagram for the peritectic FeCo binary system. Compositions greater than 15 at% Co show double recalescence

Figures 1 and 2 show that in order to access the metastable phase there must be some critical undercooling achieved. During formation of the mushy-zone the temperature of the undercooled liquid rises to the intermediate plateau temperature. After a brief delay, the stable phase nucleates within the mushy-zone and the temperature rises again. This process is known as double recalescence [11] and all residual metastable phase is consumed through engulfment, melting, and resolidification into the stable phase.

Both the critical undercooling and the transformation delay are influenced by melt convection [2]. Understanding how convection influences nucleation within the mushy-zone is the unifying theme for the LODESTARS and ELFSTONE projects.

4. Nucleation kinetics

The delay between recalescence events is commonly known as the incubation time and this represents how long it takes to develop nuclei of the stable phase. Convection significantly influences the incubation time during which the fcc phase clusters form on the bcc scaffold and grow to critical size to thus become stable. In classical nucleation theory [12-14], the steady-state nucleation rate, I_S, and time dependent nucleation rate, I, are related by the equations

$$I = I_S \exp\left[-\frac{\tau}{t}\right] \qquad \text{for } I_S = I_O \exp\left[-\frac{\Delta G^*}{k_B T}\right] \tag{2}$$

where τ is a characteristic incubation time, t the observed delay time, I_O a pre-exponential factor, ΔG^* the Gibbs free energy for formation of a critical nucleus of n* atoms of a pure material, and T the transformation temperature. The Boltzmann constant has a value of $k_B = 1.38 \times 10^{-23}$ J/atomK. Steady-state nucleation does not become appreciable until $\Delta G^* > 60\ k_BT$ while transient nucleation does not become appreciable until $t >> \tau$. Turnbull estimated that for condensed phases the incubation time was a function of the size of the cluster and the rate at which atoms cross the interface, β*, between the surrounding matrix and the cluster [15].

$$\tau = \frac{(n^*)^2}{\beta^*} \tag{3}$$

The attachment rate β* has variously been related to the jump frequency [15] and the lattice diffusivity [14, 16, 17]. By invoking the principle of time reversal, where the statistical fluctuations in cluster size follow the same path during growth and decomposition, Feder *et al* [18] determined the incubation time to be

$$\tau = \frac{-4\, k_B T}{\beta^* \left.\dfrac{\partial^2(\Delta G)}{\partial n^2}\right|_{n^*}} \tag{4}$$

Russell [19] evaluated condensed phase nucleation for binary systems to account for the influence of solute partitioning and found that clusters approaching the critical size were surrounded by an enriched solute shell, contrary to expectation, and that diffusivity of the slower moving species controls the attachment rate with an interchange frequency β'

$$\beta' = \frac{x}{a_o^2}\left(\frac{2\, D_A D_B}{D_A + D_B}\right) \qquad \text{for } x = C_A\left(\frac{A_S}{a_o^2}\right) \tag{5}$$

where x is the number of atoms jumping a jump distance a_o and the subscripts on the diffusivity D represent solute A and solvent B. The number of atoms jumping can be evaluated from the surface concentration C_A, the interface area A_S, and the area per atom. A key finding for multi-component systems is that the rate controlling step for linked flux evaluations is replacement of shell atoms of the controlling species and not the interfacial jump frequency. Thus solute partitioning is critical to understanding nucleation delays.

Unfortunately, classical theory does not include a mechanism whereby melt convection influences the thermodynamics or kinetics of nucleation. Convective solute mass transfer does not significantly enhance diffusivity. For diffusive transport alone the flux may be estimated as

$$J = D\frac{\partial C}{\partial x} = D\frac{C_L\,(1-k)}{\Omega\, a_o} \tag{6}$$

where J is the atomic flux to the cluster interface, D is the diffusivity, C_L the concentration of the key chemical species in the liquid, k the partitioning coefficient, Ω the atomic volume, and a_o the jump distance. Similarly for convective mass flux in a velocity field with average flow V,

$$J = \frac{V}{\Omega} \tag{7}$$

Using typical values for each property, convection contributes on the order of only an additional 30% to the total system flux and thus diffusive transport will always dominate. Thus, the role of convection is not to influence nucleation kinetics but rather it influences heterogeneous site development. The delay is thus comprised of two parts – the time to develop a nucleation site and, once the site exists, the time to grow a critical nucleus. Since dendrite fracture is not possible [20] and fragmentation is too slow [21], dendrite collision is the preferred site development mechanism [22].

Three crucial attributes are needed in order to accurately assess the influence of convection on phase selection. First, a single sample must be used for all testing. This ensures that chemical composition is constant and the influence of trace elements, such as oxygen which significantly influences thermophysical properties, is precluded. Second, multiple runs must be achieved at different convection conditions, preferably at a single undercooling. Third, the thermophysical properties of the sample must be known in order to provide accurate model predictions. Note that two models rely on these properties. Magnetohydrodynamic flow modelling must be used to understand and control flow conditions within the droplets [23] during the experiment. Later, solidification models require accurate thermophysical properties in order to track development of the metastable mushy-zone [3] and subsequent stable phase nucleation.

5. The NASA ESL facility for ground-based levitation testing

NASA operates an Electrostatic Levitation facility (ESL) at Marshall Space Flight Center (MSFC) in Huntsville AL. This facility provides critical ground-based thermophysical property measurement capability. Of particular importance to the LODESTARS and ELFSTONE projects is the ability to levitate, melt, undercool and test samples to obtain measurements of surface tension, viscosity and density as a function of temperature and composition [24]. Phase selection studies are also carried out in the absence of convection since the melt is quiescent under electrostatic levitation conditions. This baseline data set is used to anchor space results.

Figure 3 shows a comparison between electrostatic and electromagnetic phase selection test results.

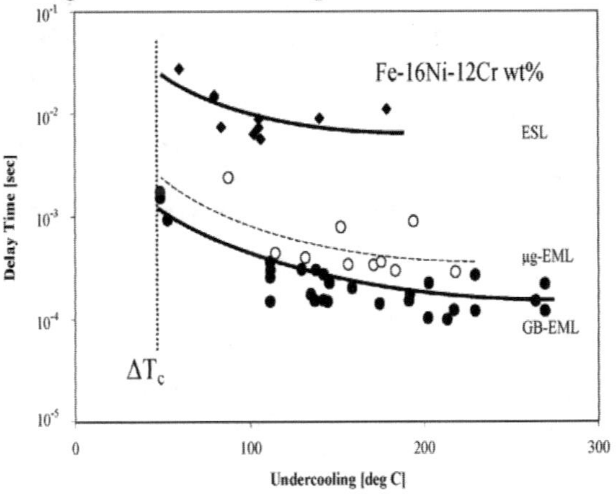

Figure 3. Comparing the nucleation delay times under conditions characterized by no flow (ESL), low flow (μG –EML) and high flow (GB-EML for a ternary steel alloy.

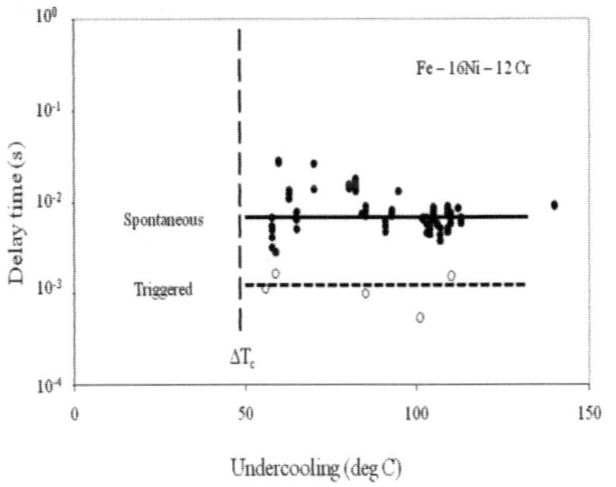

Figure 4. Nucleation stimulation creates surface forces which influence nucleation delay times even under no flow (ESL) conditions. Triggering must thus be avoided.

Nucleation delays are an order of magnitude slower with no induced flow as compared to results at high turbulent flow. Laminar flow tests during a previous space mission [25] show intermediate results. Ground-based results were obtained by spontaneous heterogeneous nucleation from the melt while the space results were obtained using a nucleation trigger to select the desired undercooling. Figure 4 shows that use of the nucleation stimulation trigger influences the delay time in ESL tests, most probably due to surface deformation with localized melt flow. Triggering is to be avoided in future space missions.

Ground-based ESL testing is enabling in that a wide variety of sample compositions and melt temperatures may be investigated using multiple samples. Unfortunately, all tests must be run without significant melt stirring and thus the influence of convection on phase selection cannot be evaluated using these techniques.

6. The MSL-EML facility for microgravity levitation testing

In ground-based electromagnetic levitation, gravity pulls the sample down into the levitation field causing significant melt convection. Turbulent conditions dominate. By testing in microgravity the need for positioning forces is significantly reduced and stirring may either be minimized or enhanced. Unlike ESL testing, microgravity EML has the ability to access a wide range of melt convection conditions on either side of the laminar/turbulent transition making it the ideal technique for phase selection investigations.

Several attributes are important to attaining the scientific goals of the LODESTARS and ELFSTONE projects. First, ISS microgravity experiment durations are sufficient to allow positioning, melting, undercooling, and phase selection testing on a single sample. Multiple tests may be run in series to provide coverage of the full range of convection conditions. Second, collaborations between the solidification modeling topical team and the thermophysical property evaluation topical team ensure proper experiment control. Recirculation flows within the droplets must be known for each test condition and flow models require accurate properties. Real-time measurement of critical properties, such as viscosity, can be accomplished on the same sample as is used for phase selection tests thus removing uncertainty created by variability introduced by trace alloy elements (such as oxygen content). The three crucial requirements of a single sample, multiple runs at different flow conditions, and know thermophysical properties can only be met using MSL-EML aboard the ISS.

7. Conclusions

The synergy provided by collaboration between the Thermophysical Property and Solidification International Topical Teams organized by ESA and supported by NASA under the LODESTARS and ELFSTONE programs is enabling for evaluating the influence of convection on phase selection. Microgravity testing provides all requirements necessary to provide for controlled convective conditions unattainable in any other environment. Results of these projects will provide new insight toward microstructural control using computerized simulation of solidification phenomena occurring in the mushy-zone for two dramatically different classes of phase transitions.

8. Acknowledgements

This article was produced as a part of the activities of the European Space Agency's International Topical Teams on *Solidification of Containerless Undercooled Melts (SOL-EML)* and *Thermophysical Properties of Liquid Metallic Alloys (ThermoLab)*. Work was partially supported by NASA under contracts NNX08AL21G and NNX10AV27G.

9. References

[1] Herlach DM, Cochrane RF, Egry I, Fecht HJ, Greer AL, 1993 *Int. Mat. Rev.* **38**[6] 373
[2] Matson DM, Hyers RW, Volkmann T, 2010 *J. Jap. Soc. Micrograv. Appl.* **27**[4] 238.
[3] Matson DM, Hyers RW, 2006, *Phil. Mag* **86**[24] 3795.

[4] Matson DM 1999, *Materials in Space – Science, Technology, and Exploration*, MRS Symposium Proceedings Vol. 551, Hepp AF, Prahl JM, Keith TG, Bailey SG, and Fowler JR, eds., (Materials Research Society, Warrendale PA) 227.
[5] Koseki T, Flemings MC, 1995 *Metall. Mater. Trans.* **26A**[11] 2991.
[6] Koseki T, Flemings MC, 1996 *Metall. Mater. Trans.* **27A**[10] 3226.
[7] T. Volkmann, W. Löser, and D.M. Herlach, 1997 *Metall. Mater. Trans.* **28A** 861.
[8] Hermann R, Löser W, Lindenkreuz G, Diefenbach A, Zahnow W, Dreier W, Volkmann T, Herlach D, 2004 *Mat. Sci. Engr.* **A375-377** 507.
[9] Barth M, Eckler K, and Herlach DM, 1991 *Mat. Sci. Engr.* **A133** 790.
[10] Shuleshova O, Woodcock TG, Lindenkreuz HG, Hermann R, Löser W, Büchner B, 2007 *Acta Mat.* **55** 681.
[11] Matson DM, Shokuhfar A, Lum JW, Flemings MC, 1996 *Solidification Science and Processing*, Ohnaka I and Stefanescu DM, eds., (TMS Warrendale PA) 19.
[12] Volmer M, Weber A, 1926 *Z. Phys. Chem.* **119** 277.
[13] Becker R, Döring W, 1935 *Ann. Phys.* **24**[8] 719.
[14] Turnbull D, Fisher JC, 1949 *J. Chem. Phys.* **17** 71.
[15] Hollomon JH, Turnbull D, 1953 *Prog. Met. Phys.* **4**, 333.
[16] Becker R, 1938 *Anal. Phys* **32**[1-2] 128.
[17] Kantrowitz A, 1951 J. *Chem. Phys.* **19**, 1097.
[18] Feder J, Russell KC, Lothe J, Pound GM, 1966 *Adv. Phys.* **15** 111.
[19] Russell KC, 1968 *Acta Met.* **16**, 761.
[20] Hilling J, Hellawell A 1996 *Met. Trans.* **27A** [1] 229.
[21] Herlach DM, Eckler K, Karma A, Schwarz M 2001 *Mat. Sci. Engr.* **304-306** 20.
[22] Hanlon AB, Matson DM, Hyers RW 2006 *Phil Mag. Lett.* **86**[3] 165.
[23] Hyers RW, Matson DM, Kelton KF, Rogers JR, 2004, *Ann. NY Acad. Sci* **1027** 474.
[24] Bradshaw RC, Warren ME, Rogers JR, Rathz TJ, Gangopadhyay AK, Kelton KF, Hyers RW, 2006 *Ann. NY Acad. Sci* **1077** 63.
[25] Team TEMPUS 1995 *Materials and Fluids Under Low Gravity*, Ratke L, Walter H, Feuerbacher B, eds., (Springer Publications NY) 233.

Journal of Physics: Conference Series **327** (2011) 012010 doi:10.1088/1742-6596/327/1/012010

Effect of cooling rate on solidification of Al-Ni alloys

A Ilbagi[1], P Delshad Khatibi[1], H Henein[1], R Lengsdorf[2], D M Herlach[2]
[1] Department of Chemical and Materials Engineering
University of Alberta
Edmonton, Alberta, CANADA T6G 2G6

E-mail: ilbagi@ualberta.ca

[2] Institut für Materialphysik im Weltraum
Deutsches Zentrum für Luft- und Raumfahrt
51147 Köln, Germany

Abstract. Particles of Al-Ni alloys with different compositions (Al–50 wt-% Ni and Al–36 wt-% Ni) were produced using a drop tube-impulse system, known as Impulse Atomization. The microstructure of these rapidly solidified particles was compared with those solidified in a DSC at low cooling rates (0.083 and 0.33 K/sec). Also, the microstructure of the sample solidified in microgravity on-board of the TEXUS 44 sounding rocket was analyzed. Neutron diffraction was used to investigate the phases formed during different solidification processes. From SEM micrographs and neutron diffraction it was found that the inner parts of the TEXUS sample and the sample that was cooled at 0.083 K/sec contain almost no eutectic structure. The outer rim of the TEXUS sample showed the highest amount of Al_3Ni and lowest amount of Al_3Ni_2 Increasing the cooling rate from 0.083 to 0.33 K/sec increased the Al_3Ni/Al_3Ni_2 ratio. Opposite trend was observed in the impulse-atomized particles, where increasing the cooling rate decreased the Al_3Ni/Al_3Ni_2 ratio.

1. Introduction

The materials characteristics are greatly influenced by microstructure evolution during the solidification, which is a function of processing parameters during production. A common solidification reaction in industrial alloys, such as Fe-C steels and Al-, Cu- and Ti-based alloys is the peritectic reaction [1]. Aluminum alloys with nickel are practically significant due to their high temperature strength and heat resistance [2] and also for their catalytic capability [3]. However, it has been shown that the final properties of the alloy depend on its constituent phases formed during the solidification process. For example, the activity of Raney nickel catalysts is strongly affected by the amount and morphology of the Al_3Ni [3-4].

Recently, modeling of the solidification of Al-rich Al-Ni alloys has attracted the attention of many scientists who try to understand and simulate this important and complex system [5-8]. On the other hand, many different experimental results have been published using different techniques to understand the effect of various processing parameters on the final phase fractions and the evolved microstructure [9-11]. Bao et al. [9] analyzed Al–Ni powders produced by gas atomization having different compositions on the Al-rich side of the phase diagram. They used neutron diffraction (ND) and X-ray diffraction (XRD) to identify the existing phases at the surface and in the bulk of the particles. Their results show a significant dependence of phase selection during the solidification of different sized gas atomized particles. Using impulse atomization technique, it has been shown that in

Al-50 wt-%Ni, by increasing cooling rate, the ratio of Al_3Ni to Al_3Ni_2 decreases, while opposite behavior was observed in Al-36 wt-%Ni [10-11].

Although atomization techniques offer containerless undercooling in conjunction with high degrees of cooling rates, the measurement of the whole history of nucleation and growth in these techniques is difficult. Therefore, electromagnetic levitation has been extensively used to containerlessly undercool bulk samples to study the solidification of a freely suspended droplet but at small cooling rates. The unique possibility of undercooling bulk samples due to the avoidance of heterogeneous nucleation on container walls coupled with accessibility of the solidifying droplet for direct observation and temperature measurement make the electromagnetic levitation technique a powerful tool for solidification studies [12]. This technique was used to study the effect of melt undercooling prior to solidification on the dendrite growth velocity of various Al-Ni alloys. The results showed that while on the Ni-rich side, growth velocity increases with increasing undercooling, Al-rich alloys show an unusual decrease in growth velocity with increasing undercooling [13]. Since it was already known that the melt stirring due to electromagnetic forces can generate pronounced forced convection [14], which in turn would affect the solidification, it was decided to perform the experiment in microgravity. Microgravity conditions provide the advantage of reduced electromagnetic stirring which is essential for the investigation of nucleating kinetics and growth in the absence of convection. The experiment was performed during the sounding rocket mission TEXUS 44 of the European Space Agency (ESA) and the German Aerospace Center (DLR). TEXUS sounding rocket missions provide an excellent quality of reduced gravity of the order of 10^{-4}g [15]. An Electro-Magnetic Levitator (EML) for use in reduced gravity environment (TEMPUS-EML) was utilized to process the sample in reduced gravity [16].

In this paper, solidification of Al-50 wt-%Ni and Al-36 wt-%Ni produced using different techniques will be studied. Solidification of both alloys in slow cooling rate condition in a furnace, high cooling rate condition during free-fall in a drop tube and high undercooling condition in reduced gravity on-board of the TEXUS 44 rocket will be discussed.

2. Experimental

2.1. Impulse atomization

Two different compositions of Al-Ni alloys (36 and 50 wt-% Ni) were produced by melting 99.9% pure Al and Ni (Alfa Aesar) in an induction furnace located at the top of 4 meters high drop tube. Each alloy was held for 30 minutes at 100 K above its liquidus temperature. Then, the liquid was pushed through small nozzles placed at the bottom of the crucible. In this technique, known as Impulse Atomization (IA), a liquid jet is generated when the molten metal emerges from the nozzles. Then, it breaks up into spherical droplets due to Rayleigh instability. The falling droplets lose heat to the surrounding stagnant gas (helium) and solidify before reaching the bottom of the atomization vessel. The details of this technique can be found elsewhere [17]. The solidified particles were washed and sieved into different size ranges, from 100 to 1000 μm, according to Metal Powder Industries Federation (MPIF) standard 05 [18].

Neutron diffraction (ND) was used to characterize the phases formed during solidification. The experiments were conducted using a neutron beam of 1.33 Å wavelength at National Research Council of Canada - Canadian Neutron Beam Centre (NRC-CNBC) in Chalk River, ON, Canada. To study the microstructure, Scanning electron microscopy was then performed using Zeiss Evo MA15 SEM with 20 keV electron beam energy.

2.2. Solidification at Low Cooling Rates

In order to compare the microstructure of IA samples to those solidified in near equilibrium conditions, 30 mg of IA particles of Al-36wt-%Ni and Al-50wt-%Ni with diameter of 750 μm were completely melted again by increasing the temperature to 1773 K. This was done in a Setaram Labsys Evo Differential Scanning Calorimetry (DSC). The DSC had been calibrated for temperature and heat measurement for the entire temperature range using standard samples of Zn, Sn, Al, Ag, Au and Ni. The molten samples solidified at two different cooling rates of 0.083 K/sec and 0.33 K/sec. The microstructure of as solidified particles was studied using SEM.

Image analysis was performed on the SEM micrographs of the impulse-atomized particles and those solidified using DSC to find out the volume fraction of each phase. This analysis was performed by ImageJ, a public domain Java image processing program [19].

2.3. Solidification under Microgravity

The TEMPUS facility, which is designed for electromagnetic levitation (EML) in reduced gravity [16], has been integrated into a sounding rocket. TEXUS sounding rockets provide about 320 seconds of reduced gravity time. The crystallization front velocity of Al-50wt-%Ni in the undercooled liquid phase was measured during the TEXUS44 sounding rocket flight [9]. Three undercooling and solidification cycles were obtained. The sample was heated by more than 200 K above the liquidus temperature of this alloy in order to reduce and even eliminate Al-Oxides at the surface of the liquid drop, which can act as heterogeneous nucleation sites of high catalytic potency. The sample undercooled during the subsequent solidification cycles to ΔT_1=185 K and ΔT_2=220 K, respectively. The TEXUS 44 sample was then studied using SEM and neutron diffraction. Image analysis using the software ImageJ was also used to find the volume fraction of the phases in the TEXUS sample.

3. Results and Discussion

3.1. Solidification at slow cooling rates

Figure 1 and 2 show the microstructure of Al-50 wt-%Ni particles solidified at cooling rates of 0.083 K/sec and 0.33 K/sec, respectively. It can be observed that in these figures the microstructure is slightly refined at the higher cooling rate.

In both figures, Al_3Ni_2, the light grey phase, and Al_3Ni, the dark grey phase, were identified using EDX. No sign of the primary AlNi phase was observed in these samples. This will be further discussed in 3.2. It was also noticed that in the particles solidified under the lower cooling rate (0.083 K/sec), visually no eutectic structure can be found. On the other hand, in the particle that solidified at 0.33 K/sec cooling rate, eutectic microstructure can be seen (Figure 2). It must be noted that the equilibrium phase diagram does not predict any eutectic transformation for this alloy. This will be discussed in the section "Phase quantification". On the other hand, microstructural analysis of Al-36 wt-%Ni particles solidified with cooling rates of 0.083 K/sec and 0.33 K/sec also showed that the particle cooled at higher cooling rate has finer microstructure. At both cooling rates, solidified particles contain the primary phase Al_3Ni_2, the peritectic phase Al_3Ni and the eutectic. It is clear that for this composition even at such low cooling rates the peritectic reaction of Liquid+$Al_3Ni_2 \rightarrow Al_3Ni$ does not go to completion. This is not surprising since peritectic reactions need diffusion in solid and therefore behave very sluggish [20].

3.2. Solidification at high cooling rates

It is known that impulse atomization technique produces rapidly solidified particles [17]. Microstructure of an impulse-atomized particle of Al-50 wt-%Ni with the diameter of 328 μm is shown in Figure 3. Clearly, the microstructure is significantly finer than those in Figures 1 and 2. EDX analysis of the phases observed in Figure 3 showed that the white phase, light grey phase and dark grey phase are indeed Al_3Ni_2, Al_3Ni and eutectic structure, respectively and no sign of AlNi was observed. Shuleshova et al. [21] by in situ X-ray measurements on highly undercooled levitated droplets showed that in Al-50 wt%Ni even under non-equilibrium conditions the primary phase is AlNi. It obviously decomposes on further cooling as it is not observed in the final microstructure. The observed microstructure will be discussed in the section "Phase quantification".

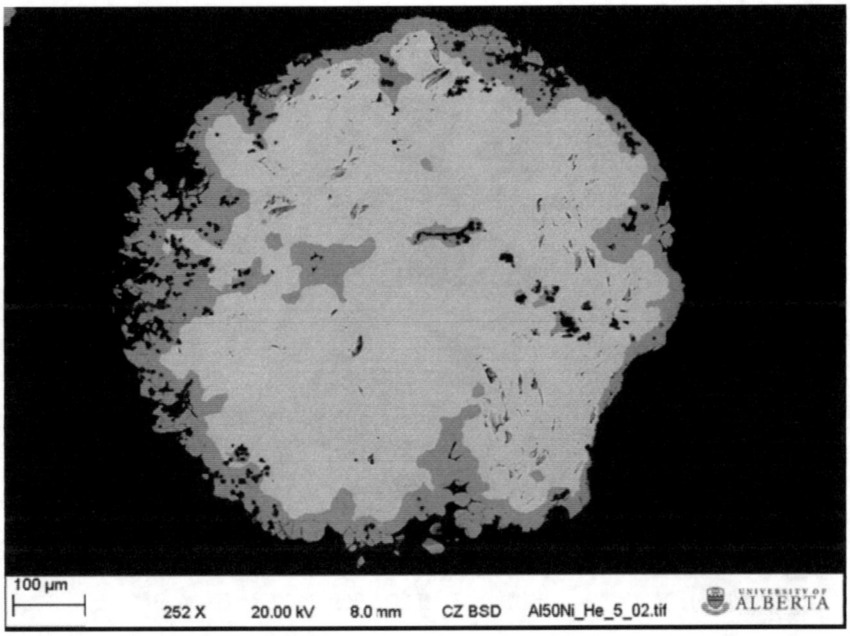

Figure 1. SEM micrograph of Al-50 wt-%Ni particles solidified under cooling rates of 0.083 K/sec.

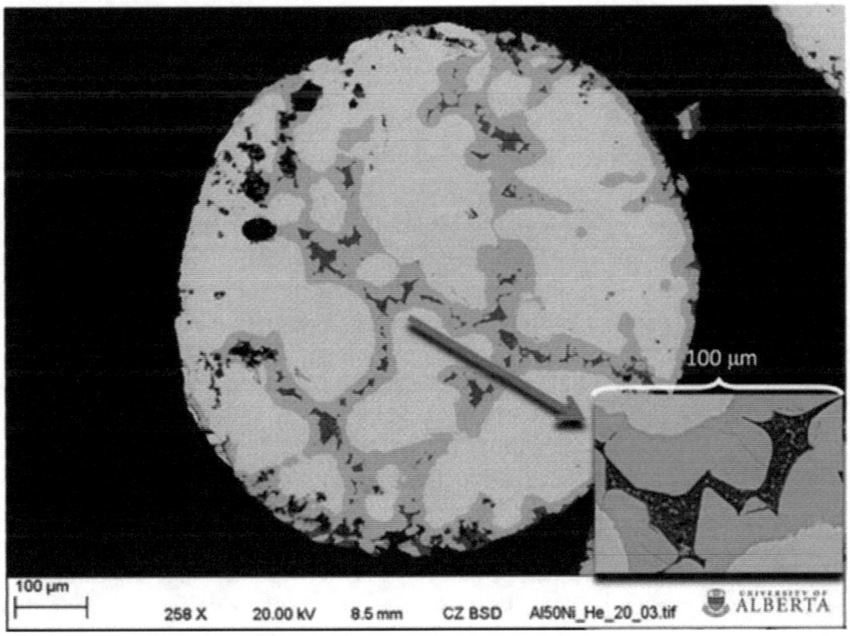

Figure 2. SEM micrograph of Al-50 wt-%Ni particles solidified under cooling rates of 0.33 K/sec.

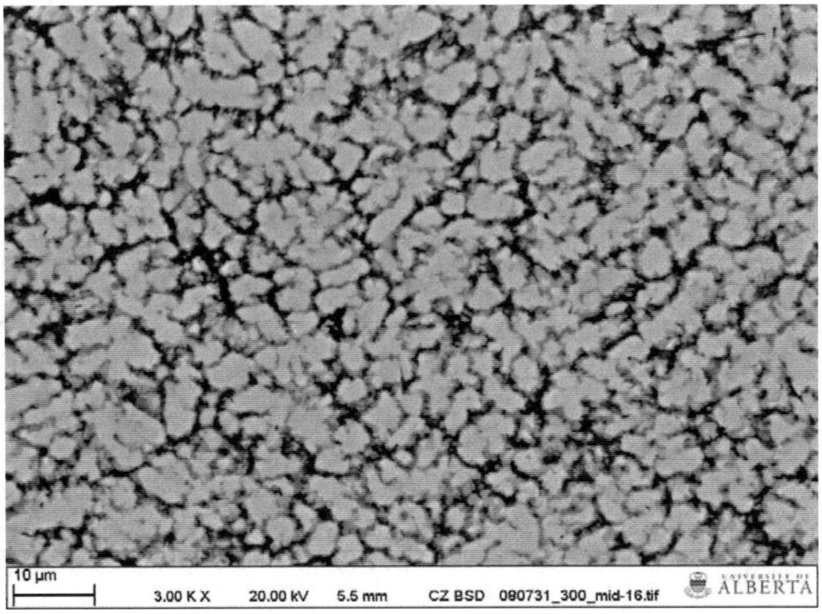

Figure 3. SEM micrograph of Al-50 wt-%Ni impulse-atomized particle D=328 µm.

3.3. Solidification in reduced gravity

The temperature-time profile of the TEXUS sample, Al-50 wt-%Ni alloy, is shown in Figure 4. The spikes on the temperature curve at the beginning of the experiment are due to oxides that were still on the sample's surface. Since oxides show in general a higher emissivity than metals the pyrometer signal changes abruptly each time when an oxide layer on the surface of the rotating sample moves thru the observation window of the pyrometer. After oxides were eliminated by increasing the temperature, the oscillations of the temperature signal vanished and the sample was cooled until the first recalescence was completed. Then the sample was completely melted again for the second cooling cycle. During this cycle, and shortly after the first recalescence, the reduced gravity period was finished. The primary phase undercooling achieved was ΔT_1=185 K and ΔT_2=220 K for the first and second cycles, respectively. The cooling rate achieved by the sample before the first recalescence of the second cycle was inferred from the temperature-time profile as 14 K/sec, assuming that the slope of the cooling curve was constant in the last 200 K prior to the start of solidification. The cooling rate was further decreased after the first recalescence to be 10 K/sec.

The growth velocity of the solidification front in the undercooled liquid was also measured. The result of this measurement was presented elsewhere [13]. In brief, performing the experiment on Al-50 wt-%Ni on-board of the TEXUS 44 sounding rocket resulted in observation of two significant and important differences between the experiments in normal gravity and in reduced gravity. First, the growth velocity measured in reduced gravity is substantially smaller than the data taken under 1g conditions. Second, in reduced gravity, it was apparent that the growth velocity increases with increasing undercooling as usual for a great variety of alloys. In this section the solidification microstructure of the TEXUS 44 specimen will be discussed.

Microstructural analysis of the solidified specimen showed a distinct difference between the inner part and outer rim of the particle. The microstructure of the outer rim of the TEXUS sample is shown in Figure 5a and 5b. The outer rim of the sample contains Al_3Ni_2 primary dendrites, which are surrounded by the peritectic phase Al_3Ni and an interdendritic Al-rich eutectic.

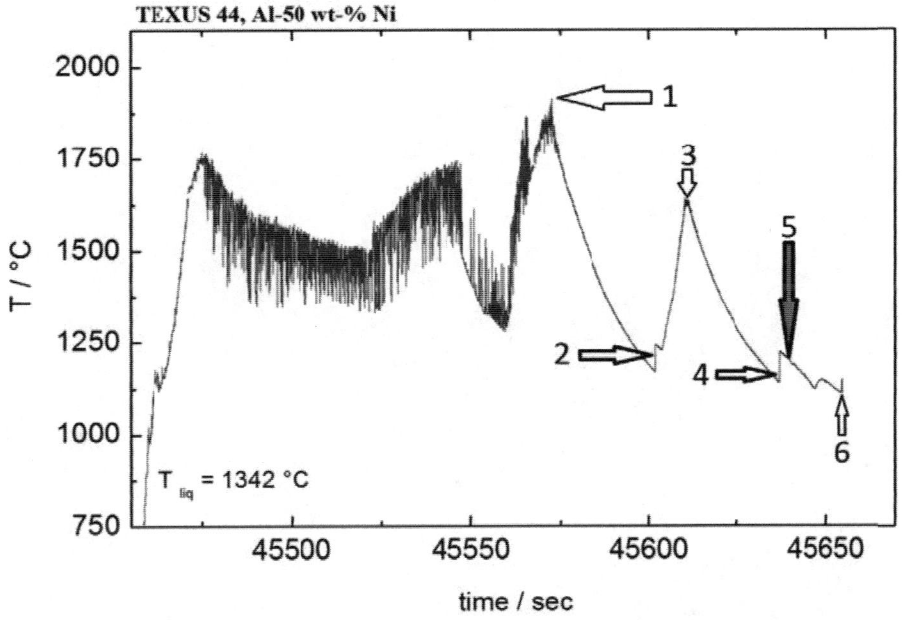

Figure 4. The temperature-time profile of the TEXUS 44 specimen. The arrows are: 1. Start of first cooling cycle, 2. First recalescence of the first cycle, 3. Start of second cooling cycle, 4. First recalescence of the second cycle, 5. End of microgravity, 6. End of temperature measurement period.

Figure 6 shows an SEM micrograph of the inner part of the Al-50 wt-%Ni sample processed during TEXUS 44 flight (in the following denoted as TEXUS sample). The inner part is characterized by a large amount of porosity, the absence of the Al-rich eutectic and dominant occurrence of the Al_3Ni_2 phase.

Neutron diffraction was used to investigate the phases formed within the impulse-atomized particles and the TEXUS sample. Figure 7 shows a small range of the diffractograms from all samples superimposed on each other. All peaks found in the diffractogram correspond to the phases Al, Al_3Ni and Al_3Ni_2.

From Figure 7 it is evident that the TEXUS sample is missing the characteristic (111) peak of Al solid solution phase, a component of the eutectic, at 2.68 $Å^{-1}$. Given the high depth of penetration of neutrons in the sample and the absence of Al phase peak in the diffraction pattern, clearly if any Al phase is present in the sample, it is below the ND detection limit, which is about 0.5 vol.% [22].

3.4. Phase quantification

SEM micrographs from the samples solidified in the DSC, impulse-atomized particles and the TEXUS samples were further investigated using ImageJ to find the volume fraction of consisting phases. For this purpose, from each of the samples solidified in the DSC, ten particles were used for image processing. The average of the phase fractions were then taken and reported as the volume fraction of each phase.

For impulse-atomized samples, twelve images from each particle size were used to measure the volume fraction of phases in each particle size. Then, the secondary dendrite arms spacing was measured and the cooling rate achieved for each particle size was estimated [10-11] using the coarsening model proposed by Kurz and Fisher [23].

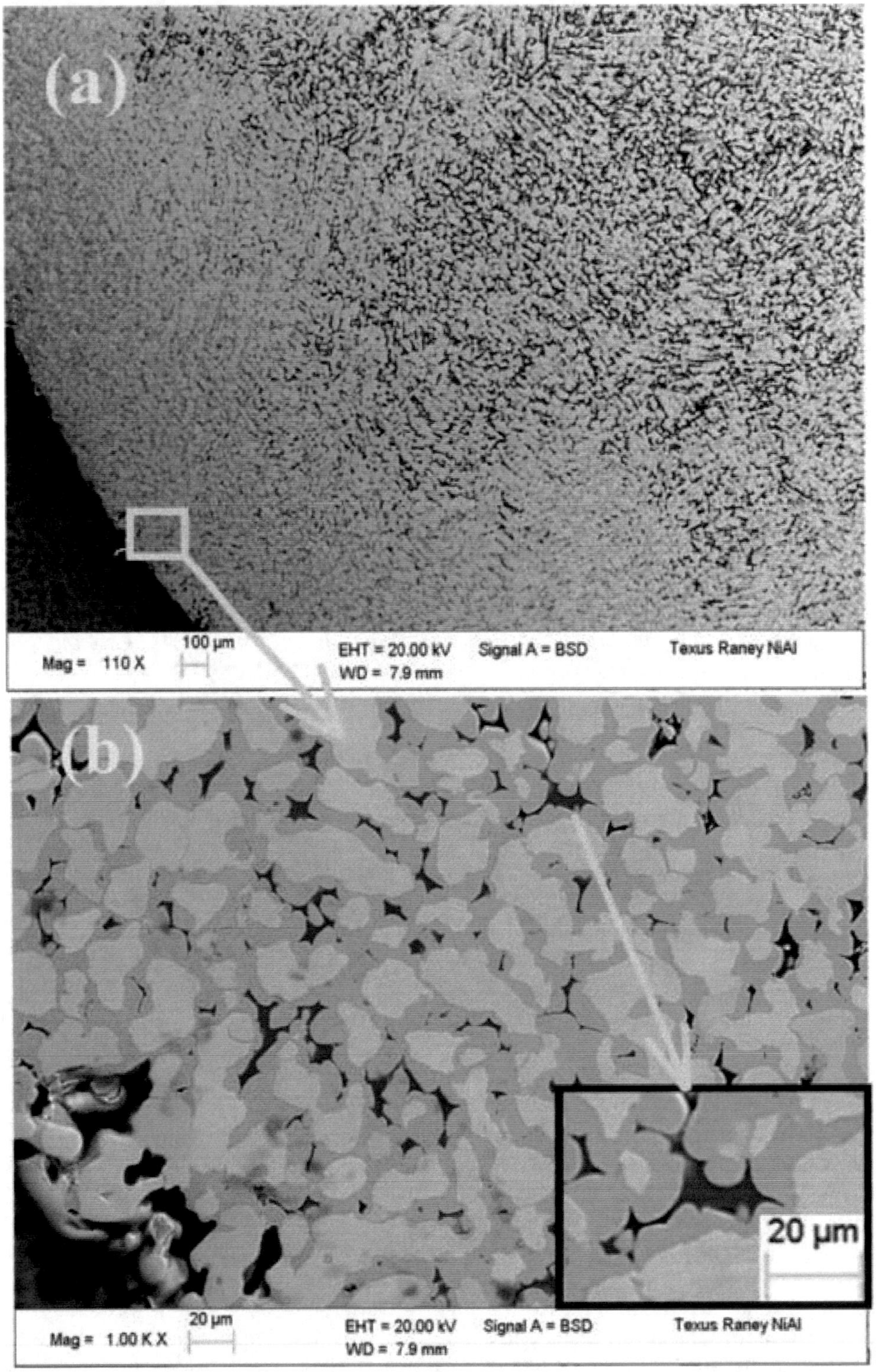

Figure 5. a) SEM backscattered electron image of the TEXUS Al-50 wt-%Ni specimen. Outer rim and inner part. b) Outer rim of the specimen shown in Figure 5a in higher magnification.

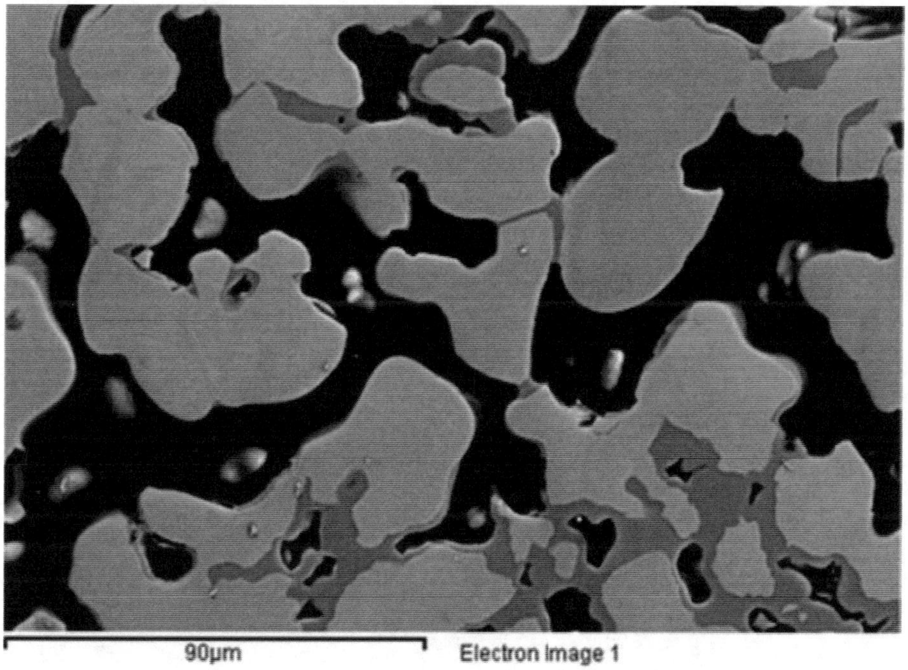

Figure 6. SEM backscattered electron image of the inner part of the TEXUS solidified Al-50 wt-%Ni specimen exhibiting the Al_3Ni_2, light grey area, and Al_3Ni phase, dark grey area. The black areas represent voids.

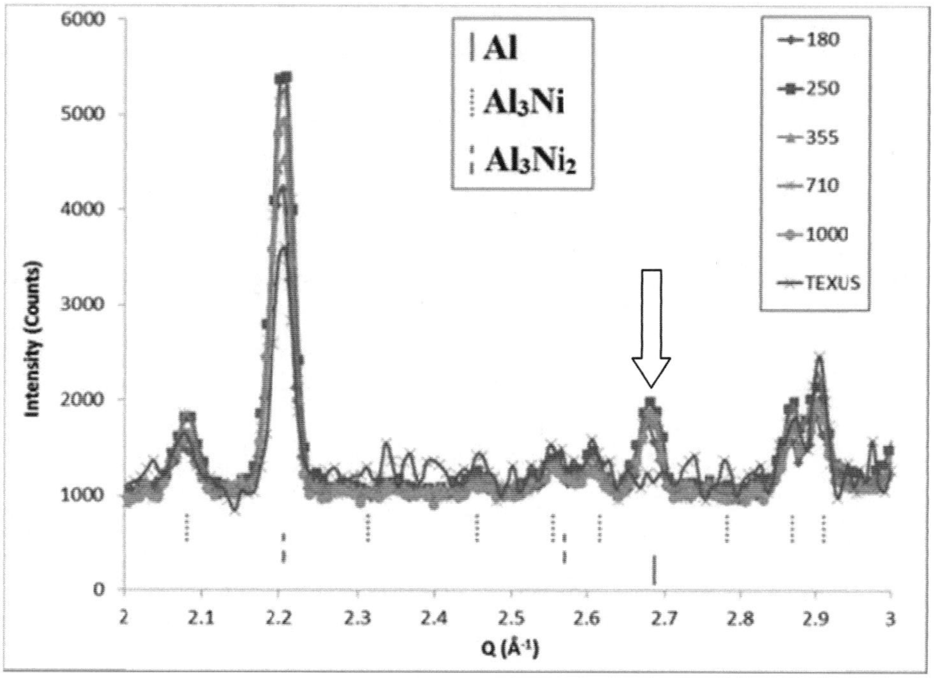

Figure 7. Part of neutron diffractogram of various particle sizes of impulse-atomized Al-50 wt-%Ni and the TEXUS sample, showing the absence of the Al peak in the TEXUS sample.

The secondary dendrite arm spacing was related to the cooling rate using Eq. 1 and Eq. 2 for Al-50 wt%Ni and Al-36 wt%Ni, respectively [11].

$$\lambda = 75.38\dot{T}^{-0.33} \tag{3}$$

$$\lambda = 69.58\dot{T}^{-0.33} \tag{4}$$

The volume percent of the phases in the inner part and outer rim of the TEXUS sample were also measured from the SEM images of each part. For the cooling rate, from Figure 4, it was found that the maximum cooling rate was achieved at the beginning of the solidification, 14 K/sec, which then declined to 10 K/sec after the first recalescence. It is expected that this cooling rate goes further down as more solid forms and releases its latent heat; and also, as the temperature gradient between the droplet and the environment decreases.

Figure 8 shows the Al_3Ni to Al_3Ni_2 ratio that was found from image analysis and was plotted as a function of cooling rate. The data for TEXUS sample is drawn with a straight line for each of the inner parts and the outer rim. Both lines were cut off at 14 K/sec since that was the maximum cooling rate achieved. Also on this figure, the results reported by Patchett and Abbaschian [26] are shown by circles along with the technique that was used to produce the sample.

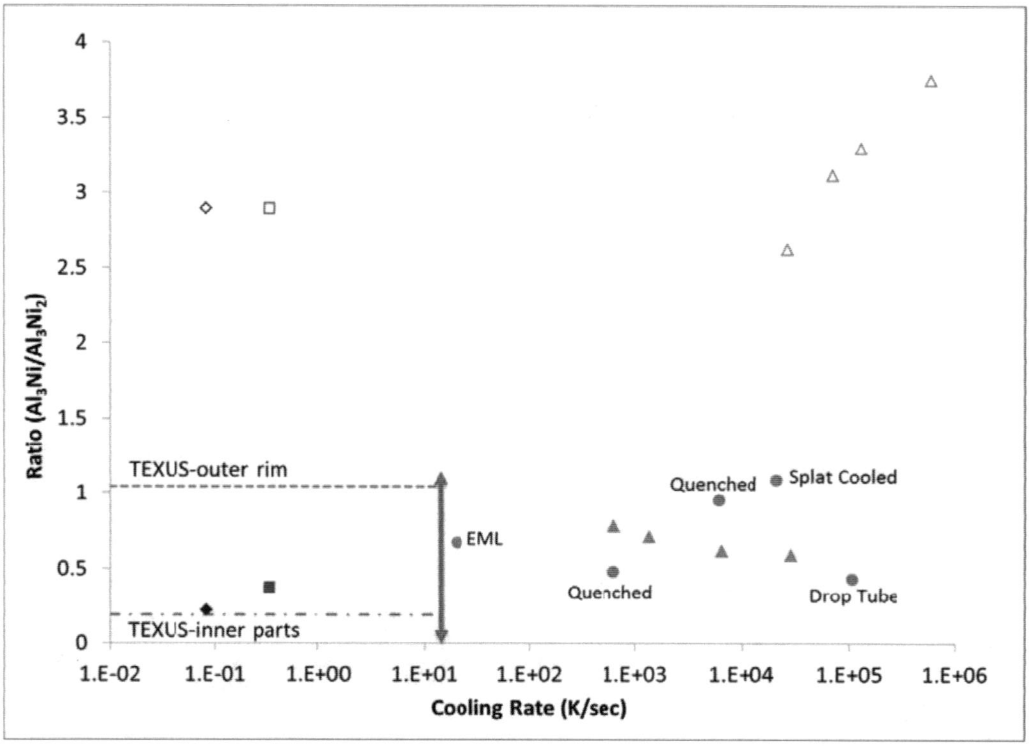

Figure 8. The Al_3Ni/Al_3Ni_2 ratio as a function of cooling rate from the image analysis on samples of Al-50 wt-%Ni (filled symbols) and Al-36 wt-%Ni (empty symbols). Different symbols represent different production techniques: DSC (at two cooling rates of 0.08 K/sec, shown by diamonds, and 0.33 K/sec, shown by squares), impulse atomization, shown by triangles, and EML during TEXUS 44 campaign, shown by dashed lines, (only Al-50 wt-%Ni), as well as from the literature, shown by circles. The line with arrows represents the cooling rate achieved by the TEXUS sample before solidification.

The first striking feature of Figure 8 is the similarity of the Al_3Ni/Al_3Ni_2 ratio in the sample that was cooled at 0.083 K/sec cooling rate in the DSC and that of the inner parts of the TEXUS sample. Also, as it was mentioned in the previous sections, in both samples the eutectic was not observed. Understanding the similarity of the Al_3Ni/Al_3Ni_2 ratio in the inner parts of the TEXUS sample to that of the slow cooled sample requires better understanding of the solidification of both samples. Possible solidification scenarios are now discussed.

The sample that was cooled in the DSC, experienced prolonged time at high temperatures resulting in extended coarsening of the primary phase. This leaves a little surface area for the nucleation of the peritectic phase Al_3Ni. On the other hand, due to extensive coarsening of the primary phase there is not enough liquid for peritectic reaction to proceed further. These hypotheses can explain why the ratio of Al_3Ni/Al_3Ni_2 is much smaller than that predicted by the equilibrium phase diagram (~0.89). Increasing the cooling rate, by almost one order of magnitude in the DSC, results in less coarsening and finer structure, which leaves more surface area for Al_3Ni nucleation. That can explain the increase in the Al_3Ni/Al_3Ni_2 ratio. However, larger cooling rate does not allow for further continuation of Al_3Ni growth, as the temperature reaches the eutectic temperature. Therefore, the remaining liquid solidifies into eutectic microstructure (Figure 2).

In the TEXUS sample, after it experiences 220 K primary phase undercooling, solidification starts from the surface, since the heat transfer is highest on the surface, and solid grows towards the inner parts. The initial growth rate is high because of high level of undercooling. However, this high growth rate decreases rapidly as the recalescence increases the sample temperature and decreases the cooling rate. Tourret et al. [5] showed that in Al-42 wt%Ni less than 10 vol.% of the solid formed by the end of the recalescence period in which the growth rate is the fastest, while more than 60 vol.% of the solid formed after the recalescence and before the peritectic reaction $L+Al_3Ni_2 \rightarrow Al_3Ni$. The growth rate in this stage is much slower than that during the recalescence. It is expected that the amount of solid formed before the peritectic reaction to be higher in Al-50 wt%Ni alloy. It is argued that the decreasing cooling rate as a result of increasing solid formation and poor heat extraction from the large droplet (~6mm diameter) in the EML chamber [24-25], resulted in the formation of large amount of Al_3Ni_2. Therefore, similar to the sample solidified in the DSC, there are not enough surfaces available for the nucleation of Al_3Ni. That is why the ratio of Al_3Ni/Al_3Ni_2 in the inner parts of this sample is low. It is also expected that most of the remaining liquid in the small interdendritic regions to be consumed by the peritectic reaction, and thus, no eutectic region is seen in this sample.

From Figures 8 it is clear that the value of Al_3Ni/Al_3Ni_2 ratio in the outer rim of the TEXUS sample is significantly higher than that of other samples studied in this research. As it was discussed, it is expected that the solidification starts from the surface at high growth rate. As a result, the microstructure in this region is refined and more surface area is available for the Al_3Ni nucleation. However, this cannot justify the high volume fraction of Al_3Ni in this region. It is known that the heat treatment of the Al-Ni samples results in an increase in Al_3Ni fraction [26]. It is argued that since this region acted as the heat passage for the entire volume of the specimen, a phenomenon similar to heat treatment has occurred to the phases in the outer rim.

On the other hand, Figure 5 shows that the outer rim of the TEXUS sample contains eutectic structure. During solidification, slow growth of Al_3Ni phase coupled with the higher cooling rate in the outer rim does not allow for further continuation of Al_3Ni growth, which in turn can result in remaining of liquid to transform to eutectic structure.

Impulse atomization on the other hand can effectively extract the heat from the solidifying droplet. Extensive refinement of the microstructure as a result of high cooling rate and undercooling due to containerless processing provides massive surface area for the Al_3Ni to nucleate. However further increasing the cooling rate results in reduction of Al_3Ni/Al_3Ni_2 ratio. This can be explained by the slow growth rate of Al_3Ni phase. It should be noted that the effect of cooling rate on the phase fractions is two-fold; (1) it affects the nucleation and growth rate of each phase and (2) it refines the microstructure of the primary phase, which in turn increases the interface area between solid and liquid and influences the heterogenous nucleation of the peritectic phase [27]. These competing

Journal of Physics: Conference Series **327** (2011) 012010 doi:10.1088/1742-6596/327/1/012010

phenomena can define the ratio of the phases within the droplets.

Comparing the results of the techniques used in this research with those reported by Patchett and Abbaschian [27] confirms the importance of the processing parameters in the phase fractions of a solidified specimen. While the Al_3Ni/Al_3Ni_2 ratio of the two samples that they studied in a containerless system such as EML and drop tube follows similar trend to that of the TEXUS and impulse-atomized samples, the results from the quenched and splat cooled samples are significantly different. This may be due to the constrained growth of the solid in the quenched and splat cooled samples, in which the released latent heat is extracted through the solid. In containerless techniques such as EML, impulse atomization or drop tube, however, the latent heat is extracted through the undercooled liquid. This can result in different levels of undercooling (primary phase, peritectic and eutectic) for the solidifying phases. Considering the complexity of the Al-Ni system with two peritectic and one non-equilibrium eutectic transformations it is not surprising that the different levels of undercooling achieved in different techniques resulted in different trend of changes in Al_3Ni/Al_3Ni_2 ratio. Also, the possibility of an anisotropic structure coupled with measurements using microscopy techniques may lead to errors and large scatter in measurements. Further work using 3D quantitative characterization techniques, such as neutron diffraction, on samples solidified using processes that result in constrained growth is needed.

In the case of Al-36 wt-%Ni, increase of the cooling rate from 0.083 K/sec to 0.33 K/sec did not have a significant effect on the Al_3Ni/Al_3Ni_2 ratio. Increasing the cooling rate to that of impulse-atomized droplets seems to initially decrease the ratio. However, further increasing the cooling rate resulted in higher ratio. In the lower cooling rate regime, such as those in the DSC, increasing the cooling rate refined both Al_3Ni_2 and Al_3Ni, which did not affect the ratio significantly. Further increasing the cooling rate may result in the reduction of the ratio as the growth of Al_3Ni is sluggish. However at high cooling regime, such as in impulse-atomized droplets, increasing the cooling rate results in suppression of the primary Al_3Ni_2. It is also shown that the cooling rate required for suppression of primary Al_3Ni_2 decreases as the composition moves toward the peritectic liquidus temperature [27].

4. Summary

Two different compositions of Al-Ni alloys (36 and 50 wt-% Ni) were solidified under different conditions. Slow cooling rate in a DSC, impulse atomization in a drop tube. Also, Al-50 wt-%Ni was solidified on-board of TEXUX 44 sounding rocket using the TEMPUS EML facility for containerless processing of liquid metals in reduced gravity.

From the SEM micrographs it was found that both the inner parts of the TEXUS sample and the sample that was cooled at 0.083 K/sec in a DSC contain almost no eutectic. Neutron diffraction confirmed this observation.

The outer rim of the TEXUS sample showed the highest amount of Al_3Ni and lowest amount of Al_3Ni_2. This observation was attributed to the reheating of Al_3Ni_2 in that zone during post-recalescence period.

Increasing the cooling rate from 0.083 K/sec to 0.33 K/sec in DSC experiments resulted in the formation of some eutectic microstructure and it increased the Al_3Ni/Al_3Ni_2 ratio. Opposite trend was observed in the impulse-atomized particles at higher cooling rates, where increasing the cooling rate decreased the Al_3Ni/Al_3Ni_2 ratio.

In Al-36 wt-%Ni, increasing the cooling rate from 0.083 K/sec to 0.33 K/sec did not significantly change the volume fraction of the respective phases. However in the impulse-atomized sample the Al_3Ni/Al_3Ni_2 ratio was increased.

5. Acknowledgements

This research was supported by Canadian Space Agency (CSA) and by Natural Sciences and Engineering Research Council of Canada (NSERC). In addition, the authors would like to thank National Research Council of Canada – Canadian Neutron Beam Centre, Chalk River, ON, Canada for their support in conducting the ND experiments.

This work is part of NEQUISOL project supported by ESA within contract 15236/02/NL/SH and CSA within contract number 9F007-08-0154 and SSEP Grant 2008. The authors thank Stefan Schneider for assistance in conducting the TEXUS experiments.

6. References

[1] Kerr H W and Kurz W 1996 *Int. Mater. Rev.* (US) **41** 129
[2] Povarova K B, Lomberg B S, Filin S A, Kazanskaya N K, Shkol'nikov D Y and Bespalova M D 1994 *Izv. Ross. Akad. Nauk, Ser. Met.* **3** 77
[3] Devred F, Reinhart G, Iles G N, Van D K, Adkins N J, Bakker J W, Nieuwenhuys B E 2011 *Catal. Today* **163**(1) 13
[4] Devred F, Gieske A H, Adkins N 2009 *Appl. Catal., A* **356**(2) 154
[5] Tourret D, Gandin C -A, Volkmann T 2011 *Acta Mater.* **59**(11) 4665
[6] Tourret D and Gandin C -A. 2009 *Acta Mater.* **57**(7) 2066
[7] Tourret D and Gandin C -A 2009 *Proc. the 12th Int. Conf. on Modeling of Casting, Welding, and Advanced Solidification Processes* 603
[8] Siquieri R, Doernberg E, Emmerich H and Schmid-Fetzer R 2009 *J. Phys. Condens. Matter* **21**:464112 6
[9] Bao C M, Dahlborg U, Adkins N and Calvo-Dahlborg M 2009 *J. Alloys Compd.* **481**(1-2) 199
[10] Ilbagi A, Henein H and Phillion A B 2010 *J. Mater. Sci.*, **46** 6235
[11] Ilbagi A, Delshad Khatibi P, Swainson I P, Reinhart G and Henein H 2011 *Can. Metall. Q.*, **50**(3) 295
[12] Herlach D M, Galenko P and Holland-Moritz D 2007 *Metastable Solids form Undercooled Melts* (Pergamon Materials Series, New York)
[13] Lengsdorf R, Holland-Moritz D and Herlach D M 2010 *Scr. Mater.* **62**(6) 365
[14] Reutzel S, Hartmann H, Galenko P K 2007 *Appl. Phys. Lett.* **91**(041913)
[15] Kuhl R, Friker A, Friedrichs D, Gritzner C 2009 *Proc. 19th ESA Symp. on European Rocket and Balloon Programmes and Related Research* (Bad Reichenhall, Germany, 7–11 June 2009) (ESA SP-671, September 2009)
[16] Piller J, Knauf R, Preu P, Herlach D M, Lohöfer G 1986 in: *Proc. 6th Eur. Symp. on Materials Sciences under Microgravity* ESA SP-256(Bordeaux, 1986) 437
[17] Henein H 2002 *Mater. Sci. Eng. A* **326**(1) 92
[18] Standard test methods for metal powders and powder metallurgy products 1993 Metal Powder Industries Federation (Princeton, NJ)
[19] Abramoff M D, Magelhaes P J, Ram S J 2004 *Biophotonics Int.* **11**(7) 36
[20] St John D H, Hogan L M 1987 *Acta Metall.* **35**(1) 171
[21] Shuleshova O, Holland-Moritz D, Loser W, Reinhart G, Iles G N and Buchner B 2009 *Europhysics Letters (EPL)*, **86**(3) 19
[22] Grosse M, Niffenegger M and Kalkhof D 2001 *J. Nucl. Mater.* **296**(1-3) 305
[23] Kurz W and Fisher D J 1989 *Fundamentals of solidification* Aedermannsdorf (Transaction Technical Publication) 88
[24] Liu R P, Volkmann T and Herlach D M 2001 *Acta Mater.* **49**(3) 439
[25] Filip O and Hermann R 2006 *J. Optoelectron. Adv. Mater.* **8**(2) 504
[26] Yuasa E, Morita K, Kondoh K and Aizawa T 2002 *J. Jpn. Soc. Powder Powder Metall.* **49**(11) 1036
[27] Patchett J A, Abbaschian G J 1986 In: Mehrabian R, Parrish PA (eds) *Proc. of the fourth conf. on rapid solidification processing: principles and technologies* (University of California, Santa Barbara, CA, USA 1986)

Versatile compact X-ray radiography module for materials science under microgravity conditions.

F Kargl[1], M Balter[1], Ch Stenzel[2], Th Gruhl[3], N Daneke[3] and A Meyer[1]

[1] Institut für Materialphysik im Weltraum, Deutsches Zentrum für Luft- und Raumfahrt (DLR), 51170 Köln, Germany
[2] Astrium - Space Transportation, Physical Sciences (TO 5), 88039 Friedrichshafen, Germany
[3] Viscom AG, Carl-Buderus-Straße 9-15, 30455 Hannover, Germany

E-mail: `florian.kargl@dlr.de`

Abstract. A versatile compact microfocus X-ray radiography facility is presented. The facility serves as a technology demonstrator showing the applicability of X-ray radiography to experiments in space. It has been designed as an insert fully compatible with requirements of the Materials Science Laboratory aboard the International Space Station. The facility consists of a microfocus X-ray source delivering up to 20 W X-ray power at 100 kV acceleration voltage and a 49.2×49.3 mm RadEye2 sensor with a Scint-X scintillator at 48 μm per pixel resolution with a 14 bit dynamic range. The total device weight including sample chamber is 43 kg. The facility is classified as a fully protected radiography equipment according to German radiation safety laws. The capabilities of the facility for research in materials sciences are demonstrated in ground-based experiments.

1. Introduction

In recent years in-situ studying of processes became increasingly important for various fields of science. In materials sciences the in-situ study of materials and processes enables to discover not only new phenomena but also to revise previously obtained results sometimes subject to a larger error. In-situ methods hereby greatly benefit from the improved and novel experimental equipment developed in recent years.

In materials science and here in particular in the area of research on metal-based materials the advent of fast high resolution area detectors and highly brilliant synchrotron sources enabled to study of phenomena in the area of solidification on real samples [1, 2, 3] and not only on the previously used more easily accessible transparent analogue materials [4]. With the advent of microfocus X-ray sources in-situ studies are also possible in the laboratory as demonstrated for studying diffusion [5] and even under microgravity as shown for foaming of metals [6, 7, 8]. The latter studies are part of the X-ray monitoring (XRMON) project set-up as part of the ESA-MAP programme. The project focuses not only on the investigation of foaming but on a number of other processes in metallic melts that greatly benefit from in-situ analysis like e. g. diffusion and solidification.

In metal research experiments under microgravity are essential to study processes without being influenced by unwanted effects caused by the presence of the gravitational force as e. g. buoyancy-driven convective flow or sedimentation. Hereby, various possibilities exist to achieve

Figure 1. Left: Photograph of assembled fully MSL compatible X-ray radiography insert. The access port to the X-ray source compartment is opened with the turbomolecular pump visible at the top and the X-ray driving electronics at the bottom. Water cooling pipes are shown at the top cooling the X-ray front end. Cable feed through ports on the sample chamber are not mounted with the access holes covered by lead shielded blanks. Right: Photograph of the X-ray insert installed in the MSL equivalent processing chamber. Water cooling pipes connected to the detector are visible.

microgravity conditions. Amongst those are drop tower, parabolic flight, sounding rocket, and satellite experiments as well as experiment on the International Space Station (ISS). The order of the platforms is according to increasing microgravity time available which ranges from a few seconds to days.

Radiation safety requirements initially precluded the use of X-ray radiation in microgravity research due to bulkiness and weight constraints of the source as well as the heavy metal shielding required. However, as demonstrated in parabolic flight campaigns [9] but more recently also aboard sounding rockets (MASER-11 May 2008 [10], MAXUS-8 March 2010, and MASER-12 November 2011) microfocus X-ray sources are sufficiently compact to enable operation on these platforms. Sealed-tube microfocus X-ray sources have been used so far. However, these sealed-tube sources show a finite lifetime and a pre-defined target-substrate combination. Therefore, they might not be an optimal choice considering research in the MSL aboard the ISS, where long-time usability and flexibility of such a device is a key factor.

Therefore, the development of a compact but flexible microfocus X-ray facility that meets requirements of research under microgravity conditions and in particular demands by the MSL aboard ISS is mandatory to enable increased future use in in-situ studies of materials under microgravity conditions. Here, such a facility that meets not only scientific requirements in terms of source power, resolution, and image depth but also technical and safety requirements for operation aboard ISS is presented. The facility was developed by Astrium GmbH together with Viscom AG and partners.The previously mentioned key factors are met by using an actievely pumped X-ray tube as part of the facility. The here discussed facility has a total weight of only 43 kg and can be safely operated in any laboratory. The facility is sufficiently flexible to allow for operation on different other microgravity platforms with minor adaptations.

2. Facility Parameters

Figure 1 left shows a photograph of the X-ray radiography insert. Figure 1 right shows the insert integrated into the MSL compatible processing chamber. The insert consists of three

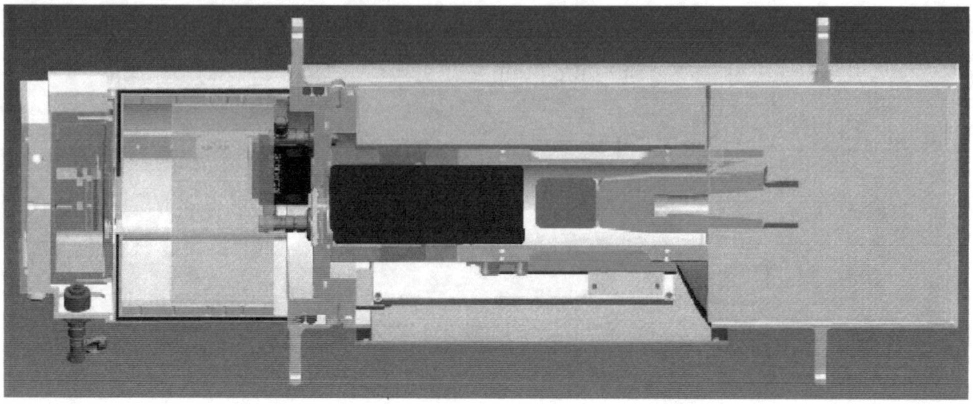

Figure 2. 2D section view of the X-ray radiography insert. Cathode (red), X-ray focusing optics (blue) and target (green) on the right of the figure. The X-ray driving electronics is represented in light blue and the high voltage generator in light green, whereas the turbomolecular pump is represented in yellow. The sample compartment is surrounded by lead shielding (thin dark-red layer). The volume occupied by the sample environment is given by the semitransparent green area. The detector module is shown to the left with the detector being the dark/light grey structure close to the beam entrance window.

independent entities: the microfocus X-ray source to the right, the sample environment chamber in the center, and the detector module to the left. The drawing of a 2D section of the insert is shown in figure 5. The 2D representation gives a better view on the inner parts of the insert not visible in the photograph.

The X-ray source entity consists of an outer housing acting both as electrical as well as radiation protection. The housing is vacuum tight, thereby ensuring ambient conditions required for safe operation of the X-ray tube. Communication, power supply, and safety interlock cabling as well as a pre-vacuum line are fed trough a vacuum flexible tube which can be connected via a standard KF25 flange to a flange adapter at the inside of the MSL chamber. Thereby the X-ray housing is protected from the MSL chamber atmosphere may it be under vacuum or partially pressurized. Operation outside the MSL chamber does not require this connection. The housing contains the X-ray tube, its high voltage generator, and the X-ray tube driving electronics. An adapted version of the standard X9100 tube by Viscom is used. The X-ray tube is fixed to the housing from the detector side. To the back of the tube the high voltage generator is connected. The only connection required is a communication cable and a power supplying cable delivering power at a voltage of 40V to the X-ray tube electronics board. The target currently in place consists of a 5 μm thin tungsten layer on a 200 μm thin aluminium window. It can be easily replaced by other target materials or by a Beryllium substrate depending on required source characteristics. The target is located behind a circular aperture matching the X-ray cone-beam size to the detector size. Thereby unwanted X-ray radiation disturbing either image quality or requiring more substantial shielding is avoided. Further shielding is included inside the housing at the radiation safety relevant positions. Thereby X-ray emission in backward direction through the housing is suppressed far below required limits. The tube is actively pumped via a compact turbomolecular pump in combination with a diaphragm pump. For continuous operation active cooling is required. This is guaranteed by thermal connection of the critical parts to the front plate of the X-ray housing which is actively cooled by chilled-water. The X-ray source of the insert delivers radiation at a maximum acceleration voltage of 100 kV and at a maximum total tube power of 20 W. Continuous operation of the source is possible up to 80 kV at maximum tube power. At 100 kV and maximum tube power the source can be operated for up to 20 min before it is switched off required for cooling of the X-ray driving electronics. The electron-beam size ultimately defining the X-ray beam size is continuously adapted according to the X-ray power supplied by the source. For low power operation (< 4 W) the smallest beam-size is below 3 μm and therefore the best resolution can be achieved. At increasing power the electron-beam size is increasing at the expense of resolution. With the current source detector configuration power values beyond 15 W lead to a marked increase in electron beam size negatively impacting on image resolution. Increasing of the electron beam size with increasing power is required to not overheat and damage the target. Target deterioration over time can be accounted for by rotating the target around its surface normal. With the electron beam slightly off-center with respect to the target rotation axis fresh parts of the target can be brought into the electron beam. Two independent X-ray safety interlock switches are mounted besides the X-ray beam aperture in the X-ray tube housing front end. The matching interlock parts are installed to the inside of the sample environment chamber. Hereby, safe operation of the source is ensured with the beam path fully enclosed by the sample environment chamber and the detector module as shown in figure 5. Most importantly, the insert is according to German radiation safety laws classified as a fully protected radiography equipment, i. e. the insert can be safely operated in the lab without any further radiation shielding. Hereby, the acceleration voltage is limited to a maximum of 80 kV at maximum tube power. Within the MSL chamber operation is safely possible up to the maximum acceleration voltage of 100 kV at maximum tube power of 20 W. Any staff operating the facility is in both operation conditions, in the lab or in the MSL chamber, not required to wear any radiation badges.

The detector is housed inside a vacuum tight aluminium chamber. Vacuum tightness is achieved by mounting to the sample environment chamber. The detector chamber is clad by a thin layer of lead (3 mm thickness) at the backplate. The shielding absorbs the primary X-ray radiation transmitted into the chamber through the rectangular carbon-fibre window. The minimal selectable thickness of the carbon-fibre window is 0.3 mm. The detector consists of two RadEye^{TM} 100 image sensors mounted side by side in RadEye2 configuration. To the CMOS sensors a protective fibre optics face-plate (FOFP) is mounted. The FOFP serves as radiation protection for the sensor. Onto the FOFP a Scint-X scintillator is glued. The detector provides an active detection area of 49.2 x 49.3 mm at 48 μm pixel size resulting in a total of 1024x1024 pixels. The detector is actively cooled by a Peltier element. Active cooling to a temperature of 16 °C significantly reduces the detector noise. The hot-end of the Peltier element is actively cooled by connecting the detector chamber to the chilled-water circulation. The detector housing has two vacuum tight feed-throughs. One feed through is responsible for supplying the power for the Peltier elements and the detector. The other feed through is required for communication with the detector. Detector communication occurs via a USB 2.0 read-out interface connected to the detector and housed inside the same aluminium chamber. A continuous frame rate of 2 frames per second with a depth of 14bit is achieved with this connection. The detector housing is mounted to the sample environment chamber back-end with six nuts that can only be opened with a special tool. The detector housing whilst mounted to the sample environment chamber is vacuum tightly sealed by means of an O-ring.

Currently two sample environment chambers are available. Both chambers are made of stainless-steel and are shielded by a layer of lead of 3 mm thickness. Both chambers have a wide opening of diameter 190 mm at the front-end enabling easy integration of the sample environment. The chamber back-ends are sealed by a carbon-fibre window of 1 mm thickness. Currently available source to detector distances are 150 mm. This means that with 1 cm distance between sample and source a maximal magnification of 15 and therefore the resolution limit given by the minimal X-ray spot-size can be achieved. Correct positioning of the sample environment chambers to the X-ray tube housing is facilitated by two ground locating pins and bushings

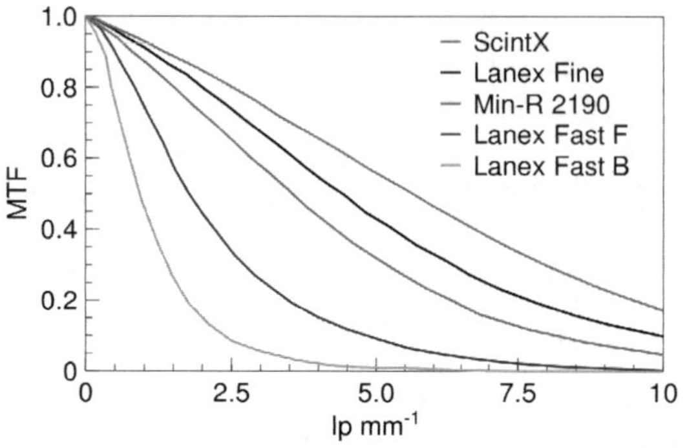

Figure 3. Modulation transfer function for the RadEye2 sensor equipped with the Scint-X scintillator as well as with standard scintillators (Lanex and Min-R) delivered by RadEye. The latter data were obtained by digitizing a figure supplied with the RadEye2 information sheet. The data shows that the Scint-X scintillator outperforms the standard scintillators.

made of quenched steel. The MSL sample environment chamber enables the installation of two lead-clad labyrinth like cable ducts at the back-end of the chamber. These cable-ducts are required to supply the respective sample environment installed inside of the chamber. The on-bench chamber is equipped with five welded ports equipped with KF flange connectors.

The total combined weight of the MSL-compatible version of the insert consisting of tube housing, MSL-compatible sample environment chamber, and the detector module amounts to only 43 kg.

The X-ray source is operated by the VxC software of Viscom. The software enables setting of acceleration voltage and beam current within the source operation limits. X-ray tube conditioning, filament warm-up, and beam-centering procedures can be carried out. For image recording the software XMC is used. This software allows in-situ correction of the recorded images for background and differences in pixel detection sensitivity. Recording time per image, averaging of images, and a number of other standard options for image manipulation and recording can be set.

3. Evaluation Experiments

Different test procedures were conducted to evaluate the performance of the equipment.

3.1. Modulation Transfer Function

The modulation transfer function shown in figure 3 of the detector with the Scint-X scintillator was evaluated by Viscom. Evaluation was carried out by using a tungsten edge. Hereby the testpattern was directly mounted to the detector conducting the test in magnification $M = 1$ configuration. Hence, the per pixel resolution of the detector rated to be about 10 linepairs per mm is probed. The X-ray source setting was chosen such that the X-ray beam size at the origin is less than the detector pixel size. The MTF for the present scintillator can be compared with the standard scintillator MTFs usually delivered with the RadEye2 sensor (cf. figure 3). It clearly shows that the present scintillator outperforms the standard scintillators.

To achieve such high resolution with high contrast is only possible due to the scintillator used. The scintillator consists of a honeycomb patterned silicon substrate filled with the scintillator material. Only light in forward direction within a relatively tight angle of acceptance is transmitted through the honeycomb pattern. Pixel-cross talk is thereby avoided to a large extent. The downside is the approximately 50% reduction in transmission. However, due to the short detector-source distance of 150 mm only, this is a minor problem compared with the major gain in image sharpness. Further improvement of the scintillator technology is expected over the years to come increasing transmission efficiency.

3.2. Sample Measurements

An interdiffusion experiment was carried out on a AlCu10at% vs. AlCu15at% diffusion couple at a processing temperature of 700 °C using the standard long-capillary diffusion X-ray radiography setup described elsewhere [5, 11]. The diffusion couple was mounted inside the graphite furnace. For this test detector and source were operated in separated configuration inside a lead box in order to be able to use the standard capillary setup. Hence, the source detector distance was increased from the nominal 150 mm in flight configuration to 390 mm for the test. The capillary was mounted such that a magnification factor of $\sim$ 1.9 was achieved. Effective pixel resolution amounts therefore to $\sim$26 μm.

For the experiment the sample was pre-heated to 510°C in several steps. The sample was kept at pre-heat temperature for about 5 min. It was then heated to a temperature of700 °C with about 7 K s^{-1} at which the diffusion experiment was conducted. This temperature was stabilized. The sample was kept at this temperature for $\sim$10 min followed by cooling to ambient temperature by switching of the heating power.

Recordings were taken from the start of heating to the diffusion temperature until the sample had fully solidified upon cooling. Images were taken with a frame rate of 1 Hz. The X-ray source was set to 80 kV and 170 μA beam current.

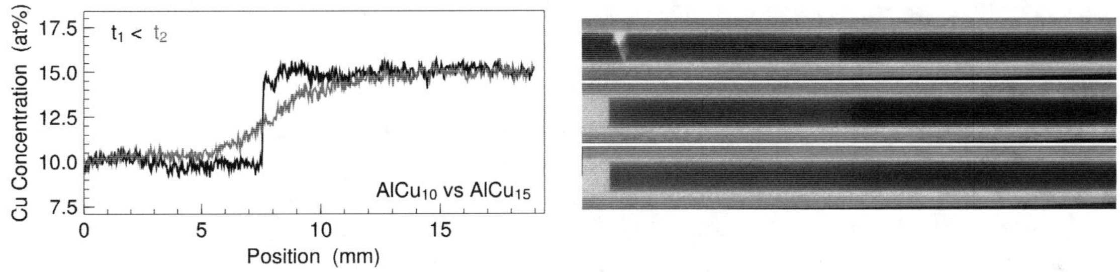

Figure 4. Left: Cu concentration as a function of position in the diffusion couple at different times: t_1 solid and t_2 later stages of diffusion. Right: Recorded images of the diffusion capillary for the same times (top t_1 and bottom t_2) and during melting (middle).

Images of the still solid diffusion couple, the diffusion couple during early stages, and during later stages of the diffusion anneal are shown in figure 4 on the right. The images were subjected to standard correction procedures consisting of background subtraction and pixel gain correction. From the recorded images grey value profiles were derived by integrating radially over the diffusion pair. The axial grey value distribution can be transformed to Cu concentration following procedures outlined in Ref. [5]. The concentration profiles corresponding to the recorded images for the solid sample and during diffusion annealing are shown in figure 4 left. Comparing the diffusion profiles with profiles obtained using a RadEye detector with standard scintillator and without cooling it turns out that though the X-ray intensity being detected is lower contrast resolution is increased by 50% and contrast to noise ratio is about similar.

Besides the described interdiffusion experiment monitored in real time images of two different solid samples have been recorded. A solidified Al-Cu alloy subjected to equiaxed growth was provided by N. Mangelinck-Noel and A. Bogno of the IM2NP-CNRS in Marseille. The specimen was 30x5x0.1 mm in dimension. Recordings with various source settings showed that images can be recorded in real time with 2 Hz repetition rate providing at the same time better resolution than recordings taken with a standard detector. In this sample the dendritic growth pattern was clearly visible. For the high magnification low tube power setup it might even be possible to follow Cu concentration from the dendrite into the surrounding liquid with an effective per pixel

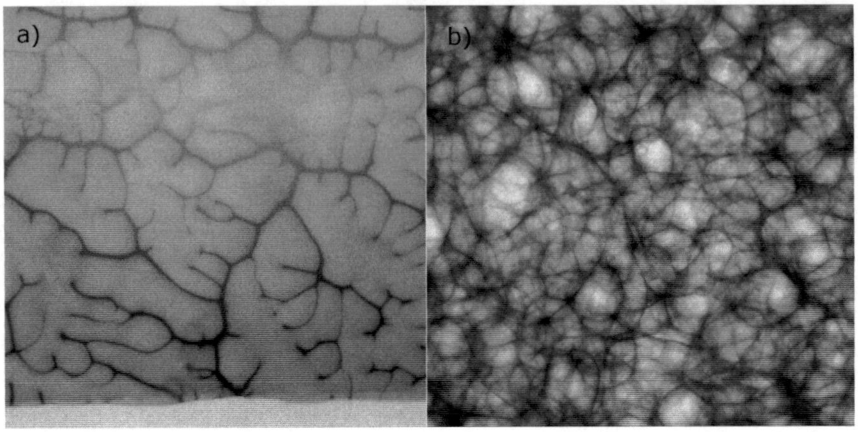

Figure 5. X-ray radiograms taken with a X-ray source setting of 60 kV and 170 μA at a recording time of 2 s for a) the solidified Al-Cu equiaxed growth sample and b) the solidified Al-Cu foam sample. A magnification of approximately 10 was used for recording of a) and approximately 2 for b), resulting in an effective pixel size of approximately 5 μm and 25 μm, respectively.

resolution of $\sim 4\,\mu$m. Further, images were taken for a solidified Al-Cu foam sample provided by F. Garcia-Moreno from Helmholtz Center Berlin, Germany. Again it can be shown that similar image quality at a slightly better resolution than with a conventional detector can be achieved at 2 Hz image repetition rate. Edges of bubbles in the foam sample are clearly visible. The entire sample of dimensions of 20x20x1 mm can be recorded at a magnification factor of about 2.5 resulting in an effective pixel size of between 20 μm and 24 μm.

In future, for both, the solidified AlCu dendrite and the solidified foam sample, intensity and therefore the contrast could be increased by using a different target, as e. g. molybdenum on aluminium thereby increasing the X-ray intensity at lower X-ray energies. Direct-converting detectors might also be an option provided they reach image stability and quality currently achieved with scintillator-based detectors and a roughly comparable total detector area. In this case the signal would be increased by at least a factor five enabling either low-power high-resolution operation of the X-ray source at better image quality or faster imaging rates.

4. Conclusion

In conclusion a novel microfocus X-ray demonstrator insert for the Materials Science Laboratory aboard the International Space Station has been presented. Characteristic features of this facility were presented in detail and results of first tests of operation conditions and imaging capabilities of this device were discussed. The X-ray source can be operated at acceleration voltages up to 100 kV at a maximal power of up to 20 W. Images of 1024×1024 pixel at a pixel resolution of 48 μm and with a depth of 14 bit per pixel can be taken at up to two frames per second. The detector provides a field of view of 49.2×49.3 mm. The detector uses a RadEye2 image sensor in combination with a Scint-X scintillator. The CMOS sensor is shielded by a fibre optics face plate enabling irradiation with higher energy X-rays. Image resolution is increased compared with the available standard scintillators. First radiography results of an interdiffusion experiment on an Al-Cu alloy have been presented. The detector showed an increased contrast resolution at a decreased contrast to noise ratio. It has been demonstrated that this novel equipment is able to provide similar results as previously obtained in laboratory experiments using a less integrated piece of equipment. Improvements in detector technology over the coming years are expected to enable faster recordings with higher intensity as currently encountered. The insert is currently slightly modified to enable its use aboard the DLR sounding rocket MAPHEUS in 2012 together with the recently designed and tested compact shear-cell furnace [12] for studying diffusion in metallic melts.

Acknowledgments

We thank the German Ministry for Economy and Technology (BMWi) for funding of this project through the Economic Stimulus Package I. We thank Nathalie Mangelinck-Noel and Abdul Aziz Bogno of the IM2NP-CNRS in Marseille for providing us with a solidified AlCu sample as well as Francisco Garcia-Moreno of the Helmholtz Center in Berlin for providing us with a Al-Cu metal foam sample. We thank Henning Weis for critically reading our manuscript.

References

[1] Mathiesen R H, Arnberg L, Mo F, Weitkamp T and Snigirev A 1999 *Phys. Rev. Lett.* **83** 5062
[2] Thi H N, Jamgotchinan H, Gastaldi J, Hartwig J, Schenk T, Klein H Billia B, Baruchel J and Dabo Y 2003 *J. Phys. D Appl. Phys.* **36** A83.
[3] Yasuda H, Ohnaka I, Kawasaki K, Suglyama A, Ohmichi T, Iwane J and Umetani K 2004 *J. Cryst. Growth* **262** 645
[4] Noel N, Jamgotchinan H and Billia B 1997 *J. Cryst. Growth* **181** 117
[5] Zhang B, Griesche A and Meyer A 2010 *Phys. Rev. Let..* **104** 035902
[6] Moreno F G, Fromme M and Banhart J 2004 *Adv. Eng. Materials* **6** 416
[7] Garcia-Moreno F, Holm P and Banhart J 2007 *ESA Spec. Publ.* **647** 389
[8] Holm P, Houltz Y, Lofgren O, Andersson P and Garcia-Moreno F 2007 *ESA Spec. Publ.* **647** 385
[9] Garcia-Moreno F, Jimenez C, Mukherjee M, Holm P, Weise J and Banhart J 2009 *Coll. Surf. A: Physchem. Eng. Aspects* **344** 101
[10] Garcia-Moreno F, Jimenez C, Mukherjee M and Banhart F 2009 *ESA Spec. Publ.* **671**
[11] Griesche A, Zhang B, Solórzano E and Garcia-Moreno F 2010 *Rev. Sci. Instrum..* **81** 056104
[12] Neumann Ch, Sondermann E, Kargl F and Meyer A 2011 *J. Phys.: Conf. Series* (submitted)

Journal of Physics: Conference Series **327** (2011) 012012 doi:10.1088/1742-6596/327/1/012012

Investigation of gravity effects on solidification of binary alloys with *in situ* X-ray radiography on earth and in microgravity environment

H Nguyen-Thi[1, 2], A Bogno[1, 2], G Reinhart[1, 2], B Billia[1, 2], R H Mathiesen[3], G Zimmermann[4], Y Houltz[5], K Löth[5], D Voss[6], A Verga[6] and F de Pascale[6]

[1)] Aix Marseille Université & [2)] CNRS, IM2NP UMR 6242, Campus Saint-Jérôme, Case 142, 13397 Marseille Cedex 20, France
[3)] Dept. of physics, NTNU, N-7491 Trondheim, Norway
[4)] ACCESS e.V., Intzestrasse 5, 52072 Aachen, Germany
[5)] Swedish Space Corporation, P.O. Box 4207, SE-171 04 Solna, Sweden
[6)] European Space Research & Technology Centre, Keplerlaan 1 Postbus 299, 2200 AG Noordwijk, the Netherlands

E-mail: henri.nguyen-thi@im2np.fr

Abstract. As most of the phenomena involved during solidification are dynamic, *in situ* and real-time X-ray imaging should be retained as the method of choice for investigating the solidification front evolution of metallic alloys grown from the melt. On Earth, natural convection in the melt is well known to be the major source of various disturbing effects which can significantly modify or mask important physical mechanisms. Microgravity environment is an efficient way to eliminate buoyancy and convection to provide benchmark data for the validation of models and numerical simulations. In addition, a comparative study of solidification experiments carried out on Earth and in space can also enlighten the effects of gravity. In the frame of the ESA - MAP programme entitled XRMON, an experimental set-up has been developed to perform directional solidification in microgravity conditions with *in situ* X-ray radiography observation. In the first part of this paper, we will present a brief review of some effects induced by gravity on the solidification process and investigated by mean of synchrotron X-ray radiography at ESRF (European Synchrotron Radiation Facility). In the second part of this paper, we will describe some results obtained with a prototype of the XRMON-Gradient Furnace set-up. These preliminary results show the large capabilities of the experimental set-up in terms of thermal behaviour, as well as X-ray observation.

1. Introduction

Structural material properties are directly related to their solidification microstructures, so that a precise control of growth process is crucial in engineering. As most of the phenomena involved during solidification are dynamic, *in situ* and real-time X-ray imaging should be retained as the method of choice for investigating the solidification front evolution of metallic alloys grown from the melt [1-3]. Elsewhere, an important issue to clarify in the solidification process is the role of convection on macrosegregation and on microstructure formation. The coupling between fluid flow and solidification has been the subject of a great deal of experimental, theoretical and numerical works [4,5]. The main conclusion of these studies is that natural convection in the melt is the major source of various disturbing effects which can significantly modify or mask important physical mechanisms on Earth (1g). It has been shown that microgravity (μg) environment is an efficient way to eliminate buoyancy and convection to provide benchmark data for the validation of models and numerical simulations [6-10]. In addition, a comparative study of solidification experiments at 1g and μg can also enlighten the effects of gravity. Therefore, in the frame of the ESA - MAP programme entitled XRMON (*In situ* **X**-**R**ay **MON**itoring of advanced metallurgical processes under microgravity and terrestrial conditions), an experimental set-up was recently developed by SSC (Swedish Space Corporation) to perform directional solidification in microgravity conditions with *in situ* X-ray radiography observation. The first solidification experiment in microgravity using this facility is scheduled during MASER 12 sounding rocket mission, in autumn 2011.

This paper is divided is two parts. In the first part, we will present a brief review of some effects induced by gravity on the solidification process and investigated by mean of synchrotron X-ray radiography. Experiments were carried out at ESRF (European Synchrotron Radiation Facility) in thin Aluminium-based alloys. In the second part of this paper, we will describe the breadboard tests carried out in Stockholm (Sweden) with a prototype of the XRMON-Gradient Furnace set-up. Preliminary results show the large capabilities of this experimental set-up in terms of thermal behaviour, as well as X-ray observation.

2. Analysis of the influence of convection on solidification process by synchrotron X-ray radiography

The synchrotron experiments are carried out at the European Synchrotron Radiation Facility (ESRF). Solidifications are performed inside a Bridgman furnace, which was already described in details [11]. It has been shown [12,13] that this furnace can be used either for directional solidification (i.e. with a vertical temperature gradient) or for equiaxed growth (i.e. in nearly isothermal conditions). *In situ* and real time observation of the solidification process is possible by the use of synchrotron X-ray radiography. The principle of this technique is quite simple: The main surface of the sheet-like sample (40 x 6 mm^2) is set perpendicular to the incident monochromatic X-ray beam, with an energy in the range 14-18 keV. The beam energy is chosen as a function of the alloy composition (Al – 4 wt% Cu or Al – 10 wt% Cu alloys) and the sample thickness (200 μm). The transmitted beam is recorded by a CCD–based (FReLoN) camera placed behind the sample. Absorption is the main source of the image contrast and depends on the atomic number of the elements and the solute content. In all radiographic pictures, the solid aluminium microstructures appear in white and the copper-rich liquid in dark grey. The optics is chosen to obtain a good compromise between a large field of view (15 x 6 mm^2) and a sufficient resolution (pixel size: 7.46 μm). For all experiments, the image quality was improved by applying an image processing described in [14].

2.1. Convection effects on solid–liquid interface during the initial transient of solidification

It is well known that even when solidification is performed under both thermal and solutal stabilizing configuration, convection in the melt can nevertheless be strong due to residual transverse gradient, in particular at low growth rates [9,15]. Indeed, in real experimental configurations there are generally small departures of the initial interface from planarity, even after a long thermal stabilisation phase. As solidification proceeds, the solute is rejected at the solid–liquid interface. Due to gravity, the denser Cu-enriched liquid flows towards depressed regions of the front, which results in solute accumulation in those regions. According to the phase diagram, solute accumulation in depressed regions retards solidification and further deepens the solid–liquid interface modulations. This leads to a large

Journal of Physics: Conference Series **327** (2011) 012012 doi:10.1088/1742-6596/327/1/012012

macroscopic deformation of the solidification front. In addition, the transverse solute gradient gives rise to a transverse microstructure gradient, as already shown by *post-mortem* analysis [9].

Figure 1 shows a sequence of four images recorded at successive times during a solidification experiment at a cooling rate of 0.3 K/min and an average temperature gradient G of 35.5 K/cm. In this experiment an initially slightly tilted smooth interface (figure 1a) is observed. A crude estimation of the transverse temperature gradient Gy can be obtained from the expression $Gy = (\Delta z_0/d)G$, where Δz_0 is the distance between the highest and the lowest levels of the curved front at t = 0 (before solidification starts) and the corresponding transverse distance d = 5 mm. In the presented experiment, $\Delta z_0 = 0.45$ mm, which gives Gy = 3.2 K/cm. When the cooling of the sample begins the growth rate gradually increases as solidification proceeds, leading to the destabilisation of the interface first on the left-hand side of the sample (figure 1b). The propagation of the morphological instability towards the right-hand side of the sample is visible in figure 1c and figure 1d, as well as the transverse gradient of microstructure. Concomitantly, macroscopic distortion of the solid–liquid interface becomes larger and larger with time, due to fluid flow in the melt. The amplitude of the front distortion increases from $\Delta z_0 = 0.45$ mm at the beginning to nearly 4 mm at the end of the experiment (figure 1d).

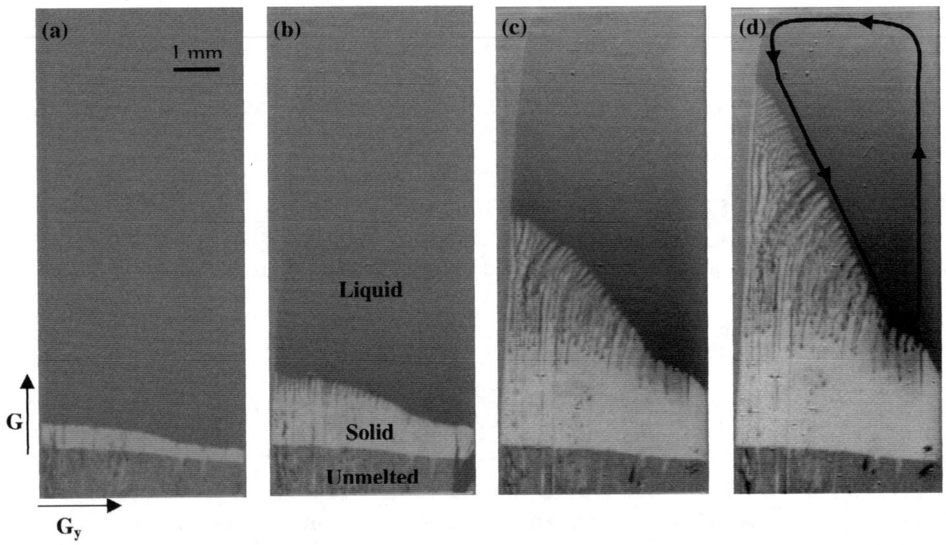

Figure 1. Sequence of four images showing the increasing macroscopic deformation of the solid-liquid interface during the initial transient of directional solidification of Al – 4 wt% Cu (cooling rate = 0.3 K/min; G = 35.5 K/cm. (a) t ≈ 17 min; (b) t ≈ 34 min; (c) t ≈ 70 min; (d) t ≈ 88min.

In figure 1c and figure 1d, it can be seen that the liquid phase on the right-hand side of the sample just ahead of the solid–liquid interface becomes darker as solidification proceeds, which confirms an increase in the liquid solute concentration there. At the same time, the solute amount in the liquid phase on the left-hand side of the sample also increases but at a slower rate. Recently we developed a novel image analysis technique which enables measurement of the long-range solute concentration profiles in the liquid phase [14]. By using this procedure, solute concentration were measured at three different y-abscissae during the initial transient. Figure 2 shows the time evolution of the concentration at the solid–liquid interface for the three different abscissae $y_1 = 0.7$ mm, $y_2 = 2.6$ mm and $y_3 = 4.0$ mm (y-axis reference was chosen as the left-hand side of the sample). It can be seen that in the early stages of solidification (t < 8 min) the solute concentration variation at the solid–liquid

Journal of Physics: Conference Series **327** (2011) 012012 doi:10.1088/1742-6596/327/1/012012

interface along the y-axis $\Delta Cint(t) = C_{right} - C_{left}$ is close to 0, indicating that the liquid phase is almost homogeneous at the beginning of the solidification. $\Delta Cint(t)$ then gradually increases during the initial transient and finally reaches a value of 2.0 wt.% at the end of the experiment (t = 70 min).

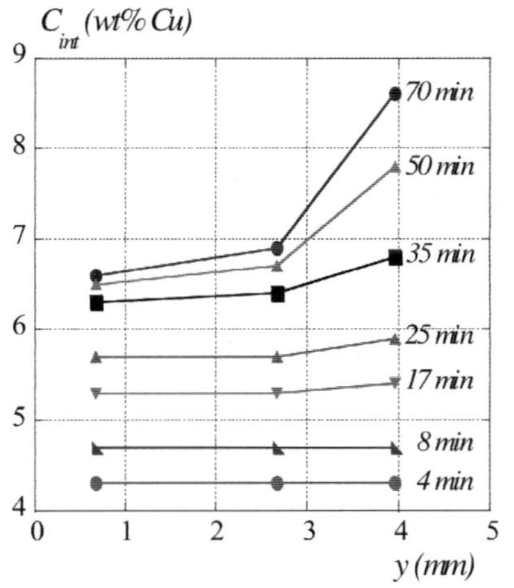

Figure 2. Evolution of the concentration at the solid–liquid interface for the three different abscissae y_1 = 0.7 mm, y_2 = 2.6 mm and y_3 = 4.0 mm during the initial solidification transient of Al – 4 wt% Cu, G = 35.5 K/cm, cooling rate = 0.3 K/min.

2.2. Self-poisoning effect during equiaxed growth

In a typical equiaxed growth experiment, the temperatures of both heater elements are adjusted in order to achieve nearly isothermal conditions all over the sample. After a stabilization period of two hours, solidification is triggered by applying simultaneously the same cooling rate on both heaters [12,13]. For the experiment presented in this paper the cooling rate applied was R = 0.5K/min. *In situ* and real time observation of the solidification process was made possible by the use of synchrotron X-ray radiography, from $t_0 = 0$ corresponding to the application of the cooling rate to the end of experiment when lengthening of the dendrite arms was no more detectable with naked eye.

Length measurements are performed for the four arms of a single grain (figure 3), growing roughly at right angles to each other along the <100> direction. As can be seen in figure 3d, the behaviors of the four dendrite arms are not identical: the initial slopes (or growth velocities) of the four curves as well as the values of their respective plateaus are different. For instance, at the beginning *dendrite arm 3* has the highest growth rate, while the longest length is achieved by *dendrite arm 4*. The difference of the arm growth velocities at the beginning is attributed to a "self – poisoning" of the grain, already revealed in the case of the facetted growth of an equiaxed grain [16]. In fact, the grain is initially isolated from the surrounding grains (figure 3a) and we can assume that there is no influence from other grains. As solidification proceeds, the rejected solute (Cu) is carried downwards by gravity driven fluid flow, leading to the velocity decrease of the dendrite arms situated aside and below. Then, the disparity in the asymptotic values of the arm lengths can be attributed to the interaction with neighboring grains in the late stage of solidification (figure 3b-3c). Owing to the grain interaction, the growth is hindered and is finally stopped, although there is no mechanical impingement of the dendrites as it will be shown later. In addition, the grain interaction is highlighted on the curve representing the growth of the *dendrite arm 4,* where a plateau is distinctly visible following the initial slope of the curve (figure 3d). This plateau indicates that the arm has stopped growing for this short period of time, due to the influence of another grain falling in its vicinity.

 doi:10.1088/1742-6596/327/1/012012

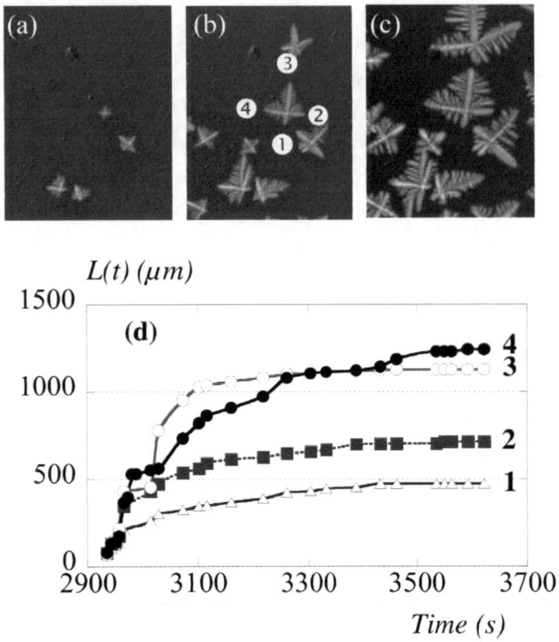

Figure 3.
(a) Equiaxed grain of Al – 10 wt% Cu initially isolated from the surrounding grains at the early stages of solidification (t = 2957sec, R = 0.5 K/min).
(b) The grain having its four dendrite arms growing at right angle to each other (t = 3037sec).
(c) The same grain surrounded by other grains at the final stage of solidification (t = 3362 sec).
(d) Time variation of dendrite arms lengths. The curves are referenced by the numbers given on the right hand side of the graph.

2.3. Sedimentation and Columnar-to-Equiaxed transition

Sedimentation and rotation of the equiaxed grains are two key concepts to understand the final grain arrangement and can only be analyzed by *in situ* observation [17-20]. For instance, the gravity-settling behaviour of few equiaxed grains is shown in figure 4, with a sequence of radiographs recorded during AlCu growth. In the beginning of the experiment, nucleation of small grains occurs in different area of the field-of-view, as well as in the region just above the field-of-view. Most of them are stuck on crucible wall, but one grain falls down due to gravity (figure 4a to 4d). In addition, live monitoring also shows that the equiaxed grain rotate because of uneven lengthening of their dendrite arms. Finally, the grains stop to move as soon as their size is comparable to the crucible thickness. It should be stressed that the falling down of equiaxed grains is enhancing grain packing at the bottom of the sample. At the end of the experiment presented, another grain moves upward, in a opposite direction of the first grain, as shown in figure 4e to figure 4h. The reason of this movement could be: *(i)* either there is a convective roll in this region of the sample or *(ii)* the grain becomes buoyant with respect to the fluid owing to local segregation in this region of the sample. Indeed, since the liquid become denser owing to the solute rejection, the relative difference of density between the grain and the surrounding liquid gradually changes. For Al-Cu alloys, the solid formed is denser than the bulk liquid below a nominal composition of about 10 wt% Cu, whereas it is lighter for higher concentrations. This explains the change in the direction of the movement between the two grains.

The grains settlement could be critical in the Columnar-Equiaxed Transition. In experiments dedicated to CET carried out on refined Al–3.5 wt% Ni alloy, we first applied a low pulling rate to establish dendritic columnar microstructure. After that, the pulling rate is at reference time t_0 jumped to a higher value to trigger CET. Rapidly, equiaxed grains become visible in the solute-rich boundary layer adjacent to the columnar structure (figure 5a). Under gravity the falling down of equiaxed grains is enhancing grain packing above the columnar dendrites and eutectic front. According to this, CET is expected to occur earlier on earth than in microgravity. Columnar growth is thus stopped and equiaxed growth regime installs and propagates (figure 5c), according to scenario proposed by Martorano [21].

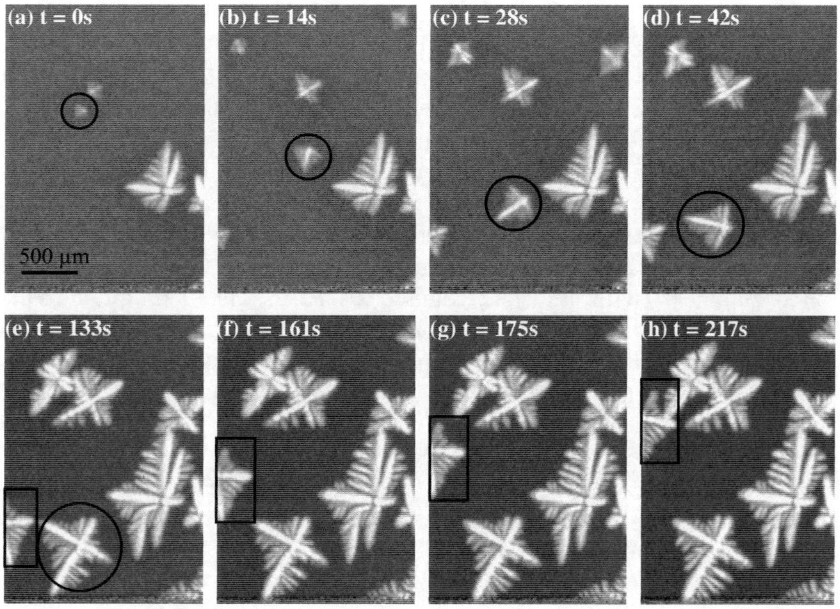

Figure 4. Sequence of radiographs recorded during Al-10wt% Cu equiaxed solidification showing the sedimentation and floatation of equiaxed grains.

Eventually, equiaxed grains get trapped in the eutectic phase, as it happens for the first fallen grains in figure 5b and 5c, giving the false impression of a mixed columnar–equiaxed structure. This sedimentation effect can obviously be misleading in *post mortem* analysis of the grain structure, as well as in the feedback to modeling of the solidification sequence. After the CET, equiaxed growth is self-sustaining, and a fully equiaxed mushy zone formed by closely packed equiaxed grains (figure 5d). Equiaxed growth proceeds by the propagation of the forefront of this mushy zone with the velocity of leading dendrite tips nearing the pulling rate. In the melt just above, new grains repeatedly nucleate, grow and fall down, screening the current leading grains. This operating cycle is maintained till the end of the sample (figure 5d).

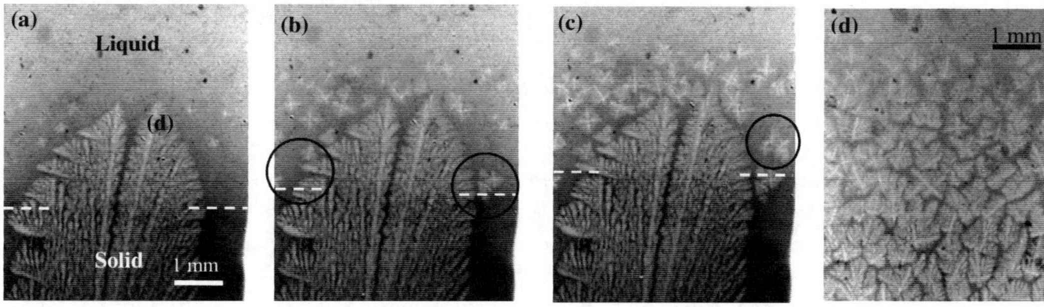

Figure 5. Sequence of radiographs of refined Al-3.5wt% Ni solidification, recorded during the CET induced by a sharp increase of the pulling rate from 1.5 to 15 μm/s at $t = t_0$, G = 30 K/cm. (a) $t_0 + 63$ sec, (b) $t = t_0 + 87$ sec, (c) $t = t_0 + 111$ sec and (d) propagation of the equiaxed microstructures.

3. XRMON – GF project

3.1. XRMON-GF device

The conceptual design of the breadboard setup is given in figure 6. It consists of a gradient furnace system for solidification of AlCu-based alloys and an attached high-resolution X-ray diagnostic system. The most challenging tasks for the development came from the distance between the source and the sample, which must be about 5 mm. This short distance required special thermal insulation to avoid heat leakage from the sample to the cooled target. In addition such a thermal insulation had to be transparent to X-rays as much as possible to guarantee maximum intensity.

The metallic samples had a sheet-like geometry with a length of 50mm, a width of 5mm and a constant thickness of less than 0.2mm along the sample. Each sample was first mechanically polished down to the desired thickness (with a surface rugosity of 1μm), and then spray coated with boron nitride (BN). Then it was sandwiched between two rectangular glass plates, welded together. The BN coating was required because of its combination of properties: *i)* it prevents chemical reaction between the Al-Cu sample and the glass plates and *ii)* it is transparent to X-rays in the range of energy of our interest.

To investigate the dynamics of solidification processes, the field of view for the X-ray diagnostics was about 5 x 5 mm^2 with a spatial resolution of 3-5 μm and a temporal resolution of 2-3 Hz. The X-ray source was based on a microfocus transmission target in molybdenum using polychromatic radiation with a peak at 17.5 keV.

(a)
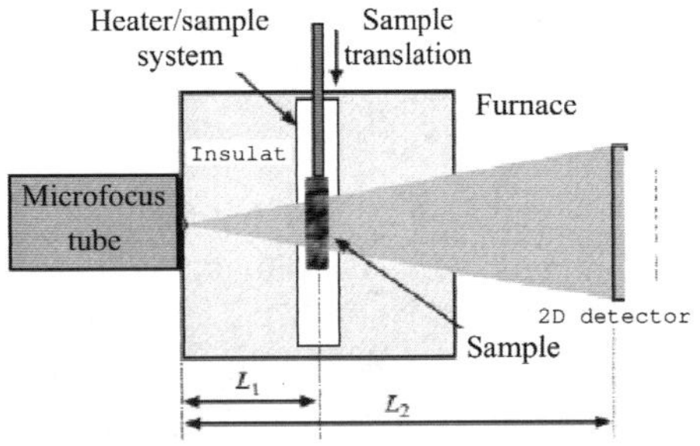

(b)
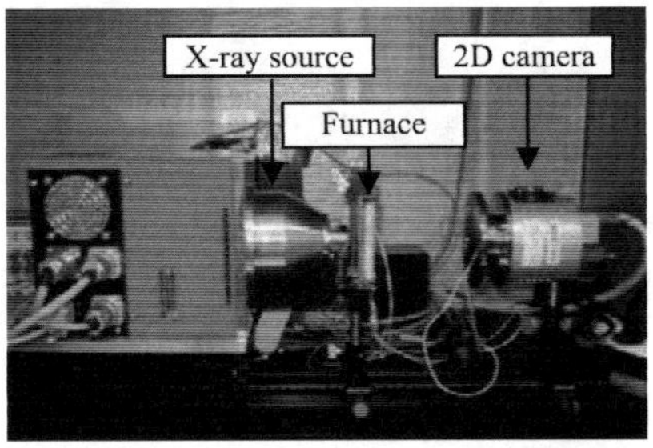

Figure 6. (a) Conceptual design of the breadboard set-up, (b) picture of the breadboard set-up showing the X-ray source (left), the furnace (very close to the microfocus source) and the detector (right).

3.2. Initial transient during directional solidification of Al - 10 wt% Cu alloy

Figure 7 shows a sequence of six images taken during the initial solidification transient of an Al - 10 wt% Cu alloy for a temperature gradient of 43 K/cm and a cooling rate of the hot heater element at 6 K/min. This figure shows the time evolution of the interface pattern, from the breakdown of the planar interface to the subsequent developments of the microstructure. For this experiment, the solidification microstructure was photographed every 2 seconds with an exposure time of 2 seconds. In figure 7a ($t = t_0 + 150s$, with t_0 the reference time, just before the beginning of the solidification) the first disturbances are visible along the whole interface, following the planar front breakdown induced by the Mullins–Sekerka instability [22]. One can already see in figure 7a that side branches begin to grow on most of the perturbations, which indicates the inception of the formation of dendrites. At this stage, it is possible to determine an average wavelength λ_i for the microstructure, $\lambda_i \approx 180$ μm. Further development of the pattern occurs with a progressive increase of the amplitude of the disturbances (figure 7b, $t = t_0 + 170s$), while the liquid ahead of the columnar front becomes darker and darker, due to the solute rejection during the liquid-solid transformation. In the subsequent stages, both the amplitude and lateral size of dendrite increase concomitantly, with solute screening causing a strong decrease of the growth of the neighbouring dendrites. In our experiments, a steady state was never achieved for several reasons: *(i)* the power-down technique used to trigger the solidification, *(ii)* the low value of the partition coefficient ($k \approx 0.14$) which implies a long transient regime, and *(iii)* the short sample length (about 50 mm). Nevertheless, the general shape of the microstructure does not change drastically in the latest stages of solidification and is mainly composed of dendrites protruding markedly into the liquid phase with an average final primary spacing $\lambda_f \approx 1.260\mu m$ (figure 7f).

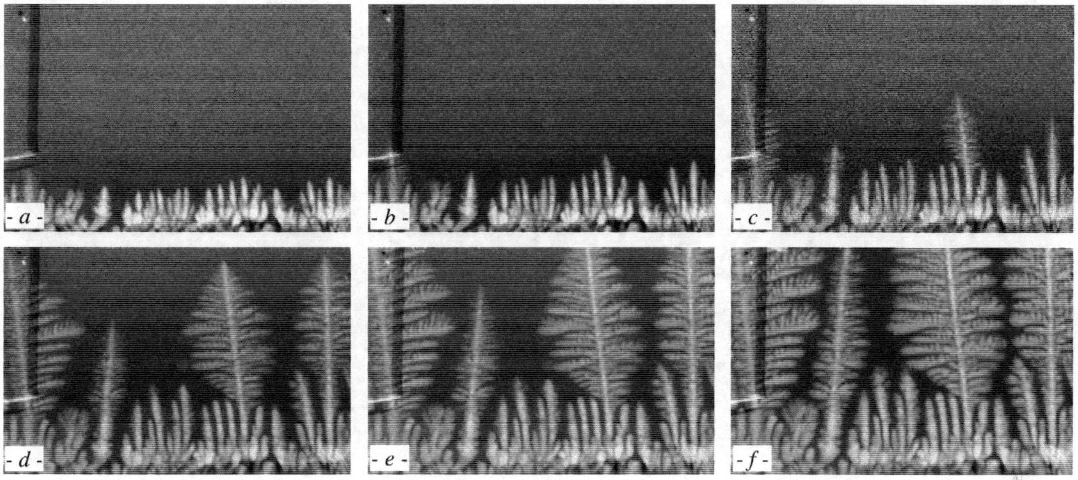

Figure 7. Sequence of radiographs showing the formation of a columnar structure of Al-10wt% Cu (cooling rate = 6K/min; G ≈ 43 K/cm) (a) $t=t_0+150s$; (b) $t=t_0+170s$; (c) $t=t_0+200s$; (d) $t=t_0+260s$; (e) $t=t_0+300s$; (f) $t=t_0+400s$. t_0 is the reference time, just before the beginning of the solidification. Image width=4.5 mm.

3.3. Fragmentation phenomenon during solidification of Al – 20 wt% Cu alloy

One mechanism for the formation of equiaxed grains during the Columnar-to-Equiaxed Transition [23] is the dendrite fragmentation, which is believed to be at the origin of the central equiaxed core region in casting processes. Fragmentation occurs when dendrite branches are detached from the main primary trunks. If these fragments are transported ahead of the columnar front by buoyancy forces or convection, they can continue to grow and then form equiaxed grains that can stop the advancing columnar front. Fragmentation is favoured by the dendrite ripening process but also by the initiation of local remelting due to the local pile-up of solute within the partially solid-liquid sample region usually called the "mushy zone" [24-26]. Fragmentation is an essential feature of the CET in non-refined alloys; however, the details of the fragmentation phenomenon are insufficiently understood and controlled. Thus, it is compulsory to confirm the predominance of the fragmentation phenomenon in CET and also to improve the characterisation and understanding of its mechanism, in particular the role of gravity at each step of CET.

To achieve this goal, an efficient approach is to perform a comparison between solidification experiments carried out at 1g and in µg, in both cases with *in situ* characterization by means of X-ray radiography. Several experiments were performed on a single non-refined Al – 20 wt% Cu sample, for various high temperature gradients (200 - 400 K/cm) and cooling rates in the range 1 – 20 K/min. For the experiment presented in figure 8, the high nominal solute concentration lead to a higher contrast than in previous experiment with an Al – 10 wt% Cu. Therefore, the exposure time was reduced to one second, and the evolution of the solidification microstructure was recorded every second. In figure 8a one can see the gradual establishment of columnar growth, as described in the previous section. In this experiment, an equiaxed grain nucleated ahead of the columnar front, on a small heterogeneity. This grain remained fixed and simply continued to grow in interaction with the columnar front. In figure 8b and figure 8d, the black circles pinpoint the presence of several dendrite fragments. For Al – 20 wt% Cu, the solid (mainly composed of aluminium) being much lighter than the surrounding liquid (Cu - enriched), dendrite fragmentation was promoted by buoyancy forces acting on secondary dendrite arms. After the fragmentation, the dendrite pieces were immediately carried away to the upper part of the sample. In this experiment, fragmentation occurred continuously during the growth process. Radiographs made it possible to have a rough estimation of the fragments velocity, about 60µm/s in this experiment. It is worth noticing that these dendrite fragments could not promote CET, because they were carried up far into the liquid where they were re-melted. Furthermore, the fact that all Al-enriched dendrite fragments were transported by buoyancy forces into the upper part of the sample was at the origin of a strong segregation along the sample as the solidification front advanced.

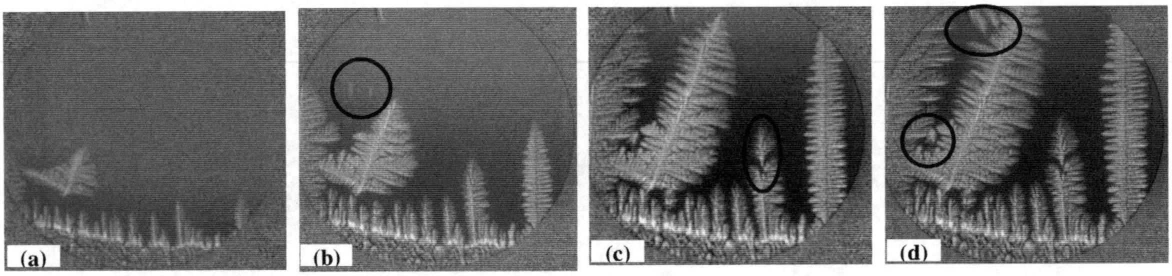

Figure 8. Sequence of radiographs during columnar growth of Al – 20 wt% Cu showing several dendrite fragmentations (cooling rate = 20K/min; G = 400 K/cm) (a) $t=t_0+50s$; (b) $t=t_0+100s$; (c) $t=t_0+150s$; (d) $t=t_0+168s$. t_0 is the reference time, just before the beginning of the solidification. Image width=5 mm.

In figure 8c, the fragmentation of a dendrite primary trunk can be noticed, indicated by a black ellipse. Its size being large compared to the sample thickness, the primary trunk fragment could not move far into the liquid phase. This phenomenon was rarely observed compared to the secondary dendrite arm fragmentation, certainly because of the difference in diameter between primary trunk and secondary arms. It has recently been observed by using combined radiography and topography that the rotation of some tenth of degrees around the growth axis of the entire primary trunk during the growth process can occur [27]. This rotation has been interpreted as a consequence of a torque induced by shear stress that builds up with the growth [28]. This suggests that stress accumulates there to finally provoke the trunk to crack. In our experiments, it was also observed that this phenomenon occurred mainly for a specific crystallographic dendrite direction <110>, which suggests an effect of dendrite morphology.

4. Conclusions

This paper reports on a selection of results obtained during metallic alloy solidification experiments characterized *in situ* and in real time by X-ray radiography. Experiments were carried out with both a Synchrotron source (ESRF) and a microfocus tube, with a dedicated novel experimental set-up developed by SSC in the frame of the ESA-MAP XRMON. These results emphasize the great impact of gravity and convection on the dynamics of the solidification front by showing the influence of convection on the solid-liquid interface, the solutal self-poisoning during equiaxed growth, the sedimentation/buoyancy of grains, the bending/fragmentation of dendrite secondary arms. The next step of the project is to perform an experiment in microgravity environment to obtain purely diffusive transport conditions and then to provide benchmark data for the validation of models and numerical simulations. In addition, a comparative study of solidification experiments at 1g and μg will also enlighten the effects of gravity. This is the goal of the solidification experiment scheduled during MASER 12 sounding rocket mission, in autumn 2011. For the future, several research projects could also benefit greatly from this novel experimental setup, which enables X-ray characterization in microgravity experiments.

Acknowledgments
This research work is supported by the XRMON project (AO-2004-046) of the MAP program of the European Space Agency (ESA) and by the French National Space Agency (CNES).

References

[1] Mathiesen, R. H., Arnberg, L., Ramsokar, K., Weitkamp, T., Rau, C.and Snigirev, A. 1999 *Phys. Rev. Lett.* **83**, 5062
[2] Yasuda, H., Ohnaka, I., Kawasaki, K., Sugiyama, A., Ohmichi, T., Iwane, J.and Umetani, K. 2004 *Journal of Crystal Growth* **262**, 645
[3] Schenk, T.et al 2005 *Journal of Crystal Growth* **275**, 201
[4] Davis, S. H. 1990 *J. Fluid Mech.* **212**, 241
[5] Hurle, D. T. J. *Interactive Dynamics of Convection and Solidification*; Kluwer Academic: Pays-Bas, 1992
[6] Giummarra, C., LaCombe, J. C., Koss, M. B., Frei, J. E., Lupulescu, A. O.and Glicksman, M. E. 2005 *Journal of Crystal Growth* **274**, 317
[7] LaCombe, J. C., Koss, M. B.and Glicksman, M. E. 1999 *Physical Review Letters* **83**, 2997
[8] Koss, M. B., LaCombe, J. C., Tennenhouse, L. A., Glicksman, M. E.and Winsa, E. A. 1999 *Metallurgical and Materials Transactions a-Physical Metallurgy and Materials Science* **30**, 3177
[9] Nguyen Thi, H., Dabo, Y., Drevet, B., Dupouy, M. D., Camel, D., Billia, B., Hunt, J. D.and Chilton, A. 2005 *J. of Crystal Growth* **281**, 654
[10] Drevet, B., Nguyen Thi, H., Camel, D., Billia, B.and Dupouy, M. D. 2000 *J. Crystal Growth* **218**, 419
[11] Buffet, A.et al 2008 *Phys. Stat. Sol.* **204**, 2721
[12] Bogno, A., Nguyen-Thi, H., Bergeon, N., Mangelinck-Noël, N., Schenk, T., Billia, B., Boller, E.and Baruchel, J. 2010 *Nuclear Instruments and Methods in Physics Research Section B: Beam Interactions with Materials and Atoms* **268**, 394
[13] Bogno, A., Nguyen-Thi, H., Billia, B., Bergeon, N., Mangelinck-Noel, N., Boller, E., Schenk, T.and Baruchel, J. 2009 *Transactions of the Indian Institute of Metals* **62**, 427
[14] Buffet, A., Nguyen Thi, H., Bogno, A., Schenk, T., Mangelinck-Noël, N., Reinhart, G., Bergeon, N., Billia, B.and Baruchel, J. 2010 *Materials Science Forum* **649**, 331
[15] Burden, M. H., Hebditch, D. J.and Hunt, J. D. 1973 *J. Cryst. Growth* **20**, 121
[16] Nguyen Thi, H.et al 2006 *Phys. Rev. E* **74**, 031605
[17] Appolaire, B., Albert, V., Combeau, H.and Lesoult, G. 1998 *Acta Materialia* **46**, 5851
[18] Ramani, A.and Beckermann, C. 1997 *Scripta Materialia* **36**, 633
[19] Badillo, A., Ceynar, D.and Beckermann, C. 2007 *Journal of Crystal Growth* **309**, 216
[20] Mangelinck-Noël, N.et al 2005 *J. Phys. D: Appl. Phys.* **38**, 28
[21] Martorano, M. A., Beckermann, C.and Gandin, C.-A. 2003 *Metallurgical and Materials Transactions A* **34A**, 1657
[22] Mullins, W. W.and Sekerka, R. F. 1964 *J. Appl. Phys.* **35**, 444
[23] Spittle, J. A. 2006 *International Materials Reviews* **51**, 247

[24] Jung, H., Mangelinck-Noel, N., Nguyen-Thi, H., Bergeon, N., Billia, B., Buffet, A., Reinhart, G., Schenk, T.and Baruchel, J. 2009 *Int. J. Cast. Metals Res.* **22**, 208

[25] Ruvalcaba, D., Mathiesen, R. H., Eskin, D. G., Arnberg, L.and Katgerman, L. 2007 *Acta Materialia* **55**, 4287

[26] Mathiesen, R. H., Arnberg, L., Bleuet, P.and Somogyi, A. 2006 *Metall. And Mater. Trans. A* **37A**, 2515

[27] Reinhart, G.et al 2008 *Metallurgical and Materials Transactions a-Physical Metallurgy and Materials Science* **39A**, 865

[28] Billia, B., Bergeon, N., Nguyen Thi, H.and Jamgotchian, H. 2004 *Phys. Rev Lett.* **93**, 126105

Journal of Physics: Conference Series **327** (2011) 012013 doi:10.1088/1742-6596/327/1/012013

Fluid-Flow Effects on Phase Selection and Nucleation in Undercooled Liquid Metals

Hyers, R.[1]; Matson, D.M.[2]; Kelton, K.F.[3]; Holland-Mortiz, D.[4]; Volkmann, T[4]

[1]University of Massachusetts, Amherst, MA, UNITED STATES; [2]Tufts University, Medford, MA, UNITED STATES; [3]Washington University, St. Louis, MO, UNITED STATES; [4] Institut für Materialphysik im Weltraum, DLR, Cologne, GERMANY

[1]hyers@ecs.umass.edu

Abstract. Fluid flow can have a profound effect on phase selection and nucleation in undercooled liquid metals.

In phase selection experiments, the sample does not solidify directly to the thermodynamically stable phase, but instead first forms a metastable phase, then transforms to the stable phase after some delay. Previous experiments and modeling on phase selection in near-eutectic Fe-Cr-Ni stainless steels have shown that the lifetime of the metastable phase is limited by the incubation time for formation of a critical nucleus of the stable phase, and that for some conditions, fluid flow can change that incubation time by an order of magnitude or more. New microgravity experiments in this area will be performed on new materials, including peritectic alloys, in the Materials Science Laboratory Electromagnetic Levitator (MSL-EML) as a part of projects PARSEC and THERMOLAB. The models that predict the effect of fluid flow on phase selection will be extended to these new experiments.

In another class of experiments, nucleation in quasicrystal- and glass-forming alloys, testing a new theory on nucleation kinetics requires control of the shear rate in fluid flow to prevent interference of the solute fields around the sub-critical nuclei. These theories will be tested in microgravity using the MSL-EML under the projects ICOPROSOL and THERMOLAB.

1. Introduction

Fluid flow has been shown to have an effect on a variety of solidification and nucleation processes. Effects which have been observed include distortion of solutal and thermal diffusion fields, gross fluid flow leading to macrosegregation, and even mechanical effects such as the bending or breaking of dendrites.

Two groups of experiments are discussed in this paper. Each group is being developed by international teams assembled under the European Space Agency's announcements of opportunity. The first group of experiments deals with phase selection in metals. Eutectic alloys including Fe-Cr-Ni alloys are being studied under the European Thermolab project, with the American partners funded under NASA's project LODESTARS. Related experiments on peritectic alloys including Fe-Co and Ti-Al-Ta are under ESA's PARSEC and NASA's ELFSTONE. These projects are described in more detail in [1].

The second group of experiments discussed here is looking at nucleation in quasicrystal- and glass-forming alloys. This group also includes samples under ESA's Thermolab projects, and ESA's ICOPROSOL / NASA's QUASI projects. These experiments are described in more detail in [2].

2. Phase Selection in Eutectic and Peritectic Alloys

In phase selection experiments, the sample does not solidify directly to the thermodynamically stable phase, but instead first forms a metastable phase, then transforms to the stable phase after some delay. Previous experiments and modeling on phase selection in near-eutectic Fe-Cr-Ni stainless steels have shown that the lifetime of the metastable phase is limited by the incubation time for formation of a critical nucleus of the stable phase, and that for some conditions, that incubation time may change by an order of magnitude or more (Fig. 1). All of the experiments shown in Fig. 1 had spontaneous nucleation.

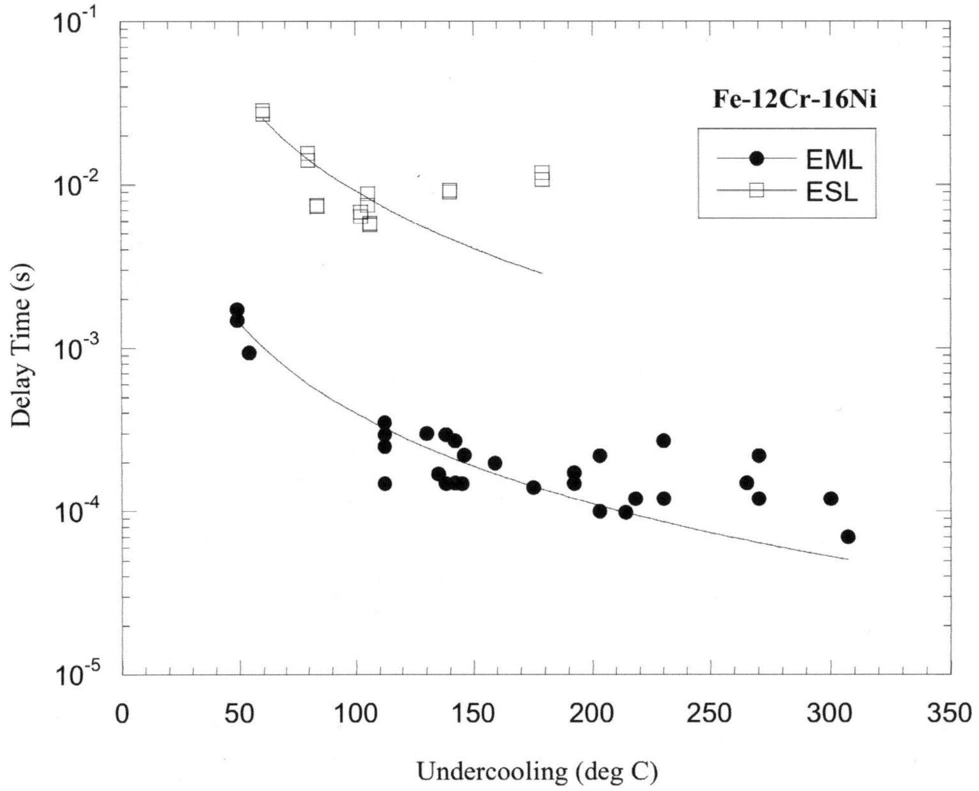

Figure 1. Lifetime of the metastable phase (Delay Time) vs. undercooling for steel samples processed in 1-g in Electromagnetic Levitation (EML) and Electrostatic Levitation (ESL) [3].

Fair, *et al.*, investigated a number of potential causes of this change in incubation time [4]. They concluded that none of those mechanisms was a plausible explanation for the observed behavior of the samples. One hypothesis that remained was examined by Hanlon, *et al.*, [5]: that flow in the liquid part of the sample was causing deflection of the dendrite arrays sufficient to cause collision of the secondary arms, leading to crevice nucleation (Fig. 2). They used computational fluid dynamics to determine the drag force on the tips of the primary arms (Fig. 3), and then evaluated analytical models to make quantitative predictions for different mechanisms of deflection including creep and elastic and plastic bending. They determined that elastic deformation of the growing dendrites would be sufficient to cause collision when the dendrites reached a critical length, and determined the time to reach that length from measurements of the growth velocity [6]. The resulting time to collision predicted by the model matched the experimental delay time within the scatter of the experimental data [5].

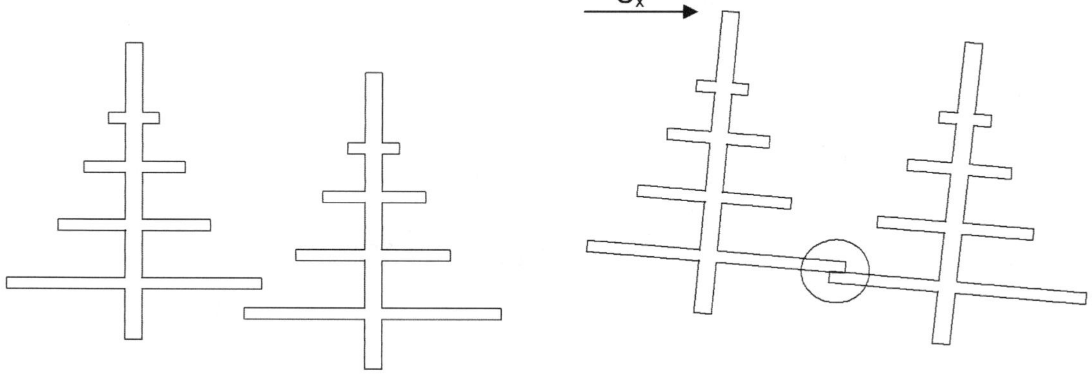

(a) **(b)**

Figure 2. Hypothesized mechanism of nucleation of the stable phase due to the influence of fluid flow on the growing metastable dendrites [5].

This model requires a significant amount of information about the microstructure of the metastable phase, including diameters of the primary arms of the dendrites and spacing of the secondary arms. This information is collected from micrographs of samples quenched between the formation of the stable and metastable phases, e.g. [7]. The micrographs of the quenched structure show that at a time within fractions of a millisecond before the nucleation of the stable phase, the dendrites have the approximately parallel arrangement assumed in the model.

The input data necessary for the models were available for two other compositions, one with a very low thermodynamic driving force for transformation (the difference in free energy between the metastable and stable phases), and one with an intermediate driving force. The sample with an intermediate driving force showed fair agreement with the model, indicating that the same mechanism is plausible for that alloy as well. For the smallest driving force, there was no agreement, indicating that this mechanism is not active. For a transformation with a small driving force, it is plausible that the transformation might not happen immediately on establishment of the crevice.

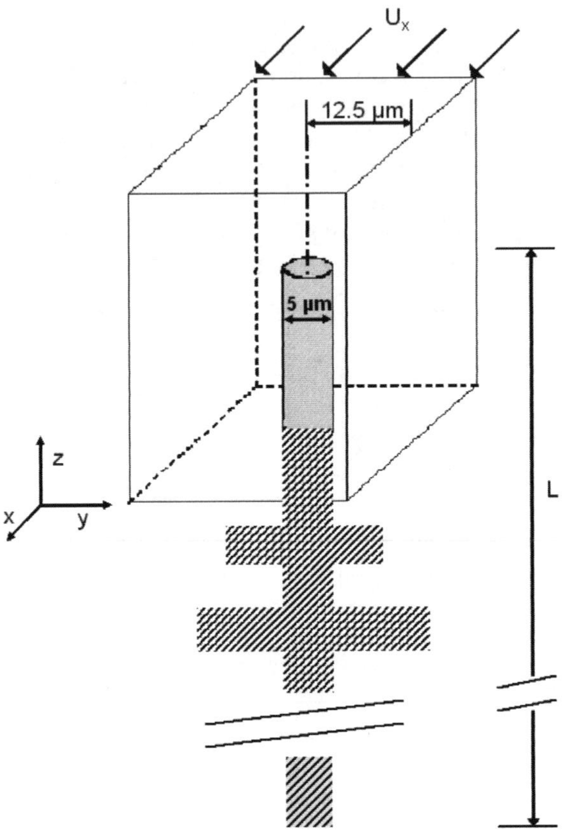

Figure 3. Computational domain for determining the drag forces exerted on the growing dendrites by fluid flow [5].

For application of Hanlon's model to the new eutectic and peritectic systems to be tested under THERMOLAB/LODESTARS and PARSEC/ELFSTONE, respectively, a number of data are required. These include:

- Properties of the liquid including density, viscosity, electrical conductivity. These properties will be measured on the flight samples as a part of the experiments in MSL-EML.
- Properties of the solid at its melting point, including elastic modulus and yield strength. The delay time predicted by the model is only weakly influenced by these properties, so extrapolations from lower temperatures will be used.
- Properties of the metastable dendrite array, including primary and secondary arm spacing and the diameter of the primary arms. These are the most difficult to obtain of all the needed data. Hanlon used micrographs by Kosecki [7] of samples quenched in the metastable state. Replicating this measurement is an important part of the ground support program for these experiments.

Once these data are obtained, the chain of models will allow determination of the range of convection where fluid-flow effects are expected to be important, and from there, the parameters needed to test for the presence of this mechanism in phase selection in the new alloys.

 doi:10.1088/1742-6596/327/1/012013

3. Nucleation in Undercooled Metals

The coupled-flux theory of nucleation has been well established for oxide glasses [8]. Testing this theory in metals is experimentally very difficult, in part because convection tends to distort the diffusion profiles around the subcritical nuclei since the viscosity is many orders of magnitude smaller than in the silicates. Microgravity experiments provide the possibility of removing the convective distortion, but models are needed to quantify convective flow and relate it to controllable experimental parameters.

The simple model for the relevant phenomena is depicted in Fig. 4. Two nuclei, each surrounded by its diffusion field, are carried along with the moving fluid. However, since the velocity of the fluid cannot be uniform throughout an EML droplet, nuclei on nearby streamlines move at different velocities. If the centers of the nuclei are close enough in the direction normal to the flow, then the diffusion fields will eventually interact. If the centers are farther apart than one diffusion field diameter, then the nuclei will pass without interacting.

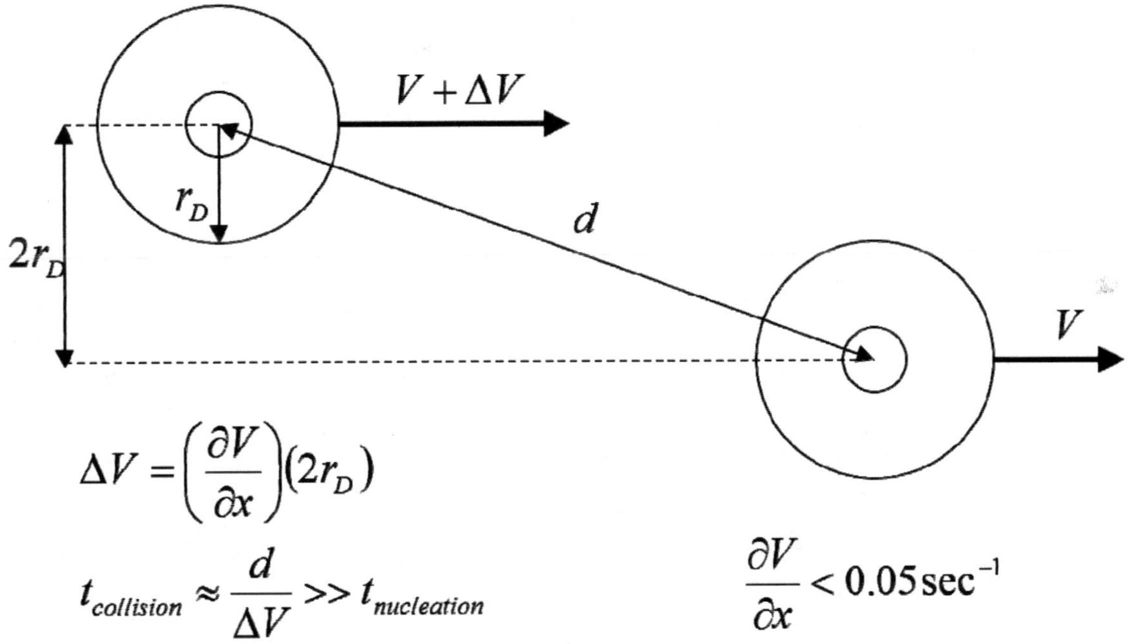

Figure 4. Model for the effect of fluid flow on the diffusion fields around subcritical nuclei [9].

The case illustrated in Fig. 4 provides the shortest possible time before the nuclei interact. Knowing the dimensions of the diffusion field and the number density of nuclei, it is possible to calculate a constraint on the allowable shear rate to prevent collisions during the time of the experiment. For a cluster separation d of 1 – 0.1 μm, a diffusion fields of radius r_D of about 10 nm, and a cooling rate of about 10 °C s^{-1}, the result is that the velocity is not explicitly constrained, but the shear rate must be less than 0.05/s throughout the drop [9].

Other effects must also be considered. First, the diffusivity of momentum (kinematic viscosity) is perhaps four orders of magnitude greater than the diffusivity of solute in these alloys, so momentum transfer will happen much faster and at much longer range than the diffusion fields around the nuclei.

However, the nuclei are very small, only nanometers in diameter, so the Stokes number is very close to zero (of the order 10^{-13}!). That means that the nuclei follow the path of the surrounding fluid very precisely, and do not distort the streamlines. Accordingly, each differential volume of fluid behaves the same whether the next differential volume is liquid or crystalline, and the nuclei are transported as in the model in Fig. 4.

Second, the shear in the fluid affects each nucleus directly. Because of fluid drag, the top of each nucleus in Figure 4 must move faster than the center, and the bottom must move slower, so each nucleus must rotate. The rotation rate in radians per second must approximately equal to the shear rate in inverse seconds. Rotating spheres in a flow experience lift due to the Magnus effect, which causes baseballs and tennis balls to curve. The lift due to the Magnus effect is proportional to the cross product of the vorticity and the relative velocity between the sphere and the fluid. In this case, as already established, the relative velocity is zero, so the Magnus forces are also zero.

Finally, the rotation itself will cause some distortion of the fluid flow as the nucleus rotates like a centrifugal pump. This effect will cause an additional radial outflow in the plane of Figure 4, and a compensating inflow toward the nuclei from planes above and below the figure. However, the rotation rate is very low, of the order 0.05 rad/sec, and the impeller radius is also very small, of the order nanometers. The resulting flow is too small to have a significant effect on the diffusion fields, again confirming the simple picture in Figure 4.

Finally, now that the constraints on the flow field have been established, magnetohydrodynamic models are used to determine the corresponding constraints on the operational parameters such as coil currents in the MSL-EML. These models use properties measured in the ground support program, and will be updated with properties determined on the flight samples as a part of the experimental program.

4. Conclusions

Fluid flow is important in both classes of experiments described above, but through different mechanisms. In one case, the shear rate in the fluid must be kept everywhere below a particular value to test a theory about nucleation of crystals in a liquid, while in the other the fluid's velocity must be varied within a range to test a theory about nucleation of the stable phase in a mixture of liquid and metastable solid. Computational models of the magnetohydrodynamic flows in electromagnetically levitated droplets, combined with accurate thermophysical properties, allow the use of flow as an experimental parameter.

5. Acknowledgments

The authors gratefully acknowledge support from ESA under AO-2009-0898, AO-2009-1020, and AO-2009-0959, and NASA under NNX10AR71G. Dr. Hyers also wishes to thank Prof. Jan Fransaer for fruitful discussions.

References

[1] D.M. Matson R.W. Hyers, T. Volkmann, and H.-J. Fecht, this volume.

[2] K. Kelton, this volume.

[3] Matson D M, Fair D J, Hyers R W and Rogers J R 2004 Contrasting electrostatic and electromagnetic levitation experimental results for transformation kinetics of steel alloys Ann. NY Acad. Sci. 1027 435–46

[4] Fair D J, Venkatesh R, Boghosian B and Matson D M 2005 Role of sample size in the nucleation kinetics of phase transformations in steel alloys Microgravity Sci. Technol. : Int. J. Microgravity Res. Applic. 16 5, and D. J. Fair, R. Venkatesh, B. Boghosian and D. M. Matson, "Role of Sample Size in the Nucleation Kinetics of Phase Transformations in Steel Alloys", *Proceedings of the 2nd International Symposium on Physical Sciences in Space ISPS-2*, 2004.

[5] A. B. Hanlon, D. M. Matson, R. W. Hyers, "Internal Convective Effects on the Lifetime of the Metastable Phase in Undercooled Fe-Cr-Ni Alloys", *Philosophical Magazine Letters*, 86(3), 2006, 165 – 174.

[6] D.M. Matson, W. Löser, and M.C. Flemings, "Phase Selection and Rapid Solidification of Undercooled Fe-Cr-Ni Steel Alloys in Microgravity", in *Solidification 1999*, W. H. Hofmeister, J. R. Rogers, N. B. Singh, S. P. Marsh, and P. W. Voorhees, *eds*., TMS Warrendale PA, (1999), pp. 99-106

[7] T. Koseki and M.C. Flemings, "Solidification of undercooled Fe-Cr-Ni alloys: Part III. Phase selection in chill casting", Met. Mat. Trans. A 1997 28A(11) 2385-2395.

[8] K.F. Kelton, "Crystal Nucleation in Liquids and Glasses", Solid State Physics 45 1991 75-177.

[9] Kelton K F, Gangopadhyay A K, Hyers R W and Holland-Moritz D 2003 Studies of nucleation and growth, specific heat and viscosity of undercooled melts of quasicrystals and polytetrahedral-phase-forming alloys Science Requirements Document QUASI-RQMT-0001, NASA.

In-situ characterization of droplets during free fall in the drop tube-impulse system

P Delshad Khatibi[1*], A Ilbagi[1], D Beinker[2], H Henein[1]

*Email: delshadk@ualberta.ca

[1]University of Alberta, Edmonton, Canada
[2]Universitat Bremen, Germany

Abstract. Powders of copper were produced using a drop tube-impulse atomization technique. In this system, molten metal is pushed through orifices, forming ligaments, which eventually break down and spherodize into droplets. A 3-D translation stage was designed, constructed and installed in the drop tube to allow for measurements of velocity and droplet size in flight using a Shadowgraph and radiant energy using DPV-2000. A mathematical model of the evolution of droplet velocity and temperature for different sized copper droplets at various heights was developed. The experimental results from the Shadowgraph and the DPV-2000 are compared to the model's results. In addition, the extent to which microgravity prevails during flight and droplet solidification was investigated by using the model and the Shadowgraph results. It was found that the acceleration of falling droplets near the melting point is close to gravitational acceleration and as a result the falling droplets do not reach their terminal velocity at their melting point. The results of in-situ measurements during the atomization of copper showed that the larger droplets have higher radiant energy than that of the smaller ones. Correlation between experimentally measured radiant energy and predicted temperature of falling droplets will be investigated. The current work is part of the NEQUISOL project supported by ESA within contract number 15236/02/NL/SH and CSA within contract number 9F007-08-0154 and SSEP Grant 2008.

1. Introduction

Solidification of alloys is a complex phenomenon arising in many modern experimental techniques and industrial technologies related to casting and surface processing. The variation of different solidification conditions (such as undercooling or cooling rate) provides the possibility to control the morphology and size of crystal structures, which substantially influences physical and chemical properties of alloys; more specifically, deep undercooling of alloys below the equilibrium liquidus and eutectic temperature results in rapid solidification and yields materials with improved mechanical, magnetic and electrical properties. Deep undercooling results in metastable phases or in high growth velocities which in turn leads to a finer microstructure.

A number of models have been reported for predicting or estimating the undercooling temperature of droplets in gas and single fluid atomization systems as well as drop tubes [1-3]. There is no feasible means available yet to directly validate these models. An indirect method of validation was presented by Prasad et al [4]. In this approach, microtomographic images are analyzed in order to locate the region of initial growth in a particle. The volume of this region is measured and compared with predictions obtained using a microsegregation model incorporating a variable undercooling of the primary phase [5, 6]. This approach has led to estimations of primary phase undercooling of about 15 to 20°C in Al-Cu droplets. Eutectic undercooling has also been estimated by measuring the fraction of eutectic in the solidified structure either using SEM images or neutron diffraction [5, 7]. The measured quantity is compared with the respective phase diagram of the system under study. Solidus and liquidus lines are extended until agreement is found between measured eutectic fraction and the extended lines on the phase diagram. The eutectic temperature corresponding to this agreement is taken to be the nucleation temperature of the eutectic. Gandin et al [8] reported temperature measurements of a droplet levitated in an electromagnetic field using a high speed pyrometer. They clearly report measurements on an Al-24 Cu droplet of 6 mm diameter of primary phase and eutectic undercooling.

A systematic effort on the investigation of thermal history of falling droplets in-situ during atomization is rare.[9] In addition to the thermal history, in-situ measurement of the droplet diameter, initial velocity and instantaneous velocity is crucial in validating solidification models. This study reports on an effort to make such measurements on droplets in flight and in situ during the atomization of copper. Two measuring instruments were used: a Shadowgraph to measure droplet velocity and diameter and a DPV-2000 (a two color pyrometer) for droplet radiation measurements.

One of the important parameters affecting the thermal behavior of falling droplets during atomization is the amount of microgravity achieved. It has been reported that the level of microgravity achievable using drop tube technique is relatively high, i.e. up to 1×10^{-5} of the earth's gravity [10]. The problem involved with the drop tube system is that the duration of microgravity happens in the drop tubes is short and many of the samples cannot solidify during their free fall in an evacuated drop tube. For example, the temperature of a 2 mm droplet of aluminum with the superheat of $0.1T_m$ (melting point) decreases only 19K during 50 meters falling height [11]. Adding gases to the tube ensures us that the solidification of the droplets happens in the reasonable falling height. But, this can deteriorate the low gravity condition. In this situation, smaller droplet size, higher gas pressure and faster initial falling velocity result in lower microgravity condition. Drop-tube-Impulse Atomization is a method in which atmospheric gas pressure is used and it provides a containerless solidification environment for the droplets to solidify with a high cooling rate.

In this study, a correlation between radiant energy and temperature of falling droplets using experimental and a mathematical code will be studied in order to investigate the sensitivity of DPV-2000 for measuring undercooling during droplet solidification. Furthermore, the microgravity condition of the falling droplets in the drop-tube impulse atomization system will be investigated using the same code and the experimental velocity results.

2. Experiments

Copper shots with 99.99% purity (manufactured by Alfa Aesar) were melted in an induction field of a drop tube-impulse atomization technique to produce copper powder. In this system, molten metal is pushed through orifices, forming ligaments, which eventually break down and spherodize into droplets according to Rayleigh instability [12]. After heating the copper shots to superheat of 1400°C inside a graphite crucible using an induction furnace, atomization was then started. During atomization, the liquid copper is pushed through 37 orifices with the diameter of 400 μm on a nozzle plate at the bottom of the crucible via a vibrating plunger. Falling droplets were cooled in argon atmosphere having a maximum oxygen content of 90 ppm and lose heat to the surrounding gas.

A 3-D translation stage was designed, constructed and installed in the drop tube to allow for measurements of radiant energy using a DPV-2000 (Tecnar Automation Ltée) and velocity and droplet size in flight using a Shadowgraph (Sizing Master Shadow from LaVision GmbH in Gottingen, Germany).

DPV-2000 is an optical sensing device based on a patented technology developed by the National Research Council of Canada. DPV-2000 uses a dual slit optical sensor that can measure the radiant energy of up to 800 droplets per second with the depth of field of 1.9 mm, and it can measure radiant energy of droplets at two different wavelengths [13].

Sizing Master Shadow from LaVision is using the shadowgraph technique (backlighting) to visualize droplets for image analysis. A light source, in this case a pulsed laser combined with the diffuser optics, illuminates in-flight droplets 5 times per second. Backlight of droplets inside the measurement volume of 6×6×6 mm is captured by a high resolution imaging system consists in a high-speed camera. It is possible to investigate droplet sizes down to 5μm using the shadowgraph device. Figure 1 shows a schematic of the shadowgraph system [14].

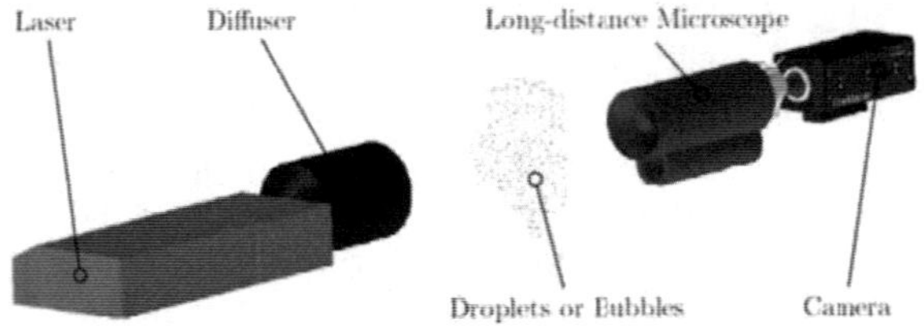

Figure 1. General view of a shadowgraph system.

Particle size distribution of falling droplets (D50 and D90) and the velocity were measured via the shadowgraph device. Also, using the 3-D stage, the droplet size, velocity and radiant energy can be measured at different distances from the nozzle plate. In this study, these measurement instruments were located at 10 cm below the nozzle plate. The focal point of both shadowgraph and DPV-2000 was set to an imaginary vertical line going through the center of the nozzle plate; therefore, any unfocused droplets would be rejected by the criteria set by the operator in the respective software of each instrument.

3. Modeling

In order to verify the experimental results, a mathematical model of the evolution of droplet velocity and temperature for different sized copper droplets at different heights was developed. The model calculates the droplet velocity and temperature as a function of time and distance traveled. For the purposes of this paper, the cooling and velocity of droplets up to their melting point is considered.

Droplets begin their downward trajectory with an initial velocity. The subsequent trajectory of the droplets depends on this initial velocity, and the forces of gravity, buoyancy and drag as shown schematically in Figure 2.

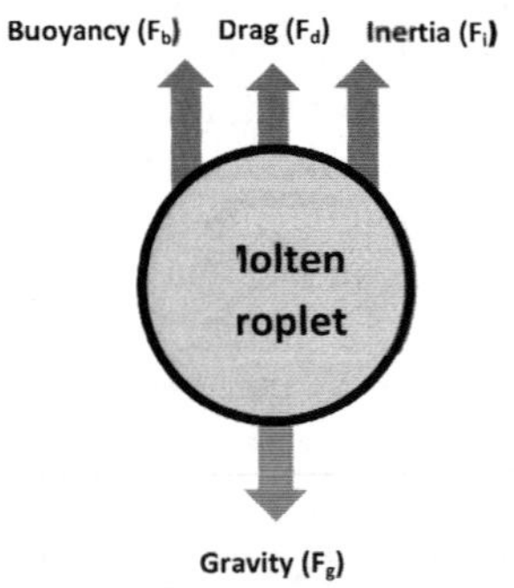

Figure 2. Schematic of the forces acting on a falling droplet [1].

Applying Newton's Second Law on the droplet, the instantaneous acceleration can be found as follows [15].

$$\frac{dv}{dt} = g - \frac{3}{4}\frac{\rho_g}{\rho_p}\frac{c_d}{d} . v^2 \quad (1)$$

where v is the relative velocity between the droplet and the atomization gas, (which in the case of Impulse Atomization v_{gas}=0) and ρ_p and ρ_g are the densities of the droplet and atomization gas, respectively. The gravitational acceleration is given by g and d is the droplet diameter. The discharge coefficient, C_d, for droplets in laminar flow is given by: [16]

$$C_d = \frac{18.5}{Re^{0.6}} \quad (2)$$

The Reynolds number, *Re*, in Equation (2) is given by:

$$Re = \frac{\rho_g . v . d}{\mu_g} \quad (3)$$

where μ_g is the gas viscosity.

To calculate the thermal history of a falling droplet, an equation describing the rate of heat energy lost from the surface of the droplet to the surrounding gas is used and is given by:

$$q = h_{eff} A(T_m - T_0) \quad (4)$$

where h_{eff} is the effective heat transfer and consisting of the additive contribution of convection, conduction and radiation heat transfer mechanisms, A is the surface area of the droplet, T_m is the droplet surface temperature and T_0 is the gas temperature. It is assumed that the surface temperature of the droplet represents the entire droplet temperature, since the internal temperature gradient within the droplets are negligible (Biot <0.1) [17]. It is also assumed that the temperature increase due to surface oxidation is negligible (see appendix 1). The effective heat transfer coefficient is mostly dominated by convection in low temperature alloys but for the high temperature alloys, radiation heat transfer can be significant.

One approach to quantify the convective component (H_c) of h_{eff} has been through the use of semi-empirical equations (e.g. Ranz-Marshall or Whitaker) [18, 19] where the Nusselt number *(Nu)* is averaged over the entire droplet surface (Equation 5).

$$Nu = \frac{H_c . d}{k_g} = A + B . Pr^m . Re^n \quad (5)$$

where the Prandtl number, *Pr*, in Equation (5) is given by

$$Pr = \frac{C_p \mu_g}{k_g} \quad (6)$$

and d is the droplet diameter, k_g ,C_p , ρ_g and μ_g are the conductivity, heat capacity, density and viscosity of the gas, respectively, and all other parameters are constants.

Wiskel et al. [17] modified the Whitaker correlation in a way that it takes into account the variability of the gas conductivity across the boundary layer. Equation 7 shows the modified Whitaker correlation:

$$Nu = \frac{h_C D_s}{k_g} = 2\frac{C}{k_s(m+1)} \cdot \frac{\left(T_s^{m+1} - T_g^{m+1}\right)}{\left(T_s - T_g\right)} + \left(0.4\,Re^{1/2} + 0.06\,Re^{2/3}\right)Pr^{0/4}\left(\frac{\mu_g}{\mu_s}\right) \tag{7}$$

where from the variation of gas conductivity with temperature (k_g=C×T^m) for argon C = 1.86×10^{-4} and m = 0.7915 [19], the model was able to closely predict the range of time and distance in which the droplets completely solidified under the condition that k_s be evaluated at the metal droplet surface temperature and the *Re* and *Pr* numbers evaluated at the free stream gas temperature. Table 1 lists the properties of pure copper and argon which were used in the model.

Table 1. Thermophysical properties of pure copper and argon at 1400°C and 40°C, respectively [20]

	Temperature (°C)	**Thermal conductivity (W/K.m)**	**Specific heat (J/kg.K)**	**Density (kg/m^3)**	**Viscosity (m^2/s)**
Copper	1400	325	532	7722	-
Argon	40	0.01853	20.80	0.287	0.021

4. Results and Discussion

4.1. Online measurements

The mean droplet diameter (D50=565μm) and the standard deviation (σ=1.52) were measured using shadowgraph. The velocity versus droplet size of the falling droplets as measured by the shadowgraph is shown in Figure 3. For this measurement, the shadowgraph was continuously collecting droplet size and velocity data at 5Hz frequency. A total number of 6100 droplets were measured during the entire atomization time of two minutes. From Figure 3, each droplet size can have a range of velocities, with larger droplets having a smaller range than smaller ones.

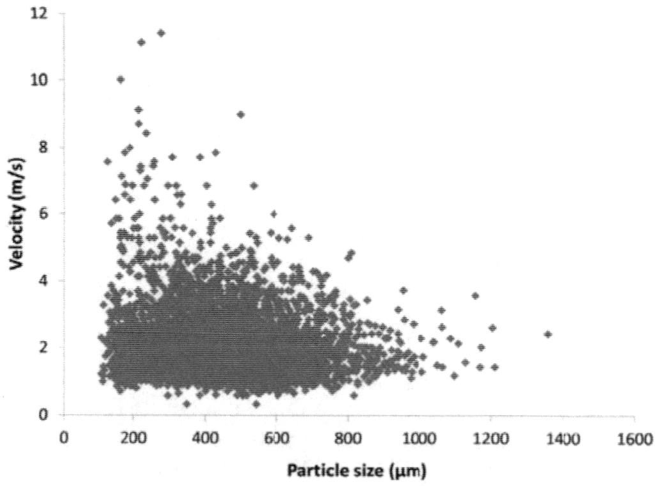

Figure 3. Velocity as a function of droplet size of falling droplets collected by shadowgraph system at 10cm below the nozzle plate.

While the shadowgraph was collecting the velocity and diameter of falling droplets at the center of the plume, the DPV-2000 was measuring the radiant energy of the droplets from the same location. Figure 4(a) shows the signal counts of each droplet that DPV-2000 measured at two wavelengths. The area under the curves shown in Figure 4(a) represents the radiant energy and as such it is a dimensionless value. The radiant energies measured for different droplet sizes at two different wavelengths are shown in Figure 4(b). The measurement was done at 10cm below the nozzle plate. It can be seen that larger droplets have higher radiant energy.

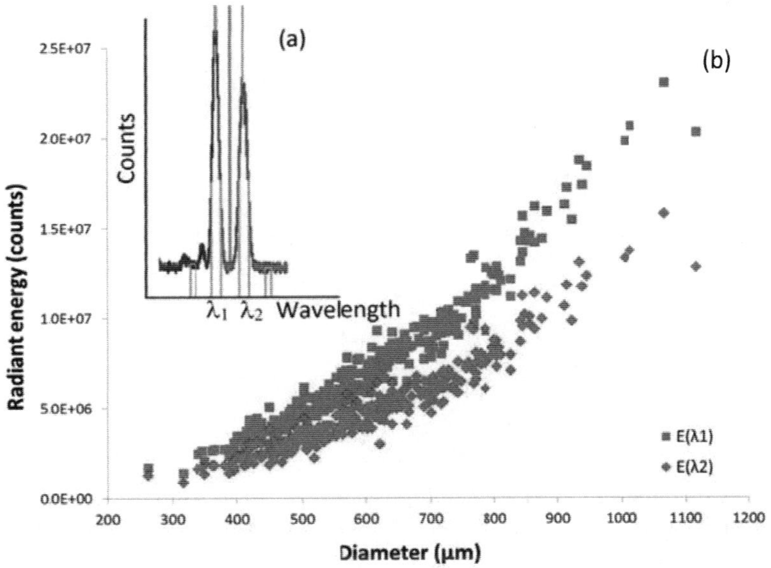

Figure 4. (a) The signal counts measured at two different wavelengths for a single droplet, (b) Radiant energy vs. droplet diameter measured by DPV-2000 at 10cm below the nozzle plate.

4.2. Droplets temperature calculations

To model the temperature of a given droplet size at different distances from the nozzle plate, the droplet's initial temperature (i.e. atomization temperature) and initial velocity (i.e. the exit velocity from the nozzle plate) are required. The initial temperature in this experiment was 1400°C, and the initial velocity of the droplets can be found using the instantaneous velocities measured by shadowgraph (Figure 3) and the model discussed in section 3. The mean droplet diameter (D50=565µm) is used as an example to show the process of droplet temperature calculations.

A range of velocities was measured for each droplet size, as shown in Figure 3. Running the model under different initial velocities reveals the initial velocities which result in the measured values at the 10 cm distance for a 565 µm droplet size (Figure 5). Thus, the initial velocity for this droplet size ranges from 0.52 to 1.36 m s^{-1}. For a given droplet size, based on the frequency of velocity measurements obtained by the shadowgraph (Figure 3), a mean initial velocity and a standard deviation is determined.

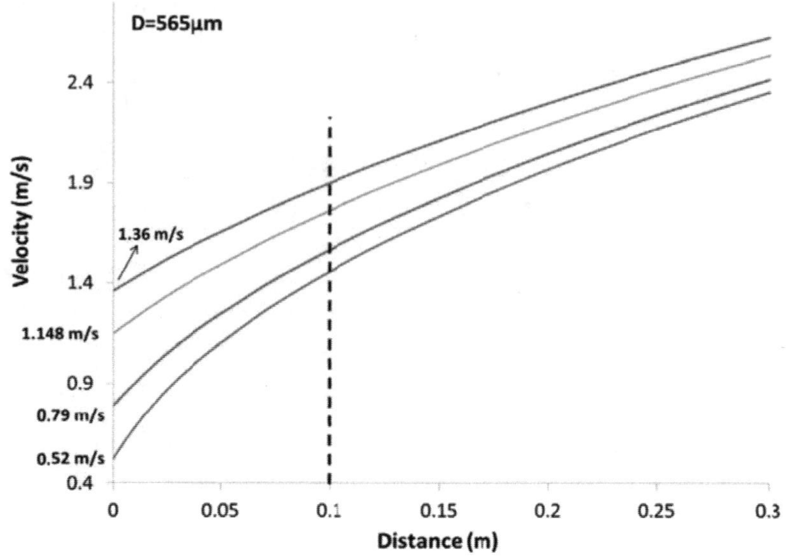

Figure 5. Velocity as a function of distance of droplets from the nozzle plate for the 565µm droplet size, generated from the model.

Using the range of calculated initial velocities and the initial temperature of melt in the model, the temperature of a 565µm droplet at different distances from the nozzle plate can be determined. Figure 6 shows the temperature-distance relationship of a 565µm droplet with different initial velocities. It is evident that different initial velocities result in different temperature at a given distance. The average of these different temperatures will be used to represent the temperature of each droplet size.

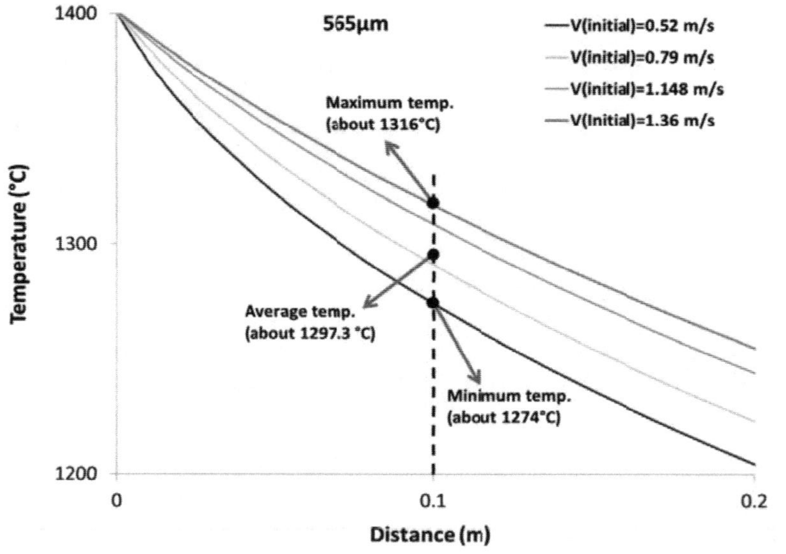

Figure 6. Temperature vs. distance from nozzle plate (plotted using the model). Dashed line shows distance from the nozzle plate.

4.3. Droplets acceleration calculations

In order to investigate the level of gravity of falling droplets during solidification, the acceleration of the droplets must be calculated. For this purpose using the average initial velocity, the velocity-time profile of a 565μm droplet is plotted in Figure 7. Furthermore, the temperature-time profile of the same droplet as a function of time is included. At the melting point of copper (1084°C),the corresponding time (0.26s) to reach this equilibrium solidification temperature is determined. At this time of the trajectory, the derivative of the velocity results (shown in Figure 7) is used to calculate the acceleration of the droplet. This calculation was performed on different droplet sizes at their respective times to reach the solidification temperature. The results are shown in Figure 8 and compared with the gravitational acceleration.

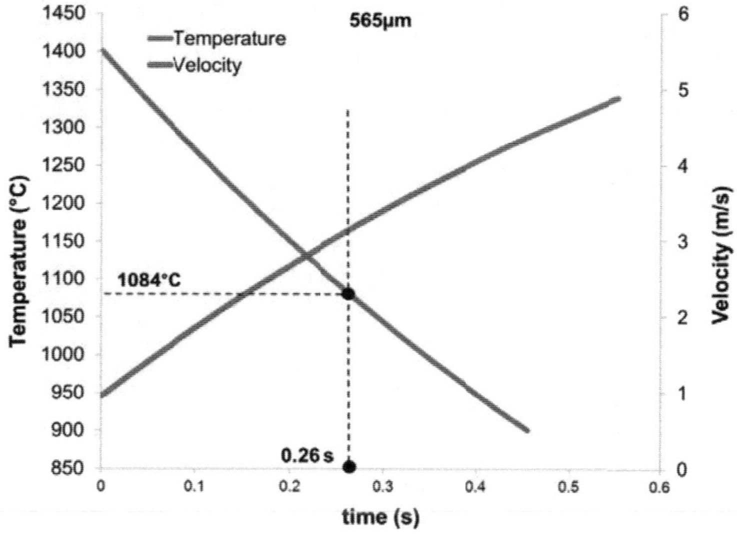

Figure 7. Temperature and velocity as a function of droplet traveling time, for 565μm droplet.

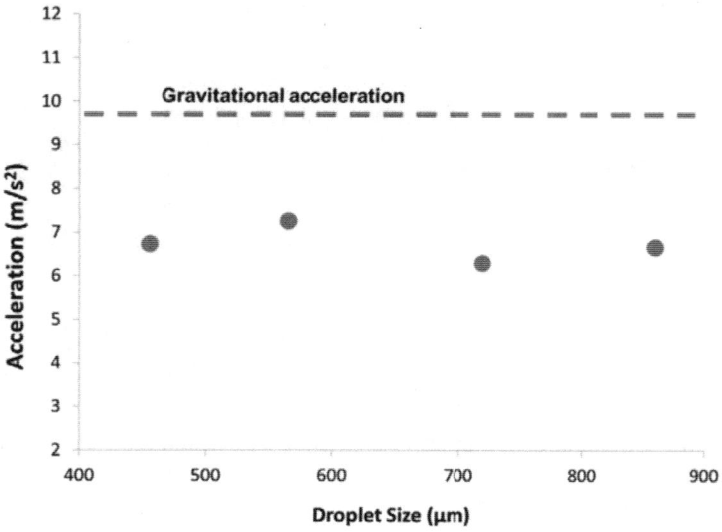

Figure 8. Acceleration of different sized droplets at a time in which the droplets temperatures reach to the melting point. Dashed line shows the gravitational acceleration.

Journal of Physics: Conference Series **327** (2011) 012014 doi:10.1088/1742-6596/327/1/012014

The variation in the amount of acceleration experienced by each droplet size in Figure 7 is due to different frequencies of velocities for each particle size. It can be seen that the droplets acceleration are smaller than the gravitational acceleration. This is due to the buoyancy and drag forces applied from the surrounding gas atmosphere on the falling droplets. Buoyancy and drag forces resist the droplets falling and decrease the effective acceleration on the droplets applied by the gravitational acceleration. Also, the droplet acceleration of about 7 m/s^2 shows that the droplets do not experience their terminal velocities at the time of the solidification. This means that the gravity is still the dominant force acting on the droplets until they reach their equilibrium solidification temperature.

4.4. Calculated temperature and measured radiant energy correlation

As was shown in section 4.1, the radiant energy measured by DPV-2000 is a dimensionless value and is dependent on a number of variables in the system, including emissivity, view factor, and the nature of the droplet surface. It is therefore important at this stage of the study to determine if there is a correlation between the calculated droplet temperatures and the measured radiant energies measured by the DPV-2000. Figure 9 shows the preliminary results of temperature-energy values achieved at 10 cm below the nozzle plate for different droplet sizes. It can be seen that at a 10 cm distance from the nozzle plate, larger droplets have higher temperature than the smaller ones. The measured radiant energy also follows the same trend. Ongoing research is directed at investigating the sensitivity of the DPV-2000 in measuring of the rate of change of the radiant energies at the two wave lengths and comparing them to the predicted cooling rate of liquid droplets. Subsequent efforts will focus on exploring the ability of the DVP 2000 to detect the undercooling of the droplets.

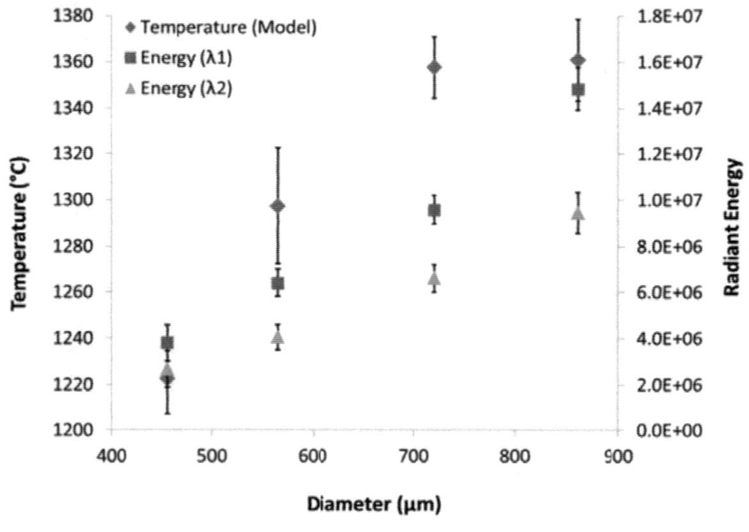

Figure 9. Temperature and Energy as a function of droplet diameter, in 10cm below the nozzle plate.

5. Summary

1- Particle size, particle size distribution and velocity of Impulse Atomized droplets have been measured using a Shadowgraph device.
2- Radiant energy of molten copper droplets was measured with DPV-2000 at 10 cm below the nozzle plate.
3- Acceleration of falling droplets near the melting point is close to gravitational acceleration.

4- Predicted droplet temperature at 10cm from nozzle plate decreases with decreasing the droplet size. Measured $E(\lambda_1)$ and $E(\lambda_2)$ also decreases with decreasing droplet size.

6. Acknowledgments

The authors would like to thank Natural Sciences and Engineering Research Council of Canada (NSERC) and Canadian Space Agency (CSA) for the financial support and Nils Ellendt for the modeling.

7. Appendix

In the model, it was assumed that the temperature increase due to oxidation is negligible. Here, the validity of this assumption will be investigated.

As mentioned in the experimental section, the amount of oxygen in the atomization tower at the start of the run is about 90ppm. The oxidation of the droplets during atomization under this condition is thermodynamically favorable and is exothermic. Thus, it may result in a temperature increase of the droplet. An estimate of the maximum increase in the droplet temperature is performed.

Liquid copper oxidizes according to the following reaction: [21]

$$2\mathrm{Cu}_{(l)} + \frac{1}{2}\mathrm{O}_2 = \mathrm{Cu}_2\mathrm{O}_{(s)} \quad , \Delta H = -188.3 \text{ kJ/mole} \tag{8}$$

Volume of the atomization tower is about $1.67\mathrm{m}^3$ and total oxygen content of the tower atmosphere is 90ppm. This gives 6.14×10^{-3} moles of O_2 available to oxidize 900g of Cu used for atomization. This amount of Cu would produce about 1.2×10^6 droplets of 565μm (D50) (a monosized droplets distribution is assumed).

Considering the total number of droplets, total available oxygen in the tower and the oxidation reaction, the average thickness of oxide layer on each droplet may be calculated to be about 240nm. This amount of oxidation was found to increase the temperature of a single Cu droplet by 5°C. This temperature increase is less than the error bars shown in Figure 9. Thus the assumption of neglecting surface droplet oxidation in the model is valid.

7. References

[1] Prasad A, Mosbah S, Henein H and Gandin Ch-A 2009 *ISIJ International*, **49**(7) 992
[2] Prasad A and Henein A 2008 *Journal of Material Science* **43** 5930
[3] Zeoli N Gu S, Kamnis S 2008 *International Journal of Heat and Mass Transfer* **51** 4121
[4] Prasad A, Henein H, Maire E, and Gandin Ch-A 2004 *Canadian Metallurgical Quarterly* **43**(2) 273
[5] Prasad A, Henein H and Conlon K 2006 *Metallurgical Transactions A* **37** (5) 1589
[6] Prasad A, Henein H, Gandin Ch-A and Mosbah S 2009 *ISIJ International* **49**(7) 992
[7] Henein H, Buchoud V, Schmidt R, Watt C Malahof D, Gandin Ch-A, Lesoult G and Uhlenwinkel V 2010 *Canadian Metallurgical Quarterly* **49**(3) 275
[8] Gandin Ch-A, Mosbah S, Volkmann Th and Herlach D M 2008 *Acta Materialia* **56** 3023
[9] Cui C, Fritsching U, Schulz A and Li Q 2005 *Acta Materialia* **53** 2775
[10] Van Zoest T, Gaaloul N, Singh Y, Ahlers H, Herr W, Seidel S T, Ertmer W, Asel E, Eckart M, Kajari E, Arnold S, Nandi G, Schleich W P, Walser R, Vogel A, Engstock K, Bongs K, Lewoczko-Adamczyk W, Schiemangk M, Schuldt T, Peters A, önemann T, Müntinga H, Lämmerzahl C, Dittus H, Steinmetz Hänsch T W and Eichel J 2010 *Science* **328** 1540
[11] Chen L and Luo X H 2007 *Acta Metallurgica Sinica* **43**(7) 769
[12] Henein H 2002 *Materials Science Engineering A* **326**(1) 92

[13] Tecnar Automation Ltée. *DPV-2000 Calculation Principles*, *http://www.tecnar.com /images/stories/TSS/DPV/pdf/dpv_calculation_principles.pdf*, Accessed: August 22, 2011
[14] LaVisionGmbH. 2007 *Product-Manual for DaVis*. Göttingen
[15] Ding Y, Henein H and Fallavollita, *Methods and apparatus for producing droplets*, US Patent 5,609,919 (1997); Canadian Patent 2,147,407 (2003)
[16] Wiskel J B, Navel K Henein H and Maire E 2002 *Canadian Metallurgical Quarterly* **41** (2) 193
[17] Wiskel J B, Henein H and Maire E 2002 *Canadian Metallurgical Quarterly* **41**(1) 97
[18] Ranz W E and Marshall Jr W R 1952 *Chemical Engineering Progress* **48**(3) 141
[19] Whitaker S 1972 *AIChE* **18**(2) 361
[20] Brillo J and Egry I 2003 *International Journal of Thermophysics* **24** 1155
[21] Gaskell D R *2003 Introduction to the thermodynamics of materials* (Taylor & Francis) Fourth edition

Materials Science Investigations using Electromagnetic Levitation

A Seidel[1], W Soellner and C Stenzel

Astrium GmbH, Dept. TO54, Claude-Dornier-Str., 88090 Immenstaad, Germany

Email: achim.seidel@astrium.eads.net

Abstract. EML on ISS allows levitating liquid samples both above and below their melting points for extended periods under ultra-high vacuum or ultra clean noble gas atmosphere. Various stimuli can be applied to the samples for dedicated experiment objectives. The heat input into the sample can be modulated to induce a thermal response of the sample, short heater pulses can be used to induce surface shape oscillations of the liquid sample, a custom made trigger needle can be driven into the undercooled sample to induce heterogeneous nucleation at a predefined temperature, touching of the sample by a dedicated chill cool plate or application of a forced gas flow can be used to increase the cooling rate of the sample or to simulate convection for reference experiments. Dedicated diagnostics elements are available to measure the physical properties of the sample. Sample temperature is measured by a pyrometer; two video units in orthogonal views provide both high spatial (up to 1 Megapixel and relative size resolution 2 * 10-4) and temporal (up to 30 kHz) resolution. Additional capabilities are under discussion which would allow to measure the electrical conductivity of the sample from electrical data of the rf coil system, and to determine the residual oxygen content of the process atmosphere.

1. Scientific Objectives

Electromagnetic levitation provides unique opportunities for the investigation of electrically conductive liquid metals, alloys and semiconductors (if pre-heated or doped), both above and below their melting temperatures. In particular, the undercooled regime where the sample is liquid at temperatures below its equilibrium melting temperature is accessible through electromagnetic levitation thus allowing the determination of thermo-physical properties of materials in this particular state as well as studying of nucleation and solidification phenomena.

Thermo-physical properties that can be measured comprise heat capacity, enthalpy of fusion, thermal transport coefficients in the liquid phase, viscosity, surface tension, thermal expansion coefficient, hemispherical emissivity, fraction solid and electrical conductivity. In studies of nucleation and solidification phenomena the nucleation kinetics and solidification velocity can be determined. [1], [2]

The results gathered by this technique can improve our understanding of the solidification process and of the physics of undercooled melts, and are therefore useful for both, pure science and practical applications. For example, in the industrial production a large fraction of all metallic materials is made from the melt. The solidification process in applications like casting determines the material properties

[1] To whom any correspondence should be addressed.

and product quality; therefore the results of the process are influenced by the properties of the material while it is undercooled. In order to improve the casting process and to avoid high costs due to iterations of the process, numerical simulations of the melt processing including casting and solidification is increasingly applied in industrial production. The quality and predictive power of the simulation critically relies on the availability of reliable thermo-physical data in the liquid phase, however there is a lack of such data due to the high chemical reactivity of high-temperature metals in the liquid phase with any crucible. Containerless processing can provide thermophysical data that are otherwise not available and can therefore help improve the modeling of casting processes.

In a recent industry survey conducted among European industries the need for better thermo-physical properties became evident, in particular in the industries dealing with casting, metal and alloy production and refinement [3].

2. Ingredients

In order to prepare the undercooled state and to perform experiments aiming at the determination of the thermo-physical properties without adverse impacts related to chemical reactions, heterogeneous nucleation, or lack of visibility, the following ingredients are essential:

2.1. Electromagnetic levitation

Electromagnetic levitation allows processing of the samples without contact to a container. A wide range of electrically conductive samples including pure metals, alloys and semiconductors can be processed with this technique. The sample is placed in the center of a coil system which is part of an oscillating RF circuit and generates an RF electromagnetic field. The interaction of eddy currents induced in the sample with the electromagnetic field leads to a displacement force which keeps the sample at the center of the coil system. Heating is achieved by ohmic losses of the eddy currents flowing in the sample. In ground-based experiments the required levitation force and therefore electromagnetic field strength to counteract gravity is so large that many materials are melted just by applying the field to position them. Heating and positioning of the sample are therefore not independent and the undercooled regime of samples with low melting temperatures is not accessible. Furthermore, the electromagnetic pressure exerted by the strong fields leads to strong convection and a deformation of the liquid samples which are not compatible with many experiment objectives such as the determination of viscosity and surface tension. The need for lower electromagnetic field strengths thus becomes evident which means that gravity has to be largely eliminated in order to perform experiments with undisturbed samples over a wide temperature range [4] … [8].

2.2. Ultraclean process environment:

In order to reduce the potential for heterogeneous nucleation as much as possible an ultra-high vacuum or ultra-pure noble gas process atmosphere is necessary. In particular, the amounts of reactive trace gases, like for example oxygen, need to be minimized.

2.3. Contactless diagnostics

The diagnostic instruments have to be compatible with the containerless processing principle as well, such as pyrometers for measurement of the sample temperature and various dedicated video cameras for specific purposes, e.g. cameras with high spatial or temporal resolution.

2.4. Sample handling

Even though the samples are processed without contact to a container wall they need to be carefully prepared and handled during the experiment. Typical experimental set-ups provide the possibility to store several samples, to select a dedicated sample for processing and to transfer the selected sample into the coil. Therefore, the samples are placed in so-called sample holders made of chemically inert thermal shock resistant material which are connected to a transfer mechanism and provide enough free space for the sample such that the containerless principle is respected during processing.

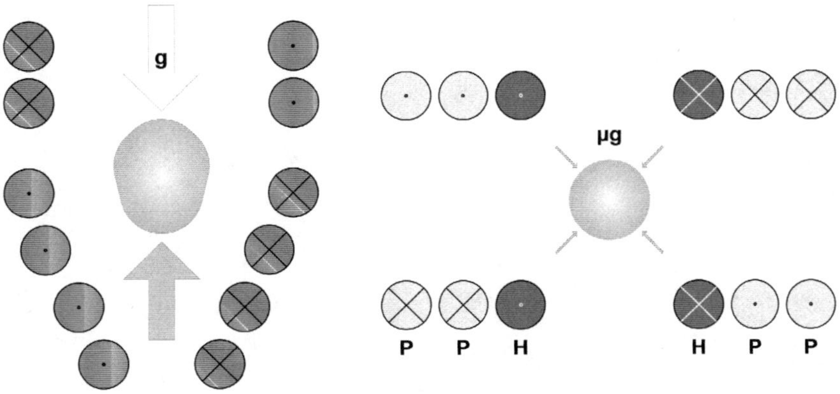

Figure 1. Principle design of levitation coils for levitation under 1g (left) and μg (right). In a 1g levitation coil the positioning force has to counteract gravity, therefore strong electromagnetic fields are needed just for the positioning of the sample, which leads to a strong heating and deformation of the sample (left). In contrast, the forces required to position a sample in a μg environment are orders of magnitude smaller, therefore the heating and deformation of the sample due the positioning field are negligible. As a consequence, heating and positioning of the sample can be controlled largely independently, e.g. by using dedicated coils for heating and positioning, respectively (right). Note that this illustrative sketch does not show the current EML flight coil geometry.

3. Undercooling at Work

Driven by the size of the RF coils which are optimized for positioning and heating efficiency under microgravity the samples are spherical with diameters typically ranging from 5 to 8 mm. The samples are manufactured by the scientists in their labs and are then stored and transferred under controlled environmental conditions into the experimental setup where they are processed. The process environment for the samples is either UHV or ultra pure noble gas (He or Ar) up to 400 mbar.

A typical experiment consists of several melt cycles. Starting from the solid state the sample is heated up above the melting temperature (overheating of up to 300 K), and then the heater power into the sample is reduced. During the subsequent cooling phase, the majority of the experimental data are gathered. Depending on its properties the sample undercools to temperatures below the melting temperature until solidification sets in rapidly and the heat of fusion is released in a very short time such that the sample temperature rises again to the melting temperature, a phenomenon known as recalescence. During the following cool down of the solid sample the cooling rate can be enhanced by additional convective cooling with the process gas being circulated through a pipe and filter system and released through a nozzle at the sample's vicinity. A typical melt cycle is shown in figure 2.

Due to the small size and related low heat content of the sample the typical duration of a melt cycle is in the order of seconds to minutes, depending on the sample material and the definition of the process, such that many such melt cycles can be performed during a the time frame available for an on-orbit experiment which is in the order of hours.

At high temperatures samples processed in ultra-high vacuum may evaporate some material which leads to a deposition of evaporated material on the inner surfaces of the experiment set-up in a physical vapor deposition process. Elements that are sensitive to deposition like the components of the optical path from the sample to the pyrometer and video cameras have to be protected from deposition by e.g. exchangeable double mirror systems. If samples are processed in a noble gas atmosphere the evaporated material does coagulate into microscopic dust particles which are toxic and need to be removed from the experiment setup by a dedicated gas cleaning system.

After processing samples can be retrieved from the facility and returned to ground for further evaluation

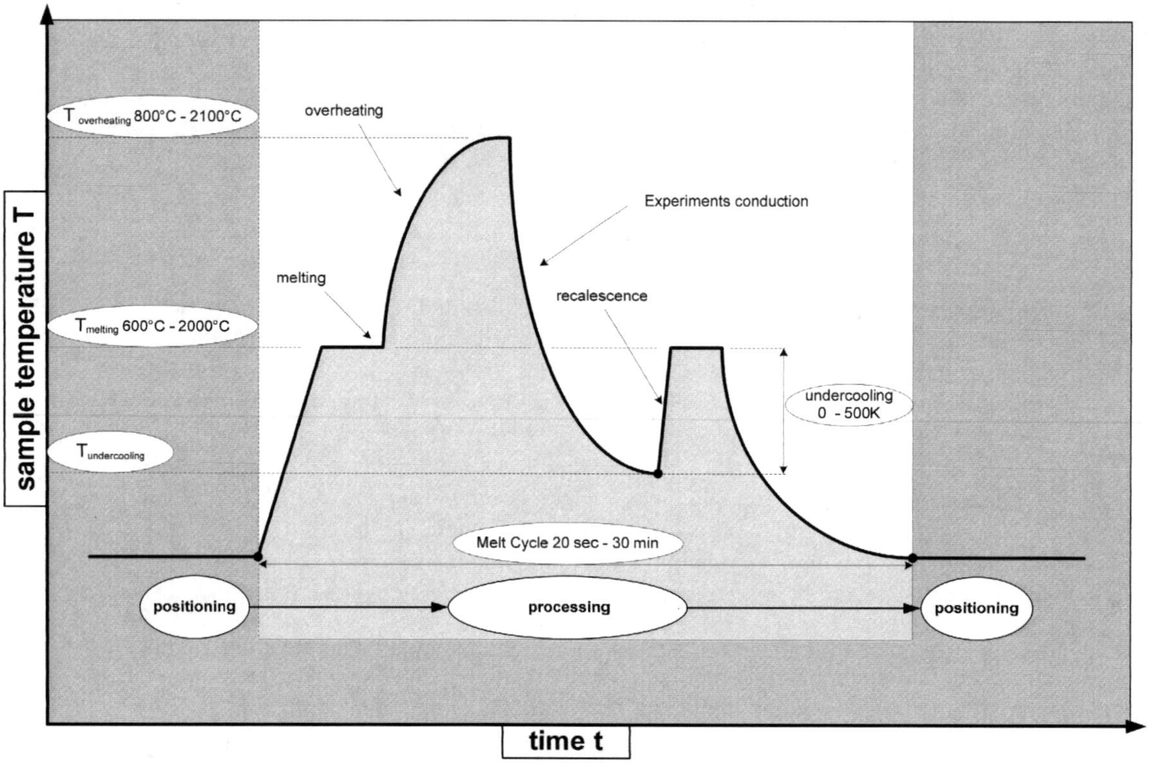

Figure 2. Principle sketch of the temperature-time profile of a typical melt cycle in the EML facility. Experiments are typically performed on the cooling flank when the sample is in the liquid state.

4. EML on ISS: Experiment Classes and Measurement Capabilities

Based on a long and successful heritage of electromagnetic levitation facilities on various carriers and missions, the electromagnetic levitation facility EML for the International Space Station ISS is currently being developed by Astrium under contracts to ESA and DLR. A comprehensive discussion of the performance and design features of EML is provided in [9].

The EML facility supports 5 classes of experiments which cover a wide range of scientific objectives. An overview of the experiment classes and the main physical parameters that are measured is provided in Table 1 below.

Journal of Physics: Conference Series **327** (2011) 012015 doi:10.1088/1742-6596/327/1/012015

Table 1. The five EML Experiment Classes.

	**Class A: Undercooling and nucleation** Solidification front speed, nucleation statistics and phase selection & morphology
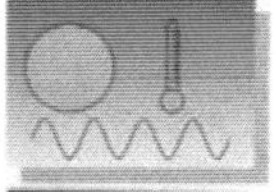	**Class B: Modulation Calorimetry** Heat capacity, heat of fusion, total hemispherical emissivity, thermal transport coefficients, fraction solid
	**Class C: Surface oscillations** Surface tension, viscosity
	**Class D: Size Measurement** Thermal expansion
	**Class E: Electrical Coupling** Electrical conductivity

4.1. Class A: Undercooling and nucleation

The scientific objectives for this experiment are to:

1. measure solidification front speed
2. perform nucleation statistics (maximum undercooling)
3. study morphology and phase selection depending on undercooling and induced fluid flow [10]

An experiment of class A type is being executed as follows: The molten sample will be cooled by turning off the heater, whereas the cooling rate may be enhanced by dedicated convective cooling per adjustable gas flow. Recalescence may occur either spontaneously or it can be triggered by touching the undercooled sample with a custom-made trigger needle at a predefined undercooling temperature. The recalescence of the undercooled sample is detected automatically by an algorithm that detects the sudden increase of the pyrometer signal upon recalescence. A high-speed video camera is used to observe the sample from a direction perpendicular to the trigger needle such that the nucleation front passing from the nucleation point (trigger needle tip) across the surface of the sample can be observed. Once recalescence has been detected the high-speed video acquisition is stopped. Pre- and post-event storage of video data is facilitated by the use of a ring memory for the video data. See also the first pictogram in table 1.

The positioning field can be varied to determine the effect on the induced fluid flow versus the Marangoni convection prior to solidification. Typical duration of one melt cycle is in the range of 30-60 seconds.

The temperature is recorded with a frequency of 100 Hz; video images are taken with up to 30.000 frames/s @ 256x256 pixel spatial resolution. An example of high-speed video recording is shown in figure 3.

The following equipment is needed for the execution of a class A experiment:

- high-speed and high resolution camera to resolve solidification process, this includes high digitalization of the camera chip to support good thermal resolution of the phases solidifying
- trigger needle to induce nucleation at desired temperatures of the undercooled liquid
- precision pyrometer

4.2. Class B: Modulation Calorimetry

The scientific objectives for this experiment class are to measure thermo-physical properties and if applicable their temperature dependence [3] as e.g.

1. Heat capacity
2. enthalpy of fusion
3. total hemispherical emissivity
4. thermal transport coefficients
5. fraction solid

The class B experiment is carried out as follows: At a constant base temperature of the molten sample the heater power is modulated sinusoidally and the sample's temperature response is evaluated. The modulation can be repeated at different sample temperatures with different modulation frequencies in order to obtain several data points within one melt cycle. The typical duration of a melt cycle ranges from 10 to 20 minutes. The temperature data will be acquired with 100 Hz; furthermore input data of RF coil system (voltage, current) are needed.

The following equipment is mandatory to perform a class B experiment:

- high precision pyrometer
- calibrated modulated power input source (amplitude or power modulation is possible)

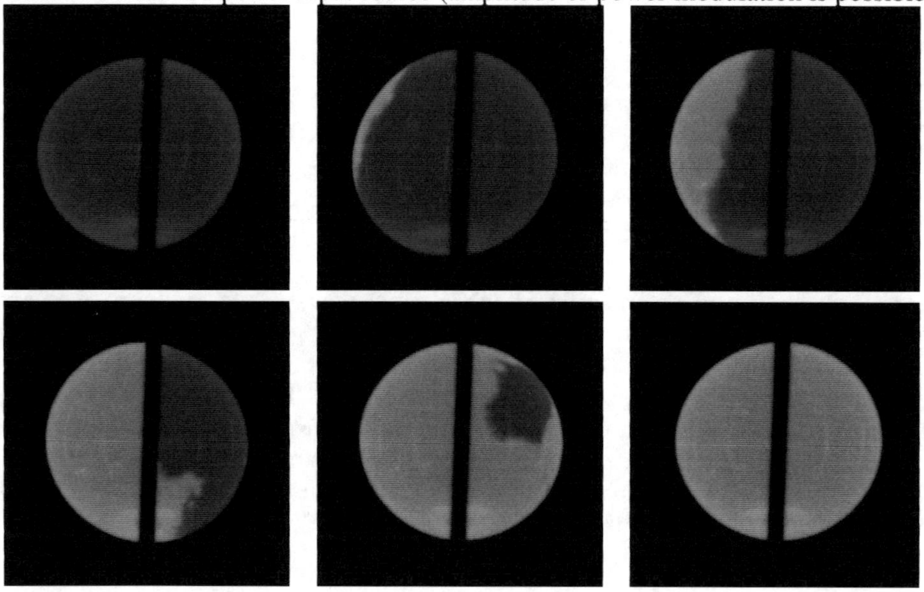

Figure 3. Class A Experiment: Sequence of a NiAl sample solidifying from the undercooled liquid captured by high-speed video camera @ 10.000 frames/s during parabolic flight. The total duration of the recalescence event was 29.1ms.

4.3. Class C: Surface Oscillations

The measurement of surface tension and viscosity and their temperature dependence are the main scientific objectives for these experiments.

The experiment starts with heating up the sample slightly above its melting point. During subsequent cooling of the molten sample a short heater pulse is applied which excites surface oscillations due to the dipole nature of the heating field, and the radial force of the field squeezing the sample. The surface oscillations are imaged by two video cameras in orthogonal views. From the

Journal of Physics: Conference Series **327** (2011) 012015 doi:10.1088/1742-6596/327/1/012015

oscillating frequency the surface tension can be derived and from the oscillations decay the viscosity can be derived since this is the only damping mechanism involved for these oscillations. The temperature as well as the stimulus (pulse amplitude & shape) can be altered to achieve various measurement points during one melt cycle. For increasing viscosity the amplitude can be increased to observe a high enough oscillation signal. Instead of an excitation pulse also a frequency sweep can be applied to find the resonance frequency of the sample's surface oscillations. The typical duration of a melt cycle is in the range of 5-15 min.

The video images are usually taken with the radial camera @ 200 frames/s and 600x600 pixel spatial resolution and with axial camera @ 150 frames/s and 350x350 pixel spatial resolution. An example is shown in figure 4. The temperature is recorded with 100 Hz.

To perform a class C experiment the following equipment must be provided:

- fast and high resolution video cameras to resolve shape oscillations such that offline Fourier analysis on the sample diameter can be performed to derive the oscillation frequency and decay time
- high precision pyrometer
- free programmable power input source to apply either short pulses or a user defined signal shape as e.g. frequency sweep

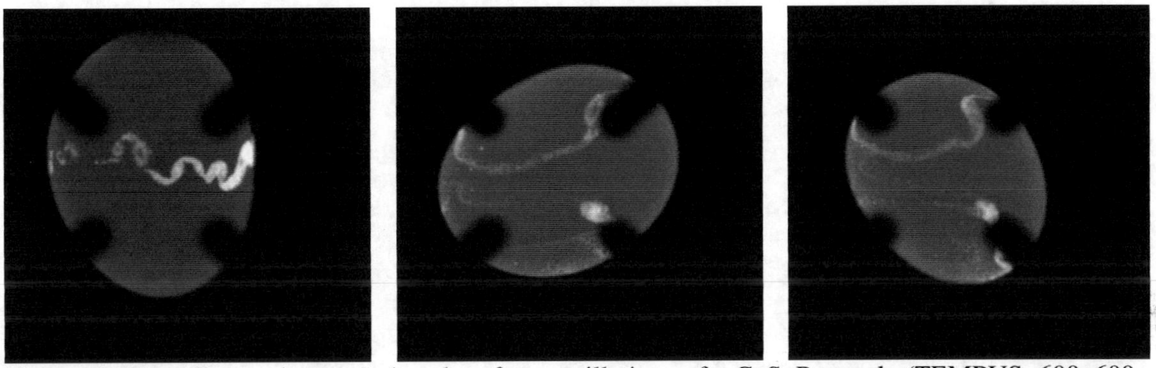

Figure 4. Class C experiment: Induced surface oscillations of a CuSnP sample (TEMPUS, 600x600x pixels @ 200 fps)

4.4. Class D: Size Measurement

The scientific target for this class of experiments is to measure the thermal expansion coefficient and its temperature dependence. It is carried out as follows: The diameter of the solid and molten sample is precisely determined by the two video cameras in orthogonal view along the changing temperature of a melt cycle. For this purpose a high spatial resolution of the camera is necessary which will be even enhanced by a sub-pixel resolution algorithm which is applied to the video images off-line during evaluation. The algorithm requires the image of the sample's edge to be smeared out over a small number of camera pixels such that the algorithm can fit a 50% intensity value, as assumed sample edge, with a precision higher than the pixel resolution. The slightly blurred imaging of the sample's equator is achieved by setting the camera's focal plane slightly in front of it still achieving a sharp picture of the sample's surface. The typical duration of a melt cycle is in the range of 1-5 min. Video images captured by the axial camera are taken with typically 25 frames/s at up to 1000x1000 pixel resolution, whereas the radial camera offers 600x600 pixel spatial resolution. The selected frame rate depends on the temperature gradient and the desired temperature resolution of individual measurement points. The temperature data are recorded with 100 Hz.

The following equipment is needed for the execution of a class D experiment:

- high resolution video cameras to resolve sample diameter

- high precision pyrometer

4.5. Class E: Electrical coupling

The measurement of the electrical conductivity and its temperature dependence are the scientific targets for these investigations.

A class E experiment is carried out as follows: The electrical data of the RF oscillating circuit are used to derive the electrical conductivity of the sample since the sample itself is changing slightly the characteristics of the RF oscillating circuit by its inductivity, which is determined by its geometry and electrical conductivity. Some calibration measurements on the empty coil have to be performed to distinguish between the coil's inductivity and the one introduced by the sample. The relevant housekeeping data are measured by the levitation power supply which uses them mainly for internal control of the high RF power output. The basic resolution and accuracy can be further enhanced by using dedicated measurement equipment for this task- the so-called SCE which is described later.

The measurement can be performed together with other experiment classes since it requires only the high precision pyrometer as diagnostics, as all the other experiments too, and the RF oscillating circuit data which are part of the regular EML housekeeping data anyway. The acquisition frequency for the temperature data is 100 Hz, the RF oscillating circuit data (voltage, current, frequency) are recorded with 10 Hz.

5. Experiment preparation

Experiment preparation involves all aspects of transforming a scientific objective or idea into a fully developed flight experiment. Since EML provides a wide range of functions related to the definition of the sample processing including diagnostics, a deep understanding of the associated systems, their capabilities, functions, interrelations and constraints is needed in order to develop a successful flight experiment. Support to the scientists for the preparation of their experiments is given in the frame of the Ground Support Program GSP by a joint team consisting of Astrium and the DLR Microgravity User Support Center MUSC. Thereby the experience made in former TEMPUS missions is fully provided to the scientific community. A brief outline of the support tasks is given hereafter.

From the EML facility operation point of view the experiments are organized into melt cycles which in turn are decomposed into process steps. Each process step is characterized by a number of EML facility settings such as heater power, camera speed, etc. These settings are called Experiment Parameters or EP. In order to perform an experiment in the EML facility all EPs belonging to this experiment are needed since the EML facility is being operated using these parameters.

One of the major tasks of the GSP is therefore to translate an experiment idea into an EP set, a task that involves multiple steps and activities (see also figure 5):

- A rough outline of the experiment is developed in standardized form which is already organized with the detailed implementation in mind, i.e. with defined melt cycles, desired temperature-time profiles, diagnostics requirements in each phase etc. The result is captured in so-called "theater scripts". This work is done in close collaboration with the scientists and is supported by a simulator which provides a first indication on how the desired experiment can be implemented.
- In order to design an EML experiment with a sample and to define the heater and positioner settings the coupling of the sample to the rf field, the emissivity of the sample and the cooling of the sample by the gas atmosphere must be known precisely. These parameters are determined by the GSP in dedicated ground facilities using ground reference samples that are provided by the scientists.
- Based on the information contained in the theater scripts and the results of the sample characterization measurements, the detailed experiment parameters for each melt cycle and process phase can now be developed. This step is again performed in close collaboration with the scientists and is supported by the simulator which is now used to determine the heater and positioner settings needed to achieve the desired temperature-time profiles using the results of

the sample characterization measurements as inputs to the underlying thermo physical model. The result of this step is an EP set which is essentially a table of EML facility settings.

- In a next step the EP sets are verified in a representative EML facility. The objective of this step is to verify that the EP set is compatible with the scientific objectives and the EML facility such that sufficient confidence is achieved that the actual flight experiment in the EML facility on ISS can actually be run based on this EP set. After successful verification the EP sets are ready to be loaded into the EML facility.
- The preparation of the flight samples by the scientists is supported by the GSP by providing requirements for the flight samples and support to the verification of these requirements. This aspect includes for example dimensional requirements, verification of the sample composition, and transportation of the flight samples to the integration site in a controlled environment.
- After delivery to the project the flight samples will be integrated into their sample holders and then into the Sample Chamber in a clean environment. The fully integrated Sample Chamber will then be shipped to the launch site, uploaded to the ISS, attached to EML, and is then ready for the execution of the experiments.

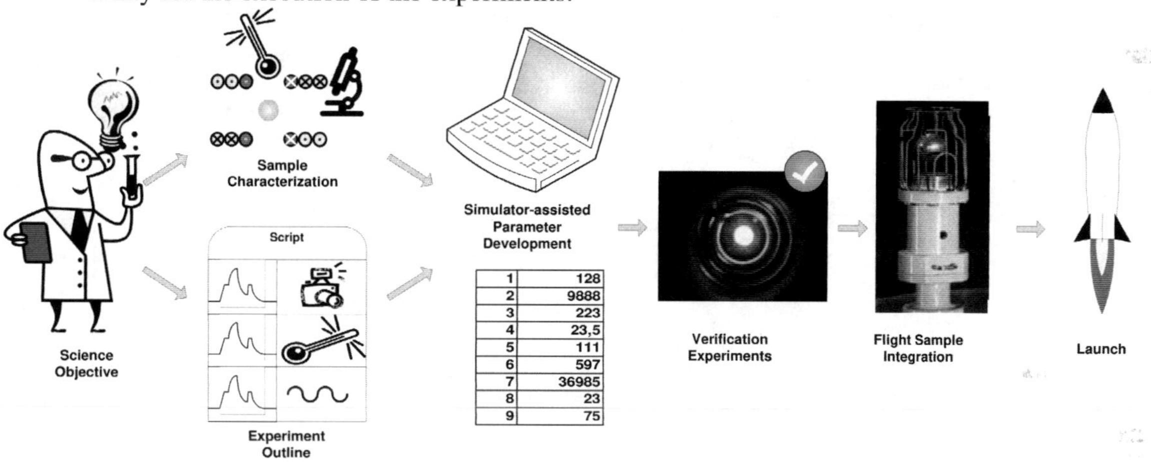

Figure 5. Preparation tasks for conducting experiments in EML onboard the ISS.

6. Advanced experimental techniques

The scientific outcome of the experiments as described in Chapter 4.1 may be significantly improved by implementing the following two upgrades of the EML facility [1], [9]:

- Oxygen control system
- Sample coupling electronics (SCE).

Oxygen represents a serious and potentially harmful contaminant to many materials at elevated temperatures due to its high chemical reactivity. Especially for experiments on the determination of the viscosity and surface tension of levitated metal melts, which are part of the EML scientific programme, the presence of oxygen in the atmosphere might lead to a contamination of the sample with oxygen leading to the formation of a surface oxide layer or to dissolution of oxygen into the liquid sample. This could significantly alter the physical parameters of the sample to be determined and finally the experimental results.

Furthermore, the sample itself may act as an oxygen source or sink, depending on the preparation procedure. Hence, processing the sample in the liquid state could lead to a release or absorption of oxygen to or from the process atmosphere over several orders of magnitude. In both cases, a non-steady condition concerning the process atmosphere is apparent which makes the interpretation of the measured data quite difficult and ambiguous.

For these reasons the precise knowledge of the residual oxygen in the process atmosphere and its control is essential. The oxygen control system as introduced in [11] is suited to measure the partial pressure of oxygen and to control this value within short time constants. Hence, the oxygen concentration in the process atmosphere could be kept constant during the entire experiment within a range of a few percent and a broad range of partial pressures could be adjusted ($pO_2 = 1$ -10^{-19} bar), see also figure 6 for reference. This would enable systematic investigations on the surface tension and/or viscosity of samples versus partial pressure of oxygen at a constant temperature [12].

The second major target for expanding the scientific potential of EML experiments in an upgraded facility would be to measure the electrical conductivity of the sample at a much higher precision than possible with the EML baseline feature using the high power RF supply as diagnostic tool. With the help of the dedicated diagnostics SCE the parameter electrical conductivity would complete the list of thermo-physical data for materials in the under-cooled state which are relevant for processing these industrial metallic alloys in terrestrial applications. The measurement principle of the SCE is contactless using induction where the sample changes by its properties the characteristics of the oscillating circuit. Hence the SCE would directly connected to a dedicated interface of the RF oscillating circuit in order to pick up the subtle changes to the field by its sensitive measurement circuitry. Moreover the SCE device as described in [13] allows also measuring the surface tension and viscosity due to its high sensitivity and fast response time such that even sample shape oscillations can be detected. Thus the SCE is a perfect complement to the optical diagnostic tools.

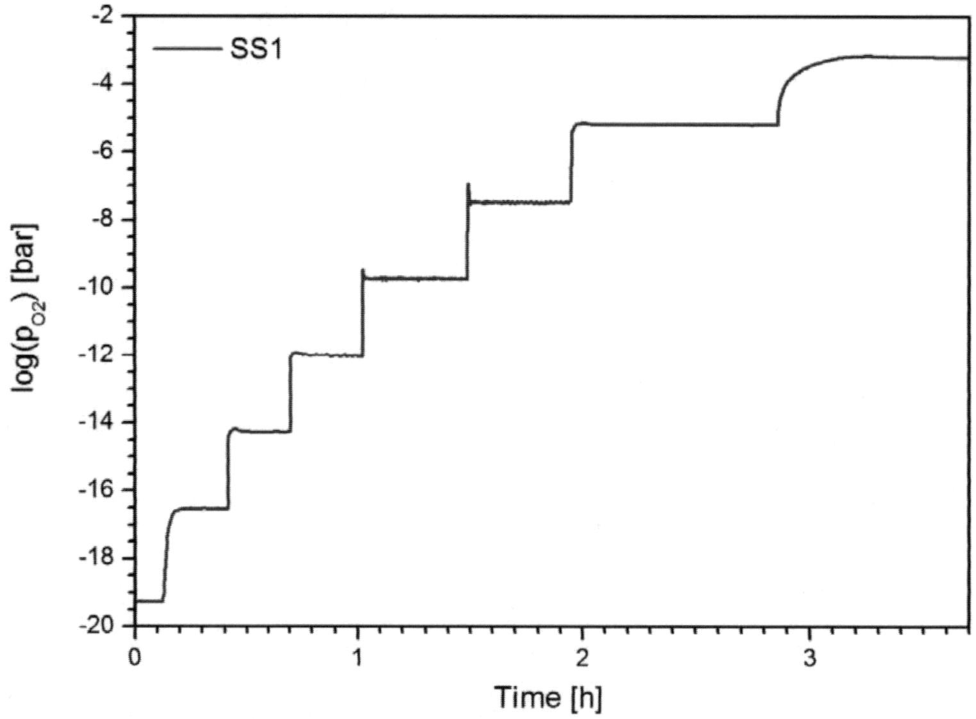

Figure 6. Range of oxygen partial pressure of oxygen as controlled by oxygen control system.

Acknowledgement
TEMPUS and EML are developed by Astrium GmbH under programs of and funded by the European Space Agency (ESA) and by the Space Agency of the German Aerospace Center (DLR e.V.) with funds of the German Federal Ministry of Economics and Technology.

References

[1] Soellner W, Seidel A, Stenzel C, Dreier W and Glaubitz B 2010 EML - Containerless Processing Facility for Material Science Research onboard The ISS *J. Jpn. Soc. Microgravity Appl.* 27 No. 4 183
[2] Egry I and Voss D 2010 Present Activities fo the Investigators Working Group (IWG) for the Electromagnetic Levitator (EML) on ISS - A Status Report *J. Jpn. Soc. Microgravity Appl.* 27 No. 4 176
[3] Fecht H J, Wunderlich R, Ricci E, Etay J, Egry I, Seetharaman S and Battezzati L 2010 The Thermo-Lab Project: Thermophysical Property Measurement in Space for High-Temperature Alloys J. *Jpn. Soc. Microgravity Appl.* 27 No. 4 190
[4] Team TEMPUS 1995 In Materials and Fluids Under Low Gravity: *Proc. IXth Symp. on Gravity-Dependent Phen. in Physical Sciences* ed L Ratke et al. (Berlin: Springer)
[5] Piller J et. al. 1986 Electromagnetic Positioning and Inductive Heating under micro-g: *6th European Symposium on Materials Sciences under Microgravity (Bordeaux)*
[6] Knauf R, Piller J, Seidel A and Dreier W 1992 TEMPUS - An Electromagnetic Levitation Facility for Containerless Processing under Micro-g in Spacelab: *Proc VIIIth European Symposium on Materials and Fluid Sciences in Microgravity (Brussels, Belgium)* ESA SP-333
[7] Geiger R and Thomas J 1993 Resonant High Power Supply for Electromagnetic Levitation and Melting of Metallic Samples *European Space Power Conference (Graz)*
[8] Knauf R et. al. 1994 Containerless Processing Facility TEMPUS for Materials Science Research onboard Spacelab: *6th International Symposium on Experimental Methods for Microgravity Material Science (San Francisco)*
[9] Seidel A, Soellner W and Stenzel C 2011 EML - An Electromagnetic Levitator for the International Space Station, *this proceedings*
[10] Matson D, Hyers R and Volkmann T 2010 Peritectic Alloy Rapid Solidification with Electromagnetic Convection *J. Jpn. Soc. Microgravity Appl.* 27 No. 4 238
[11] Study data package of ESTEC TRP study "Oxygen Sensing and Control" - ESTEC contract no. 19963/07/NL/PM
[12] Ozawa S, Suzuki S, Hibiya T and Fukuyama H 2011 Influence of oxygen partial pressure on surface tension and its temperature coefficient of molten iron *J. of Appl. Phys.* 109 014902
[13] Lohöfer G 2009 EML Sample Coupling Electronics Test Model Final Report *EML/SCE-TM-Rep 1* (DLR, Cologne)

Diffusion Limited Silicon Dissolution into Germanium Melt

Neil Armour and Sadik Dost

Crystal Growth Laboratory, University of Victoria, Victoria, BC Canada V8W 3P6

sdost@uvic.ca

Abstract. The dissolution of silicon into germanium melts has been investigated for application to solution crystal growth. Small diameter samples were processed to weaken flow structure in the melt making the transport mechanism diffusion dominated. 8mm diameter samples were processed at three different temperatures. The temperature dependence of the dissolution rate of silicon was found to be not measurably significant in the experiments presented here. The orientation of the sample, with respect to gravity, had a profound effect on the dissolution rate. A small number of 4mm diameter samples were also processed and the same effects were apparent.

1. Introduction

There has been recent interest in the silicon germanium alloy (Si_xGe_{1-x}) as an emerging semiconductor material since it possesses full miscibility across its composition range allowing for adjustment of the band gap and lattice parameter [1-2]. Bulk single crystals of Si_xGe_{1-x} have application in photodetection, photovoltaics, thermal imaging, and as a substrate for Si_xGe_{1-x} epitaxial layers [3]. The bulk Si_xGe_{1-x} substrates can be lattice-matched to the device layer, reducing strain. Among device structures of interest, the heterobipolar transistor has been most studied due to its applications in telecommunications. As for other important applications for Si_xGe_{1-x}, modulation doped field effect transistors and 1.3μm optoelectronics can be identified [4].

There are a number of techniques used to grow single crystal Si_xGe_{1-x} alloys. The Czochralski (Cz) crystal growth technique is currently the most efficient method for producing bulk Si_xGe_{1-x} crystals. However, since the liquidus and solidus curves in the Si_xGe_{1-x} phase diagram are widely separated, producing compositionally uniform material requires that silicon be replenished during growth from the melt [5]. Similarly in Bridgman growth of Si_xGe_{1-x} alloys, the transport of the species and the changing composition of the melt are again important for producing compositionally uniform crystals [6-7].

Silicon germanium crystals have been produced by the liquid phase diffusion (LPD) crystal growth technique [8]. This technique avoids the miscibility gap issues with melt growth. The rate limiting step in this solution growth technique is the transport of silicon to the growth interface. It has been shown that silicon transport proceeds in a diffusion dominated manner in this growth technique [9]. In order to further advance this technique, proper understanding of the transport mechanisms at work is necessary.

In previous work, dissolution has been shown to be influenced by flow structure in the germanium melt [10]. The dissolving silicon stabilizes the melt against convection in the configuration where silicon is dissolving into the melt from above. This is due to the lighter silicon reducing the density of the melt as the silicon concentration increases. This helps to weaken any convective flow structure.

However, we have seen that the weak flow structure present still has an influence on dissolution. In order to explore the diffusion behavior of silicon in the melt, we have reduced the diameter of the crucibles to 8mm in order to further weaken the flow structure in the melt.

2. Experimental Procedure

The experiments were conducted in a three zone vertical DC resistance tube furnace. An isotherm was maintained over the crucible length at temperatures of 1000°C, 1050°C or 1100°C. The isotherm is maintained within 3 degrees axially and 1 degree radially. The materials were contained in a quartz crucible. The silicon used was single crystal optical grade, 5N, material. The germanium used was 6N material. The silicon and germanium were both cleaned and etched prior to loading into the crucible. The cleaning, etching and loading took place in a clean room. Once the materials were loaded, the crucible was evacuated to approximately $1x10^{-3}$ Pa.

The crucible was hung in the furnace inside a quartz ampoule. It was first preheated for approximately 1 hour above the hot zone at approximately 800°C, i.e. below the melting point of Ge. This quickens the time to melt the sample and begin dissolution into the melt. To start the experiment, the crucible was dropped into the isothermal area of the furnace from the preheating position. It was allowed to remain there for the experiment time, 60 minutes. At the conclusion of this time, the crucible was pulled from the furnace and allowed to cool in air. The samples were then sectioned in half. One section was differentially etched to reveal structure.

Two different material arrangements were used. The first was a silicon source floating on top of the germanium melt. In this case the silicon source covers the melt free surface. This arrangement is similar to that used in the Liquid Phase Diffusion growth system for SiGe [8-9]. This was used for the majority of experiments. For comparison, two samples were processed at 1050°C with the silicon source at the bottom of the crucible.

3. Results

The samples processed with the silicon dissolving from the top all exhibited almost the same level of dissolution over the temperatures used. A dissolution height of around 0.5mm was measured across all samples processed as seen in figure 1. This implies that the temperature dependence of the dissolution rate and consequently the diffusion coefficient is not overly significant in this range of temperatures. It would appear from results that there is a temperature dependence but it is not resolvable with this experimental setup. The length of experiments was one hour. It is possible that a longer experiment time could allow the dependence to be better resolved.

In previous work, these types of experiments were conducted in a 25mm diameter crucible [10]. The experiments were run for shorter periods of time and only at 1100 °C. In those experiments the dissolution height was up to 1mm. This highlights the increased mixing present in the melt with the larger diameter due to the contribution of stronger convection in the melt [11]. The decreased diameter crucibles used in the current experiment work do help to weaken convection in the melt. Experiments on 4mm and 2mm diameter crucibles were performed with little success as surface tension often prevented the silicon from fully wetting with the melt and voids formed separating the melt into segments. The crucible would require modification so pressure could be exerted to ensure proper wetting between all elements.

Numerical modelling did not show significant change in flow structure between the 8mm and smaller diameter crucibles. As seen in figure 2, the composition profiles after the one hour experiment time are virtually identical. The flow structure was computed with the effects of solutal buoyancy and similar weak flow is present in both crucibles. All numerical work was completed using the commercial ANSYS CFX package.

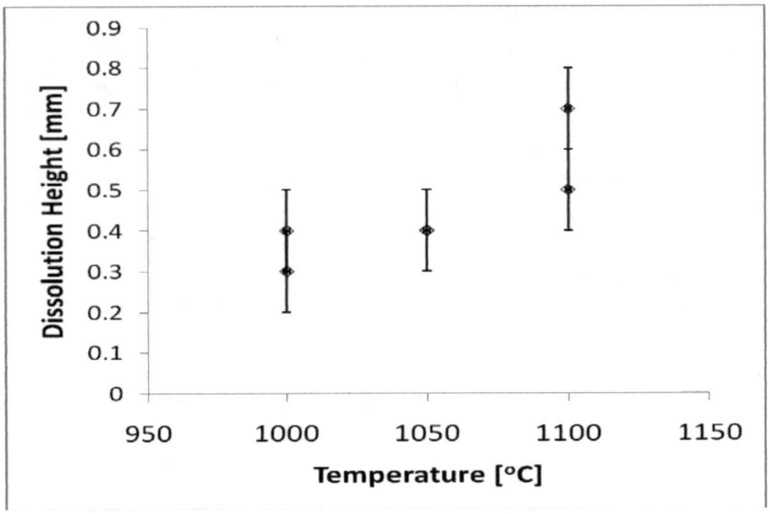

Figure 1. Plot of dissolution height of processed samples versus temperature. The lack of differential between temperatures with respect to error is apparent.

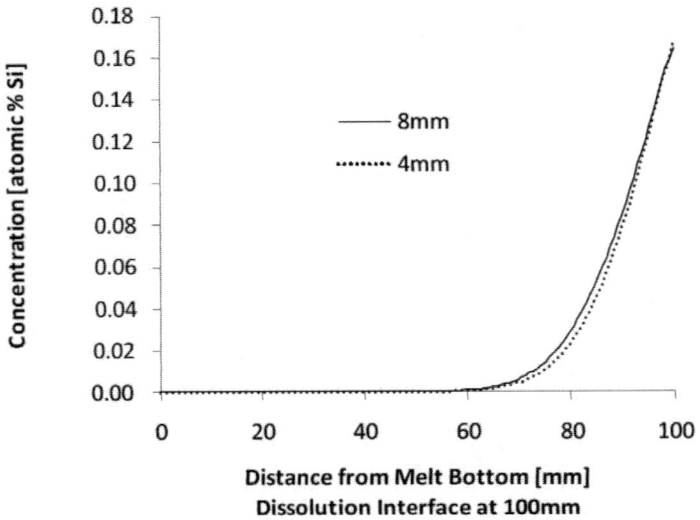

Figure 2. Computed concentration profiles for dissolution experiments. The solid line is for an 8mm diameter crucible and the dotted line is for a 4mm diameter crucible. Little difference in the computed concentration profiles is visible. The experiment conditions are 1000°C for one hour. The diffusion coefficient was 2.8×10^{-8} m^2/s

The samples processed with the silicon at the bottom of the melt exhibited a drastically different result. In the cases of a ten minute and thirty minute experiment times, the silicon source material is completely dissolved. This is dissolution of 5mm of silicon. As can be seen in figure 3, no silicon is visible in the sample with the silicon dissolving from the bottom while most of the seed remains when the silicon is dissolving from the top. This is dramatically faster transport than what was seen in the opposite orientation of the dissolution interface. This highlights the significant effect of gravity (convection) on the dissolution process.

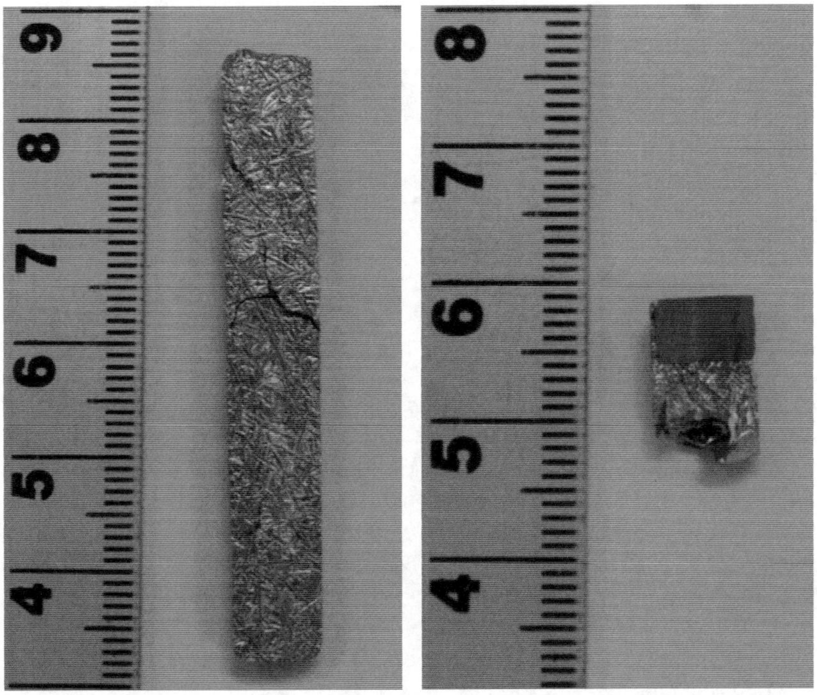

Figure 3. On the left is a sample processed with the silicon dissolving from the bottom. There is no silicon source material left. The needle structure apparent over the cross section is indicative of the presence of silicon in the material. The sample on the right is the top of a sample processed at 1100°C with the silicon dissolving from the top. The darker material is the left over silicon source material. A small amount of the needle structure is apparent around the dissolution interface.

In previous results, a difference in transport rates between orientations was observed similar to that seen here but not as dramatic [10-11]. A key difference in this set of experiments is that the ratio of germanium to silicon was increased. Therefore from the previous work, the dissolution of silicon from the bottom of the melt was limited by the saturation of the germanium melt with silicon. The rate at which silicon dissolves in this configuration seems limited only by the saturation level of the melt. Enough silicon is used in the experiment to fully saturate the melt. The approximate composition of the melt at saturation at 1000°C is 18% silicon by mole. As the melt approaches this threshold, the transport away from the dissolution interface will slow and therefore the dissolution rate will slow.

The longer column of germanium was used so that useful composition information could be later extracted to directly estimate the diffusion coefficient. However, due to the low concentration of silicon present in the melt there is significant expansion of the melt during solidification. The presence of silicon in the melt reduces the amount the melt expands on solidification. Due to this expansion during quenching, the melt column is significantly disturbed by cracking and leaking. This unfortunately makes it impossible to reliably correlate the position in the column to a composition. The use of shear cell would seem necessary to extract this type of information accurately.

4. Conclusions

Dissolving silicon into a germanium melt in an 8mm crucible resulted in a more diffusion dominated process than previously observed. However, we were unable to observe any temperature dependence of the dissolution height. This indicates that the dissolution rate and therefore the diffusion coefficient are not significantly affected by temperature over the range of temperatures examined here. In addition, it was noted that the flow structure with the silicon dissolving from the bottom is extremely strong leading to extremely fast dissolution when compared to the opposite orientation.

5. References

[1] Kasper E 1995 *J. Crystal Growth* **150** 921
[2] Seidenberg P 1997 *Facets: New perspectives on the history of semiconductors* ed A Goldstein and W Aspray (New Brunswick:IEEE Press) pp 36
[3] Usami N, Kitamura M, Obara K, Nose Y, Shishido T and Nakajima K 2005 *J. Crystal Growth* **284** 57
[4] Schuppen A and Dietrich H 1995 *J. Crystal Growth* **157** 207
[5] Yonenaga I 2005 *J. Crystal Growth* **275** 91
[6] Volz M P, Schweizer M, Kaiser N, Cobb S D, Vujisic L, Motakef S and Szofran F G 2002 *J. Crystal Growth* **237** 1844
[7] Volz M P, Walker J S, Schweizer M, Cobb S D, Szofran F R, 2005 *J. Crystal Growth* **282** 305
[8] Yildiz M, Dost S and Lent B 2005 *J. Crystal Growth* **280** 151
[9] Yildiz M and Dost S 2005 *Int. J. Engng. Sci.* **43** 1059
[10] Armour N, Dost S and Lent B 2007 *J. Crystal Growth* **299** 227
[11] Armour N and Dost S 2007 *J. Crystal Growth* **306** 200

Acknowledgements

The financial support provided by the Space Science Enhancement Program (SSEP) of the Canadian Space Agency (CSA), the Canada Research Chairs (CRC) Program, and the Natural Sciences and Engineering Research Council (NSERC) of Canada is gratefully acknowledged.

Journal of Physics: Conference Series **327** (2011) 012017 doi:10.1088/1742-6596/327/1/012017

Homogeneous SiGe crystal growth in microgravity by the travelling liquidus-zone method

K Kinoshita[1], Y Arai[1], Y Inatomi[1], H Miyata[2], R Tanaka[2], T Sone[3], J Yoshikawa[3], T Kihara[4], H Shibayama[4], Y Kubota[4], T Shimaoka[5], Y Warashina[6], K Sakata[1], M Takayanagi[1] and S Yoda[1]

[1]Japan Aerospace Exploration Agency, 2-1-1, Sengen, Tsukuba, 305-8505, Japan
[2]Advanced Engineering Services Co. Ltd, 1-6-1, Takezono, Tsukuba, Japan
[3]Japan Manned Space Systems Corporation, 1-1-26, Kawaguchi, Tsuchiura, Japan
[4]Space Engineering Development Co. Ltd., 1-12-2, Takezono, Tsukuba, Japan
[5]Japan Space Forum, 2-2-1, Otemachi, Chiyoda-ku, Tokyo, Japan
[6]Inet Corporation, 5-19-6, Shikatebukuro, Minami-ku, Saitama, Japan

E-mail: kinoshita.kyoichi@jaxa.jp

Abstract. Homogeneous SiGe crystal growth experiments will be performed on board the ISS "Kibo" using a gradient heating furnace (GHF). A new crystal growth method invented for growing homogeneous mixed crystals named "travelling liquidus-zone (TLZ) method" is evaluated by the growth of $Si_{0.5}Ge_{0.5}$ crystals in space. We have already succeeded in growing homogeneous 2mm diameter $Si_{0.5}Ge_{0.5}$ crystals on the ground but large diameter homogeneous crystals are difficult to be grown due to convection in a melt. In microgravity, larger diameter crystals can be grown with suppressing convection. Radial concentration profiles as well as axial profiles in microgravity grown crystals will be measured and will be compared with our two-dimensional TLZ growth model equation and compositional variation is analyzed. Results are beneficial for growing large diameter mixed crystals by the TLZ method on the ground. Here, we report on the principle of the TLZ method for homogeneous crystal growth, results of preparatory experiments on the ground and plan for microgravity experiments.

1. Introduction

Growth of homogeneous mixed bulk crystals has been investigated by many researchers and variety of zone melting methods have been invented [1-4]. Among them, the travelling liquidus-zone (TLZ) method is the most promising method for growing homogeneous mixed bulk crystals [4-6]. Validity of the method has been proved by the growth of InGaAs platy crystals and 2mm diameter SiGe crystals on the ground by confining melts in thin layers or capillary tubes for suppressing convection [4-11]. However, large diameter cylindrical crystals are difficult to be grown on the ground due to convection in melts. We plan crystal growth experiments by the TLZ method in microgravity for proving usefulness of the TLZ method by the growth of larger diameter crystals; 10mm diameter $Si_{0.5}Ge_{0.5}$ cylindrical crystals will be grown and radial compositional homogeneity as well as axial one will be evaluated. In evaluation, measured compositional profiles will be compared with our two-dimensional TLZ growth model equation and conditions for growing large diameter homogeneous mixed crystals will be established.

Here, we report on the principle of the TLZ method for homogeneous crystal growth, results of preparatory experiments on the ground and plan for microgravity experiments.

2. TLZ method

The TLZ method is a kind of zone melting method but is different from conventional zone melting method in forming a liquidus-zone in which solute is almost saturated. Such liquidus-zone is possible when temperature gradient is low, zone width is narrow and low melting temperature material is rich in the zone. Sample configuration, concentration profile in the sample and its relation to the phase diagram is shown in **Figure 1** in the case of $Si_{0.5}Ge_{0.5}$ crystal growth.

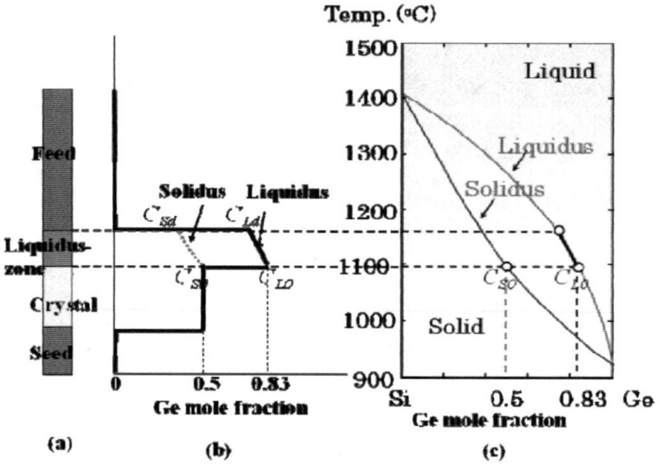

Figure 1. Principle of the TLZ method, (a) sample configuration, (b) Ge concentration profile in the sample and (c) Si-Ge phase diagram.

Ge concentration profile as shown in Figure 1 is obtained by heating a sample in which Ge is sandwiched by a Si seed and a Si feed at around 1100°C and at a temperature gradient of about 10°C/cm. At the freezing interface, a melt coexists with a grown crystal and therefore Si concentration is saturated. At the dissolving interface, the melt coexists with a solid feed and Si concentration is saturated, too. If temperature gradient is low and melt zone width is narrow, Si concentration in the melt is almost saturated in the whole as shown in Figure 1. Then, we named such a solute saturated melt zone as a liquidus-zone.

Crystal growth occurs spontaneously due to inter diffusion of Si and Ge because Ge and Si concentration difference is formed in the liquidus-zone. When Ge is transported away from the freezing interface, Si becomes richer than its equilibrium concentration at the corresponding temperature and crystal growth proceeds. At the crystal growth, Ge is segregated and the segregated Ge is transported to the dissolving interface (opposite side of the freezing interface). At the dissolving interface, Ge concentration increases due to transported Ge and its concentration exceeds equilibrium concentration and excess Ge dissolves a solid Si feed and the dissolving interface becomes equilibrium again. By the continuation of this process, crystal growth proceeds spontaneously.

We estimated crystal growth rate as described in the references [6, 9, 11]. The obtained equation is as follows.

$$-R = \frac{D}{(C_{LO} - C_{SO})}\left(\frac{\partial C_L}{\partial Z}\right)_{Z=0} = \frac{D}{(C_{LO} - C_{SO})}\left(\frac{\partial C_L}{\partial T}\right)\left(\frac{\partial T}{\partial Z}\right)_{Z=0} \quad (1)$$

Here, R is freezing rate, C_{LO} and C_{SO} are liquidus and solidus concentrations at the freezing interface, D is interdiffusion coefficient between solute and solvent. Since solute is saturated, $(\partial C_L/\partial Z)_{Z=0}$ is expressed by the reciprocal of the slope of the liquidus line $(\partial C_L/\partial T)$ multiplied by temperature gradient $(\partial T/\partial Z)$. Very important point in equation (1) is that freezing rate R is controlled by temperature gradient. Then it is very easy to translate a sample device synchronizingly in the opposite direction to the freezing rate and to keep freezing interface temperature constant. Relation between R and $(\partial T/\partial Z)$ in equation (1) was confirmed by measuring temperature gradient in a melt and growth rate R, and by comparing the measured growth rate with the calculated one [6].

3. Experimental procedures

A Ge zone forming material was sandwiched by a Si seed and a Si feed and was inserted into a boron nitride crucible. The crucible was vacuum sealed by a quartz ampoule. This sample device was heated in a temperature gradient furnace. Sample configuration and its relation to temperature profile in the furnace is shown in **Figure 2**. Growth experiments were performed on the ground using a laboratory furnace or a ground model furnace. According to equation (1), temperature gradient of 7°C/cm in a melt gives growth rate of 0.1 mm/h for the composition $Si_{0.5}Ge_{0.5}$. Therefore, temperature gradients between 7 and 14°C/cm were selected for $Si_{0.5}Ge_{0.5}$ growth experiments. Crystals with 2mm diameter and 10mm diameter were grown for elucidating effects of convection in melts. In 2mm diameter crystals, convection in melts is suppressed due to small bores of crucibles while convection in melts will occur in 10mm diameter crystal growth on the ground.

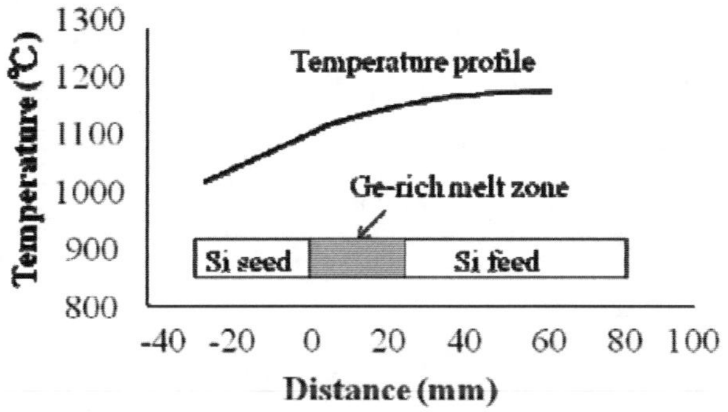

Figure 2. Sample configuration and its relation to a temperature profile.

Grown crystals were cut parallel to the growth axis and were mirror polished. Concentration profiles were measured by an EPMA (electron probe micro-analyzer) and crystallinity was evaluated by EBSP (electron back scattering pattern) and by XRD (X-ray diffraction).

4. Results and discussion

4.1. Confirmation of the TLZ growth model

Since the TLZ growth requires diffusion-controlled mass transport, principle of the TLZ method is studied by the growth of 2mm diameter crystals with suppressing convection in melts [6, 9]. **Figure 3** shows an example of axial concentration profiles for a TLZ-grown 2mm diameter crystal. A 13mm long compositionally uniform $Si_{0.49}Ge_{0.51}$ crystal was obtained. In the figure, Ge concentration profile of a Bridgman-grown crystal is depicted for comparison. No homogeneous part is observed in a Bridgman-grown crystal. Superiority of the TLZ method in compositionally uniform mixed crystal growth is clear. In the growth, measured temperature gradient was 8°C and sample translation rate was 0.11mm/h. This result agrees well with equation (1) if we use Si-Ge inter diffusion rate 9.5 x 10^{-8} m^2/s which is estimated by Adachi *et al.* [9].

4.2. Large diameter crystal growth

Since we succeeded in growing compositionally uniform small diameter $Si_{0.5}Ge_{0.5}$ crystals by the TLZ method, we tried 10mm diameter crystal growth. It was shown that radial compositional inhomogeneity increased as crystal growth proceeded. **Figure 4** shows a result of two dimensional Ge concentration analysis. Freezing interface can clearly be seen because Ge concentration in the crystal is different from that in the liquidus zone. It can be said that the freezing interface is concave towards

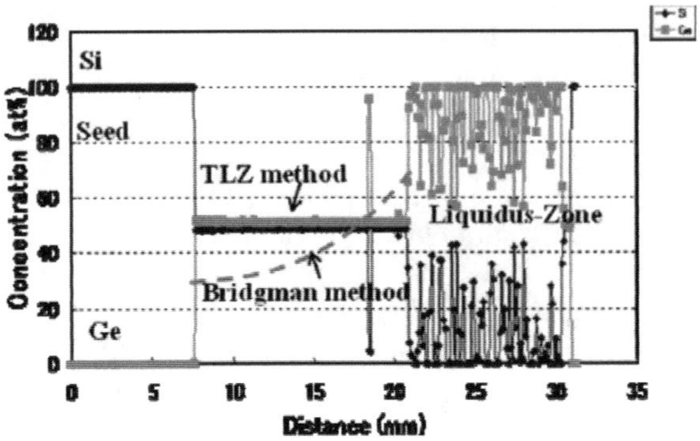

Figure 3. Axial compositional profiles of a TLZ-grown 2mm diameter crystal along with Ge concentration profile of a Bridgman-grown crystal..

the liquidus zone. Seed/crystal interface (the initial freezing interface) is almost flat but radius of curvature of the freezing interface is about 20mm at the growth length of about 10mm. Therefore, the interface curvature degrades radial compositional uniformity. **Figure 5** shows radial Ge concentration profile near the freezing interface which is about 10mm away from the initial interface. About 1 at% higher at the center of the crystal compared to the peripheral region.

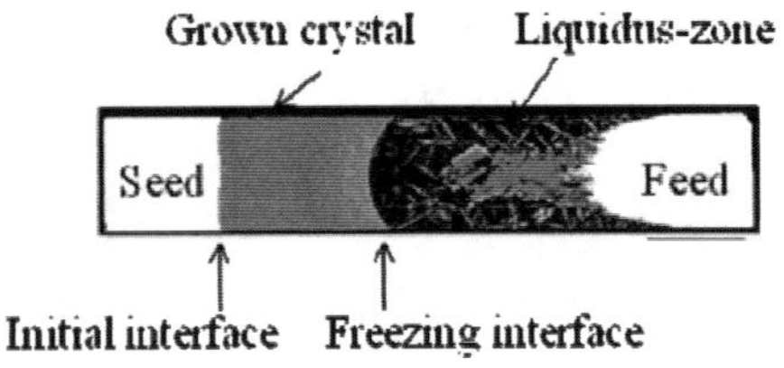

Figure 4. Two-dimensional Ge concentration map.

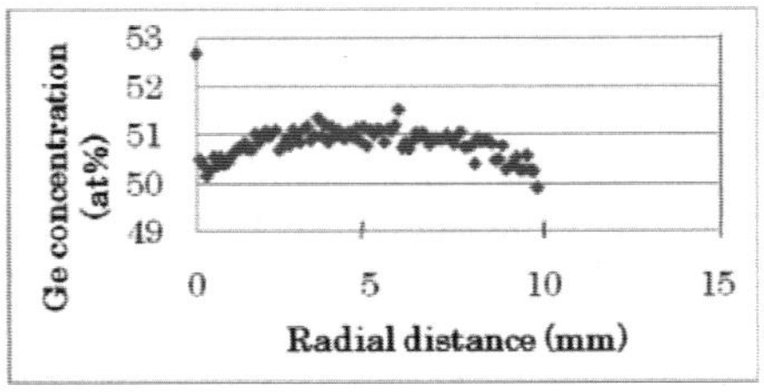

Figure 5. Radial Ge concentration near the growth interface.

4.3 Two-dimensional model

As described above, when crystal diameter is increased, radial composition degrades and two-dimensionality should be considered. We therefore proposed a two-dimensional TLZ growth model [12]. In the model, radial temperature gradient term is added to the equation (1), and we obtain equation (2).

$$\frac{\partial f}{\partial t}=-\frac{D}{(C_L-C_S)}\frac{\partial C_L}{\partial T}\left(\frac{\partial T}{\partial Z}-\frac{\partial T}{\partial r}\frac{\partial f}{\partial r}\right)_{Z=0} \qquad (2)$$

Where, the z coordinate of the freezing interface is defined by equation (3). $\partial f/\partial t$ in equation (2) expresses interface translation rate and coincides with R in equation (1)**.** It should be noted that equation (2) coincides with equation (1) if the second term in the parenthesis of right side of equation (2) is zero.

$$z=f(r,\ t) \qquad (3)$$

We want to compare freezing interface position of large diameter $Si_{0.5}Ge_{0.5}$ crystals with our two dimensional growth model but interface shapes are not always reproductive for terrestrially grown crystals. A variety of interface shape were observed . In **Figure 6** two examples of interface shapes are shown. Regular interface shape as shown in figure 4 is very rare and irregular interface shapes as shown in figure 6 are often observed. Origin of such irregularity may be caused by convection in a melt because density difference between Si and Ge is considerable; density of Si in a liquid phase is about 2.5 whereas density of Ge is about 5.6 at 990°C. Thus, two-dimensional growth model equation cannot be applied to terrestrially grown crystals. One aim of our space experiments is to obtain interface curvature dependence on growth rate (on temperature gradient) and on growth length without convection in a melt and compare interface curvature with our two-dimensional model. If we succeed in quantitative evaluation of equation (2), we can calculate effect of radial temperature gradient on radial compositional profile and we can estimate critical radial temperature gradient for establishing compositional uniformity in the radial direction. Such estimation will be useful for growing large homogeneous crystals by the TLZ method.

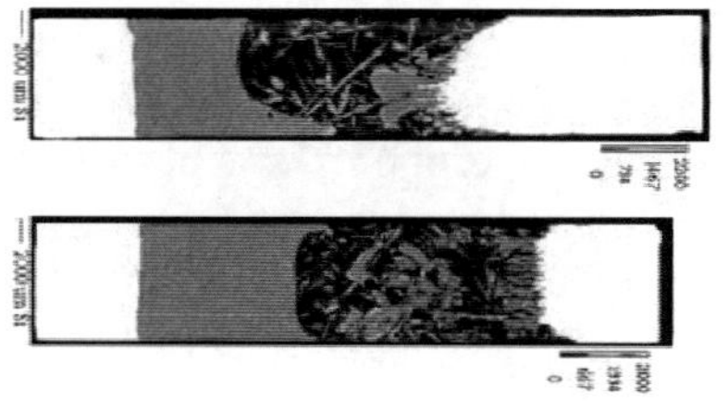

Figure 6. Two examples of interface curvature of terrestrially TLZ-grown crystals.

4.4 Preparatory experiments on the ground

Since heater length of the GHF is shorter than the usual laboratory furnace, maintaining a constant temperature gradient is difficult. Therefore, growth conditions which were obtained using laboratory furnace cannot be directly applied to a flight model of the GHF. We investigated growth conditions for a flight model of the GHF using a ground model of the GHF. At first, temperature profiles were measured using a dummy sample which is composed of solid Si as a seed, solid boron nitride as a zone and solid Si as a feed. By the translation of heater zones, temperature profiles were changed. Therefore, we changed heating conditions so that the temperature profile is not varied by the translation of heater zone. In the second step, we grew SiGe crystals by the TLZ method. In the first run, initial composition and temperature gradient were deviated from the expected values as shown in **Figure 7**. However, adjustment of heating conditions are easy because composition reflects interface temperature and growth rate reflects temperature gradient in the TLZ method. After several experiments, we obtained appropriate growth conditions for the flight model furnace. **Figure 8** shows one example of adjusted compositional profiles obtained on the ground. It is shown that Si and Ge concentrations are 50±1 at% in a grown crystal for the length of more than 15mm.

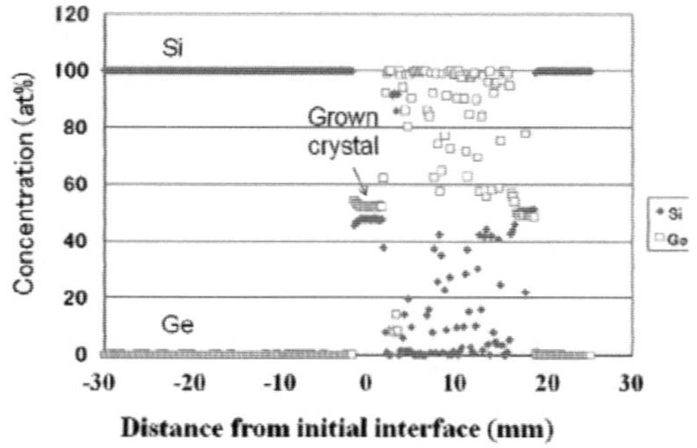

Figure 7. Si and Ge concentration profiles in the first run crystal.

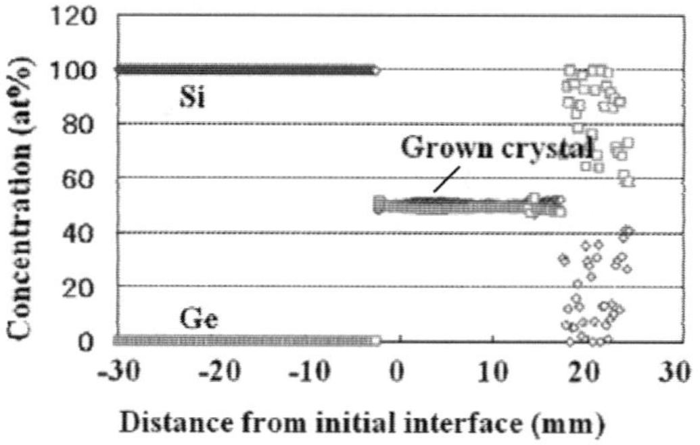

Figure 8. Si and Ge concentration profiles in a crystal after several times adjustment of growth conditions.

5. Plan for microgravity experiments

We are planning to obtain reproducible interface curvatures in the absence of convective flow in a melt and we will compare two dimensional TLZ growth model equation with experimentally obtained interface curvatures. Data obtained by microgravity experiments are used to obtain large sized homogeneous $Si_{1-x}Ge_x$ crystals which are useful as substrates of high speed CMOS devices. The GHF furnace for crystal growth experiments has already been on board the ISS "Kibo" and is being checked on its feasibility. After checking out, growth experiments will be performed.

6. Summary

A new crystal growth method named the TLZ method has been invented for growing homogeneous mixed crystals. Axial compositional homogeneity has been evaluated by the growth of small diameter crystals on the ground because convection in a melt is suppressed. However, evaluation of radial compositional homogeneity in large diameter crystals is not so easy because convection causes local compositional inhomogeneity on the ground. Therefore, microgravity experiments are planned for controlling radial compositional inhomogeneity and for evaluating a two dimensional TLZ growth model. Some results of preparatory experiments on the ground and plan for space experiments are described.

References

[1] Sell H E and Mueller G 1989 *J. Cryst. Growth* **97** 194
[2] Azuma Y, Usami N, Ujihara Fujiwara T, Sazaki G, Murakami Y and Nakajima K 2003 *J. Cryst. Growth* **250** 298
[3] Yildiz M, Dost S and Lent B 2005 *J. Cryst. Growth* **280** 151
[4] Kinoshita K, Kato H, Iwai M, Tsuru T, Muramatsu Y and Yoda S 2001 *J. Cryst. Growth* **225** 59
[5] Kinoshita K, Hanaue Y, Nakamura H, Yoda S, Iwai M, Tsuru T and Muramatsu Y 2002 *J. Cryst. Growth* **237–239** 1859
[6] Nakamura H, Hanaue Y, Kato H, Kinoshita K and Yoda S 2003 *J. Cryst. Growth* **258** 49
[7] Kinoshita K, Ogata Y, Koshikawa N, Adachi S, Yoda S, Iwai M, Tsuru T and Muramatsu Y 2005 *Int. J. Mater. & Product Tech.* **22** 95
[8] Kinoshita K, Ogata Y, Adachi S, Yoda S, Tsuru T, Miyata H and Muramatsu Y 2006 *Ann. New York Acad. Sci.* **1077** 161
[9] Adachi S, Ogata Y, Koshikawa N, Matsumoto S, Kinoshita K, Yoshizaki I, Tsuru T, Miyata H, Takayanagi M and Yoda S 2005 *J. Cryst. Growth* **280** 372
[10] Miyata H, Adachi S, Ogata Y, Tsuru T, Muramatsu Y, Kinoshita, K Odawara O and Yoda S 2007 *J. Cryst. Growth*, **303** 607
[11] Kinoshita K and Yoda S 2011 *J. Cryst. Growth* **318** 1026
[12] Kinoshita K, Adachi S, Inatomi Y, Miyata H, Tanaka R, Arai Y and Yoda S 2010 *Adv. Astro. Sci.* **138** 647

Morphological Transition in Crystallization of Si from Undercooled Melt

Kazuki Watanabe, Katsuhisa Nagayama and Kazuhiko Kuribayashi
Shibaura Institute of Technology, Toyosu, Koto, Tokyo 135-8548, Japan

E-mail: k_kuribayashi@nifty.com

Abstract. Using CO_2 laser equipped electro-magnetic levitator, we carried out the crystallization of Si at undercoolings from 0 K to 200 K. From the point of the interface morphologies, the relationship between growth velocities and undercoolings was classified into two regions, I and II, respectively. In region I where the undercooling is approximately less than 100 K, needle-like thin plate crystals whose interface consists of faceted plane were observed. In region II, the morphology of growing crystals changed to massive dendrites. Although the interface morphologies look quite different between region I and II, the growth velocities are expressed by two dimensional (2D) nucleation-controlled growth model, and at undercoolings larger than 150 K, the growth velocities asymptotically close to the analysis of the mono-parametric linear kinetics growth model. In this stage, the kinetic coefficient of 0.1 m/sK is equivalent to that derived by the diffusion-controlled growth model. This result means that with increase of undercooling, the rate-determining factor changes from 2D nucleation on the faceted interface to random incorporation of atoms on the rough interface.

1. Introduction

The rapid increase in the demand for a polycrystalline Si solar cell has caused a shortage in Si raw materials. This has brought a great deal of attention to a drop-tube process for manufacturing spherical crystals with a diameter of approximately 1mm. The first trial of drop-tube processing for spherical crystallization of Si was performed by McKee in 1982 [1]. In drop-tube process, raw materials of Si are melted in a crucible and then ejected into a drop-tube to be spherically crystallized during free fall. Although this process is very simple, the quality of as-dropped samples was much less than that of single crystals grown by conventional crystal growth techniques such as Czochraiski (CZ) and Bridgeman methods. In order to improve the quality of as-dropped samples, remelting and regrowing treatment has to be introduced as the post drop-tube process. The cost-up associated with this extra process incurs the disadvantage of spherical crystals for use to solar cell. In fact, although many venture companies in Japan intend to produce the single-crystalline sphere without any extra processes [2-5], the method hasn't been established.

In order to elucidate the crystallization behavior of Si droplets during free fall, Aoyama and Kuribayashi [6] (referred to hereafter as AK) carried out containerless crystallization of Si using a CO_2 laser equipped electro-magnetic levitator (EML). They revealed that at low undercooling less than 100 K (region I) plate-like facet crystals grow and at undercooling larger than 100 K (region II) the morphology of the crystal-melt interfaces changes to a dendritic microstructure. Nagashio and Kuribayashi [7] (referred to hereafter as NK) suggested that the plate-like facet crystals are to be grown by the preferential incorporation of atoms into the re-entrant corners formed at the edge of two parallel twin planes (Twin-related Lateral Growth: TrLG), while the dendritic crystals are grown by random incorporation of atoms on the non-faceted rough interface (Twin-free Continuous Growth: TfCG).

In crystal growth of Si, the concept of TrLG was first proposed as one of morphologies of facet dendrites more than half a century ago [8, 9]. However the related experiments carried out recently on containerless solidification into the undercooled melt have been analysed based on the simple dendrite growth model incorporating kinetic undercooling [10, 11]. It is well known that the dendrite growth model proposed by Lipton, Kurz and Trivedi (hereafter referred to as LKT) [12] is principally based on the assumption that the crystal-melt interface is rough so that atoms can be incorporated at any sites of the interface. However, in the case of Si, plate-like crystals formed at low undercoolings suggest that the crystal growth is controlled by either the screw dislocation-assisted spiral growth or the two-dimensional (2D) nucleation [13].

Until now, however, a lack of precise data for growth velocities at low undercoolings has prevented the mechanism of the crystal growth of spherical Si to be clarified. Hence, in the present investigation, the growth velocities particularly at relatively low undercooling are measured and the mechanism of spherical crystallization of Si is discussed.

2. Experimental

Figure 1 shows a schematic illustration of the used electro-magnetic levitator (EML). Undoped 5N Si spheres with a diameter of 8 mm were melted by a CO_2 laser were levitated in a 5N Ar gas atmosphere. The sample, after heating to approximately 1800 K, was cooled to predetermined temperatures by blowing of 5N He gas. Nucleation was triggered by touching the side of levitated sample with a molybdenum needle.

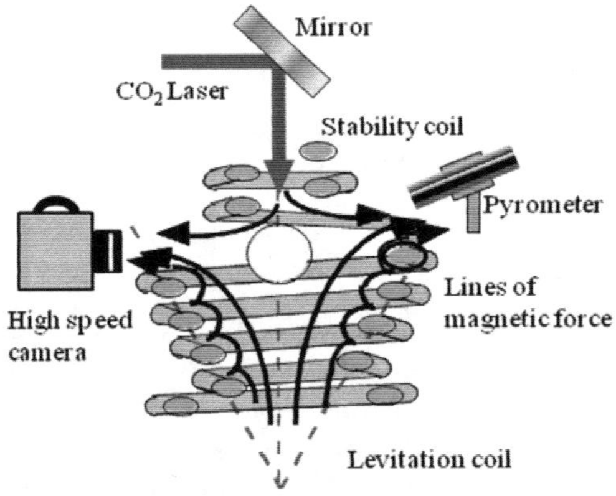

Figure 1. Schematic illustration of the electro-magnetic levitation (EML) furnace. In order to apply EML to semiconductive material, a CO_2 laser serves as a pre-heater.

Figure 2 shows the typical temperature *vs.* time profile measured with a pyrometer with operating wavelength: 900 nm and 1,550 nm. After recalescence, the temperature oscillates in a definite range. In semiconducting Si, since the emissivity ε of the solid phase is much larger than that of the liquid phase, this temperature oscillation can be attributed the fluctuation of the location of the measuring point from liquid to solid phase in the levitated sample. The undercooling was determined by adjusting ε so that the temperature just after the recalescence represents the equilibrium melting temperature of Si. The growth velocities of samples were measured using a colour High-Speed Video (HSV) with a maximum sampling rate of 640,000 frames/s.

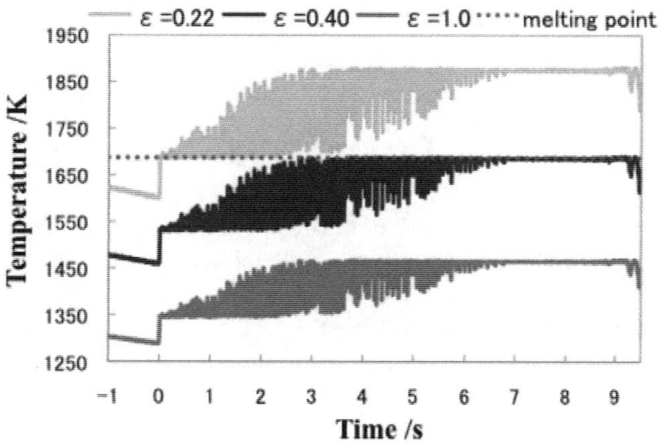

Figure 2. Typical profile of temperature *vs.* time measured with a pyrometer. The temperature of the samples was determined by adjusting the emissivity so that the temperature just after the recalescence is the equilibrium melting temperature of Si.

3. Results

The selected HSV images taken during recalescence are shown in Fig. 3, where ΔT is the undercoolings. The dark and bright area respectively show the undercooled melt and the solidified region. The release

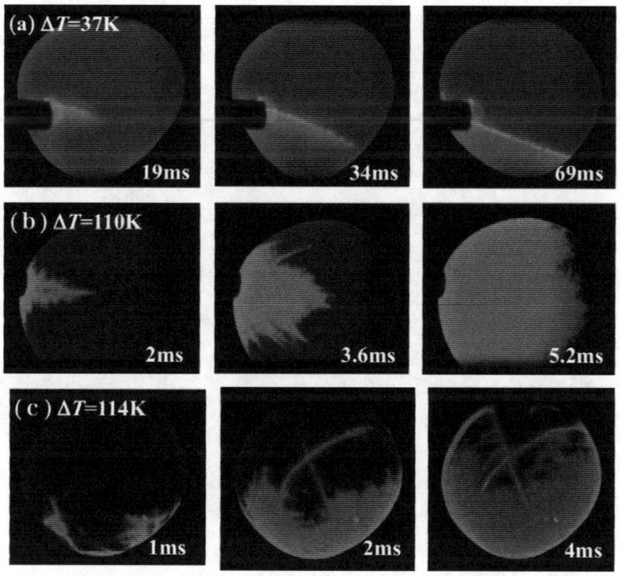

Figure 3. HSV images of samples taken during recalescence. Dark and bright area respectively show the undercooled melt and the solidified region. (a) HSV images taken at low undercooling, ΔT = 37 K, showing anisotropic line-shaped crystal that forms a circumferential ring on the surface of the sample. (b) HSV images taken at medium undercooling, ΔT = 110 K, showing a facetted dendrite. (c) Mixed-mode of facetted dendrites and line-shaped crystals. The solidification front initially advances massively, and subsequently line-shaped crystal protrudes from the facetted dendrite showing a mixed-mode of both morphologies. The velocity of line-shaped crystal is much higher than that of facetted dendrites.

of latent heat as well as the difference of emissivities between the solid and the liquid phases enhances the brightness of the solidified region.

In this figure, (a) is typical HSV images taken successively at low undercooling, ΔT = 37 K, showing anisotropic line-shaped crystal that forms a circumference ring on the surface of the sample. Note that this ring is discontinuous, suggesting that the line-shaped crystal is not a "line" crystal located on a surface but a plate-like crystal that penetrates the sample. Facetted dendrites with a similar shape appear when the undercooling becomes higher than 100 K. At medium undercooling, however, not only a facetted dendrite but also a mixed mode of facetted dendrites and line-shaped crystals are observed. The solidification front initially advances massively, and subsequently line-shaped crystals protrude from facetted dendrites showing a mixed mode of both morphologies. The velocity of the line-shaped crystal is much higher than that of facetted dendrites.

According to the LKT model, the bulk undercooling ΔT for pure material is given by

$$\Delta T = \Delta T_t + \Delta T_r + \Delta T_k, \tag{1}$$

where ΔT_t is the thermal undercooling given by

$$\Delta T_t = \frac{\Delta H_f}{C_p} \mathrm{Iv}(P_t), \tag{2}$$

for the case in which the shape of the growth front can be approximated as an elliptical paraboloid. In this equation, ΔH_f is the enthalpy of fusion, C_P is the specific heat of the melt at constant pressure, and $\mathrm{Iv}(P_t)$ is the Ivantsov function given by

$$\mathrm{Iv}(P_t) = P_t \exp(P_t) \int_{P_t}^{\infty} \frac{\exp(-P_t)}{P_t} dP_t, \tag{3}$$

where $P_t = VR/2a_l$ denotes the thermal Peclet number, V the growth velocity, R the radius of the growth front, and a_l the thermal diffusivity.

In Eq. 1, ΔT_r, the curvature undercooling due to the Gibbs-Thomson effect, is expressed by

$$\Delta T_r = \frac{2\Gamma}{R}, \tag{4}$$

where $\Gamma = \gamma/\Delta S_f$ is the Gibbs-Thomson coefficient.

The kinetic undercooling ΔT_k is related to growth velocity depending on the interfacial kinetics. If the interface is assumed to be rough, ΔT_k is expressed by

$$\Delta T_K = \frac{V}{\mu} \tag{5}$$

where μ is the kinetic coefficient. According to the marginal stability analysis [14], the radius of the growth front that is assumed to be equal to the wavelength of a critical perturbation of a planar interface λ_i is derived as

$$R = \frac{\Gamma/\sigma^*}{P_t \frac{\Delta H_f}{C_P}\left[1 - \left\{1 + \left(\sigma^* P_t^2\right)^{-1}\right\}^{-\frac{1}{2}}\right]}, \tag{6}$$

where σ^* a stability constant, was approximately derived to be 0.025 using the marginal stability criterion.

The growth velocities of samples measured as a function of undercoolings are plotted in Fig. 4, where squares and triangles correspond to those of the plate-like crystal and the faceted dendrite, respectively [15]. In this figure, the growth velocities denoted by Tr were thoese of the samples where the nucleations were triggered by touching with molybdenum needles, and the others are of the samples where the nucleation occurred spontaneously. In this figure, the solid and dotted lines correspond to the cases where the kinetic coefficient, μ, is assumed to be 0.1 m/sK and 0.02 m/sK, respectively. The physical parameters used in the calculation are listed in Table 1.

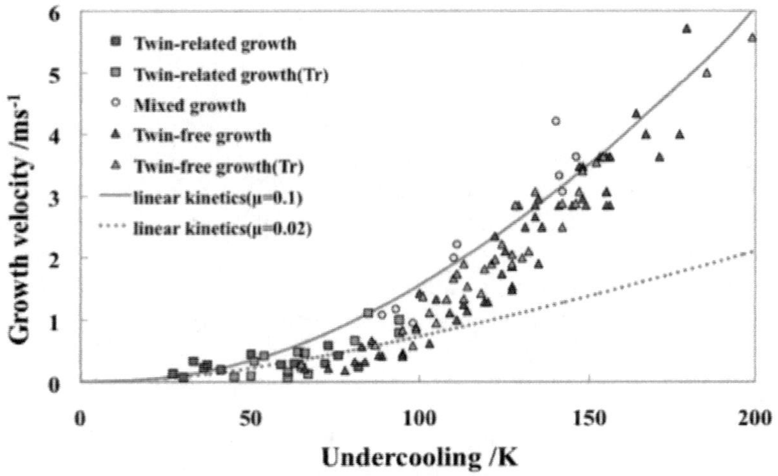

Figure 4. Growth velocities as a function of undercoolings. Squares and triangles correspond to those of the plate-like crystal and the faceted dendrite, respectively. Solid and dotted lines, which are calculated using the LKT model with linear kinetics, correspond to the cases that the kinetic coefficient μ is assumed to be 0.1 m/sK and 0.02 m/sK, respectively. Data denoted by Tr correspond to those where the nucleations were triggered with molybdenum needles, and the others are where the nucleation occurred spontaneously.

Table 1. Physical properties used in the calculation

Parameter			
Melting temperature	T_m	1687	K
Entropy of fusion	ΔS_f	29.99	J/molK
Heat of fusion	ΔH_f	5.06×10^4	J/mol
Surface energy	γ	0.438	J/m^2
Thermal diffusinity	a_l	2.35×10^{-5}	m^2/s
Specific constant	C_p	25.6	J/molK
Gibbs Thomson coefficient	Γ	9.61×10^{-11}	m

4. Discussions

In Fig. 4, V *vs*. ΔT for plate-like crystals can be fitted by $\mu = 0.02$ m/sK at relatively low undercoolings – less than 100 K. However, at medium and high undercoolings larger than 100 K, V *vs*. ΔT can be fitted by $\mu = 0.1$ m/sK. Although this result seems to show that the kinetic coefficient is enlarged with increasing undercoolings, it is to be unable to express V *vs*. ΔT with a mono-parametric μ, suggesting a change in the mechanism of the interface kinetics.

With regard to other interface kinetics, two types of mechanisms have been put forward [13]. The first one is quadratic kinetics given by

$$V = \phi\left(\Delta T_K\right)^2, \tag{7}$$

where ϕ is the kinetic coefficient for quadratic kinetics. The second one is exponential kinetics given by

$$V = \beta \exp\left(-\frac{E}{k_B \Delta T_K}\right), \tag{8}$$

where β and E are the kinetic coefficient for the exponential kinetics and the energy barrier for forming a critical nucleus, respectively. These two models describe screw-dislocation assisted spiral growth and 2D nucleation controlled growth, respectively. It has been well-known that at relatively low undercooling the quadratic kinetic or 2D nucleation model controls the crystal growth, whereas at relatively high undercooling linear kinetics is the rate determining factor. AK previously reported that the relationship between growth velocities and undercoolings in Si can be well expressed by LKT model with the mono-parametric interface linear kinetics [6]. As mentioned previously, the interface linear kinetics, which describes the random attachment of thermally activated atoms onto the interface, is well applied to the metallic materials having a rough interface. However, for {111} in Si that is an atomically smooth interface, atoms are incorporated to preferential sites of interface such as a kink and/or a step. If the plate-like crystal which is observed at region I was grown with the twin-plane re-entrant corner mechanism, the crystal growth kinetics may be controlled not by interface attachment linear kinetics but by either the screw dislocation assisted spiral growth mechanism or the 2D nucleation model. In the case of a diamond lattice, however, the lattice distortion at the core of the screw dislocation with Burgers vector along <110> direction is too large for the covalent bonds to be formed there [16]. This implies that the rate-determining process at the growth of the plate-like crystal is not the interface attachment kinetics but the 2D nucleation model.

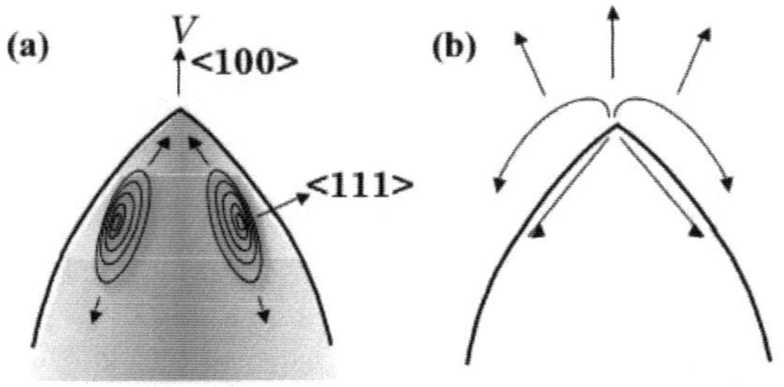

Figure 5. (a) Schematic illustration of the tip of a facetted dendrite: In FCC and DC lattices, if we can assume a non-isothermal interface, the level of the undercooling is highest at the tip where the interface attachment kinetics controls the growth velocity. The undercooling decreases along the slope that continues away from the tip. (b) Arrows indicate heat flow: Compared with the case of isothermal interface where the heat flows perpendicularly from interface to melt, the faster growth velocity can be expected.

On the facetted dendrite, Maurer et al. [17], using NH_3Br as a sample for crystal growth, reported that facetted dendrite shape strongly depends on the growth velocity. Namely, facet planes become visible at the trail of the tip forming a facetted dendrite at decreased growth velocities, while at large velocities the shape of the crystal appears unfacetted. For this shape, they determined a parabola which fits the interface over distances appreciable compared to its own tip radius of curvature, and then obtained the relationship between the length of facet area, λ, and the growth velocity, V, as

$$\lambda \propto V^{-0.5}. \tag{9}$$

Journal of Physics: Conference Series **327** (2011) 012018 doi:10.1088/1742-6596/327/1/012018

Furthermore they evaluated the product between a square of tip radius, ρ^2, and V, and observed that at high growth velocity the value oft $\rho^2 V$ is large and at low growth velocity small. This result implies that the stability constant σ^* of facet interface is larger than that of dendrite. In fact, Dougherty and Gollub [18] experimentally obtained the stability constant of NH_3Br as 0.081±0.02, despite that the dependency on the growth velocity is small than expected.

On the other hand, for facetted dendrite Tiller provides us with a highly suggestive hint in his literature. Based on the hierarchic model of terrace-ledge-kink (TLK), he proposed a novel idea to solve the conflict which is inherent in understanding facetted dendrite [19]. Figure 5 shows the concept of his idea. That is, if we can assume the non-isothermal interface, the level of the undercooling is highest at the tip where the interface attachment kinetics controls the growth velocity. The undercooling decreases toward the sloping region that continues away from the tip. At the sloping region, instead of linear kinetics, spiral growth or 2D nucleation controls the growth velocity, forming a facet plane (Fig. 5(a)). In this case the direction of heat flow is denoted with arrows as shown in Fig. 5(b). Compared with the case of isothermal interface growth where the heat flows perpendicularly from interface to melt, a faster growth velocity can be expected. According to this hypothesis, the plausible geometrical configuration of the dendrite which meets the conditions is a fourfold <100> dendrite, because there is maximum number of heat sink ({111} planes) adjacent to the tip. His suggestion implies that even in facet dendrite rate-controlling process is not interface attachment kinetics but 2D nucleation on {111} slope area.

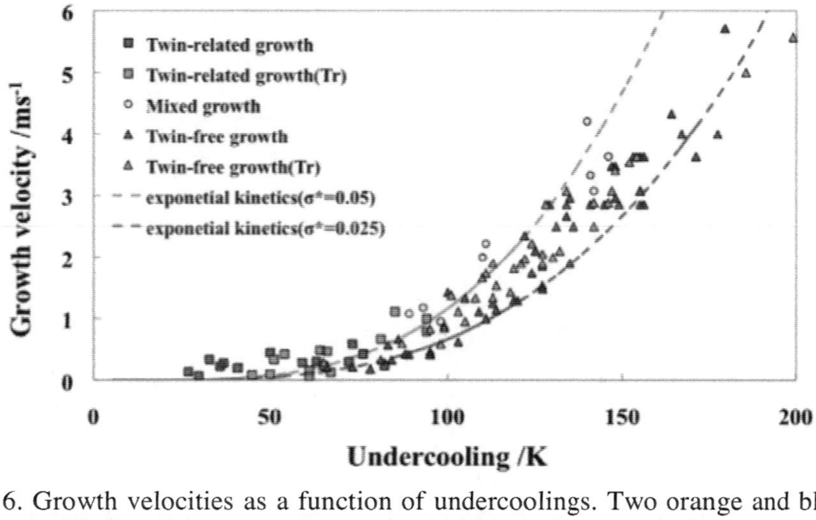

Figure 6. Growth velocities as a function of undercoolings. Two orange and blue lines are calculated using LKT model with an exponential kinetics. Parameters β and E/k_B are optimized to 7.0×10^5 and 3.5×10^2 K, respectively. The stability constant σ^* is set to 0.05 for plate-like crystal, including the mix-mode case, and 0.025 for facetted dendrites in region II. Each of these two lines qualitatively agrees with the respective relations between growth velocities and undercoolings.

On the basis of this idea, instead of applying linear kinetics we incorporated the exponetial kinetics given by Eq. 8 into the LKT model to fit the experimental data. Figure 6 shows the results for these two cases of σ^*, where β and E/k_B are optimized to 7.0×10^5 and 3.5×10^2 K, respectively. Each of these two lines qualitatively agrees with the observed relationship between growth velocities and undercoolings. Note that we must relax our interpretation of the LKT model when applied to crystallization of faceted material, particularly in respect to use of the Ivantsov function. Here, the narrow area around the tip must be approximated by a paraboloid of revolution even for a faceted material.

As clearly shown in this figure, the exponential kinetics provides a fit to the experimental data. However, at relatively high undercoolings typically $\Delta T > 150$ K, the behaviour is better asymptotically by the linear interface kinetics - the coefficient of which is that of Wilson-Frenkel model [20, 21]. At high undercoolings, dendrite breakdown due to the Gibbs-Thomson effect causes poly-crystallization of the droplet, where fragmented dendrite can change the rate-controlling process from 2D nucleation to interface attachment kinetics, due to the formation of seed crystals. Hence, it can be summarized that the crystal growth at relatively low and medium undercoolings is controlled by the 2D nucleation not only in twin-related <110> facet dendrite but also in twin-free <100> case, and at relatively high undercolling by diffusion-controlled linear kinetics.

5. Conclusion

The recent approach to containerless crystallization of semiconductive material was presented including the historical background for production of spherical Si for solar cell. Although the pioneer work on the crystallization into the undercooled melt was carried out more than half a century ago, the related containerlress experiments are now analysed in terms of modified dendrite growth models. The plate-like crystals formed at low and medium undercoolings suggest that the growth mechanism is not the simple heat-rejection-controlled model but the interface kinetic-controlled model. In this paper, precise observation of microstructure allows identification of the growth mechanism is the twin-plane reentrant corner mechanism; the rate-controlling process is the 2D nucleation not only in plate-like crystal but also in faceted dendrite.

Acknowledgement

This work was financially supported by a grant-in aid for Scientific Research from the Ministry of Education, Culture, Sports, Science and Technology.

References

[1] McKee W R 1982 *IEEE Trans. Components, Hybrids, and Manufacturing Tech.* **5** 336.
[2] Minemoto T, Okamoto C, Omae S, Murozono M, Takakura H and Hamakawa Y 2005 *Jpn. J. Appl. Phys.* **44** 4820.
[3] Taira K, Kogo N, Kikuchi H, Kumagai N, Kuratani N, Inagawa I, Imoto S, Nakata J and Biancardo M 2005 *Tech. Dig., PVSEC-15* 202.
[4] Masuda S, Takagi K, Kang Y-S and Kawasaki A 2004 *J. Jpn. Soc. Powder Metallurgy*, **51** 646.
[5] W. Dong, K. Takagi, S. Masuda, and A. Kawasaki, *J. Jpn. Soc. Powder Metallurgy*, 2006, **53,** 346-351.
[6] Aoyama T and Kuribayashi K 2000 *Acta mater.* **48** 3739.
[7] Nagashio K and Kuribayashi K 2005 *Acta mater.* **53** 3021.
[8] Billig E 1955 *Proc. Roy. Soc.* **A 229** 346.
[9] Hamilton D R and Seidensticker R G 1960 *J. Appl. Phys.* **31** 1165.
[10] Li D and Herlach D M 1997 *J. Mater. Sci.* **32** 1437.
[11] Aoyama T, Takamura Y and Kuribayashi K 1999 *Metall. Mater. Trans. A* **30A** 1333.
[12] Lipton J, Kurz W and Trivedi R 1987 *Acta mater.* **35** 957.
[13] Saito Y 1996 *Statistical Physics of Crystal Growth*, World Scientific, Singapore 86.
[14] Langer J S and Muller-Krumbhaar 1978 *Acta Metall.* **26** 1681.
[15] Watanabe K, Nagayama K and Kuribayashi K (in press) *J. Jpn Soc. Microgravity Appl.*
[16] Hornstra J 1959 *J. Phys. Chem. Solids* **5** 129.
[17] Maurer J, Bouissou P, Perrin B and Tabeling P 1988 *Erophys. Lett.*, **6** 609.

[18] Dougherty A and Gollub J P 1988 *Phys. Rev. A*, **38** 3043.
[19] Tiller W A 1991 *The science of Crystallization: microscopic interfacial phenomana*, Cambridge University Press, Cambridge.
[20] Wilson H A 1900 *Phil. Mag.* **50** 238.
[21] Frenkel J 1932 *Phys. Z. Sowjetunion*, **1** 498.

Metastable Phase Formation from Undercooled Melt of Oxide Material

Kazuhiko Kuribayashi[1,2], M. S. Vijaya Kumar[2]
[1] Shibaura Institute of Technology, Toyosu, Koto, Tokyo 135-8548
[2] Institute of Space and Astronautical Science, JAXA, Yoshinodai, Sagamihara, Kanagawa 252-5210, Japan

E-mail: k_kuribayashi@nifty.com

Abstract. Undercooling a melt often facilitates a metastable phase to nucleate preferentially. Although the classical nucleation theory shows that the most critical factor for forming a metastable phase is the interface free energy, the crystallographic stability is also indispensable for the phase to be frozen at ambient temperature. In compound materials such as oxides, authors have suggested that the decisive factors for forming a critical nucleus are not only the free energy difference but also the difference of the entropy of fusion between stable and metastable phases. In the present study, using $REFeO_3$ (RE: rare-earth element) as a model material, we investigate the formation of a metastable phase from undercooled melts with respect to the competitive nucleation and crystallographical stabilities of both phases.

1. Introduction

A metastable phase is a phase that does not exist in thermal equilibrium state and, although thermodynamically unstable, can temporarily exist when some conditions are fulfilled. Research on the metastable phase began with Ostwald's prediction that a phase formed first from supersaturated liquid is not always thermodynamically stable but is close to liquid in energy [1]. This prediction is called "step rule". More than thirty years later, Stranski and Totomanov [2] suggested that the step rule is a consequence of preferential formation of a critical nucleus of the metastable phase. That is, the activation energy required to form a critical nucleus, ΔG_n^*, controls the nature of the process. Regarding this point, the classical nucleation theory [3] states that ΔG_n^* can be understood in terms of the interfacial free energy, γ, between the liquid and solid phases. Turnbull [4] and Spaepen [5], assuming that γ of a simple metal is related not to the enthalpy change but to the entropy change at the solid–liquid interface, formulated the dimensionless solid-liquid interfacial energy, α. Furthermore, Spaepen and Meyer [6], considering the geometrical configuration of hard-spheres at the solid-liquid interface, derived α as 0.86 for fcc or hcp crystals and 0.71 for bcc structures, respectively. The α–factors, which strongly depend on the structure of both solid and liquid phase, are to be a critical parameter to determine ΔG_n^*. However, the recent numerical calculations of α based on the molecular dynamics or density functional theory show less dependence on the structure, being scattered around 0.5 [7]. On the basis of this fact, authors proposed one hypothesis that in ionic crystals such as oxides, where the close-packed tetrahedron of anions are the structural unit both in the solid and liquid phases, an entropy of fusion is to be a dominant factor in the determination of γ [8, 9]. In accordance with this idea, using $REFeO_3$ (RE: Rare-earth element) as the model material, they investigated the containerless solidification from the undercooled melt. As a result, they elucidated that in addition to the entropy of fusion, oxygen partial pressure, Po_2, is to be another important factor which controls phase selection: reduction of Po_2 in RE-Fe-O system enlarges the instability of the perovskite structure and then facilitates the metastable phase to be formed [10, 11]. This result implies that the critical factor for forming a metastable phase is not only interfacial free energies but also the relative stabilities of stable and metastable phases. Then, in the present investigation, changing Po_2 from 10^5 to 10^3 Pa and at Ar environment, Po_2 of which s approximately 10^{-1} Pa, we carried out the rapid solidification from the undecooled melt, and discussed the criterion for formation of the metastable phase from the point of the phase stabilities.

2. Experimental

Spherical samples of $REFeO_3$ were prepared from high purity (99.99%) RE_2O_3 and Fe_2O_3 powders. Details of the sample preparation are shown elsewhere [12]. Levitation and melting of samples were carried out by an aerodynamic levitator, ADL, which was designed in order to solidify undercooled melts under the precisely controlled Po_2, as shown in Fig. 1. The ADL chamber was initially evacuated down to 10^{-1} Pa and backfilled with a mixture of 6N-Ar and O_2 gas at a predetermined Po_2. The Po_2 was monitored using a ZrO_2 oxygen sensor, which was connected at the outlet of the ADL

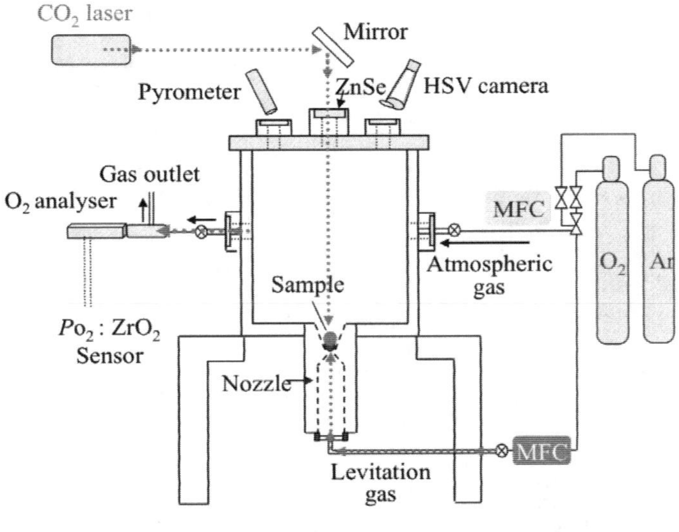

Figure 1. Schematic diagram of an aerodynamic levitation (ADL) chamber combined with a zirconia oxygen sensor. The conical nozzle angle is 60°, and orifice is $1mm^{\phi}$.

chamber.

The $REFeO_3$ sphere was levitated and then completely melted by the CO_2 laser irradiation and held at approximately 1850 K for 5 minutes to equilibrate the melt with the chamber atmosphere. Then the sample was solidified by turning off the laser. The surface temperature of the levitated sample was measured using a two-colour pyrometer at a sampling rate of 1 kHz with central wavelengths of 0.9 and 1.55 μm and a spot size of 1 mm in diameter. The recalescence behaviour was monitored using a colour high-speed video (HSV) camera, at a sampling rate of 1 kHz with a spatial resolution of 256 × 240 pixels.

The spontaneously solidified samples under various Po_2 were subjected to systematic characterization. The constituent phases in as-solidified samples were identified using X-ray diffractometry (XRD). The bulk composition was analyzed by inductively coupled plasma atomic emission spectroscopy (ICP-AES). The surface morphologies and cross-sectional microstructures of the as-solidified samples were observed by scanning electron microscopy (SEM), and the chemical compositions of the phases were analyzed by energy-dispersive spectroscopy (EDS).

3. Results

Figure 2 shows the typical cooling curves of the $REFeO_3$ (RE: (a) Lu and (b) Er) samples solidified under various Po_2. In $LuFeO_3$, the undercooled samples spontaneously nucleated below the melting temperature T_M of the $LuFeO_3$ perovskite (~1900 K) and recalescence was observed due to the release of latent heat during solidification. The nucleation (T_N) and post-recalescence temperatures ($T_{Post\text{-}recal}$) are commonly defined as the temperatures at the onset and at the end of the recalescence, which are approximately 1810 K and 1868 K, respectively. These temperatures decreased with decreasing Po_2, from 10^5 to 10^3 Pa, showing the sudden decrease in nucleation temperatures at $Po_2=3\times10^3$ Pa. If $T_{post\text{-}rec}$ is assumed to be nearly equal to the melting temperature of the phase formed during recalescence, the undercooling level ($T_{Post\text{-}recal}$ -T_N) looks independent of Po_2.

In $ErFeO_3$, which was used to study the effect of ionic radii of RE elements on the formation of metastable phases, because the ionic radius of Er^{3+} (0.1062 nm) was slightly larger than that of Lu^{3+} (0.1032 nm). In $ErFeO_3$, although T_N and $T_{Post\text{-}recal}$ also gradually decreased with decreasing Po_2 down to 10^3 Pa, the undercooling level is quite different from those of $LuFeO_3$. That is, in $LuFeO_3$, the undercooling levels are approximately 60 K independently of Po_2, whereas in $ErFeO_3$ the undercooling levels strongly depend on Po_2, more than 200 K at $Po_2=1\times10^5$ and less than 60 K at $Po_2=1\times10^3$.

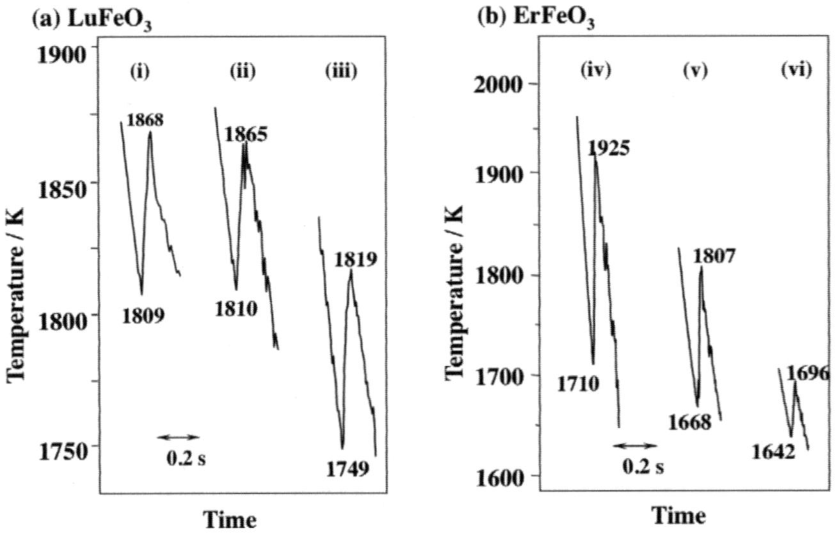

Figure 2. The typical cooling curves of the $REFeO_3$ (RE: (a) Lu and (b) Er) samples solidified under various Po_2, (i) 1×10^5, (ii) 1×10^4, (iii) 1×10^3, (iv) 1×10^5, (v) 3×10^3 and (vi) 1×10^3 Pa, respectively.

Figure 3 shows the XRD patterns of $ErFeO_3$ samples processed at 10^5 Pa and 10^3 Pa of Po_2. The XRD patterns were recorded using monochromatic radiation with a wavelength of 0.154 nm for Cu $K\alpha$. In the sample solidified at 10^5 Pa, the peaks of the stable orthorhombic $ErFeO_3$ (o-$ErFeO_3$) phase with a space group of *Pbnm* were identified. However, decreasing Po_2 to 1×10^3 Pa caused the peak intensities of the o-$ErFeO_3$ to decrease, and metastable phases appeared. In our previous study on the containerless processing of $LuFeO_3$ using a similar experimental apparatus, we already reported the matastable hexagonal phase with a space group of $P6_3cm$ appeared even at 10^5 Pa of Po_2 [10]. In

$ErFeO_3$, although the most characteristic peak of the space group $P6_3cm$ at approximately 15° is absent, the secondary characteristic peak at approximately 29° is obvious, suggesting that the hexagonal metastable phase with a space group of $P6_3cm$ was formed. In addition, another metastable phase such as h-$Er_3Fe_2O_7$ can be appreciated though the intensity is weak.

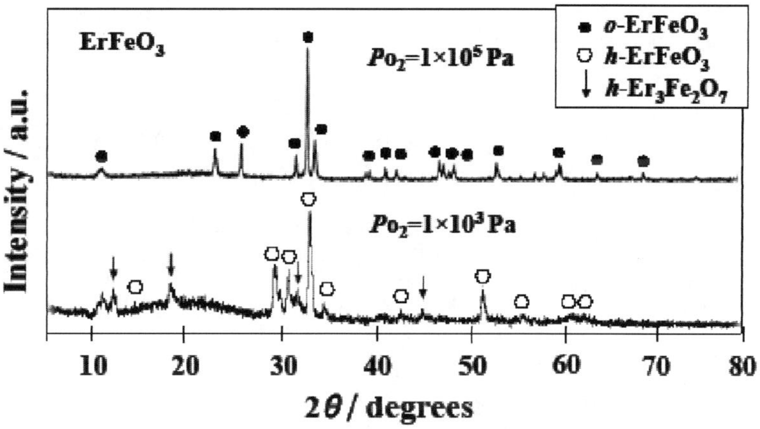

Figure 3. XRD patterns of the $ErFeO_3$ samples processed at $Po_2=10^5$ Pa and $Po_2=10^3$ Pa. In the sample solidified at 10^5 Pa, the peaks of the stable orthorhombic $ErFeO_3$ (o-$ErFeO_3$) phase with a space group of $Pbnm$ were identified. However, decreasing Po_2 to 1×10^3 Pa caused the peak intensities of the o-$ErFeO_3$ to decrease, and metastable phases appeared.

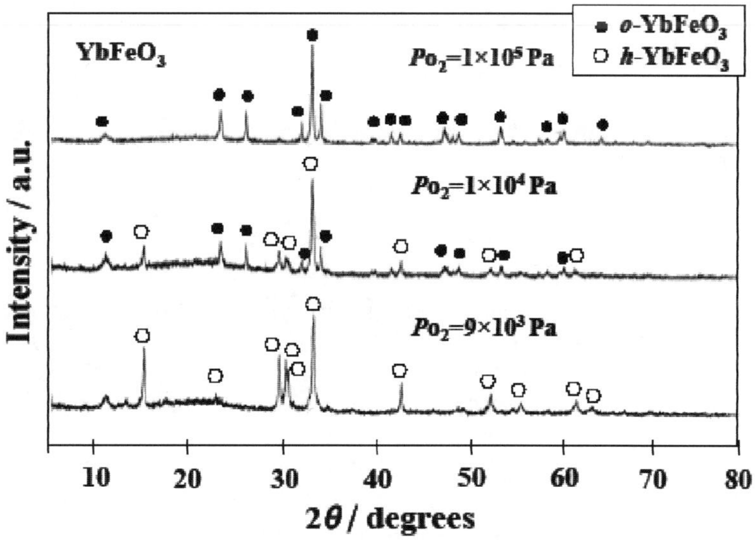

Figure 4. XRD patterns of the $YbFeO_3$ samples processed at $Po_2=10^5$ Pa, $Po_2=10^4$ Pa and $Po_2=9\times10^3$ Pa, respectively. At 10^5 Pa of Po_2, the stable orthorhombic phase (o-$YbFeO_3$) was formed as in the case of $ErFeO_3$. At 10^4 Pa of Po_2, however, the metastable hexagonal phase (h-$REFeO_3$) appeared, forming the dual phase with o-$YbFeO_3$, and at 9×10^3 Pa, the o-$YbFeO_3$ phase thoroughly disappeared.

Journal of Physics: Conference Series **327** (2011) 012019 doi:10.1088/1742-6596/327/1/012019

In $YbFeO_3$, where the ionic radius of Yb^{3+} is also slightly larger than that of Lu^{3+} but less than that of Er^{3+}, the influence of Po_2 on the phase selection between stable and metastable phases is more obvious. Figure 4 shows the XRD patterns of $YbFeO_3$ samples processed at controlled Po_2. At 10^5 Pa of Po_2, the stable orthorhombic phase (o-$YbFeO_3$) was formed as in the case of $ErFeO_3$. At 10^4 Pa of Po_2, however, the metastable hexagonal phase (h-$REFeO_3$) appeared, forming the dual phase with o-$YbFeO_3$, and at 9×10^3 Pa, the o-$YbFeO_3$ phase thoroughly disappeared.

These results suggest that the decrease of Po_2 facilitates the undercooled melt to solidify metastable h-$REFeO_3$ phase rather than the stable o-$REFeO_3$ phase, particularly in samples with RE^{3+} of relatively small ionic radius.

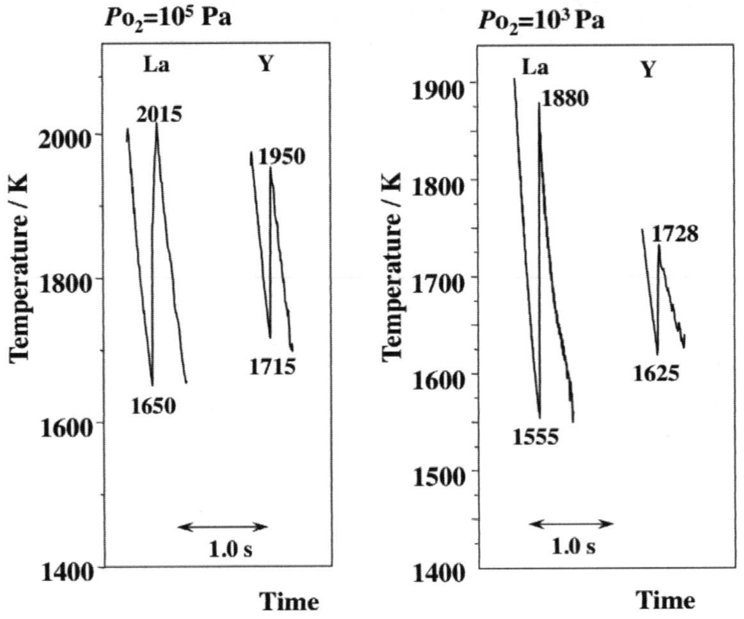

Figure 5. The typical cooling curves of the $REFeO_3$ (RE: La and Y) samples solidified at 1×10^5 Pa and 1×10^3 Pa, respectively.

Figure 5 shows the cooling curves of the $REFeO_3$ samples having relatively larger ionic radius of RE element such as La (0.1216 nm) and Y (0.1075 nm) with Po_2 of 10^5 and 10^3 Pa. As shown in this figure, the undercooling level of $LaFeO_3$ is higher than 300 K independently of Po_2. In $YFeO_3$, however, the undercooling level is similar to that of $ErFeO_3$, that is, decreasing Po_2 the undercooling level decreased from 235 K at 10^5 Pa to 103 K at 10^3 Pa.

Figure 6 shows typical images taken sequentially during recalescence in samples of $LaFeO_3$ and $YFeO_3$, each of which is processed at Po_2's of 10^5 Pa (Fig. 6a) and 10^3 Pa (Fig. 6b), respectively. The elapsed time indicated in each image was set to 0 sec for one frame before the nucleation. At 10^5 Pa, single recalescence that can be ascribed to the phase transition from undercooled melt to equilibrium perovskite phase was observed in $LaFeO_3$ samples. On the other hand, double recalescence, which could not be identified in the temperature-time profiles measured with the pyrometer, was observed in $YFeO_3$ samples. After the primary phase solidified from the undercooled melt, the secondary phase with higher brightness was initiated at the interface between melt and the primary phase. The dark and bright regions of the photograph represent the undercooled melt and the growing solid, respectively. The solid-liquid interface is clearly indicated by the large difference in brightness between the solid and liquid. The high brightness implies that the melting temperature of the secondary phase was much higher than that of the primary phase.

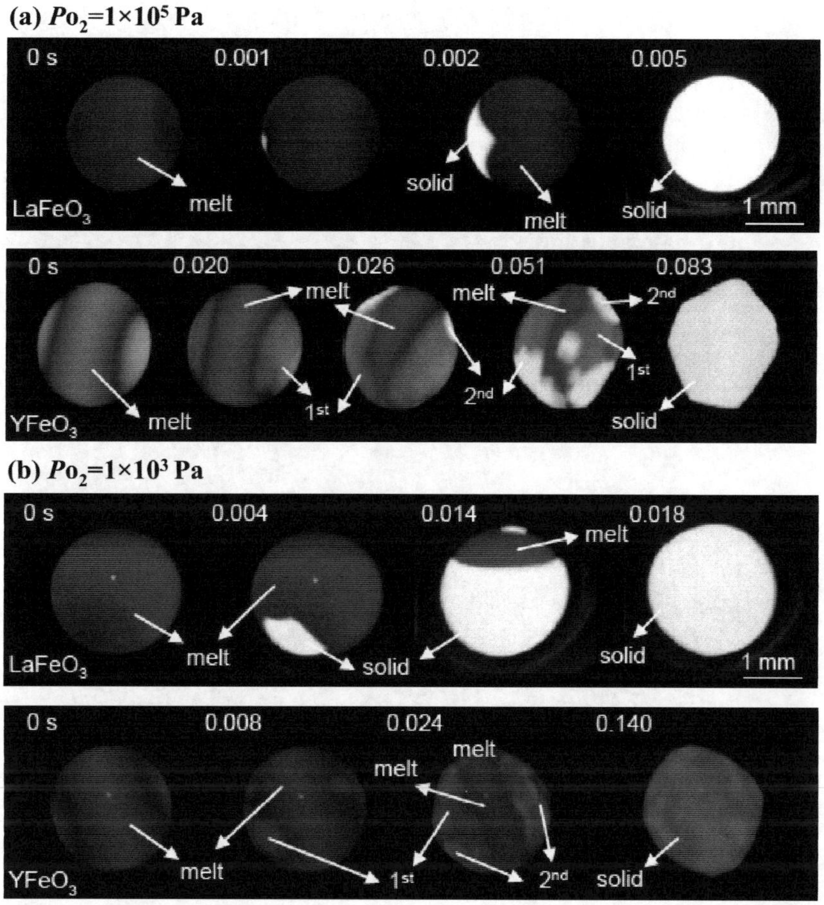

Figure 6. Sequence photographs of HSV images taken during recalescences in the $REFeO_3$ (R = La and Y) samples processed at (a) $Po_2=10^5$ Pa, and (b) $Po_2=10^3$ Pa. Although single recalescence was observed even at 10^3 Pa of Po_2 in $LaFeO_3$, double recalescence indicating the formation of metastable phases was observed even at 10^5 Pa of Po_2 in $YFeO_3$.

Previously, Nagashio et al. reported the same kind of recalescence in an undercooled $YFeO_3$ melt [13]. Later, time-resolved XRD observation on an undercooled $YFeO_3$ melt revealed that metastable hexagonal $YFeO_3$ ($P6_3cm$) was formed as the primary phase and this phase was subsequently transformed into the stable orthorhombic perovskite (*Pbnm*) structure during recalescence [14]. Therefore, in the present study, we conclude that the primary phase was the metastable hexagonal $REFeO_3$ (h-$REFeO_3$) and the secondary phase was the stable orthorhombic $REFeO_3$ (o-$REFeO_3$).

If we summarize the aforementioned experimental results, we can obtain the following conclusion: In $REFeO_3$ oxide system, decrease of Po_2 facilitates the undercooled melt to solidify the metastable h-$REFeO_3$ phase rather than the stable o-$REFeO_3$ phase, particularly in samples with relatively small ionic radius of RE^{3+}. In $LaFeO_3$, where the ionic radius of RE^{3+} is maximum in this series, the undercooling level required to form the critical nucleaus of the absolutely stable o-$REFeO_3$ phase is as large as approximately 400 K. On the other hand, in $LuFeO_3$, where the ionic radius of RE^{3+} is minimum in this series, the metastable h-$REFeO_3$ phase can be formed under relatively low undercooling level, typically less than 100 K. This result means that in $LuFeO_3$, even though the thermo-equilibrium phase is *o*-$LuFeO_3$, in the deeply undercooled melt the activation energy for forming the critical nucleus of the metastable *h*-$LuFeO_3$ phase is lower than that of the stable *o*-$LaFeO_3$ phase, suggesting that the interfacial free energy of *h*-$LuFeO_3$ is lower than that of *o*-$LuFeO_3$ if the catalytic potencies for heterogeneous nucleation of both phases are assumed to be equal. In $YFeO_3$, where double recalescence was observed, the undercooling level is medium typically 200 K, suggesting that the secondary phase can be heterogeneously nucleated at the interface between primarily solidified metastable phase and the melt.

Figure 7 shows the relation between the recalescence results and Po_2 as a function of the ionic radii of RE^{3+}, in which Shannon ionic radii for CN=9 were used. Decrease of Po_2 extends the range of *TF* for metastable *h*-$REFeO_3$ phase to be formed.

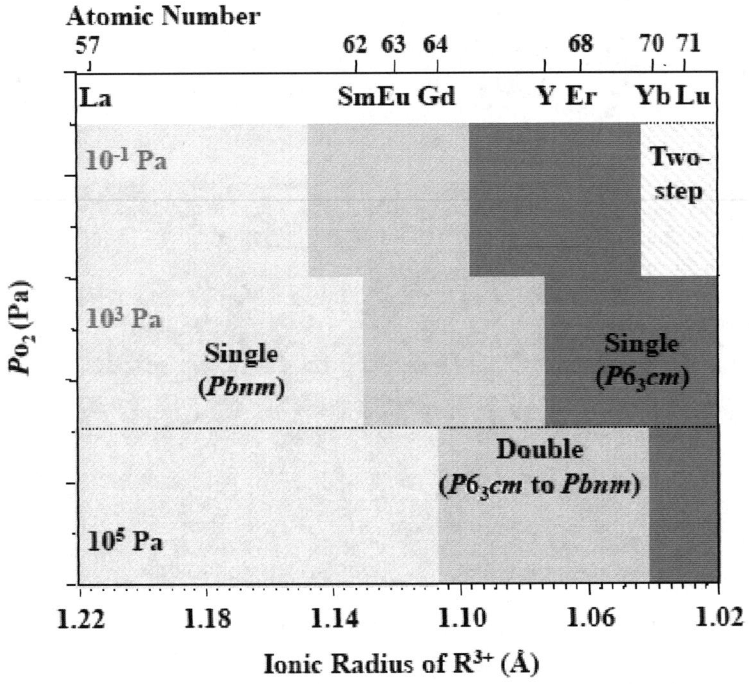

Figure 7. Relation between the recalescence results and Po_2 as a function of the ionic radii of RE^{3+}, in which Shannon ionic radii for CN=9 were used.

4. Discussion

Bertaut *et al.* [15] and Yakel *et al.* [16] have first reported the two hexagonal modifications in the ABO_3 systems, the space groups of which are $P6_3/mmc$ and $P6_3cm$, respectively. In the *h*-$REMnO_3$ system, the $P6_3cm$ type modification was formed as a low temperature phase for RE^{3+} with small ionic radius (Ho-Lu, Y and Sc), whereas the $P6_3/mmc$ type modification was reported as a high temperature phase [17]. In our experiment, although the space group of the metastable *h*-$REFeO_3$ phase belonged to $P6_3cm$, the high temperature phase can be deduced to belong to the $P6_3/mmc$ space group because the ionic radius of Fe^{3+} is as same as that of Mn^{3+} (0.0645 nm for 6 coordination). Hence, in this investigation, the geometrical analysis of the atomic configuration in *h*-$REFeO_3$ is developed on the assumption that the space group of the primary phase is $P6_3/mmc$.

The atomic configuration of the $P6_3cm$ modification in ABO_3 system can be described as a dense oxygen-ion packing (ABCACB) with B^{3+} ions having coordination number CN=5 (five-fold distorted trigonal bipyramidal coordination), and A^{3+} with CN=7 (seven-fold monocapped octahedral coordination), forming a noncentrosymmetric structure. On the other hand, centrosymmetric $P6_3/mmc$ is assumed to be described simply with B^{3+} ions of undistorted CN=5 and A^{3+} of CN=6 (octahedral coordination).

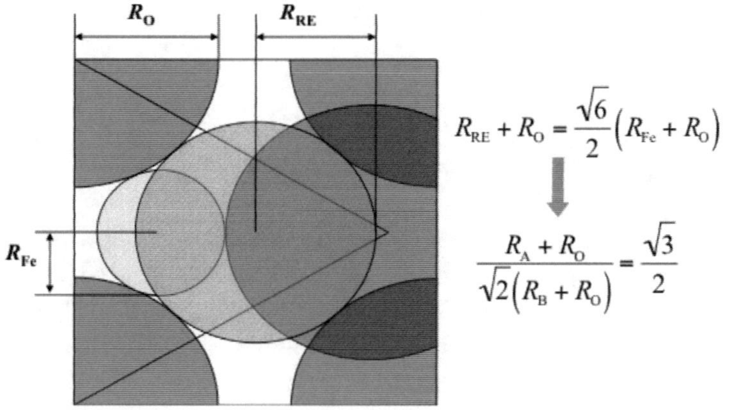

Figure 8. Geometrical configuration among RE^{3+} (CN=6), Fe^{3+} (CN=5), and O^{2-} in a space group of $P6_3/mmc$, in which the constituent ions are packed with no space between neighboring ions. The relation among the ionic radii of constituent ions of h-$REFeO_3$ is expressed as

$$R_{RE} + R_O = \sqrt{6}/2\left(R_{Fe} + R_O\right).$$

The h-$REFeO_3$ phase will be the ideal when the next equation is fulfilled,

$$TF = \sqrt{3}/2 \approx 0.87.$$

Goldschmidt [18] discussed the stability of the perovskite structure using the tolerance factor, *TF*:

$$TF = \frac{R_A + R_O}{\sqrt{2}\left(R_B + R_O\right)}, \tag{1}$$

where R_A and R_B are ionic radii of rare-earth element and iron, respectively, and R_O is the ionic radius of oxygen. From the systematic investigation, he summarized that the perovskite structure is stable at $TF > 0.8$ and contrary unstable at $TF < 0.8$. Using Shannon ionic radii from La (0.1216 nm) to Lu (0.1032 nm), *TF*'s for the $REFeO_3$ system were calculated to be 0.905 for $LaFeO_3$ to 0.841 for $LuFeO_3$. Therefore, the perovskite structure is expected to be stable in the $REFeO_3$ system.

Figure 8 shows the geometrical configuration among RE^{3+}(CN=6), Fe^{3+}(CN=5), and O^{2-} in a space group of $P6_3/mmc$, in which the constituent ions are packed without any spacing between neighboring ions. As shown in this figure, the relation among the ionic radii of constituent ions of h-$REFeO_3$ is expressed as

$$R_{RE} + R_O = \frac{\sqrt{6}}{2}\left(R_{Fe} + R_O\right). \tag{2}$$

Therefore, the h-$REFeO_3$ phase will be ideal when the next equation is fulfilled,

$$TF = \frac{\sqrt{3}}{2} \approx 0.87 \tag{3}$$

Figure 9 shows the schematical presentation for stabilities of o-$REFeO_3$ and h-$REFeO_3$ phases. If *TF* is deviated from the ideal value of 0.87 as expressed by Eq. 2, the stability of h-$REFeO_3$ phase

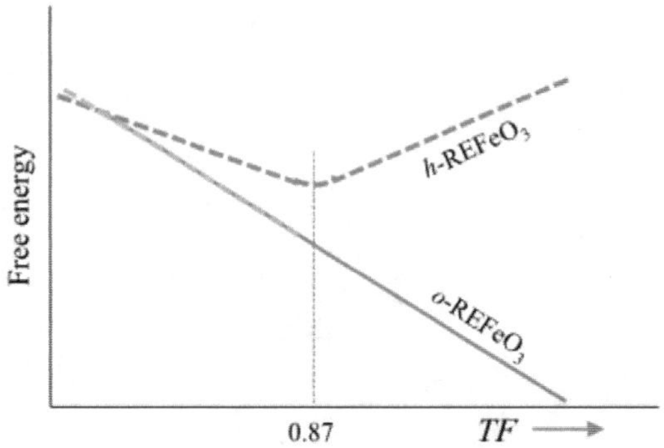

Figure 9. Schematical presentation for stabilities of o-$REFeO_3$ and h-$REFeO_3$ phases. If *TF* deviates from the critical value, the stability of h-$REFeO_3$ phase decreases. However, in the case that $TF < 0.87$, h-$REFeO_3$ phase can be formed, whereas unformed in the contrary case, because the perovskite phase is absolutely stable there.

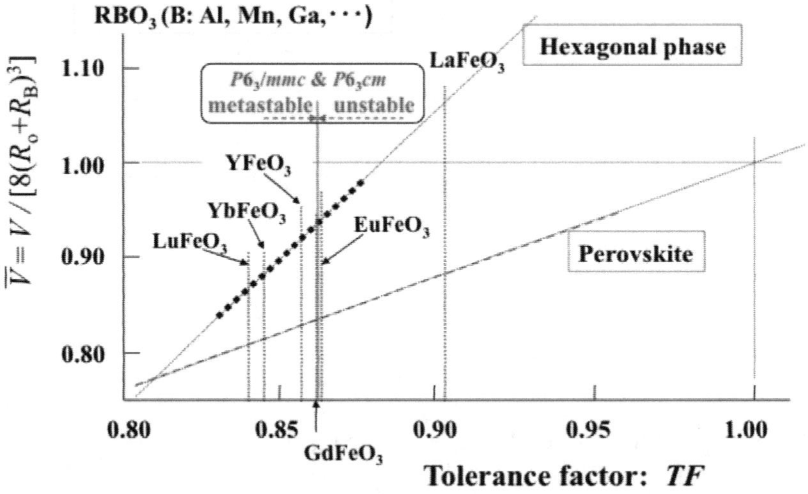

Figure 10. Relation between renormalized volume of unit cell and *TF* for both perovskite and hexagonal series [18].

decreases. However, in the case that $TF < 0.87$, h-$REFeO_3$ can be formed as a metastable phase through a phase transition from $P6_3/mmc$ to $P6_3cm$, provided that the criterion for forming the critical nucleus is fulfilled, whereas unformed in the contrary case, because the perovskite phase is absolutely stable. This result fulfilled the criterion for forming the critical nucleus of the metastable phase. In fact, our previous experiment of real-time measurement of X-ray diffraction during rapid solidification into undercooled melt of $REFeO_3$ (RE: Y and Lu) using the synchrotron orbital radiation (SOR) showed that the diffraction peak peculiar to the $P6_3cm$ phase has never been observed at the recalescence stage, although the solidified phase is evidently $P6_3cm$ [14]. Considering that the definitive difference in XRD patterns of $P6_3cm$ from that of $P6_3/mmc$ is the existence of the diffraction peak with high intensity at d~0.6 nm. Therefore, the initial phase formed during recalescence is expected to be not the $P6_3cm$ phase but the $P6_3/mmc$ phase. The experimental results showed that the h-$GdFeO_3$ phase is formed because TF of which is 0.867. Whereas, the h-$EuFeO_3$ is not formed because of slightly high TF of 0.871. This suggests that the aforementioned condition expressed by Eq. 3 is to be the criterion for the metastable h-$REFeO_3$ phase to be formed.

Decrease of Po_2 extends the range of TF for metastable h-$REFeO_3$ phase to be formed as shown in Fig. 7. Decrease of Po_2 increases the amount of Fe^{2+} (0.078 nm for CN=6) with larger ionic radius than that of Fe^{3+} (0.0645 nm for CN=6), resulting in decrease of TF.

Figure 10 shows the relation between renormalized volume of unit cell and TF for both perovskite and hexagonal series [18]. Nearly linear relationship was obtained. This means that TF remains a relevant parameter for the hexagonal structures.

5. Concluding remarks

Using $REFeO_3$ (RE: rare-earth element) as a model material, containerless solidification for forming a metastable phase from undercooled melts was carried out as a function of Po_2 (oxygen partial pressure). Based on the geometrical consideration on the ionic radii of constituent ion, RE^{3+}, Fe^{3+} and O^{2-}, tit was derived that the criterion for metastable hexagonal phase is also expressed by tolerance factor, TF:

$$TF < 0.87.$$

Experimental result well agreed with this criterion under reduced Po_2 as well as in ambient one.

Acknowledgement

This work was financially supported by a grant-in aid for Scientific Research from the Ministry of Education, Culture, Sports, Science and Technology.

References

[1] Ostwald W 1897 *Z. Phys. Chem.* **22** 282.
[2] Stranski I and Totomanov D 1933 *Z. Phys. Chem.* **A 163** 399.
[3] Kelton K F 1991 *Solid State Physics* **45** 75.
[4] Turnbull D and Cech R E 1950 *J. Appl. Phys.* **21** 804.
[5] Spaepen F 1975 *Acta Metall.* **23** 729.
[6] Spaepen F and Meyer P B 1976 *Scripta Metall.* **10** 257.
[7] Marr D W and Gust A P 1993 *J. Chem. Phys.* **99** 2024.
[8] Kuribayashi K and Ozawa S 2006 *J. Alloy Compd.* **408-412** 266.
[9] Kuribayashi K, Nagashio K, Niwata K, Vijaya Kumar M S and Hibiya T 2007 *Mater. Sci. Eng.* **A 449-451** 675.
[10] Vijaya Kumar M S, Nagashio K, Hibiya T and Kuribayashi K 2008 *J. Am. Ceram. Soc.* **91** 806.
[11] Vijaya Kumar M S, Kuribayashi K and Kitazono K 2009 *J. Am. Ceram. Soc.* **92** 903.
[12] Nagashio K, Takamura Y, Kuribayashi K and Shiohara Y 1999 *J. Crystal Growth* **200** 118.
[13] Nagashio K and Kuribayashi K 2002 *J. Am. Ceram. Soc.* **85** 2550.
[14] Nagashio K, Kuribayashi K, Vijaya Kumar M S, Niwata K, Hibiya T, Mizuno A, Watanabe M and Katayama Y 2006 *Applied Physics Letters* **89** 241923.
[15] Bertaut E F and Mareschal J 1963 *Compt. Rend.* **257** 867.
[16] Yakel H L, Koehler W C, Bertaut E F and Forrat E F 1963 *Acta. Cryst.* **16** 957.
[17] Goldschmidt V M 1926 *Die Gesetze der Krystallochemie* Math -Naturv. Kl. Oslo, Norway, **8** 1.
[18] Bosak A A, Dubourdieu C, Senateur J-P, Gorbenko O Yu and Kaul A R 2002 *Crystal Engineering* **5** 355.

Temperature Dependence of Surface Tension of Molten Iron under Reducing Gas Atmosphere

S Ozawa[1], S Takahashi[1], H Fukuyama[2] and M Watanabe[3]
[1] Department of Aerospace Engineering, Tokyo Metropolitan University, 6-6 Asahigaoka, Hino, Tokyo 191-0065, JAPAN
[2] Institute of Multidisciplinary Research for Advanced Materials (IMRAM), Tohoku University, 2-1-1 Katahira, Aoba-ku, Sendai 980-8577, JAPAN
3 Department of Physics, Gakushuin University, 1-5-1 Mejiro, Toshima-ku, Tokyo 980-8577, JAPAN

E-mail: shumpei.ozawa@tmu.ac.jp

Abstract. Surface tension of molten iron was measured under Ar-He-5vol.%H_2 gas by oscillating droplet method using electromagnetic levitation furnace in consideration of the temperature dependence of oxygen partial pressure, Po_2, of the gas. For comparison, the measurement was carried under Ar-He atmosphere to fix the Po_2 of the inlet gas at 10^{-2}Pa. The surface tension was successfully measured over a wide temperature range of about 780K including undercooling condition. When Po_2 is fixed at 10^{-2} Pa, the surface tension increased and then decreased with increasing temperature like a boomerang shape. When the measurement was carried out under the H_2-containing gas atmosphere, the temperature dependence of the surface tension shows unique kink at around 1810K instead of liner relationship due to competition between the temperature dependence of the Po_2 and that of the equilibrium constant of oxygen adsorption reaction. The relationship between the calculated $\ln K_{ad}$ with respect to inverse temperature using Szyszkowski model was different between the atmospheric gases.

1. Introduction

An accurate surface tension and its temperature dependence are strongly required to improve and optimize various high temperature melt processes involving a free surface such as welding and casting through a numerical calculation because the surface tension difference is a driving force of the Marangoni convection [1]. The Marangoni convection is characterized by the following dimensionless number,

$$Ma = \frac{\partial\sigma}{\partial T}\cdot\frac{\Delta TL}{\mu\alpha} \qquad (1)$$

where σ is the surface tension, T is the temperature, L is the characteristic length, μ is the viscosity, and α is the thermal diffusivity.

---

[1] To whom any correspondence should be addressed.

The surface tension of molten metals has often been measured under a reducing atmosphere such as H_2 containing gas to suppress oxidation of the melt surface [2-5], in which the Po_2 of the ambient atmosphere becomes lower by condensing the H_2O formed from the following reaction,

$$H_2 + \frac{1}{2}O_2 \rightleftarrows H_2O \qquad (2)$$

The Po_2 varies depending on the sample temperature due to the temperature reliance of the equilibrium constant of K_{H_2O} in the reaction (2). However the influence of the temperature dependence of the Po_2 on surface tension of molten metal has not been considered carefully in many cases though surface tension is strongly affected by the Po_2.

In this article, we would like to report the surface tension of molten iron measured by oscillating droplet method using electromagnetic levitation (EML). This paper is intended as an investigation of the effect of the Po_2 and atmospheric gas on surface tension measurement of molten metal.

2. Experimental Procedure

The experimental facility is exhibited elsewhere [5].The high purity of electrolytic iron placed onto a quartz sample holder was positioned in the center of levitation coil. The chemical composition of the iron is depicted in table 1. The sample was electromagnetically levitated and then melted under flow condition (2 L/min) of the Ar-He-5vol.%H_2 mixed gas with a moisture content of 2.66 ppm to lower the Po_2. For comparison, the sample was also melted under flow condition of a mixed gas of high purity commercial argon and helium to fix the Po_2 of 10^{-2}Pa regardless of temperature. The Po_2 of the inlet gas was confirmed by zirconia oxygen sensors operated at 1008K.

Table 1 Chemical composition of iron sample (mass %)

Mn	Ni	Cr	Si	Mo	P
<0.0001	<0.0001	<0.0001	<0.002	<0.0001	<0.001
Cu	Al	C	O	S	Fe
0.0001	<0.0001	<0.0004	0.0013	0.001	Bal.

The oscillation behavior and the temperature of the droplet were monitored from above using a high speed video camera and a single color pyrometer using a half mirror. The temperature of the droplet was controlled by changing the flow rate of atmospheric gas.

The frequencies of the surface oscillation of the $m = 0, \pm1$, and ±2 for the $l = 2$ mode and motion of the center of gravity of two dimensional image were analyzed through fast Fourier transformation (FFT) and maximum entropy method (MEM) from the time-sequential data of the observed images in consideration of the influence of two types of droplet rotations, i.e. real rotation and apparent rotation [6]. The surface tension of the molten iron was calculated from these frequencies by the following equations in which the Rayleigh equation [7] is calibrated by Cummings and Blackburn [8],

$$\sigma = \frac{3}{8}\pi M\left[\frac{1}{5}\sum_{m=-2}^{2} \nu_{2,m}^2 - \nu_t^2\left\{1.9 - 1.2\left(\frac{z_0}{a}\right)^2\right\}\right], \qquad (3)$$

$$z_0 = \frac{g}{8\pi^2\nu_t^2}, \qquad (4)$$

$$a = \sqrt[3]{\frac{3M}{4\rho\pi}}, \qquad (5)$$

where v_m is the surface oscillation, M is the sample mass, ρ is the density, and g is the gravitational acceleration. The density of the molten iron was determined from data reported by Beutel et al. [9],

3. Results and discussions

When the Ar-He-5vol.%H_2 gas was used for the surface tension measurement, it is important to consider that the Po_2 varies depending on temperature because the following equilibrium constant of reaction (2) shows temperature dependence.

$$K_{H_2O} = \frac{P_{H_2O}}{P_{H_2} \cdot P_{O_2}} \quad (6)$$

The Po_2 of the Ar-He-5vol.%H_2 gas was evaluated as a function of temperature using the following standard Gibbs energy of formation of H_2O and the Po_2 of 2.0×10^{-23}Pa of the inlet gas measured by the oxygen sensor maintained at 1008K,

$$\Delta G^{\circ} = -246535 + 54.94T \text{ [J·mol}^{-1}\text{]}. \quad (3)$$

The result is exhibited in figure 1. This confirms that Po_2 of the Ar-He-5 vol.%H_2 gas increases with rising temperature.

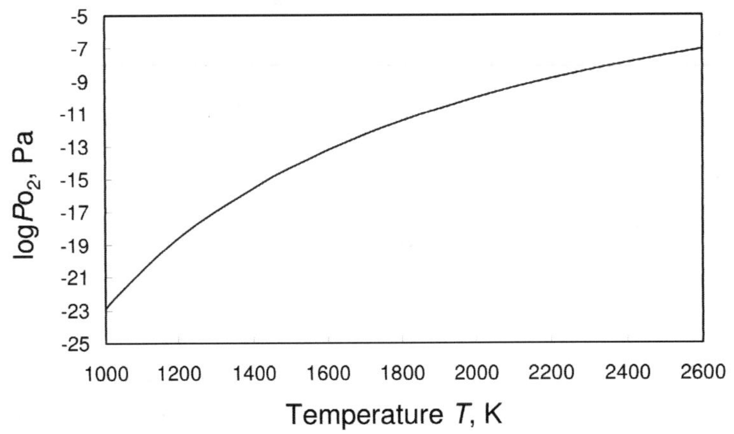

Figure 1. Temperature dependence of oxygen partial pressure (Po_2) of Ar-He-5vol.%H_2 gas evaluated through the standard Gibbs energy of formation of H_2O

Figure 2 depicts the surface tension of molten iron measured by oscillating droplet method using EML under the flow condition of mixed gases of Ar-He-5vol.%H_2 (□) and Ar-He (◆). Uncertainty for the measurement plot was evaluated based on the GUM (ISO Guide to the Expression of Uncertainty in Measurement) [10] using the coverage factor of K_p=2. The qualitative magnitude of the equilibrium constant of oxygen adsorption reaction, K_{ad}, is expressed by a color gradient of the background in this figure. Surface tension of molten iron was successfully measured over a wide temperature range of 780K including undercooling state regardless of the atmosphere gas.

When Po_2 is controlled at 10^{-2} Pa, the FeO is formed in solid state below 1650K and in liquid state above this temperature from simple thermodynamic calculations using the Gibbs energy of FeO formation [11]. However, no appreciable FeO was detected in the droplet images and temperature profile by high speed camera and pyrometer. If FeO is formed on the droplet, the droplet image becomes bright due to its higher emissivity, and it can be detected in the time sequential data of droplet image and emissivity. Since Fe-O system has a miscibility gap in the liquid state between Fe and FeO, the liquid FeO may be mixed as an immiscible state inside the droplet. The strong

Journal of Physics: Conference Series **327** (2011) 012020 doi:10.1088/1742-6596/327/1/012020

electromagnetic stirring would promote the mixing of the electromagnetically levitated droplet. Even in this case, apparent emissivity of the droplet becomes higher than that of the liquid iron. Furthermore, the apparent emissivity of the droplet image should be enhanced because the volume fraction of liquid FeO increases in the droplet with time.

Since the small amount of evaporation for the iron droplet was observed during the containerless levitation of the droplet, a metallic iron vapor would work as an oxygen getter pump in the vicinity of the melt surface as is the case for molten tin [12, 13].

When the Po_2 of the measurement atmosphere is fixed at 10^{-2}Pa regardless of sample temperature using high purity Ar-He gas, boomerang shape temperature dependence of surface tension is observed; the surface tension increases as the sample temperature rises up to about 2150K, and then it decreases above this temperature. When Po_2 is high during surface tension measurement, oxygen adsorption induces a significant decrease of the surface tension. However it is desorbed from the melt surface as the sample temperature rises because the equilibrium constant of oxygen adsorption reaction becomes small. As a result surface tension temperature coefficient becomes positive value. When the Po2 becomes less effective at about 2150K, the surface tension temperature coefficient is changed from a positive value to a negative value.

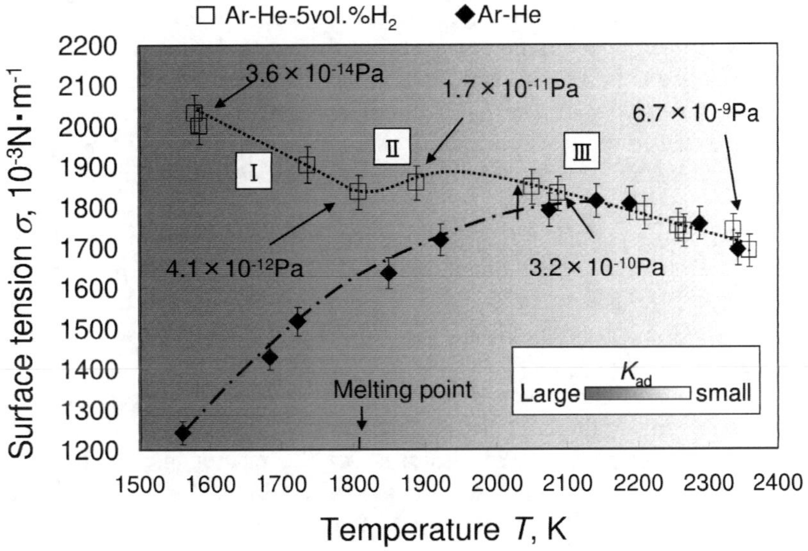

Figure 2. The surface tension of molten iron measured by oscillating droplet method using EML under the Ar-He-5vol.%H_2 and Ar-He gases.

When the measurement is carried out under the Ar-He-5vol.%H_2 gas atmosphere, the temperature dependence of the surface tension shows a peculiar kink at around 1810K; the surface tension decreases and then increases once when the sample temperature rises up to 1950K. When the sample temperature exceeds about 1950K, the surface tension decreases with increasing temperature again. Since the variation of the surface tension at this kink is beyond the uncertainty of the measurement, it is not the scattering of the measurement originating from the measurement accuracy. This unique kink in the temperature dependence of the surface tension can be explained when the temperature dependence of the Po_2 under the Ar-He-5vol.%H_2 gas is considered [5]. The Po_2 increases from 3.6×10^{-14}Pa to 4.1×10^{-12}Pa as the sample temperature rises from 1580K to 1810K (see figure 1). Higher Po_2 usually induce a lower surface tension of molten metal due to oxygen adsorption as observed at the region I.

Note that equilibrium constant of oxygen adsorption reaction becomes small due to the temperature elevation simultaneously. Therefore there is competition between the temperature dependence of Po_2 and that of equilibrium constant of oxygen adsorption reaction to decide the amount of the oxygen adsorption. If the amount of oxygen adsorption becomes small due to the temperature elevation even at high Po_2, the surface tension of molten iron would increase to approach the surface tension at pure state as observed at region II. Eventually pure surface tension is observed at high temperature as exhibited at region III because the Po_2 becomes less effective.

von Szyszkowski empirically derived the following equation about the influence of the surfactant on surface tension of a dilute solution [14],

$$\sigma = \sigma^{P} - RT\Gamma_{O}^{sat}\ln(1 + K_{ad}a_{O}) \quad (7)$$

where σ^P is the pure surface tension without any contamination [N· m^{-1}], σ is the surface tension contaminated by surfactant, R is the gas constant [J· K^{-1}· mol^{-1}], T is the temperature [K], Γ_O^{sat} is the full-coverage oxygen adsorption at the liquid surface [mol· m^{-2}], and a_O is the activity of oxygen in the liquid. This equation is theoretically formulated by March and Tosi from the combination of Langmuir and Gibbs adsorption isotherms [15]. Belton also derived this equation for the surface tension of liquid Fe-S system theoretically [16]. Our group successfully expressed the surface tension of molten silver as functions of both Po_2 and temperature using this model. When the experimental results are applied to this model, the $\ln K_{ad}$ can be calculated as a function of inverse temperature. The validity of the measurement result was confirmed by the second-law method using the equation (7). The enthalpy of the oxygen adsorption reaction on the melt surface is considered as constant within the short temperature range, which can be calculated from the slop of the $\ln K_{ad}$ and inverse temperature (van't Hoff plot). If a good linearity of the $\ln K_{ad}$ with respect to the $1/T$ is confirmed, the validity of the measurement result is confirmed. The result is shown in figure 3. Although a good linearity of the $\ln K_{ad}$ with respect to the $1/T$ is confirmed, the absolute values of the $\ln K_{ad}$ deviate between the atmospheric gases used in the measurement. As mentioned in the previous section, metallic iron vapor would act as an oxygen getter pump in the vicinity of the melt surface as is the case for molten tin [12, 13]. In this case, the calculated $\ln K_{ad}$ would exhibit the discrepancy between the atmospheric gases because the Po_2 in the vicinity of the melt surface becomes lower than that of the bulk gas used in the calculation of $\ln K_{ad}$. Furthermore, another possibility of the discrepancy in the calculated $\ln K_{ad}$ is oxygen adsorption reaction under H_2-containing gas is different from that under Ar-He gas. A continuous investigation of the influence of oxygen partial pressure of atmospheric gas on surface tension of molten metals would clarify the discrepancy of the calculated K_{ad} between the atmospheric gases.

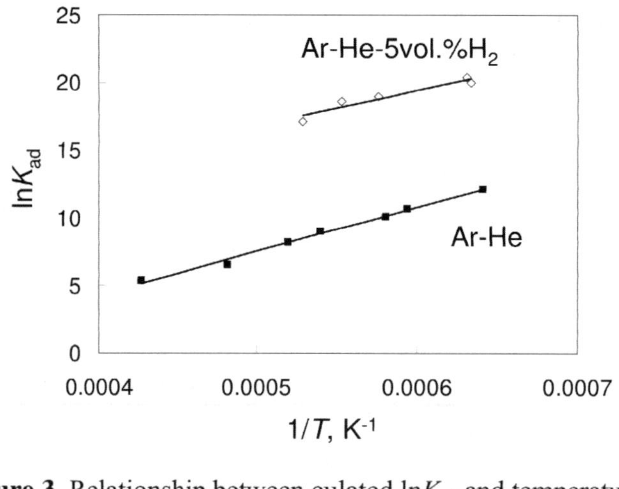

Figure 3. Relationship between culated $\ln K_{ad}$ and temperature

4. Conclusion

The surface tension of molten iron was measured over a very wide temperature range of over 780K under Ar-He and Ae-He-5vol.%H_2 gas atmosphere by an oscillating droplet method using EML. When the Po_2 is fixed at 10^{-2}Pa using high purity Ar-He gas, the boomerang shape temperature dependence of surface tension was observed. When the measurement was carried out under Ae-He-5vol.%H_2 gas atmosphere, the temperature dependence of surface tension exhibited a peculiar kink at around 1810K instead of a usual liner relationship due to the competition between the temperature dependence of Po_2 and that of equilibrium constant of oxygen adsorption reaction. Although the $\ln K_{ad}$ showed a good liner relationship against $1/T$, the absolute values of it was different between the atmospheric gases.

Acknowledgments

This research was supported by Grant-in-Aids of Japanese Science and Technology (JST) Agency. This work was partially supported by the Sasakawa Scientific Research Grant from The Japan Science Society. We also thank Prof. T. Hibiya (Keio Univ.), Prof. K. Mukai (Kyushu Inst. Tech.) for their helpful discussions.

References

[1] C. Marangoni, *Il Nuovo Cimento Series* **3** (1878) 97.
[2] G. Wille, F. Millot and J. Rifflet, *Inter. J. Thermophysics* 23, 1197 (2002).
[3] J. Brillo and I. Egry, *Journal of Mater. Sci.* **40** (2005) 2213.
[4] H.-K. Lee, M. G. Frohberg and J. P. Hajra, *Steel Research* **64** (1993) 191.
[5] S. Ozawa, S. Suzuki, T. Hibiya, and H. Fukuyama: *J Appl Phys.*, **109** (2010) 014902.
[6] S. Ozawa, T. Koda, M. Adachi, K. Morohoshi, M. Watanabe and T. Hibiya, *J. Appl. Phys.* **106** (2009) 034907.
[7] Load Rayleigh, *Proceedings of the Royal Society of London* **29** (1879) 71.
[8] D. L. Cummings and D. A. Blackburn, *J. Fluid Mech.* **224** (1991) 395.
[9] M. Beutl, G. Pottlacher and H. Jager, *Inter. J. Thermophysics* **15** (1994) 1323.
[10] Guide to the Expression of Uncertainty in Measurement, (International Organization for Standardization, Geneva, 1995).
[11] NIST-JANAF Thermochemical Tables, Fourth Edition, Part II, JANAF edited by Malcolm W. Jr. Chase, (The American Chemical Society, Washington, DC and the American Institute of Physics, New York, 1998) p. 1238.
[12] Z. Yuan, K. Mukai, K. Takagi, and M. Ohtaka, *J. Japan Inst. Metals* **65** (2001) 21.
[13] E. Ricci, E. Arato, A. Passerone and P. Costa, *Advances in Colloid and Interface Science* **117** (2005) 15.
[14] B. von Szyszkowski, *Z. Phys. Chem.*, **64** (1908) 385.
[15] N. H. March and M. P. Tosi, Atomic Dynamics in Liquids, (MacMillan, London, 1976), p.275.
[16] G. R. Belton, *Metall. Trans. B*, **7B** (1976) 35.

Journal of Physics: Conference Series **327** (2011) 012021 doi:10.1088/1742-6596/327/1/012021

The effect of rotation on resonant frequency of interfacial oscillation of a droplet using electrostatic levitator

Rui Tanaka[1], Satoshi Matsumoto[2], Akiko Kaneko[1] and Yutaka Abe[1]

[1]Department of Engineering & Mechanics Energy, University of Tsukuba, 1-1-1, Tennoudai, Tsukuba, Ibaraki, 305-8573, Japan
[2]Institute of Space and Astronautical Science, Japan Aerospace Exploration Agency, 2-1-1, Sengen, Tsukuba, Ibaraki, 305-8505, Japan

E-mail : rtanaka@edu.esys.tsukuba.ac.jp

Abstract. Under the microgravity environment, liquid can be treated without container due to weightlessness. Therefore physical properties of a material are expected to be measured with a high accuracy. In surface tension measurement of levitated droplet, it is calculated by using the formula derived under the assumption of linearity of the resonant frequency. However, it is indicated that the resonant frequency of a droplet changes due to oscillation and rotation. And mechanism of frequency shift has not been revealed yet. The final goal of this study is to make clear the mechanism of frequency shift. A droplet is levitated by using electrostatic levitator, and induced to oscillate and rotate. Resonant frequency, amplitude, angular velocity and aspect ratio are measured from droplet motion. It is confirmed oscillation and rotation effect on resonant frequency. And the frequency shift becomes zero under the appropriate combination of amplitude and angular velocity. It is experimentally clarified that the resonant frequency is strongly related with time average of aspect ratio.

1. Introduction

It has become possible to conduct various experiments under the microgravity environment in space. Under the microgravity condition, the buoyant convection and the phase separation of mixing liquid due to density difference does not occur. So, it is expected to establish the new methods of manufacturing new materials and of measuring thermo-physical properties of molten materials utilizing the microgravity condition. Understanding the nature of a levitating liquid drop is indispensable for the containerless processing. Recently, there are a lot of studies with levitating technique under gravity environment on the ground. For example, it has become possible to measure thermo-physical properties of extreme high-temperature molten materials exceeding 3000 K with the electrostatic levitation technique [1-3]. On the measurement of physical properties, the axisymmetric oscillation is applied to a levitating drop, and the surface tension is calculated from the resonant frequency by the oscillating drop method derived by Rayleigh [4]. In this method, the surface tension is calculated by equation (1):

$$\sigma = \frac{\rho R_0^3 f_0^{\ 2}}{8} \qquad (1)$$

where σ, R_0, f_0 are surface tension, droplet radius, initial resonant frequency, respectively. However, the equation was derived under linear approximations; inviscid, very small deformation only in the radius direction, potential flow. Therefore, there is a possibility that a linear theory is not applicable in the case of larger deformation [5]. Figure 1 shows the surface tension measurement of propylene carbonate by the oscillating drop method. Difference between actual measurement value 42.6 mN/m and calculated surface tension is small in small oscillation amplitude. But the difference become large with increase of amplitude. On the measurement of thermophysical properties, the horizontal rotation often has to be applied for stable levitation of sample. However, it is indicated the resonant frequency change due to rotation [6]. Watanabe [7] analyzed the oscillation behavior of rotating droplet numerically. He examined how oscillation amplitude, rotational velocity, and aspect ratio of droplet effect on resonant frequency, respectively (see Figures 2 and 3). It was indicated that resonant frequency is not suffered from the frequency shift if the time-averaged aspect ratio of droplet was maintained unity. However, mechanism of frequency shift has not been revealed yet, and there are only a few experimental data for oscillation of rotating levitated droplet.

The final goal of this study is to make clear the mechanism of frequency shift. In this paper, it is examined the effect of oscillation and rotation on resonant frequency and droplet shape.

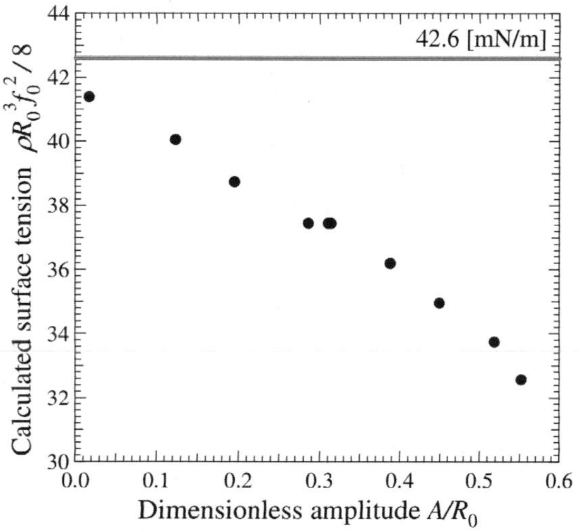

Figure 1. Effect of oscillation amplitude on surface tension measurement.

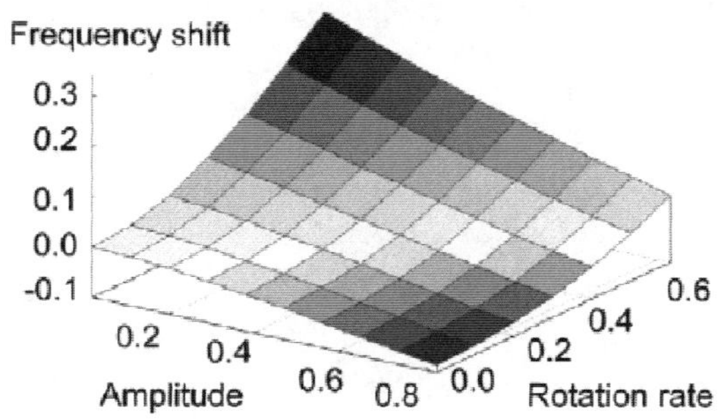

Figure 2. Dependency of frequency shift on amplitude and rotational velocity (Numerical analysis by Watanabe [7]).

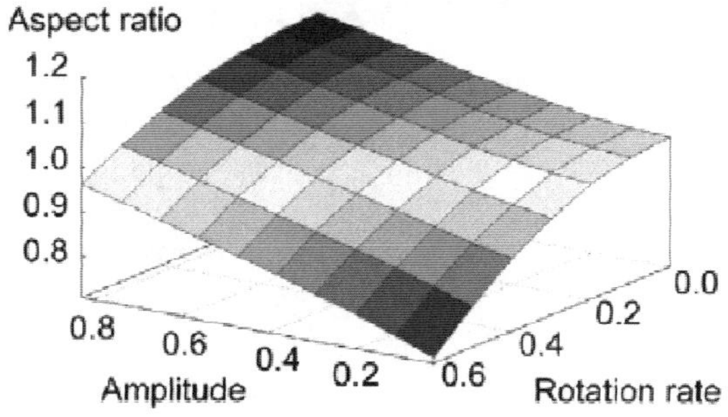

Figure 3. Dependency of aspect ratio on amplitude and rotational velocity (Numerical analysis by Watanabe [7]).

2. Experimental approach

Figure 4 shows the conceptual diagram of the electrostatic levitation. Electrostatic levitation apparatus is composed of a pair of cylindrical electrodes, high voltage amplifiers, He-Ne laser and photo detector used to control the droplet position. At first, the sample is placed on the tip of needle located in the bottom electrode. Secondly, positive voltage is applied to the bottom electrode, then the surface of sample charges positively. Thirdly, negative voltage is applied to the upper electrode. So, the electrostatic force lifts up the charged sample in the opposite direction of gravity. In this method, there is no restoring force on the levitating droplet; hence it is necessary to control the sample position. It is performed by the detecting sample position with photo detector, and by adjusting the voltage of the upper electrode through the PID algorithm.

The experimental apparatus in the present study is shown in figures 5 and 6. Experiments were performed in normal pressure at room temperature. The test fluids are shown in table 1. Propylene carbonate and Ethylene glycol are safe liquids, low viscosity and possible to levitate. Droplets of form 1.8 to 2.6 mm in diameter were levitated in the air. The voltage of 2 up to 7 kV was applied to the upper electrode, which was high-speed-feedback-controlled via PID algorithm. It changed due to levitating droplet size and droplet surface charge.

The deformation behaviors of the levitating liquid droplet were observed by using high speed video camera and charge-coupled device camera set on the side of the acrylic chamber. The vertical length of the levitating droplet was measured by a digital measurement system. The resonant frequency was calculated by the spectrum analysis from time series data of the vertical length. However, it is necessary to excite a droplet oscillation for determining the resonance frequency. In the present study, the oscillation in a perpendicular direction was excited by applying sinusoidal voltage to the bottom electrode with a voltage amplifier controlled by a function generator. The amplitude of the droplet oscillation was varied by changing the applied voltage.

The method to induce the rotation of levitating droplet was based on the theoretical prediction by Busse & Wang [8]. In this method, the rotation torque was generated by two orthogonal acoustic standing waves. In the present study, these waves were generated with two speakers set up on the side of the chamber. The phase difference between two waves was set $\pi/4$ rad, so acoustic torques was generated in the chamber and rotation in the horizontal direction was excited to a levitating droplet. The angular velocity of the droplet can be changed by adjusting the sound pressure level. The angular velocity of levitating droplet is measured by taking the motion of immiscible fine tracer particle with high speed video camera.

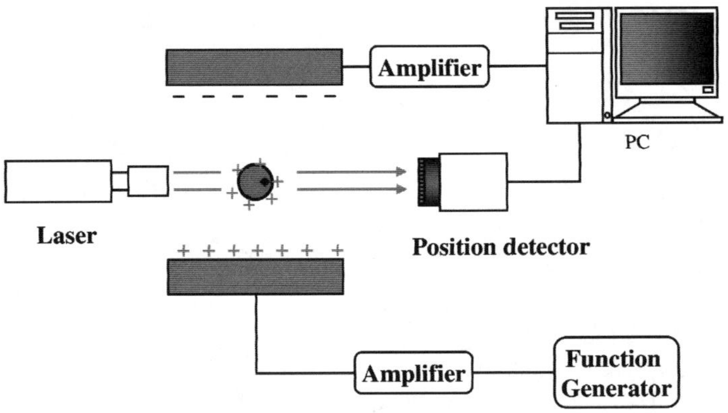

Figure 4. Conceptual diagram of the electrostatic levitation.

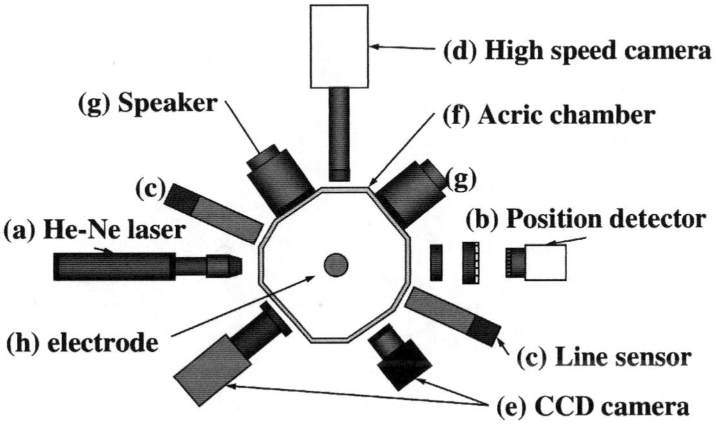

Figure 5. Schematic of Experimental apparatus (Top view).

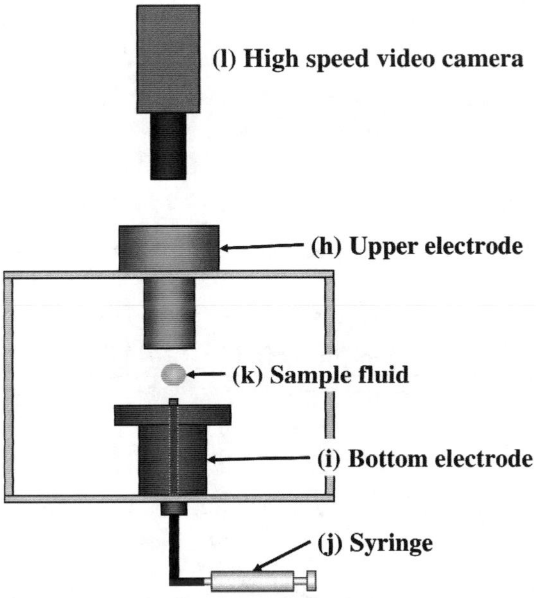

Figure 6. Schematic of Experimental apparatus (Side view).

Table 1. Physical properties of test fluids.

	Density ρ [kg/m^3]	Surface tension σ [mN/m]	viscosity μ [mPa·s]	Kinematic viscosity ν [mm^2/s]
Propylene carbonate	1205.7	42.6	2.77	2.30
Ethylene glycol	1113.2	48.4	22.8	20.5

3. Experimental Results

3.1 Oscillation of levitated droplet

Oscillation of levitated droplet is shown in figures 7 and 8. The droplet was propylene carbonate, 1.82 mm in diameter, and record speed was 2250 fps. T and A are oscillation period and amplitude, respectively. A/R_0 is dimensionless amplitude divided by droplet radius R_0. The droplet shape deformed widely with increase of amplitude. Time series of dimensionless amplitude of droplet are shown in figure 9. The oscillation period of $A/R_0 \approx 0.5$ is longer than that of $A/R_0 \approx 0.1$, it is confirmed that the resonant frequency decreases due to increase of amplitude. Examination result of resonant frequency shift of both propylene carbonate and ethylene glycol with increase of amplitude is shows in figure 10. The frequency shift is defined as follow:

$$\frac{\delta f}{f_0} = \frac{f - f_0}{f_0} \qquad (2)$$

where f is the resonant frequency with certain amplitude, f_0 is initial resonant frequency when only little oscillation is given. The line in figure 10 is result of numerical analysis by Tsamopoulos and Brown [5]. They analyzed 2nd mode axial oscillation of inviscid and incompressible droplet. The frequency shift decreased with increasing of amplitude, experimental result corresponded to numerical one by Tsamopoulos et al. qualitatively.

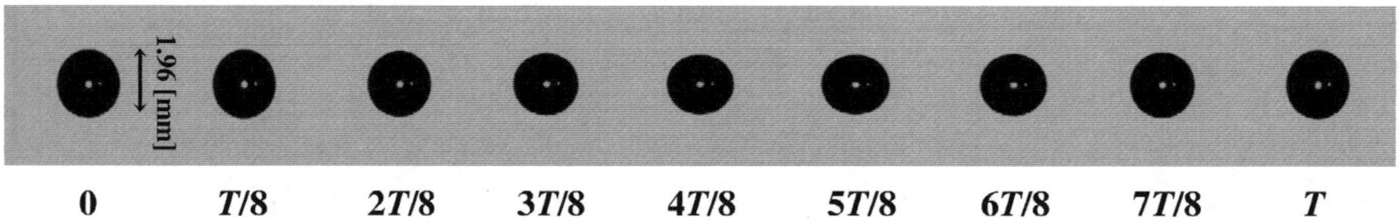

Figure 7. Observation result of droplet shape deformation in one period T ($A/R_0 \approx 0.1$).

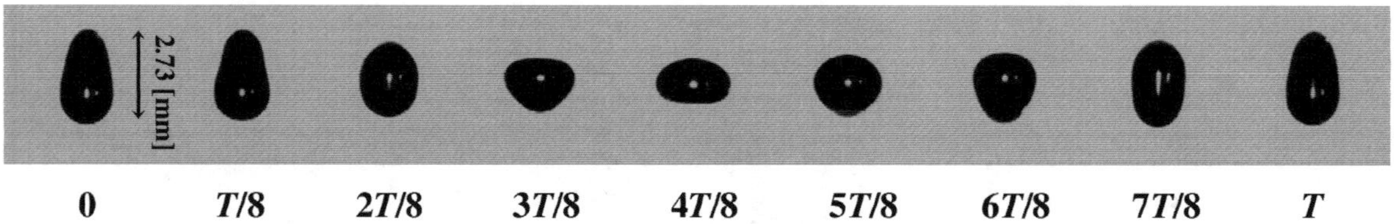

Figure 8. Observation result of droplet shape deformation in one period T ($A/R_0 \approx 0.5$).

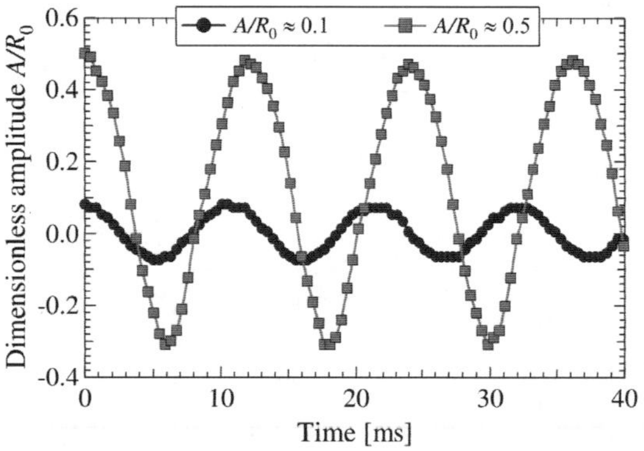

Figure 9. Time series of dimensionless amplitude of droplet.

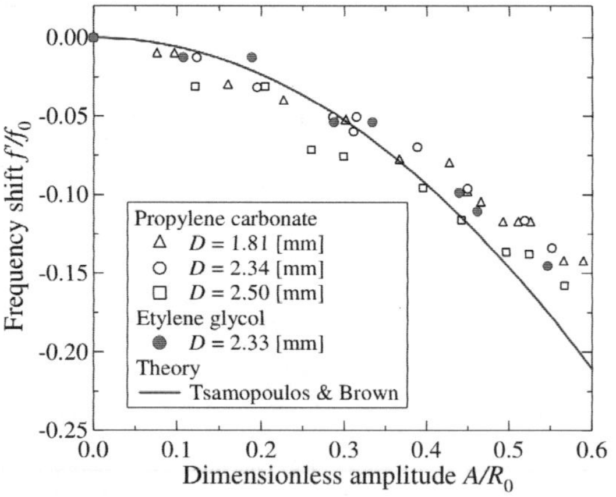

Figure 10. Effect of oscillation on frequency shift.

3.2 Oscillation of levitated droplet with rotation

Oscillation of rotating levitated droplet is shown in figures 11 and 12. The droplet was propylene carbonate of 2.15 mm diameter, and record speed was 3000 fps. Ω and Ω/ω_0 is angular velocity and dimensionless angular velocity divided by initial resonant frequency ω_0 ($=2\pi f_0$), respectively. Figures 11 and 12 show rotation of $\Omega/\omega_0 \approx 0.1$, 0.4 with oscillation of $A/R_0 \approx 0.1$, respectively. The droplet shape did not deform so much in vertical direction compared with in horizontal direction because the centrifugal force pulled droplet outward. Time series of dimensionless amplitude of rotating droplet are shown in figure 13. The oscillation period of $\Omega/\omega_0 \approx 0.4$ is shorter than that of $\Omega/\omega_0 \approx 0.1$, it is confirmed that the resonant frequency increases due to increase of angular velocity. Examination result of resonant frequency shift with increase of angular velocity is shows in figure 14. The line is result by Busse [8]. He analyzed 2 mode axial oscillation of inviscid and incompressible rotating droplet. The frequency shift increased due to increase of angular velocity, experimental results agreed with result by Busse qualitatively.

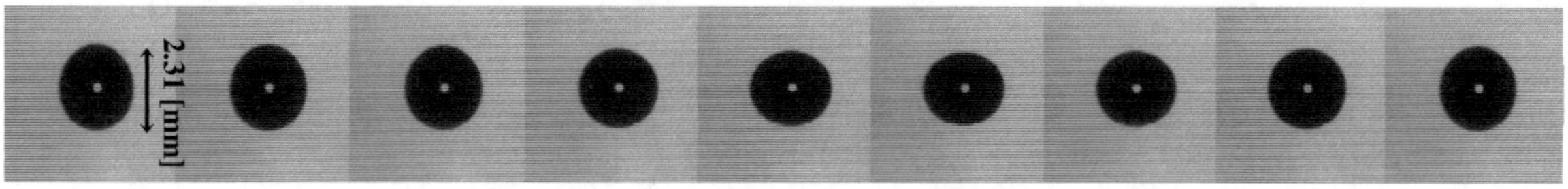

0 ***T*/8** **2*T*/8** **3*T*/8** **4*T*/8** **5*T*/8** **6*T*/8** **7*T*/8** ***T***

Figure 11. Observation result of droplet shape deformation in one period T ($\Omega/\omega_0 \approx 0.1$).

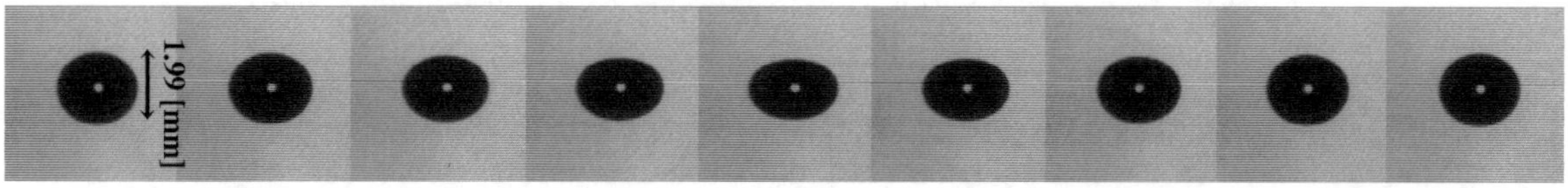

0 ***T*/8** **2*T*/8** **3*T*/8** **4*T*/8** **5*T*/8** **6*T*/8** **7*T*/8** ***T***

Figure 12. Observation result of droplet shape deformation in one period T ($\Omega/\omega_0 \approx 0.4$).

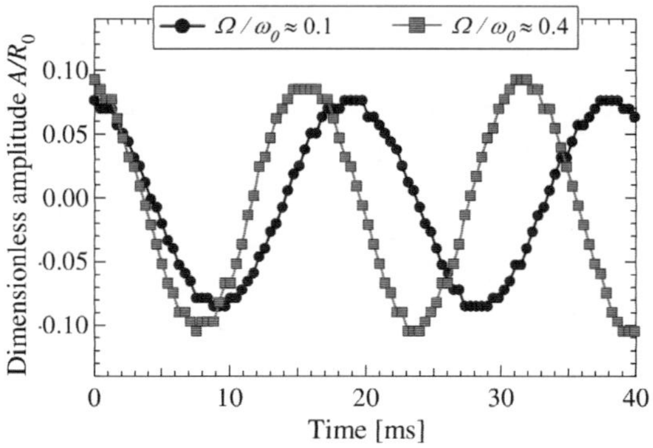

Figure 13. Time series of dimensionless amplitude of rotational droplet.

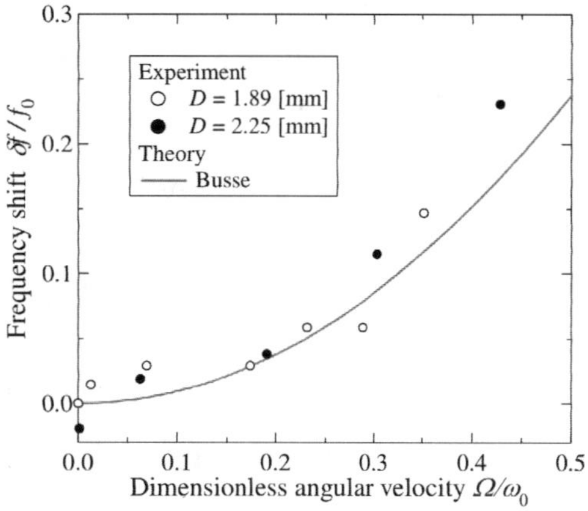

Figure 14. Effect of rotation on frequency shift.

3.3 Effect of oscillation amplitude and angular velocity on resonant frequency

Figure 15 shows experimental result of frequency shift of oscillation droplet with rotation. Frequency decreased with increasing amplitude and with decreasing angular velocity. Hence, it is confirmed that the effect of amplitude and rotation on frequency shift is opposite in this experiment in common with numerical analysis (Figure 2).

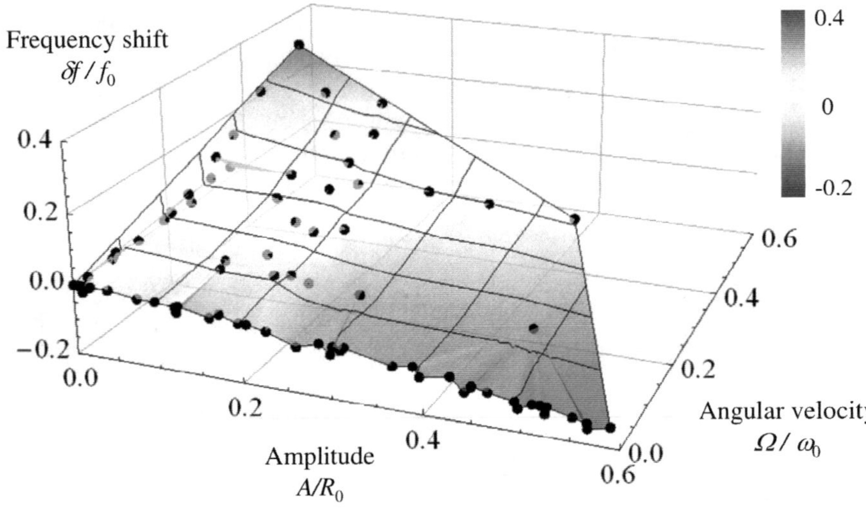

Figure 15. Effects of amplitude and angular velocity on frequency shift.

The surface tension which is estimated from frequency is shown in figure 16. It is confirmed that there are measurement errors due to frequency shift. Measurement errors of calculated surface tension, which is in $\Omega / \omega_0 \approx 0.0, 0.1, 0.2, 0.3, 0.4$, is shown in figure 17, and the line is actual measured value. It is shown that the frequency shift directly effects on accuracy of surface tension measurement. Relation between oscillation and rotation makes calculated surface tension be within ±5 % error is in figure 18. Taking into account of effect of both amplitude and rotation, the appropriate combination which makes frequency shift zero would exist. This might contribute to improve accuracy of measurement.

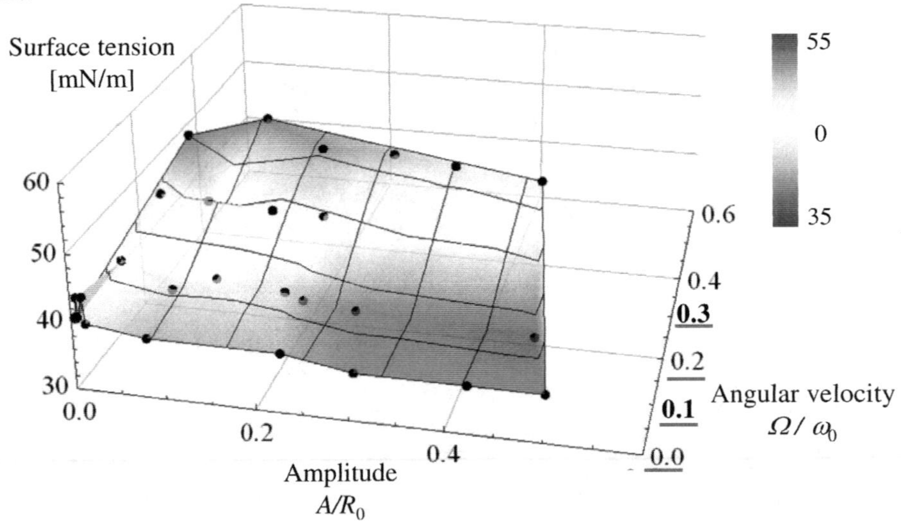

Figure 16. Effects of amplitude and angular velocity on surface tension.

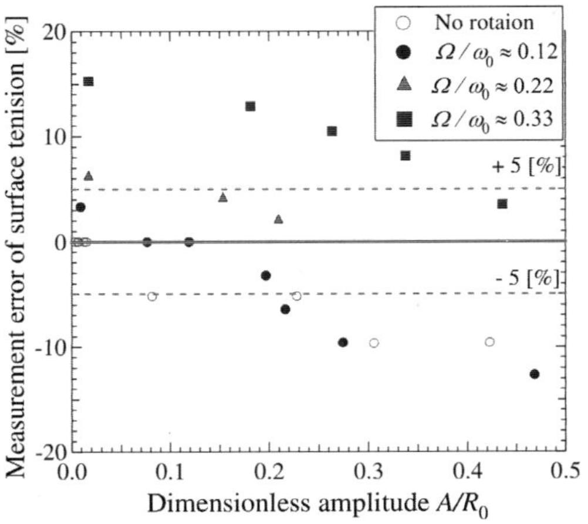

Figure 17. Measurement error of calculated surface tension.

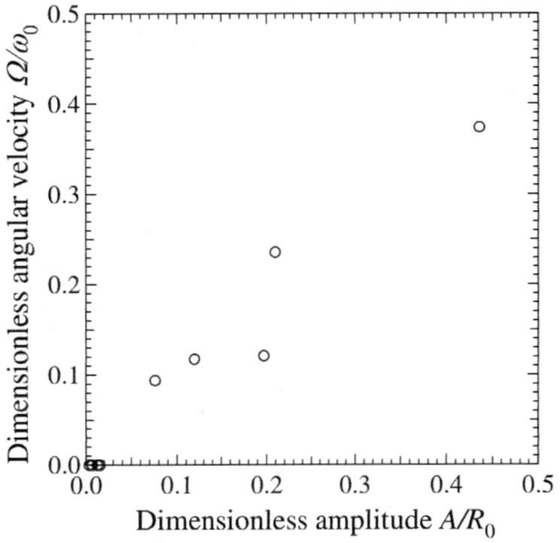

Figure 18. Relation between amplitude and angular velocity within ±5 % of measurement error of surface tension.

3.4 Relation between droplet shape and frequency shift

Figure 19 shows time series of aspect ratio of oscillating droplet without rotation. Aspect ratio is defined as vertical length b divided by horizontal length a. The line indicates aspect ratio of unity which means spherical shape. Aspect ratio greater than 1.0 is appeared for long time when dimensionless amplitude A/R_0 is 0.5. Figure 20 shows time series of aspect ratio of rotating droplet with dimensionless amplitude $A/R_0 \approx 0.1$. Aspect ratio became smaller than 1.0 due to increase of angular velocity. Figure 21 shows percentage time spent in prolate shape of oscillating and rotating droplet, and the line is numerical analysis result of non-rotating droplet by Tsamopoulos & Brown. The prolate shape indicates aspect ratio is grater than 1.0. Time spent in prolate shape increased when amplitude became larger. In the case of rotating droplet, time spent in prolate shape is smaller than rotating droplet. Thus it is possible time spent in prolate shape shift has something to do with frequency shift. Figure 21 shows time-averaged aspect ratio of oscillating droplet with no rotation and rotation. Averaged aspect ratio As is calculated from average of three oscillation cylcles. Average aspect ratio increased with increasing amplitude. In contrast, increase of angular velocity made average aspect ratio decrease. Figure 22 shows aspect ratio of oscillation droplet with no rotation and rotation. From Figures 15 and 22, relation between average aspect ratio and frequency shift is in figure 23. It is suggested that average aspect ratio correlates with frequency shift.

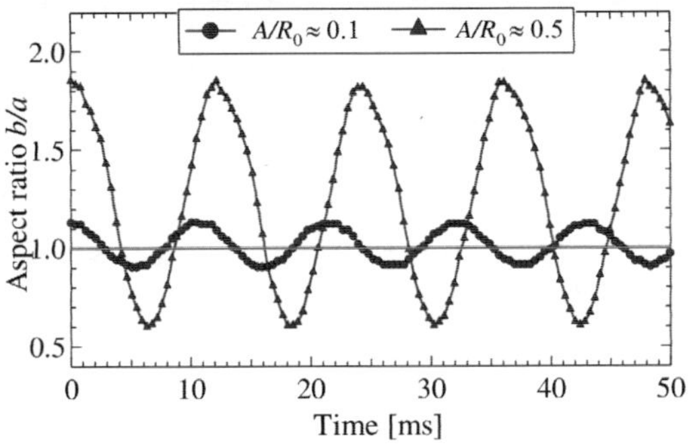

Figure 19. Time series of aspect ratio of droplet.

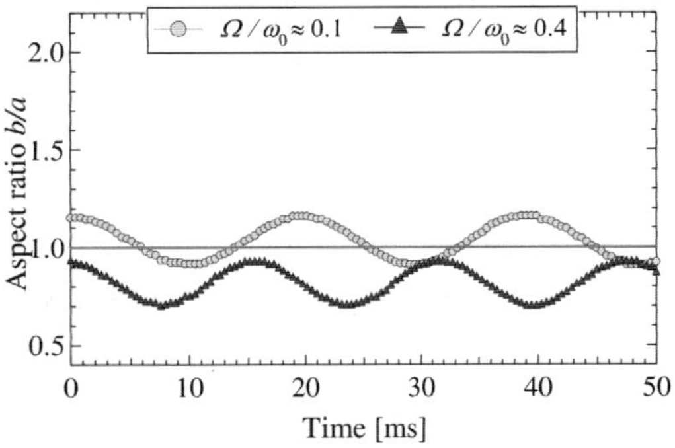

Figure 20. Time series of aspect ratio of rotational droplet.

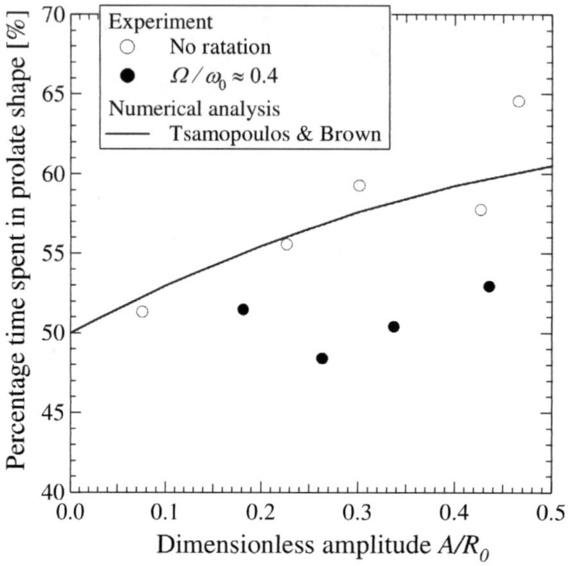

Figure 21. Percentage time spent in prolate shape of oscillating and rotating droplet.

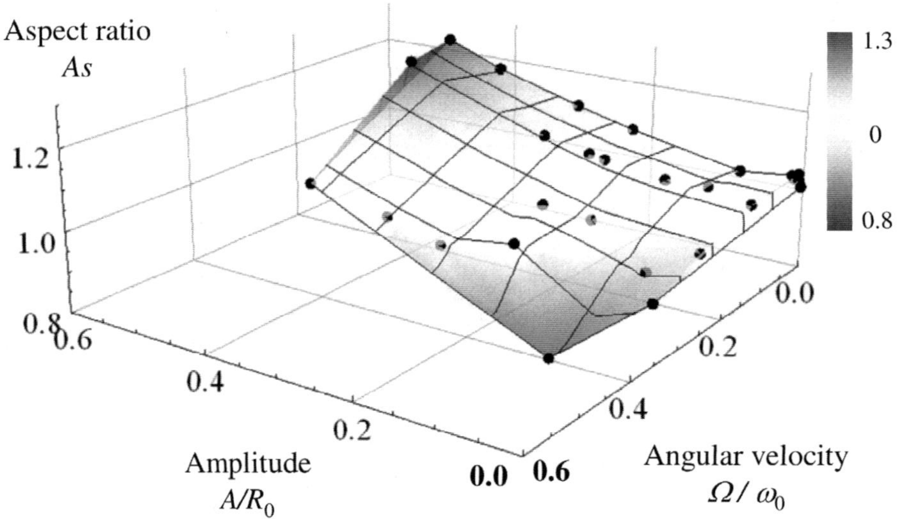

Figure 22. Effects of amplitude and angular velocity on aspect ratio.

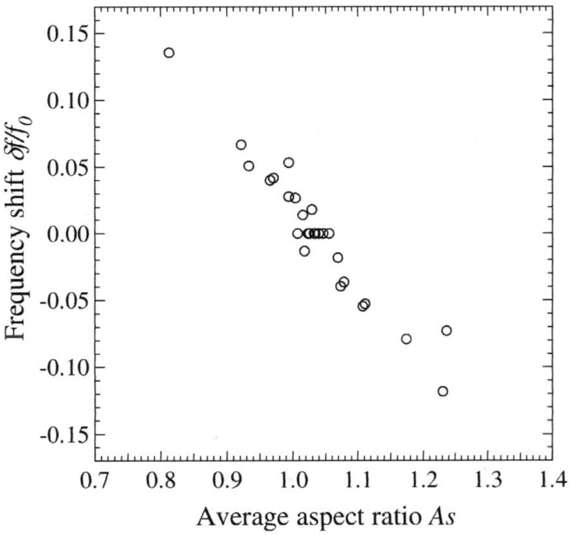

Figure 23. Relation between average aspect ratio and frequency shift.

4. Conclusion

We observed dynamics of levitated droplet by using electrostatic levitator. Behaviour of droplet shape was analysed when both oscillation amplitude and angular velocity changed. The concluding remarks are obtained as follows:

- The resonant frequency of levitated droplet was measured. It is confirmed that the resonant frequency decreases due to increase of amplitude, increases due to increase of angular velocity.
- It is suggested appropriate combination of oscillation and rotation makes frequency shift 0 in experiment.
- It is suggested that average aspect ratio correlates with frequency shift in experiment.

References

[1] Ishikawa T, Paradis P F and Yoda S 2004 *Applied Physics Letters* **85** 5866
[2] Paradis P F; Ishikawa T, Fujii, R, Yoda S 2005 *Applied Physics Letters* **86** 41901
[3] Ishikawa T, Okada J T, Paradis P F and Watanabe Y 2010 Int. *J. Thermophysics* **31** 388
[4] Rayleigh L 1882 *Phil. Mag.* **14** 184
[5] Tsamopoulos J A, Brown R A 1983 *J. Fluid Mech.* **12** 7519
[6] Busse F H 1984 *J. Fluid Mech.* **142** 373
[7] Watanabe T 2009 *Physics Letters* **A 373** 867
[8] Busse H, and Wang T G 1981 *J.Acoust.Soc.Am.* **69** 1632

Acknowledgement

The authors would like to thank Dr. T. Watanabe from Japan Atomic Energy, and Prof. K. Nishinari from of University of Tokyo for their helpful discussion.

Journal of Physics: Conference Series **327** (2011) 012022 doi:10.1088/1742-6596/327/1/012022

The viscosity of eutectic Pd-Si alloys

I. Egry

Institut für Materialphysik im Weltraum, German Aerospace Center, DLR, 51170 Köln, Germany

E-mail: ivan.egry@dlr.de

Abstract. The viscosity of eutectic PdSi alloys has been measured by different techniques. The conventional oscillating cup viscosimetry was used in the ground-based laboratory, while the oscillating drop technique was applied in a microgravity experiment. Different working equations were utilised to derive the viscosity from the damping of the oscillations, and the results thus obtained were compared to each other and available literature data. Whereas systematic deviations exist, an overall satisfactory agreement was found.

1. Introduction

The viscosity η of a liquid is one of its most fundamental thermophysical properties, both from an academic and an engineering point of view. It determines mass transport in fluids, governs the onset of convection and affects to a large extent the glass forming ability of a given alloy. Viscous flow is an activated process, and, therefore, the temperature dependence of the viscosity is often given in an Arrhenius form:

$$\eta = \eta_0 \exp\left(\frac{E_A}{kT}\right) \tag{1}$$

where E_A is the activation energy and T the absolute temperature.

Among the binary metallic alloys, Pd-Si has one of the highest reported viscosities at its eutectic compositions [1]. According to its phase diagram, two eutectics exist at very similar compositions: $Pd_{82.2}Si_{17.8}$ and $Pd_{85}Si_{15}$. Before the discovery of the bulk metallic glasses [2], these alloys were considered the best starting point for easy glass formers. Their glass forming ability could be further improved by adding a ternary component in the form of $Pd_{76}Cu_6Si_{18}$. Due to their relatively simple composition, these alloys are ideal candidates for benchmark experiments. There exist a few published data on these alloys [3, 4, 5]; their results differ by more than two orders of magnitude. Steinberg and coworkers [3] used an oscillating cup viscometer, Nishi *et al.* [4] used a vibration technique, and Lee *et al.* [5] used a capillary flow method.

During the STS-83 Spacelab mission, Egry and coworkers [6] measured the viscosity of the $Pd_{76}Cu_6Si_{18}$ alloy on electromagnetically levitated specimen, applying the oscillating drop technique, while Hyers *et al.* [7] measured the viscosity of the $Pd_{82.8}Si_{17.2}$ alloy on the same mission. The microgravity data agree well with each other, and with the terrestrial data of Lee *et al.* [5].

The evaluation of the microgravity experiments was based on Lamb's treatment of an oscillating viscous liquid sphere [8]. When the radius R of a viscous liquid drop performs small (linear) damped oscillations of the form

$$R = R_0\left(1 + \delta\cos(\omega t)\exp(-\Gamma t)\right) \quad (2)$$

then the viscosity η can be calculated from the damping Γ as:

$$\eta = \frac{3}{20\pi}\frac{m}{R_0}\Gamma \quad (3)$$

where m is the mass of the drop.

Lamb's treatment neglects any effects of surface viscosity, which may arise either from an oxide layer covering the liquid drop, or even from an intrinsic effect due to a varying density profile of the conduction electrons close to the surface [9]. When surface viscosity is taken into account, Lamb's formula, eqn (3) is no longer valid. Miller and Scriven [10] have derived an expression for the damping of a liquid drop covered by an inextensible surface layer. Their result reads:

$$\eta = \frac{54}{\pi}\frac{m}{R_0}\frac{\Gamma^2}{\omega} \quad (4)$$

As can be seen, the surface viscosity itself does not appear in the above expression; however, the viscosity now scales with the damping squared. The physical reason for this qualitatively different behaviour is due to different boundary conditions in both treatments. With surface layer, not only the normal, but also the tangential pressure balance need to be satisfied. As shown by Lyubimova and coworkers in a recent paper [11], Miller and Scriven's result may also be derived by treating a small surface viscosity as a perturbation to the original problem of Lamb.

In view of these controversial results, it seemed therefore appropriate to re-investigate the viscosity of Pd-Si again. This paper reports on measurements of the viscosity of the $Pd_{82.8}Si_{17.2}$ alloy in microgravity on board a sounding rocket during the TEXUS 46 campaign. The results obtained are compared with those of previous microgravity experiments and ground based values obtained using the new high-temperature viscometer available at DLR [12]. The new TEXUS data were evaluated by both, the Lamb and the Miller-Scriven equation.

In the next chapter, the terrestrial results will be presented and discussed, whereas chapter 3 deals with the microgravity experiment. The final chapter is devoted to the discussion of the results and some conclusions.

2. Terrestrial experiments

2.1. The instrument

All experiments have been carried out in an oscillating cup viscometer described in detail in [12]. In short, a vacuum furnace with a graphite heater is used to heat the sample. The temperature inside the furnace is controlled by a pyrometer, calibrated to the melting points of pure metals (Al, Ag, Cu, Ni, Fe). The sample inside the furnace is placed in a cup made of a temperature resistant and chemically inert material. The sample is then set in torsional oscillations, which are detected by a reflected laser beam on a position sensitive device (PSD). The oscillation period T_{osc} and the logarithmic decrement δ are extracted from the oscillation and inserted into the Roscoe-equation [13]

$$\eta Z^2(\eta) = \left(\frac{I\delta}{\pi R_{cup}^3 H}\right)^2 \frac{1}{\pi\rho T_{osc}} \quad (5)$$

for calculating the viscosity. In equation (5) η is the viscosity, Z is a known function of η, I is the moment of inertia of the whole oscillating system, R_{cup} is the inner radius of the cup, H is the height of the liquid sample inside the cup, and ρ is the density of the liquid sample. This implicit equation is then solved numerically.

In the experiments cups of yttria (99,9% Y_2O_3, Cesima) with an inner diameter of 16.5 mm, a height of 50 mm, and a wall thickness of 2 mm were used.

2.2. Results

Both alloys could be measured close to their melting points and up to 1800 K ($Pd_{82.8}Si_{17.2}$) and 1950 K ($Pd_{85}Si_{15}$), respectively, covering a temperature range of at least 700 K. Except for temperatures very close to the melting point, an Arrhenius fit describes the data correctly. The corresponding Arrhenius-fits are given by:

$\eta_{85\text{-}15} = 0.3391 * \exp(4866/T)$ [mPa s] (6a)

$\eta_{83\text{-}17} = 0.5730 * \exp(4557/T)$ [mPa s] (6b)

where the temperature T is in K.

The resulting activation energies and melting point viscosities are listed in Table 1.

Table 1. Composition, melting points, viscosities at T_L and activation energies of the investigated Pd-Si alloys, obtained from Arrhenius fits.

Composition (at%)	T_L (K)	η_L (mPa s)	E_A (kJ/mole)
$Pd_{82.8}Si_{17.2}$	1089	37.62	37.8
$Pd_{85}Si_{15}$	1094	28.97	40.5

The two alloys have similar viscosities. As expected, the deeper eutectic ($Pd_{82.8}Si_{17.2}$) has the higher viscosity by about 30%.

3. Microgravity experiments

3.1. The mission

The experiments were carried out in the framework of the TEXUS 46 sounding rocket mission using the TEXUS-EML electromagnetic levitation facility. This mission provided about 6 minutes of microgravity, to be shared by two experiments accommodated in the EML facility. The experiment time allotted to the Pd-Si experiment was 170 s. The EML has two separate high-frequency current circuits: one for positioning the sample against microgravity disturbances, providing a quadrupolar electromagnetic field, and one for heating the sample through a homogeneous dipolar electromagnetic field. More details of the EML can be found in [14].

The sample for the microgravity experiment had a nominal composition of $Pd_{82.2}Si_{17.8}$ and was prepared from the same material as used in the terrestrial viscosity measurement. The sample mass was 1.599 g. The experiments were carried out under argon atmosphere. Pyrometer and video signals were acquired with a frame rate of 200 Hz. The pyrometer signal T_p was converted to true temperatures T by adjusting its reading to the known melting temperature [15]:

$$\frac{1}{T} = \frac{1}{T_p} + \left(\frac{1}{T_L} - \frac{1}{T_{L,p}}\right), \quad \left(\frac{1}{T_L} - \frac{1}{T_{L,pyro}}\right) = 8.3\ 10^{-5} \tag{7}$$

Here, T_L = 1089 K is the known liquidus temperature, and $T_{L,Pyro}$ is the pyrometer reading at the melting temperature.

Two melt cycles were performed, both consisting of a heating and cooling phase. The heating phases heated the sample above 1400 K, in order to destroy potentially existing Pd_3Si clusters. In the cooling phase, sample oscillations were triggered every 6 seconds by applying a short pulse through the heater coil, leading to a compression of the liquid sample near the equatorial plane. No attempt was made to provide isothermal conditions, i.e. to perform the oscillation analysis at constant temperature. The reason for this is twofold: first, the experiment time is not sufficient for achieving thermal equilibrium, and second, isothermal conditions require heating of the sample by the electromagnetic field. This would lead to turbulent flow in the sample, disturbing the viscosity measurements.

The temperature profile of the experiment is shown in Figure 1. The melt plateau at 1089 K is clearly visible. The spikes on the cooling curve correspond to the heater pulses, used for triggering the sample oscillations.

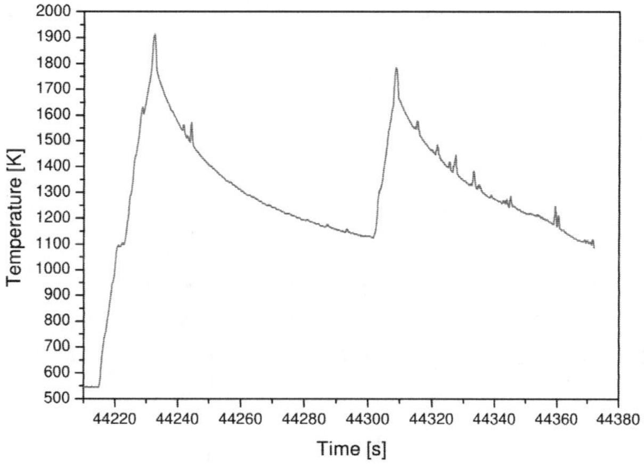

Figure 1. Temperature profile of the TEXUS 46 microgravity experiment.

Two video cameras provided axial (=top) and radial (=side) view. From the video images it was possible to extract the relevant geometrical parameters (radii in two perpendicular directions, cross section) of the drop. Due to some microgravity disturbance, the sample did not show the expected oscillation behaviour during melt cycle #1. Consequently, only cycle #2 was further analysed. The corresponding video signal is shown in Fig. 2.

3.2. Data analysis

For the analysis,all radius and area data were normalized to zero average by subtracting a linear fit from the original data. In addition, the absolute time was replaced by a relative time, starting at the beginning of cycle #2. This avoids dealing with large numbers and improves the accuracy of the fits and Fourier transforms. The damping constant Γ, eqn (2), was determined in two different ways:

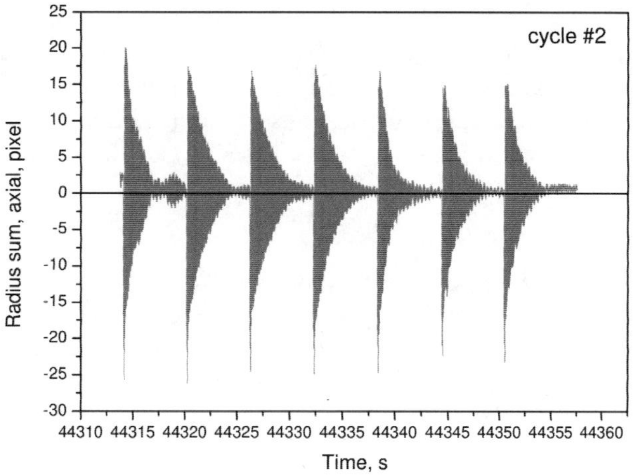

Figure 2. Sum of two perpendicular radii of oscillating drop, mean value subtracted. Axial view, cycle #2.

Method1:

The absolute value of the radius (or area) was taken, and its maxima were determined: $S(t_i) = Max\{|R(t)|\}$. The resulting data set was fitted by a simple exponential $A\exp(-\Gamma t)$.

Method 2:

The radius (or area) signal was squared, and its logarithm taken:

$$\ln\left(R^2(t)\right) = \ln\left(A^2\sin^2(\omega t)\right) - 2\Gamma t$$

The maxima were determined and fitted by a linear function, whose slope equals -2Γ.

Both method yielded consistent results.

The temperature attributed to each pulse corresponds to the temperature at the beginning of the oscillation. This overestimates the temperature, but is reasonable, since the data points at the beginning of the oscillation, where the amplitude is large, dominate the data analysis.

3.3. Results

The damping constants, determined as described above, are shown in Table 2 together with the associated temperatures..

Table 2. Damping constants and frequencies as function of temperature.

T, K	Γ, s^{-1}	ω, s^{-1}
1447	0,510	175,4
1381	0,515	175,9
1328	0,641	176,8
1252	0,781	177,6
1223	0,775	177,5

For the application of eqns (3) or (4), the mass and radius of the sample are required. For the mass $m = 1.599\ g$ was used. This value was determined before flight, a post-flight analysis was not possible. Consequently, possible mass losses due to evaporation are not considered. The radius was determined indirectly from the mass and density of the drop, assuming a spherical shape. For the liquid density a temperature-dependent value of $\rho(T) = 10.0 - 0.001(T - 1089K)\ g\,cm^{-3}$ was used, as determined before flight [16].

In eqn (4), the frequency ω of the oscillation is required in addition. This was obtained by Fourier transforming the time signal of each pulse individually and determining the peak position.

The viscosities, derived from the damping constants, by either using Lamb's or Miller's formula, are shown in Figure 3 together with our ground-based result and the earlier microgravity result of Hyers [7], evaluated according to Lamb.

4. Results

The results obtained with the oscillating cup method on ground and those obtained under microgravity show reasonable agreement, especially in the high-temperature region. Surprisingly, the evaluation using Miller's equation, eqn (4), yield higher viscosities than Lamb's equation, eqn (3). This is unexpected, since eqn (4) has an additional term Γ/ω, which in the sense of a perturbation expansion is a small quantity of the order of 10^{-3}. However, this is compensated by the different numerical prefactors, whose ratio has the same order of magnitude.

The data obtained with Lamb's formula yield viscosities somewhat smaller than those measured on ground, but with a very similar temperature dependence, whereas the evaluation according to Miller & Scriven agrees better with the ground-based data, but yields a different temperature dependence. The microgravity results from TEXUS46 and MSL-1, evaluated according to Lamb's formula show good agreement.

In summary, the results obtained do not allow to clearly decide which equation works better. In order to do so, the parameter Γ/ω needs to be varied, such that the two formulae should give results which differ by at least an order of magnitude. This is possible by varying the mass of the sample, as damping and frequency depend differently on sample radius. Such experiments are planned for the future.

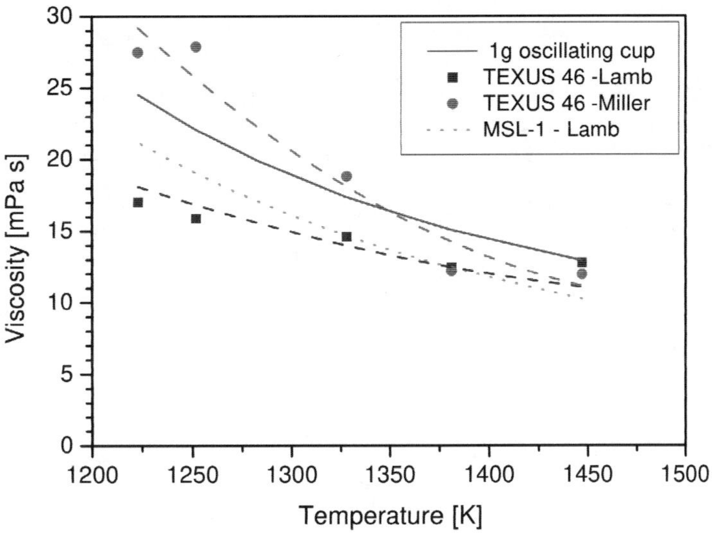

Figure 3. Viscosity of $Pd_{82.2}Si_{17.8}$, as obtained by different methods. Data points correspond to the TEXUS 46 results, full line is our ground based result, and dotted line is Hyers' result [7].

Acknowledgements
The microgravity experiments would not have been possible without the excellent support of many people involved, namely the ESRANGE launch team, the MORABA team, the EADS teams from Bremen and Friedrichshafen, and the MUSC team from DLR Cologne. Their help and advice and their professional attitude during extreme work load is highly appreciated. Last, but not least, many thanks go to the DLR Agency in Bonn for providing this flight opportunity.

References
1 L. Battezzati and L. Greer, *Acta Met.* **37** (1989), 1791
2 A. Inoue, T. Zhang, and T. Masumoto, *Mater. Trans. JIM*, **31** (1991), 425.
3 J. Steinberg, S. Tyagi, A. Lord Jr., *Appl. Phys. Lett.* **38** (1981), 878
4 Y. Nishi, N. Kayama, S. Kiuchi, K. Suzuki, T. Masumoto, *J. Japan Inst. Metals* **44** (1980), 1336
5 S. Lee, K. Tsang, H. Kui, *J. Appl. Phys.* **70** (1991), 4842
6 I. Egry, G. Lohöfer, I. Seyhan, S. Schneider, B. Feuerbacher, *Appl. Phys. Lett.*, **73**, (1998), 462
7 R. Hyers, G. Trapaga, M. Flemings, in: *Solidification 1999*, W. Hofmeister, J. Rogers, N. Singh, S. Marsh, P. Voorhees eds., TMS Warrendale, 1999, p.23
8 H. Lamb, *Proc. London Math. Soc.* **13**, (1881) 51

9 J. Earnshaw, in: *Marangoni and Interfacial Phenomena in Materials Processing*, E. Hondros, M. McLean, K. Mills, eds., IOM Communications Ltd, London, 1998, p.41.
10 C. Miller and L. Scriven, *J. Fluid Mech.*, **32** (1968), 417
11 D. Lyubimov, V. Konovalov, T. Lyubimova, I. Egry, *J. Fluid Mech.*, doi:10.1017/jfm.2011.76, (2011)
12 Kehr M, Hoyer W, Egry I. *Int. J. Thermophys.* **28** (2007) 3
13 R. Roscoe, *Proc. Phys. Soc.* **72** (1958)
14 W. Soellner, A. Seidel, Ch. Stenzel, W. Dreier, B. Glaubitz, *J. Jpn. Soc Microgravity* **27** (2010), 183.
15 G. Nutter, in: *Theory and Practice of Radiation Thermometry*, D. DeWitt, G. Nutter, eds., Wiley, New York, 1989, p.231.
16 D. Dorow-Gerspach, *DLR internal report*, 2009

Journal of Physics: Conference Series **327** (2011) 012023 doi:10.1088/1742-6596/327/1/012023

Non-equilibrium fluctuations on earth and in micro-gravity. The GRADFLEX experiment.

A Vailati[1], R Cerbino[2], S Mazzoni[3], M Giglio[1,5], C J Takacs[4] and D S Cannell[4]

[1] IFN-CNR and Dipartimento di Fisica, Università degli Studi di Milano, 20133 Milano, Italy
[2] IFN-CNR and Dipartimento di Chimica, Biochimica e Biotecnologie, Università degli Studi di Milano, Italy
[3] ESA/ESTEC, Keplerlaan 1, 2200AG Noordwijk, The Netherlands
[4] Department of Physics and ITST, University of California at Santa Barbara, Santa Barbara, CA 93106, US
[5] corresponding author: marzio.giglio@fisica.unimi.it

We present the results of measurements of giant nonequilibrium fluctuations in a single component fluid (density fluctuations) and a mixture (concentration fluctuations) driven by applied temperature gradients, both on Earth and in space. Flight data were obtained during the September 2007 FOTON M3 mission. Spatial power spectra obtained using the shadowgraph method, during flight, confirm that the asymptotic $|\nabla T|^2 q^{-4}$ behaviour extends to such low wave vector q, as to be limited by the sample thickness. Quantitative comparison with theory is provided, and is generally quite good. Temporal sequences of shadowgraph images for the mixture, both on Earth and during flight will be presented to emphasize the dramatic differences. Fluctuation lifetimes of thousands of seconds were observed during flight. The rugged but sensitive shadowgraph scattering method (phase fluctuations below 10 milliradians measured to within a few percent absolute accuracy) will be described briefly.

1. Introduction

Equilibrium fluctuations are usually small and limited to molecular length scales. A notable exception occurs for fluids near second-order phase transitions where correlations readily extend to the micron scale as evidenced by the impressive phenomenon of critical opalescence [1]. While critical phenomena was one of the most studied areas in the sixties and seventies, a great amount of work was also carried out during that period in connection with two quite diverse physical systems, the laser, and the Rayleigh-Bénard (RB) convective instability (a liquid slab heated from below). These are examples of a general class of instabilities in open systems exhibiting a threshold value for the control parameter, above which macroscopic structures are formed. It was shown that the behaviour near the threshold of these instabilities closely mimics mean-field behaviour of critical systems, in the sense that system properties (order parameter, amplitude of fluctuations, relaxation times) present a power-law dependence on the control parameter much like that observed close to a classical critical point [2-4].

The literature on this topic is too vast to be quoted here. We single out however a UC Santa Barbara paper where large fluctuations below the onset of convection were first observed and measured [5]. It was shown that fluctuations below threshold peaked around the wave vector of the convective pattern that eventually materialized above threshold, in keeping with critical fluctuations scaling arguments. The amplitude was also measured and agreed with theory to within experimental accuracy of about 50%. We quote Ref. 5 here because it is the first reported utilization of quantitative shadowgraph methods to study large-scale non equilibrium fluctuations.

This paper presents results obtained using the shadowgraph method to measure long-range non-equilibrium fluctuations in both a fluid mixture and a single component fluid. The measurements were made both on Earth and during a flight experiment, the 13-day Foton-M3 mission. None of these measurements could have been performed without quantitative shadow methods.

The fluctuations we will discuss occur quite readily without the necessity of being close to a generalized critical point. Very much like the critical fluctuations, these are non propagating fluctuations that scatter into the central Rayleigh component of the spectrum of the scattered light. The relevant theoretical work [6-9] was developed over a few years, without immediate reaction from experimentalists. It showed that subjecting a fluid or fluid mixture to a stabilizing temperature gradient by heating from above should result in non-equilibrium fluctuations with amplitudes and spatial ranges many orders of magnitude greater than those of equilibrium fluctuations. Incidentally, non equilibrium propagating fluctuations under a temperature gradient were studied in Refs. 10 and 11 slightly earlier than references above, the effect of the gradient being of generating an asymmetry between the frequency shifted Brillouin components. For the non propagating fluctuations we describe here, the hallmark of such gradient-driven fluctuations is the asymptotic behaviour for the structure factor of an unbounded sample in the absence of gravity. On Earth, gravity quenches this divergence at a well-defined wave vector determined by the fluid properties, while in the absence of gravity, the finite sample thickness was predicted to be the only limiting factor. Physically, the fluctuations are due to the coupling of thermal velocity fluctuations, with the quantity exhibiting a gradient (temperature or concentration, e.g.). While they have no effect on a homogeneous system, when they occur along a density or a concentration gradient, they can generate large fluctuations by displacing parcels of fluid having different density or concentration along the gradient. The limitation of such gradient-driven fluctuation on Earth occurs because very long wavelength fluctuations are suppressed by buoyancy which acts as a restoring mechanism for the displaced fluid parcels. In fact, such a restoring force can even result in oscillatory motion for sufficiently long wavelength fluctuations. Because of buoyant forces, $S(q)$ saturates to a constant below a critical wave vector q_c, which for a concentration gradient ∇c is given by $q_c = (\beta g \nabla c / (\nu D_C))^{1/4}$, where β is the solutal expansion coefficient, g is the acceleration of gravity, ν is the kinematic viscosity and D_C is the concentration diffusion coefficient. For most common mixtures q_c is of order few hundred cm^{-1}.

When the predictions were made they raised the issue of why not even a hint of the existence of these anomalously large fluctuations had ever been observed, despite the fact that solutions and mixtures subjected to gradients had been studied for many years. For example, using free diffusion methods to determine diffusion coefficients, or analytical centrifuges to obtain molecular weight through schlieren-sedimentation profiles [12], or while measuring Soret coefficients optically [13].

The reasons why they escaped observation are quite diverse and to some extent rather fortuitous. First, the common techniques use optical methods with the axis orthogonal to the applied gradient, while theory indicates that effects are maximized when the optical axis is aligned with the gradient, a geometry that is inappropriate (and very unpractical) for all the above methods. Second, because of the weak dependence of q_c on density mismatch, it turns out that for any choice of single-component liquid and any mixture or suspension, gradient driven fluctuations on Earth remain confined to length-scales of the order of ten to a hundred microns, and this means that they were unlikely to be observed by visual inspection. Finally, the optical distortion introduced is rather modest in comparison with the critical opalescence effects one would expect for a critical fluid with the same correlation length-scale. Critical opalescence implies that scattering is so strong, that it usually develops into strong multiple scattering, which in turns implies that a plane wave front will suffer random phase distortions large compared to π in passing through the sample. Here, as we will show, wave front distortions remain a small fraction of π, and consequently instrumentation of high sensitivity is required to detect them. The stronger disturbance caused by critical opalescence is due to the fact that the structure function $S(q)$ decays approximately as q^{-2} rather than as q^{-4}, and thus integration over wave vector results in large total scattering cross-sections.

The first observation of non-equilibrium fluctuations in a simple liquid subjected to a (stabilizing) temperature gradient is due to the Sengers group at the University of Maryland [14, 15]. The technique employed was a form of dynamic light scattering in which the scattering was optically mixed with stray light. The scattering due to the non-equilibrium fluctuations was very small compared to the enormous stray light contribution from optical imperfections and other sources. As dynamic measurements were performed, it was possible to cleanly separate the true signal from the stray light, and the q^{-4} dependence was clearly observed, as well as the $|\nabla T|^2$ behaviour. The accessible scattering angles were much too large however, to permit the detection of the gravitational cutoff, and the Authors concluded that it was unlikely that it would be detected in the future. Ref. 14 and 15 are groundbreaking papers proving that experiments could be successfully executed in this area.

Subsequently, ultra low angle scattering and shadowgraph methods were employed by the Milan group to investigate non-equilibrium concentration fluctuations in Soret–driven and free diffusion processes [16, 17]. The cutoff at low q due to the presence of gravity was observed, and found to be in excellent agreement with theory. Particularly important for the history of this project is a paper in Nature [17], describing an experiment in which no applied thermal gradient was involved, but instead the driving mechanism was the creation of a pure concentration gradient via the separation of two miscible solutions and subsequent free diffusion process. In the associated comment in the News and Views section, it was suggested that an experiment in microgravity conditions would be highly desirable, as this would give the possibility of observing fluctuations at length scales comparable with the physical size of the sample.

It was then decided to join forces between the Milan group and the UCSB group, proposals were submitted to ESA and NASA, and eventually the GRADFLEX (GRAdient Driven FLuctuation EXperiment) project was launched. It was decided that two experiments had to be flown, one on a single-component fluid (SINGLE), to be primarily the responsibility of the UCSB group, and one on a mixture (MIXTURE), for which the Milan group assumed primary responsibility. Both experiments employed a temperature gradient as the stimulus. It was applied to a 3-mm thick carbon disulphide sample in SINGLE to study density fluctuations and to 1-mm thick sample consisting of a 1.8 wt.% solution of M_w = 9,100 polystyrene dissolved in toluene for MIXTURE. The polymer solution, which has a very large Soret effect, developed a strong concentration gradient in response to the applied temperature gradient. Two publications have appeared recently describing the two experiments and the flight results, that are briefly reported here [18, 19]. As a general reference to the subject of non-equilibrium fluctuation we suggest the exhaustive and beautifully written book by Ortiz de Zarate and Sengers [20]. The present work is the first report containing results from both MIXTURE and SINGLE.

2. Experimental setups

Each setup consisted of a thin cylindrical sample cell with associated equipment to impose a stable temperature difference between the plates confining the sample. In both cases, the optical axis of a shadowgraph was oriented parallel to the gradient, and therefore optical windows that conduct heat much better than do the samples were required. Sapphire was chosen, as it has a thermal conductivity hundreds of times larger than that of typical fluids. Because the cells are rather different, they will be described below separately. The SINGLE cell (and associated optics) was about 3 times larger than those of MIXTURE because the gravitationally induced crossover wave vector is about that much larger for the mixture than for CS_2. For a single-component fluid, $q_c = \left(\alpha g \nabla T / (\nu D_T)\right)^{1/4}$, where α is the thermal expansion coefficient, and D_T is the thermal diffusivity. Typically for single-component organic fluids, q_c is of the order of 70 cm^{-1} as compared to a few hundred cm^{-1} for mixtures. As the gravitational crossover is absent in microgravity, fluctuations are limited only by the finite sample thickness, and the cross over then occurs at a much lower value near $q'_c = 2\pi/L$, where L is the sample thickness. Thus a dramatic change in the low q scattering was anticipated for MIXTURE even with a sample thickness of only 1 mm.

Both experiments utilized the quantitative shadowgraph technique. As this method proved to provide exceptional performance, but is very little known, we present below a brief description of the physical principles on which it is based and the origin of the rather complex instrumental transfer function T(q). The approach is unorthodox, no rigor will be claimed, but we hope it will be effective.

2.1. Scattering cells, choice of samples, stimuli and general layout

2.1.1. SINGLE

SINGLE deals with large-scale gradient driven density fluctuations in a single component liquid, CS_2, which was chosen to give a large signal. The apparatus was a flight version similar to the prototype developed to demonstrate the feasibility of measuring signals from a single-component fluid heated from above, something that had not been achieved previously. A 75-mm diameter 3 mm thick sample, at 1 atmosphere, was confined between the sapphire window and a silicon mirror. The window was coated with transparent conducting Indium Tin Oxide (ITO) on the surface in contact with the CS_2, and the ITO was used to generate heat. Heat was removed from the mirror by means of 4 Peltier elements separated from the mirror by a thermal diffuser. The ITO and Peltier currents were regulated to impose gradients and were also used to maintain the mean temperature at 30° C. Light of wavelength 680 nm from a super-luminous diode coupled into a mono-mode fibre, was used for illumination. The light diverged in a nearly Gaussian beam from the polished end of the fibre and passed through a beam splitter and then a lens mounted just upstream from the sample window. The lens collimated the light into a diffraction-limited beam, after which it passed through the sample and was reflected by the mirror. The beam splitter diverted 50% of the returning beam to a 1024 by 1024 CCD sensor, which recorded images at an effective distance $z = 310$ cm from the sample. By dividing images pixel by pixel, it was possible to detect the fractional intensity variation $\delta I(x, y, t)$ caused by interference between the beam and light scattered by the fluctuations. Fourier transformation of the central 512 by 512 portion of such ratio images served to separate the fluctuations by wave vector. In the end the apparatus proved to be extremely sensitive and we were able to measure signals with applied temperature differences as small as 1 K. In taking data during the mission, nominal temperature differences of 1, 2, 5, 10, 15, 20 and 30 K were applied, and the same values were used while taking data on Earth after the mission.

2.1.2. MIXTURE

MIXTURE deals with large-scale fluctuations of concentration, and to induce them, a concentration gradient must be established. A repeatable and controlled way to do so is to utilize the Soret effect [13]. Applying a temperature gradient generates a Soret mass flow, until the gradient reaches steady state where back diffusion balances the Soret flow. The resulting steady-state concentration profile is governed by $\Delta c = -S_T c(1-c)\Delta T$. Here $S_T = 6.49 \times 10^{-2}$ K^{-1} is the Soret coefficient, and c = 1.8 wt.% is the concentration. In contrast with the sample choice for SINGLE, sample choice for MIXTURE offered a broad spectrum of choices, as S_T varies substantially with solute molecular weight and species. Unfortunately, the mass diffusion coefficient typically scales in the opposite way, and has to be chosen carefully in order to make time constants compatible even with a two-week mission. Ultimately, a low molecular weight polystyrene-toluene solution was used (M_w = 9,100 g mol^{-1}), as it guaranteed a large Soret-driven concentration gradient and reasonably large mass diffusivity D_c to keep time constants within reason. The magnitude of the concentration difference can be readily altered, and the experiment repeated as desired. The thermal-gradient cell was a flight-engineered version of a prototype used for ground-based tests [21], and was similar to designs developed previously [16, 17].

The sample was a 25-mm diameter by 1.00 mm thick layer of polymer solution confined between two parallel, 12 mm thick, sapphire windows that were temperature controlled to better than ± 0.01 K. The use of sapphire windows allowed the application of a relatively uniform temperature difference, while permitting optical access to the sample in a direction parallel to the imposed gradient. The sample was illuminated with collimated, 680 nm wavelength, light from a super-luminous LED coupled to a mono-mode optical fibre. Interference between the beam and the light scattered by the diffusing fronts resulted in small but measurable variations of the intensity in the images collected by a Charge-Coupled Device (CCD) with 1024x1024 square pixels. The central 512 by 512 portion of differences of succeeding images were Fourier decomposed to reveal the mean-squared amplitude of concentration fluctuations *vs*. the scattering wave vector q.

A measurement cycle involved an initial equilibration phase at a uniform temperature of 30.0° C for 260 minutes. Toward the end of this phase, a set of 539 reference images was acquired to characterize the optical background and the camera noise. This phase was followed by the rapid imposition of a temperature difference across the sample, which started the diffusion process. The temperature differences utilized were 4.35, 8.70 and 17.40 K. The time required for the formation of a linear temperature profile was about 100 s. In contrast, the time to create a steady-state concentration profile ($\tau_0 = h^2/\left(\pi^2 D_c\right)$) was about 500 s; here $D_c = 1.97 \times 10^{-6}$ cm^2/s is the diffusion coefficient, and h = 1.00 mm the sample thickness.

2.2. The shadowgraph method

The structure function of fluctuations $S(q)$ is usually determined by measuring the angular distribution of scattered light $I(q) \propto S(q)$, the scattering angle θ being related to the wave vector of the fluctuations by $q = 2k\sin(\theta/2)$, where k is the wave vector of the light in the sample. In the simplest arrangement, sensors at various scattering angles are placed well clear of the transmitted beam. When fluctuations become spatially large, scattering angles must be reduced, and separating the scattered light from the beam becomes exceedingly difficult, as stray light scattered by imperfections in every optical surface encountered by the beam overwhelms the true scattering. It is in this regime that the power of light scattering via the shadowgraph is most valuable.

A shadowgraph instrument can be used as a scattering instrument [22], where one actually exploits the existence of the transmitted beam as a local oscillator that mixes with the weak scattered field. This mixing results in small amplitude time- and spatially-dependent fluctuations in the beam intensity which may be quantified using a digital camera. The sensor is normally so close that light scattered at any angle of interest falls onto the detector. A special scattering condition, the Raman Nath (R-N) condition (see below) is normally satisfied. Before proceeding further, one should stop for a second, and ponder the extreme simplicity of this method: a good quality plane wave, the scattering cell, and a sensor comprise the entire setup. Because of the interference between the beam and the scattered light, the low contrast intensity variations are proportional to the scattered field, rather than to the scattered intensity. Consequently, taking the Fourier transform of the intensity distribution gives the angular spectrum of the scattered radiation [22, 23]. So one could actually determine both the amplitude and phase of the scattered plane waves in all directions. Indeed for a sample consisting of spheres, such as the calibration sample discussed below, one can generate the images of individual particle via a z scan [24]. However, as we are interested in the scattered intensity, measuring the power spectrum of the detected intensity distribution $P(q)$ is more useful, rather than the overly-detailed Fourier transform. Of course phases are lost, and positions of particles are consequently also lost. From the above naïve discussion, one might be inclined to think that $P(q) \propto S(q)$. Unfortunately, the matter is more complicated, as a transfer function $T(q)$ is generally involved, and $P(q) = T(q)S(q)$. Actually, $T(q)$ is the product of a number of transfer functions, each accounting for contributions due to various phenomena, such as temporal space coherence, averaging over pixels, optical resolution and windowing caused by taking finite-sized images, etc. Some of these effects are subtle, and will not be discussed here. Others, like finite time coherence will be briefly mentioned below.

The most evident feature of the power spectra of shadow images is the typical oscillatory, fringe-like behaviour, that is a consequence of the Raman Nath condition. It is rather simple to calculate $T(q)$ due to R-N condition. By definition, it is the response to white noise signal, that is to light being scattered uniformly as a function of q, as is the case for very small particles, that have $S(q) = S(0) =$ constant. The interference at distance z for a small particle in the origin is given by $I(x,y) = \frac{2|S(0)|}{kz}\cos\left[\frac{k(x^2+y^2)}{2z}+\phi\right]$ where z is the distance and φ the scattered wave phase delay. The associated power spectrum is $S(q_x,q_y) = \frac{4\pi^2}{k^2}|S(0)|^2\sin^2\left[\frac{q^2 z}{2k}-\phi\right]$ and it coincides with the transfer function $T(q)$. This implies that the random intensity distribution is made up by identical, infinite, constant amplitude interference fringes. The extra phase ϕ accounts for size dependent phase delays, as discussed below [25]. Let us see when this holds true. As the fringe peak positions in q depend on distance, this implies that for all the q values of interest, the actual sample thickness must be so small that distance variations between $z - L/2$ and $z + L/2$ will not introduce appreciable changes for the positions of the outermost fringe like oscillations. It can be easily proven that this is the R-N condition $q^2 L/(2k_0) \le \pi$. So, if $S(q)$ decays over the wave vector range where R-N holds, it will depress accordingly the infinite sequence of fringe-like oscillations of $T(q)$. So, in a sense, it is partially true that $P(q)$ and $S(q)$ are proportional, but in "average" over the $T(q)$ oscillations, the case q=0 being pathological because of the wide minimum, that requires special attention.

In concluding, thinking in terms of circular interference fringes as the source of the power spectra can be very profitable, because one learns how to get a gross idea about additional effects like limited time coherence of the superluminous diode that has been used in both setups. As $\lambda/\Delta\lambda$ is typically of the order of 40, additional depression of the fringe like oscillations should be expected at the 40th fringe, as limited coherence adds its share in depressing the oscillations, and it must be accounted for.

While one could figure out one by one the contributions to $T(q)$, it was found much profitable to derive experimentally $T(q)$ through a calibration procedure. As often done with scattering method, a colloidal suspension was chosen. Unfortunately very small particles, say 100 nm in diameter, otherwise ideal to guarantee uniform scattered intensity and zero scattered wave phase delay, do scatter too little, and finally 2 micron polystyrene spheres in isopropyl alcohol were chosen.

Below, the scheme used for the mixture experiment is reported, the one for single being almost identical. The transfer function can be quite generally written as $T(q) = A(q)\sin(q^2 z/2k_0) + C(q)$, where the envelopes of the minima and maxima are equal to $C(q)$ and $A(q)+C(q)$, respectively. Calibrating the optical system means basically to determine the functions $A(q)$ and $C(q)$.

To carry out the calibration, 320 images were taken with the cell filled with filtered isopropyl alcohol, to determine a background signal $S_B(q)$. The cell was then filled with the colloidal suspension, and 3500 images were taken. Between each image, the sample was stirred using an automatic piston, to ensure statistically independent images. After subtracting $S_B(q)$, the signal from the spheres is given by $S_{Sp}(q) = A(q)\sin^2(q^2 z/2k_0 + \phi_{Sp}) + C(q)$, where ϕ_{Sp} is an additional phase delay that accounts for the non-vanishing optical thickness of the 2 micron spheres in isopropyl alcohol [25]. We found the phase delay to be 1.78 radians, in very good agreement with Mie theory [26]. The functions $A(q)$ and $C(q)$ were determined by fitting the data for $S_{Sp}(q)$, and this in turn allowed us to determine the instrumental transfer function $T(q)$.

3. Experimental results

3.1. Mixture

To show the striking comparison between ground and flight data, we show in Figure 1 a sequence of false colour two dimensional shadow images taken both on ground and in flight with the polymer solution and an applied temperature difference of 17.40 degrees. Images were taken 0, 500, 1000 and 2000 s (left to right) after the imposition of a 17.40 K temperature difference. The side of each image corresponds to 5 mm. Colours map the deviation of the intensity of shadowgraph images with respect to the time-averaged intensity.

The figure shows that in a gravity-free environment, fluctuations are boosted in amplitude and size. Images were acquired at a constant frame rate of 0.1 Hz while a temperature gradient was kept constant for 42 h. Estimates of the mean square amplitude of the fluctuations at various q is performed by taking the azimuthal average of the two dimensional power spectrum.

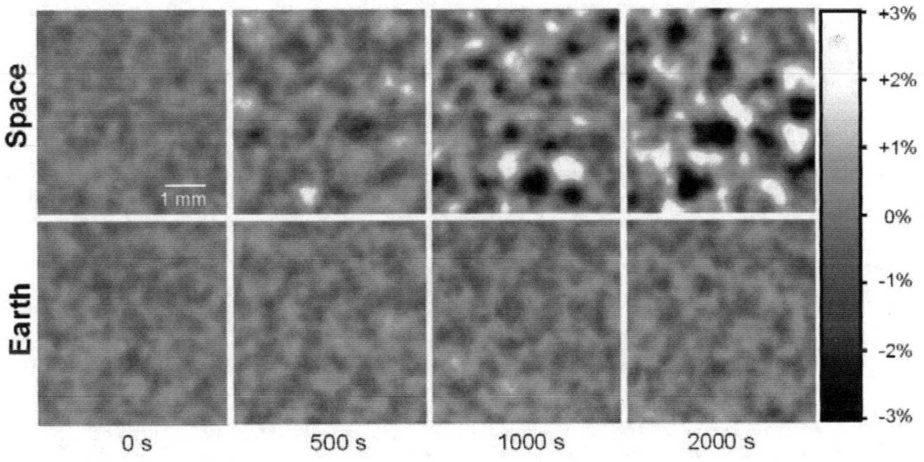

Figure 1. Shadowgraph images of nonequilibrium fluctuations in microgravity (top panels) and on Earth (bottom panels). The sample is a 1.00-mm-thick solution of polystyrene in toluene. Images were taken 0, 500, 1000, and 2000 s (left to right) after the imposition of a 17.40 K temperature difference at t=0. The side of each image corresponds to 5 mm. The false-colours map the deviation of the intensity of shadowgraph images relative to the average intensity

To isolate the signal of the concentration fluctuations from the contribution of nonequilibrium temperature fluctuations, as well as from other sources of noise (e g electronic noise from the CCD sensor), a dynamical analysis has been performed (see [18] for details). The power spectrum was finally divided by the instrument transfer function $T(q)$ to compensate for the R-N oscillations, for finite time coherence and other effects not discussed here.

The resulting $S(q)$ obtained for temperature differences of 4.35, 8.70 and 17.40 K are shown in the top part of figure 2. The curves show the striking asymptotic behaviour $S(q) \propto q^{-4}$ for large values of q, while tend to saturate at low q due to the finite-thickness of the sample (1 mm).

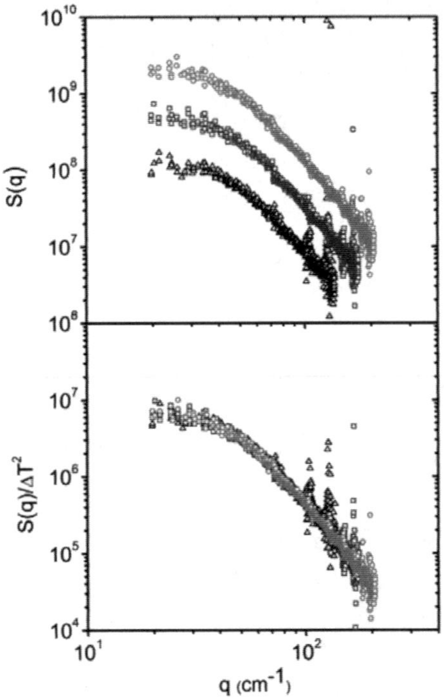

Figure 2. Top panel: experimental results obtained in microgravity in the presence of temperature differences of 4.35 K (black triangles), 8.70 K (blue squares) and 17.40 K (red circles). Bottom panel: experimental results divided by $|\Delta T|^2$

From the actual values of the fractional intensity fluctuations (see false colour scale) one can derive an absolute estimate for the rms of the phase fluctuations. It amounts to less the 10^{-2} radians. So path length changes of few Angstrom are actually evaluated. Similar conclusions can be immediately reached by noticing that the scattering phase delay ϕ is very small. So the one millimetre size giant fluctuations have an optical thickness smaller than a 100 nm polystyrene colloidal particle in water [25]. Incidentally, this also implies that the main beam fractional power removal by the fluctuations is a small fraction of 0.1 %. The proportionality of the overall amplitude of non-equilibrium fluctuations to $|\nabla T|^2$ is confirmed in the bottom part of the figure 2.

Experimental data were compared with existing numerical predictions. In the absence of gravity, and with the radius of gyration of the polymer chains much larger than the size of solvent molecules, a theoretical model [20, 27] based on a single-mode Galerkin approximation predicts that the mean-squared amplitude of fluctuations should scale onto a universal curve independent of mixture properties or applied concentration difference which scales as at large wave vectors and saturates to a constant value at small wave vectors. Figure 3 shows the experimental data plotted with the theoretical prediction, in the absence of adjustable parameters. This result provides a straightforward confirmation of the scale invariance of the fluctuations up to wavelengths comparable to the sample thickness, and of the overall increase of amplitude of fluctuations when compared to data acquired on earth (dashed lines). Above a length scale comparable to the sample thickness, the results confirm that the scale invariance of the fronts of diffusion is frustrated by the finite size of the container, as predicted by theory [27] and by two-dimensional simulations [28]. Quite interestingly, very recent simulations show that the presence of such fluctuations affects the mass transfer during a diffusion process [29, 30].

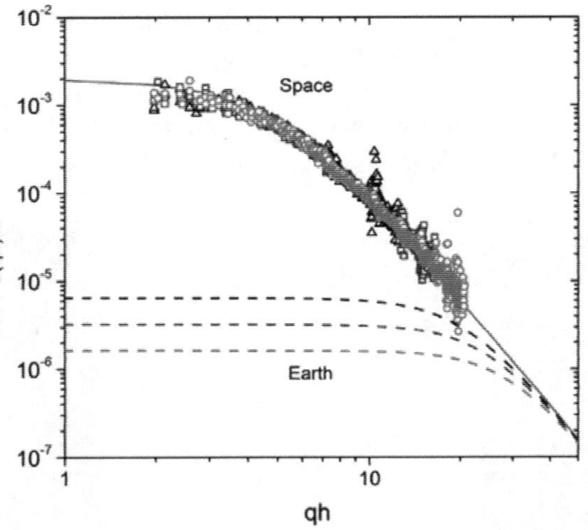

Figure 3. Comparison of the experimental results with the theoretical predictions. The solid line is the theoretical prediction for microgravity. The dashed lines are the theoretical predictions for fluctuations on Earth.

3.2. Single

We show in figure 4 a typical result for the azimuthal average of the two-dimensional power spectrum of the fluctuations for CS2 on ground, vs. the dimensionless wave vector qL, for an applied temperature gradient of 17.9 K/cm.

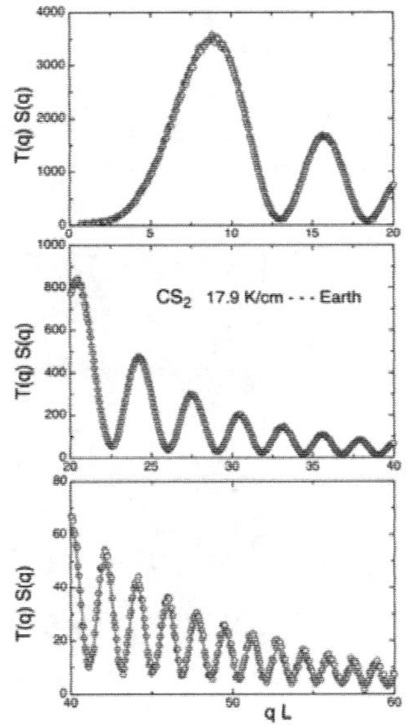

Figure 4. The product of the measured transfer function $T(q)$ and the calculated structure factor $S(q)$ vs. the dimensionless wave vector qL for a 3 mm thick CS_2 sample subjected to a temperature gradient of 17.9 K/cm on Earth

The data are shown as open circles, and the theoretical result is shown as the solid line. It should be emphasized that no parameters have been adjusted in making this comparison, including the amplitude. Thus the theory clearly does a remarkable job of describing the data, both in terms of absolute amplitude and wave-vector dependence. Of course, the theory actually predicts the structure factor $S(q)$, while what is shown for direct comparison with the data is the product of the measured transfer function $T(q)$ and $S(q)$. The oscillatory behaviour associated with the transfer function is clearly evident. Similar results were observed for all applied gradients.

SINGLE exhibited some unexpected behaviour during flight, and to date we have no explanation for what was observed, however we have devised a reasonably robust method for extracting useful data despite the difficulties. Figure 5 shows shadowgraph images taken with no gradient applied (left) and shortly after application of a nominal 30 K temperature difference across the sample (right).

Figure 5. Shadowgraph image taken during flight with no temperature difference applied (left) and with a nominal 30 K difference applied (right). Note the presence of several white spots of various intensities in the image on the right, as well as the distortions present near the periphery. The speckled region at the bottom is of no consequence.

We observed that applying a temperature gradient during flight resulted in the sudden (within seconds) appearance of several white spots in the image, as though light were being focused in those regions. These bright regions appeared in exactly the same place every time a gradient was applied, and remained fixed in location as long as the gradient was maintained. Normal fluctuations were observed in those regions however. In addition to the white spots, it is evident that there are other perturbations present in the outer portions of the cell, and these perturbations evolved slowly over the 5-hour duration of a data run, suggesting that they may have been caused by weak lateral temperature gradients. The extent of this evolution was marked, and is shown by the two images of figure 6.

Figure 6. Shadowgraph images taken during flight with a nominal temperature difference of 30 K. The image on the left was taken immediately after the warm-up period, and the one on the right about 5.3 hours later, immediately after the end of the data-taking period.

The central region from which the data were obtained seemed little affected, however, the results obtained for $S(q)$ showed additional fluctuation power at low q, and this additional power varied with what region was chosen for analysis and even from run to run. Time-domain spectral analysis of images taken in rapid succession (as rapidly as 30 Hz) revealed that presence of an additional noise source contributing very low temporal frequency power, but only for wave vectors lower than about 30 cm^{-1}.

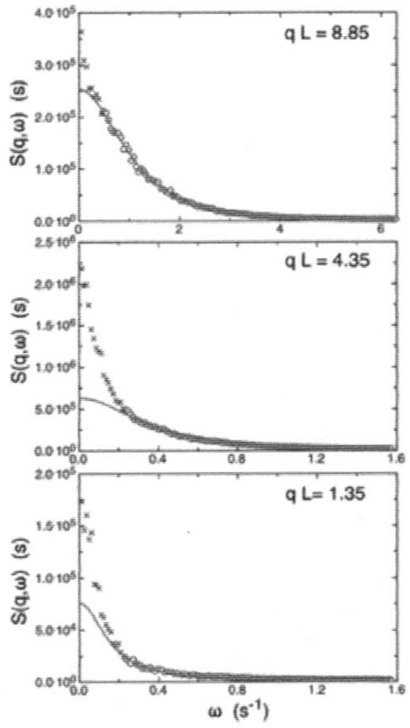

Figure 7. Time-domain spectral power for several small wave vectors with an applied gradient of 101 K/cm. Data images were taken at 2 Hz for the upper panel and at 0.5 Hz for the lower two. The theoretical form (solid line) was fit to the data shown as open circles, while the data shown as x's was excluded from the fit. The integral of the fitted function equals $T(q)S(q)$.

By fitting the higher frequency data using the predicted theoretical form and integrating the resulting spectrum it was possible to extract results for the product $T(q)S(q)$. Examples of the time domain power spectrum $S(q,\omega)$ for several very small wave vectors are shown in figure 7. As can be seen, the additional noise (visible in the data shown as x's) is significant only for angular frequencies below about 0.2 rad/s (~0.03 Hz), and is thus a very low frequency phenomenon, allowing it to be separated from the data by fitting. Clearly, the form predicted by theory provides an excellent description of the dynamics of the fluctuations, as shown by the good agreement between the open circles and the solid lines. By automating the fitting process is was possible to analyze the flight data to obtain $T(q)S(q)$ for many different q values, and the results are shown in figure 8, below.

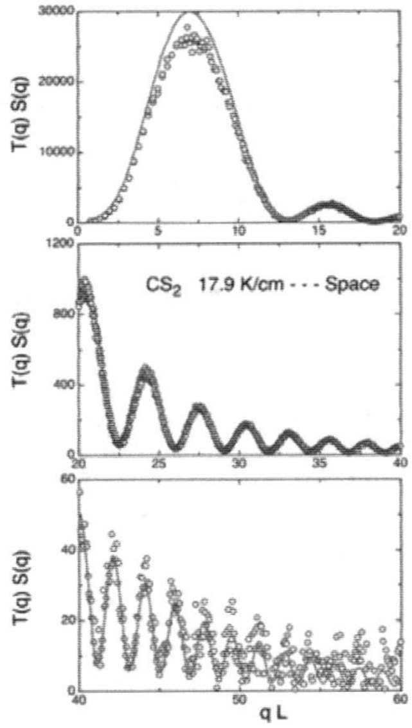

Figure 8. Data for $T(q)S(q)$ vs. the dimensionless wave vector qL taken in the absence of gravity with an applied gradient of 17.9 K/cm. The solid lines are the predicted results for $S(q)$ after multiplication by the measured transfer function. As can be seen the theoretical prediction exceeds the data by about 15%, an effect we observed for all applied gradients.

Comparison of the vertical axis scale with that of figure 4 shows that removing gravity has enhanced the signal by about a factor of 8, and that the position of the lowest peak has shifted to lower wave vector. This occurs because removing gravity eliminates the gravitational suppression of the fluctuations, and their amplitude continues to diverge as $q \to 0$ until limited by the finite sample thickness of 3 mm.

To display the behaviour more generally we may divide the data by the measured transfer function (bearing in mind that near the minima this involves dividing by small numbers!). Doing so produces experimental results for the structure factor $S(q)$ in absolute units, and the results are shown in figure 9 for several applied gradients, both on Earth and in the absence of gravity.

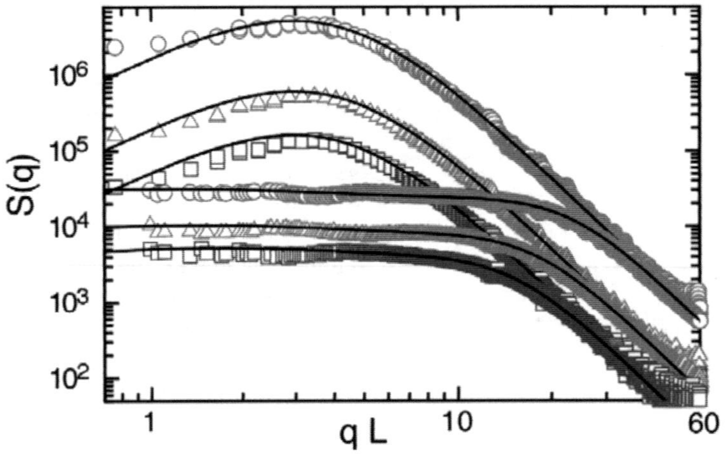

Figure 9. Data for the dimensionless structure factor $S(q)$ obtained by dividing the raw data by the q-dependent transfer function $T(q)$, vs. the dimensionless wave vector qL, for applied gradients of 17.9 K/cm (squares), 34.5 K/cm (triangles) and 101 K/cm (circles). The upper curves show the data obtained during flight and the lower ones show data obtained on Earth after the mission. The solid lines are the theoretical predictions without any adjustable parameters. From [19]

Note that the wide flat plateau region observed on Earth is replaced by a broad maximum in the absence of gravity. As predicted, this maximum occurs very near $qL = \pi$, showing that in the absence of gravity, the only thing limiting the range and amplitude of the fluctuations is the finite sample thickness. Considering that these fluctuations are thermally driven, this is certainly a remarkable observation.

Acknowledgements
We thank M. Potenza for discussions and help with the calibration procedure and F. Giavazzi for useful insights about the analysis of the shadow data. We also thank A. Verga, O. Minster, F. Molster. N. Melville, W. Meyer, R. Greger. C. Pereira and B. Hirtz for their contribution to the GRADFLEX project. Ground-based activity was supported by ESA and NASA. Flight opportunity sponsored by ESA

[1] Stanley H E 1971 Introduction to Phase Transitions and Critical Phenomena (New York: Oxford University Press)
[2] Degiorgio V and Scully M O 1970 *Phys. Rev. A* **2** 1170
[3] Zaitsev V M and Shliomis M I 1970 *Zh. Eksp. Teor. Fiz.* **59** 1583 [1971 *Sov. Phys. JETP* **32**, 866]

[4] Degiorgio V 1976 *Physics Today* **29** 42
[5] Wu M, Ahlers G and Cannell D S 1995 *Phys. Rev. Lett.* **75** 1743
[6] Kirkpatrick T R, Cohen E G D and Dorfman J R 1982 *Phys. Rev. A* **26** 995
[7] Schmitz R and Cohen E G D 1985 *J. Stat. Phys.* **38** 285
[8] Schmitz R and Cohen E G D 1985 *J. Stat. Phys.* **40** 431
[9] Ronis D and Procaccia I 1982 *Phys. Rev. A* **26** 1812
[10] Procaccia I, Ronis D and Oppenheim I 1979 *Phys. Rev. Lett.* **42**, 287
[11] Beysens D, Garrabos Y and Zalczer G 1980 *Phys. Rev. Lett.* **45**, 403
[12] Tanford C 1961 Physical Chemistry of Macromolecules (New York: Wiley)
[13] Tyrrell H J W 1961 Diffusion and Heat Flow in Liquids (London: Butterworths)
[14] Law B M, Gammon R W and Sengers J V 1988 *Phys. Rev. Lett.* **60** 1554.
[15] Segrè P M, Gammon R W, Sengers J V and Law B M 1992 *Phys. Rev. A* **45** 714
[16] Vailati A and Giglio M 1996 *Phys. Rev. Lett.* **77** 1484
[17] Vailati A and Giglio M 1997 *Nature (London)* **390** 262
[18] Vailati A, Cerbino R, Mazzoni S, Takacs C J, Cannell D S and Giglio M 2011 *Nat. Commun.* **2**:290 doi:10.1038/ncomms1290.
[19] Takacs C J, Vailati A, Cerbino R, Mazzoni S, Giglio M and Cannell D S 2011 *Phys. Rev. Lett.* **102** 244502
[20] Ortiz de Zárate J M and Sengers J V 2006 Hydrodynamic Fluctuations in Fluids and Fluid Mixtures (Oxford: Elsevier).
[21] Vailati A, Cerbino R, Mazzoni S, Giglio M, Nikolaenko G, Takacs C J, Cannell D S, Meyer W V and Smart A E 2006 *Appl Opt* **45** 2155
[22] Trainoff S P and Cannell D S 2002 *Phys. Fluids* **14** 1340
[23] Goodman J W 1996 Introduction to Fourier optics (Englewood, CO: Roberts and company)
[24] Sheng J, Malkiel E and Katz J 2006 *Appl. Opt* **45** 3893
[25] Potenza M A C, Sabareesh K P V, Carpineti M, Alaimo M D and Giglio M 2010 *Phys. Rev. Lett.* **105** 193901
[26] Van de Hulst H C 1981 Scattering of light by small particles (New York: Dover)
[27] Ortiz de Zárate J M, Peluso F and Sengers J V 2004 *Eur. Phys. J. E* **15** 319
[28] Chappa V C and Albano E V 2004 *J. Chem. Phys.* **121** 328
[29] Donev A, de la Fuente A, Bell J B, and Garcia A L 2011, *Phys. Rev. Lett.,* 106:204501
[30] Donev A, de la Fuente A, Bell J B, and Garcia A L 2011 *J. Stat. Mech.,* P06014

Foam Stability in Microgravity

N Vandewalle[1], H Caps[1], G Delon[1], A Saint-Jalmes[2], E Rio[5], L Saulnier[5], M Adler[3], A L Biance[3], O Pitois[3], S Cohen Addad[3], R Hohler[3], D Weaire[4], S Hutzler[4], D Langevin[5]

[1] GRASP, University of Liege, BELGIUM
[2] Institut de Physique, Universite Rennes, FRANCE
[3] LPMDI, Universite Paris Est, Marne-la-Vallee, FRANCE
[4] Foams and Complex Systems, Trinity College, Dublin, IRELAND
[5] LPS, Universite Paris Sud, Orsay, FRANCE

E-mail: nvandewalle@ulg.ac.be

Abstract. Within the context of the ESA FOAM project, we have studied the stability of aqueous and non-aqueous foams both on Earth and in microgravity. Foams are dispersions of gas into liquid or solid. On Earth, the lifetime of a foam is limited by the free drainage. By drainage, we are referring to the irreversible flow of liquid through the foam (leading to the accumulation of liquid at the foam bottom, and to a global liquid content decreases within the foam). When the liquid films become thinner, they eventually break, and the foam collapses. In microgravity, this process is no more present and foams containing large amounts of liquid can be studied for longer time. While the difference between foaming and not-foaming solutions is clear, the case of slightly-foaming solutions is more complicated. On Earth, such mixtures are observed to produce unstable froth for a couple of seconds. However, these latter solutions may produce foam in microgravity. We have studied both configurations for different solutions composed of common surfactant, proteins, anti-foaming agents or silicon oil. Surprising results have been obtained, emphasizing the role played by gravity on the foam stabilization process.

1. Introduction

Foams occur widely in nature and in important industrial processes and commercial applications [1]. Foams are made of closely packed gas bubbles in a liquid, typically water with an added surfactant. The liquid volume fraction φ may vary from less than one 1% (dry foam) to around 35% (wet foam). At low volume fractions, the bubbles are deformed into polyhedra. Most of the liquid is confined in the bubble edges called "Plateau borders", connecting three films and which are themselves interconnected by "nodes".

Gravity plays an important role in the formation of foam and its subsequent evolution. Its primary effect is to cause excess liquid to drain rapidly away. When the foam is stable enough, it becomes dry and the gravitational force is balanced by a vertical capillary-pressure gradient in the liquid and hence a vertical profile of liquid fraction is established. The addition of liquid to such a dry foam at the top results initially in a solitary wave with an approximately constant profile during its downward passage. Subsequently, a state of uniform drainage, with a constant volume fraction, is established. A sample undergoing such "forced" drainage can be used to study homogeneous wet foams, but only up to about $\varphi = 20\%$ liquid fraction, beyond which various dynamic instabilities (primarily convection) occur. In the case of rheology, a very interesting transition occurs at $\varphi_c \approx 35\%$ where the foam changes from liquid-like (disconnected bubbles) to solid-like (finite shear modulus). This is the "jamming transition" also encountered in other assemblies of randomly packed objects, such as emulsions, sand, clays, etc. When $\varphi > \varphi_c \approx 35\%$, the bubbles are spherical and move independently. Unfortunately, in the case of foams, the 20%-35% range, which extends to the wet foam limit at which individual bubbles separate, remains inaccessible on Earth. This limits present experiments to stable dry foams, and indeed the idealized theoretical models are largely confined to the dry foam limit as well.

A zero gravity study of wet foam hydrodynamics allows one to overcome the limits imposed by various instabilities experienced under normal gravity [2, 3, 4]. This broader experimental characterization and corresponding insight will provide a scientifically valid alternative for the necessarily conservative empiricism currently employed to estimate the operational window and design for foam handling in industrial processes (such as gas/liquid contacting, flotation and pumping).

Microgravity experiments on foams represent therefore a big challenge for scientists. Within our collaboration, parabolic flight campaigns were conducted for testing foaming processes and more recently the FOAM-S mission took place on ISS for studying the stability of a foam. Some of the results of the latter experiment are detailed in the present paper. Unexpected behaviours will be emphasized.

2. FOAM-S mission

FOAM-S benefited from a flight opportunity to the ISS in 2009 during the stay of astronaut Frank DeWinne. The experiment consisted in shaking various solutions both on Earth and in the ISS in order to study the influence of gravity on foamability and stability of foams. For that purpose, a small rack, illustrated in Figure 1, has been specially built. It contains 12 cylindrical transparent cells (polycarbonate, height = 40 mm, inner diameter = 12 mm) containing a liquid solution. The liquid volume fraction is close to 30% for all samples. This is our choice for studying foams near the critical liquid fraction φ_c. Each cell contains also a spherical ceramic bead used for the foaming process. The diameter of the bead (11 mm) is slightly smaller than the tube inner diameter. When the rack is hand shaken by the astronaut during 10 seconds, the bead motion provides enough shear stress along the tube wall and foam is created within a few oscillations of the bead. After shaking, the tubes are placed in front of a light source and images are recorded thanks to a high resolution video recorder during 30 minutes.

A total of 60 closed cells containing various foaming solutions were embarked in 5 racks. Various samples have been tested. Among others, solutions were pure liquids, water surfactant mixtures including protein based agents. Antifoaming agents were also tested as well as for a chocolate foam. The samples were sent on ISS in 2009 (see Figure 2). Images have been returned to scientific teams during winter 2010. In order to compare to normal gravity conditions, ground based experiments have been performed using the same apparatus. First results are described in the next section.

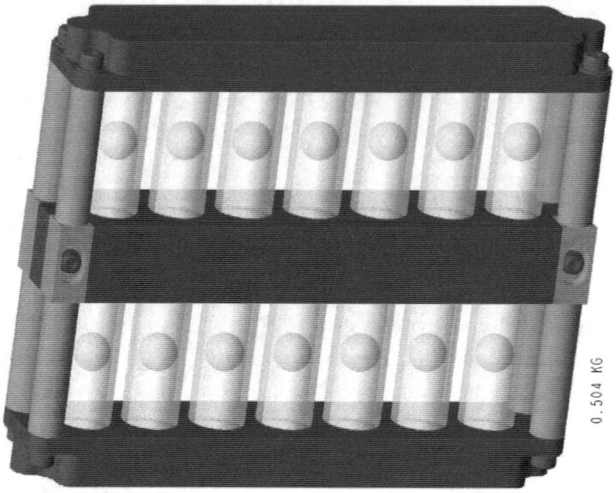

Figure 1. Sketch of the experimental setup which consists in a small rack containing 12 samples in partially filled cylindrical cells. When the system is shaken by hand, the motion of a bead in each cell will provide enough shear stress along cylinder wall, and will therefore imply the mixing of air with the liquid. The stability of this fresh foam is recorded by a camera.

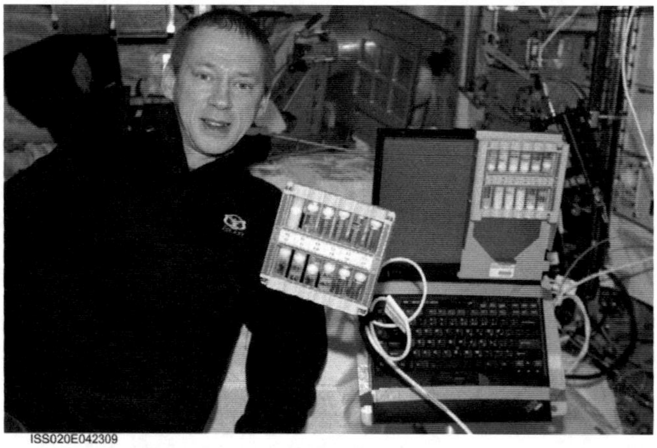

Figure 2. Astronaut Frank deWinne on the ISS. A rack is free floating in the center of the picture. A second rack is placed in front of a computer screen for backlight illumination during video recording.

3. Experimental results

The pictures of the 60 samples recorded on the ISS have been collected and analyzed. Typical pictures of foam created in cylindrical cells are shown in Figure 3 for four different solutions both in zero gravity and Earth conditions: SDS, TTAB, casein and anti-foaming agent Rodorsil TF1 © with SDS. One observes clear differences between 0g and 1g, as well as strong differences between samples. For example, the last sample of Figure 3 corresponds to an antifoam agent mixed in a SDS solution. The amount of foam created in micro-gravity seems more important

Journal of Physics: Conference Series **327** (2011) 012024 doi:10.1088/1742-6596/327/1/012024

than observed on Earth. However, the foam is stable when the system is shaken in microgravity. This behavior was completely unexpected since anti-foaming agents are meant to avoid foam creation and stabilisation [5].

In order to measure the foamability of each solution [6], the amount of foam Φ, ranging between 0% and 100%, created by shaking is measured on the pictures by image analysis. Foamability Φ is the volume fraction occupied by the foam only. It does neither consider the presence of the bead nor the presence of the remaining liquid. A rectangular zone is selected on the middle of a tube, in order to avoid the effect of the tube curvature. Such a rectangular zone is illustrated in Figure 3 by a black frame. A binarization threshold is applied to distinguish the foam from the rest of the tube: the foam appears in black. Then, the total area of the black region, representing the amount Φ of foam in the tube, is measured. The collected data are averaged over different runs. Table 1 summarizes our results. Among others, solutions shown in Figure 3 are denoted by a star "*". Note that in most cases, the foamability is almost two times larger on the ISS than on Earth. One exception is the solution of silica beads ©which was already very stable in Earth conditions [7]. A striking result is that a non-foaming solution on Earth exhibits a significant foam column in space ! This behaviour was unexpected.

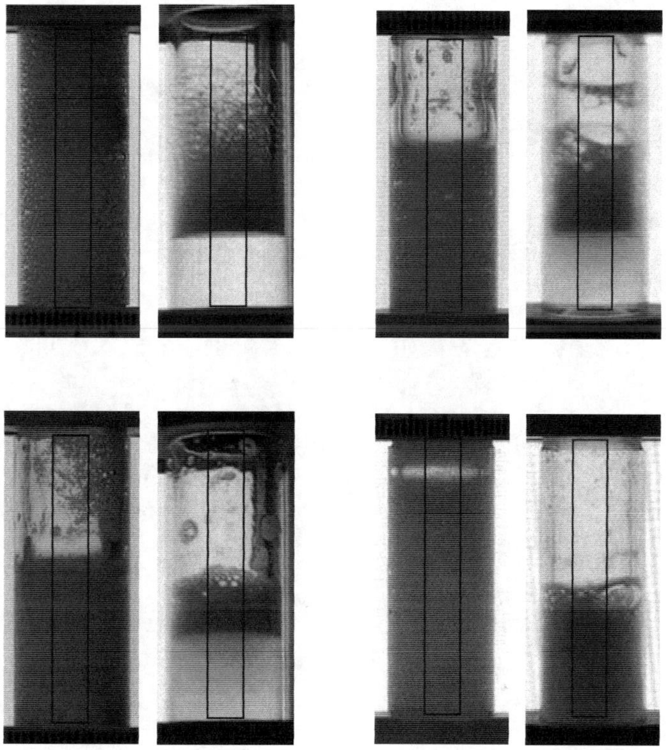

Figure 3. Snapshots of samples in respectively 0g and 1g after shaking : (top left) SDS $2g/L$ (top right) TTAB $0.34g/L$, (bottom left) caseine-based $1g/L$, (bottom right) anti-foaming agent Rodorsil with SDS $2g/L$. The rectangular black frames drawn on the pictures illustrate the zones where Φ is measured.

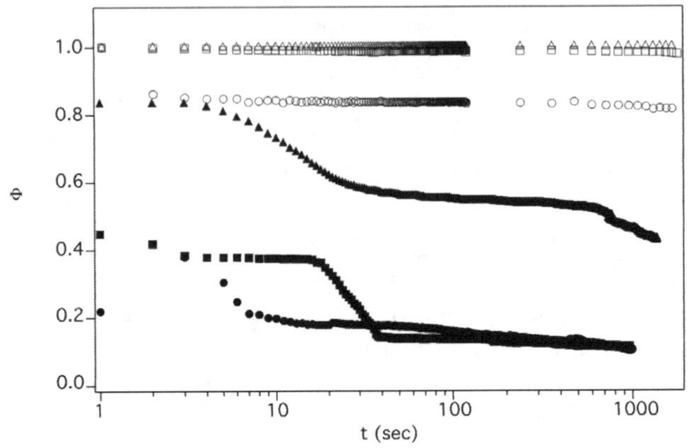

Figure 4. Semi-log plot of the foam volume fraction Φ in the cell as a function of time t, for the solution of SDS $0.05g/L$(circle and solid circle respectively at 0g and 1g), the solution of SDS $2g/L$ (triangle and solid triangle respectively at 0g and 1g), the solution of anti-foaming agent Rodorsil TF1 ©with SDS $2g/L$ (square and solid square respectively at 0g and 1g).

Image analysis provides also relevant information on the foam stability when applied to image series. Figures 4, 5 and 6 present a semi-log plot of the foam fraction Φ in cell as a function of time (after hand shaking at time $t = 0$). Different dynamical behaviors should be emphasized. In 1g, a typical foam column collapses significantly within 30 seconds before starting a slow decrease for the remaining foam. The characteristic time scales for this decay depends mainly on the foaming agent since various physico-chemical processes are involved : drainage, film ruptures, bubble reorganisations. One can note a significative difference between foams made at 0g and 1g as already noted in Figure 3. In most cases, the quantity of foam is larger in 0g (the tubes are almost totally filled with foam made with the SDS). This is true for solutions under and above the CMC, and also for solutions with anti-foaming agent Rodorsil TF1 ©. Moreover, these foams are showing a remarkable stability: the dynamics of Φ present a nearly flat curve in 0g. The amount Φ decreases rapidly in less than one minute before reaching a minimum. Whatever the observed solution on Earth, a characteristic time can be extracted, which is always around a few tens of seconds, whereas it is not the case in 0g. After ten seconds, the 0g foams are stabilized and no evolution is observed.

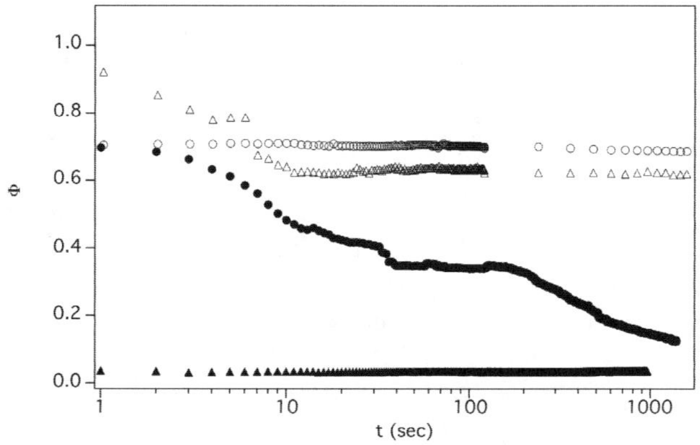

Figure 5. Semi-log plot of the foam volume fraction Φ in the cell as a function of time t, for the solution of TTAB $0.34g/L$(circle and solid circle respectively at 0g and 1g) and for the solution of TTAB $0.024g/L$ (triangle and solid triangle respectively at 0g and 1g).

Figure 5 presents the results for TTAB solution above and below the CMC. The 0g foam shows a greater stability, and the quantity of foam is significatively more important than under gravity except for the solution of 0.34g/L of TTAB for which the quantities of foams are the same with and without gravity. After ten seconds, the 0g-foams are stabilized and do not evolve further. Figure 6 shows the evolution of foam for solution of casein. The foams made in the ISS present a great stability even if a weak decay can be observed during all the experiments and fill the tube four times more than foam produced on Earth. One should also note that sudden fluctuations are seen in one of the series around $t = 40$ s. This is due to a moving object or astronaut near the experiment modifying the lightning. Such an experimental artefact is hopefully very rare.

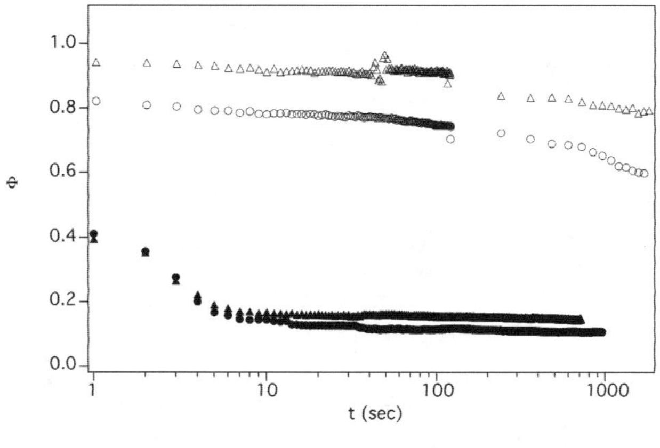

Figure 6. Semi-log plot of the foam volume fraction in the cell as a function of time, for the solution of casein $0.1g/L$(circle and solid circle respectively at 0g and 1g) and for the solution of casein $1g/L$ (triangle and solid triangle respectively at 0g and 1g).

solution	$\Phi(0g)$	$\Phi(1g)$
SDS 0.05g/L	0.569	0.295
SDS 2 g/L *	0.889	0.873
TF1 + SDS *	0.938	0.465
H2O/gly (30/70) +SDS	0.258	0.234
TTAB 0.024 g/L	0.71	0.026
TTAB 0.34 g/L *	0.697	0.351
Carboxymethylcellulose + TTAB	0.621	0.093
Tween 20 0.012 g/L	0.653	0.04
Tween 20 0.12 g/L	0.41	0.06
glass beads	0.42	0.435
Lyzozyme	0.227	0.051
Lactoglobuline	0.4951	0.238
Casein 0.01g/L	0.511	0.247
Casein 1g/L *	0.803	0.267
Egg	0.519	0.327
Chocolate foam ingredients	0.627	0.055

Table 1. List of selected samples and behaviours in 0g and 1g.

4. Discussion

Microgravity experiments allowed us to explore a new sort of foams with liquid fractions around $\varphi = 30\%$. On Earth, such materials are unstable and most of the time not called foams but bubbly liquid. The liquid fraction is indeed so high that the bubbles can keep their spherical shape allowing to minimize their free energy. In contrary, at smaller liquid fractions, bubbles are connected to others bubbles and deformed: the surface of bubbles is flatten due to the gravity driven liquid flow through Plateau borders. The main process which controls the stability of a foam on Earth is the free drainage driven by gravity. Film thinning in the foam leads to a global decrease of the cell srtucture as observed in all data for 1g, even when the foaming agent is present. The thinner films indeed allow a faster gas transfer and, thus a faster coarsening. Moreover, they tend to reach a critical thickness at which they become unstable and burst. In 0 g, drainage is suppressed such that liquid films remains thick. Because $\varphi = 30\%$, the foam remains wet. Bubble coalescence events are nevertheless seen, explaining a decrease of Φ on some plots. However, after some time, bubble motions become rare events such that the foam is more stable.

Note that the critical micellar concentration (CMC) is usually used to describe roughly the foamability. It is actually well known that a minimum concentration (around the CMC) is needed to create foam on Earth. Moreover, it has been observed that foamability is maximal around the CMC. This is actually due to surfactants remobilization. At large concentrations, the liquid indeed contains a large quantity of surfactants even after filling the surface. Hence, a small surface gradient is rapidly compensated by surfactants exchanges with the bulk and the surface elasticity is drastically decreased leading to smaller stability [10, 11]. As a result, on Earth, low concentration solutions do not foam or lead to unstable foams. Note that the considerations concerning surface rheology are still valid in the ISS but the foamability is also good for small concentrations. Then, it seems that in 0g, the CMC is no more a good order of magnitude for the frontier between stable and unstable foams. Indeed, surface rheology becomes important as soon as the films become very thin and can burst [8]. Concerning the results different on Earth and in space, there are also proteins which behave differently. In this case, it is worth noticing that they need quite a long time to reach the interface [10]. Actually, on Earth, foams stabilized by proteins (as well as particles) are very stable once generated but are very difficult to generate [11]. Obviously, on the ISS, the drainage is reduced giving time to the proteins to reach the interface. The various physical ingredients behind this unexpected stability have to be identified. We believe that the liquid fraction is an important parameter as tested in earlier parabolic flights.

5. Conclusion

From the FOAM-S experiment, we discovered the possibibilty to create super stable aqueous foams in zero gravity conditions. Even on Earth, coarsening and film ruptures are always present for a solution with foaming agents. In zero gravity, the foam evolves certainly but the amount of foam Φ does not appear to change significantly. Surprisingly, antifoaming agents have a reduced effect in microgravity, and this particular foam appears to be stable. This result raises new fundamental questions that should be investigated in future works.

In the FOAM-S experiment, cells have been filled with the same quantity of liquid (around 30% close to the rigidity loss transition). New experiments are needed to study the effect of this parameter on the amount of generated foam.

Acknowledgements

The project FOAM-S was financially supported by ESA. People at Verhaert are gratefully acknowledged.

References

[1] D.Weaire and S.Hutzler, *The physics of foams*, (Oxford University Press, USA, 2000)
[2] H.Caps, H.Decauwer, M.L.Chevalier, G.Soyez, M.Ausloos, N.Vandewalle, Eur. Phys. J. B, **33**, 115, (2003).
[3] A.Saint-Jalmes, S.Marze, H.Ritacco, D.Langevin, S.Bail, J.Dubail, G.Roux, L.Guingot, L.Tosini, P.Sung, Phys Rev Lett. **98**, 058303 (2007).
[4] D.G.T.Barrett, S.Kelly, E.J.Daly, M.J.Dolan, W.Drenckhan, D.Weaire and S.Hutzler, Microgravity Sci. Technol. **20**, 17 (2008).
[5] M.Safouane, D.Langevin, B.P.Binks, Langmuir **23**, 11546 (2007).
[6] H.Caps, N.Vandewalle, G.Broze and G.Zocchi Appl. Phys. Lett. **90**, 214101 (2007).
[7] A.C.Martinez, E.Rio, G.Delon, G, A.Saint-Jalmes, D.Langevin, B.P.Binks, Soft Matter **4**,1531 (2008).
[8] N.Vandewalle, J.F.Lentz, S.Dorbolo, F.Brisbois, Phys. Rev. Lett. **86**, 79 (2001).
[9] A.-L.Biance, S.Cohen-Addad, R.Höhler, Soft Matter **5**, 4672 (2009).
[10] D.Varade, D.Carriere, L. R.Arriaga, A.-L.Fameau, E.Rio, D.Langevin, W.Drenckhan, Soft Matter **7**, 6557 (2011)
[11] S.Tcholakova, N. D.Denkov, A.Lips, Phys. Chem. Chem. Phys., **10**, 1608 (2008).

Journal of Physics: Conference Series **327** (2011) 012025 doi:10.1088/1742-6596/327/1/012025

Foam Generation and Sample Composition Optimization for the FOAM-C Experiment of the ISS

Rodrigo Carpy[a], Gerold Picker[a], Benjamin Amann[a], Hans Ranebo[b], Sébastien Vincent-Bonnieu[b], Olivier Minster[b], Josef Winter[b], Jan Dettmann[b], Luigi Castiglione[b], Reinhard Höhler[c,e], Dominique Langevin[d]

[a]ASTRIUM Space Transportation, Department of Fluid Physics, Friedrichshafen, Germany
[b]European Space Agency, Human Spaceflight, ESTEC, The Netherlands
[c]Laboratorie de Physique des Matériaux Divisés et des Interfaces FRE 3300 CNRS-Université Paris-Est, France
[d]Laboratorie de Physique des Solides, Université Paris Sud, Orsay, France
[e]Institut des NanoSciences de Paris, UMR 7588 CNRS-UPMC Université Pierre et Marie Curie - Paris, France

FOAM-C Science Team Members
Université Paris Sud (FR), Université Pierre et Marie Curie (FR),Trinity College (IR), Université de Liège (BE), Université Paris-Est (FR), University of Pennsylvania (US), University Kyoto(JP), Moscow University (RU), MPI Golm (DE), IFP (FR), IENI Genova (IT), Unilever (UK), Nestlé (CH), Twente University (NL), St Petersburg University (RU), UTC (FR), Florence University (IT), Lafayette College(US), Alberta University (CA), Marseille University (FR), University Madrid UCM (ES), Thessaloniki University (GR), Loughborough University(UK), DLR (DE), MPI Goettingen (DE), Duke University(US)

Abstract: End of 2009 and early 2010 a sealed cell, for foam generation and observation, has been designed and manufactured at Astrium Friedrichshafen facilities. With the use of this cell, different sample compositions of "wet foams" have been optimized for mixtures of chemicals such as water, dodecanol, pluronic, aethoxisclerol, glycerol, CTAB, SDS, as well as glass beads. This development is performed in the frame of the breadboarding development activities of the Experiment Container FOAM-C for operation in the ISS Fluid Science Laboratory (ISS). The sample cell supports multiple observation methods such as: Diffusing-Wave and Diffuse Transmission Spectrometry, Time Resolved Correlation Spectroscopy [1] and microscope observation, all of these methods are applied in the cell with a relatively small experiment volume <3cm^3. These units, will be on orbit replaceable sets, that will allow multiple sample compositions processing (in the range of >40).

1. Introduction

Foaming processes are used on a large scale in the mining industry, for fire fighting and food production, but they can also be limiting factors for improved efficiency in many chemical industries like food and cosmetics production, but also in the oil industry. The experiment FOAM-C (Foam and Optics Mechanics - Coarsening) studies the coarsening process of aqueous wet foam as a function of liquid volume fraction and liquid characteristics. Microgravity allows investigating highly wet foam, which cannot be studied in detail in the earth laboratory due to drainage. Previous experiments have been performed in the frame of the FOAM project, such as parabolic flight campaigns, the Foam-S on the ISS (focused on foams stability) or the experiments in the MAXUS 6 rocket [2]. Now an extensive foam coarsening experiment under microgravity conditions for several types of probes and measurement periods as long as 24hrs for each sample experiment foam is foreseen to be performed on the ISS.

For the experiment on the ISS, where the storage times before experiment execution can be of several months, one of the crucial development points is the foam generation and quality of the generated foam. The foaming device shall be capable of generating wet foams with different surfactants and with a homogeneous initial size of foam bubbles (<100 μm), as described later in the materials and methods section.

More than 40 samples are foreseen to be processed for the first experiment mission to ISS. A concept of on-orbit replaceable sealed samples has been developed, avoiding the need of a fluid waste reservoir or on orbit sample cleaning.

The foreseen experiment sequence for the on-orbit execution is: 1) ground sample preparation with pre-defined filling procedures, concentrations and liquid fraction, 2) on-orbit foam generation, 3) foam coarsening observation with overview camera, diffusing wave spectroscopy in reflection (DWS) and transmission (DTS) and time resolved correlation spectroscopy (TRC), 4) repeat steps 2 and 3 for all samples is present.

The summary of breadboarding tests performed to demonstrate the foam generation capabilities of the system with the different experiment liquids is presented within the following chapters, additionally the initial system characterization to verify compatibility of the optical diagnostics concept to the foam coarsening observation methods have been performed.

2. Materials and Methods

The materials used for the experiment are: Dodecanol, increasing the surface viscoelasticity of the bubbles and probably decreasing the coarsening rate; Glycerol increasing the bulk viscosity and decreasing the coarsening rate; TTAB ionic surfactant, chemically very stable; and SDS as backup of the TTAB. Additionally, the following chemicals were tested: C12G2, as non-ionic surfactant in backup of TTAB; Pluronics a polymeric surfactant that adsorbs irreversibly at the air-water surface; Aethoxisclerol a pharmaceutical product for varicose veins; and Glass beads to study how a network of particles, linked by attractive capillar bridges, affects coarsening and bubble rearrangement dynamics. All of these chemicals, except Aethoxisclerol were tested as mixtures with different pure water concentrations. For the Diffusing Wave Spectroscopy Backscattering and Transmission measurements, Gillette shaving foam was used.

The setup for the foaming tests consists of a polycarbonate cell with thickness of 2mm and outer dimensions 53 x 25 x 20 mm (see Fig. 1), which contains a moving piston. The cell is cleaned with ethanol and an ultrasonic bath prior to filling. After the cleaning, the cell is filled with the desired concentration and liquid fraction and sealed. The sealing of the sample with a viton o-ring and screws, has three main advantages, which are: the liquid volume and concentration, are well defined and kept known during all phases of the experiment. The tests can be repeated with the same sample, as often as needed; and additionally there is no liquid waste, which would limit the experiment life on-orbit, due to limited available volume for liquid waste.

The foam is generated by a shearing force, occurring at the gap (1mm on each side) between the moving piston and the sample cell body. The piston, which contains a set of magnets, within its body, is magnetically coupled to the outside of the cell. The piston is driven with a frequency of 10Hz in all cases. The foam sample thickness L=1.1cm, height is 1.1cm and width 1.4cm. All experiments are carried out under normal gravity conditions at a temperature of 21±0.5°C.

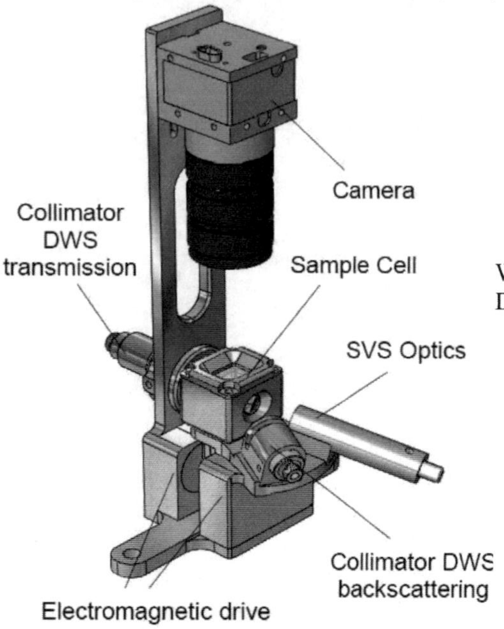

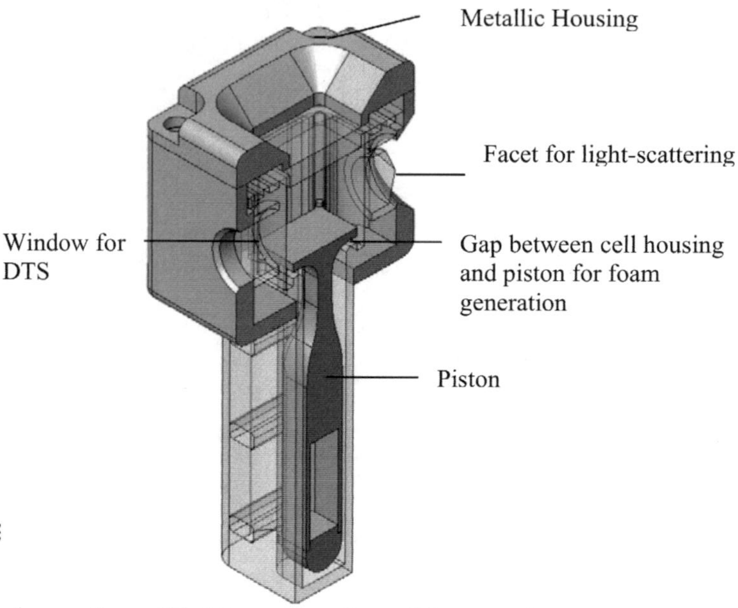

Fig. 2 (Left) Test Setup of the sample under observation. (Right) cross section of the experiment sample

The methods used for the observation (see **Fig. 2**) of the foam coarsening are, a CCD-camera for top-view observation of the first layer of bubbles, in the upper window of the sample cell, with six LEDs accurately placed below the sample, to provide both enough contrast and illumination for bubble observation. The additional light-scattering measurements (DWS, DTS and SVS) were illuminated with an expanded laser light beam (0.8mm on the foam surface), with a wavelength λ=532nm. The wavelength has no influence for the extrapolation ratio z_e, or the transport mean free path, l^* for aqueous foams as known from [3].

3. Results

The foams were prepared in different concentrations, such as: Dodecanol 2% of total mass, Glycerol 25% volume fraction, SDS 8g/l, TTAB 4.5g/l, C12G2 0.2g/l, Pluronics 17.5g/l, and the glass beads 15% of fraction of liquid. This means, a total of 13 sample compositions were tested. For each composition, different liquid fractions (LF) were tested with increments of 1% of liquid fraction for the 5-10% range and increments of 10% for 10-50%. Each test has been repeated 3 times to verify reproducibility. This sums up 390 measurements.

The success criterion was based on overview pictures of the foam sample cell (**Fig. 3**) for the study of the coarsening process, the required initial sample characteristics, have been as follows: an average bubble size <100 μm, maximum bubble size <5*average diameter, and a standard deviation between 0.3-3times the average bubble size.

Fig. 3 (Left) Example of homogeneous foam, (Right) Non-homogeneous foam. TTAB+Dodecanol+H_2O.

The samples with SDS provided the widest range of liquid fractions for which the foam generation is possible, in some cases with liquid fractions as low as 5%. Note that SDS is not as chemically stable as the TTAB (hydrolysis is present), which may be a problem if the solutions are studied only weeks after preparation. Tests carried out for the samples with TTAB have provided also high foaming properties and compatibility to the success criteria.

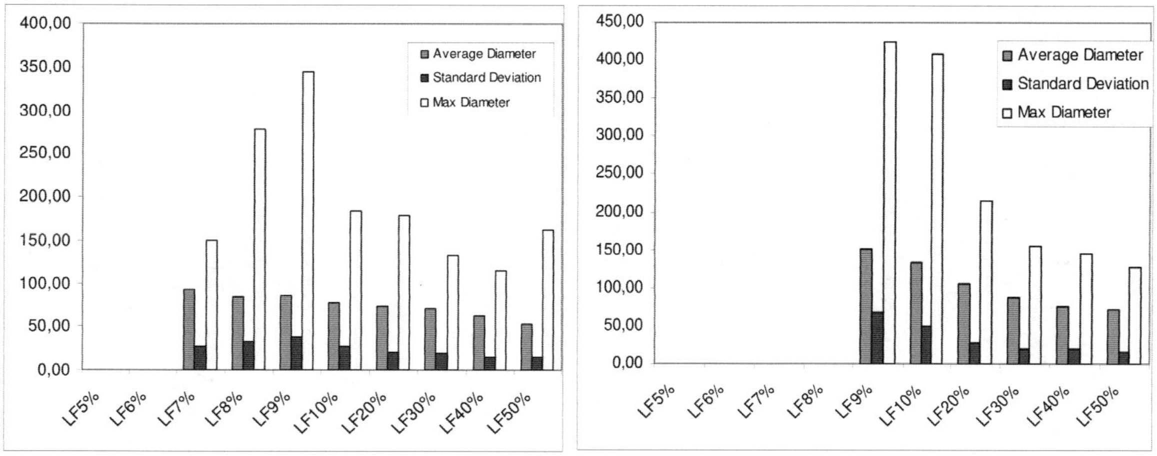

Fig. 4 Characteristics immediately after foaming of (left) TTAB + Water, (right) TTAB + Dodecanol + water.

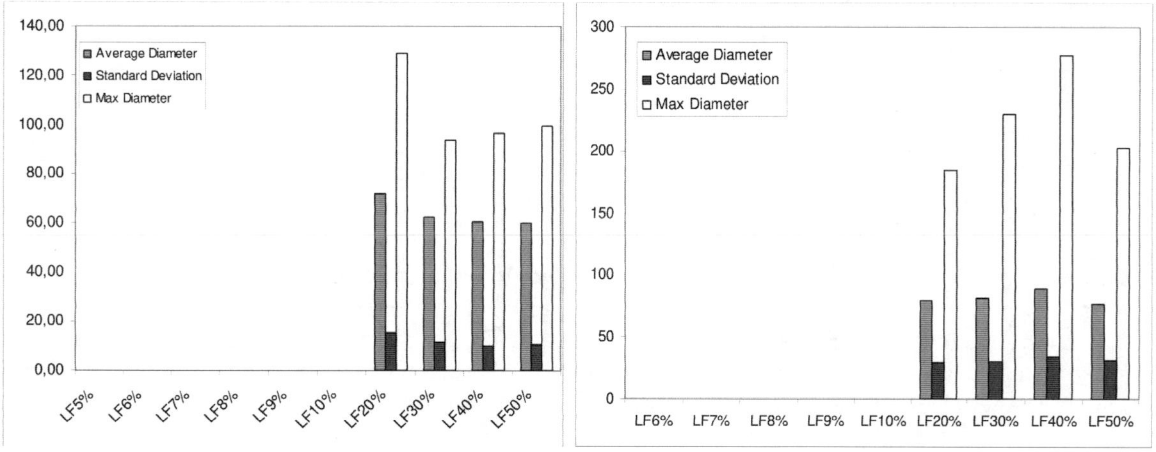

Fig. 5 Characteristics immediately after foaming (left) TTAB + Dodecanol + glycerol + water, (right) TTAB + glass beads+ water.

The results obtained with solutions containing C12G2, C12G2+Dodecanol + glycerol + water, pluronic surfactants and Aethoxisclerol were similar.

After the experiments performed to identify the liquid compositions and liquid fractions where foams of the required characteristics could be obtained, additional experiments with the setup, were executed for the additional light scattering methods. In this case, measurements under normal gravity conditions have been performed (Fig. 6) similar to [4], where comparison with other validated test setups and calculation of the average bubble size diameter using Gillette shaving foam has been performed.

From the output of the multiple tau autocorrelation the Siegert equation is used, then the formulas for $g1(\tau)$ for backscattering and transmission are used in line with [5] and [7].

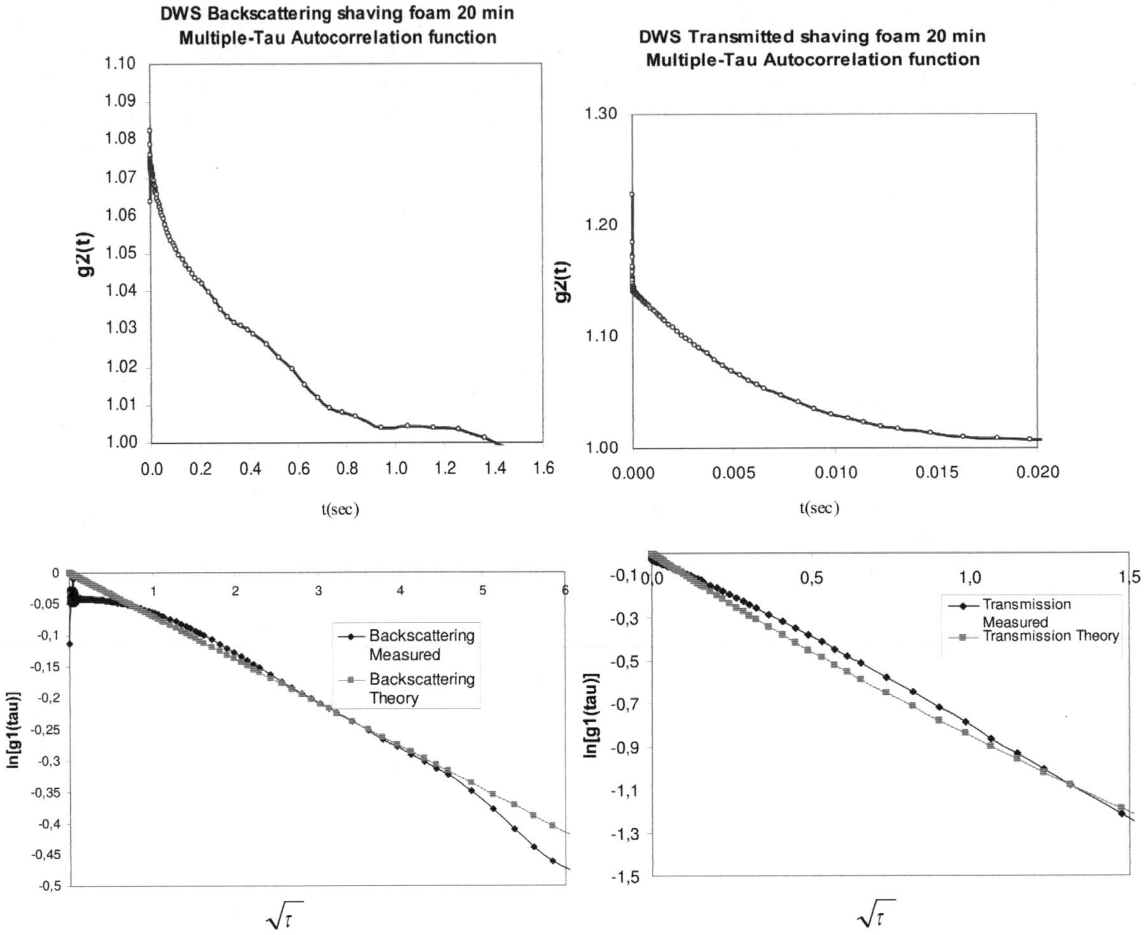

Fig. 6 Multiple-Tau Autocorrelation function of Gillette foam after 20mins for Diffusing Wave Spectroscopy Backscattering (upper-left) and Transmission (upper-right) and comparison. The logarithm of the correlation function g1 roughly scales linearly with delay time in the case of transmission, and as the square root of delay time in the case of backscattering. This is consistent with previous light scattering experiments with coarsening foams. The sample is Gillette shaving foam, at an age of 20min.

Following the reference measurements, following tests have been performed for functionality of optic and light scattering diagnostics. In this example, the coarsening of a probe of SDS+water with a liquid fraction of 20% has been evaluated. Following on the simplification applied in [6] where $T \approx 5l^*/3L$ the sample shows a linear behavior between the transport mean free path of light (l^*) versus the average bubble diameter (Fig. 7). The relationship for this probe was $l^*=(1.2\pm0.05)$d. The coefficient of this law will be refined in further analyses, taking into account a more accurate description of the optical boundary conditions. Out of this graphs, a clear short-time coarsening of the foam SDS+water (within 5 minutes), compared with other foams where times above 1000min [7] is observed. The bubble growth illustrated in Fig. 7 is different (linear) from the parabolic law usually observed in coarsening foams, possibly due to drainage effects. The growth rate of the bubbles under microgravity where drainage will not be present is one of the parameters to be evaluated during the ISS Foam-C experiment. The results are comparable to similar measurements performed within [8] where a bubble size of ≈140μm is measured for a SDS sample with LF15% after 120 seconds.

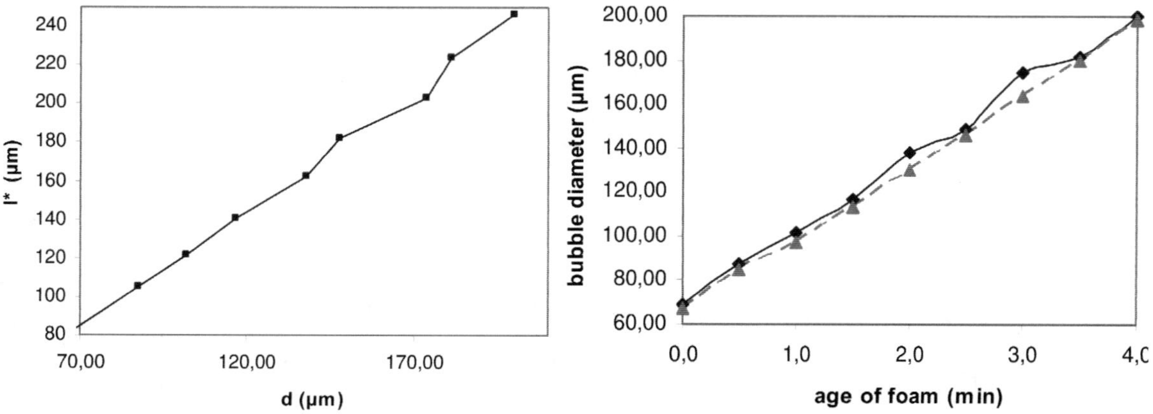

Fig. 7 (Left) Transport mean free path vs. average bubble diameter obtained by viewing first plane of bubbles for SDS+water LF 20%. (Right) Continues line shows the diameter of bubbles over time observed on the first plane of bubbles, dashed line shows the calculated diameter from measured static light transmission. Note: All measurements under gravitiy conditions.

4. Conclusions

In summary, we have described series of measurements for foam generation within a sealed volume, with different samples varying the composition, concentrations and liquid fractions with an objective of generating homogeneous foams with an average bubble diameter below 100µm. At the beginning of the measurements, it was not possible to define, which of these samples composition could be used for foam generation to fulfill the target foam homogeneity and size. After this campaign, an envelope containing the ranges and compositions of fluid, for which wet foams can be generated with the described test setup, has been defined and will be used as a baseline for the future development of the Experiment Container Foam-C for the International Space Station. In this sense, the results of these investigations have been successfully finished. The optic diagnostics have provided as well initial measurements and compatibility with previous theories or measurements with e.g. shaving cream.

An additional parabolic flight campaign is foreseen in spring 2012 to continue the investigations of the generation of aqueous foams under microgravity, as well as, the use and calibration of multiple light-scattering methods, in view of the Foam-C mission preparation for the ISS.

Acknowledgements

The industrial activities of the FOAM-C project are funded, and the science team is supported, by ESA. We are thankful to the support of Laboratoire de Physique des Matériaux Divisés et des Interfaces for the development of the optical diagnostics methods.

References

[1] Cipelletti -et al J. Phys Cond Mat 15, 257(2003).

[2] Saint-Jalmes, A., Marze, S., Safouane, M., Langevin, D., Cox, S., Weaire, D. Proceedings of the 17th ESA symposium on European rockets, Sandefjord, Norway, ESA SP-590 573.

[3] Moin U.Vera, Arnaud Saint-James, and Douglas J. Durain, Applied Optics 40, 4210 (2001)

[4] S. Cohen-Addad and R. Höhler, Phys Rev. Lett. A86, 4700 (2001)

[5] D.J. Pine, D.A. Weitz, J.X. Zhu, E. Herbolzheimer, J. Phys (Paris) 51, 2101 (1990)

[6] D. J. Durian, D.A. Weitz, and D.J. Pine, Science 252, 686 (1991)

[7] D.J. Durian, D.A. Weitz, and D. J. Pine, Phys Rev. A 44, 7902 (1991)

[8] A. Saint-Jalmes, M.L. Peugeot, H. Ferraz, D. Langevin, Surfaces A, **263**, 2 (2005)

Droplet collisions after liquid jet breakup in microgravity conditions

Francesc Suñol and Ricard González-Cinca

Departament de Física Aplicada, Barcelona Tech-Universitat Politècnica de Catalunya
Esteve Terradas 5, 08860 Castelldefels (Barcelona), Spain

E-mail: francesc@fa.upc.edu, ricard@fa.upc.edu

Abstract. The droplet dynamics and collisions after a liquid jet breakup have been experimentally studied in low gravity conditions. An experimental setup was designed in order to be used at the I.N.T.A. Drop Tower, which allows for 2.1 seconds of microgravity. The dynamics of distilled water jets injected into a rectangular tank was recorded by means of a high-speed video camera. Observations of the droplet trajectories showed a conical shape of the liquid jet caused by droplet collisions just after detachment from the liquid jet. The detached droplets initially follow straight paths at constant velocity in the direction of injection. Deviation of these trajectories is a consequence of the collision between two droplets with an impact parameter slightly different from zero. The collision between two droplets can give rise to coalescence or bouncing between droplets depending on the droplet velocity difference and impact parameter. At low values of the relative velocity, the collision leads to coalescence between droplets, while at higher values the collision results in bouncing between droplets.

1. Introduction

Liquid jets are encountered in a very large variety of natural and technological environments. The structure of liquid jets has been widely studied in the last century, due to its fundamental importance and practical applications in different fields. Nevertheless, the improvement of computer processing power, high-speed photography, and other laboratory techniques has recently attracted increasing attention to the subject.

On the analytical side, studies on liquid jets started with the work of Lord Rayleigh [1] and have been conducted in order to describe the breakup mechanisms. The fundamental tool is the linear stability analysis around the cylindrical base state, although later research refined this analysis to include the effects of viscosity, outside fluid effects, nozzle geometry, etc. [2–4]. In addition, many non-linear effects are dominant in the breakup process. Although these effects are extremely difficult to implement numerically, promising advances have been recently carried out [5–8].

From an experimental point of view, advances in numerous techniques have made possible the detailed study of the breakup process [9–13] and the structure of the resulting fragments [14–16]. In particular, the bouncing and coalescence of the resulting droplets [17–19]. Hence, a considerable amount of information can be provided for comparison with theory. However, experimental results have generally been obtained in normal gravity, with the jet injection direction aligned with the gravity force. The gravity force can be predominant in the dynamics of such flows, and it can mask the effects of inertia and surface tension. For this reason, a

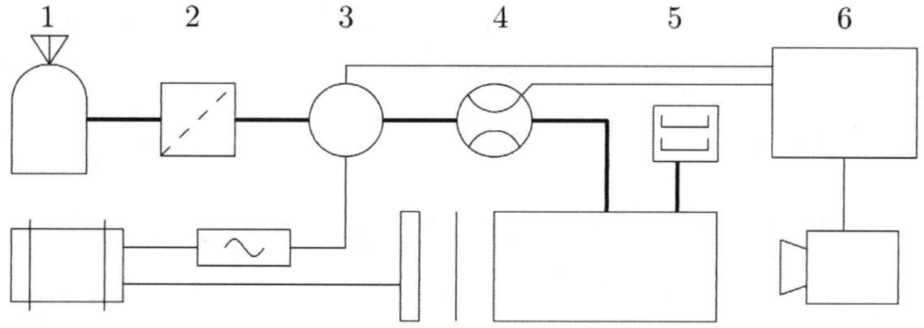

Figure 1. Sketch of the experimental setup. Thin lines: Electric connections. Thick lines: Hydraulic connections. 1: Liquid reservoir. 2: Filter. 3: Liquid pump. 4: Liquid flow meter. 5: Residual tank. 6: PC. 7: Batteries (power supply). 8: DC/AC inverter. 9: LED matrix. 10: Diffuser sheet. 11: Test tank. 12: High-speed camera.

microgravity environment is required to investigate the effects of the inertial and surface forces without the masking effects of gravity.

Umemura et al. [12] and Tsukiji et al. [13] studied the atomization regimes of a liquid jet in microgravity. The jet emerged into a gas whose temperature and pressure exceeded the critical values of the injected liquid, identifying a new hydrodynamic-assisted capillary instability. The parameter range in their investigations differs notably from the one used in the present work.

In this study, we focus on the structure of a jet after breakup by the Rayleigh mode in a low gravity environment. The breakup mechanism is determined by a direct competition between the surface and the inertial forces. The surface tension between fluids amplifies the jet surface perturbations until the radius of the perturbations equals the radius of the initial jet. At this instant, a droplet is created. This breakup mechanism occurs at low velocities and leads to the formation of droplets with a larger diameter than de initial jet diameter. On ground, the generated droplets follow a rectilinear path in a two-dimensional space initially in the direction of injection, with an increasing velocity due to gravity. The dynamics of droplets in microgravity is different. The collision between droplets after detachment gives rise to a resulting jet of conical shape, whilst the droplets follow rectilinear paths between collisions in a three-dimensional space.

A description of the experimental setup is provided in Section 2. The obtained results and discussion is presented in Section 3. Conclusions are finally drawn in Section 4.

2. Experimental setup

With the aim of studying the droplet dynamics and collisions after liquid jet breakup in a microgravity environment, an experimental setup was designed in order to be used at the I.N.T.A. Drop Tower (Torrejón de Ardoz, Spain), which allows for 2.1 seconds at 10^{-4}-10^{-5} g.

Liquid jets were injected into a stainless steel rectangular tank (with a length of 160 mm, width of 200 mm and height of 250 mm) provided with two methacrylate windows which allow the visualization of the interior of the tank. The working liquid was distilled water, with a density of $\rho = 998$ kg/m^3, viscosity $\mu = 1 \cdot 10^{-3}$ Pa·s, and surface tension $\sigma = 7.28 \cdot 10^{-2}$ N/m. Water was directed to the test tank from a pump (Ismatec MCP-Z Standard) connected to a nozzle, and the liquid flow rate Q_L was measured using a Bronkhosrt L30 liquid flow meter. Liquid flow rates were increased from 5 ml/min up to 15 ml/min. The nozzle internal and external diameter were $d = 0.7$ mm and $D = 1.8$ mm, respectively. The experiments were conducted at room temperature (around 20° C) and at ambient pressure. A RedLake MotionXtra HG-SE high-speed camera recorded the jets at framerates between 500 and 1000 frames per second

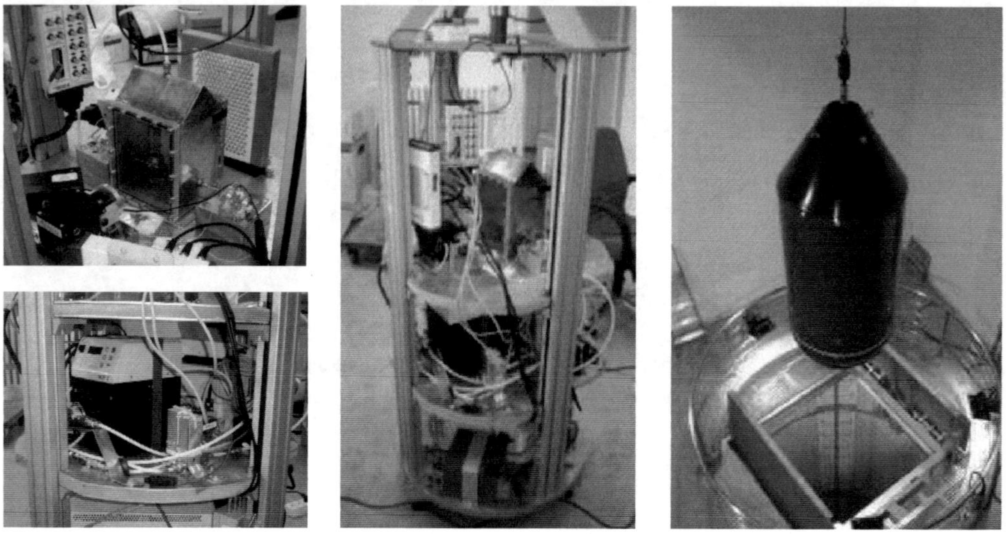

Figure 2. Pictures of the experimental rack (left and middle) and the capsule containing it (right) at the I.N.T.A. drop tower. Left: detail of the first and second floors of the rack; middle: whole rack.

(recording images at 640×512 and 640×256 pixels, respectively), depending on the experiment run. The high-speed camera had a built-in battery and an internal memory, and was activated by means of a trigger signal. The resolution of the acquired images was 0.15 mm per pixel. The illumination system consisted of a matrix of 280 ultrabright LEDs whose light was homogenized by a diffuser sheet. The liquid pump and the data acquisition system were controlled remotely using LabView software. A sketch of the experimental setup is presented in Fig. 1.

In Fig. 2, some pictures of the experimental setup in the capsule rack are shown. Equipment was distributed in three floors. Batteries were placed in the first floor. Test tank, illumination system and high-speed camera were placed in the second floor. In the third floor, there were the liquid pump, liquid flow meter, DC/AC inverter, liquid reservoir and residual tank.

3. Results and discussion

3.1. Droplet trajectories

In order to measure the droplet trajectories, an automatized method for tracking from the recorded videos the droplet centers evolving in time has been used.

In Fig 3, an x-y map (being x the injection direction) of the droplet trajectories for the case of $Q_L = 15$ ml/min is presented. It can be observed that the trajectories with the highest slope do not cross the origin (being the origin the average breakup point). In the video analysis one observes that the droplets created after the jet breakup maintain a horizontal trajectory following the direction of injection. The deflected trajectories are a consequence of droplet collisions, most of them taking place near the jet breakup point. Droplet collisions occur when a drop moving in the injection direction is followed by a slightly faster drop moving in the same direction with a low impact parameter different from zero. After the collision, the trajectories of the droplets are deviated from the injection direction. The collision process is shown in Figure 4: two droplets with impact parameter $b \neq 0$ are approaching each other with $|\vec{v}_1| < |\vec{v}_2|$. Although the initial velocities are in the injection direction (time t_1 in Fig. 4), after the collision both velocities have a non-zero component perpendicular to the direction of injection (time t_3 in Fig. 4).

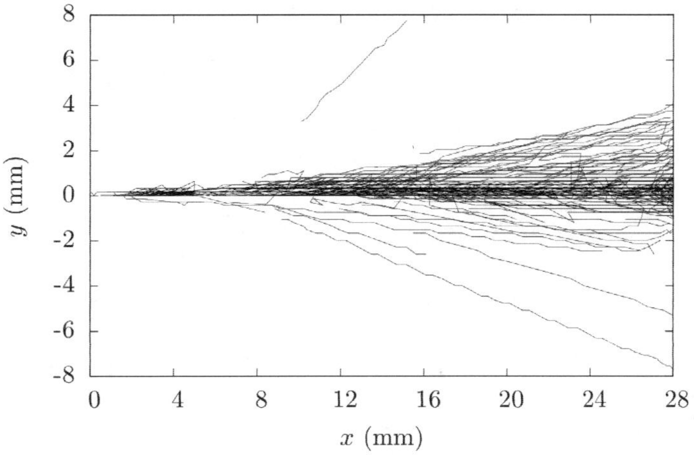

Figure 3. Map of droplet trajectories for $Q_L = 15$ ml/min.

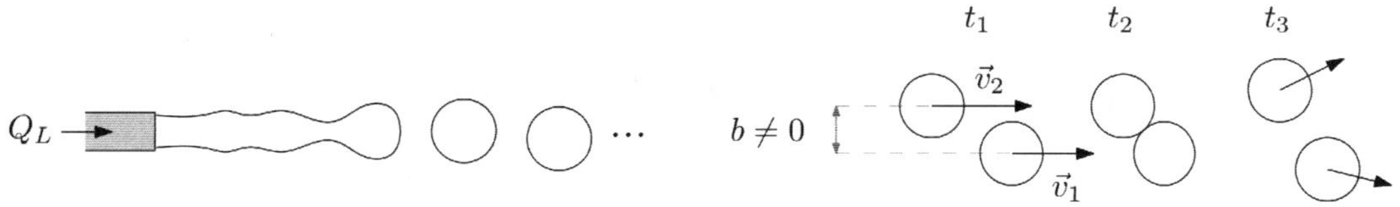

Figure 4. Collision process.

The deviation of the droplet trajectory can also be the result of multiple collisions or bounces. In Figure 5(a), a double bouncing is illustrated. At $t = 0$, a droplet is generated with a velocity v_1 in the injection direction. The following generated droplet (with a slightly smaller diameter) has a higher velocity v_2 in the direction of injection ($|v_2| > |v_1|$), and collides with the first droplet at $t = 8$ ms. At this point, v_2 is abruptly reduced, and a third droplet with velocity v_3 in the injection direction ($|v_3| > |v_2|$) collides with the second droplet at $t = 20$ ms. Since the second and third droplets have an impact parameter b different from zero, their trajectories are deviated with respect to the injection direction. This phenomenon can only be observed in microgravity conditions, since in normal gravity the acceleration of the droplets after detachment prevents the droplets from collisions.

In Figure 5(b), the time evolution of two colliding droplets is presented. In this case the collision gives rise to a coalescence phenomenon at $t = 3$ ms. Thus, in the range of droplet sizes and velocities of this work, the impact between two droplets can result in either bouncing or coalescence.

3.2. Droplet collisions: bouncing and coalescence

After a collision, droplets can bounce or coalesce. Considering that inertia and surface tension are key factors in these phenomena, the Weber number turns out to be an appropriate parameter of study. Taking into account the relative velocity $v_{\text{rel}} = v_i - v_j$ between droplets i and j prior to collision, a modified Weber number We* is introduced,

$$\text{We}^* = \frac{\rho v_{\text{rel}}^2 d^*}{\sigma}, \tag{1}$$

Journal of Physics: Conference Series **327** (2011) 012026 doi:10.1088/1742-6596/327/1/012026

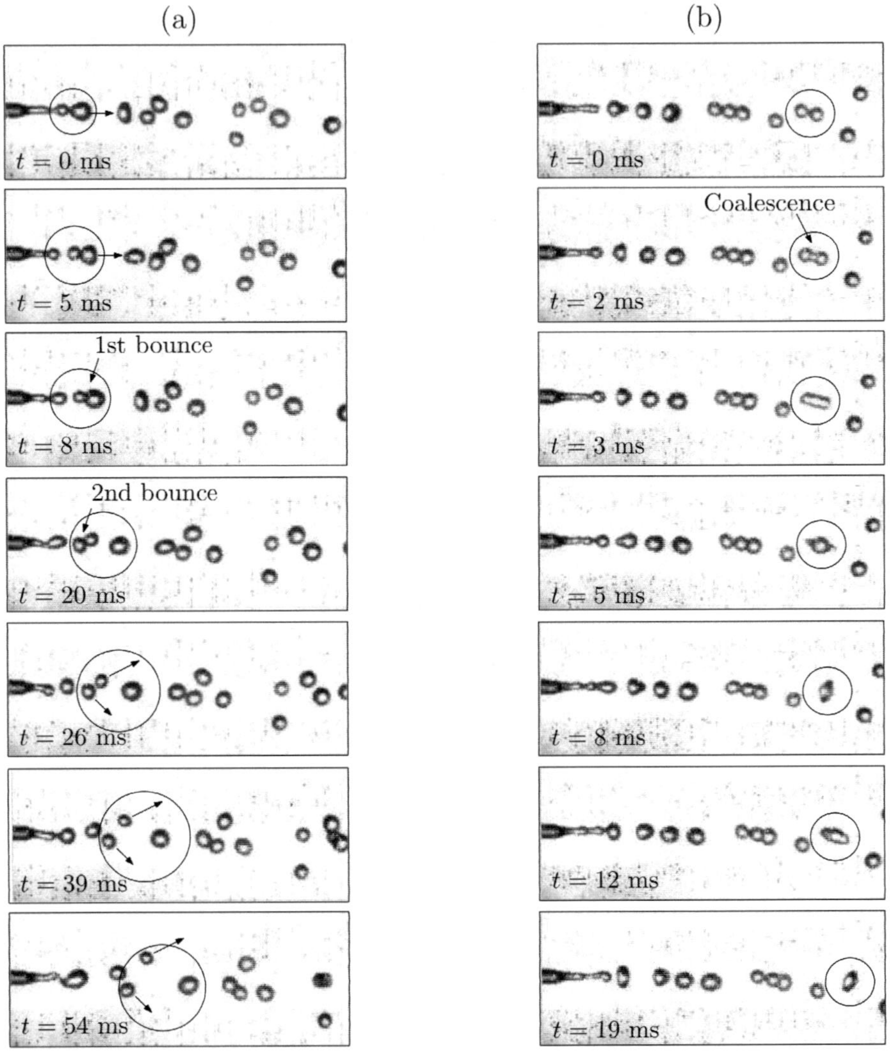

Figure 5. Time evolution of (a) Double bouncing process, (b) Coalescence process.

where $d^* = \sqrt{d_i d_j}$, being d_i and d_j the equivalent diameter of droplet i and j, respectively.

In Fig. 6, a map of the observed phenomena for given normalized impact parameter and modified Weber number is presented. Only four values of b/d (which correspond to 0, 1, 2 and 3 pixels) could be considered, due to the actual spatial resolution of the images.

Despite the few data points available, one can observe a tendency of coalescence taking place at low modified Weber numbers, while bouncing occurs more frequently at high values of We*. This may seem to be in contrast with the results obtained by Gao et al. [14] and Ko and Ryou [15] using ethanol and water droplet collisions, or the results from Chen and Chen [16] using diesel oil and water droplet collisions. In their works, bouncing is observed at lower We* and coalescence at higher values. However, the values of We* in their studies were larger than in the present work. Furthermore, in investigations of Orme, Willis and Passandideh and coworkers [17–19] an additional regime of coalescence at low values of the modified Weber number was found.

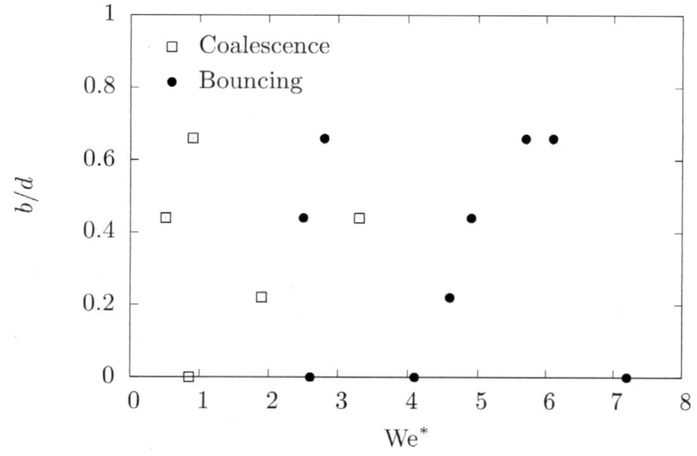

Figure 6. Observed phenomena (coalescence or bouncing) for given normalized impact parameter b/d and modified Weber number.

4. Conclusions

An experimental setup for the study of droplet dynamics and collisions in microgravity has been presented. Several experiments were run with this setup at the I.N.T.A Drop Tower. The most remarkable results obtained from the experiments can be summarized in the following points:

- A conical-shaped jet is observed in the region where droplets detach and move away from the initial jet. The conical shape is neither due to the rectilinear trajectories of the droplets detaching from the nozzle nor to the breakup phenomenon. The shape of the droplet jet is obtained from the bouncing process between colliding droplets just after detachment. This conclusion has been obtained from the trajectories of individual droplets, observing that the vertex of the cone has a positive offset in the injection direction, with respect to the jet breakup point.
- Droplet collisions at low values of the relative velocity result in coalescence. On the contrary, at higher values of the relative velocity, collisions result in bouncing between droplets.

Acknowledgments

We are thankful to I.N.T.A for their assistance and support in the payload integration. We acknowledge *Ministerio de Ciencia e Innovación* (Spain) for financial support (Project number AYA-2009-11493). F. S. also ackowledges *Comissionat per a Universitats i Recerca del Departament d'Innovació, Universitats i Empresa de la Generalitat de Catalunya i del Fons Social Europeu.*

References

[1] Rayleigh L 1878 On the instability of jets *Proc. of the London Math. Soc.* **10** 4.
[2] Teng H, Kinoshita C M, and Masutani S M 1995 *Int. J. Multiphase Flow* **21** 129.
[3] Lin S P and Reitz R D 1998 *Annu. Rev. Fluid Mech.* **30** 85.
[4] Eggers J and Villermaux E 2008 *Rep. Prog. Phys.* **71** 036601.
[5] Ménard T, Tanguy S and Berlemont A 2002 *Int. J. Multiphase Flow* **33** **(5)** 510.
[6] Fuster D, Bagué A, Boeck T, Le Moyne L, Leboissetier A, Popinet S, Ray P, Scardovelli R and Zaleski S 2009 *Int. J. Multiphase Flow* **35** **(6)** 550.
[7] Lebas R, Ménard T, Beau P A, Berlemont A and Demoulin F X 2009 Int. J. Multiphase Flow **35** **(3)** 247.
[8] Delteil J, Vincent S, Erriguible A and Subra-Paternault P 2011 *Computers and Fluids* Accepted manuscript - doi:10.1016/j.compfluid.2011.05.010.

[9] Reitz R D and Bracco F V 1982 *Phys. Fluids* **25** 1730.
[10] Liu Z and Reitz R D 1997 *Int. J. Multiphase Flow* **23 (4)** 631.
[11] Lasheras J C and Hopfinger E J 2000 *Ann. Rev. of Fluid Mech.* **32** 275.
[12] Umemura A and Wakashima Y 2002 *Proc. of the Comb. Inst.* **29** 633.
[13] Tsukiji H, Umemura A and Hisida M 2004 *Proc. of the Conf. on Aerospace Prop* 97.
[14] Gao T C, Chen R H and Lin T H 2005 *Experiments in Fluids* **38** 731.
[15] Ko G H and Ryou H S 2005 *Int. Journal of Multiphase Flow* **31** 723.
[16] Chen R H and Chen C T 2006 *Experiments in Fluids* **41** 453.
[17] Orme M 1997 *Prog. Energy Combust. Sci.* **23** 65.
[18] Willis K and Orme M 2003 *Experiments in Fluids* **34** 28.
[19] Passandideh-Fard M and Roohi E 2006 *14th Annual (International) Mechanical Engineering Conference* Isfahan University of Technology (Iran).

Numerical Modelling of Liquid Droplet Dynamics in Microgravity

S. Easter, V. Bojarevics, K. Pericleous

University of Greenwich, Park Row, London SE10 9LS, UK

V.Bojarevics@gre.ac.uk

Abstract. Microgravity provides ideal experimental conditions for studying highly reactive and under-cooled materials where there is no contact between the sample and the other experimental apparatus. The non-contact conditions allow material properties to be measured from the oscillating liquid droplet response to perturbations. This work investigates the impact of a strong magnetic field on these measurement processes for weakly viscous, electrically conducting droplets. We present numerical results using an axisymmetric model that employs the pseudo-spectral collocation method and a recently developed 3D model. Both numerical models have been developed to solve the equations describing the coupled electromagnetic and fluid flow processes. The models represent the changing surface shape that results from the interaction between forces inside the droplet and the surface tension imposed boundary conditions. The models are used to examine the liquid droplet dynamics in a strong DC magnetic field. In each case the surface shape is decomposed into a superposition of spherical harmonic modes. The oscillation of the individual mode coefficients is then analysed to determine the oscillation frequencies and damping rates that are then compared to the low amplitude solutions predicted by the published analytical asymptotic theory.

1. Introduction

The experiments with liquid metal droplets levitated in AC magnetic field show difficulties related to confinement instability and a need for complex correction functions to establish a correlation between the measurements and the droplet material properties [1-3]. A very intense internal fluid flow is visually observed, apparently being in the turbulent regime for earthbound conditions. The combination of AC and DC magnetic fields was recently recognised as an efficient tool for the electromagnetic processing of materials and for thermophysical property measurements without a contact to contaminating walls [4, 5]. The intense AC magnetic field required to produce levitation, along with the buoyancy and thermo-capillary forces, results in turbulent large-scale toroidal recirculation within the droplet, which prevents accurate measurements. The use of a homogenous DC magnetic field allows the toroidal flow to be damped. However the intense recirculation generates turbulence which could make the effective viscosity behave in a non-linear fashion depending on the DC and AC magnetic field intensity [5]. The flow in a typical droplet is approaching the conditions with laminar viscosity and heat transfer when a uniform DC magnetic field reaches about 4-5 T. Even a purely DC magnetic levitation, using para- and dia-magnetic properties of the materials, can be used for advanced material research [6, 7]. Due to the low values of magnetic susceptibility for typical liquid materials the fields are large (typically > 10 T), requiring superconducting coils with an associated cooling system. The vertical field gradient permits to compensate gravity along the central axis, while the radial variation opposes motion of the droplet and acts to centre it on the axis for stable levitation. However, in practice a small vibration of the droplet as a whole remains even in carefully conducted experiments [7].

The oscillations in a high DC magnetic field are quite different for an electrically conducting droplet, like liquid metal. In a recent publication [8] an asymptotic solution for very high magnetic field shows damping of the even axisymmetric modes, but the odd modes are not damped or damped moderately. The asymptotic linear theory predicts a considerable shift of the oscillating droplet frequencies from the non-magnetic case. The transition of the droplet behavior from the low to high magnetic field is the subject of investigation in this paper. Numerical models of the flow coupled with the moving free surface give an insight to the dynamics and complexity of the events within levitated droplets of various sizes and magnetic field intensities. For simplicity we will restrict the study to uniform DC magnetic field, leaving the more general cases of realistic gradient fields, as in the solenoidal coils of superconducting magnets, for a future work.

2. Droplet oscillation mode decomposition

The general theoretical approach assumes that the free surface shape of the droplet is defined by a small deviation from a sphere of radius R_0. The surface position can be represented by:

$$r_s = R_0\left[1+S(\theta,\phi,t)\right]. \tag{1}$$

The function defining the deviation from a sphere (S) varies in time and space and can be represented in spherical polar coordinates as a series of spherical harmonics:

$$S(\theta,\phi,t)=\sum_{L=0}^{L=\infty}\sum_{M=0}^{M=L} A_L^M(t)P_L^M(\theta)\cos(M\phi)+A_L^{-M}(t)P_L^{-M}(\theta)\sin(-M\phi), \tag{2}$$

The $Y^0_0 = P^0_0$ mode is included to account for conservation of mass (volume). The time-dependent surface coefficients are assumed to be harmonic with exponential damping:

$$A_L^M(t)=\cos\left(\omega_L^M t\right)\exp\left(-\gamma_L^M t\right) \tag{3}$$

The mode dependent oscillation frequency and damping constants are generally derived by following a perturbation approach. The zero order result for the frequency was calculated by Rayleigh [9], and is given for a droplet with density ρ and surface tension Γ by:

$$\omega_L^M=\sqrt{\frac{L(L-1)(L+2)\Gamma}{\rho R_0^3}} \tag{4}$$

The zero order result for the damping coefficient for a droplet of viscosity ν was determined by Lamb [10]:

$$\gamma_L^M=\frac{(L-1)(2L+1)\nu}{R_0^2} \tag{5}$$

It is interesting to note that the zero order approximation for both the frequency and the damping constant are independent of the azimuthal number M. These formulae provide a good first approximation for low amplitude oscillations of droplets in microgravity conditions where the effects of any external forces used to position the droplet are negligible compared with surface tension and viscous forces. However it is worth noting that these formulae are restricted and are only valid in the limit that the amplitude of oscillation and also non-linear effects tend to zero. Modifications to these lowest order formulae have been derived by Tsamopoulos & Brown [11]; they showed that for 'moderate amplitude', the frequency decreases with increasing amplitude and that a coupling of the spherical harmonic modes appears at the second order.

Performing experiments in conditions where any external positioning forces are negligible (drop shafts, parabolic flights, space) is limited due to expense and also by the amount of time in which experimental measurements can be made. The oscillating drop technique has been used in terrestrial conditions where the forces used to balance gravity and position the droplet are in the form of some sort of levitation, acoustic, aerodynamic, electromagnetic, or electrostatic. The use of a positioning field results in a more significant impact on the droplet dynamics and adjustments to the theoretical formulae are required. In the case of electromagnetic levitation correction formulae have been derived by Cummings & Blackburn [2], in which the external field is shown to remove the degeneracy of the L=2 modes resulting in a splitting of the frequency spectrum. Other levitation forces have also been considered by Suryanarayana & Bayazitoglu [3].

Perhaps the most promising method of confinement for conducting measurements using the oscillating droplet technique is diamagnetic levitation in a superconducting magnet. In this case the weak diamagnetic properties exhibited by many materials can be harnessed by a large magnetic field and used to balance the force of gravity without the more invasive Lorentz forces associated with AC electromagnetic levitation. The resulting total magnetic force on the droplet allows gravity to be compensated, although there is some variation in the force and a component that acts towards the axis, which means that it is a good approximation to microgravity conditions but not completely equivalent. Frequency measurements using the oscillating drop technique have been conducted for water droplets by Beaugnon et al. [6], where estimates were made as to the frequency modification due to the magnetic field. More recent experiments have been conducted by Hill & Eaves [7], in which a derivation of the frequency modifications due to the magnetic field are made and compared with the experimental results.

In a recent paper Priede [8], considers the effect of a constant high intensity magnetic field on the oscillation frequency and damping rate of an electrically conducting drop. The magnetic field is shown to significantly alter the dynamics of the droplet with some interesting results. There are two distinct cases; the longitudinal modes when (*L-M*) is odd and the transversal modes when (*L-M*) is even. The frequencies for the odd and even modes respectively are given by:

$$\omega_L^M = \sqrt{\frac{(L-1)(L+2)\Gamma}{\rho R_0^3}} \tag{6}$$

$$\omega_L^M = M\sqrt{\frac{(L-1)(L+2)}{L(L+1)-M^2}\frac{\Gamma}{\rho R_0^3}} \tag{7}$$

The damping constant consists of two components, damping due to the magnetic field B_0 and damping due to viscosity ν. The damping contribution of the magnetic field for the odd and even modes respectively are given by:

$$\gamma_L^M = \frac{1}{6}(L-1)(L+2)\left((L-1)(L+2)+2M^2\right)\frac{\sigma B_0^2 R_0^3}{\Gamma} \tag{8}$$

$$\gamma_L^M = \frac{(L-1)(L+2)(L^2-M^2)\left((L+1)^2-M^2\right)\left(3L(L+1)-2M^2\right)}{6\left(L(L+1)-M^2\right)^2}\frac{\sigma B_0^2 R_0^3}{\Gamma} \tag{9}$$

The contribution to the damping constant due to viscous forces is given by:

$$\gamma_L^M = C_L^M \frac{\nu}{R_0^2} \tag{10}$$

Where (C_L^M) are coefficients determined by evaluation of an integral without a closed form solution. In this paper we are only concerned with the low order mode with coefficients as given in [8]:

$$C_2^1 = \frac{5}{2} \qquad C_2^2 = 5 \qquad C_3^0 = \frac{35}{3} \tag{11}$$

In the following section an outline is given of an axisymmetric and a 3D numerical model that will be used to investigate these effects. Following results are presented that demonstrate quantitatively the theoretical predictions and the gradual transition with the increase of the magnetic field.

3. Numerical Model

The 3D numerical model uses a grid point formulation of the spectral collocation method with the Chebyshev grid for the radial direction and Fourier in the angular directions. The model uses a coordinate transformation for the free surface, which allows the problem to be solved on a unit sphere. The equations solved by the numerical model are the momentum and mass conservation equations with the Poisson equation for the modified pressure, $P_{\text{mod}} = P + \rho g z - \dfrac{\chi_v |B|^2}{2\mu_0}$:

$$\frac{\partial \mathbf{V}}{\partial t} + (\mathbf{V}\cdot\nabla)\mathbf{V} = -\frac{1}{\rho}\nabla P_{\text{mod}} + \nu\nabla\cdot\left(\nabla\mathbf{V} + \nabla\mathbf{V}^T\right) + \frac{\mathbf{J}\times\mathbf{B}}{\rho};\quad \nabla\cdot\mathbf{V} = 0; \tag{12}$$

$$\frac{1}{\rho}\nabla^2 P_{\text{mod}} = -\nabla\cdot\left[(\mathbf{V}\cdot\nabla)\mathbf{V} - \frac{1}{\rho}\mathbf{J}\times\mathbf{B}\right]. \tag{13}$$

The total force due to the magnetic field consists of two components, the Lorentz force due to the conducting properties, which is added to the momentum equations as a body force and the diamagnetic force, which is potential and is implemented in the model as boundary condition to the pressure equation along with the gravitational force, which is also potential. The full boundary conditions for the equations (12) and (13) relate the normal stress to the three forces (surface tension, gravity, and diamagnetic):

$$\mathbf{e_n}\cdot\Pi\cdot\mathbf{e_n} = \Gamma K + \rho g z - \frac{\chi_v |\mathbf{B}|^2}{2\mu_0}, \tag{14}$$

the continuity and the tangential stress conditions:

$$\nabla\cdot\mathbf{V} = 0, \tag{15}$$

$$\mathbf{e_n}\cdot\Pi\cdot\mathbf{e}_{\tau 1} = 0, \tag{16}$$

$$\mathbf{e_n}\cdot\Pi\cdot\mathbf{e}_{\tau 2} = 0, \tag{17}$$

where $\boldsymbol{e}_n, \boldsymbol{e}_{\tau 1}, \boldsymbol{e}_{\tau 2}$ are the unit vectors normal and tangential to the free surface, Π is the stress tensor, K is the surface curvature, and χ_v is the volumetric magnetic susceptibility.

The free surface shape is defined by its deviation from a sphere of radius R_0 by an amount S given as a function of time and the angular coordinates:

$$r_s = R_0\left[1 + S(\theta,\phi,t)\right] \tag{18}$$

The new surface position at each time step is evaluated using the kinematic condition:

$$\frac{\partial r_s}{\partial t}\cdot\mathbf{e_n} = \mathbf{V}\cdot\mathbf{e_n}. \tag{19}$$

3.1. Electromagnetic Forces

The magnetic field from a solenoid is modelled as a superposition of the magnetic field generated by axisymmetric coil filaments of radius a. The axisymmetric magnetic field in cylindrical coordinates, $\mathbf{B}(r,0,z) = (B_r, 0, B_z)$, is obtained from the analytical formulae [12]:

$$B_r = \frac{\mu_0 I}{2\pi} \frac{z}{r\sqrt{(a+r)^2 + z^2}} \left[-K + \frac{a^2 + r^2 + z^2}{(a-r)^2 + z^2} E \right] \tag{20}$$

$$B_z = \frac{\mu_0 I}{2\pi} \frac{1}{\sqrt{(a+r)^2 + z^2}} \left[K + \frac{a^2 - r^2 - z^2}{(a-r)^2 + z^2} E \right] \tag{21}$$

Where, a = coil radius, I = electric current, K, E are the complete elliptic integrals of the first and second kind respectively with argument, $k^2 = 4ar\left[(a+r)^2 + z^2\right]^{-1}$.

The stable levitation conditions can be established by adjusting the electric current in the coils to form a minimum in the magneto-gravitational potential (per unit volume):

$$U_{vol} = \rho g z - \frac{\chi_v |\mathbf{B}|^2}{2\mu_0}. \tag{22}$$

The Lorentz body force requires an additional equation to be solved for the electric potential. The electric current density is given by:

$$\boldsymbol{J} = \sigma\left(-\nabla\phi_E + \boldsymbol{V} \times \boldsymbol{B}\right). \tag{23}$$

The equation for the electric potential is obtained from the charge conservation condition:

$$\nabla \cdot \boldsymbol{J} = 0. \tag{24}$$

Taking the divergence of the electric current density gives:

$$\nabla^2 \phi_E = \nabla \cdot \left(\boldsymbol{V} \times \boldsymbol{B}\right). \tag{25}$$

This equation is solved subject to the boundary condition:

$$\boldsymbol{J} \cdot \boldsymbol{e}_n = 0. \tag{26}$$

The magnetic field in this paper is mostly prescribed as the constant field given in cylindrical coordinates by:

$$\boldsymbol{B}(r,\phi,z) = (0,0,B_0). \tag{27}$$

4. Results & Discussion

The following section contains the results of numerical simulations made using both the 2D axisymmetric [13] and 3D modes compared with theoretical asymptotic results from [8].

The material chosen for the numerical test cases is either molten silicon having the material property values for surface tension coefficient, density, kinematic viscosity, volumetric magnetic susceptibility, and conductivity, respectively: Γ = 0.865 N m^{-1} , ρ = 2510.0 Kg m^{-3} , ν = 3.75e-7 m^2 s^{-1} , χ_v = -4.2e-6, σ = 1.3e6 S m^{-1}; or for the molten nickel: Γ = 1.778 N m^{-1} , ρ = 7995 Kg m^{-3} , ν = 6.2e-7 m^2 s^{-1} , χ_v = 1.6e-6, σ = 1.18e6 S m^{-1}

4.1. Freely Oscillating Droplets, Axisymmetric Mode (2,0)

The first case considered is the axisymmetric mode Y^0_2 for the droplet of radius (R_0 = 5.0 mm), initial amplitude (A^0_2 = 0.01), and the velocity V=0. The constant magnetic field has strength (B_0 = 2.5T). The initial amplitude corresponds to a polar deformation of 1% of the equilibrium droplet radius. A

Journal of Physics: Conference Series **327** (2011) 012027 doi:10.1088/1742-6596/327/1/012027

small volume correction is applied in the form of a (A^0_0) coefficient, with the magnitude determined numerically.

Figure 1 shows the flow for a droplet moving in the (Y^0_2) mode with the external magnetic field. Maximum deformation occurs when t = 0, then monotonously decreases to reach the spherical shape as shown in the Figure 2 for various field magnitudes.

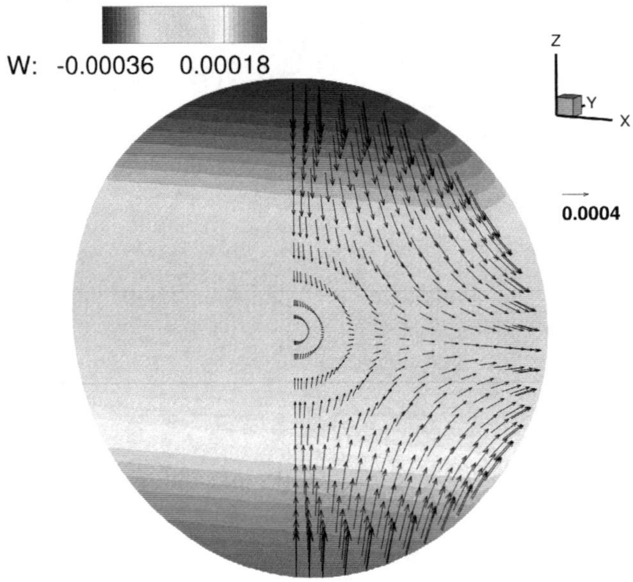

Figure 1. Oscillation mode (2,0), instantaneous velocity vectors and vertical velocity contours for Si droplet in 2.5 T magnetic field.

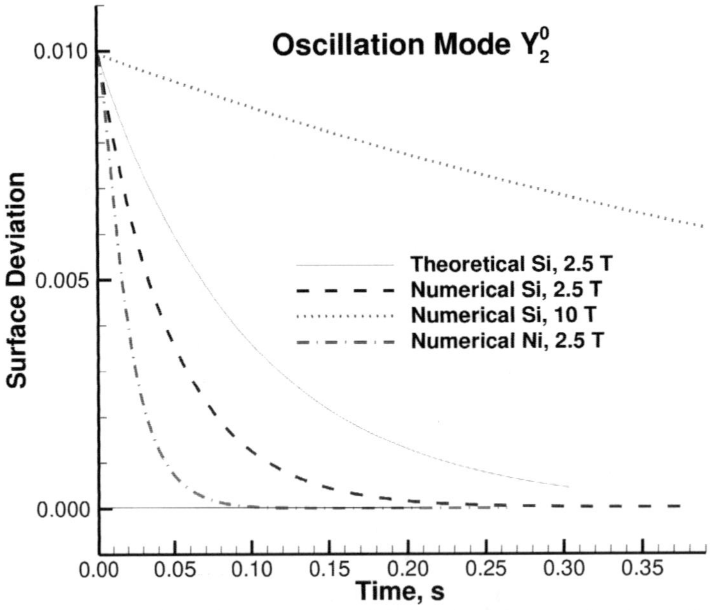

Figure 2. Surface deviation $(R_{top}-R_o)/ R_o$ function evaluated at the top point for Si and Ni droplets in various magnetic fields.

Theoretical asymptotic analysis for high B shows that the mode is over-damped by the magnetic field, the oscillation frequency is zero and the magnetic damping constant is given by:

$$\gamma_2^0 = 12\frac{\sigma B_0^2 R_0^3}{\Gamma}. \tag{28}$$

The reason for this result is clear when observing the flow profile in the Figure 1. The flow component in radial direction u is perpendicular to the uniform magnetic field along the z-axis. Due to the symmetry of the flow it is not possible to satisfy mass continuity with a potential flow only in the z-direction. Figure 2 shows the results of a numerical simulation compared with the theoretical asymptotic result. In this case the numerical model shows a higher level of damping than the theoretical because the result (28) does not include the viscous damping for the axisymmetric mode at the finite value of the magnetic field. The numerical results show the dependence on the material properties and the field intensity. For the very high field (10 T) the damping leads to a very slow motion, and the droplet reaches the spherical shape in a considerably longer time than without field.

4.2. Freely Oscillating Droplets, Axisymmetric Mode (3,0)

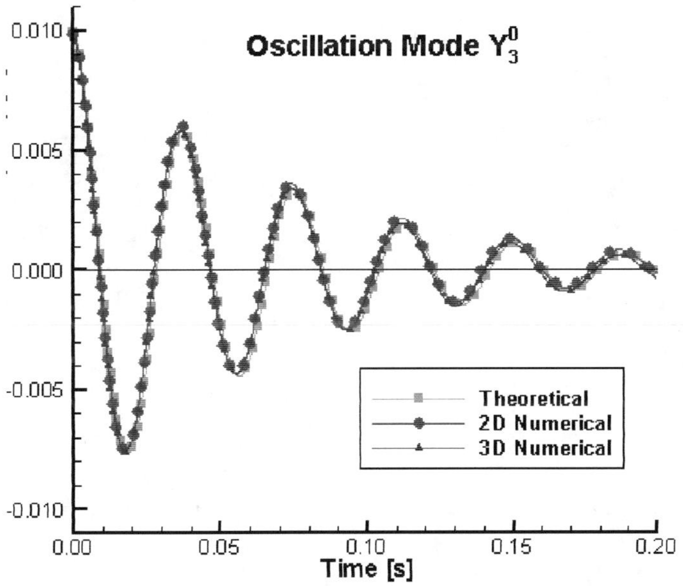

Figure 3. Surface deviation function, $(R_{top}-R_o)/R_o$, evaluated at the top point for B_o =2.5 T.

The next case considered is the axisymmetric mode (Y^0_3) droplet perturbation, the radius, amplitude, and magnetic field strength are all the same as in the previous case. An initial volume correction is again made to preserve the volume of the sphere with the radius R_o. Theoretical analysis shows that the oscillation frequency and the magnetic and viscous damping constants are given, respectively, by:

$$\omega_3^0 = \sqrt{\frac{10\Gamma}{\rho R_0^3}} \tag{29}$$

$$\gamma_3^0 = \frac{50}{3}\frac{\sigma B_0^2 R_0^3}{\Gamma} + \frac{35}{3}\frac{\nu}{R_0^2}. \tag{30}$$

Journal of Physics: Conference Series **327** (2011) 012027 doi:10.1088/1742-6596/327/1/012027

Figure 3 shows the results of the numerical simulations with both the 2D and 3D modes compared with the theoretical result. There is reasonably good agreement with all 3 results.

Figure 4 shows a Ni droplet oscillating in the (Y^0_3) mode with a high external magnetic field B_o =10 T. The radial flow is damped by the magnetic field, however the symmetry and continuity restrictions are not present here and the droplet continues to oscillate in the (Y^0_3) mode. The internal flow is quite different to the case without the magnetic field (Figure 5). In the presence of the strong DC magnetic field in the z-axis direction there is practically no radial flow, and the fluid oscillates in columns along the direction of the external magnetic field.

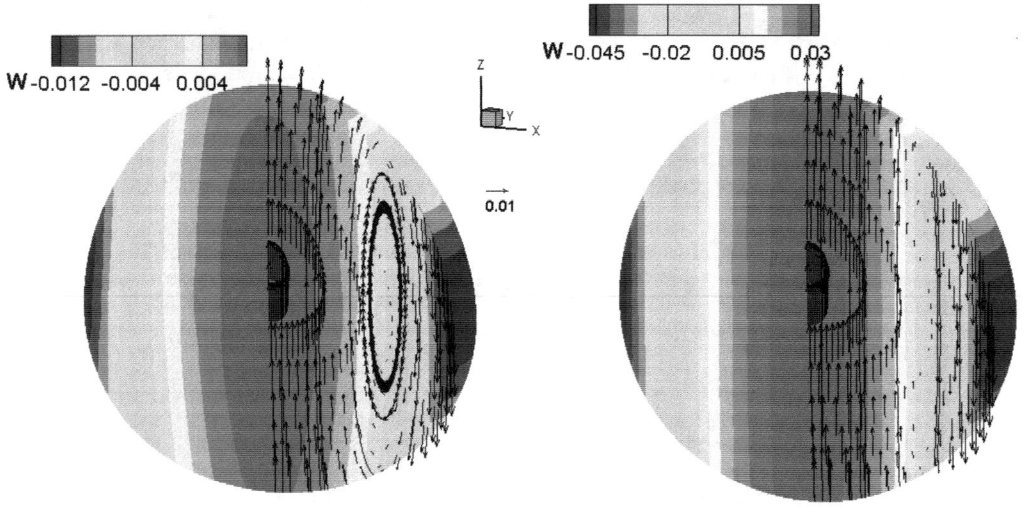

Figure 4. Oscillation mode (3,0) in B_o =10 T magnetic field, velocity field at two moments: 0.226 s, and 0.232 s for the liquid Ni droplet.

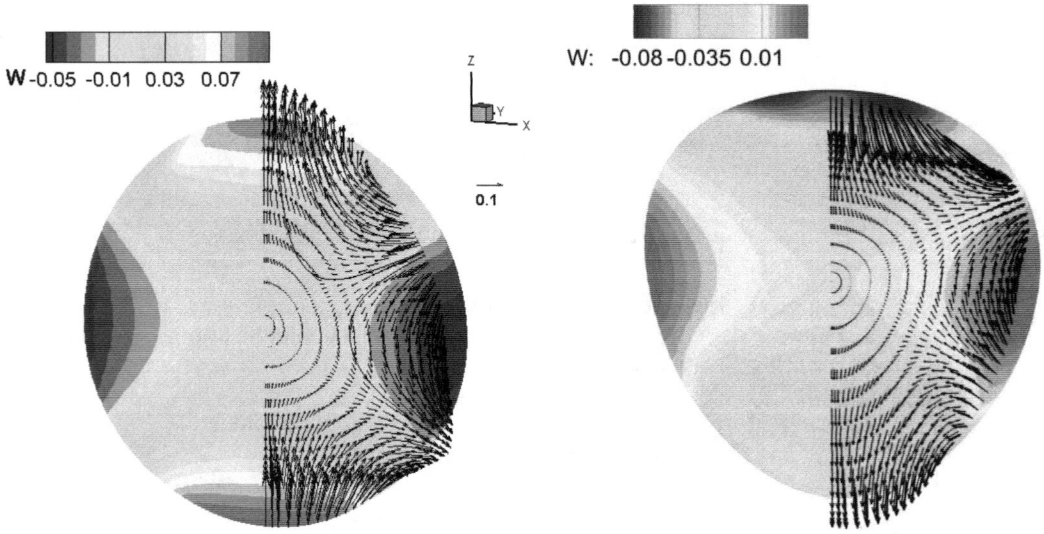

Figure 5. Oscillation Mode (3,0) without the magnetic field, velocity field at two moments for the liquid Ni droplet.

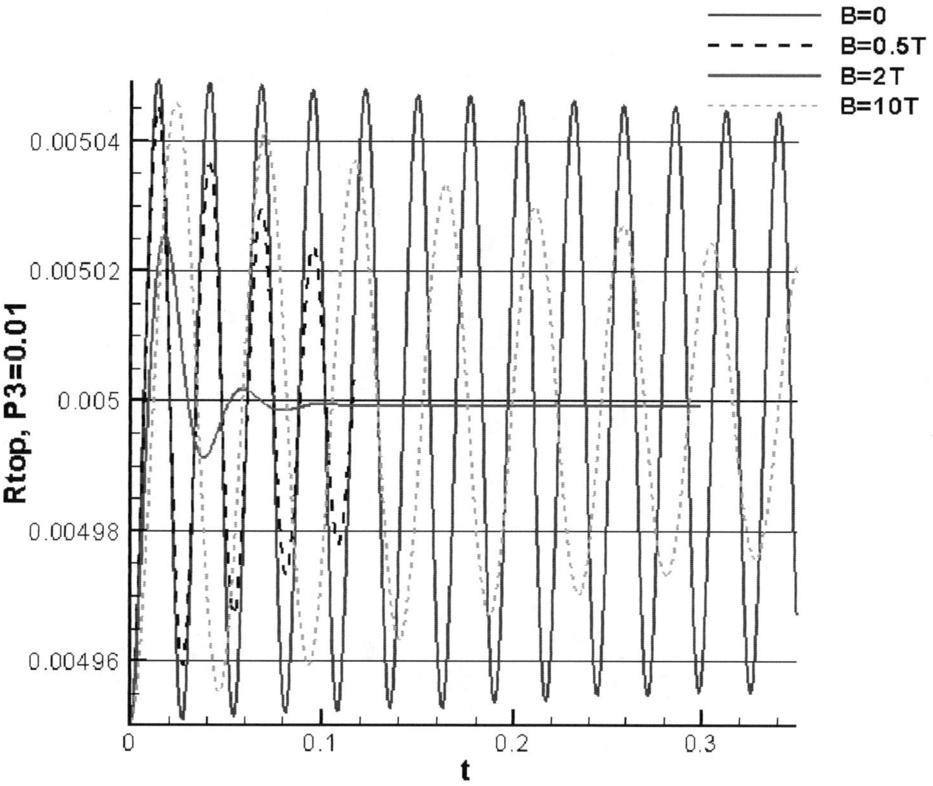

Figure 6. Oscillations of the top position for Ni droplet obtained in axisymmetric model for the (3,0) mode in different intensity uniform Bz field.

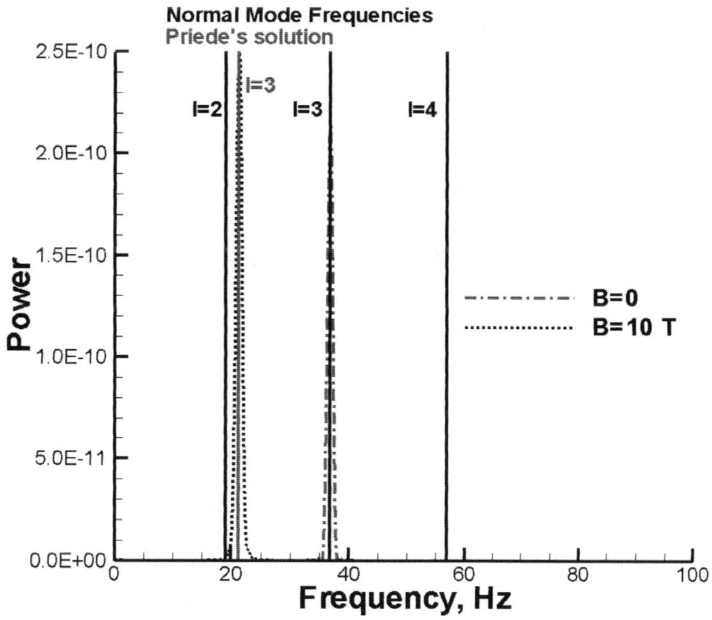

Figure 7. Spectra for the high magnetic field 10 T compared to the nonmagnetic case and the asymptotic solution [8] for the liquid Ni droplet.

It is of interest to investigate the oscillation change with the gradual increase of the magnetic field magnitude. Rayleigh normal mode frequency remains practically unchanged for a moderate magnetic field 0.5 T, however with the increased damping rate (Figure 6). Oscillations are close to overdamping at about 2 T. For extreme fields (10 T and more) the frequency gradually decreases to the asymptotic value given in [8]. We can conclude from these results that a DC field up to 0.5-1 T is beneficial to stabilize the oscillation, damp the turbulence while preserving the Rayleigh frequency. High DC fields (5 T and more) need updated models to interpret measurements. Figure 7 demonstrates that the asymptotic solution predicted frequency is achieved for the field of 10 T.

4.3. Freely Oscillating Droplets, Non-Axisymmetric Mode (2,1)

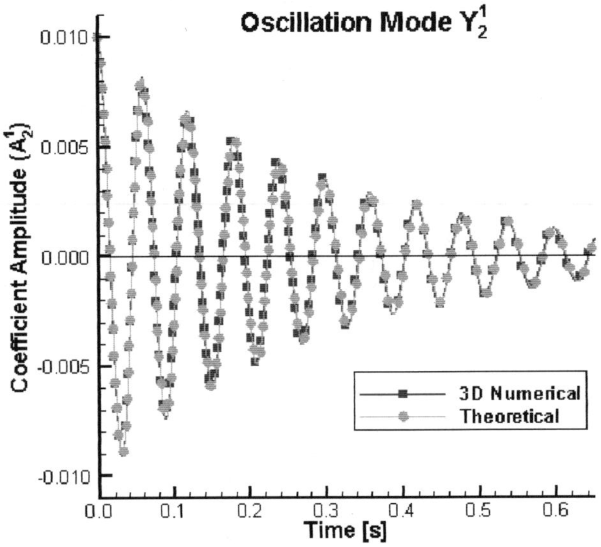

Figure 8. Time-Dependent Coefficient to ($Y^1{}_2$) mode.

The next case considered is the non-axisymmetric mode ($Y^1{}_2$), the droplet radius, amplitude, and magnetic field strength are for the Si droplet at 2.5 T. A volume correction is again made. Theoretical analysis shows that the oscillation frequency and the magnetic and viscous damping constants are given by:

$$\omega_2^1 = 2\sqrt{\frac{\Gamma}{\rho R_0^3}}, \tag{31}$$

$$\gamma_2^1 = 4\frac{\sigma B_0^2 R_0^3}{\Gamma} + \frac{5}{2}\frac{\nu}{R_0^2}, \tag{32}$$

Figure 8 shows the results of the 3D numerical simulation compared with the theoretical result. The two results show very good agreement. Figure 9 shows plots of the flow variables, flow for this mode only occurs in the x-z plane. It is clear from these contour plots that the surface oscillation mode is unchanged by the external magnetic field; however the internal flow has changed. The radial component of the flow is damped out and the fluid oscillates as columns along the direction of the magnetic field.

Journal of Physics: Conference Series **327** (2011) 012027 doi:10.1088/1742-6596/327/1/012027

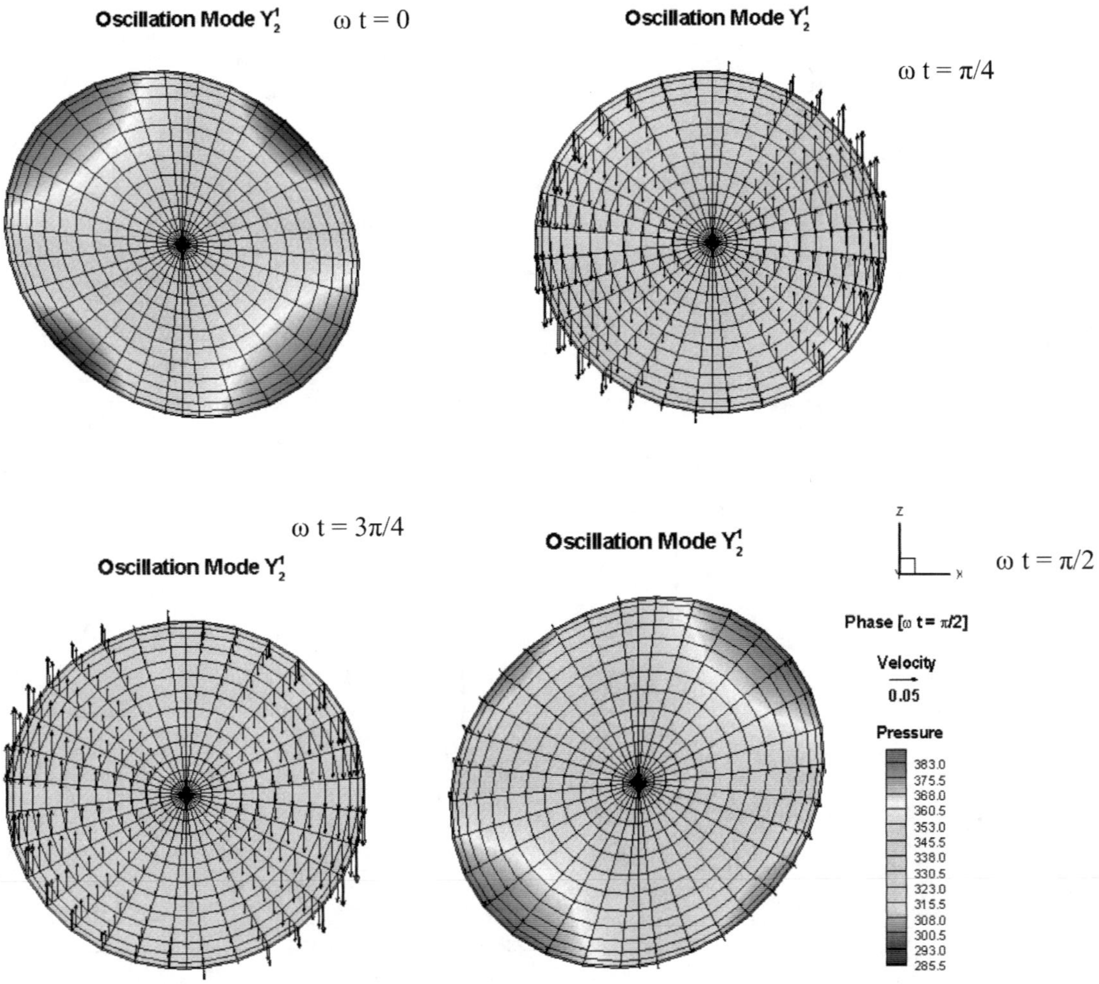

Figure 9. Oscillation mode (2,1), pressure contours and velocity vectors.

4.4. Freely Oscillating Droplets, Non-Axisymmetric Mode (2,2)

The next case considered is the non-axisymmetric mode ($Y^2{}_2$), the droplet radius, amplitude, and magnetic field strength are all the same as the previous case. A volume correction is again made. Theoretical analysis shows that the magnetic damping coefficient is zero and the oscillation frequency and viscous damping constant are the same as for the non-magnetic case:

$$\omega_2^2 = \sqrt{\frac{8\Gamma}{\rho R_0^3}}, \tag{33}$$

$$\gamma_2^2 = 0 + 5\frac{\nu}{R_0^2} \quad \text{(magnetic + viscous)}, \tag{34}$$

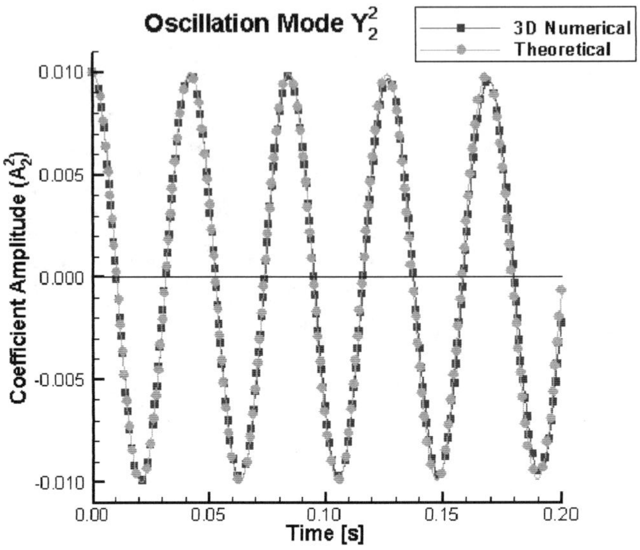

Figure 10. Time-Dependent Coefficient to ($Y^2{}_2$) mode.

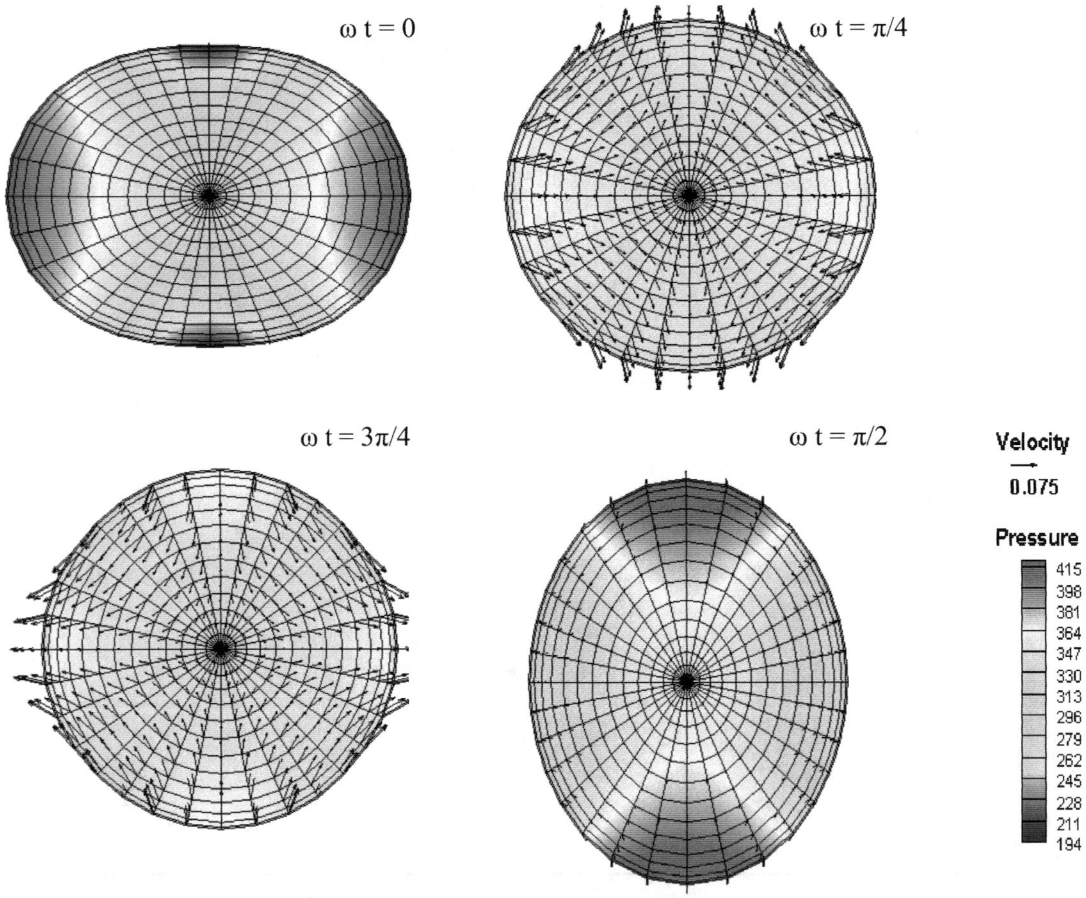

Figure 11. Oscillation mode (2,2), pressure contours and velocity vectors.

Figure 10 shows the results of the 3D numerical simulation compared with the theoretical result. The two results show very good agreement, the oscillation is unaffected by the external magnetic field. The reason for this behavior can be clearly understood from the plots of the fluid flow variables in Figure 11. Fluid flow for this mode only occurs in the x-y plane resulting in an electric field $(\boldsymbol{V} \times \boldsymbol{B})$ perpendicular to the flow but confined in the same plane. Due to the potential nature of the fluid flow, charge continuity cannot be satisfied and so an equal and opposite electric field $(-\nabla\phi)$ exists that results in zero net current and no damping of the flow by the Lorentz force.

5. Conclusions

The oscillating drop technique is a valuable method for measuring surface tension and viscosity of materials in a non-contact environment. Numerical results have been presented that confirm the asymptotic theoretical results relating surface tension and viscosity to frequency and damping rates for conducting droplets in a constant magnetic field. Further work involves use of the 3D numerical model to find qualitative and quantitative results for the oscillating drop technique in the presence of gravity forces. Comparison will also be made with experimental results where large amplitude and non-linear effects have a significant contribution to the dynamics and can impact on the measurement process.

References

[1] Egry I, Lohofer G, Seyhan I, Schneider S and Feuerbacher B 1999 *Int. J. Thermophys.* **20** (4) 1005
[2] Cummings D L and Blackburn D A 1991 *J. Fluid Mech.* **224** 395
[3] Suryanarayana P V R and Bayazitoglu Y 1991 *Phys. Fluids* A **3** (5) 967
[4] Kobatake H, Fukuyama H, Minato I, Tsukada T and Avaji S 2007 *Applied Physics Letters* **90** 094102
[5] Bojarevics V, Easter S, Roy A and Pericleous K 2009 *Proc. Int. Symp.Liquid Metal Processing and Casting*, Santa Fe, TMS, ed-s Lee P, Mitchell A, Williamson R, 319
[6] Beaugnon E, Fabregue D, Billy D, Nappa J and Tournier R 2001 *Physica* B **294-295** 715
[7] Hill R J A and Eaves L 2010 *Phys. Rev* E **81** (5) 056312
[8] Priede J 2010 *J. Fluid Mech.* **671** 399
[9] Lord Rayleigh 1879 *Proc. Royal Soc. London* A **29** 71
[10] Lamb H 1895 *Hydrodynamics*, Cambridge University Press, Cambridge
[11] Tsamopoulos J A and Brown R A 1983 *J. Fluid Mech.* **127** 519
[12] Smythe W R 1950, *Static and Dynamic Electricity*, McGraw-Hill, London
[13] Bojarevics V and Pericleous K 2003 *ISIJ International* **43** (6) 890

Journal of Physics: Conference Series **327** (2011) 012028 | doi:10.1088/1742-6596/327/1/012028

Numerical study of bubble dynamics with the Boundary Element Method

N. Méndez and R. González-Cinca

Departament de Física Aplicada, Barcelona Tech-Universitat Politècnica de Catalunya
Esteve Terradas 5, 08860 Castelldefels (Barcelona), Spain

E-mail: ricard@fa.upc.edu

Abstract.
A new Boundary Element Method software for simulating bubble dynamics is presented. This tool allows us to simulate a wide variety of scenarios, including external acoustic fields and multiple bubble interaction. We present results on the interaction between two oscillating bubbles in various configurations. Different sized bubbles under the effects of acoustic fields at different frequencies have been considered.

1. Introduction

The study of two-phase flow phenomena such as bubble dynamics is of significant interest for many terrestrial and space applications. In particular, the interaction between oscillating bubbles has attracted researchers attention since the early work of V. Bjerknes [1, 2, 3]. A good understanding of the interaction dynamics between two oscillating bubbles is crucial for the proper management of bubbly flows on ground and in space.

Two phase flow problems are an extension of the traditional fluid mechanics involving the dynamics of two different fluids in the same domain. The Navier Stokes equations combined with special considerations must be taken into account in order to describe the interaction between these two fluids. For a wide variety of problems it is convenient to simplify the model to the maximum that allows to describe the phenomenon with a reasonable accuracy. The Boundary Element Method (BEM) [4, 5, 6, 7, 8, 9, 10] provides a convenient, efficient way to describe two-phase flows in different scenarios, since it offers a low computational cost in comparison with other numerical methods, and provides an accurate representation of the interface.

In this paper we present a study on the interaction of two oscillating bubbles in different configurations by means of a BEM. We have considered bubbles of different sizes in an inviscid liquid under the effects of an external acoustic field at different frequencies. In Section 2 the BEM is introduced. The numerical procedure as well as the software package specially built for this work are presented in Section 3. The implemented model is validated by means of the comparison with previous numerical results in Section 4. In Section 5 we present results on the interaction between two oscillating bubbles in different scenarios.

2. Boundary Element Method

The boundary element formulation for three dimensional fluid dynamics is commonly used for the simulation of bubble dynamics and other problems involving mechanical stress analysis.

Journal of Physics: Conference Series **327** (2011) 012028 doi:10.1088/1742-6596/327/1/012028

The BEM has been used in the study of the dynamics of toroidal bubbles [5], and in the study of cavitation bubbles near a wall [10]. Many BEM formulations for bubble dynamics have been developed in the past years, including indirect methods [6] and completely desingularized methods [8]. In the BEM, the phenomenon of study is well defined over a domain with a set of initial and boundary conditions. This method has some limitations in fluid mechanics, since only potential flow models accomplish its requirements. However, three dimensional simulations by means of the boundary element formulation are faster than simulations with other numerical methods such as finite elements. This allows the BEM to capture more detailed information about the physical boundary than other methods with the same computational cost. In this work we have considered a BEM with inviscid liquid.

An incompressible and irrotational flow can be defined by the Laplace equation $\nabla^2\phi = 0$, ϕ being a potential defined over the fluid domain. Let us consider a bubble in an unbounded domain Ω governed by this potential. The velocity of the bubble surface can be determined from:

$$\frac{d\vec{x}}{dt} = \nabla\phi \tag{1}$$

In order to completely define the dynamics of the bubble, it is required to obtain a function that satisfies the governing equation, and to derive a set of boundary conditions for the bubble interface. The Green function G is one solution of the Laplace equation in the three dimensional domain:

$$G = \frac{1}{\left|\vec{P} - \vec{Q}\right|} \tag{2}$$

Eq. 2 relates the potential in $\vec{P}$ and $\vec{Q}$ due to a source in $\vec{Q}$ with potential of unitary value. Defining $\vec{r} = \vec{P} - \vec{Q}$, the velocity is obtained from Eq. 1:

$$\vec{v} = \vec{\nabla}G = -\frac{\vec{r}}{|\vec{r}|^3} \tag{3}$$

In order to complete the definition of the problem, the time dependence and the conservation of momentum are introduced as boundary conditions. For the velocity, the time dependence is introduced through the kinematic boundary condition in the bubble surface:

$$\frac{d\vec{x}}{dt} \cdot \hat{n} = \nabla\phi \cdot \hat{n}, \tag{4}$$

where $\hat{n}$ is the normal to the surface. Thus, normal velocities in the surface of the bubble must be continuous.

The momentum conservation is derived through the Bernoulli equation, relating the physical quantities at point of the fluid with the values at the far field:

$$\frac{p_\infty}{\rho} + \frac{1}{2}|\vec{v}_\infty|^2 + \frac{\partial\phi_\infty}{\partial t} - E_\infty = \frac{p}{\rho} + \frac{1}{2}|\vec{v}|^2 + \frac{\partial\phi}{\partial t} - E \tag{5}$$

where p is the liquid pressure, $p_\infty(t)$ is the pressure at the far field, ρ is density, and E is the potential energy of external sources such as gravity. If the far field is at rest, and the potential origin is taken at infinity, the above equation reduces to:

$$\frac{\partial\phi}{\partial t} = -\frac{p_\infty - p}{\rho} - \frac{1}{2}|\vec{v}|^2 + (E - E_\infty), \tag{6}$$

which can also be expressed in terms of the material derivative of the potential:

$$\frac{D\phi}{Dt} = \frac{\partial\phi}{\partial t} + \vec{v}\nabla\phi = \frac{1}{2}|\vec{v}|^2 - \frac{p_\infty - p}{\rho} + (E - E_\infty) \tag{7}$$

In order to determine the time evolution of ϕ, the velocity of the bubble surface must be provided. The BEM calculates the velocity from the boundary integral equation given a set of initial conditions for the velocity potential:

$$\frac{1}{2}\phi(\vec{P}) = \int_{\partial\Omega}\left(\frac{\partial\phi}{\partial n}(\vec{Q})G(\vec{P},\vec{Q}) - \phi(\vec{Q})\frac{\partial G}{\partial n}(\vec{P},\vec{Q})\right) d\Omega_Q \tag{8}$$

3. Numerical Procedure

The numerical procedure followed to implement the model is based on [6, 9]. In order to solve the equations, the interface is discretized in n triangular elements, while the potential and velocity are interpolated using linear basis functions. The discretized form of Eq. 8 is given by:

$$\frac{1}{2}\phi(\vec{P}_i) = \sum_{j=0}^{n}\left(\int_{s_j}\left(\frac{\partial\phi}{\partial n}(\vec{Q})G(\vec{P}_i,\vec{Q}) - \phi(\vec{Q})\frac{\partial G}{\partial n}(\vec{P}_i,\vec{Q})\right) ds_j\right) \tag{9}$$

Eq. 9 is integrated over all the surface patches s in order to form a linear set of equations to obtain the normal velocities $\frac{\partial\phi}{\partial n}$. We used the Bi-Conjugate Gradient Stabilized solver (Bi-CG Stab) in our code. The tangential component of the velocity is obtained by finite differences from the values of the potential at a given point. The control points are then advanced in time by means of the discretized dynamic and kinematic boundary conditions:

$$\frac{D\phi_i}{Dt} = \frac{1}{2}|\vec{v}_i|^2 - \frac{p_\infty(t)\left(1 - \left(\frac{V_0}{V}\right)^\gamma + 2\sigma\kappa_i\right)}{\rho} - g \cdot h_i \tag{10}$$

$$\frac{D\mathbf{x}_i}{Dt} = \vec{\nabla\phi_i} \tag{11}$$

where V_0 is the initial volume of the bubble, σ is the surface tension, κ_i is the principal curvature of the liquid-air interface, g is gravity, and h_i is the height of a bubble point measured from the inception point. The control points are advanced with its material velocity in order to avoid interpolation of the values for the potential at the next timestep. Numerical instabilities in our method are avoided by imposing a stability criteria limiting both the time step and the maximum displacement of a given node. Both a fourth order Runge-Kutta and a predictor-corrector schemes are implemented in the code.

Due to the intensive calculations that have to be carried out in the BEM, interpreted languages such as MATLAB or MATHEMATICA were discarded. Instead, we constructed a C code that takes full advantage of multiple CPUs, which accelerates the simulations. The designed software integrates all the calculus in one package, with the ability to export the results to other softwares such as EXCEL or MATLAB. This software was developed for the Mac OS X platform, which provides a native set of libraries (Cocoa Framework) that allowed us to build a graphical user interface, shown in Fig. 1, without much overhead. The software also benefits from parallelization with multiple CPUs via Grand Central Dispatch and Hardware acceleration of the 3D viewport via OpenGL framework. On the Windows side, the Graphical User Interface was designed with the Cocotron libraries, while the parallelization was carried out with openMP.

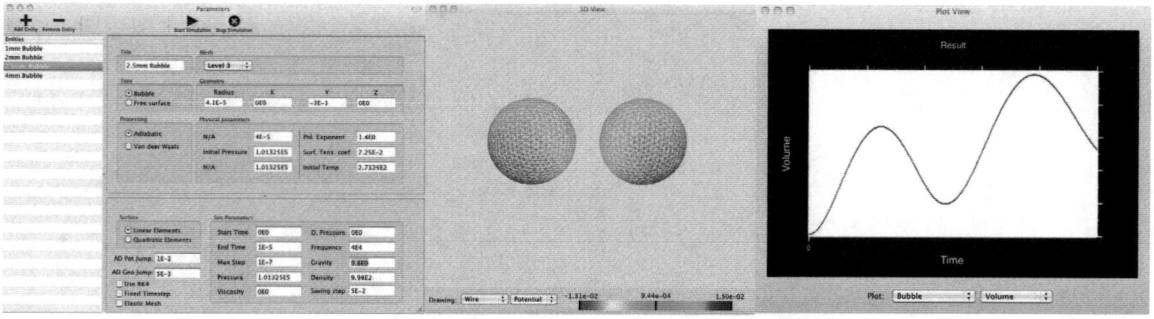

Figure 1. Main interface of the software.

4. Validation

In order to validate our BEM code, some tests have been carried out to compare results provided by the BEM simulations with existing results. We initially tested the code by simulating an oscillating air bubble in water with $R_0 = 0.1651$ and $\epsilon = 100$, where R_0 and ϵ are the dimensionless initial radius and pressure, respectively. This configuration corresponds to an overpressure of the bubble which generates its volume oscillation and shows a highly varying dynamics. Thus, it is an optimum phenomenon to validate the BEM. Fig. 2 shows the evolution of the bubble volume with time. The behaviour presented in Fig. 2 is in complete agreement with the results obtained by means of the Rayleigh-Plesset equation.

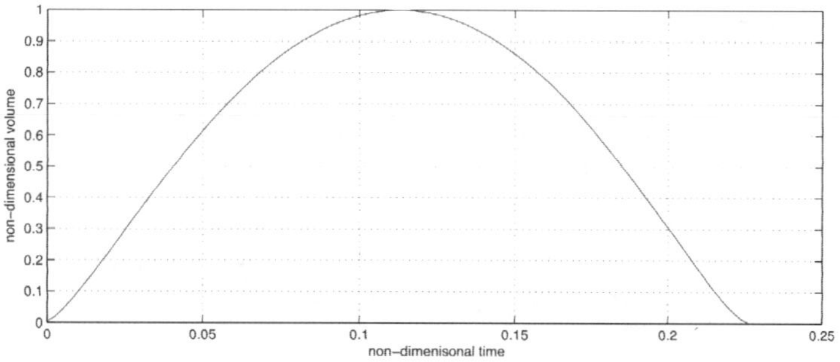

Figure 2. Time evolution of the volume of an underwater explosion bubble obtained by means of the BEM.

Our BEM code has also been validated by the comparison with results from previous studies with the BEM. Fig. 3 shows the time evolution of two in-phase bubbles under the action of gravity, as one can encounter when bubbles are generated by two underwater exploding charges. No remeshing or smoothing was required in order to advance the bubble surface in time. The behaviour of the interacting bubbles shown in Fig. 3 is in good agreement with previous results with a BEM (see Fig. 6 in [7]). Thus, our code is proofed to be an appropriate tool to reproduce scenarios with multiple bubbles.

5. Bubble interactions

We have analyzed the interaction of two bubbles under the effects of an external acoustic field with no spatial dependence. Although finite wavelength standing wave conditions have also been implemented in the code, we do not use them in this work since we aim to isolate as much as

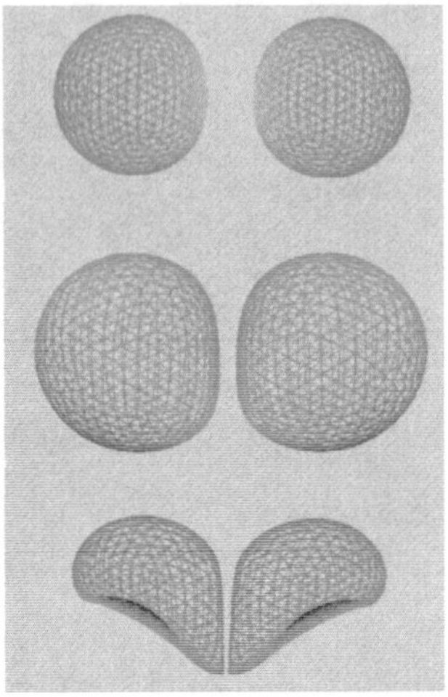

Figure 3. Time evolution of two in-phase bubbles obtained by means of the BEM.

possible the effects of the acoustic field generated by the pulsating bubbles (secondary Bjerkens force). Had we also considered standing waves of finite wavelength, additional forces would act on the bubbles. Thus, the following results show the behaviour of bubbles oscillating under the effects of an external acoustic field without spatial dependence and a spatially dependent acoustic field generated by the pulsation of the neighboring bubble.

We have considered bubbles of different size and different frequencies of the external acoustic field. Fig. 4 shows the time evolution of two bubbles with an initial $0.1mm$ radius at an initial separation between their geometric centers of $0.4mm$ under an acoustic field at 5 KHz. As a result of the attractive second Bjerknes force, each bubble approaches the other until they collide after $2.58ms$. An increasing attraction velocity with decreasing distance between bubbles can be observed.

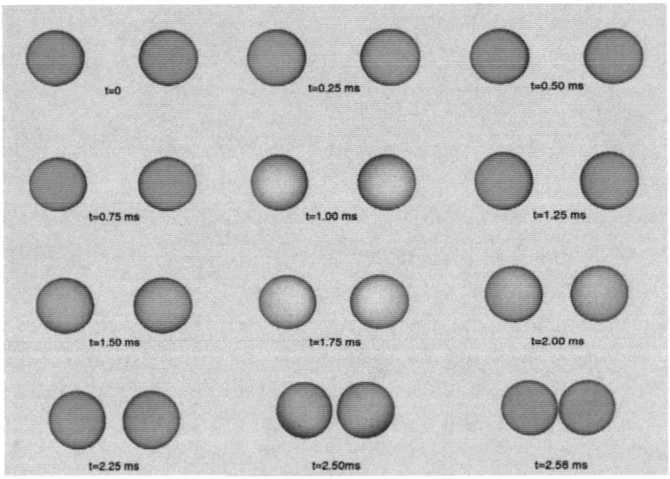

Figure 4. Time evolution of two $0.1mm$ bubbles initially separated $0.4mm$ under the presence of an external acoustic field at 5 KHz.

A similar behaviour is obtained in Fig. 5, where two bubbles of $25\mu m$ evolve under the effects of an external acoustic field with frequency 57 MHz from an initial separation distance of $0.1mm$. In this case, the deformation of both bubbles when they are very close can also be observed. One would expect that when the liquid film between the bubbles was expelled, coalescence would take place. However, the coalescence phenomenon is not included in the current version of our BEM code. Thus, additional boundary conditions should be implemented in the BEM in order to simulate coalescence.

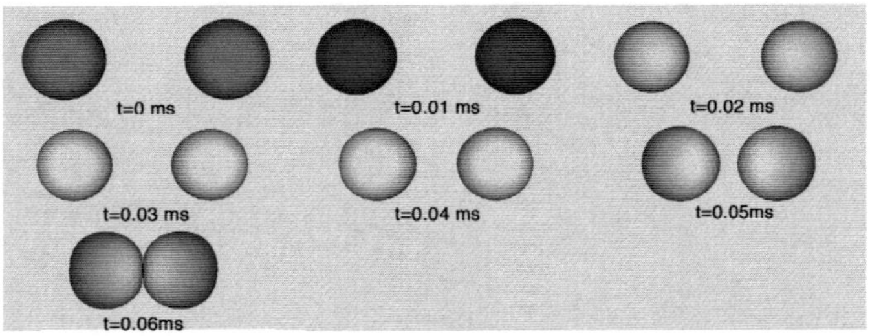

Figure 5. Time evolution of two $25\mu m$ bubbles initially separated $0.1mm$ under the presence of an external acoustic field at 57MHz.

The time evolution of the volume of one bubble as well as the velocity of the geometric center of the bubble is shown in Fig. 6, where the parameters are the same as in Fig. 4. On the one hand, the velocity is found to increase with time, as bubbles approach. On the other hand, oscillations in both volume and velocity can be observed. The bubble velocity shows two frequencies of oscillation. The lowest frequency coincides with the volume oscillation frequency, while the highest frequency of oscillation of the velocity is related to the acoustic field oscillations.

A similar oscillation pattern than the one shown in Fig. 6 is obtained with two bubbles of $25\mu m$ initially at a distance of $0.1mm$ under an acoustic field at 57 MHz (Fig. 7). The smallest frequency of the velocity coincides with the frequency of the volume oscillations, while the largest velocity frequency is related to the external field frequency. However, this behaviour is not observed in Fig. 8, where the volume and velocity time evolutions of a $25\mu m$ bubble initially at $0.1mm$ from another same sized bubble under an acoustic field at 5 KHz are presented. This behavior is not observed when the bubble collapse time is comparable with the period of oscillation, as shown in Fig. 8. In this case, the attractive velocity seems to be only modulated by the effects of the volume oscillations.

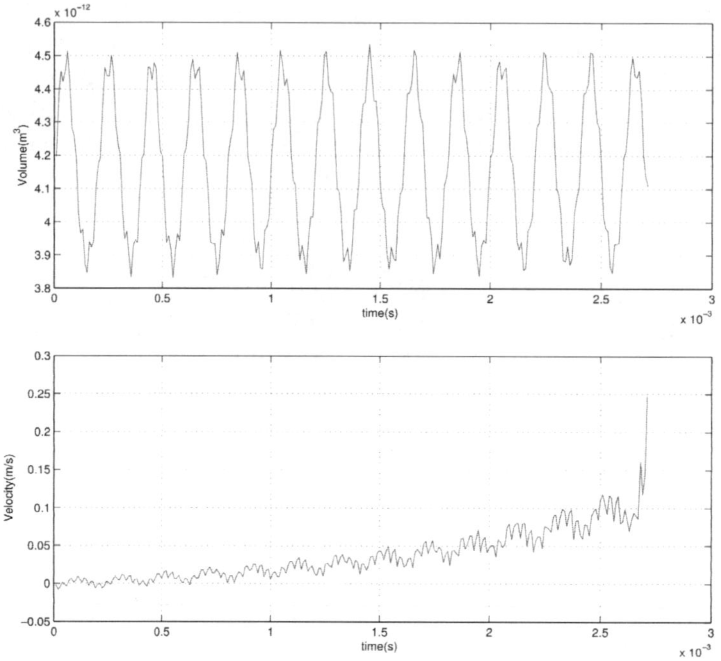

Figure 6. Time evolution of the volume and velocity of a bubble with parameters as in Fig. 4.

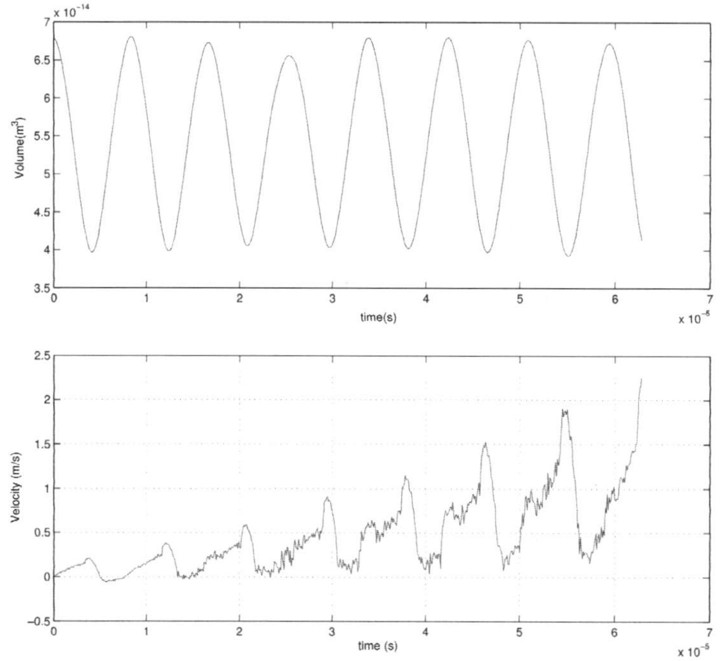

Figure 7. Time evolution of the volume and velocity of a bubble with parameters as in Fig. 5.

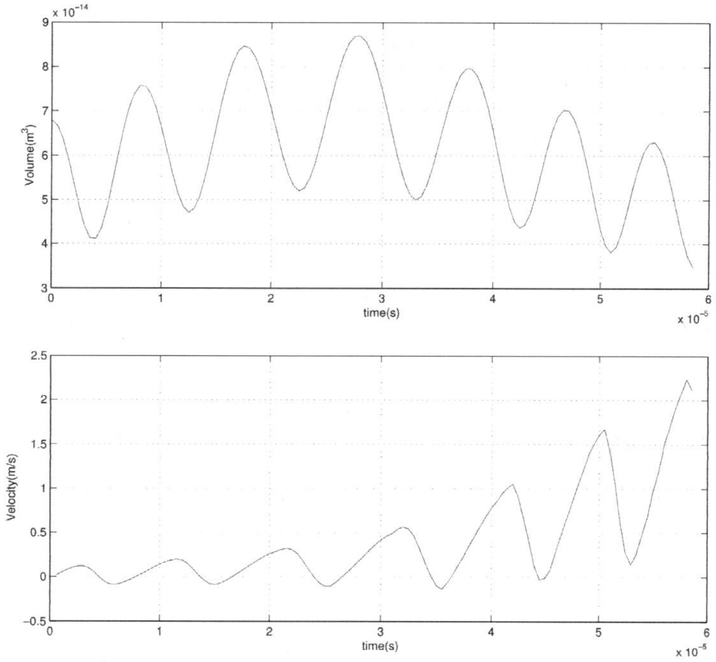

Figure 8. Time evolution of the volume and velocity of a $25\mu m$ bubble under an acoustic field at 5 KHz.

The volume oscillations in all three configurations suggest that the attraction velocity decreases when the bubble is expanding. This can be explained from the fact that the expanding bubble needs to push away the surrounding fluid, which slows down the attraction velocity of the bubbles. On the other hand, the collapse phase of the bubble accelerates the attraction as the pressure of the exterior sides of the bubble decreases and allows the bubble to advance more rapidly.

Finally, by comparing velocities in Figs. 6-8, one can observe that the attraction velocity is larger for smaller sized bubbles. This observation is in agreement with results on particle attraction.

6. Conclusions

We have presented a new embedded numerical package that allows us to study a wide range of bubble dynamics scenarios that can be defined by a BEM. This software is designed with the aim to have the maximum performance on modern computers with multiple CPUs. The software has been validated with existing numerical results. We have presented results on the interaction between two oscillating bubbles in various configurations. The effects of the secondary Bjerknes force have been analyzed.

7. Acknowledgements

This work has been financially supported by the Spanish *Ministerio de Ciencia e Innovación* (Project AYA2009-11493). R.G.C. is indebted to the Center for System Studies at the University of Alabama in Huntsville for its kind hospitality during his sabbatical leave.

References

[1] Bjerknes V 1906 *Fields of force: supplementary lectures, applications to meteorology; a course of lectures in mathematical physics delivered December 1 to 23, 1905* Publication of the Ernest Kempton Adams Fund for Physical Research (The Columbia University Press)
[2] Crum L A 1975 *The Journal of the Acoustical Society of America* **57** 1363
[3] Leighton T 1994 *The acoustic bubble* (Academic Press) ISBN 9780124419209
[4] Guerri L, Lucca G and Prosperetti A 1981 A numerical method for the dynamics of non-spherical cavitation bubbles *Proceedings of 2nd International Colloquium on Drops and Bubbles* p 175
[5] Zhang Y, Yeo K, Khoo B and Wang C 2001 *Journal of Computational Physics* **166** 336
[6] Wang C and Khoo B 2004 *Journal of Computational Physics* **194** 451
[7] A-Man Z and Xiong-Liang Y 2008 *Chinese Phys. B* **17** 927
[8] Klaseboer E, Fernandez C R and Khoo B C 2009 *Engineering Analysis with Boundary Elements* **33** 796
[9] Zhang A, Yao X and Feng L 2009 *Ocean Engineering* **36** 295
[10] Blake J, Pearson A and Otto S Boundary integral methods for cavitation bubbles near boundaries *Cavitation 2003 Symposium, Pasadena, USA*

Space experiment on the instability of Marangoni convection in large liquid bridge - MEIS-4: Effect of Prandtl number -

Taishi Yano[1], Koichi Nishino[1], Hiroshi Kawamura[2], Ichiro Ueno[3], Satoshi Matsumoto[4], Mitsuru Ohnishi[4] and Masato Sakurai[4]

[1] Department of Mechanical Engineering, Yokohama National University, 79-5 Tokiwadai, Hodogaya-ku, Yokohama, Kanagawa, 240-8501, Japan
[2] Department of Mechanics and System Design, Tokyo University of Science, Suwa, 5000-1 Toyohira, Nagano, 391-8502, Japan
[3] Department of Mechanical Engineering, Tokyo University of Science, 2641 Yamazaki, Noda, Chiba, 278-8501, Japan
[4] Japan Aerospace Exploration Agency, 2-1-1 Sengen, Tsukuba, Ibaraki, 305-8505, Japan

E-mail: nish@ynu.ac.jp

Abstract. Microgravity experiments on the thermocapillary convection in liquid bridge, called Marangoni Experiment in Space (MEIS), are carried out in "KIBO" of ISS. Three series of experiments, MEIS-1, 2, and 4, have been conducted so far. This paper reports the results obtained from MEIS-4, in which 20cSt silicone oil (Pr=207) is used to generate large liquid bridges. They are suspended between coaxial disks that are 50mm in diameter, with their maximum length equal to 62.5mm. MEIS-4 aims at (1) determining the critical temperature difference for the onset of oscillatory flow; (2) realizing high Marangoni number conditions for high Pr fluid; (3) clarifying the effects of volume ratio, heating rate, hysteresis, and cooled disk temperature; and (4) observing whether the hydrothermal wave with azimuthal mode number m=0 appears or not. The main results are presented and compared with those obtained in MEIS-1 and 2, which utilized liquid bridges of 5cSt silicone oil (Pr=67).

1. Introduction

Thermocapillary convection is the flow that is driven by the surface tension-gradient along a liquid-gas interface. In general, the surface tension, σ, negatively depends on the temperature of the working fluid. Along the liquid-gas interface with a non-uniform temperature field, the surface tension difference drives the flow from the higher temperature region toward the lower temperature region. The resultant convection is called thermocapillary convection or Marangoni convection. Thermocapillary convection becomes important in liquid films, droplets, bubbles, and liquid bridges.

The liquid bridge (LB) is the geometry that is originally seen in the floating-zone method for crystal growth, as illustrated in Fig. 1a. This method is for manufacturing high-quality single crystals of high purity. The feed material is heated by a ring heater and the melt is suspended between the feed rod at one end and the already grown crystal at the other end. The temperature difference between the melting region and the solidified region is the driving force for the thermocapillary convection. This flow configuration is modeled to a LB suspended between two coaxial disks with different temperatures, as illustrated in Fig. 1b. This simplified geometry is called a half-zone model. The temperature difference between the disks, ΔT, drives a flow that is directed from the heated disk toward the cooled disk, along the surface, and returns inside the LB. This geometry is now recognized as one of the basic flow fields for the study of thermocapillary convection, along with its instability mechanisms.

It is known that the thermocapillary convection in a LB changes from a steady, axisymmetric state to an oscillatory, non-axisymmetric one, if ΔT exceeds a certain level. The appearance of such a flow causes the deterioration of the quality in the grown crystal. This instability is related to many factors, such as, the aspect ratio of the LB, the physical properties of the working fluid, the ambient gas condition, and so on. Therefore, a large number of studies have been conducted for the instability mechanisms in half-zone LBs.

A series of microgravity experiments, called Marangoni Experiment in Space (MEIS), have been conducted as the first science experiment on the Japanese Experimental Module "KIBO" of the International Space Station (ISS). Three series of experiments, MEIS-1, MEIS-2, and MEIS-4 were carried out in 2008, 2009, and 2010, respectively. The overview of MEIS-1 and 2 was reported by Kawamura et al. [1] in 2010. The present paper reports the main results of MEIS-4. This experiment was conducted in the period from October 26 to December 22, 2010. Its purpose is to collect data on the following phenomena, in large LBs (50mm in diameter and up to 62.5mm in length), which can form only in a long-period microgravity environment on ISS:

(1) to determine the critical temperature difference for the onset of oscillatory flow,
(2) to realize high Marangoni number conditions for high *Pr* fluid,
(3) to clarify the effects of volume ratio, heating rate, hysteresis, and cooled disk temperature, and
(4) to observe if the hydrothermal wave with the azimuthal mode number m=0 appears.

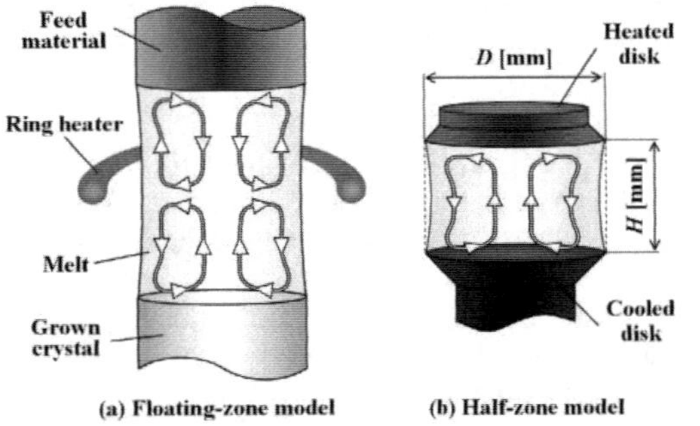

Figure 1. Thermocapillary convection (a) in the floating-zone model and (b) in the half-zone model.

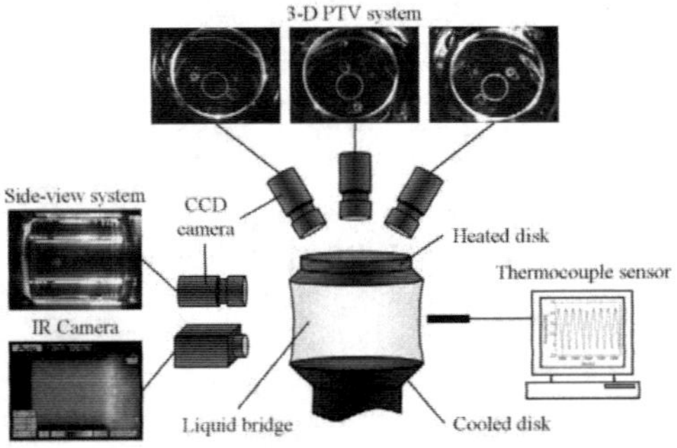

Figure 2. Schematic image of experimental apparatuses installed in the Fluid Physics Experiment Facility (FPEF).

2. Method

2.1. *Fluid Physics Experiment Facility (FPEF)*

All the experiments were performed in the Fluid Physics Experiment Facility (FPEF), which was especially designed for the present microgravity experiment. It was remote-controlled by the operation staffs and the science team members from Tsukuba Space Center (TKSC) of JAXA in Japan. Each experimental run started from 6:00 o'clock and ended at 15:00 o'clock in Japanese Standard Time (JST). This period is corresponding to 21:00 and 6:00 in Greenwich Mean Time (GMT), during which the astronauts are expected to sleep. Therefore, the g-jitters that might be caused by the astronaut activities can be minimized during experiment.

As shown in Fig. 2, several measurement apparatuses are installed in FPEF. They are (1) three top-view cameras for three-dimensional particle tracking velocimetry, (2) a side-view camera for observing the shape of the LB and the overall flow pattern, (3) an IR camera for measuring surface temperature and observing hydrothermal waves, and (4) thermocouple sensors for measuring temperature distribution and detecting the onset of oscillation. Three finger-sized B/W CCD cameras are mounted near the heated disk. Their image is 768×494 pixel at a frame rate of 30fps. The heated disk is made of sapphire so that the top-view cameras can observe the flow in the LB. An IR camera having a wavelength sensitivity of 8-14μm is used to visualize surface temperature from the same viewing direction as the side-view camera. A traversable thermocouple sensor is adopted in MEIS-4. It is to detect the onset of oscillation and can be traversed in axial and radial directions.

2.2. *Experimental conditions*

A LB of silicone oil is formed between two coaxial disks heated differently. The chamber is filled with argon gas at 94kPa and about 21°C. The geometry of the LB is illustrated in Fig. 1b. The cooled disk is made of aluminum. The disk diameter is 50mm in MEIS-4 and 30mm in MEIS-1 and 2. Each disk has 45-degree edge to prevent liquid leakage. The cooled disk is traversed in the axial direction so that a wide range of aspect ratio, Ar ($\equiv H/D$), where H and D are the LB length and the disk diameter, respectively, can be realized. The values of the maximum LB length, with the 50mm and 30mm disks, are respectively 62.5mm and 60mm, for which Ar=1.25 and 2.0, respectively. Note that the maximum length of the LB in the 1g environment is about 3mm, and the longer LB can be generated only in a microgravity environment.

Figure 3 shows one such LB with Ar=1.25; it is generated in MEIS-4. The volume ratio, Vr ($\equiv V/V_0$), defines the ratio of the LB volume, V, to the gap volume between the disks, V_0 ($\equiv D^2H/4$). The Vr=1.0 condition gives a straight cylindrical shape. The volume ratio is basically fixed at 0.95 to stabilize the LB; and, thus, to prevent fluid leakage from the edge of the disks. In some experiments, the volume ratio was changed in the range from 0.91 to 1.05 to study the effect of Vr on the onset of the oscillation.

Silicone oil with the kinematic viscosity of 20cSt is used as the working fluid in MEIS-4, while silicone oil with the kinematic viscosity of 5cSt is used in MEIS-1 and 2. The Prandtl number, Pr, of the former liquid is 207, while that of the latter one is 67, at 25°C (Shin-Etsu Chemical Co., LTD, KF-96L). The effect of Pr is, thus, examined. One of the important purposes in MEIS-4 is to observe whether hydrothermal waves with m=0 appear or not. This is motivated by the linear stability analysis of Xu and Davis [2], who analyzed thermocapillary instabilities in an infinitely long LB. They reported that the azimuthal mode number of the hydrothermal waves depends on Pr of the working fluid and that m=0 appears when Pr exceeds the value of 62.2. More recently, however, Ryzhkov [3] carried out a linear stability analysis which is similar to that reported by Xu and Davis [2] but he reported that the critical mode with m=0 for Pr higher than 62.2 is not confirmed in his analysis. The Pr in MEIS-4 is 207, which is high enough to check if the critical mode with m=0 appears for long liquid bridges.

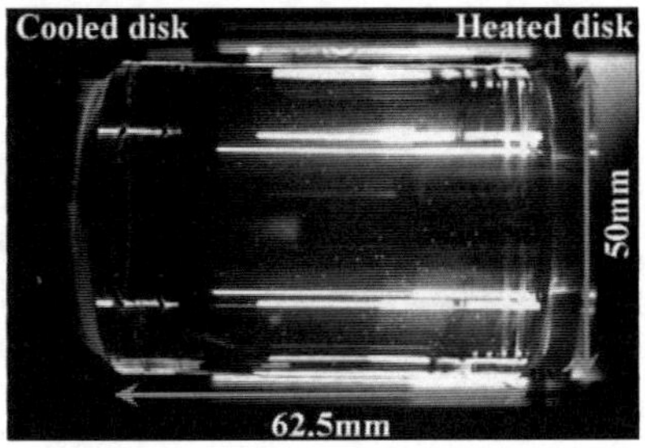

Figure 3. A liquid bridge for Ar=1.25 in MEIS-4.

Table 1. Basic physical properties of working fluid.

Working fluid	20cSt silicon oil at 25°C	5cSt silicon oil at 25°C
Prandtl number, Pr (-)	207	67
density, ρ (kg·m^{-3})	950	915
kinematic viscosity, ν (m^2·s^{-1})	20.0×10^{-6}	5.0×10^{-6}
thermal diffusivity, α (m^2·s^{-1})	9.67×10^{-8}	7.46×10^{-8}
surface tension, σ (N·m^{-1})	20.6×10^{-3}	19.7×10^{-3}
temp. conf. of σ, σ_T (N·m^{-1} K^{-1})	-6.24×10^{-5}	-6.58×10^{-5} [a] -6.23×10^{-5} [b]

[a] MEIS-1, [b] MEIS-2

The physical properties of each of the silicone oil are given in Table 1. The working fluid is seeded with the tracer particles for the flow visualization. These particles are gold-nickel-coated polymer particles, and their average diameter and density are 180μm and 1300kg/m^3, respectively. The density ratio between the tracer particle and the silicone oil is close to unity, and the fidelity of the particle to the fluid motion is checked numerically.

The critical temperature difference, ΔT_c, required for the onset of the oscillation of the flow and temperature fields, is determined by a stepwise adjustment of ΔT ($\equiv T_h - T_c$). The cooled disk temperature, T_c, is kept constant at 20°C, except for a few cases where T_c is varied to especially examine its effect. On the other hand, the T_h is always changed in a stepwise manner. The lower and higher temperature limits of the cooled and the heated disks, respectively, are 10°C and 85°C. The maximum temperature difference is about 75°C. A sufficiently long waiting time (more than 60min) is given after each of the stepwise temperature changes, ΔT. The onset of the oscillation is detected by the traversable thermocouple sensor and the motion of tracer particles in the LB.

3. Result and discussion

3.1. Instability of thermocapillary convection

In MEIS-4, ΔT_c was measured for Ar=0.225~1.25. Figure 4 shows the measured ΔT_c plot as a function of Ar. From here, it is evident that the ΔT_c decreases with Ar, showing local peaks at Ar=0.87. The results of MEIS-1 and 2 are in good agreement with each other, while the results of MEIS-4 show appreciably larger values. These results are perhaps caused by the differences in Pr and in the size of the LB (30mm vs. 50mm in diameter).

For better understanding, the ΔT_c is non-dimensionalized as the critical Marangoni number, Ma_c, as follows:

$$Ma = \frac{|\sigma_T|\Delta TH}{\rho\bar{\nu}\alpha} \tag{1}$$

where ρ, ν, α, and σ_T are the density, kinematic viscosity, thermal diffusivity and temperature coefficient of surface tension, respectively. The values of ρ, α, and σ_T for the 20cSt and 5cSt silicone oils are given in Table 1 where $\bar{\nu}$ is the mean of $\nu(T_h)$ and $\nu(T_c)$, and dependent on the disk temperature. The Ma represents the strength of thermocapillary convection, by which the onset of oscillatory flow can be specified. In microgravity, other dimensionless parameters (such as, Grashof number and Bond number), which could have potentially affected the onset of the oscillation, become zero. This is advantageous for the study of the instability mechanisms of thermocapillary convection. The obtained Ma_c is plotted as a function of Ar, in Fig. 5. The results from MEIS-1 and 2 can be seen to be in good agreement with each-other. Moreover, the results from MEIS-2 and 4 show a local peak at Ar=0.87.

Figure 6 shows the oscillation frequency, f, plotted as a function of Ar. It shows that f decreases with Ar, for MEIS-1, 2, and 4. An interesting behavior observed in MEIS-4 is the presence of the fundamental (or 1st) frequency and its 2nd harmonics, for Ar>0.63. This behavior is observed at a temperature difference that is only slightly higher than ΔT_c. Note the stepwise increase of ΔT in the determination of ΔT_c. Interestingly, the 2nd harmonics disappears when the ΔT is increased further slightly. The appearance of the 2nd harmonics at nearly ΔT_c was not observed in MEIS-1 or 2. In Fig. 7, the dimensionless oscillation frequency, $\tilde{f}$, is plotted as a function of Ar. This non-dimensionalization was proposed by Preisser et al. [4], as follows:

$$\tilde{f} = \frac{H^2}{\alpha\sqrt{Ma}} f \tag{2}$$

It is seen that the dimensionless fundamental frequency from MEIS-4 is in agreement with the dimensionless oscillation frequencies from MEIS-1 and 2, for Ar=0.67~1.25. This may indicate that the non-dimensionalization of Eq. (2) is appropriate for the LBs of different, high Pr fluids. The data from MEIS-2 exhibit a jump of $\tilde{f}$ at Ar=1.25~1.50, suggesting that the oscillation mode changes from azimuthal mode into axial mode (Schwabe [5], Yano et al. [6]). This behaviour is not confirmed in MEIS-4 due to the Ar≤1.25 condition.

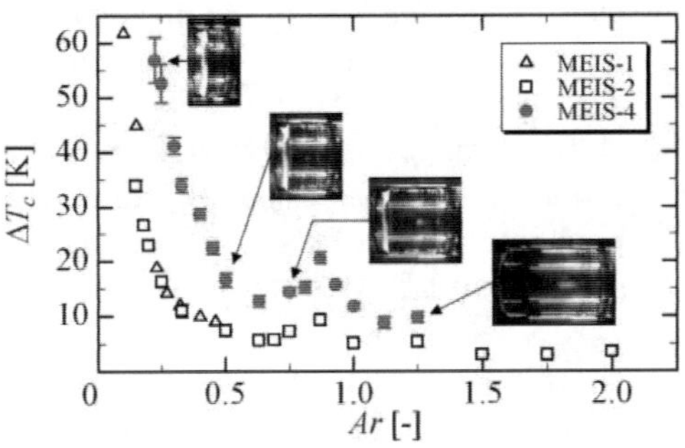

Figure 4. Plot of ΔT_c as function of Ar.

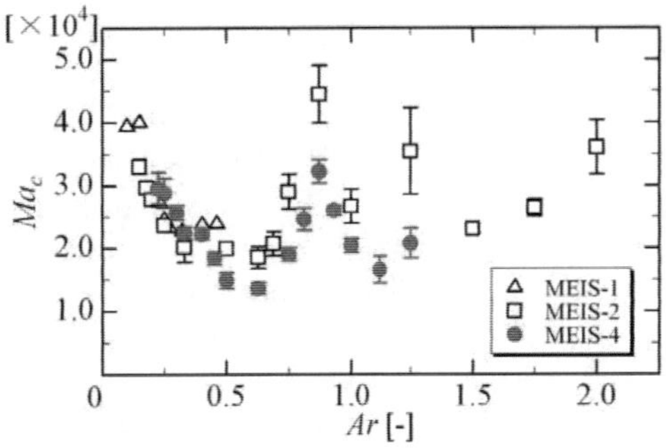

Figure 5. Plot of Ma_c as function of Ar.

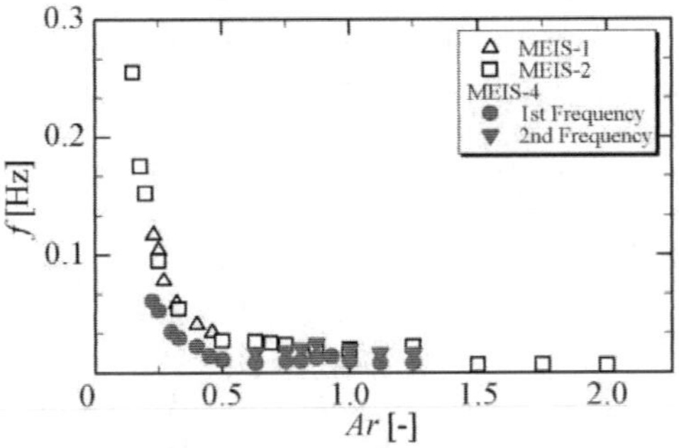

Figure 6. Plot of f as function of Ar.

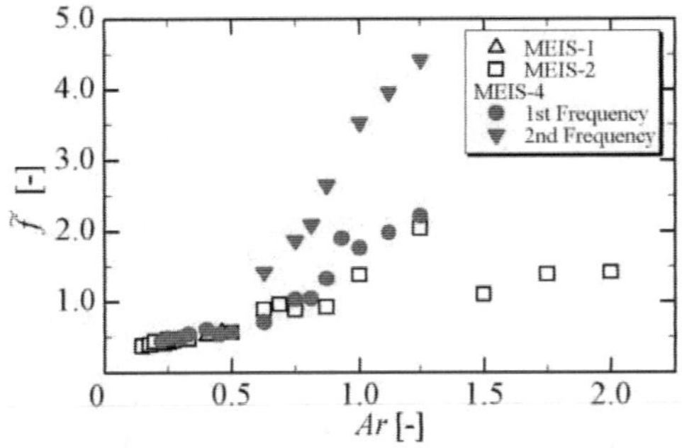

Figure 7. Plot of $\tilde{f}$ as function of Ar.

Table 2. Critical values and oscillation modes obtained from MEIS-4.

Ar (-)	Vr (-)	T_h (°C)	T_c (°C)	Ma_c (-)	$\tilde{f}$ 1st/2nd (Hz)	Oscillation mode	m (-)
0.225	0.95	76.9	20.0	2.96×10^4	0.44/-	Rotating	3
0.25	0.95	72.6	20.0	2.88×10^4	0.48/-	-	-
0.30	0.95	61.3	20.0	2.57×10^4	0.49/-	Standing	2
0.35	0.95	54.0	20.0	2.24×10^4	0.54/-	Standing	2
0.40	0.95	48.7	20.0	2.22×10^4	0.60/-	Standing	2
0.45	0.95	42.4	20.0	1.84×10^4	0.54/-	Rotating	2
0.50	0.95	36.7	20.0	1.48×10^4	0.57/-	Rotating	1
0.63	0.95	32.8	20.0	1.36×10^4	0.71/1.42	Standing	1
0.75	0.95	33.6	20.0	1.90×10^4	1.03/1.87	Standing	1
0.81	0.95	35.3	20.0	2.46×10^4	1.04/2.09	Standing	1
0.87	0.98	40.6	20.0	3.22×10^4	1.32/2.65	Standing	1
0.93	0.98	35.8	20.0	2.60×10^4	1.90/-	-	-
1.00	0.99	31.9	20.0	2.04×10^4	1.76/3.52	Standing	1
1.12	0.99	28.9	20.0	1.66×10^4	1.98/3.95	-	-
1.25	0.99	29.9	20.0	2.07×10^4	2.21/4.42	Standing	1

The modes of oscillation observed in MEIS-4 are given in Table 2, together with the value of $\tilde{f}$ and T ($\equiv f^{-1}$). The azimuthal mode, m, was determined from the motion of the tracer particles, and the oscillation with m=0 was not observed in the present experiment. This observation is in agreement with the result reported by Ryzhkov [3] who found a new branch of neutral mode with m=1 for high Pr fluids (say $Pr \geq 19.95$) for thermally insulated boundary conditions. Although not shown in this paper, the IR images acquired for Ar=1.00 and 1.25 indicate that the hydrothermal waves propagate along the LB surface in the direction from the heated disk towards the cooled disk (cf. Yano et al. [6]). This propagation direction is in opposite to that reported by Ryzhkov. The reason for this discrepancy is not clear yet and, as conjectured by Ryzhkov, the direct comparison between experiment for LB with finite length and linear stability analysis for LB with infinite length may not be possible. From the data obtained in MEIS-1 and 2, Yano et al. [6] pointed out several fundamental differences in the characteristics of oscillatory flow and temperature fields between an infinitely long LB (considered by Xu and Davis [2]) and the finitely long LBs that are experimented in MEIS. This issue will be studied in MEIS-3, which is to be conducted soon by using a 20cSt silicone oil LB with Ar=2.0.

3.2. High Marangoni number conditions

Several experiments for high Ma conditions were also carried out in MEIS-4, by exploiting the hardware capability, which allowed us to increase T_h up to as high as 85°C. As a result, ΔT=64.8°C and 61.9°C was achieved for Ar=1.0 and 1.25, respectively. These conditions correspond to Ma/Ma_c=7.1 and 8.3, respectively. Note that such high Ma conditions, relative to Ma_c, for high Pr fluids, are extremely difficult in the ground experiments where ΔT_c becomes high for short liquid bridges. Figure 8 presents the IR images of the surface temperature fluctuations at Ma/Ma_c=2.1, 5.2, and 8.3, for Ar=1.25. The mean brightness levels are subtracted to enhance the brightness levels corresponding to the surface temperature fluctuations. In these figures, higher and lower temperature regions appeared as the white and black regions, respectively. It can be observed that the lower temperature regions propagate along the axial direction, from the heated disk towards the cooled disk. Yano et al. [6] reported that such a propagation of the temperature field is associated with the propagation of the velocity field; and, therefore, this propagation is a hydrothermal wave. Initially the oscillation mode of the hydrothermal wave is a standing wave (see Fig. 8a), which latter changes into a rotating wave (see Fig. 8b). At Ma/Ma_c=8.3, in Fig. 8c, the propagation starts to exhibit chaotic features, consisting of fine and fluctuating structures in the surface temperature distributions.

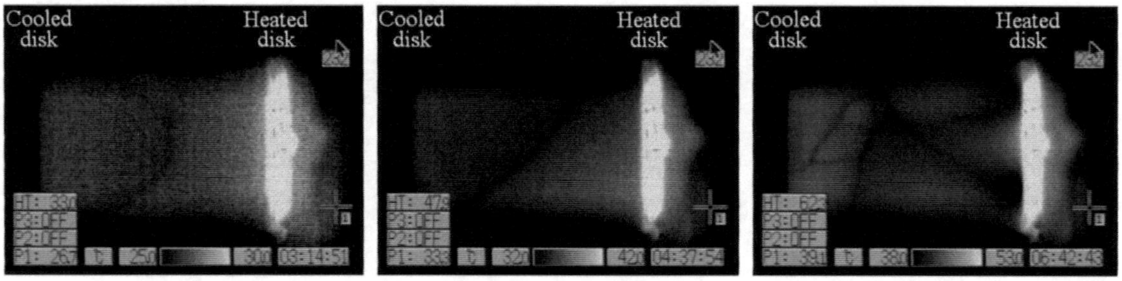

(a) Ma/Ma_c~2.1 (b) Ma/Ma_c~5.2 (c) Ma/Ma_c~8.3

Figure 8. Surface temperature fluctuation visualized by the IR camera.

4. Conclusions

This paper reports the basic characteristics of the instability of thermocapillary convection observed in large liquid bridges that are formed in long microgravity environment on the ISS. The working fluid is 20cSt silicone oil, whose Prandtl number (Pr=207) is nearly three times higher than that of the 5cSt silicone oil (Pr=67), which was used in our previous space experiments. The effect of the Prandtl number is taken into account in the comparison of these space experiments.

The critical temperature difference, ΔT_c, and the resultant $Ma_{c,}$ which is required for the onset of oscillatory flow, are determined in a wide range of *Ar* (*Ar*=0.225~1.25). As expected, the measured ΔT_c for the *Pr*=207 is substantially higher than that for the *Pr*=67, for all the *Ar* cases. However, the plots of Ma_c obtained for these Prandtl numbers, agree well with each other, with both showing a local Ma_c peak at *Ar*=0.87. The oscillation frequency is also determined for the same *Ar* range. Non-dimensional oscillation frequencies for *Pr*=207 and 67 agree well with each other, but those for *Pr*=207 reveal the presence of a fundamental (or 1st) frequency and its 2nd harmonics for *Ar*>0.67.

The mode of oscillation for each *Ar* is obtained. The hydrothermal wave with *m*=0 is not observed in the present experiment. This may indicate that the linear stability analysis, assuming an infinitely long liquid bridge, does not provide direct comparison with the experiments, whose liquid bridges have inevitably a finite length. The IR camera reveals that the surface temperature fluctuations propagate in the axial direction, from the heated disk towards the cooled disk. The oscillation mode of hydrothermal wave for, *Ar*=1.25, is standing wave, while it changes into a rotating wave, or chaotic state, with increasing *Ma*.

Acknowledgments
The authors would like to thank JAXA and the members of MEIS for their assistance in performing this study. The authors also acknowledge that a part of this study was supported by KAKENHI (Grant-in-Aid for Scientific Research (B), 21360101) from the Japan Society for the Promotion of Science (JSPS).

References

[1] Kawamura H, Nishino K, Matsumoto S and Ueno I 2010 *Proc. 14th Int. Heat Transfer Conf. (Washington, DC, USA 8-13 August 2010)* **8** 343 (also to appear in *Trans ASME*, *J. Heat Transfer*)
[2] Xu J J and Davis S H 1984 *Phys. Fluids* **27** 1102
[3] Ryzhkov I I 2011 *Phys. Fluids* **23**(8) 082108
[4] Preisser F, Schwabe D and Scharmann A 1983 *J. Fluid Mech.* **126** 545
[5] Schwabe D 2005 Phys. Fluids **17** 112104
[6] Yano T, Nishino K, Kawamura H, Ueno I, Matsumoto S, Ohnishi M and Sakurai M 2011 *Exp. Fluids*, DOI 10.1007/s00348-011-1136-9

International Symposium on Physical Sciences in Space | IOP Publishing
Journal of Physics: Conference Series **327** (2011) 012030 | doi:10.1088/1742-6596/327/1/012030

Thermocapillary and shear driven flows in gas/liquid system in annular duct

Yu Gaponenko[1], A Nepomnyashchy[2] and V Shevtsova[1]

[1] Microgravity Research Center, Université Libre de Bruxelles,
CP-165/62 av. F. D. Roosevelt, 50, B-1050 Brussels, Belgium
[2]Dept. of Math. and Minerva Center for Nonlinear Physics of Complex Systems, Technion - Israel Institute of Technology, Haifa, Israel.

E-mail: ygaponen@ulb.ac.be

Abstract. We report the results of numerical study of two-phase flows in annulus for different aspect ratios obtained in the frame of the JEREMI experiment preparation. The geometry of the physical problem is a cylindrical and non-deformable liquid bridge concentrically surrounded by an annular gas channel under conditions of zero gravity. Thermocapillary (Marangoni) convection in liquid bridge of $Pr = 68$ is analyzed in the case when the interface is subjected to an axial gas stream. The gas flow is counter-directed with respect to the Marangoni flow. The inlet gas velocity U_0^g, temperature difference ΔT between end rods of the liquid bridge and aspect ratio are the control parameters of the system. In the case when the gas stream comes from the cold side, it cools down the interface to a temperature lower than that of the liquid beneath, and in a certain region of the parameter space that cooling causes instability due to a temperature difference in the direction, perpendicular to the interface. The present study is focused on the influence of the aspect ratio on the existence and characteristic features of the oscillatory regime.

1. Introduction

The dynamics of fluid systems with interfaces remains a challenging problem in fluid physics. In such systems, the variation of surface tension due to thermal or compositional gradient along the interface can cause convective flows in the bulk fluid [1],[2]. This effect is especially important in microgravity environment, where buoyancy convection is absent and thermocapillary forces often constitute the sole cause of motion. The investigation of instabilities induced by the thermocapillary effect has been initiated by [3] who has shown that the temperature dependence of the free surface tension by itself is capable of producing an instability in a fluid layer heated from below. Interfacial flows play an important role in many natural and technological processes, such as propagation of liquid jets, motion of thin liquid films, evolution of ocean waves, etc. A wide range of important applications promoted the development of different techniques for interfacial flow control.

Thermocapillary (or Marangoni) effect essentially affects the process of crystal growth by floating zone method [4]. During this process, the feed material is slowly moved through a ring heater, so a small zone of the material near the heater is melted. The melt recrystalizes at the melt-crystal interface to form a single crystal. The liquid zone between the feed material and crystal is kept by surface tension. To study the heat and fluid flows in the melt zone, a liquid bridge (or half-zone) model is often used. In this model, the liquid is placed between two cylindrical rods (hot and cold) with a common axis. The surface tension gradient due to temperature variation along the free surface drives thermocapillary flow from hot to cold rod near the free surface and in the opposite direction at the axis. This flow is stationary for small temperature differences ΔT between the rods. When ΔT reaches some critical value, the flow becomes unstable, oscillatory in the case of high Prandtl liquids. The latter are represented by standing or axially running hydrothermal waves characterized by the azimuthal wave number m. Stability of thermocapillary flows generated by a temperature gradient directed along the liquid surface has been extensively studied in different geometries. The new flow states were recently reported [5].

The use of forced coaxial gas flow has been proposed as a way to stabilize the Marangoni convection in liquid bridges, which might have important technological applications in the floating zone technique. The planned experiment JEREMI(Japanese and European Research Experiment on Marangoni Instabilities) in Japanese KIBO module at the ISS aims at studying the influence of an external coaxial gas flow on the Marangoni convection in liquid bridges filled with silicone oil under zero gravity conditions. In the context of this experiment, some preparatory experimental and numerical studies are underway. The present study is one of the first steps on the way of the experiment preparation.

2. Problem formulation

We consider the system consisting of two co-axial cylinders, see geometry and notations in Fig. 1. The outer cylinder is a solid tube of radius R_{out}. The inner cylinder consists of two solid rods of radius R_0 and a liquid bridge between them, which is kept in its position by the surface tension force. Gas of a constant flow rate Q_{in} enters from the right or left and flows through the annular duct of a size $(R_{out} - R_0)$. The solid rods are rather long; the distance from the tube inlet is H while $H/(R_{out} - R_0) = 10$. It provides a fully developed flow in the gas, i.e. a velocity profile unchanging in the flow direction, before reaching the liquid zone. The flow structure at the entrance zone was analyzed earlier by [6], [7].

Upon reaching the liquid zone, a forced gas flow interplays with a liquid thermocapillary flow, which is present when the temperature difference is imposed between the rods. After passing the liquid zone of the length d, the gas continues to move through the annular duct over the distance H up to the tube outlet. The total length of the tube is $L = 2\,H + d$. The geometrical dimensions are given in Table 1.

In this study the gravity is not taken into account and the liquid bridge is placed horizontally in Fig. 1. Numerical study is performed in a geometry as close as possible to the laboratory set-up and with exact physical properties of the fluids. Thus the small grooves are captured by the numerical code. The physical properties of the fluids are given in Table 2. The gas is assumed to be incompressible in consistency with the low pressure drop between the tube ends.

Table 1. Geometrical scales used in calculations, unless otherwise stated.

R_{out} $10^{-3}m$	R_0 $10^{-3}m$	D_h $10^{-3}m$	H $10^{-3}m$	d $10^{-3}m$	L $10^{-3}m$
5	3	4	20	3 − 6	43

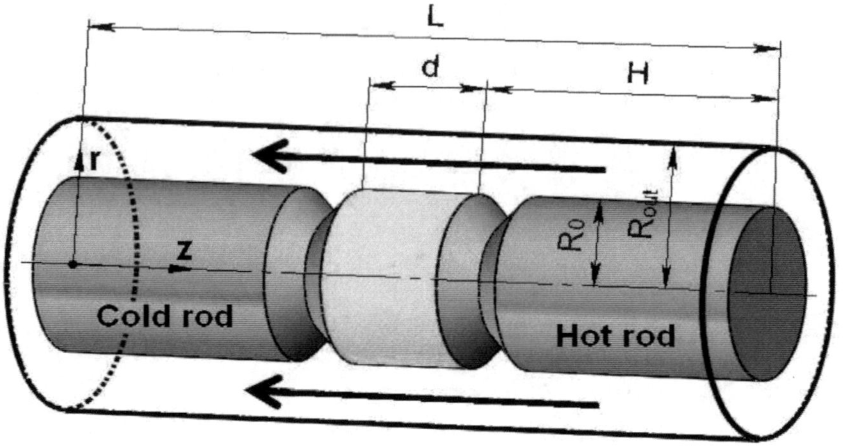

Figure 1. Geometry and scales used in numerical simulations. Gas may enter into the duct from the left or right side.

The dynamics of the system in the geometry of Fig. 1 is described by the momentum, continuity and heat transfer equations for an incompressible Newtonian fluid which are written for gas (^{g}) and liquid (^{l}) in cylindrical coordinates (r, z):

$$\begin{aligned} \partial_t \mathbf{V}^g + (\mathbf{V}^g \cdot \nabla)\mathbf{V}^g &= -\frac{1}{\rho^g}\nabla P^g + \frac{\mu^g}{\rho^g}\nabla^2 \mathbf{V}^g, \qquad \nabla \cdot \mathbf{V}^g = 0, \\ \partial_t T^g + \mathbf{V}^g \cdot \nabla T^g &= \chi^g \nabla^2 T^g, \\ \partial_t \mathbf{V}^l + (\mathbf{V}^l \cdot \nabla)\mathbf{V}^l &= -\frac{1}{\rho^l}\nabla P^l + \frac{\mu^l}{\rho^l}\nabla^2 \mathbf{V}^l, \qquad \nabla \cdot \mathbf{V}^l = 0, \\ \partial_t T^l + \mathbf{V}^l \cdot \nabla T^l &= \chi^l \nabla^2 T^l, \end{aligned} \tag{1}$$

where the velocity $\mathbf{V} = [V, U]$ includes the radial V and axial U components; T is the temperature; μ is the dynamic viscosity; χ is the thermal diffusivity and ρ is the density. The boundary conditions on the liquid–gas interface $r = R_0$ are:

a) a balance of viscous and thermocapillary stresses

$$\mu^l\, \partial_r U^l + \sigma_T\, \partial_z T^l = \mu^g\, \partial_r U^g, \tag{2}$$

where the surface tension σ is a linear function of temperature

$$\sigma(T) = \sigma(T_0) - \sigma_T(T - T_0), \qquad \sigma_T \equiv -\partial_T \sigma = const,$$

and T_0 is the reference temperature equal to the mean temperature of the system $T_0 = (T_{cold} + T_{hot})/2$;

b) axial velocities of liquid and gas are equal; radial velocities vanish (interface deformation is neglected)

$$U^l = U^g, \qquad V^l = V^g = 0; \tag{3}$$

c) heat fluxes in the liquid and gas and temperatures of liquid and gas are equal; here k is the thermal conductivity

$$k^l\, \partial_r T^l = k^g\, \partial_r T^g, \qquad T^l = T^g. \tag{4}$$

On the liquid bridge axis $r = 0$ the symmetry conditions are imposed:

$$\partial_r U^l = 0, \quad V^l = 0, \quad \partial_r T^l = 0. \tag{5}$$

No-slip conditions are imposed on the wall of the external tube $r = R_{out}$:

$$U^g = 0, \quad V^g = 0. \tag{6}$$

The temperature of the external wall can be considered as a control parameter, but it is taken equal to the mean temperature of the system if it is not stated otherwise:

$$T^g(r = R_{out}) = T_0 \tag{7}$$

No-slip conditions and constant temperatures are also imposed on the supporting rods:

$$\begin{aligned} &\text{on the left rod} \quad 0 \le z \le H \qquad && U^l = V^l = 0, \quad T^l = T_{cold}; \\ &\text{on the right rod} \quad H + d \le z \le L \qquad && U^l = V^l = 0, \quad T^l = T_{hot}. \end{aligned} \tag{8}$$

The inlet conditions at $z = 0$ or $z = L$ (depending on which side gas enters into the duct), $R_0 < r < R_{out}$ are:

$$U^g = U_0 = const, \quad V^g = 0, \quad T^g = T_{in}, \tag{9}$$

alternatively, a constant flow rate, calculated as $Q_{in} = U_0 \pi (R_{out}^2 - R_0^2)$, can be fixed.

Let us discuss the *outlet conditions* imposed at $z = 0$ (in the case where gas enters from the right), $R_0 < r < R_{out}$. The flow velocity and pressure at the outlet are not known prior to the solution of the flow problem. The "soft" conditions applied at outflow boundaries when the velocity profiles are physically appropriate for fully-developed flows are:

$$\partial_z U^g = 0, \quad \partial_z V^g = 0, \quad \partial_z T^g = 0. \tag{10}$$

The question whether these conditions allow perturbations to penetrate back inside the tube have been previously investigated by [6]. The simulations were performed with an extended outlet part to analyze artificial boundary effects. The results for the length $0 < z < L$, and an extended length $0 < z < L + H/2$ appeared to be rather similar.

The system is multi-parametric and the major control parameters are: Marangoni number Ma, Prandtl number Pr, Reynolds number of gas Re^g and aspect ratio Γ

$$Pr = \frac{\mu}{\rho \chi}, \quad Re^g = \frac{\rho^g U_0^g R_h}{\mu^g}, \quad Ma = \frac{\sigma_T \Delta T d}{\mu^l \chi^l}, \quad \Gamma = \frac{d}{R_0} \tag{11}$$

here $\nu = \mu/\rho$ is the kinematic viscosity and $R_h = 2\,(R_{out} - R_0)$ is the hydraulic radius which is a typical length scale in a pipe. For the geometry and fluid properties listed in Tables 1–2 the dimensionless parameters can be written in a way convenient for comparison

$$Pr^l = 68, \quad Pr^g = 0.75, \quad \Gamma = 2, \quad Re^g = 280\, U_0^g\, [m/s], \quad Ma = 1.91 \cdot 10^5\, \Delta T\, d\, [m].$$

The local heat flux through the interface is defined as $\tilde{q}(z) = -k^l\, \partial_r T^l(r = R_0)$. In the discussion of the results the non-dimensional heat flux q and the average heat gain/loss through the unit of the free surface q_{av} will be used

$$q(z) = \frac{\tilde{q} R_0}{k^l \Delta T} = -\frac{R_0}{\Delta T}\frac{\partial T^l}{\partial r}; \qquad q_{av} = \frac{1}{d}\int_0^d q dz \tag{12}$$

Table 2. Physical properties of silicone oil 5 cSt and air.

fluid	μ $(Pa \cdot s)$	χ (m^2/s)	k (W/mK)	ρ (kg/m^3)	$\sigma_T = d\sigma/dT$ (N/mK)	Pr
Silicone oil	$4.56 \cdot 10^{-3}$	$7.31 \cdot 10^{-8}$	0.11	912	$-6.37 \cdot 10^{-5}$	68
air	$17.9 \cdot 10^{-6}$	$1.9 \cdot 10^{-5}$	0.025	1.225		0.713

The heat flux q is positive when the liquid locally loses heat and negative for heat gain. Correspondingly, q_{av} is positive (negative) when the liquid integrally loses (gains) heat. Similarly to [8] the averaged heat flux can be considered as modified average Biot number $Bi_{av} = -q_{av}$.

Commercial solver FLUENT v.6.3 (laminar steady) was used for solving governing Eqs. (1)–(10). The problem was solved using dimensional variables in two domains for each phase separately with taking into account the boundary conditions at the interface Eqs. (2)–(4). With a user defined function, a special procedure of flows coupling was created where values for shear stresses balance on the gas-liquid interface were taken from the previous iteration time step.

3. Results and discussion

3.1. Stationary flows

In the absence of a gas stream the flow is induced by thermocapillary stresses due to the temperature gradient along the interface. The thermocapillary flow arises in the liquid for arbitrary small temperature difference ΔT between the supporting rods. The liquid moves from the hot to the cold side along the interface and the flow pattern consists of one vortex with a center shifted to the hot side. In this paragraph we compare two kinds of flows: (i) the flows in a single phase liquid bridge and (ii) two-phase flows in a system which consists of a liquid bridge and a passive gas, $U_0^g = 0$. The flow in the gas is driven by the liquid shear. The flow structure in gas consists of two small recirculation regions with the same direction of circulation (cat's eyes) embedded in a globally circulating flow over the whole duct region (see Fig. 2b). These recirculation regions correlate with two peaks on the interface velocity profile shown in Fig. 3b.

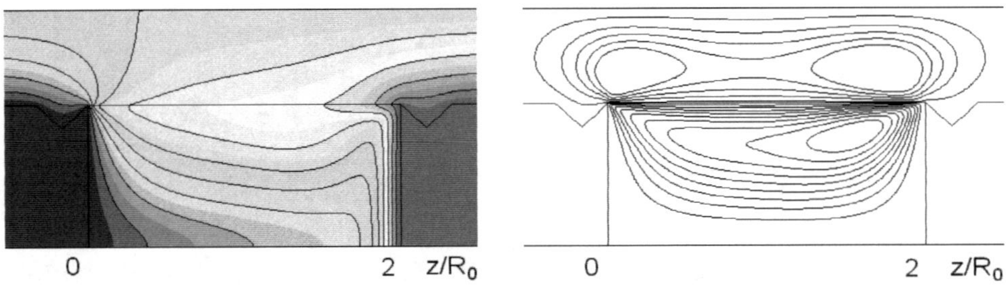

Figure 2. Temperature field and isolines of stream functions in two phases without forced gas motion, $\Delta T = 5K$, $U_0^g = 0.\,m/s$, $T^g = T_0$ and $\Gamma = 2$.

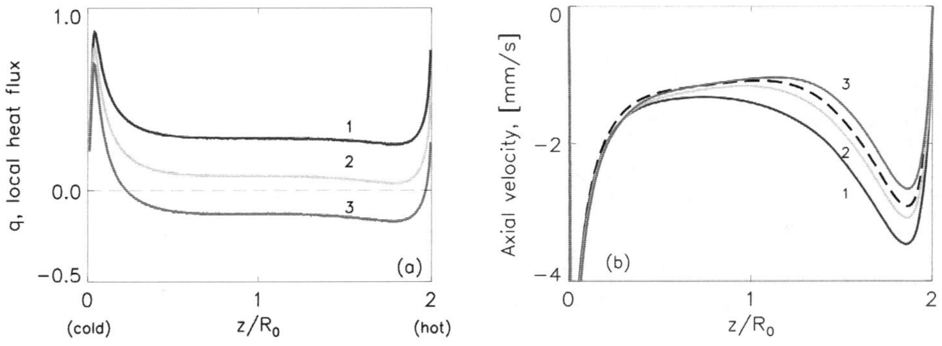

Figure 3. The local heat flux (a) and the velocity (b) along the free surface when $\Delta T = 5K$ and $\Gamma = 2$. Curves 1, 2, 3 correspond to the passive gas with temperature T_{cold}, T_{mean}, T_{hot}, respectively. Temperature of the external tube is equal to the temperature of gas. The dashed curve in graph (b) corresponds to the case when ambient gas is not considered.

Here we would like to draw the attention to the following point: the presence of ambient gas at mean temperature, which is natural for any experiment, diminishes the interface temperature, see Fig. 2a. To analyze the effect of a heat gain/loss we performed simulations for the cases when the temperature of ambient gas is equal to the mean temperature of the system and either cold or hot rod. The wall of an external tube is kept at the temperature of the gas. In the present study, the buoyancy is absent but the conductive heat transfer between liquid and gas plays an important role. For the cases mentioned above the local heat flux q is shown in Fig. 3a and velocity profiles in Fig. 3b where curves 1, 2 and 3 correspond to the ambient gas at the temperatures T_{cold}, T_{mean}, T_{hot}, respectively. It is seen without calculating an integral of $q(z)$ that the mean (as well as total) heat flux in cases 1 and 2 is positive (heat loss) while in case 3 it is negative (heat gain).

Correspondingly, the presence of ambient gas changes the temperature gradients along the interface which is crucial for the driving force. Two peaks on the velocity profile are induced by large local temperature gradients near hot and cold rods that accelerate/decelerate the liquid flow due to thermocapillary stresses, see the dashed line in Fig. 3b. The modification of the temperature profile by gas leads to the adjustment of the velocity on the interface with more pronounced changes near the hot side. The peak velocity on the cold side is too large for being shown in the graph scale, it ranges from $-5.0\,mm/s$ to $-5.9\,mm/s$ for curves 1-3. Only in the case of hot gas the flow on the free surface slows down with respect to the adiabatic case. Surrounding gas at mean or colder temperature accelerates the flow. We may conclude that the heat loss, $q_{av} > 0$, enhances the flow on the interface (curve 1 and 2 in Fig. 3b) while the heat gain slows down the flow (curve 3).

3.2. Development of the oscillatory instability under the action of a cold gas

3.2.1. Oscillatory regimes

Hereafter were consider the flow patterns when gas enters from the cold side. For majority of liquids $\partial\sigma/\partial T < 0$ and the thermocapillary flow is directed from the hot to the cold side at the interface. A gas at the mean temperature is moving against the Marangoni flow and causes a competition between the action of shear and thermocapillary stresses. When a gas stream approaches the liquid zone near the cold corner, it faces the typical sharp maximum of Marangoni velocity (see Fig. 3b), which strongly resists the incoming gas flow. For small gas velocities the shear stress is too weak to blow away this peak. Because gas is entering from

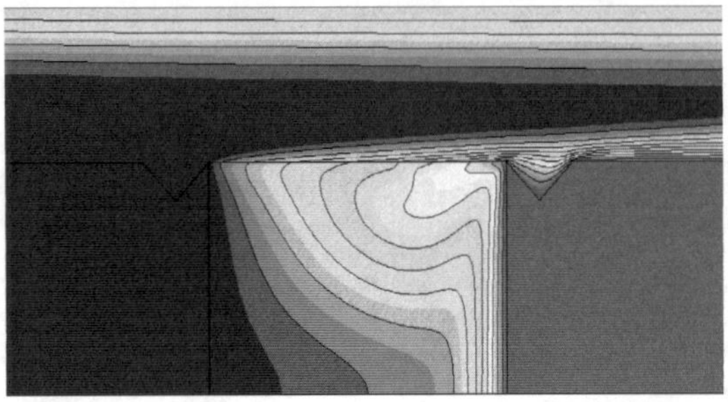

Figure 4. Snapshot of temperature field in the first oscillatory regime when $\Gamma = 1.5$, $\Delta T = 20\,K$, $U_0^g = 3m/s$.

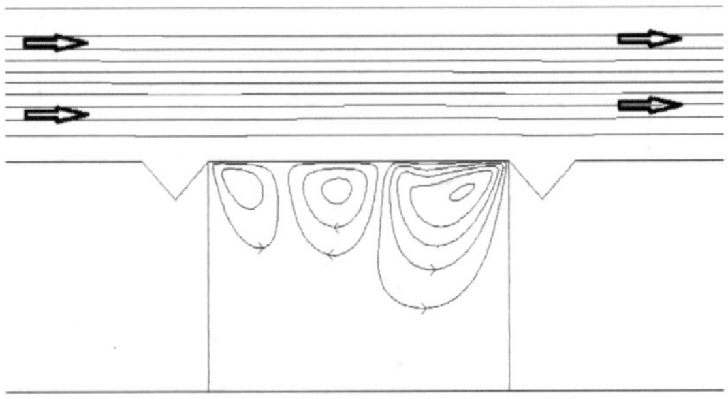

Figure 5. Snapshot of flow field in the first oscillatory regime when $\Gamma = 1.5$, $\Delta T = 20\,K$, $U_0^g = 3m/s$.

the cold side and moves along the cold rod, its temperature approaches T_{cold} on the side of the liquid. Thus, the gas affects the liquid flow not only via shear stress but also by cooling the nearby interface and reducing the temperature gradient, driving the velocity peak. The flow topology depends on the gas speed.

The shear and thermocapillary stresses act to pull the interface apart and, as a result, oscillatory instability sets in at a certain set of control parameters: U_0^g, ΔT and Γ. Here our attention is focused on the development of oscillatory regimes and their features for different aspect ratios while the instability mechanism will be described elsewhere. At the threshold of instability a strong Marangoni force acts near the hot side and significantly suppresses the oscillations in that region. The oscillations of the velocity and temperature start in the central part of the free surface. The instability occurs in a limited region of the parameter space. For each analyzed aspect ratio two different regimes of instability are identified as function (ΔT vs U_0^g).

In the *first oscillatory regime* two vortices, attached to the cold and hot ends, keep the flow direction controlled by the Marangoni force. The snapshots of temperature and flow field for this regime are shown in Figs. 4-5 for aspect ratio $\Gamma = 1.5$. At that time instant the flow pattern consists of three independent vortices. Two of them adjacent to the hot and cold walls rotate counterclockwise and the middle vortex rotates clockwise. The size and shape of the convection cells in this oscillatory regime are correlated with the changes in the gas speed. The disturbances oscillate in time, spread in the region between the vortices, and decay before attaining either the cold or the hot end.

In the *second oscillatory regime* observed at higher gas velocities. One typical vortex exists at the hot side, which is controlled by Marangoni force, and another permanent vortex sets on the cold side, being controlled by the gas stream. The snapshots of temperature and flow field for this regime are shown in Figs. 6–7 for aspect ratio $\Gamma = 1.5$. The smaller vortices attached to the central part of the interface are actively moving between the strong vortices with opposite directions of the circulation. There exists a time instant when a flow pattern consists of a sequence of four counter rotating vortices. The vortices near the solid walls tend to destroy their weaker counter-rotating neighbours and push the pair of central vortices against each other. After destroying them the basic vortices stretch from one of the walls to the center of the liquid zone, particularly this instant is shown in Figs. 6-7. Later on, the four-vortex flow state appears again. With a further increase in the gas speed the stable core of the vortex near the cold side expands drastically compressing the area where oscillations occur. After invading the major part of the interface, where the Marangoni force could be active, the instability decays.

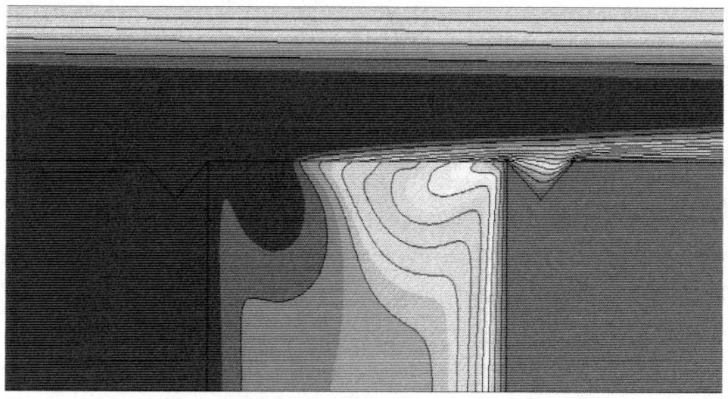

Figure 6. Snapshot of temperature field in the second oscillatory regime when $\Gamma = 1.5$, $\Delta T = 20\,K$, $U_0^g = 4m/s$.

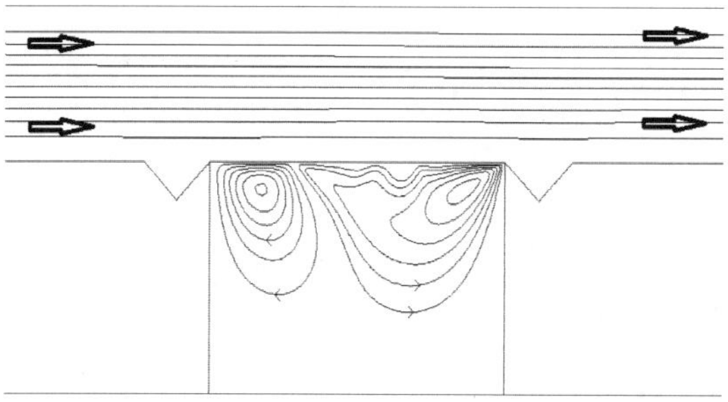

Figure 7. Snapshot of flow field in the second oscillatory regime when $\Gamma = 1.5$, $\Delta T = 20\,K$, $U_0^g = 4m/s$.

3.2.2. Influence of the aspect ratio on the stability of the flow

The described above instability occurs inside the limited range of the control parameters U_0^g, ΔT and Γ. The threshold of the instability for different ΔT only weakly depends on the gas speed. Figure 8 demonstrates that for the aspect ratio $\Gamma = 1$ the critical gas speed (or Re^g_{cr}) practically does not change when the temperature difference increases by 40%, i.e. ΔT varies from $14\,K$ to $20\,K$. Perhaps, calculations with smaller sampling in U_0^g could reveal the effect of U_0^g but it would be weak. For example, our analysis showed that for $\Gamma = 2$ the critical gas speed changes by about 20% when the temperature difference is increased by 2.5 times (not shown by plots), i.e. ΔT varies from $8K$ to $20K$. This gives a hint that the shear stress provided by gas is not a major contributor in the mechanism of instability.

But in the same time the range of gas velocities, where oscillatory regime is present, considerably depends on the length of a liquid bridge. The shorter is the liquid bridge the smaller is region of parameters where instability occurs. For short liquid bridge, when temperature gradient along the interface is larger for the same ΔT, more intensive gas flow is necessary

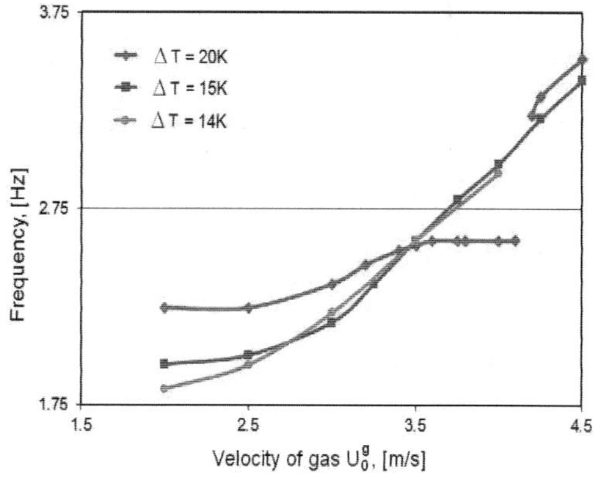

Figure 8. Dependence of the oscillations frequency on gas velocity U_0^g for different ΔT when $\Gamma = 1$.

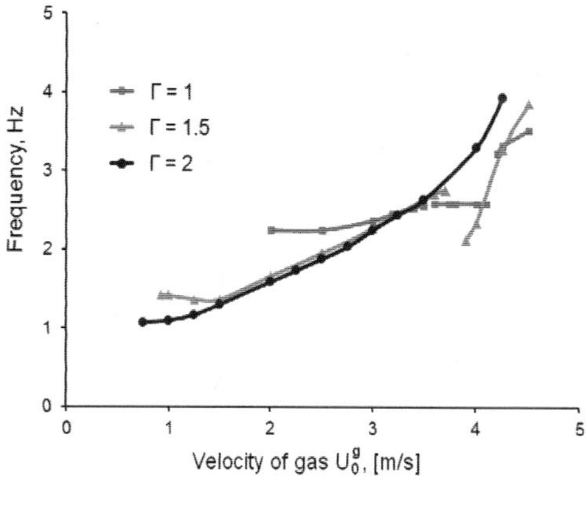

Figure 9. Dependence of the oscillations frequency on gas velocity U_0^g for different Γ when $\Delta T = 20\,K$.

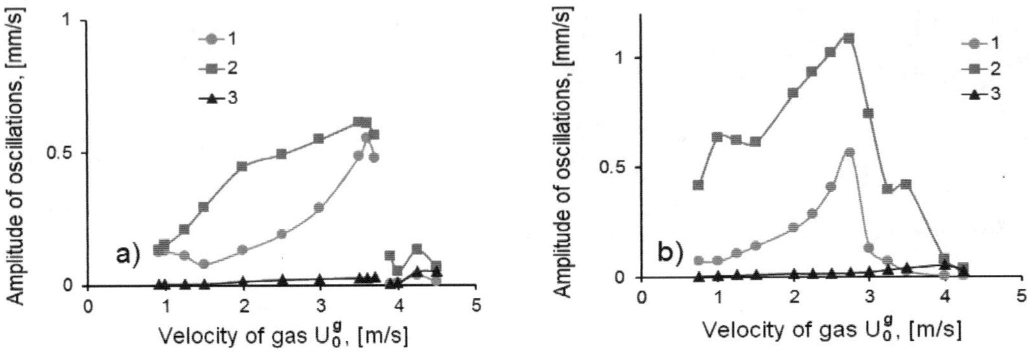

Figure 10. Amplitude of axial velocity oscillations versus U_0^g for different aspect ratios (a) $\Gamma = 1.5$, (b)$\Gamma = 2.0$ at different points of the interface: 1 - near cold rod; 2 - in the middle of liquid bridge; 3 - near hot rod;

for appearance of oscillatory regime. In addition, the solid walls impose strong constraints on the spreading of perturbations. Figure 9 shows dependence of the oscillations frequency on gas velocity U_0^g for different Γ when $\Delta T = 20\,K$. Indeed, for $\Gamma = 1$ (red curve) oscillations start at much higher U_0^g than for $\Gamma = 2$. Note that the critical temperature difference at the threshold of instability for $\Gamma = 1$ is $\Delta T_{cr} = 14K$ while for $\Gamma = 2$ is much smaller, $\Delta T_{cr} = 8K$.

As common tendency, increasing of gas velocity U_0^g leads to growing of the oscillation frequency for all considered Γ. Nevertheless the development of flow pattern depends on the length of the liquid zone. For the relatively long liquid zones, i.e. $\Gamma \geq 2$ the oscillation frequency continuously grows indicating a smooth transition between regimes described above, see Fig. 9. However in case of short liquid zones (for $\Gamma \leq 1.5$) a strong jump of the frequency occurs when gas velocity $U_0^g = 4.1\,m/s$ for $\Gamma = 1$ and $U_0^g = 3.5\,m/s$ for $\Gamma = 1.5$. This frequency jump is associated to the change of the oscillatory flow regimes. One more notable conclusion follows from comparative analysis of curves behaviour in Fig. 8. It follows that for short liquid bridges the second regime, at which a permanent vortex is created by the gas flow near the cold wall, occurs only for relative large thermal stresses, i.e. $\Delta T \sim 20\,K$ when $\Gamma = 1$.

All the transitions are clearly seen in Fig. 10 where oscillation amplitudes of axial velocity are presented for different aspect ratios and gas velocities. As a general trend the oscillations amplitude is largest in the middle of the free surface (red curves) and smallest near the hot wall, where the flow is controlled by Marangoni force (dark blue curve).

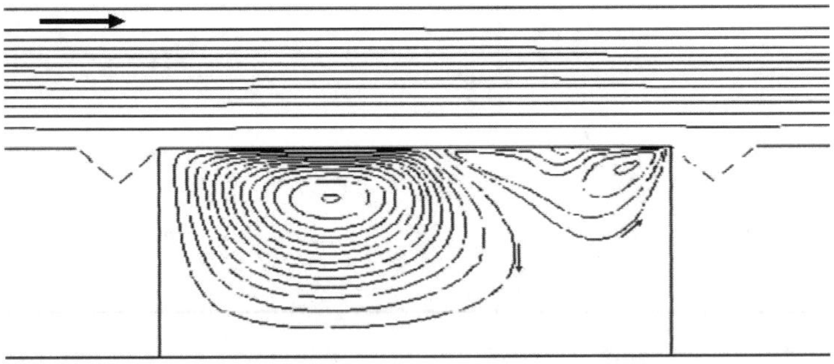

Figure 11. Flow pattern for the same case as in Fig.10b when gas velocity $U_0^g = 4\,m/s$

The variation of the amplitude in the middle of the liquid zone reflects the flow pattern transitions in the best way. For U_0^g just above the threshold of instability a little vortex moves at the central part of the interface and never penetrates deep inside the liquid. With the increase of U_0^g the vortex in the middle of the interface grows, and at a certain time moment it reaches the symmetry axis in a long liquid zones. The first small decrease of the amplitude on green (a) or red (b) curves in Fig. 10 is attributed to this transition.

Position of the maximum amplitude with respect to the gas velocity depends on the aspect ratio. For shorter liquid bridges where transition between regimes occurs via jump of frequency and amplitude (Fig. 10a), the maximum of the amplitude is achieved shortly before transitions. For long liquid bridge with continuous and smooth development of the flow the maximum of the amplitude is achieved in the second regime when vortices near the solid walls with opposite circulation attain approximately the same strength. With further increase of U_0^g the vortex on cold side becomes more active and powerful than that on the hot side. Note that with the increasing liquid bridge length the temperature gradient along the interface becomes smaller and it results in comparable magnitude of the amplitudes for different Γ.

Figure 11 shows the flow pattern for long liquid zone $\Gamma = 2$ just before decaying instability, when the gas flow is strong, $U_0^g = 4.0\, m/s$. The vortex created by the gas shear stress on cold side dominates and forces to oscillate the vortex, created by Marangoni force, which was usually steady. Correspondingly the oscillation amplitude at this point, see Fig. 10b, is the same in the middle or near the hot side.

4. Conclusions

We have performed an extensive computational study on the effects of the ambient air flow on a thermocapillary convection in a high Prandtl number liquid silicone oil 5cSt, Pr = 68. The mathematical model assumed incompressibility of the liquid and gas, linear dependence of the surface tension on temperature, with all other thermophysical properties of the fluids being taken as constant. The geometry of the physical problem is a cylindrical and non-deformable liquid bridge concentrically surrounded by an annular gas channel under conditions of zero gravity. The inlet velocity, temperature difference between end rods of the liquid bridge and aspect ratio are the control parameters of the system. We report on oscillatory instability, which occurs in the liquid when gas enters from cold side. The flow patterns, frequency and amplitude of oscillations are thoroughly analyzed for different aspect ratios.

This work is supported by the PRODEX programme of the Belgian Federal Science Policy Office, ESA and and by the Grant of Russian Foundation for Basic Research - Project N 11-01-00283-a.

References

[1] Nepomnyashchy A A, Velarde M G and Colinet P 2001 *Interfacial phenomena and convection*, Chapmann & Hall/CRC, Boca Raton, USA.
[2] Lappa M 2010 *Thermal Convection: Pattern, Evolution and Stability* John Wiley& Sons, United Kingdom.
[3] Pearson J R A 1958 *J. Fluid Mech.* **4**, 489.
[4] Kuhlmann H C 1999 *Thermocapillary Convection in Models of Crystal Growth* Springer, Berlin.
[5] Shevtsova V, Melnikov D E and Nepomnyashchy A 2009 *Phys. Rev. Lett.* **102**, 134503.
[6] Gaponenko Yu, Ryzhkov I and Shevtsova V 2010 *FDMP* **6**, 1, 75.
[7] Gaponenko Yu, Mialdun, A. and Shevtsova V 2011 *Eur. Phys. J. Special Topics* **192**, 63.
[8] Kamotani Y, Wang L, Hatta L, Wang A and Yoda S 2003 *Int. J. Heat and Mass Transfer* **46**, 3211.

Journal of Physics: Conference Series **327** (2011) 012031 doi:10.1088/1742-6596/327/1/012031

IVIDIL: on-board g-jitters and diffusion controlled phenomena

V. Shevtsova[1], T. Lyubimova[2], Z. Saghir[3], D. Melnikov[1], Y. Gaponenko[1], V. Sechenyh[1], J.C. Legros[1], A. Mialdun[1]

[1]MRC, CP-165/62, Université Libre de Bruxelles, 50, av. F.D.Roosevelt, B-1050 Brussels, Belgium

[2]Institute of Continuous Media Mechanics UB RAS, Koroleva str. 1, 614013 Perm, Russia

[3]Ryerson University, Dept of Mechanical Eng, 350 Victoria St, Toronto, ON, M5B 2K3, Canada

E-mail: vshev@ulb.ac.be

Abstract. The experiment IVIDIL (Influence of Vibrations on Diffusion in Liquids) has been performed in 2009-2010 onboard the ISS, inside the SODI instrument mounted in the Glovebox at the ESA Columbus module. 55 experimental runs were carried out and each of them lasted 18 hours. The objectives of the experiment were multi-fold and here we report results for one of them. After each space experiment there is a discussion about the role of onboard g–jitters. The attention is focused on reproducibility of the results, their accuracy and comparison with numerical simulations conducted in exact geometry and using the physical properties of the system. We shortly report on the results of six experiments which were performed in natural environment of the ISS without forced vibrations. Thermodiffusion process in the cells filled with binary mixtures was monitored by means of optical digital interferometry. Perturbations of the diffusion control processes by on-board g-jitters is not observed in nominal regime of the ISS. Perturbations of thermodiffusion process were observed in non-nominal regime of the ISS, e.g. attitude control maneuvers.

1. Introduction

Molecular diffusion occurs when a concentration gradient exists in a mixture: there is a net mass flux that tends to decrease the magnitude of this concentration gradient. To analyze the effect of temperature gradients on separation of components in liquids the thermodiffusion process, also called Soret effect, is considered. This effect is usually slow but can be quite important in the analysis of the distributions of components in oil reservoirs. The Soret coefficient $S_T = D_T/D$ measures the separation of the components of a mixture subjected to a temperature difference.

There is no universal technique that works for measuring the Soret coefficient of any binary mixture. Each technique has its own limitations. Besides, all experimental techniques require establishing a temperature gradient in the fluid in gravity field. This condition easily provokes convection in terrestrial conditions which disturbs or even erases component separation even in the case of heating from above. The effect of convection is more pronounced in the case of negative Soret when the heavy liquid goes to the hot wall (against gravity). Note that convection can also appear in theoretically stable configurations due to the non-ideality of an experimental setup, which leads to unstable temperature (density) stratification.

The microgravity environment is a unique tool for studying the behaviour of liquids and measurements of transport coefficients. Note that the benefits of microgravity environment on space platforms such as International Space Station (ISS) could be altered by the residual acceleration and vibrations (g–jitter). It can result from aerodynamic drag, non-ideal free falling, trembling of space vehicle, onboard machinery, and crews activity. The presence of high-frequency oscillations in the spectrum of g-jitters could significantly affect diffusion experiments. Such oscillations generate time–averaged convective flows which may disturb purely diffusive heat and mass transfer [1]. There is a large body of theoretical studies on the influence of g-jitters on liquid behaviour and diffusion (see e.g. [2]-[6]) in microgravity and only very few experimental studies addressed this topic.

One of the first experiments studying the impact of residual accelerations onboard a spacecraft on convection in differentially heated cylindrical cavities was performed by Naumann [7] on STS–95 and independently by Russian researchers in DACON [8]. In these experiments, the temperature was monitored at several fixed points. The results did not provide clear evidence of time-averaged flows and related heat transfer caused by periodic high-frequency g-jitter.

Experimentally the existence of averaged flows in non-uniformly heated fluids under vibrations in microgravity environment has been qualitatively indicated in the experiments on MIR station[9] and has been qualitatively demonstrated in TEVICON parabolic flight experiments in the frame of IVIDIL preparation [10]-[11].

The MEPHISTO data from the USMP-3 Space flight mission (1996) presented the first quantitative data on g-jitter effects on a directional solidification experiment [12]. However it was concluded that there is a lack of well-documented experimental results that can be used for verification of numerical modelling to provide reliable predictions of g-jitter effects.

Dedicated experiments were carried out on-board the Mir station using the Canadian MIM facility [13]. Their primary analysis indicated that diffusion coefficients in dilute binary metallic alloys depend upon the residual accelerations and the quality of microgravity. Recent thorough reanalysis of these original data claimed that very limited numbers of data were considered for post-flight analysis and the results from the liquid metal diffusion experiments conducted on MIR do not support the conclusions drawn [14].

One of the objectives of joint Russian-German experiments on-board Foton M3 (2007) was aimed at studying the influence of g-jitter effect on solidification front. However the grown crystals of Ge and GeSi were locally detached from the wall and had a characteristic shape with a "neck" due to a gas volume, which appeared at melting a semiconductor ingot . Under these conditions Marangoni convection occurs in the melt.

The latter citations clearly indicate the lack of the experimental data quantifying on-board g-jitters effects and demonstrate timeliness and the importance of the IVIDIL space experiment.

2. Experiment procedure

IVIDIL experiment was placed inside ESA SODI multi-user facility (Selectable Optical Diagnostics Instrument) which, in turn, was installed inside the MSG on the 23rd of September 2009. SODI was developed as a multi purpose Facility for conducting initially three experiments: IVIDIL, COLLOID and DSC. Hardware development was provided by QinetiQ Space NV (Kruibeke, Belgium) and optical development was provided by Lambda-X (Nivelles, Belgium). The first in the sequence, the experiment IVIDIL (Influence of Vibration on Diffusion in Liquids) started on the 5th October 2009. In total 55 experimental runs (41 original runs and 14 re-runs) were successfully completed by January 20, 2010. Each original run lasted 18 hours and all of them were controlled via telescience provided by the Spanish User Support Center (E-USOC, Madrid). Each experimental run was performed in two steps. During the first step (duration of 12 hours) a concentration gradient is established by imposing a temperature gradient across

Journal of Physics: Conference Series **327** (2011) 012031 doi:10.1088/1742-6596/327/1/012031

Figure 1. IVIDIL inside Glavbox at the ISS

the experimental cell to a uniform mixture. Due to the Soret effect, the concentration profile is slowly generated in the initially homogeneous binary mixture. After 12h the temperature gradient is removed and diffusion occurs during 6 hours. The vibrations of different amplitude and frequency were applied during these 18 hours if it was prescribed, for more details see [16]-[17]. The level of onboard g-jitters was recorded by SAMS and decoded by science team. Six hours of preheating time for thermostabilization of the optical system was required after switching off SODI for weekend or due to technical reasons.

The IVIDIL experiment included 2 cell arrays: one mixture with negative and another with positive Soret effect, water (90%)-isopropanol (10%) and water (50%)-isopropanol (50%), respectively. The physical properties of liquids are given in Table 1. The cell arrays were exchanged on 2 December 2009. Each cell array included two cells: a primary and a companion. The only difference between them is the added tracer particles in the companion cell.

3. Experimental technique

The temperature and concentration fields were monitored by measuring the variation of refractive index inside the cell by optical digital interferometry. The setup is based on the concept of Mach–Zehnder interferometer. The prototype of flight model was designed and developed in the laboratory prior to the experiment as well as corresponding software [18]. It enabled processing the control images in real time and introducing small corrections in the experimental procedure. The detailed description of optical digital interferometry as well as results from ground prototype can be found in [19]–[20].

The important part of the experimental setup is the diffusion cell. The working liquid is placed in a cubic cell with transparent walls of internal size $L = 10$ mm, see Fig. 2a. The cell is made of quartz Suprasil. The top and bottom walls are kept at constant temperatures T_{hot} and T_{cold}, respectively, by Peltier modules. In the experiments, the mean temperature $(T_{cold}+T_{hot})/2$ was fixed at 25°C, while the applied temperature difference $\Delta T = T_{hot} - T_{cold}$ was 5, 10 K or 15 K. The experimental cell is attached to the linear motor, which performs translational harmonic

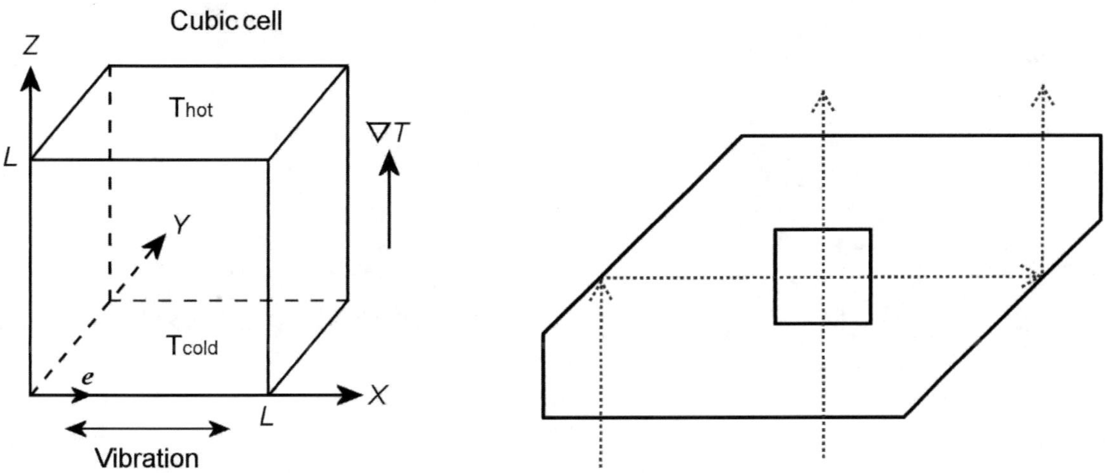

Figure 2. (a) The cubic cell and (b) top view of the experimental cell.

oscillations perpendicular to the temperature gradient. In the experiments, the frequency and amplitude were varied in the ranges 0.5–2.8 Hz and 10–61 mm, respectively. The maximum vibrational acceleration $A\omega^2 = 0.998\, g_0$ was achieved at $A = 61$ mm and $f = 2$ Hz ($\omega = 2\pi f$). The upper limit for frequency increases with decreasing the amplitude.

The external walls of the cell are shaped in the form of two prisms (Fig. 2b) to allow optical observation in two perpendicular directions[11]. An example of interference pattern in front view of the cell is shown in Fig. 3. Interferograms are reconstructed by performing 2D fast Fourier transform (FFT) of the fringe image, filtering a selected band of spectrum, performing the inverse 2D FFT of the filtered result and phase unwrapping[20]. The interferograms are the only source of the scientific information in the IVIDIL experiment along with the housekeeping data. For scientific examination of running experiments 283 digital images were downloaded for each run via telemetry. Totally 653 images per run have been recorded and they have returned on Earth on flash disks a few months later.

The knowledge of phase shift gives information about the gradient of refractive index. For a given wavelength, the variation of the refractive index is

$$\Delta n(x, z, t) = \left(\frac{\partial n}{\partial T}\right)_{T_0, C_0} \Delta T(t) + \left(\frac{\partial n}{\partial C}\right)_{T_0, C_0} \Delta C(t) \quad (1)$$

where $\Delta T(t)$ and $\Delta C(t)$ are temperature and concentration variations at point (x, z). The characteristic thermal time, $\tau_{th} = L^2/\chi$, is about 500 times shorter than the diffusion time,

Table 1. The physical properties of the mixtures at 25°C.

	ρ	β_T	β_C	ν	χ	D	S_T
	kg/m^3	10^{-4} K^{-1}		10^{-6} m^2/s	10^{-7} m^2/s	10^{-10} m^2/s	10^{-3} 1/K
H_2O(90%)-IPA(10%)	984	3.1	-0.14	1.41	1.3	7.4	-7.08
H_2O(50%)-IPA(50%)	905	7.7	-0.25	3.10	0.85	1.62	5.45

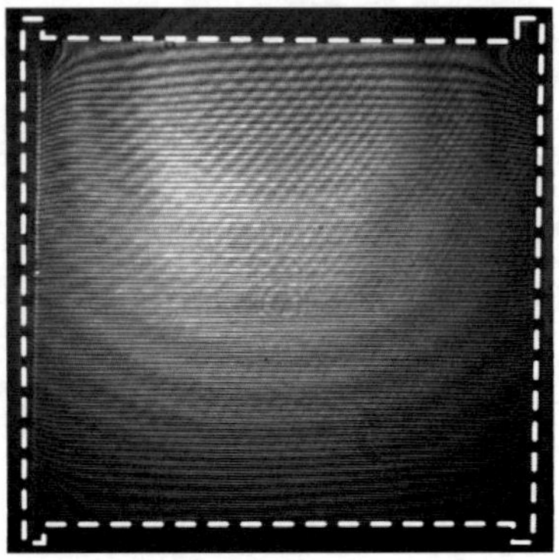

Figure 3. Interference pattern of the cell; front view. White line depicts the volume filled with liquid

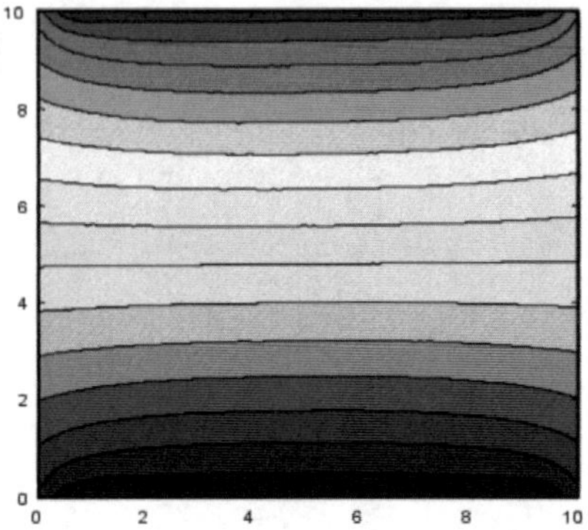

Figure 4. Temperature field restored from the interferogram; heating from above, $\Delta T = 15\,K$.

$\tau_D = L^2/D$, here χ is the thermal diffusivity and D is the diffusion coefficient. Thus, the contributions from temperature and concentration are separated in time. The typical example of the temperature field 10 min after imposing temperature gradient is shown in Fig. 4 (Run 33). Beneficial part of this technique is that the records are done over entire cross-section of the cell in two perpendicular directions during the whole diffusive process. It allows observing the evolution of the process with time and investigating convection when it exists, [18], [21].

In the absence of vibration an approximate analytical solution describing the process separation in the diffusion cell can be written as [18]

$$\Delta C(t) = -S_T C_0(1 - C_0)\Delta T \left[1 - \frac{8}{\pi^2} \sum_{n,odd} \frac{1}{n^2} exp\left(-n^2 \frac{t}{\tau_r}\right)\right]. \tag{2}$$

here C is the concentration of denser component, $\Delta C(t) = C(L,t) - C(t)$ is the concentration difference between hot and cold sides of the cell, C_0 is the initial concentration of homogeneous mixture and ΔT is the imposed temperature difference. In Eq.(2) there are two unknown parameters: the Soret coefficient S_T and the relaxation time $\tau_r = L^2/\pi^2 D$. By fitting numerous experimental (Eq.1) and theoretical (Eq.2) points for $\Delta C(t)$ from the beginning to several characteristic times the two unknown parameters S_T and τ_r are determined. The value of Soret coefficient can be also obtained from the steady state, $t \to \infty$

$$\Delta C = -S_T C_0(1 - C_0)\Delta T. \tag{3}$$

In the case of the observation of the lighter component the sign will be positive.

4. Results and discussion

The primary goals of this manuscript is to clarify the role of onboard g-jitters. For this reason only runs without imposed vibrations are considered which are listed in Table 2 in chronological order. In the second column the Run identification is given according to "Daily Operations Report" during the experiment if someone in future would like to verify results as it was the case [13], [14]. Two experiments without vibrations were performed for the liquid with negative Soret effect, H_2O(90%)-IPA(10%), and four experiments for the liquid with positive Soret effect. In the case of experiment repetition (letter "R") the starting time allows to identify what is the time lag between the experiments.

Table 2. List of the experiments without imposed vibrations.

Mixture	Run	ΔT	start (time,date)	initial condition	μg condition
H_2O(90%)-IPA(10%)	2	5	11:37, 07/10/09,	after break	nominal
H_2O(90%)-IPA(10%)	2R	5	13:58, 11/11/09	after 15R	nominal
H_2O(50%)-IPA(50%)	33	15	18:25, 10/12/09	after run32	nominal
H_2O(50%)-IPA(50%)	1N	5	10:32, 12/01/10	after run36	perturbed*
H_2O(50%)-IPA(50%)	2N	10	05:04, 13/01/10	after run 1N	nominal
H_2O(50%)-IPA(50%)	33R	15	17:20, 15/01/10	after run 3N	nominal

**One of the experiments, Run1N, has been performed during ISS activities which may imply microgravity disturbances:*
- ESP3 relocation between 14:25-15:55 GMT
- ISS attitude control maneuvers happened at 15:00 GMT (5 minutes duration)
- MT Translation moving the ESP3 platform between 17:35-19:35 GMT (thrusters were disabled during the activity).

4.1. First cell array, 90%H_2O – 10%IPA

Each time a measurement is taken, a full concentration distribution over the thermodiffusion path is determined (similar to temperature field in Fig.4). From this knowledge the concentration profiles across the cell are obtained at successive time moments; the single profile at t = 12h is shown in Fig.5 for denser component, i.e. water. The pink curve shows experimental dependence for Run2 while dark blue curve corresponds to numerical simulations performed for exact parameters of the system. From the sequence of such concentration profiles the evolution of components separation with time $\Delta C(t)$ can be obtained between two arbitrary points in the direction of temperature gradient (z-direction). The experimentally obtained transient path between hot and cold walls $\Delta C(t) = C(L,t) - C(0,t)$ over 12 hours is presented in Fig.6 for original Run2 and its repetition one month later. The smooth behavior of curves and excellent reproducibility indicates that either onboard g-jitter is not important or absolutely identical during the month.

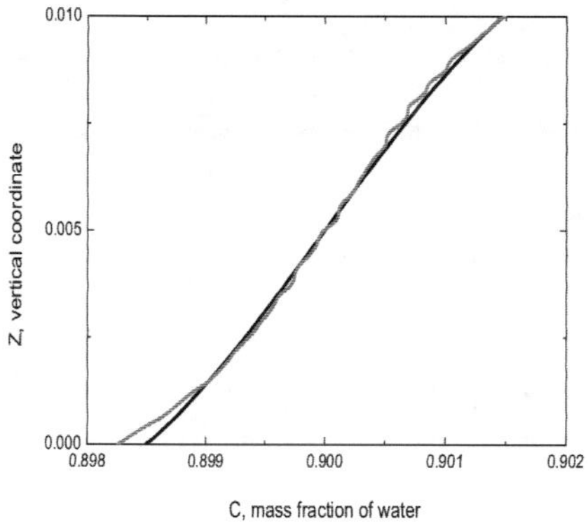

Figure 5. Concentration profiles over the height of the cell, $\Delta T = 5\,K$.

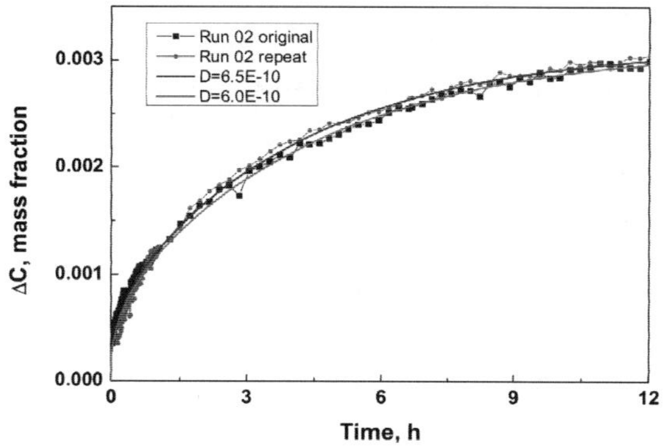

Figure 6. Separation of the components with time. The IVIDIL results and numerical simulations.

In experiments without vibrations with the mixture displaying negative Soret effect the imposed temperature difference was $\Delta T = 5\,K$ (Run2 and Run2R). As follows from Eq. 3 the small ΔT leads to the small component separation, i.e. weak signal. Indeed, while the initial concentration is $C_0 = 0.9$ ($90\% H_2O$) the measured variation is only 0.003 what is about 0.3% from initial value. Keeping in mind good repeatability, it indicates on the high accuracy of measurements. Two solid curves in Fig.6 correspond to numerical simulations for slightly different diffusion coefficients, $D = 6.0 \cdot 10^{-10} m^2/s$ and $D = 6.5 \cdot 10^{-10} m^2/s$. This excellent agreement between experimental and numerical results performed without g-jitters shows that even for weak signals effect of g-jitter on diffusion controlled phenomenon is not visible.

4.2. Second cell array, $50\% H_2O - 50\% IPA$

The behavior of the mixture in the second cell array is very different from the first one and it is not only due to the different Soret sign, $S_T > 0$. The Soret coefficient for the 2nd mixture is 1.3 times smaller but the diffusion coefficient is about 5 times smaller, see physical properties of the mixtures in Table 1.

The thermodiffusion process starts not only near the hot and cold walls but also near lateral walls because of an additional lateral temperature gradient (see Fig. 4) due to technical design of the cell, as a result of so called non-ideality of the experiment. In the second mixture (H_2O(50%)-IPA(50%)) diffusion process is too slow to eliminate these lateral concentration variations. The distribution of the concentration in the cell 10 min after imposing temperature difference is shown in Fig. 7. It is clearly seen that due to thermodiffuison (Soret effect) the concentration field changes near the lateral wall. On long time scale it will diffuse and concentration field at t=15h (Fig.8) will be similar to the temperature field (Fig. 4).

Duration of the thermodiffusion process in the IVIDIL (t=12h) is not sufficient to reach the state similar to that shown in Fig. 8 for numerical results. Thus concentration field observed in the 2nd cell of IVIDIL at t=12h still contains some variation at the central part near the lateral walls, see Fig.9. Two views show concentration fields in two perpendicular directions at the central part, not on the walls.

However the transient process of the thermodiffusion step is well reproducible and after 12h converges to the same final state shown in Fig.9. The transient behavior of run 33 and 33R

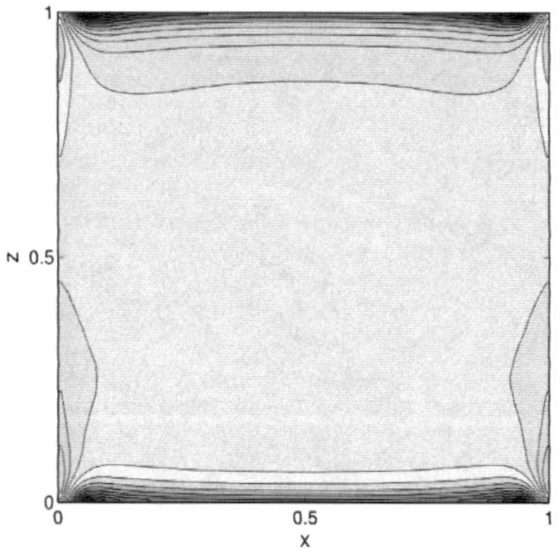

Figure 7. (Experimental results) Concentration fields 10 min after imposing $\Delta T = 15\,K$, Run 33.

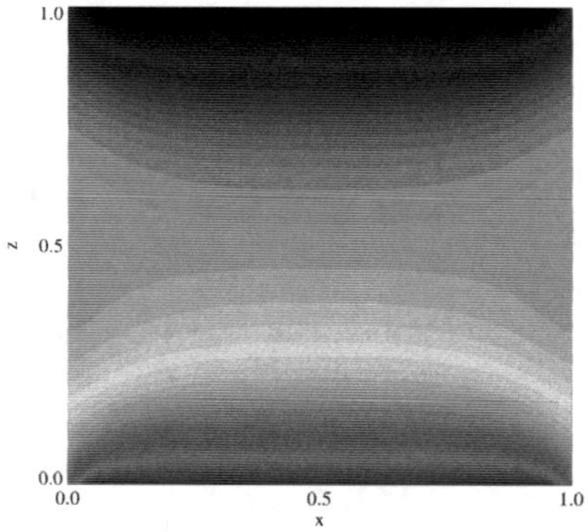

Figure 8. (Numerical results) Concentration fields 15h after imposing $\Delta T = 15\,K$, Run 33.

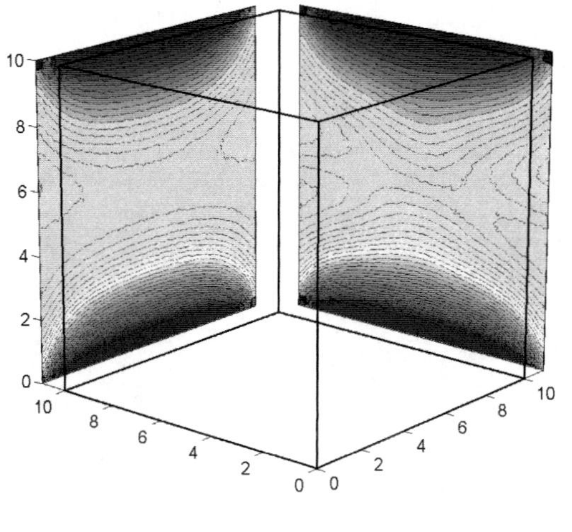

Figure 9. (Experimental results) Concentration fields in two perpendicular directions at the central part of the cell at t=12h, $\Delta T = 15\,K$; Run33.

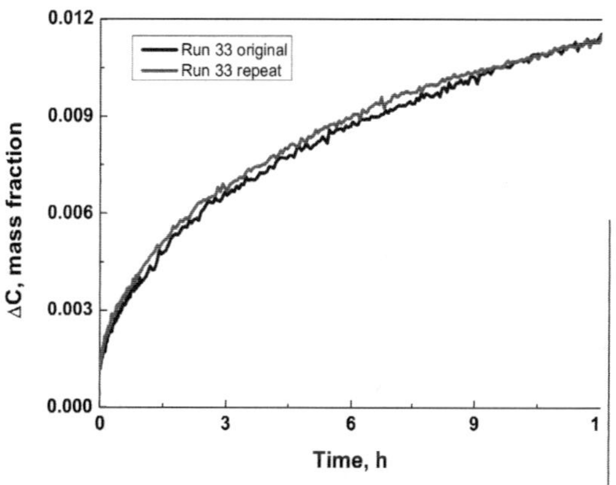

Figure 10. (Experimental results) Evolution of the component separation $\Delta C(t) = C(0,t)_2 - C(L,t)$ with time when $\Delta T = 15\,K$; Run 33 and Run33R.

performed with time lag more than one month, including total stop of the SODI and Glovebox for three weeks of Christmas Holidays, displays good agreement between curves in Fig. 10. Note, that in this case dependence $\Delta C(t) = C(0,t) - C(L,t)$ is shown because of positive Soret sign. An excellent reproducibility enables confidence to the obtained results and does not display dependence of the diffusion process upon onboard g-jitters. The transport coefficients S_T, D obtained by fitting the experimental curves in Fig.10 with the numerical solution Eq. (2) for the 2nd cell array are

$$S_T = 5.28 \times 10^{-3}\, 1/K, \quad D = 1.62 \times 10^{-10} m^2/s \tag{4}$$

The same coefficients should be recovered from all the experiment in the 2nd cell array, i.e. at

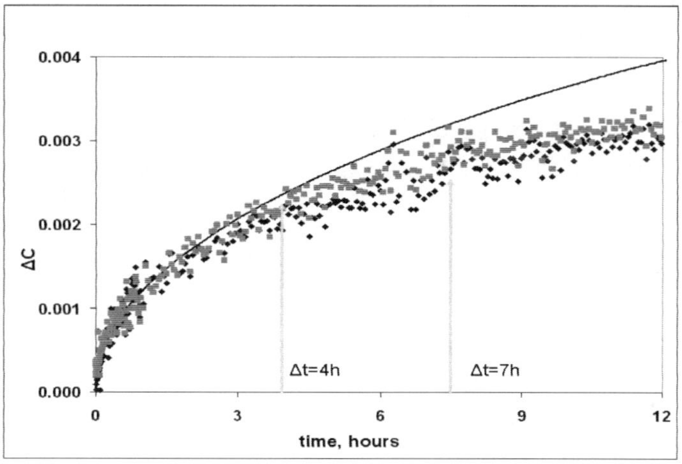

Figure 11. Evolution of the component separation $\Delta C(t) = C(0,t) - C(L,t)$ with time when $\Delta T = 5\,K$; Second cell array, Run 1N.

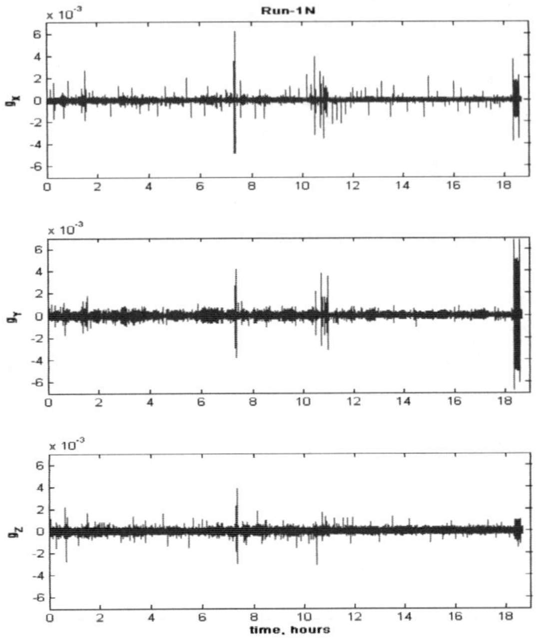

Figure 12. Residual accelerations on the ISS recorded by SAMS during the IVIDIL experiment, Run1N.

runs 1N, 2N and 33.

Looking at the perfectly reproducible results one may assume that the system under consideration is not sensitive to the microgravity perturbations. However it is not the case for strong microgravity disturbances caused by exceptional situations on the ISS. Run 1N has been performed during special activities on the ISS. Evolution of the component separation $\Delta C(t) = C(0,t) - C(L,t)$ with time for this experiment is shown in Fig. 11 where pink symbols corresponds to the front view and blue symbols corresponds to the side view. The continuous curve is the numerical solution with transport coefficients determined during the run without perturbations, i.e. by Eq. 4. Already at the first four hours experimental points display larger scattering than at nominal regime but they lie close to the numerical curve. It seems that preparatory activities by astronauts on the ISS started before the planed actions. As soon as system faced the microgravity perturbations the experimental points started to deviate from the theoretical curve and exhibit larger scattering. The strong peak perturbation at $\Delta t \approx 7h$ provides significant impact on thermodiffusion process and experimental points deviate stronger from the numerical curve.

During this run the SAMS instrument was attached to Glovebox and onboard accelerations (g-jitters) were recorded. The residual accelerations over the entire experiment in all three directions are shown in Fig. 12. The most distinctive peaks are correlated with perturbations of diffusion process. The detailed study of correspondence of the perturbations with the response of the diffusion cell is continued and will be published elsewhere.

5. Conclusions

Unprecedented experimental investigations onboard the ISS have been made to estimate the effect of g-jitters on diffusion controlled phenomena. The working liquids were binary mixtures of water and isopropanol of two different compositions. A temperature gradient was applied to the cells filled with intitially homogeneous liquid. The component separation of the mixture caused by thermodiffusion has been thoroughly analyzed during 12 hours. The evolutions of the concentration field with time over the whole cross-section of the cell were monitored by means of optical digital interferometry in two perpendicular directions. Alliance of digital interferometry and absence of convection allow to follow 3D time-dependent evolution of concentration and temperature fields.

Perturbations of the diffusion control processes by on-board g-jitter are not observed in nominal regime of the ISS. The experiments were repeated with large time lag and for different temperature gradients and they displayed an excellent reproducibility. For the runs without imposed vibrations the behavior of the experimental curves is in excellent agreement with numerical simulations when the ISS is operating in nominal regime.

Perturbations are visible in non-nominal regime of the ISS, e.g. orbit correction, docking, undocking, etc.

6. Acknowledgments

This work is supported by the PRODEX programme of the Belgian Federal Science Policy Office and ESA. The authors would like to thank everyone who has made this experiment happened.

References

[1] Gershuni G Z and Lyubimov D V, Thermal Vibrational Convection, (Wiley & Sons, 1998)
[2] Savino R, Monti R and Piccirillo M 1998 *Computers & fluids* **27**, 923.
[3] Gershuni G Z, Kolesnikov A K, Legros J C and Myznikova B I, 1997 *J. Fluid Mech.* **330**, 251.
[4] Zebib A 2001 *Phys. Fluids* **13** 1829-1832.
[5] Razi Y P, Malivan K, Mojtabi M C and Mojtabi A 2004 *Eur. Phys. J. E* **15** 335-341.
[6] Shevtsova V, Melnikov D, Legros J C, Yan Y, Saghir Z, Lyubimova T, Sedelnikov G and Roux B 2007 *Phys. Fluids* **19**, 017111.
[7] Naumann R J, Haulenbeek G, Kawamura H and Matsunaga K, 2002 *Microgravity Sci. Technol* **13 (2)** 22.
[8] Babushkin I A, Bogatyrev G P, Glukhov A F, Putin A F, Avdeev S V, Ivanov A I and Maksimova M M 2001 *Cosmic research* **39 (2)** 161.
[9] Beysens D 2006 Vibrations in space as an artificial gravity? *Europhysicsnews* **37 (3)**, 22.
[10] Mialdun A, Ryzhkov I I, Melnikov D E and Shevtsova V 2008 *Phys. Rev. Lett.* **101**, 084501.
[11] Shevtsova V, Ryzhkov I I, Melnikov D E, Gaponenko Yu and Mialdun A, 2010 *J. Fluid Mech.* **648** 53.
[12] Garandet J P, Alexander J I D, Corre S and Favier J J 2001 *J. Crystal Growth* **22** 543.
[13] Smith R W, Zhu X, Tunnicliffe M C, Smith T J N, Misener L and Adamson J 2002 *Ann. N.Y. Acad. Sci.* **974** 57.
[14] Shirkhanzadeh M, 2009 *Acta Astronautica* **64** 256.
[15] Senchenkov A 2010 *61th IAC Congress, IAC-10-A2.4.9,* Prague, Czech Republic, 27 Sept.–1 Oct., 2010.
[16] Mazzoni S, Shevtsova V, Mialdun A, Melnikov D, Gaponenko Y, Lyubimova T and Saghir Z 2010 *Europhysics News* **41** 14.
[17] Shevtsova V, Mialdun A, Melnikov D, Ryzhkov I, Gaponenko Y, Saghir Z, Lyubimova T, Legros J C 2011 *C. R. Mecanique* **339** 310.
[18] Mialdun A and Shevtsova V 2008 *Int. J. Heat Mass Trans.* **51** 3164.
[19] Mialdun A and Shevtsova V 2011 *C. R. Mecanique* **339** 362.
[20] Mialdun A and Shevtsova V 2011 *J. Chem. Phys.* **134** 044524.
[21] Mialdun A and Shevtsova V 2006 *Microgravity Sci. Technol.* **XVIII (3/4)** 146.

Journal of Physics: Conference Series **327** (2011) 012032 doi:10.1088/1742-6596/327/1/012032

Ground-based activities in preparation of SELENE ISS Experiment on Self-Rewetting Fluids

R. Savino[1], Y. Abe[2], D. Castagnolo[3], G.P. Celata[4], O. Kabov[5], M. Kawaji[6], M. Sato[7], K. Tanaka[8], J.R. Thome[9], S. Van Vaerenbergh[5]

[1]University of Naples, Italy
[2]AIST, Tsukuba, Japan
[3]Telespazio, Italy
[4]ENEA, Italy
[5]Universite' Libre de Bruxelles, MRC, Belgium
[6]University of Toronto, Canada
[7]Utsunomyia University, Japan
[8]Shibaura Institute of Technology, Tokyo, Japan
[9]EPFL, Switzerland

E-mail: [1]raffaele.savino@unina.it

Abstract. SELENE (SELf rewetting fluids for thermal ENErgy management) is a microgravity experiment proposed to the European Space Agency (ESA) in response to the Announcement of Opportunities for Physical Sciences. Main objectives of the microgravity research onboard ISS include the quantitative investigation of heat transfer performances of "self-rewetting fluids" and "nano self-rewetting fluids" in model heat pipes and validation of adequate theoretical and numerical modelling able to predict their behaviour in microgravity conditions. This article summarizes the results of ground-based research activities in preparation of the microgravity experiments. They include: 1) thermophysical properties measurements; 2) study of thermo-soluto-capillary effects in micro-channels; 3) numerical modelling; 4) thermal and concentration distribution measurements with optical (e.g. interferometric) and intrusive techniques; 5) surface tension-driven effects and thermal performances test on different capillary structures and heat pipes; 6) breadboards development and support to definition of scientific requirements.

1. Introduction

Two-phase flow devices based on evaporation and condensation, surface tension and capillary effects, are very frequently used as thermal energy management systems in many applications including electronics cooling, air conditioning, propulsion, power generation, energy recovering, chemical engineering. Research and development activities are generally concerned with development of advanced heat pipes, vapour chambers or thermal ground planes for efficient cooling of semiconductor and integrated circuit devices, since energy dissipation becomes critical to their operation. In the space applications, increase in size of space platforms and future exploration missions will require a large

[1] To whom any correspondence should be addressed.

Journal of Physics: Conference Series **327** (2011) 012032 doi:10.1088/1742-6596/327/1/012032

amount of heat to be dissipated by large space radiators. Efficient thermal control in space and the reduction of moving mechanical elements by two-phase closed-loop systems, becomes of crucial importance.
The two most important objectives in electronics or other cooling engineering applications, namely the reduction of the device maximum temperature and the minimization of temperature gradients on the device surface, can be efficiently achieved by the use of micro-grooved heat pipes and/or selecting suitable working fluids.

The research activities presented in this paper deal various processes governing heat transport capacity of the two-phase flow devices and in particular on the heat transfer performances of binary or multi-component heat transfer fluids with particular surface tension, called "self-rewetting fluids", i.e. liquids with a surface tension increasing with temperature and concentration. As pointed out in [1-9], thermophysical properties, like surface tension, wettability and thermal conductivity, at different temperatures, are of crucial importance not only for binary mixtures, but also for a number of ternary aqueous solutions with relatively low freezing point and for nanoparticles suspensions. Some of them interestingly exhibit the same anomalous positive surface tension gradient with temperature as binary self-rewetting solutions. Since in the course of liquid/vapour phase change, self-rewetting fluids behaviour induces a rather strong liquid inflow (caused by both temperature and concentration gradients) from the cold region (where liquid condensates) to the hot evaporator region, several interesting space and terrestrial applications may be envisaged [10].
In this general framework, a cooperative international project supported by ESA and JAXA is under development to contribute to the understanding of the basic fluid dynamic and physico-chemical mechanisms occurring in these multi-component two-phase systems and particularly on the interplay between heat transport and surface and bulk thermophysical properties. To pursue the above objective, the behaviour of non-isothermal self-rewetting liquids will be experimentally investigated onboard the International Space Station under microgravity conditions.
This article is organized as follows. The general scientific and technological motivations and concepts of the microgravity experiment SELENE are summarized in section 2, including main objectives, preliminary design concepts and requirements and description of the experimental procedure. Fundamental theoretical, experimental and numerical results of ground activities on the subject are summarized in the next paragraphs. In particular, laboratory results achieved on ground are presented, including thermophysical properties measurements, flow visualization experiments with transparent cells, optical and interferometric analyses. Final comments will conclude the article.

2. SELENE Experiment on ISS

SELENE (SELf rewetting fluids for thermal ENErgy management) is a microgravity experiment proposed to the European Space Agency (ESA) in response to the Announcement of Opportunities (AO 2009) for Physical Sciences. The proposal received a favourable scientific and technical review and selected for inclusion in the ELIPS research pool, approved by the ESA Programme Board for Human Spaceflight, Microgravity and Exploration. The project is carried out under a cooperative ESA-JAXA programme.
The main objectives include the quantitative investigation of heat transfer performances of self-rewetting fluids in model heat pipes and validation of adequate theoretical and numerical modelling able to predict their behaviour in microgravity conditions. Once the models have been validated, they can be extended and applied also to terrestrial applications.

Microgravity offers the opportunity to avoid gravity effects and to analyse in detail typical flow patterns occurring only in the space environment, in relatively large systems, offering the opportunity to perform systematic experimental investigations of the processes, considering a large number of experimental conditions typical of space and terrestrial heat pipes. For this reason the experiments will use transparent test cells, to allow direct flow visualization, coupled with optical and intrusive diagnostic techniques to measure thermal/compositional fields.

Main physical mechanisms that shall be investigated include:

- surface tension-driven and Marangoni flows induced by evaporation, condensation, temperature and concentration gradients
- two-phase flow and liquid-vapour distribution
- influence of the vapour temperature and pressure on the flow regime
- effect of the length of the evaporator/condenser regions
- influence of the liquid fraction
- effect of the nanoparticles type and concentration on the heat transfer performances
- effect of the heat pipe cell surface geometry (e.g. single capillary channel vs. structured surface)
- liquid film stability and the dry patch formation in each heat pipe configuration.

The SELENE experiment shall feature a heat pipe cell (ideally a pair of cells) mounted on a thermal conditioning platform (see figure 1). The heat pipe cell (pair) shall be interchangeable to allow the study of numerous cells using the same facility.

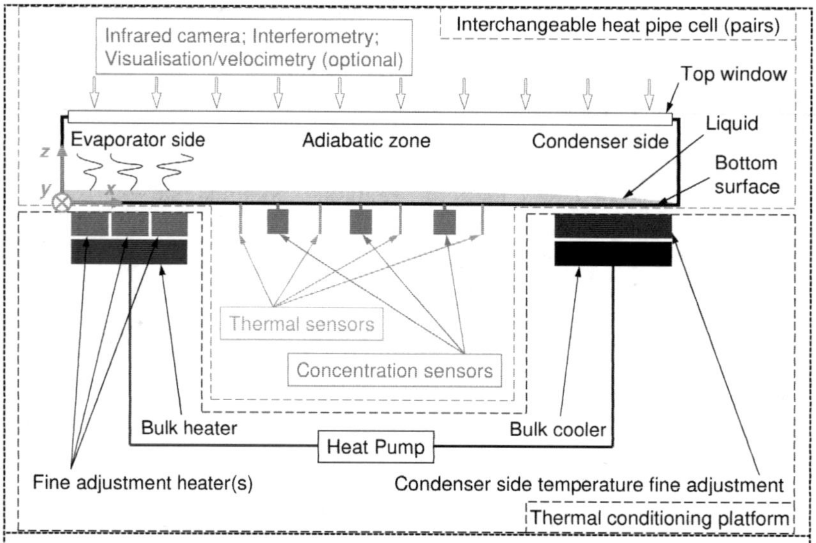

Figure 1. Schematic of the experiment concept.

The implementation of pairs of heat pipe cells could allow the parallel study of two liquids simultaneously or sequentially, but in the same configuration and settings.

Each cell represents a half-heat pipe model with one groove of appropriate cross section shape and size machined into the bottom surface. The heat pipe cells are prepared on-ground and filled with the experiment liquid. The cells may also include certain (mostly intrusive) measurement devices, such as thermal sensors. The top window of the heat pipe cell is transparent, which enables the use of various optical diagnostics (infrared observation, interferometry). Furthermore, optionally velocimetry measurements shall be possible as well. In each heat pipe cell temperature and concentration sensors shall provide local measurements of the liquid properties.

For each heat pipe, evaporation is promoted at the hot side (evaporator) with a controllable heat input while the temperature at the other end (condenser) should be controlled and kept at a lower value. For a given temperature difference, corresponding to a prescribed thermal power, the evaporated liquid will condense at the cold side where heat will be removed and the liquid condensate will be driven back to the evaporator flowing along the channel by the capillary effect and/or by the inverse Marangoni effect. The heat input (evaporator) and the temperature (condenser) can then be fine-tuned with dedicated elements. The investigation of the evaporation and condensation-driven behaviour of the liquid in this cell, when subjected to different heat stimuli is the main objective of the experimental research.

Main experiment scientific required measurements include:

- heat power provided at the evaporator section
- heat exchanged at the condenser section
- temperature distributions along the liquid-vapour interface and in the liquid at different positions along the cell, from the evaporator to the condenser
- concentrations in the liquid at different positions along the cell, from the evaporator to the condenser
- liquid distribution and in particular film thickness and shape of the liquid-vapour interface from the evaporator to the condenser using visual observation and quantitative analysis (e.g. interferometric techniques) during heat pipe operations, i.e. from relatively low power up to the dry-out limit for the different investigated working fluids
- flow visualization shall be considered also for liquid velocity measurements

A layout of a possible experimental heat pipe cell is shown in figure 2. It shall include in the "bottom surface" part a metallic surface (copper or aluminium) with a dedicated hydro- and oleophilic channel (groove) connecting the evaporator to the condenser in the nominal configuration. Nevertheless, different surface structures shall be possible as well.
The shape and size of the channel shall be designed according to the maximum thermal power and heat flux density to be representative of ordinary heat pipes.
As a baseline, the bottom surface of the groove shall be prepared with very small roughness (like an optical mirror) or eventually coated (e.g. with silver or gold) in such a way that the optical quality allows interferometric measurements in reflection mode. Nevertheless, certain samples with different surface roughness and/or structuring will be also foreseen (without interferometric measurements). The opposite surface of the cell ("top window" in figure 2) shall be made of an optically transparent material with relatively high thermal conductivity (e.g. sapphire).

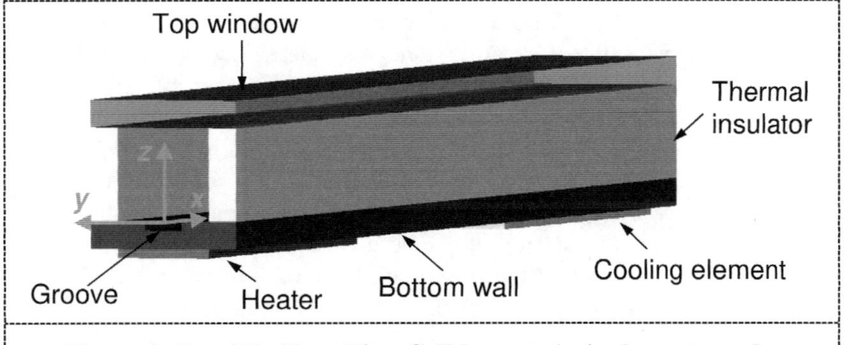

Figure 2. Possible Heat Pipe Cell layout. A single rectangular groove is shown but different surface structures shall be considered

The experiments require very long duration, therefore scientific and technological results can be obtained only on ISS using a dedicated test container.

For measurements of temperature and concentration distributions a set of temperature and concentration sensors shall be mounted on each pipe. In particular micro-sensors for measurements of liquid refractive index will be considered, in order to evaluate the liquid concentration once the temperature is known.

CCD, Infrared Camera and Interferometric analysis will complete the investigation of the basic fluid physics mechanisms occurring in the heat pipe model, to detect the liquid fraction distribution within the cell, the liquid-vapour interface shape, the thermosolutal and liquid velocity distributions.

3. Ground-based research activities

In what follows, a number of ground research activities, in preparation of the microgravity research, will be summarized.

3.1. Characterization of thermophysical properties of self-rewetting fluids and nanofluids

As discussed in [11], main thermophysical properties of interest for utilization of self-rewetting fluids in thermal energy management devices include surface tension, contact angle, latent heat of vaporization, thermal conductivity, freezing point. These properties have been investigated for a number of binary and multi-component self-rewetting solutions including water-alcohol solutions (based on Butanol, Pentanol, Exanol, Heptanol, tec.) as well as in low freezing point mixtures (brines) with relatively low concentrations of the same alcohols [11]. Comparison between different working fluids show that generally self-rewetting fluids exhibit better performances in terms of surface tension and wetting behaviour, in comparison to the base liquids. Furthermore, preliminary tests on heat pipes in different configurations pointed out better thermal performances when self -rewetting fluids or self-rewetting brines are used to replace common heat transfer fluids like water.

Particular attention has been dedicated to the study of the thermophysical properties of nanofluids based on water/alcohol solutions. They include self-rewetting nanofluids based on silver or gold nanoparticles or nanofibers [9], or nanofluids with suspended carbon nanostructures, in particular single-wall carbon nanohorns [12].

In [9] the authors have shown that "self-rewetting nanofluids" containing aqueous-butanolic solution with microwave-assisted synthesized silver nanoparticles exhibit dryout limits in heat pipes almost twice as high as that of base working fluid heat pipe and this could be explained by the enhancement of the surface tension gradients with temperature, due to the dependencies changes of silver nano-self-rewetting fluids. Since these self-rewetting nanofluids contained very dilute silver nanoparticles, other thermophysical properties such as thermal conductivity and viscosity were almost same as those of base fluid.

Another research line was dedicated to self-rewetting nanofluids prepared suspending, in butanol-water solutions, single wall carbon nanohorns (SWNH), produced by arc discharge [12]. In this case, surface tensions measurements reveal the self-rewetting behaviour of the nanofluids as the self-rewetting base solutionshave shown that the presence of the carbon nanoparticles does not affect the surface tension and contact angles properties of these fluids, in comparison to the corresponding base fluid, i.e. a binary butanol aqueous solution. On the other hand, thermal conductivity measurements performed with the parallel plate technique pointed out a significant enhancement of the equivalent thermal conductivity due to the presence of the carbon nanohorns.

Preliminary heat transfer analysis of heat pipes filled with self-rewetting nanofluids based on carbon nano-horns confirm improved performances in comparison to other self-rewetting fluids.

3.2 Flow visualization and interferometric analysis

Research activities have been focused on numerical simulations and laboratory experiments in transparent cells for the visualization and analysis with optical diagnostic systems of fluid flows in presence of thermal gradients.

Some basic fluid-dynamic mechanisms related to the behaviour of wickless transparent heat pipes with self-rewetting fluids have been investigated in [6]. In particular, the interesting behaviour of "inverse" thermocapillary bubble migration was pointed out (see figure 3) in transparent capillary tubes, showing the different Marangoni effect in self-rewetting fluids compared to ordinary liquids. This phenomenon has been discussed to show the potential advantages of the new working fluids and confirmed by the thermal characterization of wickless heat pipes filled with water and with self-rewetting fluids.

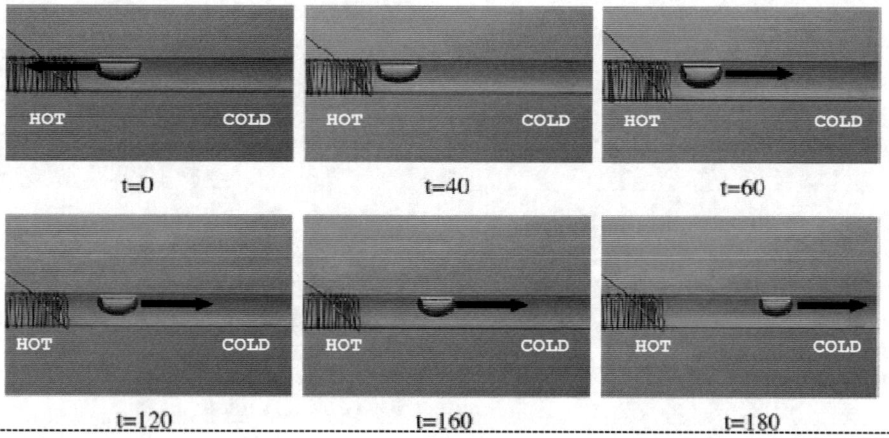

Figure 3. Inverse bubble Marangoni migration in self-rewetting fluid [6]. The capillary diametr is 6 mm. The heater is shown on the left. The bubble initially moves towards the left (hot side) and then to the right (cold) side

Experiments with binary solutions of water and long-chain alcohols have confirmed the different boiling behaviour in heat pipes filled with self-rewetting fluids in comparison to pure water [6].
For the self-rewetting fluid the evaporation region contains more liquid in comparison to water, which also implies spontaneous liquid supply to higher temperature region and vapour slug behaviour near the heating region implies a rather strong shear stress between vapour and liquid film.
In [13] thin optically transparent cuvettes with lateral heating allowed the development of a nearly 2D flow, good thermal boundary conditions and minimization of the liquid meniscus; in addition this configuration is similar to typical devices for heat transfer applicationsThe target of the experiments was to observe the Marangoni flow driven by the thermal and solutal gradients along the liquid-vapor interface. The investigated fluids include two ordinary liquids (water and ethanol) and water/alcohols mixtures (self-rewetting fluids), in particular water-butanol 5%wt, water-heptanol 0.2%wt.
As discussed in [13], for self-rewetting solutions, due to the evaporation–condensation phenomena, both thermo-capillary and solutal Marangoni effects are present. Indeed, any imposed temperature difference along a liquid-vapour interface creates a surface concentration gradient, alcohol evaporates in the hot region of the surface and condensation occurs at the cold side. For this purpose, flow visualization experiments have been accomplished using optical diagnostic systems, in particular, a parallel background illumination system to visualize tracer particles and a two wavelength Mach-Zehnder interferometer. The latter technique allows one to evaluate, once the temperature and concentration dependence of the refraction index is known at each wavelength, the temperature and concentration profiles. In this way the different contributions related to thermo-capillary and soluto-capillary effects can be independently evaluated. In particular, based on the measured surface tension derivatives for the different binary solutions investigated, the different driving actions have been predicted from the temperature and concentration distributions, for a positive mixture (like water-ethanol) and a self-rewetting water-butanol solution.
The measured temperature and concentration maps shown in figures 4 and 5 point out that, as

Journal of Physics: Conference Series **327** (2011) 012032 doi:10.1088/1742-6596/327/1/012032

expected, in pure liquids ordinary Marangoni flow is directed towards the cold side, but self-rewetting solutions exhibit a self-rewetting behaviour, i.e. liquid at the interface directed towards the hot side. In particular, the surface velocity, directed from the cold (left) towards the warm (right) side is responsible for an evident "inverse" deformation of the isotherms at the liquid-vapor interface. The numerical results are in agreement with the experimental results. Further details can be found in [13].

Figure 4. Marangoni convection in Ethanol. Interference fringes and temperature distributions, experimental and numerical [13]. The liquid layer is 1mm depth. The cell length is 4cm. Only the central part is shown. The temperatures at the hot and cold side are 60°C and 40°C

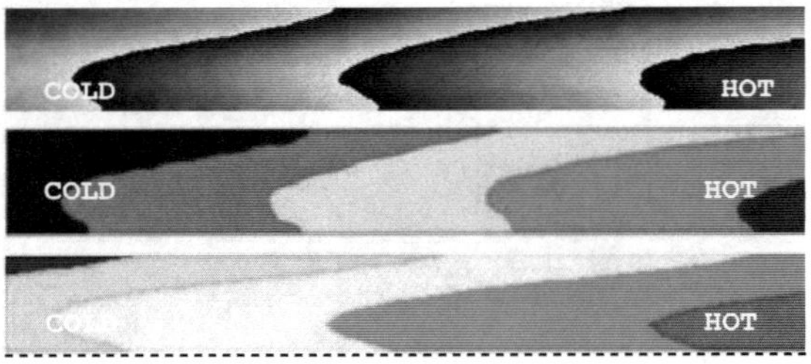

Figure 5. Marangoni convection in water-heptanol solution. Interference fringes and temperature distributions, experimental and numerical [13]. The liquid layer is 1mm depth. The cell length is 4cm. Only the central part is shown. The temperatures at the hot and cold side are 70°C and 40°C

3.3 Theoretical and numerical activities

Preliminary theoretical and numerical simulations are in progress at the University of Naples to identify the geometry and experimental conditions for the ISS experiment. In particular, the main objective is the development of analytical or numerical models able to predict the dry-out (capillary) limit of microchannels in microgravity, in order to identify the expected experimental conditions.

A model for fluid flow and heat transfer in the micro-channel of arbitrary V-shaped cross section has been developed due to evaporation and condensation. The liquid meniscus radius changes with position in a micro-channel, and the liquid pressure is given by the Young-Laplace equation.The model allows one to evaluate the dry-out limit of micro-channels with different cross section.

The model equations [14] are based on the following assumptions: (i) one dimensional steady incompressible flow along the length of heat pipe; (ii) uniform distribution of heat input; (iii) negligible heat dissipation due to viscosity; (iv) constant pressure in the vapor region in the operating range of temperature; (v) one dimensional temperature variation along the length of heat pipe; (vi) negligible shear stress at the liquid-vapor interface (zero Marangoni effect); complete wetting of the liquid (zero contact angle).

The heat pipe has three sections – the evaporator takes up dissipated heat, the condenser releases the heat to the surroundings, and the adiabatic section transports the heat from the evaporator to the condenser. Due to evaporation and condensation, the liquid meniscus radius changes with position in a micro-channel, and the liquid pressure is given by the Young-Laplace.

The pressure difference generates the capillary pumping for the liquid flow from the condenser to the evaporator. In particular, the liquid is pushed from the cold end to the hot end due to decrease in the radius of curvature caused by the intrinsic meniscus receding into the corner. The liquid film gradually becomes thinner and more curved (lower radius of curvature) as the liquid recedes further towards the apex of the corner (see figure 6).

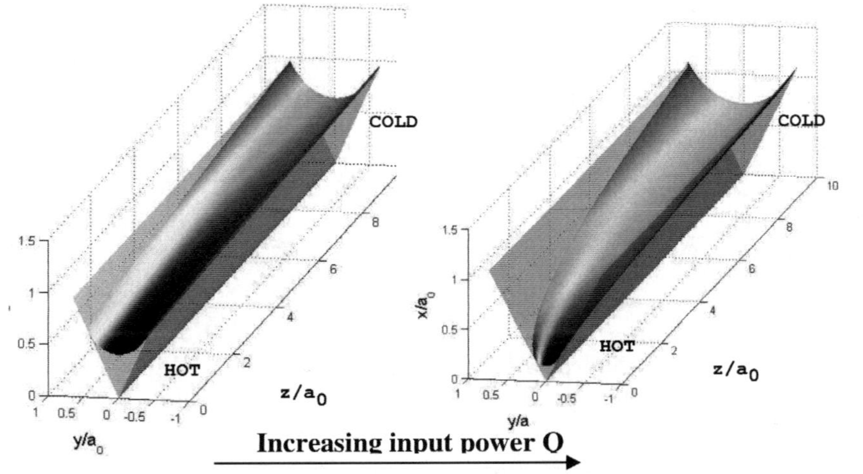

Figure 6. Qualitative distribution of liquid in a microchannel at different input powers. The left figure shows the situation at a relatively low input power. The right figure shows that, increasing the input heat, the liquid level decreases at the evaporator and increases at the condenser. a_0 and L are the initial liquid level and the length of the microchannel. The colors show the liquid level (h)

The analytical model provides the equation for the dimensionless curvature radius as function of the local coordinate and of the input power at the evaporator, if the fluid properties (viscosity, surface tension, latent heat, etc.) are known. Furthermore, the critical heat input (Q_{cr}) i.e. the maximum heat transport capacity of a micro-grooved heat pipe, without generating any hot spot in the heat pipe, can be evaluated.
In particular, according to order of magnitude scaling analysis, the maximum dimensionless heat flux at the evaporator/condenser (scaled with reference thermophysical fluid properties) is proportional to the square of the groove depth (a) to length (L) ratio:

$$\frac{\mathrm{Q}}{(\mathrm{La})(\rho \mathrm{H}_{\mathrm{v}} \sigma / \mu)} \propto \left(\frac{a}{L}\right)^2 \tag{1}$$

where Q is the heat input and Q/(La) is the heat flux density (W/m^2),ρ,H_V, σ and μ are the density, the latent heat of vaporization, the surface tension and the viscosity respectively.

Figure 7 shows the predicted evolution of the dimensionless film thickness along a microchannel with length 15 cm, cross section of width 6mm, half angle 45°, in presence of different input powers, until dry-out. It is interesting to see also the corresponding power densities in figure 8.

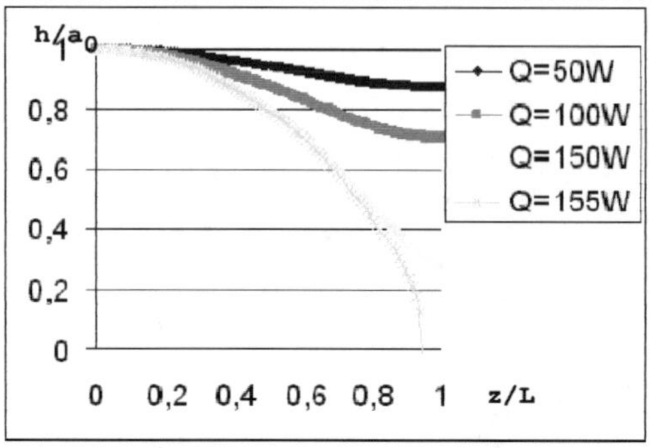

Figure 7. Computed dimensionless film thickness (h/a_0) along a microchannel for different input power [14]. z=0 and z=L correspond to the condenser (cold) and the evaporator (hot) sides.

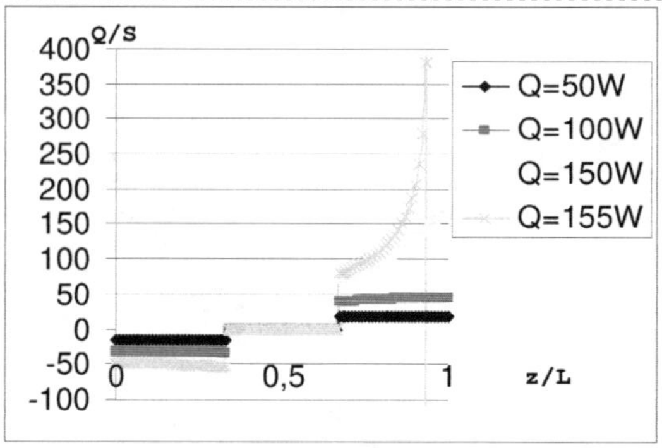

Figure 8. Computed power density (W/m^2) along a microchannel at different input power [14].

A three-dimensional numerical model has been developed to better investigate the influence of some parameters like filling ratio (i.e. ratio between the volume of the liquid and the volume of the fluid cell), and the wetting properties of the liquid on the dry-out limit.

Figure 9 shows the geometry (in particular the rectangular groove) of the investigated configuration and the corresponding three-dimensional grid. The cross section of the bottom groove is 4mm x 4 mm. The total length is 15 cm.

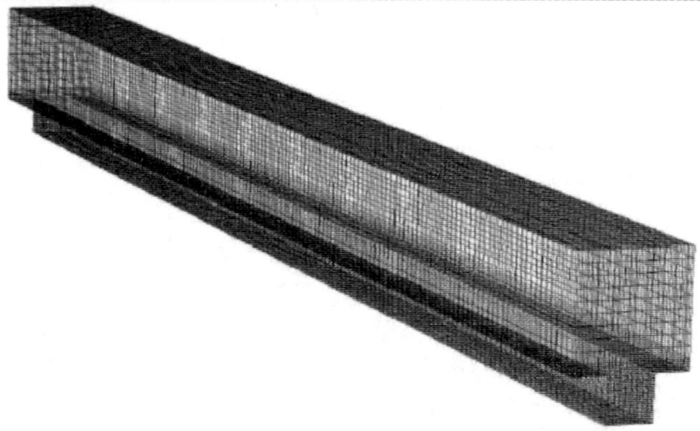

Figure 9. Geometry of the cell and computational domain.

The numerical model, based on the commercial FLUENT solver with ad-hoc user defined functions, is able to solve the energy and fluid dynamics equations for a two-phase system (liquid and vapour) in presence of evaporation and condensation.

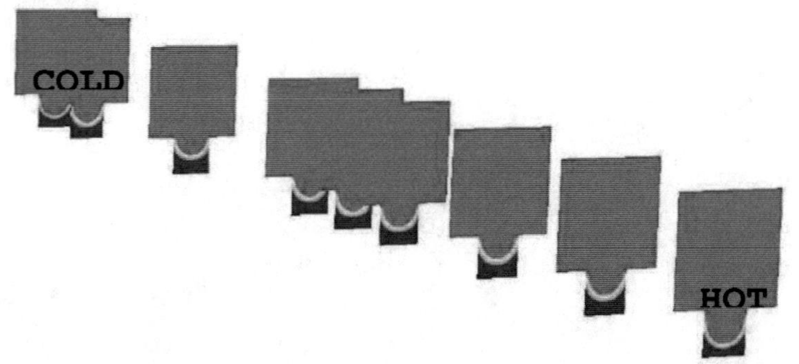

Figure 10. Computed liquid distribution for input power Q=10W and complete liquid wetting (θ=0°). The liquid phase is blue, the vapour phase is red.

Figures 10, 11 and 12 show different CFD results, considering a channel partially filled with water. It is interesting to see the different liquid distributions corresponding to different input power and different wetting behaviour. In particular figure 10 refers to a the case of a relatively low input power (Q=10W) when the liquid completely wets the bottom channel. The liquid volume decrease from the condenser to the evaporator but the capillary pressure is enough to sustain liquid return avoiding dry-out. Figure 11 shows the result for a power Q=100W. In this case the liquid content at the evaporator side is less than in the previous case, but a liquid film is still present.

For the same conditions if the liquid exhibit poor wetting properties (see figure 12) the condensated liquid is accumulated at the cold side and the evaporator becomes completely dry.

Figure 11. Computed liquid distribution for input power Q=100W and complete liquid wetting (θ=0°). The liquid phase is blue, the vapour phase is red.

Journal of Physics: Conference Series **327** (2011) 012032	doi:10.1088/1742-6596/327/1/012032

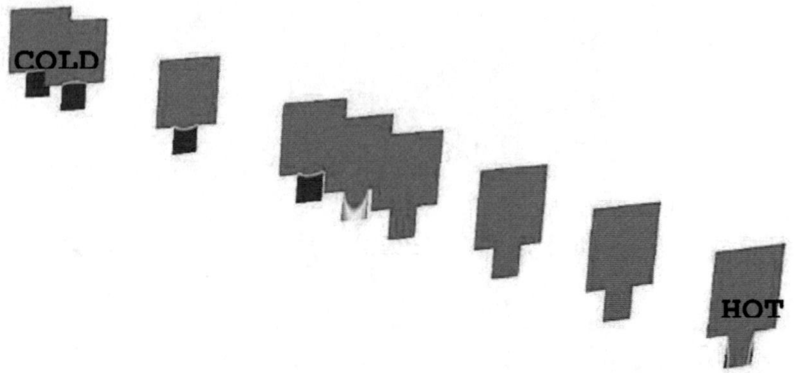

Figure 12. Computed liquid distribution for input power Q=100W and poor liquid wetting (θ=90°). The liquid phase is blue, the vapour phase is red. The dry-out conditions is established at the evaporator (hot) side

4. Experimental activities

Experimental investigations include the study of heat transfer performance of heat pipes using water/alcohol mixtures and different set up and configurations. These laboratory experiments specifically explore the use of self-rewetting fluids, self-rewetting brines and self-rewetting nanofluids. A number of results can be found, e.g. in [10-12].

In addition, in preparation of SELENE flight experiment, ground-based experimental activities will be carried out to identify the flight experimental liquids and concentrations, the requirements for the development of the flight hardware and the diagnostic systems. Different ground-based experiments are in progress (see figure 13). An experimental breadboard cell has been developed in order to study the preliminary configuration and setup, to identify possible problems and corresponding solutions.
In particular, ground-based experiments include stability of surface properties, material compatibility test between liquid and materials, diagnostics for evaluation of distribution of liquid and surface reconstruction (film shape and thickness measurements with interferometric techniques), concentration measurements with intrusive (optical probes) and optical techniques, such as-two-wavelength Mach-Zhender interferometer, nanofluids characterization including study of nanofluids stability, particles shape and size control.

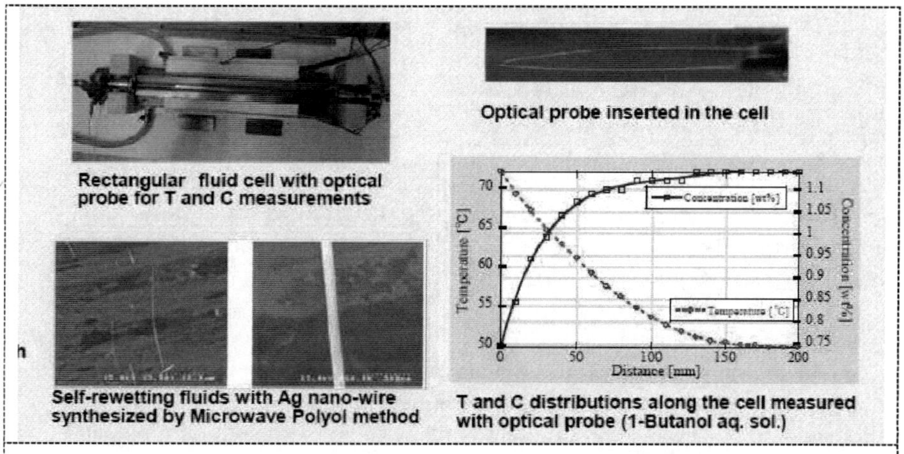

Figure 13. Examples of experimental breadboard cells and preliminary results. The volume available for liquid is a 200mm long glass tube of inner diameter 12.5mm and outer diameter 15mm. The optical probe is 100 micron diameter. The temperature is measured with a sheated thermocouple.

5. Conclusions

The scientific requirements of SELENE ISS experiment will be defined through theoretical and numerical analyses, as well as ground-based experimental activities. The main objectives include identification of experimental liquids, requirements for the flight hardware and diagnostic systems.
Current research activities include: characterization of thermophysical properties of self.rewetting fluids and nanofluids, flow visualization and interferometric analysis in relatively simple geometrical configurations to show fundamental behaviour, development of numerical tools and experiments with breadboard cells. Some of the most important results have been summarized in this article.
Other experiments are in progress, including also activities dedicated to the characterization of intrusive methods for measurements of concentration and temperature profiles. Detailed theoretical and numerical studies and comparisons with experimental results will be carried out in the future. Particular attention will be focused on the heat transfer mechanisms in grooves of different shapes and size, to properly define the flight experimental configuration, also considering requirements of quantitative investigation of the liquid distributions in microchannels using optical techniques.

6. References

[1] Savino R., Monti R. 2005 *Space Technology* **25: 59**.
[2] Tanaka K., Abe Y., Nakagawa M., Piccolo C., Savino R. 2009 *Annals of the New York Academy Science*, 1161: **554**.
[3] Savino R., Paterna D. 2006 *Physics of Fluids* **18**:**118103**.
[4] Savino R., di Francescantonio, N., Fortezza, R., Abe, Y. 2007 *Acta Astronautica* **61**: **16**.
[5] di Francescantonio N., Savino, R. & Abe, Y. 2008 *Int. Journal of Heat and Mass Transfer* **51: 6199**.
[6] Savino R., Cecere A., Di Paola R. 2009 *Int. Journal of Heat and Fluid Flow* **30**: **380**.
[7] Savino R., Abe Y., Fortezza R. 2008 *Acta Astronautica*, **63: 24**
[8] Savino R., Piccolo C. & Fortezza R., Abe, Y. 2007 *Microgravity Sci. & Tech.*, **XIX 3/4** :**75**.
[9] Sato M. et al., Proc. 17th Symposium on Thermophysical Properties, Univ. Colorado, Boulder, Boulder, CO, USA, June 21-26, 2009.

[10] Savino R., Cecere A., Di Paola R., Abe Y., Castagnolo D., Fortezza R. 2009 *Acta Astronautica*, **65**: **1582**
[11] Savino R., di Paola R., Cecere A., Fortezza R. 2010, *Acta Astronautica*, **67**: **1030**
[12] Savino R., Di Paola R., Mirabile Gattia D., Marazzi R., Vittori Antisari M. 2011, *Journal of Nanoscience and technology*, **11**: **1**
[13] Cecere A. Di Paola R., Savino R., Abe Y., Carotenuto L., Van Vaerenbergh S., 5th Conference of the International Marangoni Association, June 7-10 2010, Florence, Italy.
[14] Suman B., Savino R. 2011 Capillary Flow Driven Heat Transfer Enhancement, *accepted for publication in AIAA J. of Thermophysics and heat Transfer*

Acknowledgments
The Authors wish to acknowledge the ESA project scientist Dr. Balazs Toth for his valuable contribution to the development of the SELENE experiment scientific requirement.

Long range boundary effect of 2D intermediate number density vibro-fluidized granular media in micro-gravity

Chen Yanpei[1,2], Pierre Evesque[1], Meiying Hou[2], C. Lecoutre[3], F. Palencia[3] and Yves Garrabos[3]

[1] Lab MSSMat, Ecole Centrale de Paris,UMR 8579 CNRS, 92295 Châtenay-Malabry Cedex, France

[2] Institute of Physics, Chinese Academy of Sciences, Beijing, 100190, China

[3] ESEME, Institut de Chimie de la Matière Condensée de Bordeauz, CNRS UPR 9048, 33608 Pressac Cedex, France

E-mail: `pierre.evesque@ecp.fr` `mayhou@aphy.iphy.ac.cn`

Abstract. We present a micro-gravity experimental study of the statistical properties of intermediate number density vibro-fluidized inelastic spheres in a rectangular container. It is found that although when taking all the particles into account, the probability distributions of velocities both along and perpendicular to the vibration direction are exponential and symmetric, when dividing particles along the vibration direction into different bins, the local velocity distributions are found to deviate measurably from a symmetric distribution for the velocity component in the vibrating direction. The skewness analysis of the local distribution profiles for v_x and v_y shows that the local distribution of v_x remains symmetric, however, the skewness of the distribution profile in v_y changes nearly linear from positive to negative with skew = 0 near the center bin. This indicates a long range boundary effect of the asymmetry in v_y. We further studied the hydrodynamic profiles granular pressure p_x and p_y, and temperature T_x and T_y in positive and negative components such as p_x^+ and p_x^-, and T_x^+ and T_x^-, in accordance with the sign of velocity components. The profiles for the two components are found different along the y direction. Along vibration direction granular medium is found inhomogeneous and anisotropic not only in the particle number densities, but also in v_y, p_y and T_y. This suggests new hydrodynamical modeling is needed for such vibro-fluidized granular systems.

1. Introduction

Granular medium has caught much of physicists' attention in theoretical, simulation and experimental studies in recent years. It is deemed as an ideal model system for studying the statistical and dynamical properties of nonlinear non-equilibrium systems. Granular systems are treated by resembling the molecular gas or ordinary fluids [1] in many works although big differences may be found between them. For a nearly homogeneous granular system, it is assumed that the granular systems follow the Maxwell-Boltzmann velocity distribution $e^{-mv^2/2kT}$ and conform to equipartition of energy. From the Maxwell-Boltzmann distribution, temperature is defined by $T = 1/2m < v^2 > (k = 1)$, when there are no macroscopic flows in the system. However, an increasing number of works point out the non-Gaussian velocity distribution [2–11] in granular gases. These deviations from the Maxwell distribution find overpopulation in the high-velocity tail. Kinetic reasons of these deviations are still unclear. Among all the potential reasons, concerns on the driving mechanism [6, 12–22] have been raised. Brey [21] discussed discontinuity effects at the sawtooth boundary in direct Monte Carlo simulation being due to the particle-number conservation near the boundary. Local distribution functions f at the boundary walls are found asymmetric and discontinuous. Olaf Herbst [22] et al. also found same local velocity distributions f in simulation. However, they did not indicate whether this asymmetry and discontinuity due to the boundary effect only existed in the 'Boundary Layer' or extended to the whole cell. If it extends to the whole cell, another question to ask is how it shall vary with the distance from the boundary. For inhomogeneous granular media whether hydrodynamic descriptions hold is unclear. Breakdowns of the hydrodynamics have been shown [23, 24]. To find the suitable scope of hydrodynamics for granular systems is still a challenge and further studies, especially experimental studies, are needed.

This paper reports our experimental observations obtained in Airbus of *Novespace* (2006 Campaign) in a 2D vibro-fluidized granular system. In the experiment the particle velocity distributions are obtained in both x and y directions globally and locally. It is found that although when taking all the particles into account, the probability distributions of velocities both along and perpendicular to the vibration direction are exponential and symmetric, but when dividing particles along the vibration direction into bins, the local velocity distribution measured in each bin is found to deviate measurably from a symmetric distribution for the velocity component in the vibrating direction. Considering the asymmetric distributions of local velocities, we measure hydrodynamic profiles in positive and negative components in accordance with the sign of velocity components. We find that the hydrodynamic profiles are different for these two components. Under the influence of the vibrating walls, granular medium is found inhomogeneous and anisotropic not only in the particle number density, but also in v_y, p_y and T_y.

The outline of this paper is as follows. In Section 2, we present the experimental method. In section 3, we show the global velocity distributions are exponential and symmetric in both x and y directions, but the local velocity distributions are asymmetric in the y direction and the skewness is discussed. In section 4, the distributions of hydrodynamic parameters are investigated. Finally, a brief summary is given.

2. Experimental Method

The experiment was conducted in Airbus A300 developed by *Novespace* in 2006. A quasi-2D quadrate cell (side walls in aluminum, front and back walls in glass, V=10mm *10mm *1.4 mm) containing 47 copper spheres is driven in y direction with a sinusoidal oscillating shaker. A snapshot is shown in Fig. 1. The diameter of the copper spheres is (1.21mm $\pm$ 0.02mm) , and the area fraction is 0.54. The spheres can rotate in three dimensions but only have two dimension translational motion.

The cell is fixed on a V455 LDS shaker, which vibrates in the y direction. The vibration controller is the same as in previous works [8,25]. The movement of the beads is recorded by a fast camera (499 fps). The LEDs are mounted in the reverse side of the cell to increase the contrast of the beads vs. the background. The experiments were performed for 30 parabolic flights in a 2-hour Campaign. Each parabolic flight has about 20s for micro-gravity (0.0 g$\pm$ 0.05 g). Within the 20s of each parabolic flight only one second is recorded by the fast camera. The resolution of the cell image is 288 pixel* 288 pixel, i.e. 1 pixel= 0.035mm. The frequency range in the experiment is from 40 Hz to 130Hz, the oscillating amplitude is ranged from 0.005 to 0.912mm, and the correspondent acceleration is in the range of 0.1 g to 9 g. In this paper, we only present four sets of data with vibrating conditions shown in Table 1. We use notations (A, V_ω, Γ, f) to represent amplitude, vibrating peak velocity, acceleration and frequency, respectively.

In the table all the parameters are determined by image analysis using the software *ImageJ*.

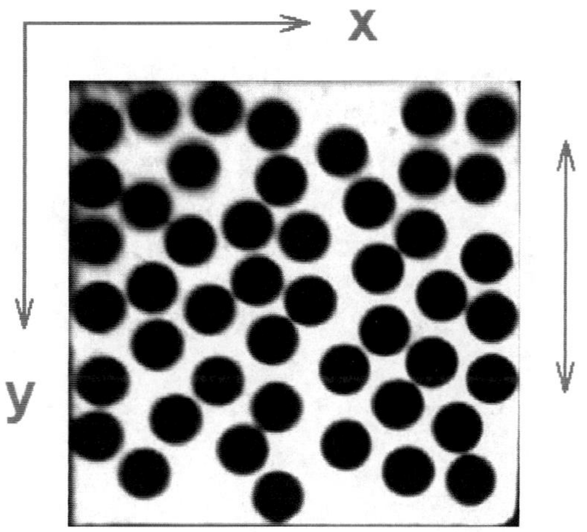

Figure 1: Snapshot of the cell particles. The cell contains 47 bronze sphere particles, driven in y direction at various vibration parameter(A, V_ω, Γ, f). The diameter of particle is 1.21 mm and cell size is (10mm*10mm*1.4mm).

Images recorded by the fast camera are firstly processed to get the positions of beads center which are calculated through the ultimate eroded points(UEPs) in the Euclidian distance map(EDM). Afterwards, particles are tracked using the program of minimum-distance algorithm. Since our system has moderate number density, the spheres can not move extensively, the minimum-distance algorithm can work well in our case. The magnitude of accuracy obtained in this way reaches $0.01mm$.

Table 1: The driving parameters in experimental. The y is the vibration direction (Amplitude A, speed $V_w = A\omega$, $\Gamma = A\omega^2$ for acceleration, frequency f).

	$Frequency(Hz)$	$A(0-peak,\ \ mm)$	$\Gamma(m/s^2)$	$V_\omega(m/s)$
1.	49	0.23	21.56	0.070
2.	97	0.11	41.28	0.067
3.	97	0.14	53.60	0.088
4.	49	0.12	11.73	0.038

3. Velocity distributions

In this section, the velocity distributions are analyzed under vibration condition 1 listed in ***Table 1***. The particle velocities are averaged for 499 image sequences within 1 second. Considering the anisotropy of the vibration, we analyze the x and y components of the velocity distributions separately. For local velocity distribution along vibrating axis, y axis, we divide the cell into seven bins and measure the velocity distribution in each bin.

3.1. The velocity distributions

The velocity distributions for components v_x and v_y are investigated as shown in Fig. 2. The velocities v_x and v_y are scaled by the characteristic velocities $v_c^x = \sqrt{\bar{v_x^2}} = \sqrt{T_x}$ and $v_c^y = \sqrt{\bar{v_y^2}} = \sqrt{T_y}$ [7], respectively. The velocity distributions in logarithmic scale are also shown in the insets of Fig. 2 (a) and (b). In the figures, the dash lines are exponential fitting $C * exp(-|B * x|)$, where A and B are fitting parameters, while the solid lines are Gaussian fitting $F * exp(-(x/G)^2)$, and F and G are fitting parameters. As can be seen in the figures, the Gaussian fitting underestimates in the parts of low and high velocities, in agreement with previous report [7]. The characteristic velocity $v_c^y = 0.057m/s$ is found larger than v_c^x $(0.041m/s)$. Anisotropy is apparent.

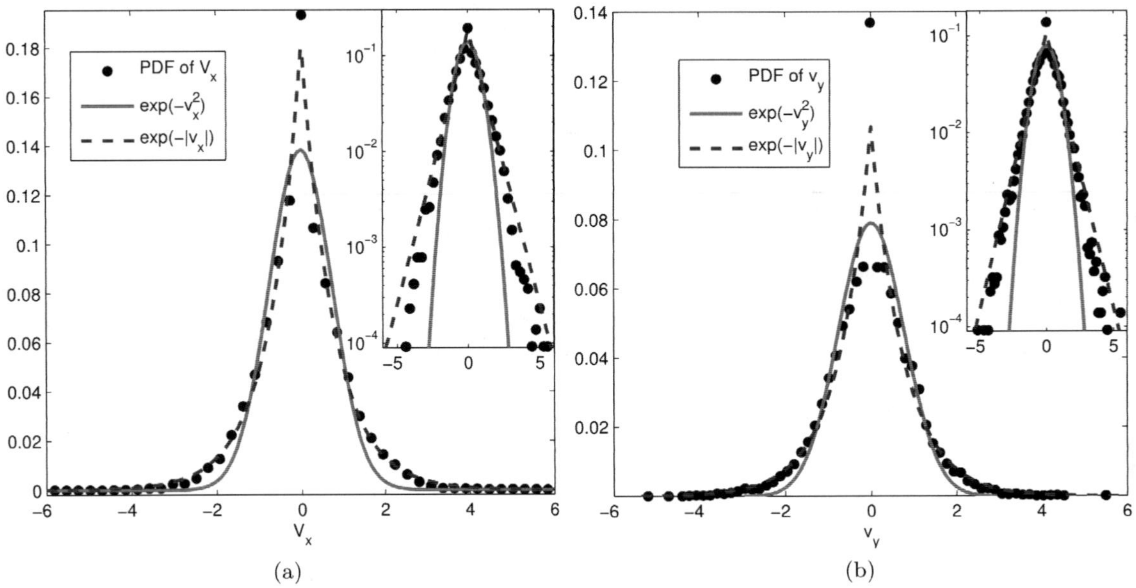

Figure 2: Distribution of the v_x (a) and v_y (b) in the experiment no. 1 of *Table 1*. The insets are the same plots on logarithmic scale. The velocities are scaled by $\sqrt{T_x} = 0.041m/s$ and $\sqrt{T_y} = 0.057m/s$. The solid lines are the Gaussian fit, where dashed lines are exponential.

3.2. Local velocity distributions

In order to understand the spatial distribution of velocities, the local velocity distributions are investigated. The cell is divided into 7 bins along the y axis, and the velocity distributions v_x and v_y are averaged in each bin. The velocity is scaled by the maximum driving velocity $V_\omega(0.07m/s)$ (condition 1 of *Table* 1).

The local velocity distribution is found anisotropic along y direction. As shown in Fig. 3, the local velocity distribution of v_y is asymmetric for each bin in y direction, while the distribution of v_x in each bin remains symmetric. The local velocity distribution of v_y in the bins near the two boundaries $\tilde{y} = 0$ or 1 $(\tilde{y} = y/L)$ shows remarkably asymmetric. While moving away from the driving wall, the asymmetry of the profile of the local velocity distribution v_y becomes less profound. In the box center, the profile becomes symmetric, while the boundary effect maybe balanced out.

For a quantitative analysis of the asymmetry, momenta technique of statistic distribution can be used and skewness is calculated for each local profile. The result is shown in Fig. 4. The

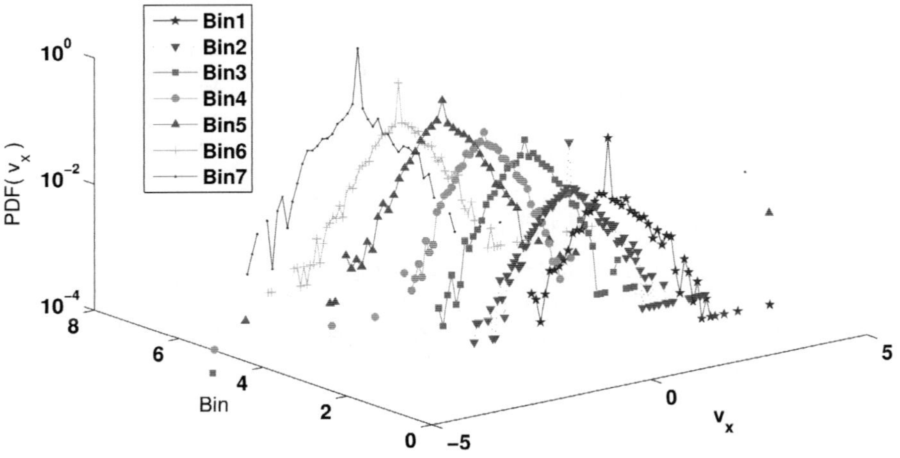

(a)

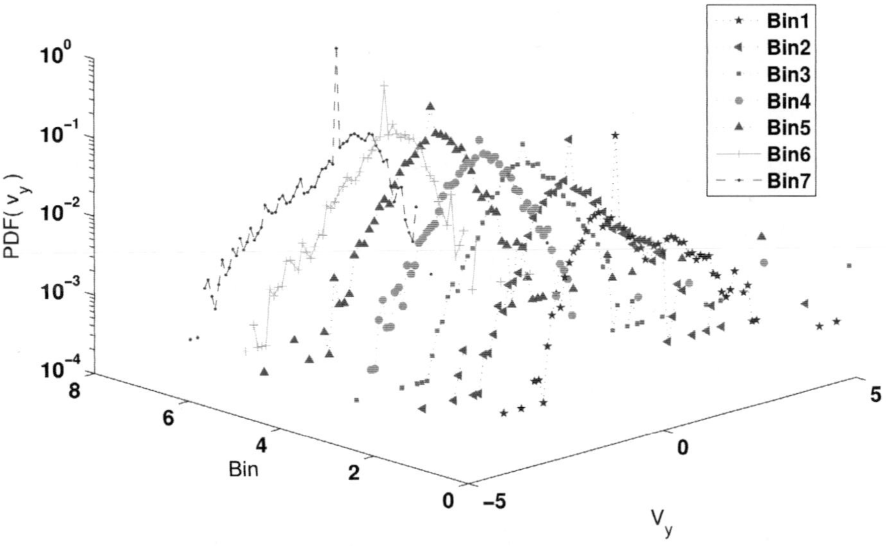

(b)

Figure 3: Local distribution functions of (a) v_x and (b) v_y in data set no. 1 of *Table 1.* There are 7 bins along the vibration direction y axis. The velocities are scaled by $V_\omega(0.07m/s)$.

skewness of n sample points x_i is defined as:

$$skewness = \frac{\frac{1}{n}\sum_{n}^{i=1}(x_i - \bar{x})^3}{(\frac{1}{n}\sum_{n}^{i=1}(x_i - \bar{x})^2)^{3/2}} \quad (1)$$

where $\bar{x}$ is the mean of all the n sample points x_i. Fig. 4 compares the skewness of profiles for v_x and v_y in different bins along y axis. The skewness is changing linearly from positive to negative and is about zero at the center of the box for v_y profiles, while skew seems flat for v_x

Journal of Physics: Conference Series **327** (2011) 012033 doi:10.1088/1742-6596/327/1/012033

profiles.

In the steady state, conservation of momentum requires [20]:

$$\int_{-\infty}^{+\infty} vf(v)dv = 0, \tag{2}$$

therefore,

$$\int_{0}^{+\infty} vf(v)dv = -\int_{-\infty}^{0} vf(v)dv. \tag{3}$$

In turn, near the boundary $y = 0$, the number of particles moving towards the wall and leaving the wall shall be conserved [20, 21], that is, the particle number leaving the vibrating wall shall be equal to the particle number moving towards the vibrating wall in a sawtooth excitation:

$$f_y^-(v_y)v_y = f_y^+(2V_\omega - v_y)(v_y - 2V_\omega) \tag{4}$$

Here, f_y^+ (f_y^-) is the velocity distribution function for the $v_y > 0$ ($v_y < 0$). The number density n near the boundaries can therefore be obtained from particle-number conservation mentioned above [21]:

$$n_y = n_y^+ + n_y^- = \int_0^{\infty} f(v_y)dv_y + \int_{-\infty}^{0} f(v_y)dv_y \tag{5}$$

At the boundary $y = 0$:

$$n_y = 2n_y^- + 2V_\omega \int dv_y \frac{f_y^-}{v_y - 2V_\omega} \tag{6}$$

It implies at the boundary $y = 0$, n_y^+ is smaller than n_y^-. The discrepancies between n^+ and n^- can be confirmed in our experimental results shown below(Fig. 7 (a) and (b)).

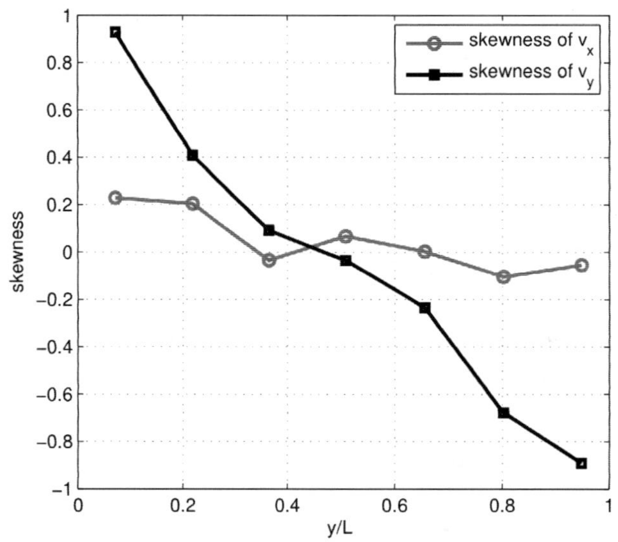

Figure 4: The skewness of v_x and v_y distribution profiles along y/L.

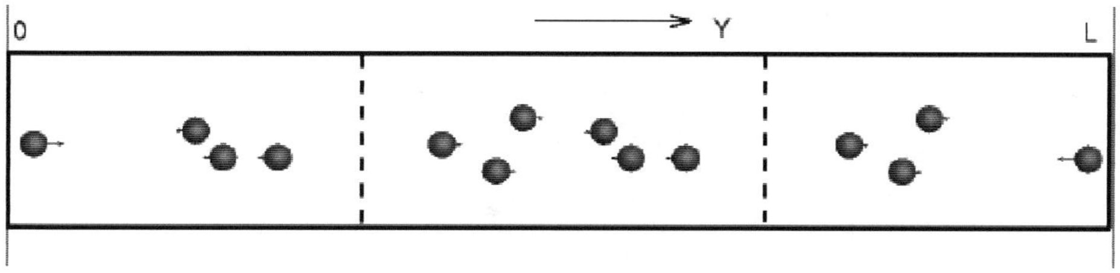

Figure 5: Sketch of the breakdown of the mean free path. On the right and left sides of the cell, the v_y distribution is asymmetric, contrary to what occurs in the cell center. As there is no net flow, it means the number density n^+ and n^- are different. This imposes to define two mean free paths through $n^- dl_c^+ = n^+ dl_c^- = 1$.

3.3. The breakdown of the mean free path

Mean free path is a basic concept in the kinetic theory. It is defined as the average distance the particle travels between two successive collisions. More importantly, the mean free path is often used to be a parameter to judge the suitable scope of the kinetic theories and hydrodynamics for granular medium [26]. The concept of mean free path is based on the assumption of particles being in random distribution, i.e., Maxwell-Bolzmann distribution is valid. However, in our system, the velocity distribution is no longer the Maxwell-Bolzmann distribution. Moreover, the local velocity distribution is asymmetric and anisotropic. This indicates the breakdown of kinetic models based on the concept of the mean free path.

At the cell wall, for instance, while the asymmetry of velocity v_y is the largest, one has to define two mean free paths, l_c^+ and l_c^-, for the two "kinds" of particles with v^+ and v^-, as demonstrated in Fig. 5. In Fig. 5 on the right and left sides of the cell, the v_y distribution is asymmetric, contrarily to what occurs in the cell center. As the average flow is zero, it means the density number n^+ and n^- are different. So this imposes the need to define two mean free paths that $n^- dl_c^+ = n^+ dl_c^- = 1$ (d is the particle diameter).

4. Hydrodynamic description

Starting from the measured positions (x, y) of particles and the velocities in four components $v_x^+ (v_x > 0)$, v_x^- $(v_x < 0)$, v_y^+ $(v_y > 0)$, and v_y^- $(v_y < 0)$, the hydrodynamic parameters p_x^+ (p_y^+), p_x^- (p_y^-), n_x^+ (n_y^+), n_x^- (n_y^-), T_x^+ (T_y^+), T_x^- (T_y^-) can be derived from velocity components using the four sets of the experimental data shown in *Table 1*. In general hydrodynamics, the pressure p is considered as a symmetric tensor; we show here that p^+ is different from p^-, that means p is not single valued, so p is not a valid concept anymore.

4.1. The spatial profile of pressure

We will use the following notation $p_x^+(y) = \sum (v_x^+)^2(y)$ to present the pressure due to the x component with sum of v_x component with sum of v_x^2 for $v_x > 0$ of each bin along y direction. The mass of the particles are assumed to be unitary, $m = 1$. The pressure is scaled by $p_0 = (N/7)k_B T_{drive}$, where $N = 47$, $T_{drive} = V_\omega^2$, and V_ω is the peak value of the vibrating velocity, shown in *Table 1*.

For the p_x components it is observed that $p_x^+(y) \approx p_x^-(y) \approx constant$ (data are not shown here). However, the y components shown in Fig. 6 is rather unexpected. Firstly, the components of pressure p_y are no long constant in steady state [19, 20]. This result is also found in some simulation [27]. Secondly, in the vibrating direction, $p_y^+(y)$, $p_y^-(y)$ are asymmetric with respect to the center of the cell, and $p_y^+(y)$ and $p_y^-(y)$ are mirror symmetric to each other. Therefore, the pressure must be (i) anisotropic ($p_x \neq p_y$), and even more (ii)$p_y^+ \neq p_y^-$. Point (ii) is linked

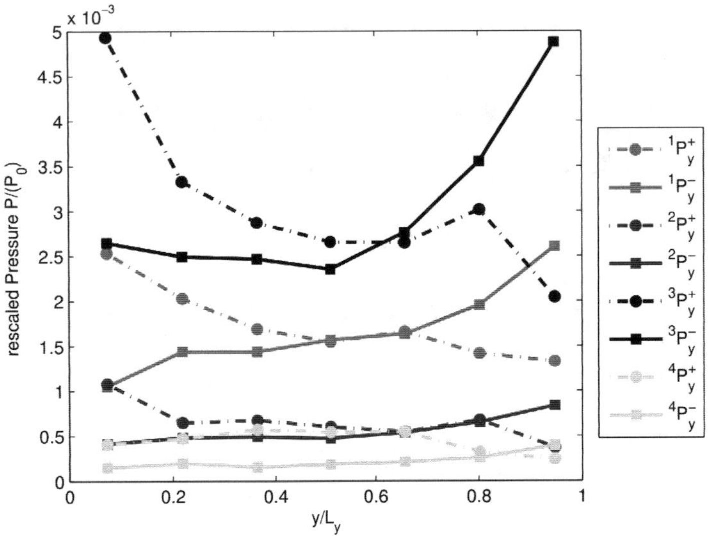

Figure 6: The scaled pressure $p_y^{\mp}(y)$ are plotted in the four sets of vibration parameters shown in *Table* 1.

to the asymmetric velocity distribution and to equation 2. This imposes $p_y^+ \neq p_y^-$ in such a system. Moreover, it is found that p_y components are always larger than p_x components.

From another view point, the pressure can be viewed as the kinetic energy in unit volume. So Fig. 6 can be seen as the energy profiles of the system. Obviously, $p_x^+(y)$ and $p_x^-(y)$ gain energy from the components in y direction to balance the energy dissipation in the x direction. In y components, $p_y^+(y)$ and $p_y^-(y)$ get energy from the two boundaries to balance the energy dissipation in the y direction and the energy transformed to the x direction. The difference $p_y^+(0) - p_y^+(L)$ as shown in Fig. 6 due to the energy dissipation along the y direction plus the energy being transformed to x direction.

4.2. Density inhomogeneity

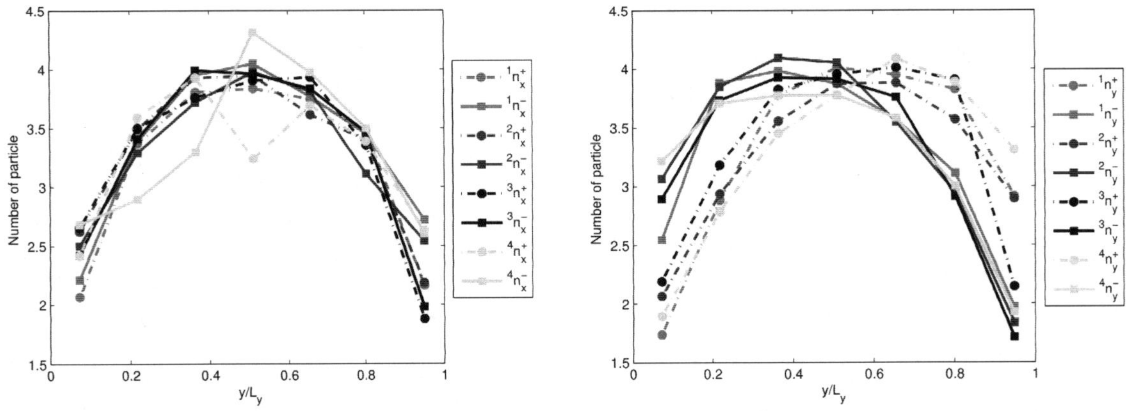

(a) (b)

Figure 7: The number density of particles (a) with velocity $v_x^{\pm}$ (b) with velocity $v_y^{\pm}$ are plotted in the four sets of vibration parameters shown in *Table* 1.

The number density with four components $n_x^+(y)$, $n_x^-(y)$, $n_y^+(y)$, $n_y^-(y)$ are shown in Fig. 7. The notation of $n_x^+(y)$ means the spatial profile of the number density of particles with velocity $v_x > 0$ along the y direction. We observe that $n_x^+(y)$ and $n_x^-(y)$ overlap and peak at the cell center as in Fig. 7(a). The curves in Fig. 7 (b) conform that the difference between $n_y^+(y)$ and $n_y^-(y)$ near the boundaries predicted in equations (3) and (4). Moreover, n_y^+ and n_y^- do not overlap. The components $n_y^+(y)$ and $n_y^-(y)$ are mirror symmetric to each other as shown in Fig. 7(b). It is apparent that $n_y^+(y) = n_y^-(y)$ is not in generally valid as assumed by the theoretician when calculating the transport parameters in granular system [1, 26].

4.3. Granular temperature

The temperature anisotropy due to the non-isotropic driving in vibro-fluid granular systems has been discussed in recent years [22, 27–30]. Here, $T_x^+(y)$ (T_x^-(y)) represents the average of v_x^2 for $v_x > 0$ ($v_x < 0$) groups of each bin along y direction. We observe that $T_x^+(y) \approx T_x^-(y)$, and $T_x^+(y)(T_x^-(y))$ are axisymmetric with central axis of the cell (it is not shown in this paper). However, $T_y^+(y)$ ($T_y^-(y)$) is asymmetric along y axis. The values of y component $T_y^+(y)$ ($T_y^-(y)$) is seen greater than that of the x components $T_x^+(y)$ ($T_x^-(y)$). $T_y^+(y)$ ($T_y^-(y)$) is greater near the boundary at $y = 0$ ($y = L$) as shown in Fig. 8. In Ref [30], an explanation is provided about the temperature anisotropy. However, in the paper, a Gaussian distribution was assumed. The non-Gaussian and asymmetric velocity distribution observed in our system imply the Gaussian distribution misleading. Our velocity distribution already verify a non-Gaussian existed. It means Gaussian distribution approximation is misleading in [30]. Temperature anisotropy is apparent, and the two components T_y^+ and T_y^- are mirror symmetric to each other.

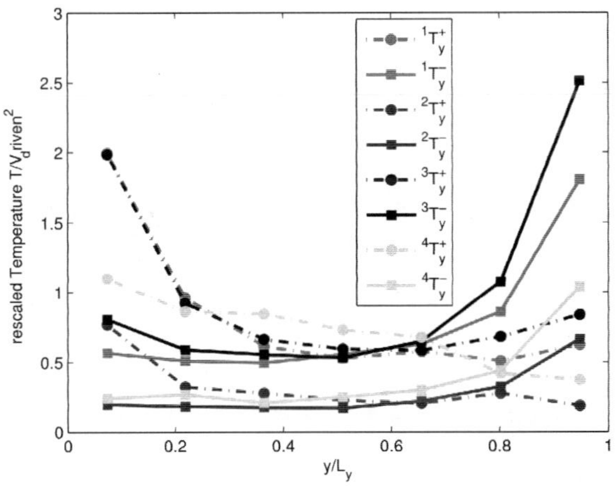

Figure 8: The scaled temperature $T_y^{\mp}(y)$ are plotted in the four sets of vibration parameters shown in *Table* 1.

5. Summary

The objective of this work is to provide an experimental study of the boundary effect in vibro-fluidized granular gas system. We have studied the velocity distribution in the vibration direction locally and globally. The velocity distribution is found anisotropic, and both its x and y components are non-Gaussian. We measure the local velocity distribution along the vibration direction and find that the v_x distributions are still symmetric, while the v_y distributions become asymmetric. The skewness of the profiles is found linear along y with skew $= 0$ at the cell center. This implies a long range boundary effect. The particles number density is found not a constant but peaked at the center. The hydrodynamics-parameter profile, $p_x^+(y)$ and $p_x^-(y)$, $T_x^+(y)$ and $T_x^-(y)$, $n_x^+(y)$ and $n_x^-(y)$ are overlapped, but the $p_y^+(y)$ and $p_y^-(y)$, $n_y^+(y)$ and $n_y^-(y)$, $T_y^+(y)$

and $T_y^-(y)$ are found mirror symmetric to each other. The asymmetry in the y component profiles can be understood as the dissipation of the energy input through collision with the boundary walls and energy transformed to x component through particles collisions. This long range boundary effect complicates the statistical properties of the confined fluidized granular gas system, and suggests further hydrodynamical studies are essential.

Acknowledgements
The authors want to thanks Novespace for helpful technical assistance and CNES for the fund of three years. This research was supported by CNES, ESA, ECP, CNRS, CAS, and CNSA for partial funding.

References

[1] Grossman, E. L. and Zhou, Tong and Ben-Naim, E. 1997 *Phys. Rev. E* **55** 4200
[2] Sergei E. Esipov, Thorsten Poeschel 1997 *J. Stat. Phys* **86** 1385
[3] Olafsen, J. S. and Urbach, J. S. 1999 *Phys. Rev. E* **60** 2468
[4] Olafsen, J. S. and Urbach, J. S. 1998 *Phys. Rev. Lett.* **81** 4369
[5] Rouyer, Florence and Menon, Narayanan F. 2000*Phys. Rev. Lett.* **85** 3676
[6] Kudrolli, A. and Wolpert, M. and Gollub, J. P., 1997 *Phys. Rev. Lett.* **78** 1383–1386 *Granular Matter*, **2**, 189
[7] Losert, W. Cooper, D. G. W. Delour, A. Kudrolli, J. P. Gollub 1999 *Chaos* **9** 682
[8] M. Hou, R. Liu, G. Zhai, Z. Sun, K. Lu 2008 *Microgravity Sci Technol* **20** 73-80
[9] Noije, T.P.C. and Ernst, M.H 1998 *Granular Matter* **1** 57
[10] Brey, J. Javier and Ruiz-Montero, M. J. and Cubero, D. 1996 *Phys. Rev. E* **54** 3664-3671
[11] Huthmann, M.,Orza, J. A., and Brito, R. 2000
[12] P. Evesque 2005 *poudres & grains* **15**,1
[13] P. Evesque 2007 *poudres & grains* **16**,3
[14] P. Evesque 2001 *Poudres & Grains* **12** 17-73, 60-82
[15] P. Evesque 2002 *Poudres & Grains* **13** 40-73
[16] P. Evesque 2004 *Poudres & Grains* **14** 8-53
[17] P. Evesque 2005 *Poudres & Grains* **15** 1-16,18-34
[18] P. Evesque 2007 *Poudres & Grains* **16** 38-62
[19] R. Liu, M. Hou, P. Evesque 2009 *poudres & grains* **17**,1
[20] P. Evesque 2010 *poudres & grains* **18**,1
[21] Brey, J. Javier and Ruiz-Montero, M. J. and Moreno 2000 *Phys. Rev. E* **62** 5339
[22] Herbst, Olaf and Muller, Peter and Otto, Matthias and Zippelius, Annette, E. 2004 *Phys. Rev. E* **70** 051313
[23] Du, Yunson and Li, Hao and Kadanoff, Leo P. 1995 *Phys. Rev. Lett.* **74** 1268
[24] Goldhirsch, I. and Zanetti, G. 1993 *Phys. Rev. Lett.* **70** 1619
[25] M. Leconte, Y. Garrabos, F. Palencia, C. Lecoutre, P. Evesque, D. Beysens 2006 *Appl. Phys. Lett.* **89** 243518
[26] Nikolai V. Brilliantov Thorsten Pöschel 2004 *Kinetic Theory of Granular Gases* (NewYork: Oxford University Press).
[27] J. Javier Brey, and D. Cubero 1998 *Phys. Rev. E* **57** 2019
[28] Barrat, Alain and Trizac, Emmanuel 2002*Phys. Rev. E* **66** 051303
[29] McNamara, Sean and Luding, Stefan 1998*Phys. Rev. E* **58** 813
[30] D. van der Meer , P. Reimann 2006 *Europhys. Lett.* **74(3)** 384-390

Directed clustering in driven compartmentalized granular gas systems in zero gravity

Yinchang Li[1], Meiying Hou[1], Pierre Evesque[2]
[1]Institute of Physics, Chinese Academy of Sciences, Beijing, China
[2]Lab MSSMat, EcoleCentrale de Paris, 92295 Châtenay-Malabry, France
E-mail: mayhou@aphy.iphy.ac.cn

Abstract. Clustering of shaken fluidized granular matter in connected compartments has been observed and studied in the laboratory. This clustering behavior in granular gas systems is related to the dissipative nature of granular system, and therefore shall not depend on gravity. This clustering phenomenon in compartmental configuration may provide a means for particle depletion and transportation in microgravity environment. In this work we propose different configurations for possible directed clustering in zero gravity. The related experiment has been planned for the Chinese satellite SJ-10.

1. Introduction

Clustering is one of the important features of granular gas systems [1–6]. It occurs spontaneously due to inelastic collisions among particles. On one hand, clustering is one kind of pattern formations commonly seen in non-equilibrium systems, which has caught much attention of physicists in recent years [7–15] using granular gases for understanding behaviors of non-equilibrium systems. On the other hand, granular matters have important industrial applications. Understanding the clustering mechanism can be economically and environmentally important.

In a vertically shaken two-compartment system, mono-dispersed particles will spontaneously cluster into one of the two compartments [16]. This clustering behavior demonstrates the "Maxwell's demon" phenomenon, in which the energy dissipation between particles acts as the demon to preferentially let particles pass only in one direction to cluster in one compartment [17,18]. This kind of clustering behavior can also be found in N connected compartments [19,20], for example in three connected compartments it is found that for sufficiently strong shaking the granular gas will be equal-partitioned, but for some moderate shaking the gas will cluster in one of the compartments. For bi-dispersed particles, competitive clustering behavior will cause new phases [21, 22], such as oscillatory clustering [23–29], degenerate oscillation state, and semi homogeneous state [30]. In a cyclic three-compartment system, a chaos-like clustering behavior, i.e., two kinds of particles cluster into compartments in a random order [31], is also observed.

In this paper, we will first review the clustering phenomena of mono-disperse granular gas in symmetric two-compartment system. We then propose and study possible directed clustering with asymmetric configurations for possible application of transporting particles under micro-gravity.

Journal of Physics: Conference Series **327** (2011) 012034 | doi:10.1088/1742-6596/327/1/012034

2. Clustering Phenomenon under Gravity

2.1. For cells with symmetric configuration: Maxwell Demon

A molecular dynamics simulation is performed for a container of two identical compartments, which are connected by an opening at the wall in between the compartments, as is shown in Fig.1. The ground area of each compartment is $26\sigma \times 26\sigma$, (σ is the diameter of the particles). The lower edge of opening is 15σ above the bottom plate and the opening size is $26\sigma \times 30\sigma$. The bottom plate of the container provides a driving velocity v_b in the unit of $\sqrt{2g\sigma}$ (g is gravity acceleration) in a saw-tooth manner. Gaining kinetic energy by colliding with the bottom plate, particles may jump from one compartment to the other through the opening. Hard sphere method is used to describe collisions between particles. Particles are considered to be perfect hard sphere that the deformation during collision is ignored. Identical particles with unit mass and unit diameter are used, and the dissipation in a particle-particle collision is characterized by the coefficients of restitution. For simplicity, only the normal coefficient of restitution is considered, thus the particle-particle collision can be described as follows:

$$\vec{v}_i' = \vec{v}_i - \frac{1+e}{2}[(\vec{v}_i - \vec{v}_j)\cdot\hat{n}]\hat{n} \tag{1}$$

$$\vec{v}_j' = \vec{v}_j + \frac{1+e}{2}[(\vec{v}_i - \vec{v}_j)\cdot\hat{n}]\hat{n} \tag{2}$$

where i, j refer to the two particles and n the normal direction of their contact, v and v' are pre- and post-collision velocities, respectively, and e is the normal coefficient of restitution, which is set to a constant value 0.9 throughout this paper. Particle-wall collisions are assumed to be elastic. Friction and rotation are ignored for all collisions. Each collision is considered to be a key event, and a standard event-driven algorithm [32] is adopted in our simulation.

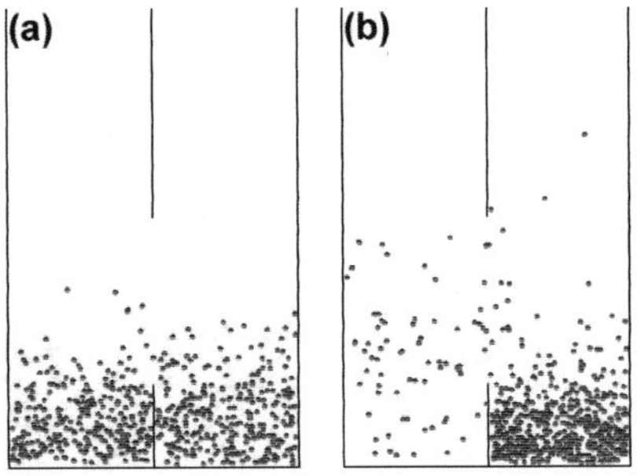

Figure 1. Homogeneous clustering state(HOM) (a) and asymmetric clustering state(ASY) (b) in mono-disperse two-compartment system by molecular dynamics simulation. Total number of particles N_{tot} is 500. Shaking velocity v_b is 0.90 (a) and 0.70 (b).

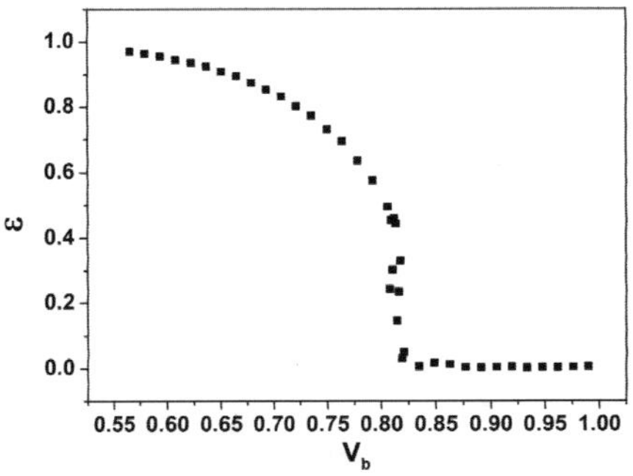

Figure 2. Asymmetric factor ε changes with shaking velocity v_b. Total number of particles N_{tot} is 500.

Fig.1(a) shows the homogeneous distribution of particles in the two compartments when shaking strength is high enough. When shaking strength is getting lower, clustering in one compartments is observed (Fig.1(b)). A parameter $\varepsilon = \langle (N_l - N_r)/N_{tot} \rangle$ is defined to describe the asymmetric clustering phenomenon, where N_l, N_r, N_{tot} are the distribution number of particles in left compartment, right compartment and the total number of particles, respectively. $\langle ... \rangle$ is averaged over time when system reaches stable state. Figure 2 shows a symmetry breaking

Journal of Physics: Conference Series **327** (2011) 012034 doi:10.1088/1742-6596/327/1/012034

transition from homogeneous clustering (HOM) to asymmetric clustering (ASY) with decreasing v_b.

2.2. Asymmetric structure: directed clustering

In the symmetric two-compartment system, the clustering can be in either compartment. By introducing an asymmetric structure, preferential directed clustering is possible [33]. One way to do it is by adding a slab into one of compartments (shown in fig. 3(a)). An experiment was performed. Our experimental setup consists of a glass cell with base size $2.5 \times 5cm^2$ and height $15cm$. The cell filled with $N_{tot} = 800$ glass beads of radius $r = 0.5mm$, is mounted on a sinusoidal shaker. The cell is separated into two equal compartments by an aluminum plate, which has an opening of $2 \times 2.5cm^2$. This opening is at a height with its lower edge $2cm$ above the base. Different slabs of thickness δh ranging from $6mm$ to $12mm$ are glued one at a time to the bottom of one of the two compartments. In this way the depth of the two compartments from the opening height become asymmetric. We refer the shallow one as compartment A and the deeper one as compartment B. The shaker frequency f is fixed at $40Hz$. Driving velocity $v_b = af$.

By changing the driving velocity, different clustering behaviors can be observed, as demonstrated in fig. 3(a)- (e). The demonstration is under an initial condition that all particles are put in compartment A (fig. 3(a)) with a slab of thickness $\delta h = 10mm$ in it. When the driving velocity is lower than a certain critical value v_c, a small percentage of beads are able to jump through the opening to compartment B, a large portion of beads remains in compartment A. This is the case shown in fig. 3(b). While the driving velocity reaches v_c, nearly all the beads initially in compartment A will cluster in compartment B, as shown in fig. 3(c). When driving velocity goes higher than v_c, beads in compartment B gradually flow back to A, and at large enough driving velocity the distribution will slowly approach homogeneous state (fig. 3(d)). When the driving velocity is reversed to a lower value, particles in compartment A will be relocated to compartment B. Instead of returning to the state as shown in fig. 3(a) even with same driving velocity, most of the particles would eventually cluster in the deeper compartment B, which is shown in fig. 3(e). But if we set the initial distribution for most of the beads originally to be located in compartment B, no abrupt phase transition would occur. The system would evolve along one phase trajectory without any bifurcation.

To get a clear view of the experimental observation, the number of particles in each compartment at any driving velocity is measured (fig. 4). After long enough time (usually several minutes after reaching a steady state), the shaker is turned off and the number of particles remaining in each compartment is counted. This procedure is performed several times with different initial conditions at each fixed driving velocity. A phase diagram of N_A (the number of particles in compartment A) vs. v_b is obtained. There are two regimes in this phase diagram as shown in fig. 4. In regime I (below v_c): for low shaking strength, particles can cluster in either compartment A or compartment B. While increasing the shaking strength, the state that most particles are in compartment A will suddenly change at a critical velocity v_c to populate in compartment B; while in another state that most particles are in compartment B, they will evolve continuously without this sudden change. In regime II (above v_c): the system evolves in a relative simple and reversible way. It will eventually approach an asymptotically homogeneous distribution at high shaking strength.

As shown above, directed clustering occurs when v_b is larger than v_c that most particles will cluster in the lower compartment no matter what initiate distribution is. Such an asymmetric structure can be extended to N-compartment, which may be used as a method for directed transportation of granular particles.

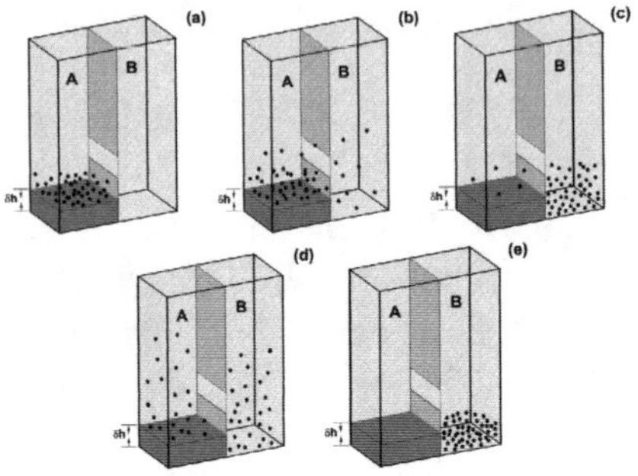

Figure 3. Illustrations of the system evolution with the initial condition that all particles are located in compartment A. (a) is the initial distribution, (b) is the distribution when v_b is lower than v_c, (c) is the distribution when v_b is right at or above v_c, (d) is the asymptotically homogeneous distribution when v_b is further increased, (e) is the distribution when v_b is lowered again to below v_c.

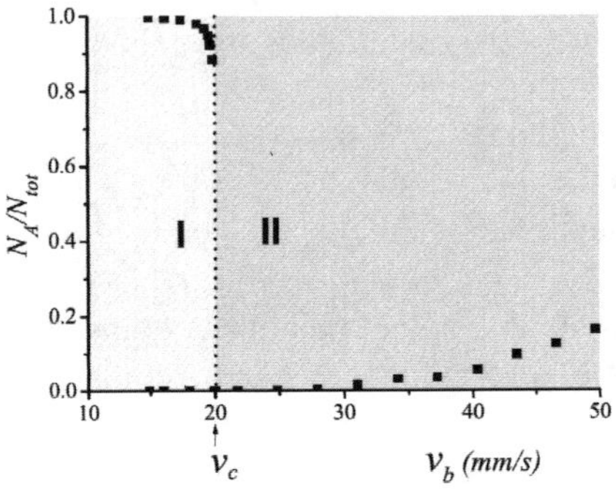

Figure 4. When δh is set to $10mm$, system's evolution was obtained. Particle populations of steady state in compartment A changes with driving velocity v_b.

3. Directed clustering in zero gravity

3.1. Symmetric structure

A 3-D molecular dynamics simulation is performed for a two-compartment cell connected by an opening on the sidewall, under zero gravity, with the top and bottom plates oscillating in saw-tooth manner. The size of each compartment is chosen to be $30\sigma \times 30\sigma \times 120\sigma$, and the connection opening at the sidewall is $30\sigma \times 30\sigma$, where σ is the particle size. In a dilute system with small number of particles, the system is found to be homogenously distributed, as shown in Fig. 5(a). This symmetric distribution state is broken when more particles are added into the system. Particles will become asymmetrically distributed in the system and cluster in the middle part of one of the compartments (Fig. 5(b)). This phenomenon is very similar to the clustering behavior under gravity. Asymmetric factor ε increases rapidly with total number of particles N_{tot} as shown in Fig. 6. In this simulation the coefficient of restitution is assumed to be independent on relative velocity for particle-particle collisions. This final clustering state can be reached with any non-zero value of v_b as long as the right N is used.

3.2. Asymmetric structure

Similar directed clustering behavior can be found in zero gravity. Asymmetric configuration can be made by adding a slab in one of the two compartments, as shown in Fig. 7. If initially all particles are placed in compartment A, most of particles will remain in this compartment when the height of slab is not large enough (Fig. 7 (a)). Directed clustering can be observed by increasing the height of slab (Fig. 7(b)). The distribution can be clearly demonstrated in figure 8. Black square points represent the final distribution state with initial population of particles in compartment A; red circle points indicates the final population of initial state of particles in compartment B. When the height of slab δh is not large enough, the final clustering state depend upon the initial clustering state (regime I). When δh is above certain threshold value,

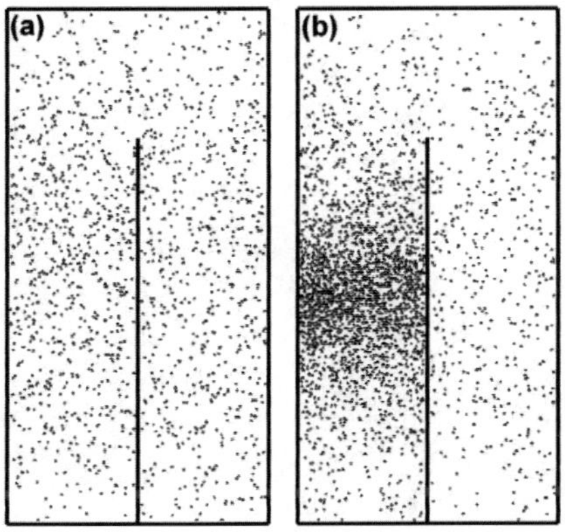

Figure 5. Simulation results of homogenously clustering state (a) and asymmetric clustering state (b) in zero gravity. The total number of particles N_{tot} in this simulation is 1500 (a) and 3000 (b).

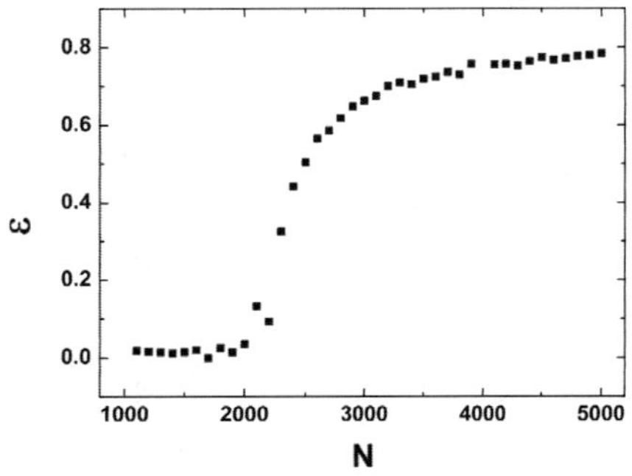

Figure 6. Asymmetric factor ε changes with N_{tot}.

directed clustering is found: all particles will cluster in the deeper compartment (regime II).

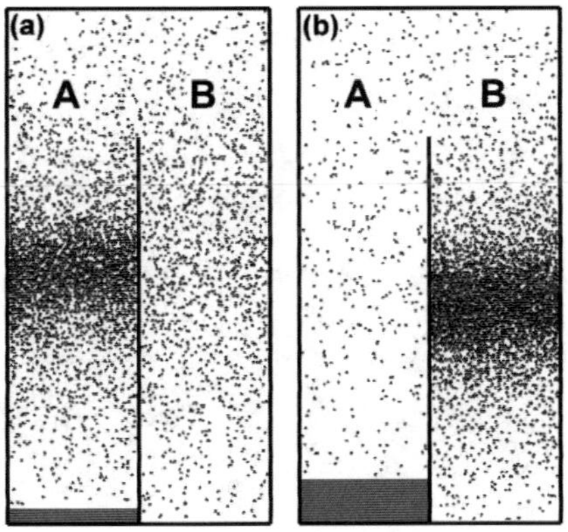

Figure 7. Simulation result of different clustering state in asymmetric two-compartment system in zero gravity. Total number of particles is 5000, the height of slab δh is (a) 5 and (b) 10 radii of the particle.

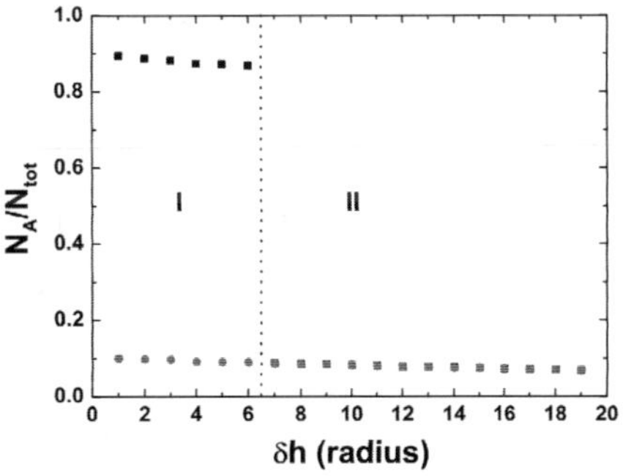

Figure 8. Number of particles remaining in compartment A vs. the height of slab δh when initially all particles are in compartment A (black square points) or in compartment B (red circle points).

3.3. N-compartment configuration

With specially designed asymmetric N-compartment configuration, particles can be transported in a desired direction. Two examples are presented in Fig. 9. As shown in Fig. 9(a), ten compartments are connected with openings located near either end boundaries in an alternative manner. Particles will cluster around non-shaking boundaries (black lines). For example, particles will be pushed by the top boundary of compartment 1 and cluster around the bottom.

Journal of Physics: Conference Series **327** (2011) 012034 doi:10.1088/1742-6596/327/1/012034

Because of large density gradient at the bottom boundaries of compartment 1, particles will diffuse into compartment 2 and then cluster around its top boundary. In this way, particles will flow from compartment 1 to compartment 10, and form a stable cluster at one end of compartment 10.

Fig. 9(b) shows another asymmetric configuration of ten compartments with two moving boundaries and opening in the middle wall. Particles initially cluster in the middle part of the compartment 1. Since the opening is set at the middle, particles will diffuse from compartment 1 into compartment 2, and through compartment 3, 4 ... and eventually to compartment 10.

The time evolution of the population can be obtained by simulation and is as shown in Fig. 10 (a) and (b). The system starts with all particles in compartment 1 (black curves), will populate through compartment 2 to 3, 4... and eventually evolve to compartment 10.

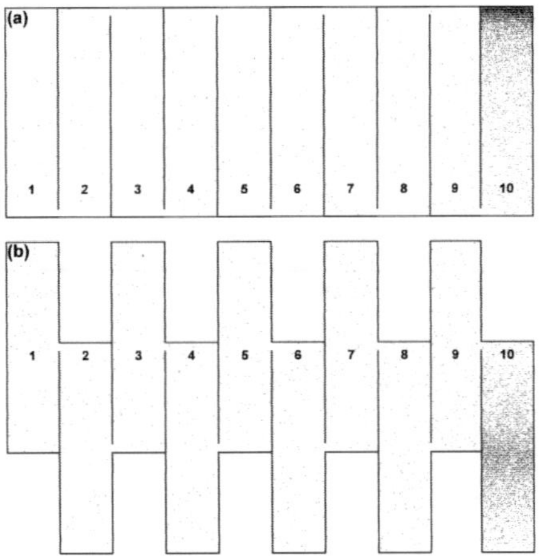

Figure 9. Two types of asymmetric N-compartment system in zero-gravity. In (a), red lines represent the moving boundaries. In (b), both top and bottom boundaries are moving. Particles are initially placed in compartment 1 and finally cluster into compartment 10.

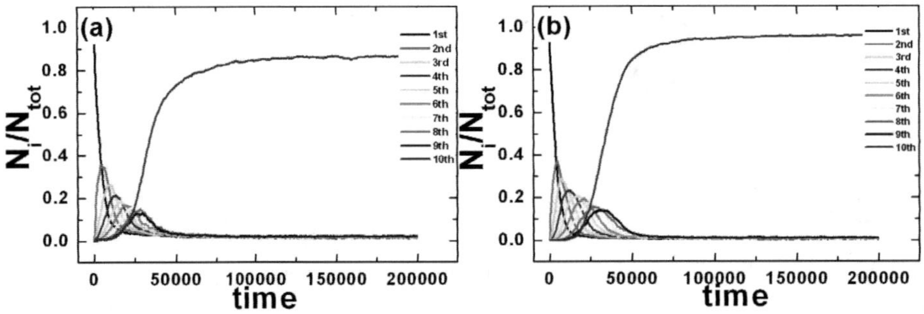

Figure 10. Time evolution of number of particles in different compartments. Numbers in different compartments are indicated by different colors. Curves in (a) and (b) correspond to time evolution of configurations shown in Fig.9 (a) and (b), respectively.

4. Conclusion

Clustering occurs in compartmentalized granular gases due to the dissipative nature of the system. By constructing asymmetric configuration the cluster can be directed in a desired way. In this work MD simulation is performed and the directed clustering in zero gravity is achieved in some asymmetric two-compartment structure and in N-compartment cells with proper shaking strength.

5. Acknowledgement

This work is supported by the National Natural Science Foundation of China (Grant Nos. 10720101074 and 10874209), the Knowledge Innovation Program of the Chinese Academy of Sciences (Grant No. KJCX2-YW-L08), and grants from CSSAR and CNES.

References

[1] Goldhirsch I and Zanetti G, *Phys. Rev. Lett.* **70**, 1619 1993.
[2] McNamara S and Young W R, *Phys. Rev. E* **50**, R28 1994.
[3] Du Y, Li H, and Kadanoff L P, *Phys. Rev. Lett.* **74**, 1268 1995.
[4] McNamara S and Young W R, *Phys. Rev. E* **53**, 5089 1996.
[5] Zhou T and Kadanoff L P, *Phys. Rev. E* **54**, 623 1996.
[6] Grossman E and Mungan M, *Phys. Rev. E* **53**, 6435 1996.
[7] Melo F, Umbanhowar P B, and Swinney H L, *Phys. Rev. Lett.* **75**, 3838 1995.
[8] Umbanhowar P B, Melo F, and Swinney H L, *Nature (London)* **382**, 793 1996.
[9] Jaeger H M, Nagel S R, and Behringer R P, *Rev. Mod. Phys.* **68**, 1259 1996.
[10] Liffman K, Metcalfe G, and Cleary P, *Phys. Rev. Lett.* **79**, 4574 1997.
[11] Ristow G H, Strassburger G, and Rehberg I, *Phys. Rev. Lett.* **79**, 833 1997.
[12] Olafsen J S and Urbach J S, *Phys. Rev. Lett.* **81**, 4369 1998.
[13] Ristow G H, Pattern Formation in Granular Materials (Springer, New York, 2000), 161.
[14] Shinbrot T, *Nature (London)* **429**, 352 2004.
[15] Aranson I S and Tsimring L S, *Rev. Mod. Phys.* **78**, 641 2006.
[16] Schlichting H Jand Nordmeier V, *Math. Naturwiss. Unterr.* **49**, 323 1996.
[17] Eggers J, *Phys. Rev. Lett.* **83**, 5322 1999.
[18] Jean P, Bellenger H, Burban P, Ponson L, and Evesque P, *Poudres & Grains* **13**, 27 2002.
[19] Weele K van der, Meer D van der, Versluis M and Lohse D *Europhys. Lett.* **53** 328 2001.
[20] Meer D van der, Weele K van der, and Lohse D, *Phys. Rev. E*, **63**, 061304 2001.
[21] Mikkelsen R, Meer D van der, Weele K van der, and Lohse D, *Phys. Rev. Lett.* **89**, 214301 2002.
[22] Mikkelsen R, Meer D van der, Weele K van der, and Lohse D, *Phys. Rev. E* **70**, 061307 2004.
[23] Costantini G, Paolotti D, Cattuto C, and Marconi U M B, *Physica A* **347**, 411 2005.
[24] Marconi U M B, Costantini G, and Paolotti D, *J. Phys.: Condens. Matter* **17**, S2461 2005.
[25] Lambiotte R, Salazar J M, and Brenig L, *Phys. Lett. A* **343**, 224 2005.
[26] Miao T, Liu Y, Miao F, and Mu Q, *Chin. Sci. Bull.* **50**, 740 2005.
[27] Viridi S, Schmick M, and Markus M, *Phys. Rev. E* **74**, 041301 2006.
[28] Hou M, Tu H, Liu R, Li Y, Lu K, Lai P, and Chan C K, *Phys. Rev. Lett.* **100**, 068001 2008.
[29] Lai P, Hou M, and Chan C, *J. Phys. Soc. Jpn.* **78**, 041001 2009.
[30] Liu R, Li Y, and Hou M, *Phys. Rev. E* **79**, 052301 2009.
[31] Liu Y, Mu Q, Miao T, and Liao J, *Europhys. Lett.* **84**, 14004 2008.
[32] Pöshel T, Schwager T, *Computational Granular Dynamics: Models and Algorithms* (Springer-Verlag Berlin Heidelberg, 2005).
[33] Zhang Y, Li Y, Liu R, Cui F, Evesque P, and Hou M, *Granular Matter* to be published.

Journal of Physics: Conference Series **327** (2011) 012035 doi:10.1088/1742-6596/327/1/012035

Dynamical Regimes of a Granular Gas in Microgravity : a Molecular Dynamics Study

E. Opsomer, F. Ludewig and N. Vandewalle

GRASP Université de Liège, Allée du 6 Août, B-4000 Liège

E-mail: `eric.oposmer@doct.ulg.ac.be`

Abstract. We propose a numerical model, based on molecular dynamics, which is able to reproduce the behavior of a dissipative granular gas in microgravity. Granular material is confined in a cubic box of side length L following a sinusoidal motion of amplitude A and angular velocity $\omega = 2\pi f$ along the z-axis. The simulations are performed using the parameters of earlier Texus experiments [15]. Our results are in excellent agreement with experimental data. Moreover, we discovered various dynamical regimes and the physical conditions for their appearance : a gaseous state, the formation of either small or large clusters, the collective motion of grains. Phase diagrams are drawn where transitions between these different granular states are emphasized. Transitions are discussed by considering Statistical Physics models.

1. Introduction

The intriguing behaviors of grains and powders fascinate scientists since the nineteenth century. Properties like Reynolds dilatancy [1], Brazil Nut effect [2, 3], heterogenous force networks [4, 5] and clustering [6, 15] have been discovered. Nowadays, granular material like coal, sand and even powders are part of all economic activity such as agriculture, construction, cosmetics and aerospace research.

Different granular states could be encountered. Solid and liquid states have been deeply investigated during the last decades by studying granular compaction [7, 8, 9] and granular flows [10, 11]. A great number of models were suggested but could not give complete answers to all fundamental questions [12]. Among all phases, gaseous are the most complex. The dissipative character of the collisions implies a constant external energy supply and microgravity to subsist. Indeed, experimentation with granular material requires parabolic flights, sounding rockets, drop tower experiments or ISS missions that are bound to high costs and long preparation times. Nevertheless, these micro-gravity studies are of an capital interest in order to prepare longtime space missions, to study the impact of asteroids [13] or to understand the dynamics of planetary rings [14].

Pioneer work has been done by Falcon *et al.* during Mini-Texus 5 experimentation [15, 16]. Clustering has been reported and refers to particles mostly remaining in the center of the box, forming a denser region in the system. Incoming particles are trapped in the cluster, while other particles leave the dense region. Different mechanisms like inelastic collapse [17] have been proposed to explain the clustering phenomenon. A criterion for the appearance of the cluster has been proposed, based on experimental observation. It is related to the number of grain layers inside the box. Low densitiy granular gases have been investigated by Evesque [18].

Journal of Physics: Conference Series **327** (2011) 012035 doi:10.1088/1742-6596/327/1/012035

Local measures of granular pressure and granular temperature as well as exponential velocity distributions were reported. Numerical simulations have been realized and sustain experimental results.

However, the behavior of granular gases, especially for extreme densities, is still not well understood. Given the great number of parameters, too few experiments have been realized and nowadays no model is able to describe completely the behavior of this material. A lot of fundamental research will be needed to illuminate the remaining questions.

2. Model

In opposition to classical Event Driven algorithm (ED) [17, 19] for which inelastic collapse is avoided by a time cutoff model [20], we are using a three-dimensional model based on Molecular Dynamics (MD) [21, 22]. A cubic container of side length L, following a sinusoidal motion of amplitude A and angular velocity $\omega = 2\pi f$ along the z-axis, is filled with N spherical particles of mass m and a radius r. At initialization, each one is given a different random position (that allows no contact between the particle and the rest of the system). Moreover, their initial linear and angular velocities are null. Contact forces are computed following two simple models.

Normal forces F_{ij}^n are composed by a repulsive (F_{ij}^{rep}) and a dissipative (F_{ij}^{dis}) component. The repulsive component follows a simple Hooke's Law

$$F_{ij}^{rep} = -k_n \delta_{ij} \tag{1}$$

where $\delta_{ij} = d_{ij} - 2r$ with d_{ij} the distance between the centers of the solids i and j. The constant k_n is the normal stiffness which is a purely numerical parameter. Indeed, this stiffness is not linked to specific physical properties of the simulated material. It's value is calculated in order to keep the static deformation of the particles lower than $10^{-4}r$. (In our simulations k_n ranges typically between 100 and 1000 kg/s^2). The dissipative component is taken into account by viscous forces according to the following law

$$F_{ij}^{dis} = -\gamma_n(k_n, \varepsilon)\frac{\partial \delta_{ij}}{\partial t} \tag{2}$$

where the viscous constant γ_n is a non-trivial function of the normal stiffness k_n and the restitution coefficient ε. This restitution coefficient is used for both, grain-grain and grain-wall collisions. In opposite to other models (like Hertz' model) using a Hooke's Law and viscous dissipation allows us to work with a constant contact duration Δt, independent on the impact velocities. In order to assure the integration of the forces, each contact should be resolved in about 100 time steps τ ($i.e.$ $\tau \approx \Delta t/100$).

Tangent forces F_{ij}^t are bounded and depend on the relative tangent velocities v_{ij}^t between the colliding solids i and j. One has

$$F_{ij}^t = -k_t v_{ij}^t \quad \text{and} \quad \|F_{ij}^t\| \leq \mu F_{ij}^n \tag{3}$$

where μ is a friction coefficient and k_t a purely numerical constant. Friction coefficient μ as wall as restitution coefficient ε can be adapted to fit the intrinsic properties of the system. A complete description of MD simulations is given by Taberlet [23].

3. Results

In our simulations, we reproduced the results of Falcon *et al* [15]. Three cubic cells of side length $L = 10$ mm are filled with bronze spheres of radius $r = 0.175$ mm. The system is oscillating with an amplitude $A = 2.5$ mm and a frequency $f = 30$ Hz. According to the materials properties, friction coefficient is set to $\mu = 0.2$, restitution coefficient to $\varepsilon = 0.9$ and volume density to $\rho = 8\ 10^3$ kg/m^3. Figure 1 presents simulations and experiments of the system with $N = 1420$, 2840 and 4510 particles at its maximum downward velocity $-A\omega$. Dense regions are observed thanks to shading effects [15].

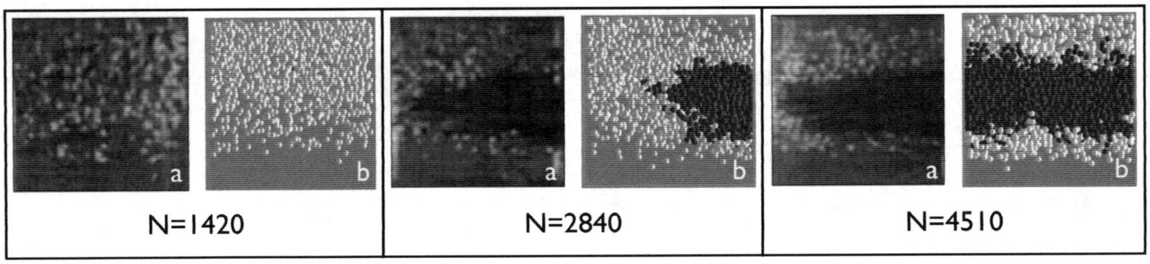

Figure 1. (a) The experimental results for respectively N=1420, 2840 and 4510 particles at the box maximum downward velocity [15]. (b) Numerical results in the same conditions. Shading of the cluster is realized by a proximity rule described in the text.

The right three pictures (b) correspond to our simulations where dense regions are blackened according to a simple proximity rule : (i) Two particles are neighbors if the distance between them is less than $2.5r$. (ii) Particles that are close enough to the system's center of mass are forming a seed. (iii) Step by step, the seed is extended by its neighbors. (iv) At saturation, the cluster is colored in black (see Figure 1). Simulations are in a qualitative agreement to Falcon's experiment presented in pictures (a). Indeed, the less dense system ($N = 1420$) of Figure 1 stays in a gaseous state and presents a density gradient decreasing towards the center of the box. For the densest case ($N = 4510$) clustering is observed around the center of the box. Moreover, the distribution is homogeneous over the xy-plane. Finally, the intermediate case ($N = 2840$) presents a new phenomenon : the cluster does not occupy the entire xy-plane. That behavior was not addressed in Falcon's experiment but can be seen on their pictures.

After the reproduction of the Texus experiments and the detection of partial clustering, the second step was to modify the different parameters. We decided to investigate the behavior of a system with a large number of grains. At this point, the densest system ($N = 4510$) had only a global density of $\eta = 0.11$. By progressively increasing the number of grains up to $N = 6800$ ($\eta \approx 0.15$) a fourth state has been found. Particles stay gathered and present a coherent motion with a velocity close to $\pm A\omega$. Figure 2 describes the different dynamical regimes of a system with respectively $N = 1420, 2840, 4510$ and 6800 grains at its maximum downwards velocity $-A\omega$. Different colors correspond to selected ranges of the vertical velocity : blue if $|v_z| < A\omega$, green if $A\omega \leq |v_z| < 2A\omega$ and orange if $2A\omega \leq |v_z|$.

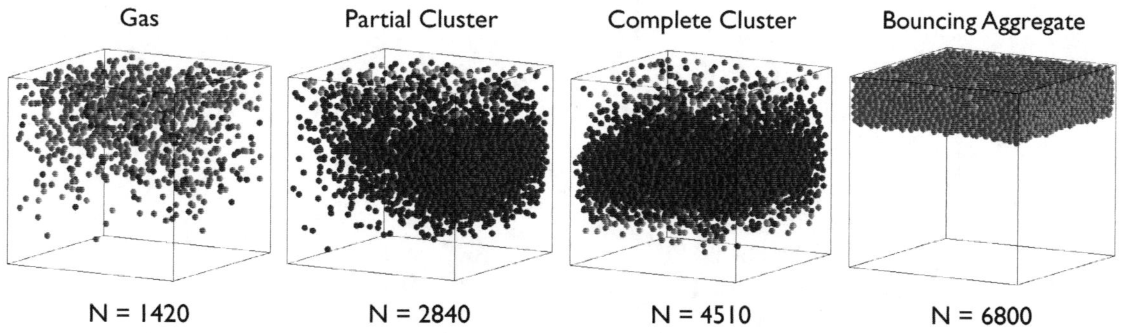

Figure 2. By increasing the grain number N, one can observe four different dynamical regimes: a gaseous phase, partial and complete clustering and an aggregate phase. Color stand for different velocity ranges: Blue for subsonic grains, green for velocities between one and two times the maximum velocity of the box and orange for even higher velocities.

Each dynamical regime is discussed in the following and introduced by two figures. On the left, the Probability Density Function (PDF) of particle positions, along the three axis, collected at the systems maximum velocities $\pm A\omega$. On the right, the PDF of U, the vertical velocities of the particles normalized by $A\omega$. Data is collected at the systems maximum downward velocity $-A\omega$.

3.1. Gas state

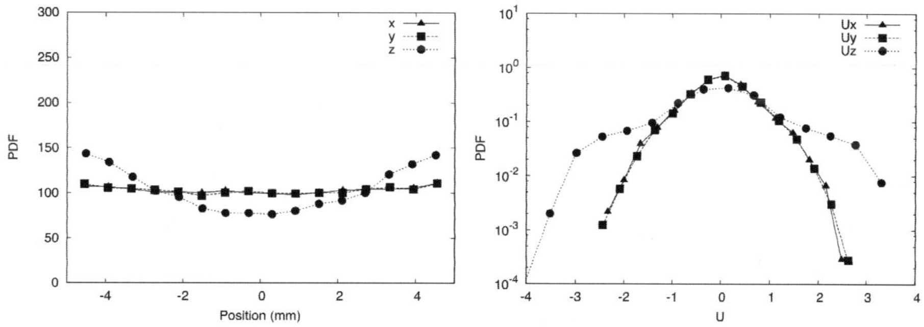

Figure 3. (Left) PDF of particle positions in the system collected every half period. (Right) Semi-log PDF plot of particle velocities in the gaseous state.

The gaseous state is found in dilute systems, the grains exhibit a broad and asymmetric distribution of vertical velocities v_z as shown in Figure 3. This asymmetry evolves according to the phase oscillation of the box. Particles can be found everywhere in the system, but follow a density gradient decreasing from one wall towards the center of the box, as shown in Figure 2 and 3. According to earlier experimentations [24, 25], we found v_x and v_y distributions of the form $P_\lambda^\gamma(v) = \lambda \exp\left(-|\lambda v|^\gamma\right), \gamma \in \{1, 2\}$ and λ as fitting parameter. Note that $\lambda \approx \frac{1}{v_0}$, where v_0 is the mean velocity. A γ value is preferred to another depending on the velocity range in the system.

3.2. Partial Cluster

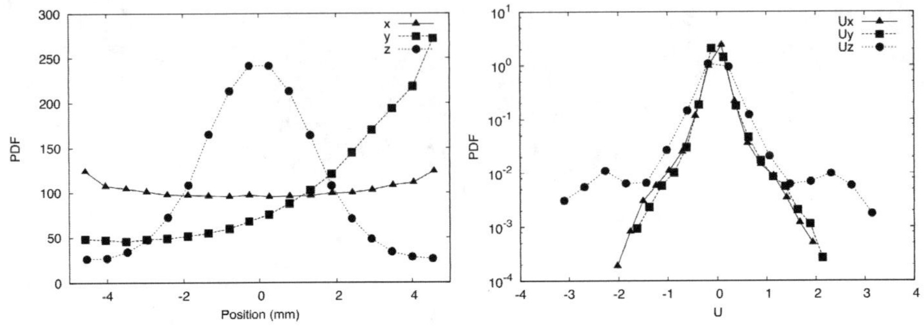

Figure 4. (Left) PDF of particle positions in the system collected every half period. (Right) Semi-log PDF plot of particle velocities in the a partial cluster.

When the number of grains increases, the system could present a partial clustering. Particles are more or less stabilized around the origin of the z-axis but stay gathered along a wall of the box. One could assumed that the system needs a typical cluster width to be energetically stable. In the distribution of velocities along z (Figure 4), a sharp peak appears around zero which is a signature for clustering. However, the distribution exhibits a broad foot since a gaseous regime coexists in the system. Multiple simulations have shown that a cluster can appear on a particular lateral wall with a probability 1/4. Moreover, simulations in a cylinder with identical volume and base section revealed that the lateral clustering is not related to corner effects. The reason of this phenomenon is yet not well understood

3.3. Complete Cluster

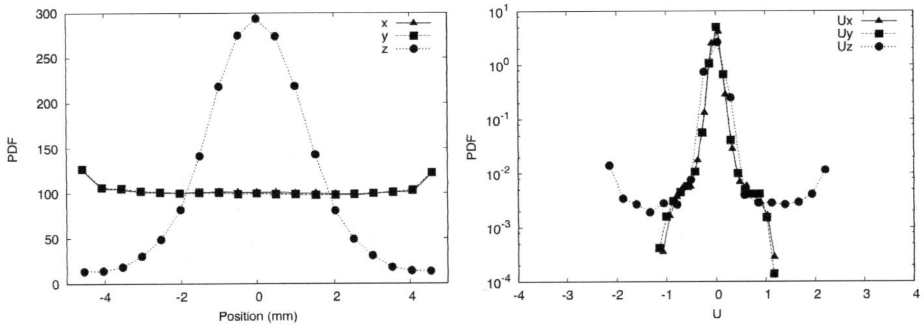

Figure 5. (Left) PDF of particle positions in the system collected every half period. (Right) Semi-log PDF plot of particle velocities in the a complete cluster.

When the density further increases, we observe that the cluster spreads over the entire xy plane. A peaked velocity distribution around zero can be observed in Figure 5. The presence of two sharp peaks at $v_z = \pm 2A\omega$ (also seen for partial clustering) is due to the granular gas trapped in between the cluster and the vibrating walls. We assist to an equilibrium between condensation and evaporation of the cluster. By tracking particles and analyzing the dynamics of clusterisation during several periods, we can conclude that the cluster is in a dynamical state with a constant renewal. Indeed, high speed particles arriving from the walls crush into the cluster, dissipate their energy and eventually become part of the cluster. The energy partially transmitted across the cluster provokes internal rearrangements and eventually ejects particles. It should be noted that the packing fraction measured inside the cluster is about $\eta = 0.30$ which is far below the jamming limit.

3.4. Bouncing aggregate

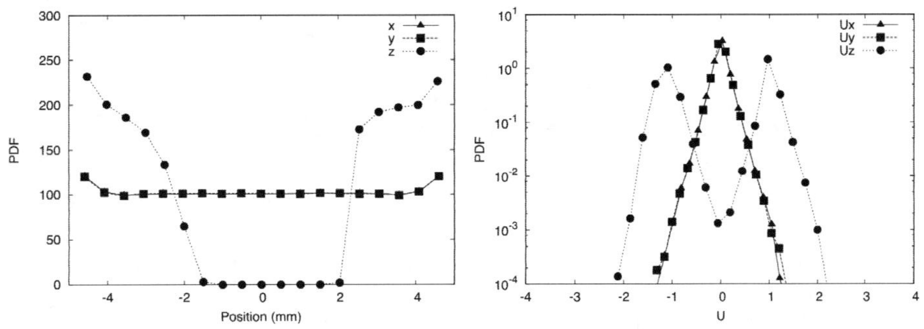

Figure 6. (Left) PDF of particle positions in the system collected every half period. (Right) Semi-log PDF plot of particle velocities in a bouncing aggregate.

For denser systems a new phenomenon is observed. The important amount of granular material confined in the system leads to higher kinetic energy than in other regimes (see figure 9). This energy cannot only be dissipated through successive collisions like in a cluster. Indeed, the local density in the system's center is high enough to assure permanent contacts between many particles, when the gaseous phase encounters this denser region all the grains a dragged towards the wall. This bouncing state is characterized by a coherent motion of all particles and peaked velocity distribution around $\pm A\omega$. Particles stay gathered in a dense arrangement which describes a movement similar to a totally dissipative bouncing ball [26]. Moreover, density distribution along the z-axis revealed crystal-like structures inside the packing. Layers are observed along the three axes.

4. Discussion

4.1. Phase Diagrams

An automated and reliable detection of the different dynamical regimes can be done by statistical means. The complete cluster detection relies a two-sample Kolmogorov-Smirnov (KS) test [27] with a significance level $\alpha = 0.01$. The KS test compares the cumulative distribution function (CDF) of a uniform distribution with the CDF of the particles distribution along z axis. The null hypothesis (H_0) of the test is the assumption that the observed distribution is uniform, hence that the system is gaseous. The alternative hypothesis is noted (H_1). We define the parameter D as the following

$$D = \sup_{|z|\leq L/2} \|F(z) - U(z)\| \tag{4}$$

where $F(z)$ is the observed CDF of the particles distribution along z-axis and $U(z)$ the theoretical uniform distribution. The KS test value is defined by $T_z = D\sqrt{n/2}$ where n is the number of classes. H_0 is rejected at level α if $T_z > K_\alpha$. The statistical threshold K_α can be found in tables. In a partial cluster regime, the system displays a strong heterogeneity in the xy-plane, therefore its detection relies on a two-sample KS test between the particles distribution along the x and y-axes with a significance level $\alpha = 0.01$. The test is denoted by T_{xy}. Finally, bouncing regime is detected by analyzing frequency fluctuations in the center of the box. If the central part of the box has a density at least two times lower than the one corresponding to a uniform distribution, the system is evolving like an inelastic body bouncing between two oscillating plates. A large number of MD simulations allowed us to obtain four typical (η, r) phase diagrams that are shown in figure 7.

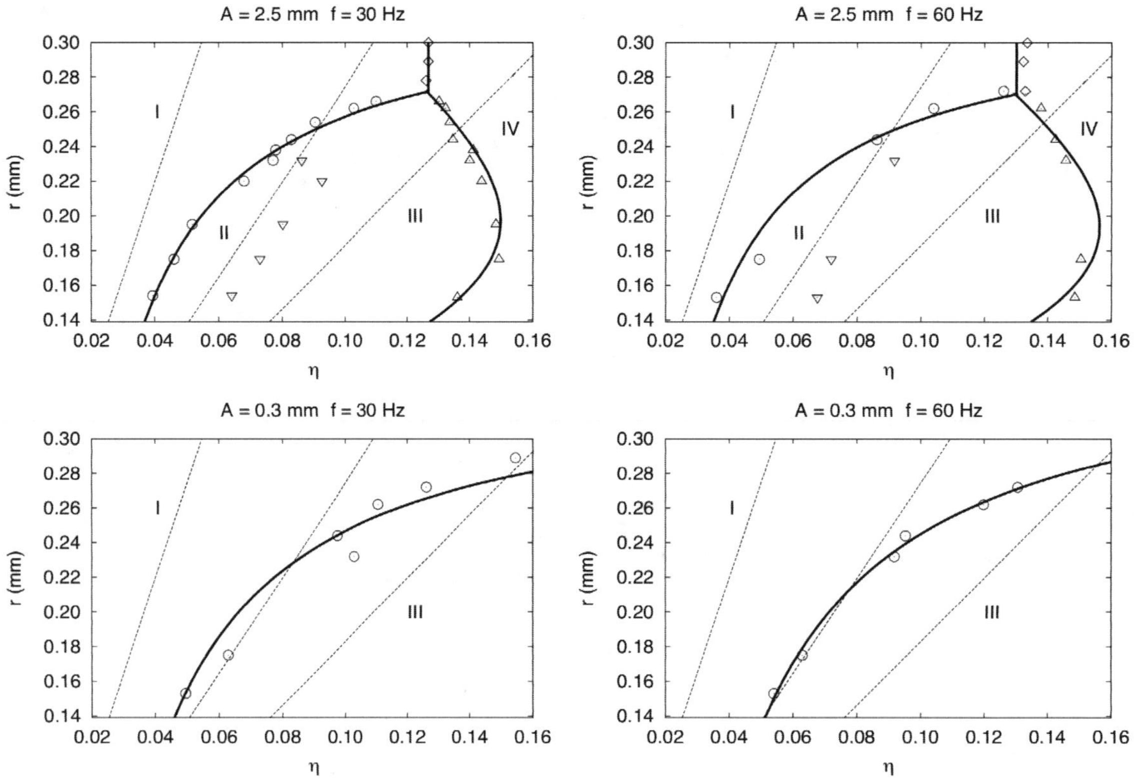

Figure 7. Simulated (r,η) phase diagrams for two different amplitudes $A = 2.5$ mm and $A = 0.3$ mm and for two different frequencies $f = 30$ Hz and $f = 60$ Hz. Curves correspond to the transitions from one dynamical regime to another. Differnt phases are denoted by numbers from I to IV.

For high amplitude ($A = 2.5$ mm), the four dynamical regimes can be observed and are denoted on figure 7 (I gas, II partial cluster, III complete cluster and IV aggregate). For low amplitude ($A = 0.3$ mm), only the gas and the complete cluster state is observed. Parameters such as η, r and A seem to have a deep impact on the systems dynamics in opposite to ω whose fluctuations have only minor effects. The emergence of a cluster for low amplitude is not trivial. While the formation of a high density zone in the case of $A = L/4 = 2.5$ mm could be triggered by geometrical reasons and confinement, the appearance of such a dynamical regime for $A = 0.3$ mm requires the presence of additional phenomena such as inelastic collapse. Moreover, measuring the standard deviation σ of the position PDF along the z-axis for 2 simulations with $N = 5000$, $r = 0.175$ mm and respectively $A = 2.5$ mm and $A = 0.3$ mm shows us that the difference of the cluster width is negligible compared to the difference of amplitude. Collecting data every period, we found in the first case $\sigma_{2.5} = 1.51$ mm and in the second $\sigma_{0.3} = 1.72$ mm. Figure 8 shows the position PDF along z-axis for both simulations. Shaded zones refer to the volume visited by the box' walls during oscillation.

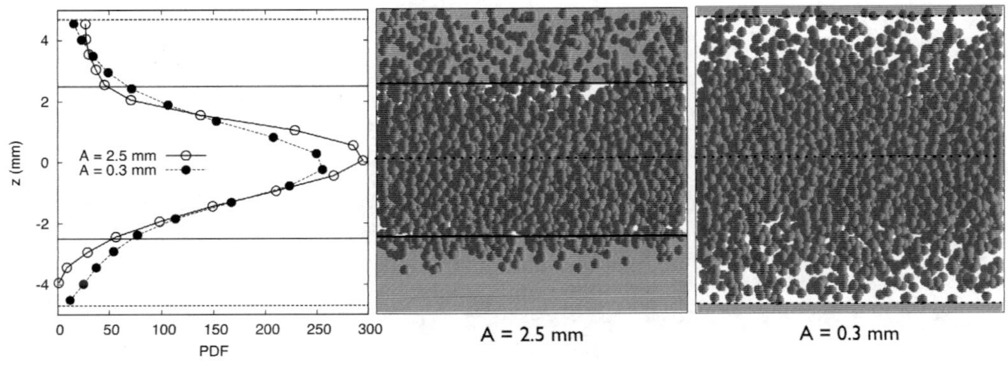

Figure 8. (Left) Distribution of the particle positions along z-axis for 2 simulations with $N = 5000$, $r = 0.175$ mm and respectively $A = 2.5$ mm and $A = 0.3$ mm. Data is collected every period. (Center) Lateral view of the system with an amplitude of $A = 2.5$ mm. (Right) Lateral view of the system with an amplitude of $A = 0.3$ mm. Shaded zones refer to the volume visited by the box' walls during oscillation.

4.2. Transition lines

For large amplitudes, the (r, η)-phase diagrams, illustrated in the first row of Figure 7, exhibit all the dynamical regimes. Different regimes are separated by transition curves that meet on triple points. Earlier studies [15] considered that cluster may be expected once there is initially more than a single layer of particles in the system. In this configuration, particle-particle collisions become frequent events which encourages dissipation and thus cluster formation. In our phase diagram, iso-layer curves correspond to dashed straight lines with an angular coefficient proportional to the inverse number of granular layers. The layer criterion clearly misses the transition from a gas to a (partial) cluster. The (◯)frontier is better described by the empirical law

$$r_{clust} - r = \delta_{clust} \left(\frac{1 - \eta}{\eta} \right) \tag{5}$$

where δ_{clust} and r_{clust} are constants. Eq. (5) supports two arguments : (i) a critical size $r_{clust} \approx 0.32$ mm for the grains above which no cluster appears, and (ii) the ratio between free volume and occupied volume becomes a relevant parameter. It should be noted that this ratio is also relevant for describing mean free path in our system.

The transition between partial and complete clustering is rather complex. One should note that partial clustering is only present for small particles and large amplitudes and. Moreover, it needs a long formation time (about 100 periods).

The bouncing aggregate regime IV appears for large packing fractions. Large particles condensate without clustering. The ◊-frontier between gaseous and condensed states is a vertical line of equation $\eta = 0.127$. Considering the whole granular assembly as a single inelastic body, this critical value can be established theoretically by adapting latest results concerning the Fermi-Box problem [28]. A punctual body is confined in a box of height H which follows a sinusoidal motion of amplitude A. For a restitution coefficient $\varepsilon = 0$, it is shown that the dynamics switch from a mono-periodical motion (similar to the diluted gaseous state) to a regime presenting sticking phenomena (similar to the condensed state) once the critical height $h = (H/A) = \pi$

is reached. In our case, the size of the particles cannot be neglected, therefore the Fermi-Box condition becomes

$$\frac{L - \left(\frac{L\eta^*}{\eta_{rlp}}\right)}{A} = \pi \tag{6}$$

where the random loose packing fraction $\eta_{rlp} \approx 0.59$. The critical value $\eta^* \approx 0.127$ is obtained in an excellent agreement with our numerical simulations.

For smaller particles, the transition between cluster and bouncing aggregate is also triggered when η increases. We have noted that the $\triangle$-frontier corresponds roughly to the relationship $r/L \sim \log N$ which consists in a Lambert-W function [29] provided by the inversion of (7).

$$\eta = \frac{4\pi}{3}\left(\frac{r}{L}\right)^3 \exp\frac{r_{ba} - r}{\delta_{ba}} \tag{7}$$

This transition curve needs deeper investigations since different bouncing modes (like period doubling) can be observed in some cases. This is left for future works.

The different values of the free fitting parameters for the various transition curves are given in table 1. In opposite to the small impact of the frequency, we can see that the variations are important when the amplitude changes.

A (mm)	f (Hz)	δ_{clust} (mm)	r_{clust} (mm)	δ_{ba} (mm)	r_{ba} (mm)
2.5	30	6.8817 10^{-3}	0.31908	6.5718 10^{-2}	0.75250
2.5	60	6.3164 10^{-3}	0.31239	6.4548 10^{-2}	0.74515
0.3	30	9.2490 10^{-3}	0.33101	-	-
0.3	60	1.1143 10^{-2}	0.34529	-	-

Table 1. Values of the free fitting parameters for the various transition curves in four diagrams

4.3. 3D Phase Diagram

In order to understand the origin of the cluster phenomenon, the internal energy of the system with amplitude $A = 2.5$ mm and $f = 30$ Hz has been investigated. By collecting the translation and the rotation velocities of the particles during one period, a three dimensional graph representing the mean kinematic energy in the system as a function of the packing fraction η and the grain radius r has been obtained. As shown in figure 9, the surface is colored by a color gradient and displays different energetic levels according to the dynamical regimes we determinate by statistical means. The corresponding transition lines shown in the previous figures are denoted by purple vertical surfaces.

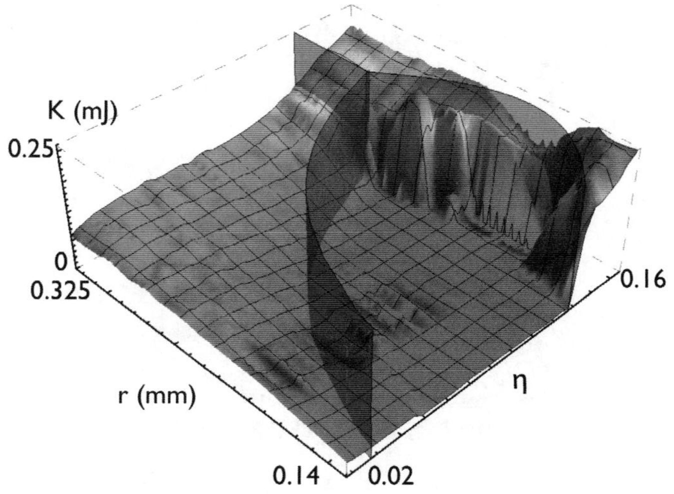

Figure 9. 3D view of the mean kinematic energy in the system with an amplitude $A = 2.5$ mm and frequency $f = 30$ Hz as a function of grain size r and packing fraction η. The corresponding transition lines shown in the previous figures are denoted by purple vertical surfaces.

5. Conclusion and Perspectives

In summary, the phase diagram is richer than expected since we have found four different dynamical regimes instead of gas/cluster regimes. Moreover, the transition curves cannot be described by simple rules or arguments, except for the behavior of large grains. Small variation of A and ω parameters will quantitatively modify the diagram but the phases and the transitions will be qualitatively recovered. We can assure that the diagram is locally stable. Our results raise fundamental questions and opens new perspectives in particular for future experiments in microgravity planned on the International Space Station.

In order to collect new relevant data, ESA is setting up "VIP-Gran", a series of experimentation expected to join the ISS in a closer future. In the experiment a certain amount of granular material is trapped in a cell volume of 30x30x70 mm between two independent oscillating walls. The main diagnostic tool is a CCD camera. Effects of the global packing fraction on the systems dynamics will be studied with a special interest in very dense systems and systems in a Knudsen regime. Different grain sizes as well as different shapes for the box will be tested. Moreover, experimentation with granular mixtures are planned. The following work has been realized in order to accomplish a first step for the calibration of VIP-Gran. Effects of parameters as grain size and packing fraction have been investigated and a granular phase diagram has been obtained.

6. Acknowledgments

We thank E. Falcon and S. Vincent-Bonnieu for fruitful discussions. This work has been supported by Prodex (Belspo, Brussels) and the European Space Agency program TT VIP-Gran. We also thank the TREX Morecar project (Feder, Wallonia) for supporting the development of our numerical model.

References

[1] O. Reynolds, Phil. Mag. Ser. **50**, 20 (1885)
[2] A. Kudrolli, Rep. Prog. Phys. **67**, 209 (2004)
[3] F. Ludewig and N. Vandewalle, Eur. Phys. J. E **18**, 367 (2005)
[4] D. Howell and R.P. Behringer, Phys. Rev. Lett. **82**, 5241 (1999)
[5] T.S. Majmudar and R.P. Behringer, Nature **435**, 1079 (2005)
[6] A. Kudrolli, M. Wollpert, J.P. Gollub, Phys. Rev. Lett. **78**, 1383 (1997)
[7] G. Lumay and N. Vandewalle, Phys. Rev. Lett. **95**, 028002 (2005)
[8] G. Lumay and N. Vandewalle, Phys. Rev. E. **74**, 021301 (2006)
[9] F. Ludewig, S. Dorbolo, T. Gilet and N. Vandewalle, Eur. Phys. Lett. **84**, 44001 (2008)
[10] P. A. Thomson and G. S. Grest, Phys. Rev. Lett. **67**, 13 (1991)
[11] G. D. R. Midi, Eur. Phys. J. E. **14**, 341(2004)
[12] P. deGennes, Rev. Mod. Phys. **71**, 374 (1999)
[13] A. M. Walsh, K. E. Holloway, P. Habdas and J. R. de Bruyn, Phys. Rev. Lett. **91**, 10 (2003)
[14] F. Spahn, J. Schmidt and M. Sremcevic, Lect. Notes in Phys. **557**, pp 507(2000)
[15] E. Falcon, R. Wunenburger, P. Evesque, S. Fauve, C. Chabot, Y. Garrabos and D. Beysens, Phys. Rev. Lett. **83**, 2, 440 (1999)
[16] E. Falcon, S. Fauve and C. Laroche, Eur. Phys. J. B **9**, 183 (1999)
[17] S. McNamara and W.R. Young, Phys. Rev. E. **50**, 1 (1994)
[18] P. Evesque, Y. Garrabos, C. Lecoutre, F. Palencia, and D. Beysens, Powders and Grains 2005 **1** (2005)
[19] S. McNamara and E. Falcon, Phys. Rev. E. **71**, 031302 (2005)
[20] S. Miller and S. Luding, Phys. Rev. E. **69**, 031305 (2004)
[21] P. A. Cundall and O. D. L. Stark, Geotechnique **29**, 47 (1979)
[22] F. Ludewig, PhD Thesis, University of Liege (2007) (URL http://orbi.ulg.ac.be/handle/2268/70899)
[23] N. Taberlet, PhD Thesis, Université de Rennes I (2005)
[24] M. Hou, R. Liu, G. Zhai, Z. Sun, K. Lu, Y. Garrabos and P. Evesque, Microgravity Sci. Technol. **20**, 73 (2008)
[25] J. J. Brey and M. J. Ruiz-Montero, Phys. Rev. E. **67**, 021307 (2003)
[26] T. Gilet, N. Vandewalle and S. Dorbolo, Phys. Rev. E. **79**, 055201 (2009)
[27] W. T. Eadie, D. Drijard, F. E. James, M. Roos and B. Sadoulet, Statistical Methods in Experimental Physics, 2nd Edition (World Scientific, Amsterdam, 2006)
[28] N. Vandewalle, T. Gilet and S. Dorbolo (preprint 2010)
[29] S. R Valluri, D. J. Jeffrey and R. M. Corless, Canadian Journal of Physics **78**, 823 (2000)

Journal of Physics: Conference Series **327** (2011) 012036 doi:10.1088/1742-6596/327/1/012036

Agglomeration of Ni-nanoparticles in the Gas Phase under Gravity and Microgravity Conditions

S Lösch[1], G N Iles[2], B Schmitz[3], B H Günther[1]

[1] Fraunhofer Institute for Manufacturing Technology and Advanced Materials (IFAM), Wiener Str. 12, 28359 Bremen, Germany
[2] European Space Agency (ESA), ESTEC Nordwijk, The Netherlands
[3] Astrium GmbH, Space Transportation, Airbus-Allee 1, 28199 Bremen, Germany

E-mail: loesch@ifam.fhg.de

Abstract. The agglomeration of metallic nanoparticles can be performed using the well-known inert gas condensation process. Unfortunately, thermal effects such as convection are created by the heating source and as a result the turbulent aerosol avoids ideal conditions. In addition, the sedimentation of large particles and/or agglomerates influences the self-assembly of particles. These negative effects can be eliminated by using microgravity conditions. Here we present the results of the agglomeration of nanoscale Ni-particles under gravity and microgravity conditions, the latter provided by adapted microgravity platforms namely the European sounding rocket MAXUS 8 and the European Parabolic Flight aircraft, Airbus A300 Zero-G.

1. INTRODUCTION

Agglomeration phenomena within metal aerosols enable many technical solutions for functional materials and generally occur under the combined effect of Van der Waals, electrostatic, and magnetic interactions [1][2][3][4]. Particularly interesting would be long chains consisting of monosized particles. For ferromagnetic materials, such as Ni, Co, and Fe, the agglomeration process is dominated by magnetic interparticle forces [5] leading to magnetic pearl necklaces, which could be useful, e.g., as functional ingredients in sensor applications. In this paper, such necklaces are synthesized from thermally produced nickel aerosol using the well-known inert gas condensation technique [6] with the clear objective to investigate the relationship between background gas pressure, agglomeration time and size/shape of the resulting Ni-agglomerates under laminar flow conditions. It is expected that the "free-flow" residence time of magnetic particles correlates to the length of the deposited particle chains. The experiments have been conducted under microgravity (μ-g) conditions in order to suppress thermally-induced convection. Several parabolic flight campaigns (PFC) as well as a sounding rocket flight (SR) were used as experimental platforms to extend the time available for Ni-particles/agglomerates on their way from the heating source to the substrate up to the minutes range.

2. EXPERIMENTAL SET-UP

The typical inert gas condensation process with its single stages is illustrated by a simple diagram as shown in figure 1. Hereby the metallic material is heated up above its melting temperature inside an adapted crucible. The inert gas, whose gas volume is variable, carries the particles originated by nucleation and coagulation, and agglomerates upstream within a defined flow chamber until reaching a deposition area where substrates can be removed after each experiment for ex-situ analysis. The left sketch of figure 2 shows a schematic of the Ni evaporation condensation vacuum source made from standard vacuum parts used for agglomeration experiments of metallic materials in laboratory and during parabolic flights. A 5g Ni ingot (purity 99,95%) is heated up to 1850°C using an induction coil rf-heater. The melted Ni is located inside an Al_2O_3 crucible. The inert gas Ar flows through the silica glass tube and carries the Ni-nanoparticles and agglomerates downstream. At the exit of the tube the Ni-agglomerates are deposited on electro-polished substrates made from non-magnetic stainless steel. More technical details and operational procedures about this set-up can be found in [7].

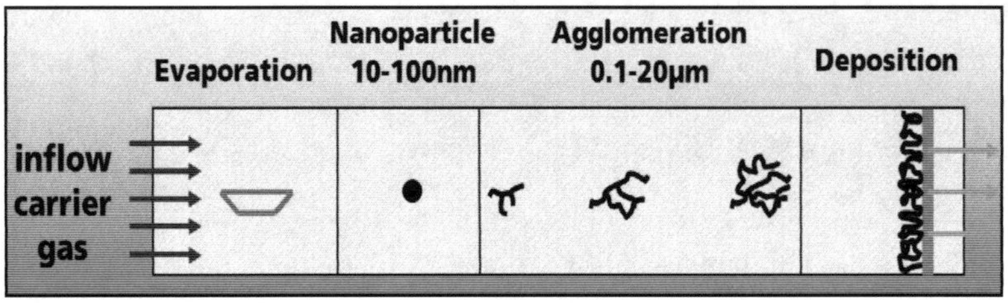

Figure 1: principle of inert gas condensation process

In gravity (1-g) turbulences do not show up at residence times below 8sec. At smaller flow rates thermally induced turbulences within the aerosol prevent a well-defined and reproducible deposition behavior. For example a very limited range of residence times can be used in 1-g, while keeping well-defined flow conditions. Unfortunately the rf-heating equipment could not be used on a sounding rocket due to restrictions in electric power and weight. Therefore the evaporator concept had to be modified as shown in the right-side sketch of figure 2. Approximately 0,1g of Ni-wire (purity 99,99%) was wound on a tungsten filament and evaporated at 1850°C. With this method a steady evaporation of ca.12min could be achieved. The distance between heating source and site of sampling is in both cases 44cm.

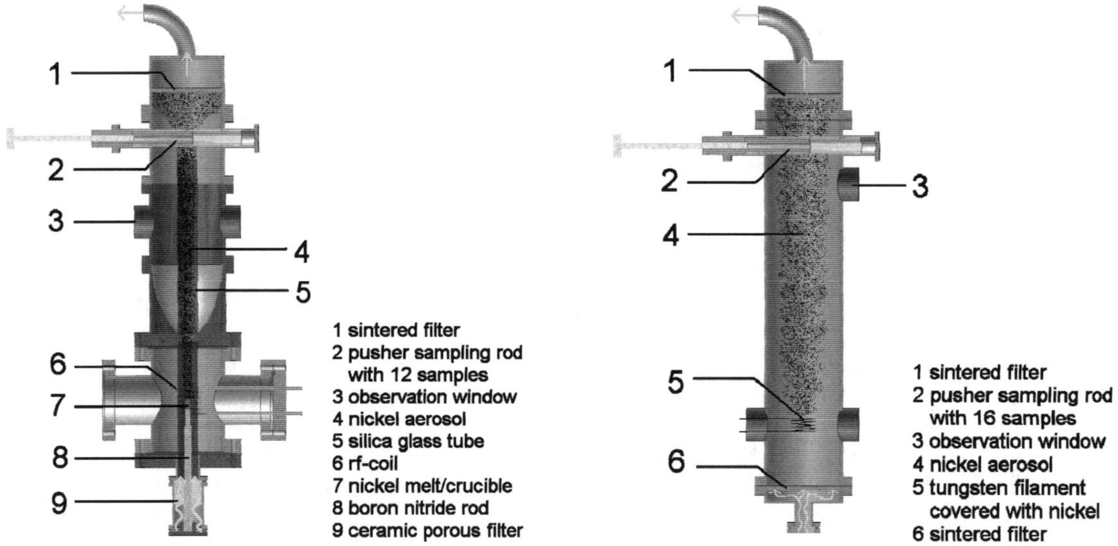

Figure 2: schematic of Ni evaporation condensation device; left: used in lab and parabolic flights, right: used in lab and on MAXUS 8

Likewise the Joule-heating set-up could not be used on parabolic flights due to the limited evaporation time of the nickel and the risk of loosing Ni-drops from the tungsten filament during the hypergravity phase of each parabola. In both set-ups Ni-agglomerates were collected in the same principle as shown in figure 3. Hereby the substrate is fixed on a sampling rod which is located in a horizontal stainless steel tube with an opening for aerosol exposure. For sampling the rod moves the deposition section of the substrate over the small opening and the aerosol comes in contact with it. After sampling for several seconds the opening becomes closed by moving the rod until defined position is reached. When in this defined position, the deposited section is protected against further deposition. After one full side of the substrate has seen aerosol exposure, the rod is turned 180°, and the same process occurs on the reverse side of the substrate which has more sampling sections. The morphology of the deposited agglomerates was observed ex-situ via Scanning Electron Microscopy (SEM) and Transmission Electron Microscopy (TEM).

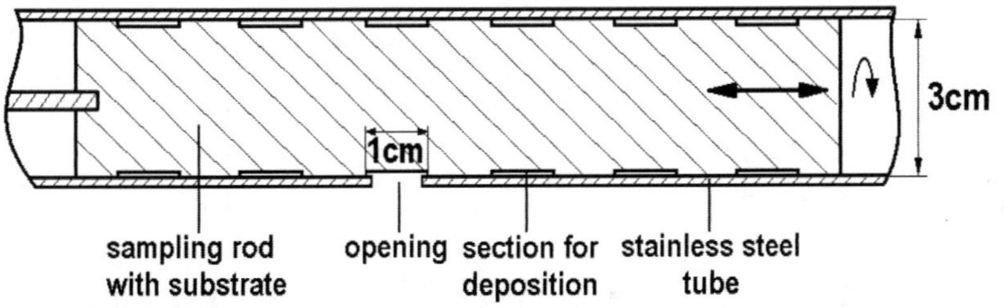

Figure 3: prinicple of sampling method

The residence time is derived from the Ar gas velocity assuming laminar flow with a parabolic velocity profile (figure 4). At Reynolds numbers less than 2300 the gas velocity can be solved by the law of Stokes:

$$v(r) = \frac{\Delta p_V * \left(\frac{d^2}{4} - r^2\right)}{4 * \eta * l} \qquad [1]$$

with the pressure difference Δp_V between entrance/exit port of flow tube with diameter d and length l and the dynamic viscosity η of the carrier gas.

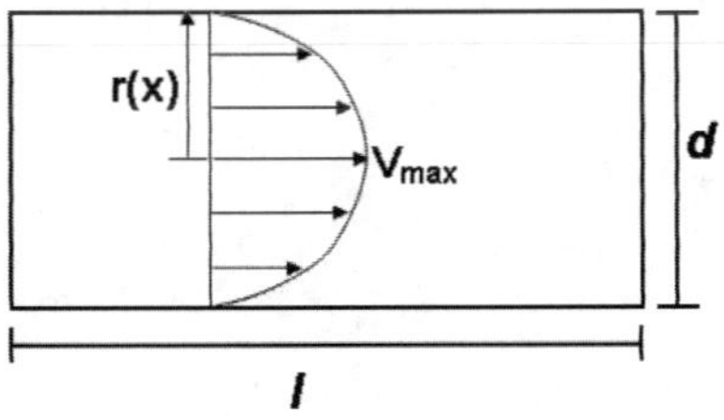

Figure 4: parabolic flow profile of an ideal gas in a flow tube

By using the law of Hagen-Poisouille Δp_V can be found:

$$\dot{V} = \int_{r=0}^{d/2} v(r) * 2\pi * r dr = \frac{\Delta p_V * \pi * d^4}{128 * \eta * l} \qquad [2]$$

Now the minimal residence time can be calculated with:

$$t_{res} = \frac{l}{v_{\max}} \qquad [3]$$

In 1-g the calculated residence time can be used at least as a "figure-of-merit" to correctly compare the agglomeration results by neglecting any effects from convection and sedimentation.
In μ-g (parabolic flights) the maximum residence time is limited to 22 sec during each parabola, where ca. 6s must be subtracted as sampling time. Extension of this value is possible by using the sounding rocket μ-g platform. With the ESA-operated MAXUS a maximum residence time of up to 12min can be realized.

3. EXPERIMENTAL RESULTS

3.1 AGGLOMERATION OF NI-PARTICLES IN NORMAL GRAVITY CONDITIONS (1-g)

Figure 5 shows typical Ni-agglomerates obtained in 1-g conditions. There is no obvious difference in morphology of particles obtained with Joule-heating and rf-heating. However, we observed an increase in agglomerate length up to 10µm at longer residence times up to 8sec (figure 6). Large argon gas flow rates resulting in short residence times result in short Ni-chains of just several 100nm.

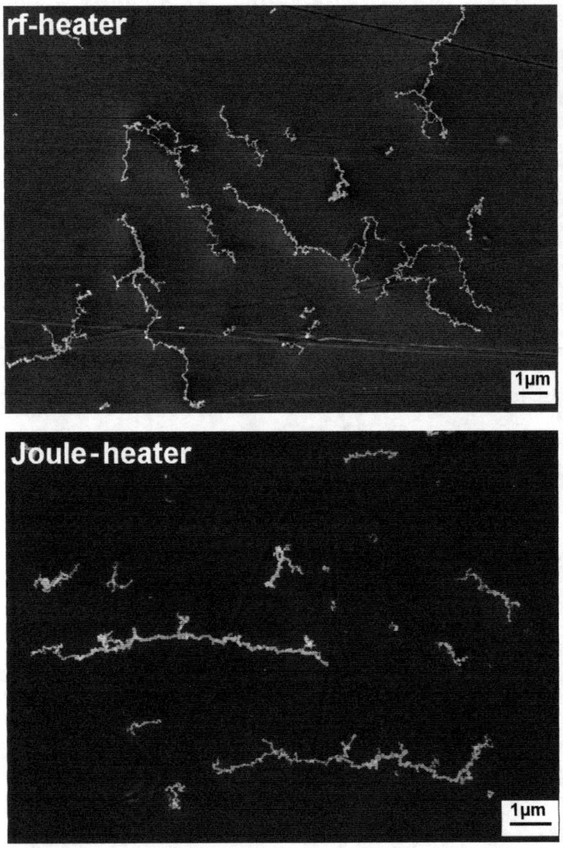

Figure 5: Ni-agglomerates obtain- ed in 1-g with rf-heater and Joule-heater (upper and lower image, resp.)

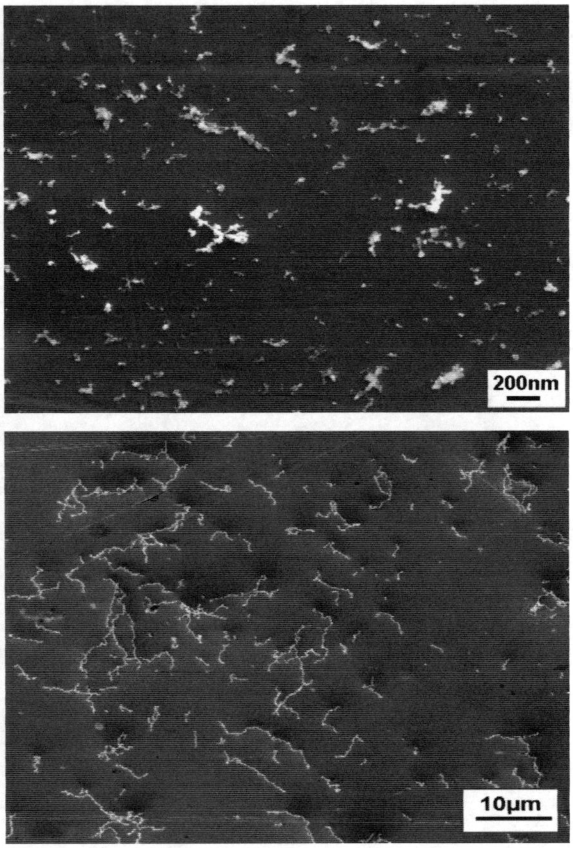

Figure 6 Ni-agglomerates obtained in 1-g at residence times <1sec and ca. 8s (upper and lower image, resp.)

To define the particle size distribution of the Ni-particles within the agglomerates, individual measurements were taken at different gas pressures and velocities. Up to 700 single particles had been used for each measurement, which was performed by an adapted computer program. The intra-agglomerate size distribution of the Ni-particles made in 1-g is rather broad, see figure 7. A plot of particle sizes, shown in figure 8, indicates that inside one agglomerate Ni-particles between 10 and 150nm can be found. This observation most likely results from convection effects close to the heating source, where a mixing of particles occurs having seen different growth histories. In general no significant dependency between the gas pressure and the particle size distribution could be detected. While with decreasing the aerosol velocity the particle size distribution becomes more normally distributed as shown in figure 8.

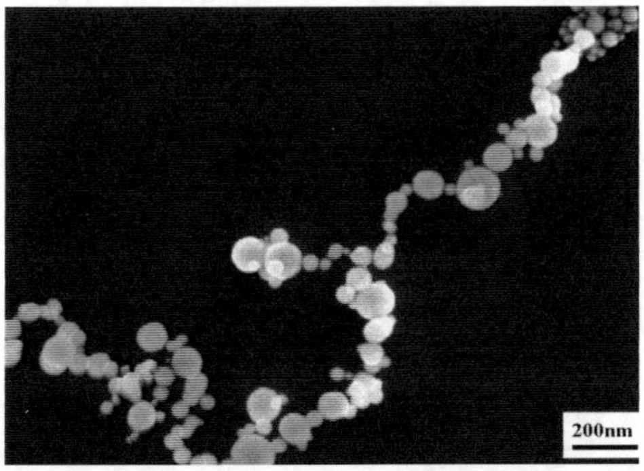

Figure 7: Ni-particles inside Ni-agglomerates

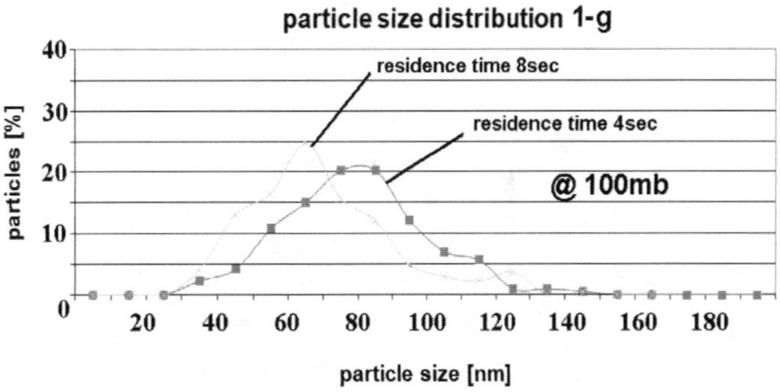

Figure 8: size distribution of Ni-particles within agglomerates made in 1-g

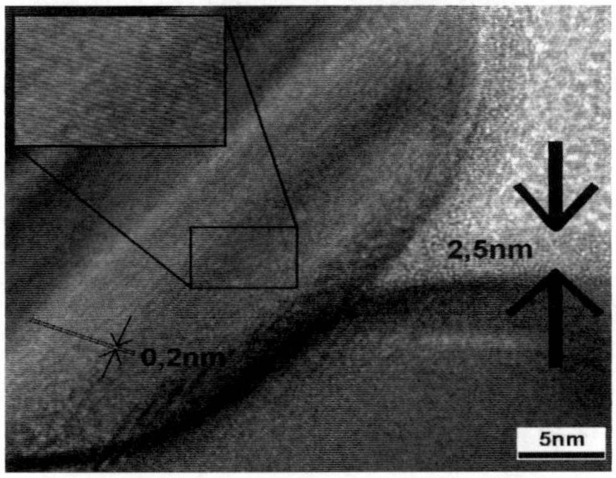

Using transmission electron microscopy, it was possible to visualize the Ni atoms. Figure 9 shows that individual particles are surrounded by an oxide layer which is to be expected once the samples are removed from the chamber. The TEM image indicates that this oxide layer is approximately 2,5nm thick. Furthermore, the Ni atoms are clearly arranged in a crystalline structure with an atomic spacing of 0,2nm.

Figure 9: TEM image of 2 Ni-particles with single-crystalline structure; the oxide layer is 2,5nm thick

3.2 AGGLOMERATION OF NI-PARTICLES IN MICROGRAVITY CONDITION (μ-g)

3.2.1 Parabolic flight platform (PF)

Typical agglomerates of Ni-nanoparticles which have been aggregated in μ-g-conditions are shown in figure 10 and figure 11. In contrast to agglomerates produced in 1-g conditions, here the particle chains are stretched much more perfect. Additionally, with increasing "free-flow" residence time the mean length of the agglomerates increases. The longest agglomerates which could be found had a length of ca. 30μm.

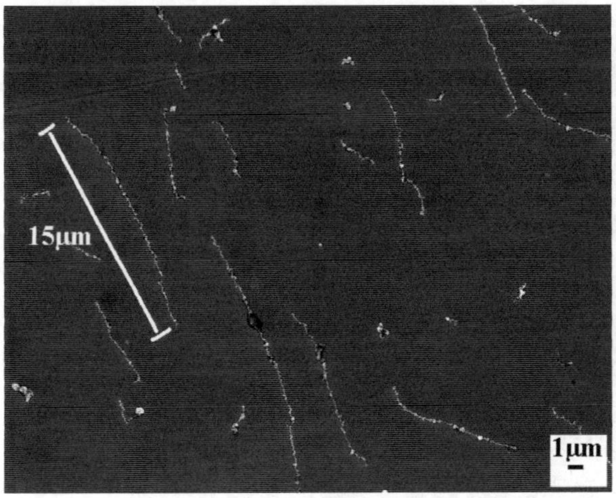

Figure 10: typical deposit of Ni-chains obtained from microgravity experiment (parabolic flight) at p=100mb, t_{res}=8sec

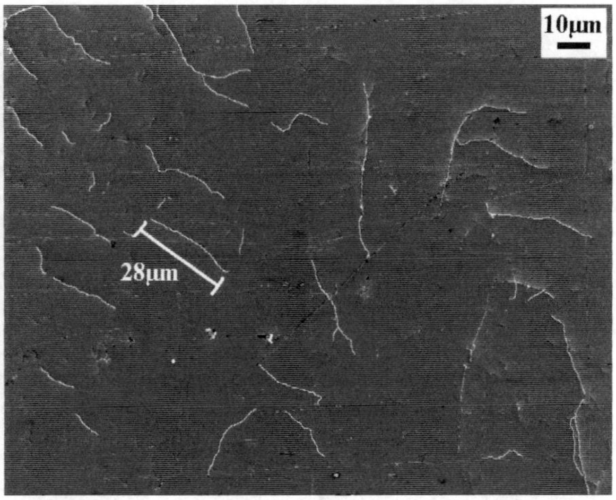

Figure 11: Ni-agglomerates from microgravity experiment (parabolic flight) at p=100mb, t_{res}=16sec

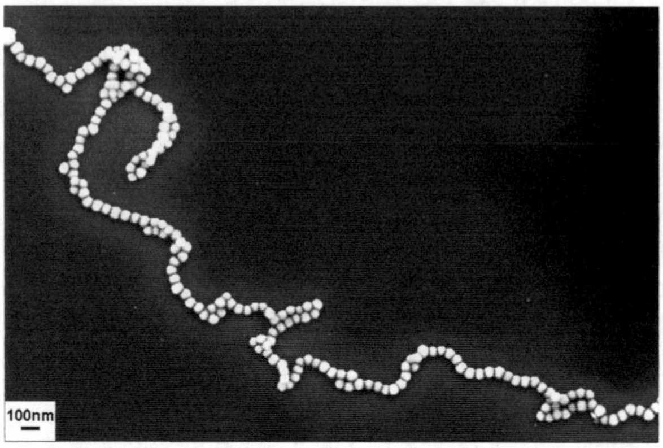

The intrachain particle size distribution, shown in figure 12, depicts how μ-g experiments can produce more consistent particles throughout the chains as compared to those obtained from 1-g produced particle chains (figure 7). This probably results from the absence of convection effects close to the heating source, where a non-mixing of particles occurs.

Figure 12: agglomerate of Ni-nanoparticles obtained from parabolic flight experiment at p=100mbar

The dependency between the mean length of the Ni-agglomerates and different residence times is shown in figure 13. Thus for each parameter, 100 Ni-chains had been measured on the SEM-images via an adapted computer program.

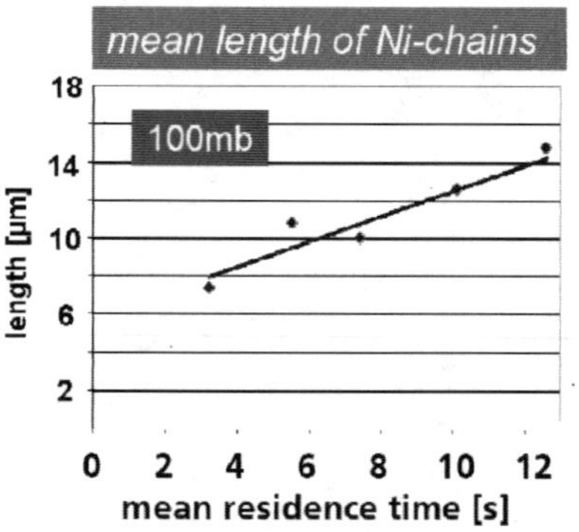

Figure 13: mean length of Ni-agglomerates obtained during parabolic flights as a function of "free-flow" residence time. The straight line is just a guide to the eye.

3.2.2 Sounding rocket campaign MAXUS (SR)

Regarding the results from the parabolic flight experiments the intention of the agglomeration experiments on-board a sounding rocket (MAXUS 8) was to increase the residence time into the minute range. There was microgravity time for running experiments with two different background gas pressures – 100 and 300mb. At each pressure 4 different residence times were set at 13, 40, 70 and 100sec. During the flight the μ-g quality was approx. 10^{-5}-g, guaranteeing ideal conditions for particle agglomeration ex-cluding any sedimentation or convection effects. This was proven by recording the aerosol with a CCD-camera. The quality of the Ni-aerosol flow at different time intervals during the 300mb experiment is shown in figure 14.

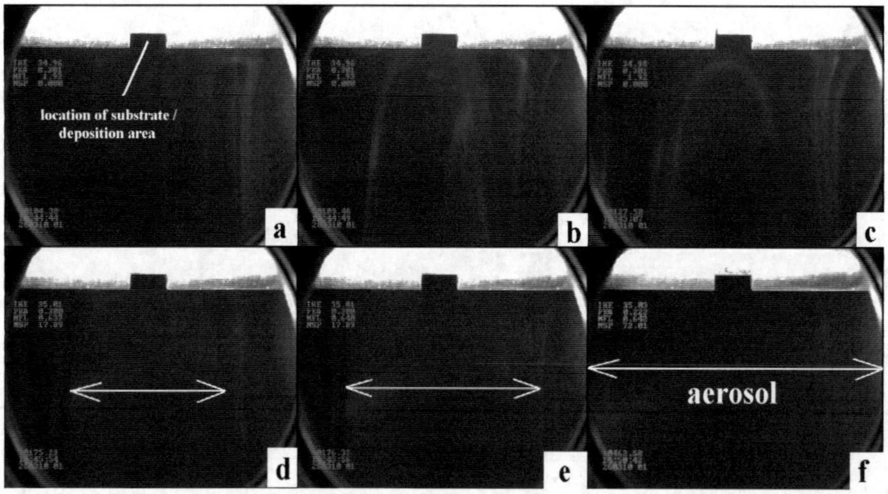

Figure 14: CCD-images of Ni-aerosol at 300mb Ar during MAXUS 8 flight taken close to sampling area; (a) t=10s – immediately after switching on the evaporator – turbulent flow due to gravitational effects, (b) t=15sec – beginning of μ-g conditions, but still turbulent flow, (c) t=22sec – steady and well-defined aerosol (d) t=80sec, (e) t=150sec and (f) t = 250sec; the arrow indicates the width of the aerosol plume; Note: the slit on the tube has a length of 8mm

Surprisingly, no chain-like agglomerates were found in the Ni-deposits. Instead the particles aggregated as fractal structures as well-known from aggregation of non-magnetic materials. In addition we could not find a tendency regarding residence time and mean size of agglomerates as expected (figure 15).

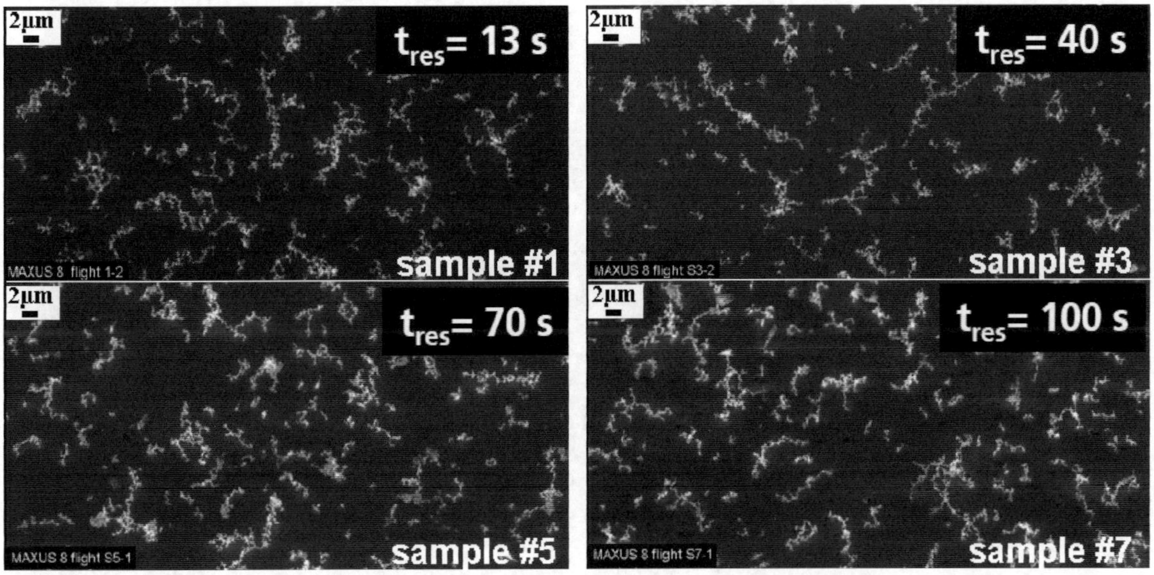

Figure 15: SEM images of Ni-agglomerates produced on sounding rocket MAXUS 8 flight at 300mb and residence times 13s, 40s, 70s and 100s for samples #1,3,5,7 resp.

Furthermore no pressure dependence of the agglomerates' morphology could be detected (figure 16 and figure 17), whereas the size distribution of the primary particles is similar to those obtained in parabolic flight experiments (figure 18, figure 19).

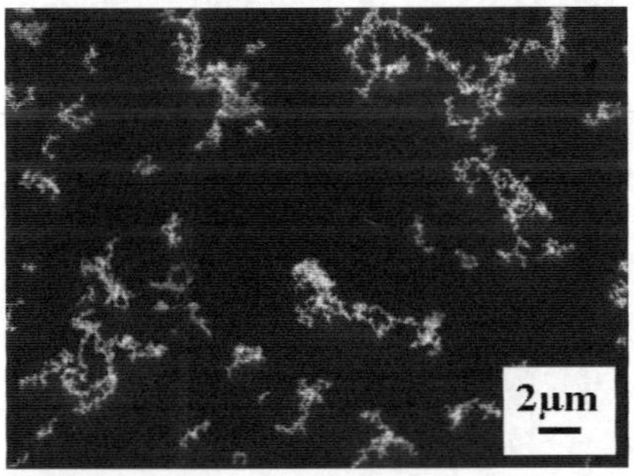

Figure 16: SEM image of Ni-agglomerates produced on sounding rocket at 300mb and t_{res}=13sec

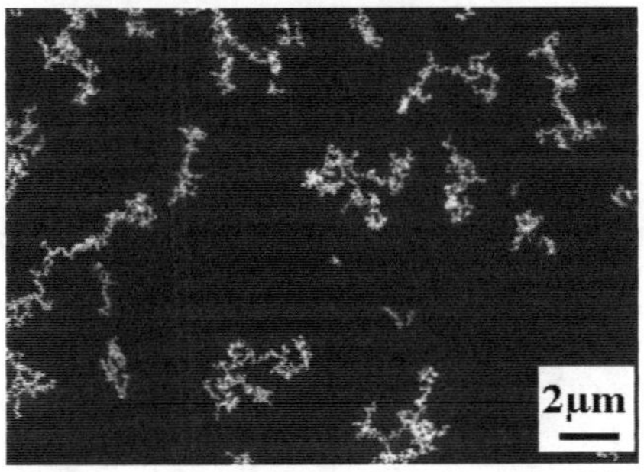

Figure 17: SEM image of Ni-agglomerates produced on sounding rocket at 100mb and t_{res}=13sec

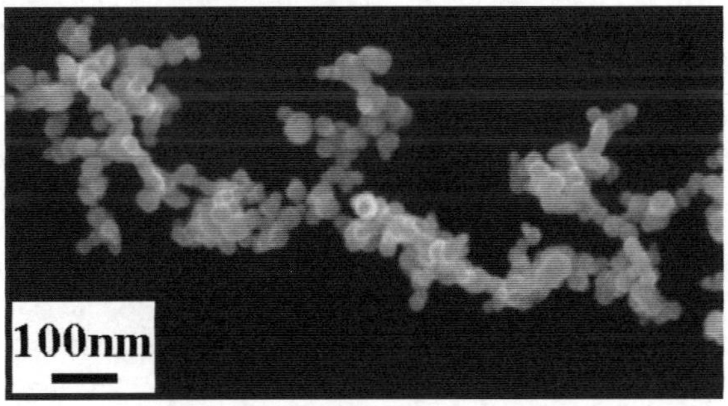

Figure 18: typical Ni-particles agglomerate obtained from MAXUS 8 experiment

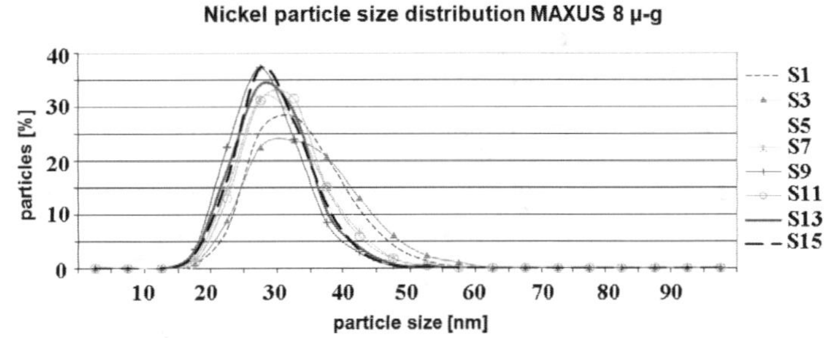

Figure 19: size distribution of Ni-particles inside agglomerates made during MAXUS 8 flight for different gas pressures and residence times, i.e., p=300mb at 13s (S1), 40s (S3), 70s (S5), 100s (S7); p=100mb at 13s (S9), 40s (S11), 70s (S13), 100s (S15). The total number for each curve is about 1000 particles

4. DISCUSSION

Performing an evaporation experiment with permanent gas flow on ground leads to turbulent aerosol conditions driven by thermal effects as shown in the left sketch of figure 20. However, the investigation of the agglomeration of magnetic nanoparticles, i.e., in the gas phase requires ideal aerosol conditions without any disturbing side effects (right sketch of figure 20). This could be provided by using suitable microgravity platforms. The formation of extended chains of nanoparticles is a well-known phenomena observed as a result of agglomeration within magnetic aerosols. Usually the magnetic dipole–dipole interaction is considered responsible for this type of self-organization (see, e.g., [5]). Li et al. [9] observed morphologies of Ni-agglomerates depending on the presence and strength of an external magnetic field (figure 21), i.e., applying stronger magnetic fields results in straighter and longer particle chains. Unfortunately in all these experiments the domain structure of the individual nanoparticles is unclear. However, it can be reasonably assumed that multi-domain particles would not align as perfectly in linear chains as single-domain particles do, if the magnetic interaction between particles dominates the agglomeration process. However, in our experimental set-up external magnetic fields are involved, too, a) due to the W-heater coil in the case of the MAXUS-setup and b) due to the Earth's magnetic field.

In total we have tested 4 experimental situations (μ-g, 1-g both with rf- and Joule-heating set-up) and only in a single case, i.e., via Joule-heating on sounding rocket platform, no chain-like aggregates have formed (figure 22). We had performed several pre-tests before the MAXUS 8 flight in order to prove if the strong impact of the payload during the reentry into the atmosphere and the landing on ground could modify the deposited Ni-agglomerates. Therefore we had deposited substrates with Ni in a lab and analyzed them via SEM. Subsequently we made vertical drop tests (ground material: ceramic; distance: 1,5m) and vibration tests (amplitude: 1mm, frequency: 50Hz) with these deposited substrates. The SEM studies did not show any changes. In addition we had analyzed the deposit (1-g, μ-g) on the sintered filter which was located at the top of the flow chamber to verify if the design of the sampling module and the procedure to sample the Ni-agglomerates can affect the shape of the agglomerates. Again, the SEM examinations showed no changes to the Ni-structures deposited on the substrates. Due to these tests we could determine the strong adhesion of the Ni-agglomerates on the substrates. To elucidate the possible reason for the non-chainlike Ni-aggregates, we need to figure out the main differences for particle interactions during the agglomeration process. These are

- strength and direction of Earth's magnetic field
- competition between thermal fluctuation of the direction of the particles magnetic moment (decreasing with particle size) and the dipole-dipole-interaction (increasing with particle size for single domain particles)
- convection effect.

We assume that Earth's magnetic field being unstable in size and direction during the agglomeration period is responsible for the fluffy morphology of the Ni-aggregates obtained during the MAXUS 8 flight. The apogee of this SR-flight was at an altitude of approx. 720km. During the ascent to and descent from the apogee combined with a slow spin (<0,1Hz) of the rocket payload the magnitude and direction of the earth magnetic field relative to the flow tube axis varies during agglomeration (figure 23).

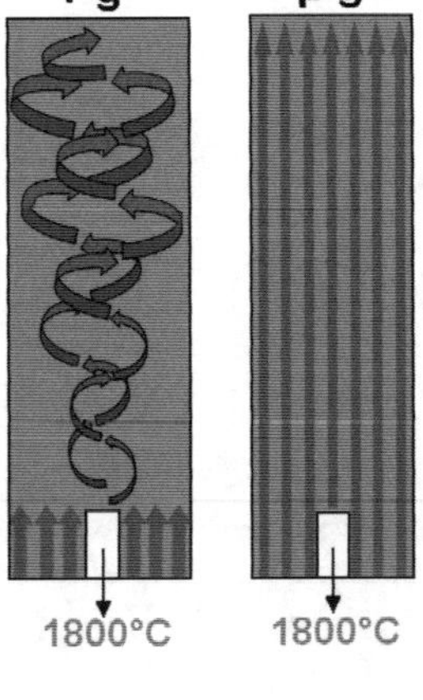

Figure 20: typical behavior of an aerosol inside a flow chamber with an evaporator on the bottom; turbulent characteristic due to convection and sedimentation effects under gravity (1-g); ideal gas flow conditions under microgravity (μ-g)

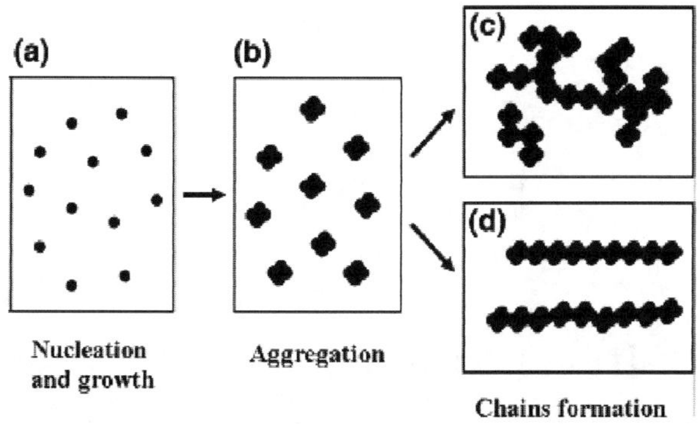

Figure 21: Schematic diagram for the growth mechanism of the Ni nanochains (taken from [8]). Processes of (a) nucleation/growth, (b) aggregation, (c) self-assembly and formation of dendritic chains, and (d) formation of axially aligned nanochains in magnetic field

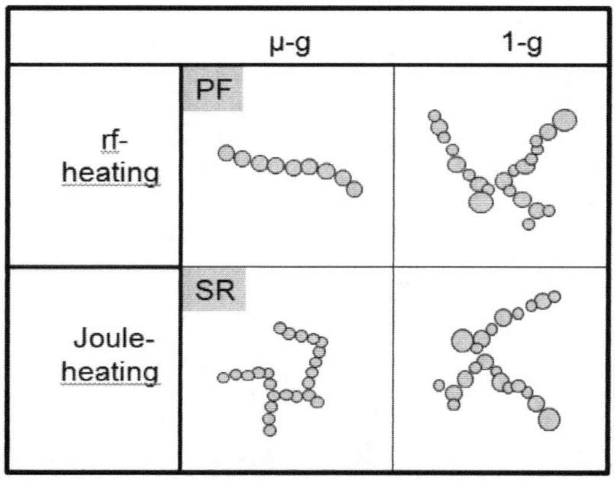

Figure 22: morphology characteristics of Ni-particles agglomerated under microgravity (μ-g) and gravity (1-g) conditions using the 2 evaporation methods

Bremen (lab)	Parabolic flights (during parabola)	MAXUS 8 (at launch)	MAXUS 8 (start of μ-g phase)	MAXUS 8 (at apogee)
48,5 μT[1]	Ca. 46,8 μT[2]	50,2 μT	54,9 μT	43,1 μT

Figure 23: earth magnetic fields |B| at different locations; [1] http://www.pimath.de/magnetfeld/gesamtfeld.html, [2][10]; values during MAXUS flight have been measured via magnetometer

5. CONCLUSIONS

With this work we have presented new results on the agglomeration behavior of Ni-nanoparticles in aerosols under earth gravity and microgravity conditions by using the inert gas condensation technique. In general, necklace-like chains are obtained with primary Ni-particles being larger than about 20 nm. Agglomerates obtained in μ-g during parabolic flights are similar in morphology with those obtained in 1-g, but tend to be much longer, if the time available for agglomeration is increased. Further the intra-chain particle size distribution is much narrower compared to those obtained in 1-g conditions. In contrast, the results from a sounding rocket campaign (MAXUS-8) do not fit into this scheme: no particle chains were found even with Ni-particles much larger than 20 nm. This might be due to a different (multi-domain) magnetic structure of the nanoparticles. This will be part of future investigation on the particles' magnetic properties.

ACKNOWLEDGEMENTS
The authors thank ESA, SSC, and Novespace for providing μ-g facilities, sounding rocket launch support, and assistance during parabolic flights, respectively. This work was partly funded under the EC-integrated project IMPRESS Contract No NMP3-CT-2004-500635.

REFERENCES

[1] Kruis F E 1998 J. Aerosol Sci. Vol. 29 No. 5/6 pp.511
[2] Parkin S S 2008 Science 320 pp.190
[3] Wang H 2008 Materials Research Bulletin 43 pp.3529
[4] Oberholz A 2003-2005 Band 2 Kapitel 9 pp.821
[5] Maceira V 2006 Journal of Magnetism and Magnetic Materials 303 pp.163
[6] Granqvist C G 1976 J. Appl. Phys. Vol. 47. No.5 pp.2200
[7] Lösch S 2009 Rev. Sci. Instrum. 80 083907
[8] Lie P 2010 Nanoscale Res Lett 5 pp.597
[9] Rost M 2006 Copernicus

Journal of Physics: Conference Series **327** (2011) 012037 doi:10.1088/1742-6596/327/1/012037

Structures and dynamics of fine particles in fine particle plasmas under microgravity and friction between two-dimensional layers of charges

Hiroo Totsuji

Graduate School of Natural Science and Technology, Okayama University, Tsushimanaka 3-1-1, Kitaku, Okayama 700-8530, Japan

E-mail: `totsuji@elec.okayama-u.ac.jp`

Chieko Totsuji

Academic Support Center, Kogakuin University, Nishishinjuku 1-24-2, Shinjukuku, Tokyo 163-8677, Japan

Abstract. Low temperature structures and dynamics of fine (dust) particles in the ambient (background) plasma with the cylindrical symmetry are analyzed by numerical simulations and theory. Fine particles are modeled as Yukawa particles and the one- and two-component cases are considered. In the one-component case, particles form well-defined concentric cylindrical shells and the structures are expressed by simple interpolation formulas to a good accuracy. When we introduce the difference into charges of particles forming a shell, particles are separated into a smaller shell of larger charges and a larger shell of smaller charges. In the two-component Yukawa system, we usually have structures composed of charge-separated shells where larger and smaller charges appear alternately in the order of radius. When we have charge-dependent forces along the symmetry axis, shells of larger and smaller charges can move relative to each other keeping their own structures and the mobility for the relative motion can be defined. We give examples obtained for some combinations of parameters. It is found that, when appropriate conditions are satisfied, concentric shells form a kind of 'fine-particle (dust) screw' and the mobility is increased.

1. Introduction

Fine particles in fine particle (dusty, complex) plasmas, mixtures of fine (dust) particles and the ambient plasma, provide us with a field where various phenomena in strongly coupled charge systems can be directly observed at the kinetic level. One of typical examples is the formation of static structures generally called dust (Coulomb) crystals[1, 2, 3, 4]. Structures are closely related to mutual interactions between particles and expected to give us basic information on the latter. In order to obtain such information and have insight into the nature of these structure formations, observations under the condition of high symmetry will be helpful.

Next to the homogeneous and isotropic systems which are difficult to realize, we have those with spherical or cylindrical symmetry or with restricted degrees of freedom. In experiments on the ground, however, we have to use the electric field in the sheath to levitate fine particles against strong influence of gravitation. Fine particles are usually distributed around the minimum of the total potential forming horizontal layers whose number is controlled by the strength of mutual repulsion relative to the strength of vertical confinement[5]. It is therefore difficult to have not only homogeneous systems but also with the ideal spherical or cylindrical symmetry. In addition, the supersonic ion flow in the sheath possibly modify the structures.

Long straight discharge chambers under microgravity seem to be one of candidates for such observations in systems with ideal cylindrical symmetry. Recently, systematic analyses of fine particle plasmas in such a chamber are in progress[6]. This apparatus is planned to succeed the highly successful parallel-plate type one now operated in the International Space Station (ISS)[7] and the structure formation has been observed on the ground[8]. We may thus expect to observe phenomena in the system with the ideal cylindrical symmetry under microgravity.

Colloid particles in charge stabilized colloidal suspensions also enable us to make kinetic level observations of strongly coupled charge systems such as colloidal (molecular) crystals[9]. Colloid particles confined between plates[10] or in two-dimensional channels[9] have been extensively investigated. They are free from the effect of gravity and their structures may be applied as microstructures in photonic or phononic crystals. Colloid particles in cylindrically symmetric systems also seem to be worth serious experimental investigations.

We here analyze the structure formation of Yukawa particles in a system with the cylindrical symmetry by numerical simulations and theory. Fine particles interacting via the Yukawa repulsion embedded in the ambient (background) plasma of electrons and ions can be regarded as a simple model of fine particle plasmas[11, 12, 13]. We assume that the background plasma is uniformly distributed in a cylinder of radius R_0 and the overall charge neutrality is satisfied. When one neglect the hard core in the Derjaguin-Landau-Verwey-Overbeek potential, Yukawa system can also work as a model of colloidal suspensions.

In various ion traps, we also have nearly cylindrical systems of ions confined by a combination of magnetic field and electrodes or by electrodes. We can regard these confinements as equivalent to the existence of the uniform background charges in a cylinder. We give a general picture for the structure formation in cylindrical systems covering these Coulombic systems where structure formations have been investigated[14, 15, 16, 17].

In Section 2, the interaction energy and the potential due to the ambient (background) plasma are described. The structure formation in the case of one-component system is shown in Section 3. Structures in the two-component system are given in Section 4 with emphasis on the separation of charges. Possible relative motion of different species is presented in Section 5. Concluding remarks are given in Section 6.

2. Interaction energy of Yukawa particles

2.1. Total interaction energy

We take z along the symmetry axis and express coordinates as $\mathbf{r} = (\mathbf{R}, z)$. The interaction between Yukawa particles with the charge $-Qe$ is given by

$$\frac{(-Qe)^2}{r}\exp(-r/\lambda), \tag{1}$$

where λ is the screening length. In the case of fine particle plasmas, the screening length is given by

$$\frac{1}{\lambda^2} = \frac{4\pi n_e e^2}{k_B T_e} + \frac{4\pi n_i e^2}{k_B T_i}, \tag{2}$$

where the number density, the charge, and the temperature of electrons and ions in the ambient plasma are denoted by (n_{e}, $-e$, T_{e}) and (n_{i}, e, T_{i}), respectively. The total interaction energy including particle-particle, particle-background, and background-background interactions is written in the form[13]

$$\int\int d\mathbf{r}d\mathbf{r}'\frac{\exp(-|\mathbf{r}-\mathbf{r}'|/\lambda)}{2|\mathbf{r}-\mathbf{r}'|}\rho(\mathbf{r})\rho(\mathbf{r}') - \frac{(Qe)^2}{2}\sum_{\mathrm{i}}\int\int d\mathbf{r}d\mathbf{r}'\frac{\exp(-|\mathbf{r}-\mathbf{r}'|/\lambda)}{|\mathbf{r}-\mathbf{r}'|}\delta(\mathbf{r}-\mathbf{r}_{\mathrm{i}})\delta(\mathbf{r}'-\mathbf{r}_{\mathrm{i}}). \tag{3}$$

Here $\rho(\mathbf{r})$ is the total charge density of Yukawa particles and the background. Since the latter charge density is $\rho_0 = (-e)n_{\mathrm{e}} + en_{\mathrm{i}}$ for $R < R_0$ and 0 otherwise, we have ($\theta(\cdot)$ being the step function)

$$\rho(\mathbf{r}) = \sum_{\mathrm{i}}(-Qe)\delta(\mathbf{r}-\mathbf{r}_{\mathrm{i}}) + \rho_0\theta(R_0 - R). \tag{4}$$

In (3), self-interactions formally included in the first term are subtracted.

2.2. Potential due to background with cylindrical symmetry

When the charge density of the ambient plasma is $\rho_b(\mathbf{r})$, its potential $\Phi(\mathbf{r})$ is given by

$$\Phi(\mathbf{r}) = \int d\mathbf{r}_1\frac{\rho_b(\mathbf{r}_1)}{|\mathbf{r}-\mathbf{r}_1|}\exp(-|\mathbf{r}-\mathbf{r}_1|/\lambda). \tag{5}$$

In our system with cylindrical symmetry, $\rho_b(\mathbf{r}) = \rho_b(R)$ and $\Phi(\mathbf{r}) = \Phi(R)$. For uniform distribution of the ambient plasma inside of R_0 with the density ρ_0, we have $\rho_b(R) = \rho_0\theta(R_0 - R)$ and

$$\Phi(R) - \Phi(R=0) = 4\pi\rho_0 R_0^2\phi(R/R_0; R_0'), \tag{6}$$

where

$$\phi(x; R_0') = \frac{K_1(R_0')}{R_0'}[1 - I_0(xR_0')] \qquad \text{for } x = \frac{R}{R_0} < 1, \tag{7}$$

$$\phi(x; R_0') = \frac{I_1(R_0')}{R_0'}K_0(xR_0') + \frac{K_1(R_0')}{R_0'} - \frac{1}{(R_0')^2} \qquad \text{for } x = \frac{R}{R_0} > 1, \tag{8}$$

$R_0' = R_0/\lambda$, and K_ν and I_ν are modified Bessel functions.

For Yukawa particles with the negative charge $-Qe$, the background works as a source of the confining potential $(-Qe)\Phi(R)$. The values of the function $-\Phi/4\pi\rho_0 R_0^2$ is shown in Fig.1. With the increase of the screening, the potential becomes shallower and flatter in accordance with the decrease of mutual repulsion between particles as in the case of uniformly filled spheres[13]. In the limit $\lambda \to \infty$, Φ reduces to the Coulombic case as

$$\Phi(R) - \Phi(R=0) = -\pi\rho_0 R^2 \qquad \text{for } R < R_0, \tag{9}$$

$$\Phi(R) - \Phi(R=0) = -\pi\rho_0 R_0^2[1 + 2\ln(R/R_0)] \qquad \text{for } R > R_0. \tag{10}$$

Since particles are distributed within the radius R_0 at low temperatures, the potential outside of R_0 has no effect on their structures.

2.3. Characteristic parameters

We denote the *linear* density of Yukawa particles by n. Due to the charge neutrality, we have

$$(-Qe)n + \pi R_0^2\rho_0 = 0 \tag{11}$$

and (6) is rewritten as

$$\Phi(R) - \Phi(R=0) = 4Qen\phi(R/R_0; R_0'). \tag{12}$$

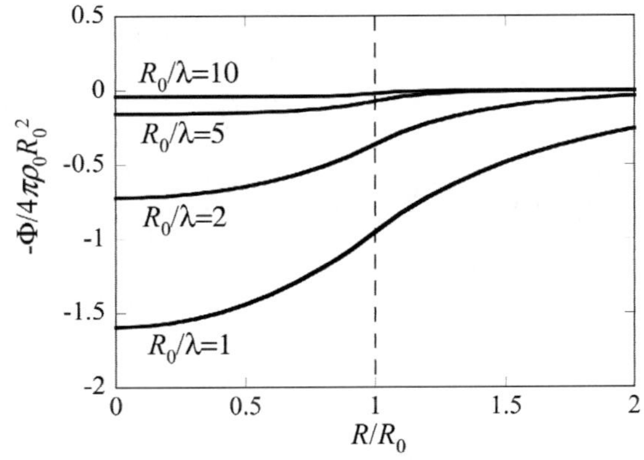

Figure 1. Potential for fine particles in the field of a uniform cylinder of neutralizing ambient (background) plasma with radius R_0.

Noting that $\rho_0/(Qe)$ is the average three-dimensional density of particles, we define the mean distance a by

$$a = \left(\frac{4\pi}{3}\frac{\rho_0}{Qe}\right)^{-1/3}. \tag{13}$$

The radius of the background R_0 is expressed by a as $R_0/a = (4/3)^{1/2}\,(na)^{1/2}$. With the mean distance a, we express the strength of screening by the parameter $\xi = a/\lambda$.

Rewriting the parameters in the function as $\phi[(R_i/a);\ (na)^{1/2},\ \xi]$, we have the potential energy of particles in the form

$$\frac{(Qe)^2}{a}\left[\frac{1}{2}\sum_{i\neq j}\frac{\exp(-\xi|\mathbf{r}_i-\mathbf{r}_j|/a)}{|\mathbf{r}_i-\mathbf{r}_j|/a} - 4na\sum_i \phi[(R_i/a);\ (na)^{1/2},\ \xi]\right]. \tag{14}$$

Our system at low temperatures is characterized by two independent dimensionless parameters and we take the set

$$(na,\ \xi), \tag{15}$$

the degrees of packing and screening. It will be shown that the structure is directly related to the former but is almost insensitive to the latter when scaled by the mean distance a.

Hitherto we have implicitly assumed that the system contains fine particles of a single species. Characteristic parameters in multi-component cases will be considered in sections 4.

3. Structures of one-component Yukawa and Coulomb systems in cylindrically symmetric background

3.1. Structure formation at low temperatures

In order to obtain low temperature structures of the system, we perform Monte Carlo simulations based on the usual Metropolice importance sampling. The periodic boundary condition is imposed along the z-axis and independent particles between $N = 200$ and $N = 1000$ are placed in the basic cell. We confirm these numbers are sufficient. In this section, we briefly summarize the results[18].

Some examples of the structures are shown in Fig.2 where particles are projected onto the xy-plane. Particles are organized into well-defined thin concentric shells. We number the shells

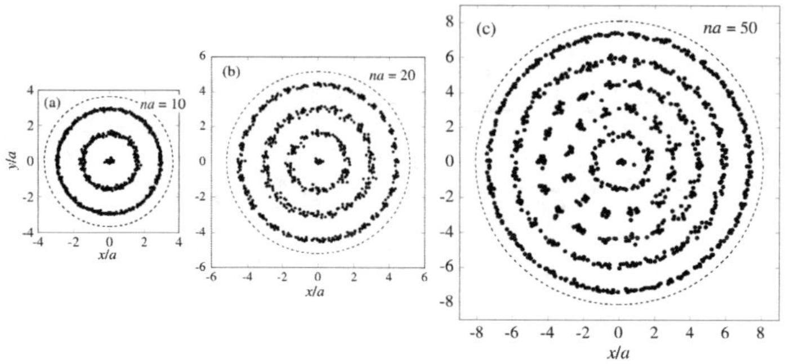

Figure 2. Examples of the distribution in the xy-plane. Dotted lines are $R = R_0$. (a) $na = 10$, $N = 800$, (b) $na = 20$, $N = 800$, and (c) $na = 49.3$, $N = 1000$.

by $m = 1, 2, \ldots$ from outside and denote their radii by R_m. On each shell, particles form a triangular lattice with defects as shown in Fig.3.

Because of the periodic boundary condition, we need to use larger number of independent particles in order to simulate the case of multiple layers accurately. We adopt the number of independent particles so that the length of the basic cell along the axis is long enough compared with the lattice constant of quasi-two-dimensional lattices on shells.

With the increase of na, the number of concentric shells increases. As is shown in Fig.4 in Ref.18, values of R_m/a ($m = 1, 2, \ldots$) linearly increase with $(na)^{1/2}$ and the spacing between them is almost constant. In addition, the value of ξ varies from 0.115 to 0.531 in these simulations but the structures are almost independent of the value of ξ. We note that the results of the Coulombic system[15] which correspond to the case of $\xi = 0$ are also covered by these ξ-insensitive values.

The surface density of particles on shells is plotted in Fig.5 in Ref.18 including the case of $\xi = 0$[15]. We observe that the surface density is almost common for various cases except for the innermost shell of radius smaller than a. The average (common) surface density for shells with radii larger than a is given by

$$\sigma a^2 \approx 0.352. \tag{16}$$

3.2. Summary of structure formation

With the decrease of the temperature, particles are organized into well-defined thin concentric cylindrical shells and, on each shell, particles form a triangular lattice with defects. Structures measured in the units of a are almost insensitive to the strength of screening. Shell radius has a linear $(na)^{1/2}$-dependence and the spacing is almost common when measured in a. The number of shells increases discretely with the increase of $(na)^{1/2}$ and the critical values for the appearance of new shells $(na)_m^{1/2}$ are linearly dependent on the number of shells. The surface density of particles on shells measured in a is almost common except for shells with small radius compared with a.

3.3. Interpolation formulas for structures

Based on the above observations, we express R_m/a in the form

$$R_m/a = c_0[(na)^{1/2} - (na)_m^{1/2}] \qquad (na)^{1/2} > (na)_m^{1/2}. \tag{17}$$

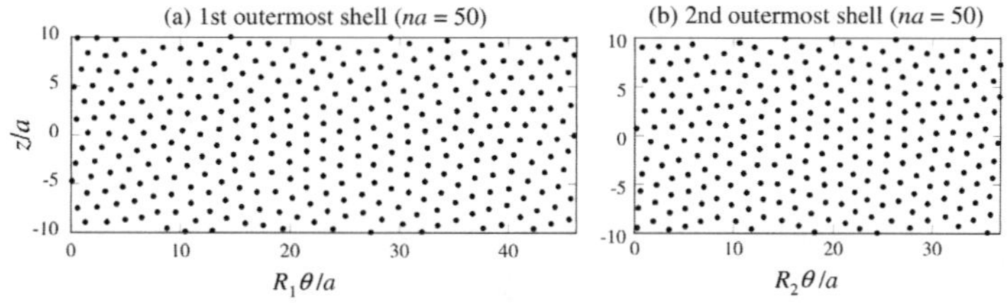

Figure 3. Examples of particle distribution in shells in the case of the structure Fig.2(c). (a) The first ($m = 1$) and (b) the second ($m = 2$) outermost shells. The abscissa is the azimuthal position, θ being the angle.

with $c_0 = 1.138$. The relative errors are within 2%. Since the critical values for the appearance of the shell m are expressed as

$$(na)_m^{1/2} = c_1 (m - c_2) \tag{18}$$

with constants $c_1 = 1.411$ and $c_2 = 0.501$, we have

$$R_m/a = 1.138\,(na)^{1/2} - 1.411\,(m - 0.501) \tag{19}$$

within relative error of 4%. By the common surface density of particle on shells, the number of particles on a shell of length L is given in the form

$$N_m = 2\pi\sigma R_m L = 2\pi(\sigma a^2)\frac{R_m}{a}\frac{L}{a} = 2\pi \cdot 0.352\,[1.138(na)^{1/2} - 1.411(m - 0.501)]\frac{L}{a}. \tag{20}$$

When we assume that the spacing between shells is a constant d, we have a heuristic formula for the positions of shells in the form

$$R_m/a = (4/3)^{1/2}(na)^{1/2} - (d/a)(m - 0.5). \tag{21}$$

Compared with this expression, the results of simulations summarized in the form (19) give 0.501 instead of 0.5, 1.138 instead of $(4/3)^{1/2} = 1.1547$, and

$$\frac{d}{a} = 1.411. \tag{22}$$

The formula (19) is thus consistent with the result of this heuristic reasoning.

As for the value of d/a, we have $(d/a)^{\rm cp} = 1.477$ in the three-dimensional closest packing (cp) structure. With this value, (21) takes the form

$$R_m/a = (4/3)^{1/2}(na)^{1/2} - 1.477(m - 0.5). \tag{23}$$

In the results of simulations (19), we have 1.411 which is close to 1.477. We observe that (23) reproduces results of simulations when m is small or in the case of outer shells as expected. The average surface density σa^2 from our simulation 0.352 is also close to the value for the lattice plane in the closest packing structure $(\sigma\, a^2)^{\rm cp} = 3^{1/6}/(2\pi)^{2/3} = 0.3527$. We may thus regard the local structure of one-component dust particles in cylindrical symmetry as being similar to the closest packing with equally spaced triangular lattices.

The structures are almost independent of the strength of screening also in the case of uniform spherical background where we have[13]

$$\frac{R_m}{a} = N^{1/3} - 1.474\,(m - 0.488), \tag{24}$$

Journal of Physics: Conference Series **327** (2011) 012037 doi:10.1088/1742-6596/327/1/012037

$$\frac{d}{a} = 1.474, \tag{25}$$

$$\sigma a^2 = 0.356, \tag{26}$$

and

$$N_m = 4\pi\sigma R_m^2 = 4\pi \cdot 0.356\,[N^{1/3} - 1.474\,(m - 0.488)]^2 \tag{27}$$

with the mean distance a assuming a uniform distribution in the sphere. The local structure is also close to the three-dimensional closest packing structure.

4. Structures of two-component Yukawa system in cylindrically symmetric background

The charge on a fine particle is approximately proportional to its radius and the electron temperature. When we have fine particles of different radii and therefore different charges, the parameter space for static structures has additional dimensions, the ratios of charges and their concentrations. (As far as the gravity has no significant effect, the static structures are generally independent of the mass of particles. In our case, the mass ratio is not independent of the charge ratio and does not compose a new dimension even with the effect of gravity.)

We denote the charge and linear number density of species α by $(-Q_\alpha e,\ n_\alpha)$. The condition of the charge neutrality is written as

$$\sum_\alpha (-Q_\alpha e) n_\alpha + \pi R_0^2 \rho_0 = \sum_\alpha (-Q_\alpha e) n_\alpha + (-e) n_{\mathrm{e}} + e n_{\mathrm{i}} = 0. \tag{28}$$

We restrict ourselves within the minimum multi-component system or two-component mixtures. Even in this case, we have three characteristic parameters, putting aside the strength of screening which is expected to be not effective, and the overall survey is difficult. We give some examples which may help us predict characteristic properties of multi-component systems.

4.1. Parameters for two-component system

We here analyze the systems of two species of particles, 1 and 2, with charges and linear densities $(-Q_1 e,\ n_1)$ and $(-Q_2 e,\ n_2)$, respectively. By the total linear density n,

$$n = n_1 + n_2, \tag{29}$$

we define the average charge on a fine particle $-\bar{Q}e$ and the mean distance a by

$$\bar{Q}e = \frac{Q_1 e n_1 + Q_2 e n_2}{n_1 + n_2}, \tag{30}$$

and

$$a = \left(\frac{4\pi}{3}\frac{\rho_0}{\bar{Q}e}\right)^{-1/3}, \tag{31}$$

respectively. Without loss of generality, we assume $Q_2 > Q_1$. Now the system has two more parameters, the charge ratio Q_2/Q_1 and the concentration ratio n_2/n_1, in addition to na and ξ. We thus have three effective characteristic parameters, na, Q_2/Q_1, and n_2/n_1.

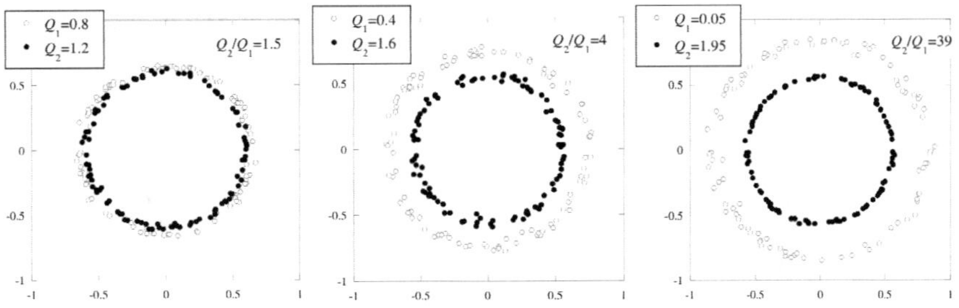

Figure 4. Examples of separation of shells by the magnitudes of charges. Larger charges form the shell of smaller radius. The ratio of densities is $n_2 = n_1$.

4.2. Separation of shells: Inner shell of larger charges and outer shells of smaller charges
We first analyze how the structure changes when the one-component system is continuously modified into the two-component system. When $na = 2.66$ ($nR_0 = 5$), the one-component system forms only one shell. Let us now increase the charges on half of the particles and decrease those of the rest keeping the average charge unchanged so that the background needs not to be changed. By large or small of charges we refers to their magnitudes irrespective of their signs.

Examples in Fig.4 indicate that larger charges move inward and smaller ones move outward forming two separate shells. The difference of the two radii depends on the ratio of charges as shown in Fig.5. The difference increases with the increase of $Q_2/Q_1 - 1$ starting from the linear dependence and, when $Q_2/Q_1 > 10$, the changes of both radii are almost saturated.

In Fig.6, we show the cases where the ratio of the number densities is changed in comparison with the one-component case composed of the average charge. The ratio n_2/n_1 is changed form 0.125 to 0.875. Larger charges are located inside of the corresponding one-component case and the radius becomes close to the latter when the ratio $n_2/n_1 > 1$. The radius of the shell of smaller charges, on the other hand, is larger than the radius in the one-component case and is almost independent of $n_2/n_1 - 1$.

Figure 7 shows examples which start from the multiple-shell structures of the one-component case where we have one, two, and three shells for $na = 2.66$ ($nR_0 = 5$), 12.3 ($nR_0 = 50$), and 19.6 ($nR_0 = 100$), respectively. Other parameters are $Q_2/Q_1 = 100$ and $n_2/n_1 = 1$. We observe that each shell in the one-component case is separated into two shells and we have alternate shells of large and small charges in the order of increasing radius. From the charge-ratio dependence shown in Fig.5, we expect to have similar structure also in the cases of smaller charge ratios.

Though our simulations are not exhaustive, we find no example where shells of the same charge are adjacent. In other words, we have always alternate shells of different charges within our exploration of the parameter space.

The results our simulations of two-component systems on static structures are summarized as follows. [A] Each shell in the one-component case is separated into the one of smaller radius with larger charges and the one of larger radius with smaller charges. [B] We have shells of one-by-one alternating charges in the order of increasing radius.

4.3. Origin of charge separation
The most important phenomenon characterizing structures of two-component systems is the separation of shells according to charges: Starting from the same shell, particles with increased (decreased) charges form a shell with smaller (larger) radius. In order to understand this separation, let us consider what happens when we introduce a small difference of charges into

Journal of Physics: Conference Series **327** (2011) 012037 doi:10.1088/1742-6596/327/1/012037

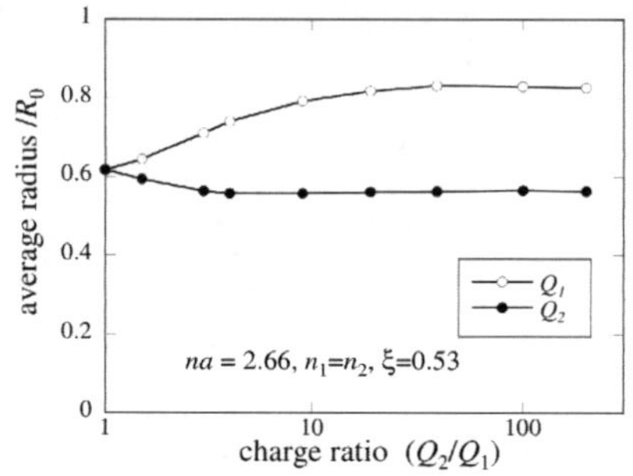

Figure 5. The separation of shells vs. charge ratio. The ratio of densities is $n_2 = n_1$.

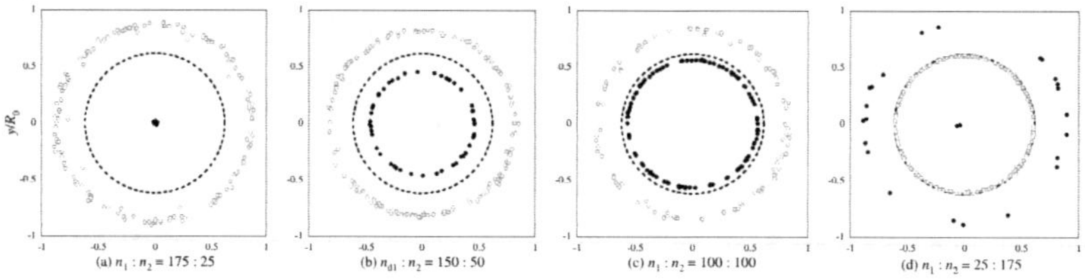

Figure 6. Separation of shells vs. number density ratio for $na = 2.66$ ($nR_0 = 5$) and $Q_2 = 100Q_1$. Thin dotted circle is the position of the shell in the one-component case.

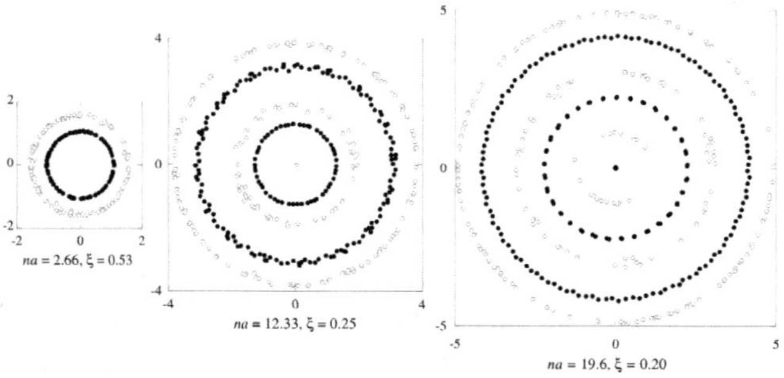

Figure 7. Separation of shells. The parameter na is changed from 2.66 to 19.6 ($n_2 = n_1$, $Q_2 = 100Q_1$).

particles which are in the equilibrium distribution of the one-component system. We slightly modify the charge of the i-th particle as

$$Q_i = -Q(1 + \Delta_i), \tag{32}$$

where

$$\Delta_i = \Delta \;\text{ or }\; -\Delta \quad (1 \gg \Delta > 0). \tag{33}$$

We assume

$$\sum_i \Delta_i = 0, \tag{34}$$

so that we do not need to change the density of the background.

Journal of Physics: Conference Series **327** (2011) 012037 doi:10.1088/1742-6596/327/1/012037

Due to these changes, the balance of forces acting on each particle will be lost. The force on the i-th particle is given by

$$\mathbf{f}_i = Q_i e(\mathbf{E}_{\Phi,i} + \sum_{j(\neq i)} Q_j e \mathbf{f}_{ij}). \tag{35}$$

Here $Q_i e \mathbf{E}_{\Phi,i}$ is due to the background and $Q_i Q_j e^2 \mathbf{f}_{ij}$ is due to the other particle j;

$$\mathbf{E}_{\Phi,i} = \mathbf{E}_{\Phi}(\mathbf{R}_i) = -\frac{\partial}{\partial \mathbf{R}_i}\Phi, \tag{36}$$

$$\mathbf{f}_{ij} = -\frac{\partial}{\partial \mathbf{r}_i}\frac{1}{r_{ij}}\exp(-r_{ij}/\lambda). \tag{37}$$

We rewrite (35) into the form

$$\mathbf{f}_i = Q_i e(\mathbf{E}_{\Phi,i} - Qe \sum_{j(\neq i)} \mathbf{f}_{ij}) + Q^2 e^2 (1+\Delta_i) \sum_{j(\neq i)} \Delta_j \mathbf{f}_{ij}. \tag{38}$$

Since we start from the equilibrium state of the one-component system, we have

$$\mathbf{E}_{\Phi,i} - Qe \sum_{j(\neq i)} \mathbf{f}_{ij} = 0 \tag{39}$$

and the first term of (38) vanishes. If the distribution of $\{\Delta_i\}$ is not mutually correlated, the second term also vanishes on the average and the replacement of charges has no effect on the radius of the shell.

Since we are considering the low temperature structures, the distribution of $\{\Delta_i\}$ and the positions of particles are determined so as to minimize the total potential energy

$$\sum_i Q_i e \Phi(R_i) + \frac{1}{2}\sum_{i\neq j} \frac{Q_i Q_j e^2}{r_{ij}} \exp(-r_{ij}/\lambda). \tag{40}$$

We consider the effect of $\{\Delta_i\}$ in two steps: We first distribute $\{\Delta_i\}$ keeping the positions of particles unchanged and then consider the effect of rearrangement of positions.

In the first step, the first term of (40) is not changed. We rewrite the second term into

$$\frac{1}{2}\sum_{i\neq j} \frac{Q^2 e^2}{r_{ij}} \exp(-r_{ij}/\lambda) + \sum_i \Delta_i \sum_{j(\neq i)} \frac{Q^2 e^2}{r_{ij}} \exp(-r_{ij}/\lambda) + \frac{Q^2 e^2}{2} \sum_{i\neq j} \frac{\Delta_i \Delta_j}{r_{ij}} \exp(-r_{ij}/\lambda). \tag{41}$$

Since the one-component system at low temperature can be regarded as a lattice,

$$\sum_{j(\neq i)} \frac{Q^2 e^2}{r_{ij}} \exp(-r_{ij}/\lambda) \tag{42}$$

is independent of i and the second term vanishes due to (34). In order to minimize the last term, the distribution of $\{\Delta_i\}$ has to be correlated so that $\Delta_i \Delta_j < 0$ for near neighbors. As a result, particles with positive Δ_i ('plus' particles) are surrounded by particles with negative Δ_i ('minus' particles) and vise versa. Since the shell has a small but finite curvature,

$$\sum_{j(\neq i)} \Delta_j \mathbf{f}_{ij} \tag{43}$$

tends to have the inward (outward) direction when i is a 'plus' particle ('minus' particle). The shell thus tends to be separated into the shell of larger charges with smaller radius and the one of smaller charges with larger radius in the first order in Δ. In the second step where positions of particles are relaxed, particles are reorganized. The changes, however, are of the order of Δ when normalized by the original positions and their effect to the results obtained in the first step is of the second order. In this way, we may understand the directions of the separation of radius of shells at least when Δ is small.

5. Two-component system: Dynamics

Fine particles in fine particle plasmas also serve as the important source of the information on dynamics of strongly coupled systems. Among many interesting dynamical aspects of strongly coupled charge systems, such as collective modes and transport coefficients, we here focus on the relative motion of different species of charges. In strongly coupled systems there exists a possibility that different components of charges keep their own structures and, at the same time, move relative to other components. This kind of motion may be characterized by the mobility, the friction coefficient, and the stopping power. The system with the cylindrical symmetry seems to fit particularly the observation of this phenomenon.

In two-dimensional systems, we can have the cases where different components are spatially separated. For example, we have systems of electrons (holes) in adjacent layers in semiconductor microstructures and the mutual friction between electrons (holes) in adjacent layers may be applied to control electronic devices. Our cylindrically symmetric system of two-component fine particles with separated shells might be regarded as a kind of two-dimensional system with spatially separated components.

The motion of fine particle against the rest of the system can be excited by irradiating the laser beam. The excitation, however, will be limited to a small group of particles and it seems to be difficult to have collective motions of some specific component in the system. In fine particle plasmas, we have systems with different species of particles. Different species of particles have different radius and therefore different charges and there occur a variety of structures even in the case of two components. We may therefore expect the mutual friction between different species may be directly observed in fine particle plasmas with multiple species.

Since fine particles are embedded in the ambient plasma, one may wonder how to excite the relative collective motion. Even if the direct (full) application of the electric field is difficult due to the screening effect by the ambient plasma, some small part of the applied electric field may penetrate and give different accelerations to different species of particles. One may also induce some flow of the neutral gas or the ambient plasma in order to give species-dependent forces. We here analyze the behavior of the two-component systems under species-dependent accelerations.

5.1. Equations of motion

The equation of motion for the particle i is written as

$$m_i \frac{d^2\mathbf{r}_i}{dt^2} = Q_i e \mathbf{E}_\Phi(\mathbf{R}_i) + Q_i e \sum_{j(\neq i)} Q_j e \mathbf{f}_{ij} + \mathbf{F}_i^{ext} - \nu_i m_i \frac{d\mathbf{r}_i}{dt}. \tag{44}$$

Here m_i is the mass, $\mathbf{F}_i^{ext}$ is the external force, and ν_i is the friction coefficient which is usually due mainly to collisions with neutral atoms. We assume the gas species is Ar and estimate the coefficient ν_i by[19, 20, 21]

$$\nu_i \sim c n_n (d_i/2)^2 \frac{m_n}{m_i} v_{th,n}, \tag{45}$$

where n_n is the number density of neutral atoms, $d_i/2$ the radius of the particle, $v_{th,n}$ the thermal velocity of neutral atoms, and c a factor of the order unity. Measuring the pressure in Pa and the temperature in eV and expressing the particle radius $d_i/2$ in μm, we have for Ar

$$\nu_i \sim 5.93 p_n[\mathrm{Pa}](d_i/2[\mu\mathrm{m}])^{-1} \left(\frac{T_n[\mathrm{eV}]}{0.03}\right)^{-1/2} \mathrm{s}^{-1}. \tag{46}$$

Here $c = 1.4 \times (8/3)(2\pi)^{1/2}$ is adopted[21].

5.2. Effect of axial electric field

Let us assume that the neutral gas is Ar with $T_n = 0.03$eV and typically $p_n \sim 10$Pa and consider a mixture with $d_1/2 = d_0/2 = 1\mu$m for the species 1 and $d_2/2 = 10(d_0/2)$ for the species 2. The mobility is inversely proportional to the neutral gas pressure and we can scale the results accordingly. At low temperatures and under appropriate conditions, particles form two separate shells, outer shell of charge 1 and inner shell of charge 2.

When the external electric field is applied along the axis, two species of particles feel the forces of different magnitudes. They also have different coefficients for the friction in the motion among the neutral gas. Since particles are strongly coupled, each species are expected to move as a solid at least when the external field is sufficiently small. When the external field is sufficiently large, on the other hand, the difference in the forces acting on different species may lead either to a relative motion between different species with each structure kept (as in the case of weak field), or a destruction of the structure of either (or both) species. We analyze the response of two-component systems by molecular dynamics simulations.

5.3. Mobility for relative motion and formation of 'dust screw'

The results of simulations are summarized in Figs.8, 9, and 10. With the increase of the axial electric field, the relative velocity between the charges 2 and 1 increases. The dependency is locally linear and we may define a kind of mobility by the ratio of the z-component of the relative velocity v_{rel} to the z-component of the electric field E_z as

$$\mu_{rel} = \frac{v_{rel}}{E_z}. \tag{47}$$

For small electric field, the mobility seems to have small dependence on the structure. The value is given by

$$\mu_{rel} \approx \frac{9.2}{p_n[\mathrm{Pa}]}\mathrm{cm}^2/\mathrm{Vs}. \tag{48}$$

When the electric field (or the relative velocity) exceeds some critical value, the mobility increases stepwise for a structure while unchanged for other structures. the increased value is given by

$$\mu_{rel} \approx \frac{18.3}{p_n[\mathrm{Pa}]}\mathrm{cm}^2/\mathrm{Vs}. \tag{49}$$

We note that these values are much smaller than those found in two-dimensional systems of electrons or holes in semiconductor microstructures such as the inversion layers used in electronic devices.

In the case of stepwise increase, we observe that, corresponding to this increase of the mobility, the structure changes into a kind of 'screw' as shown in Fig.9. Figure 10 shows the motion of particles projected onto the xy-plane. The applied electric field thus induces the rotations of charges 1 and 2 in opposite directions. It is clear that these rotations help to have larger relative velocity under a given electric field. One might be tempted to understand these rotations in opposite directions in terms of the conservation of the total angular momentum. Fine particles, however, collide with neutral atoms and the total angular momentum is not conserved in our system.

6. Concluding remarks

We have analyzed structures and dynamics of fine particles in fine particle plasmas with the cylindrical symmetry. Such a system of high symmetry is expected to be realized in a long discharge chamber under the condition of microgravity.

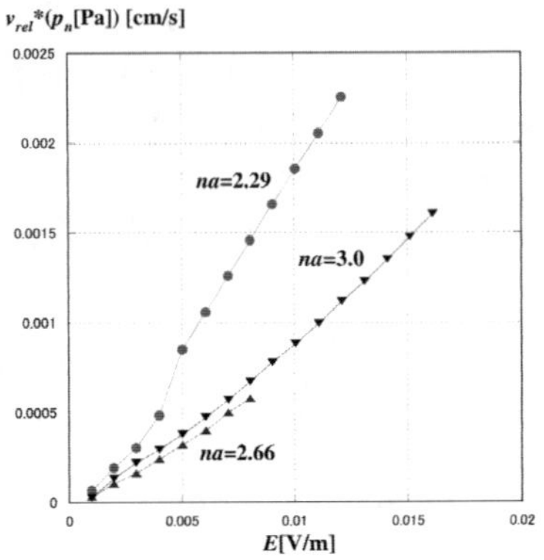

Figure 8. Relative velocity vs. electric field. The change in the slope corresponds to the structural change shown in Fig.9. Values multiplied by p_n[Pa] are shown.

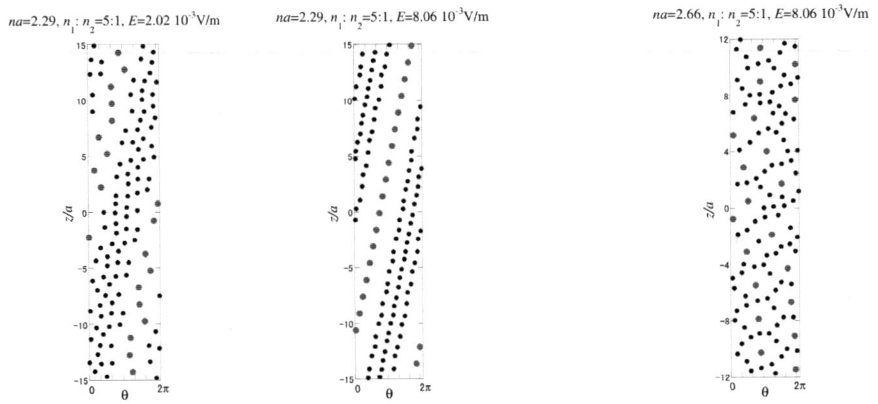

Figure 9. The structural change shown in the $(\theta,\ z)$-plane. In the case of $na = 2.29$, the 'screw' structure is formed when the electric field is increased, changing from the one in the left panel to the one in the center panel. In the case of $na = 2.66$ (and $na = 3.0$), we have no such change as shown in the right panel.

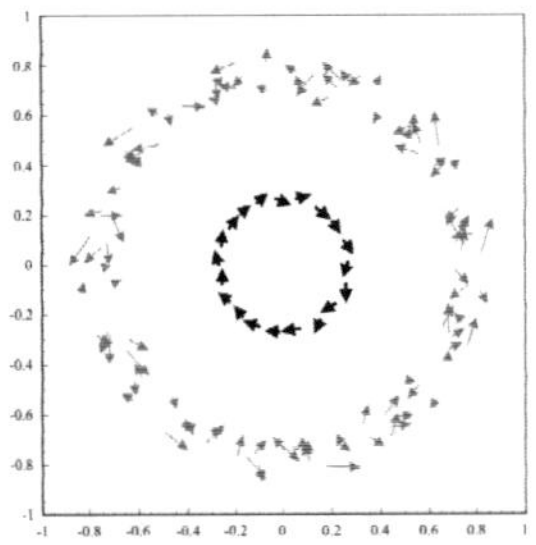

Figure 10. Velocity of particles projected onto the xy-plane. Particles in the inner shell and those in the outer shell rotate in opposite directions. The magnitudes of velocities of particle in the inner shell are magnified by the factor 4 for clarity.

In the case of one species, well-defined concentric shells are formed at low temperatures and the structures are expressed by simple but accurate interpolation formulas covering the case of Coulomb particles. In the case of two components, shells of the one-component system are separated by charges. Larger and smaller charges form shells of smaller and larger radii, respectively, and an explanation for this separation is given.

As for dynamics, the relative motion of separated shells in the two-component system is analyzed. It is shown that, by applying an appropriate force along the axis of the cylinder, the relative motion of separated shells is induced and we can define the mobility for the relative motion. An example of the formation of 'screw' by fine particles and corresponding reduction of mutual friction is shown.

In these analyses, some simplifications are made. For example, the density of the ambient (background) plasma is assumed to be uniform in a cylinder and the mutual interaction between fine particles are expressed by the Yukawa potential. We expect there exist the domain of parameters where these simplifications are allowed and the results obtained here are expected to serve as a basis of detailed investigations.

Acknowledgments
The authors thank Professor G. E. Morfill, Dr. H. M. Thomas, Dr. M. H. Thoma, Professor V. E. Fortov, Dr. V. I. Molotkov, Dr. A. M. Lipaev, and the research group at the Max-Planck Institute for Extraterrestrial Physics for a lot of useful information on microgravity experiments. Thanks are also due to Dr. S. Adachi, Dr. K. Takahashi, Dr. M. Takayanagi, and members of our Working Group of the Japan Aeronautic Exploration Agency (JAXA). This work has been partly supported by the Grant-in-Aid for Scientific Research (C) from the Japan Society for Promotion of Sciences Nos. 19540521 and 21540512.

[1] J. H. Chu and Lin I, Physica A **205**, 183(1994); J. H. Chu and Lin I, Phys. Rev. Lett. **72**, 4009(1994).
[2] H. Thomas, G. E. Morfill, V. Demmel, J. Goree, B. Feuerbacher, and D. Mohlmann, Phys. Rev. Lett. **73**, 652(1994).
[3] Y. Hayashi and K. Tachibana, Jpn. J. Appl. Phys., Part 2 **33**, L804(1994).
[4] A. Melzer, T. Trottenberg, and A. Piel, Phys. Lett. A **191**, 301(1994).
[5] H. Totsuji, T. Kishimoto, and C. Totsuji, Phys. Rev. Lett. **78**, 3113(1997).
[6] For example, V. Fortov, G. Morfill, O. Petrov, M. Thoma, A. Usachev, H. Hoefner, A. Zobin, M. Kretschmer, S. Ratynskaia, M. Fink, K. Tarantik, Yu. Gerasimov, and V. Esenkov, Plasma Phys. Control. Fusion **47**, B537(2005).
[7] H. M. Thomas, G. E. Morfill, V. E. Fortov, A. V. Ivlev, V. I. Molotkov, A. M. Lipaev, T. Hagl, H. Rothermel, S. A. Khrapak, R. K. Suetterlin, M. Rubin-Zuzic, O. F. Petrov, V. I. Tokarev, and S. K. Krikalev, New J. Phys. **10**, 033036(2008).
[8] S. Mitic, B. A. Klumov, U. Konopka, M. H. Thoma, and G. E. Morfill, Phys. Rev. Lett. **101**, 125002(2008).
[9] For example, S. Rice, Chem. Phys. Lett. **479**, 1(2009).
[10] For example, D. H. Van Winkle and C. A. Murray, Phys. Rev. A **34**, 562(1986).
[11] S. Hamaguchi and R. T. Farouki, J. Chem. Phys. **101**, 9876(1994).
[12] Y. Rosenfeld, Phys. Rev. E**49**, 4425(1994).
[13] H. Totsuji, C. Totsuji, T. Ogawa, and K. Tsuruta, Phys. Rev. E **71**, 045401(R)(2005). H. Totsuji, T. Ogawa, C. Totsuji, and K. Tsuruta, Phys. Rev. E **72**, 036406(2005). T. Ogawa, H. Totsuji, C. Totsuji, and K. Tsuruta, J. Phys. Soc. Japan **75**, 123501(2006).
[14] D. H. E. Dubin and T. M. O'Neil, Rev. Mod. Phys. **71**, 87(1999).
[15] A. Rahman and J. P. Schiffer, Phys. Rev. Lett. **57**, 1133(1986).
[16] H. Totsuji and J.-L. Barrat, Phys. Rev. Lett. **60**, 2484(1988).
[17] R. W. Hasse and J. P. Schiffer, Annal. Phys. **203**, 419(1990).
[18] H. Totsuji and C. Totsuji, Phys. Rev. E **84**, 015401(R)(2011).
[19] E. M. Lifshitz and L. P. Pitaevskii, *Physical Kinetics*, Pergamon Press, 1981, section 12.
[20] P. S. Epstein, Phys. Rev. **23**, 710(1924).
[21] S. A. Khrapak, A. V. Ivlev, and G. E. Morfill, Phys. Rev. E **70**, 056405(2004).

Effects of Soot Formation on Shape of a Nonpremixed Laminar Flame Established in a Shear Boundary Layer in Microgravity

Hui Ying WANG[a], Juan Luis Florenciano Merino[a] and Philippe Dagaut[b]

a. Institut Pprime (CNRS-UPR 3346)
Département Fluide-Thermique-Combustion
ENSMA - BP 40109
Téléport 2, 1 Av Clément ADER
F-86961 FUTUROSCOPE CHASSENEUIL Cedex, France
b. CNRS – INSTII, ICARE, 1c, Ave de la Recherche Scientifique
45071 Orléans Cedex 2, France

wang@ensma.fr

Abstract. A numerical study was performed to give a quantitative description of a heavily sooting, nonpremixed laminar flame established in a shear boundary layer in microgravity. Controlling mechanisms of three dimensional flow, combustion, soot and radiation are coupled. Soot volume fraction were predicted by using three approaches, referred respectively to as the fuel, acetylene and PAH inception models. It is found that the PAH inception model, which is based on the formation of two and three-ringed aromatic species, reproduces correctly the experimental data from a laminar ethylene diffusion flame. The PAH inception model serves later to better understand flame quenching, flame stand-off distance and soot formation as a function of the dimensionless volume coefficient, defined as $C_q=V_F/V_{ox}$ where V_F is the fuel injection velocity, and V_{ox} air stream velocity. The present experiments showed that a blue unstable flame, negligible radiative feedback, may change to a yellow stable flame, significant radiative loss with an increase of C_q; this experimental trend was numerically reproduced. The flame quenching occurs at the trailing edge due to radiative heat loss which is significantly amplified by increasing V_F or decreasing V_{ox}, favouring soot formation. Along a semi-infinite fuel zone, the ratio, d_f/d_b, where d_f is the flame standoff distance, and d_b the boundary layer thickness, converges towards a constant value of 1.2, while soot resides always within the boundary layer far away from the flame sheet.

1. Introduction

The most likely scenario of a flame incidentally initiated and spread on board of a spacecraft is a wall of combustible material adjacent to a low Reynolds number gas flow (forced by the venting system) and exposed to a critical ignition source. As a major portion of heat released from the igniting stage of combustion is transported by forced convection in the flow direction, the superior mode of spread is of co-current (forward) type. Under conditions of reduced gravity in spacecraft where the effects of buoyancy are small, due to the absence of natural convection, radiation becomes the predominant mode of heat transfer even for a small forward diffusion flame [1]. A proper evaluation of material flammability requires a detailed understanding of the effect of oxidizer flow velocity on local soot concentrations, to which the structure of the flow field in the vicinity of the flame connects. Konsur et al. [2] have attempted to describe the sooting behaviour of non-buoyant jet diffusion flame by using the Smoke-Point concept proposed originally by Markstein et al. [3]. Co-current flame spread over a solid plate could then be linked to a critical soot concentration [4] controlling the trailing edge quenching. Despite the practical utility of this approach, the local soot concentrations have not been comprehensively described. Several experimental investigations [5, 6] have demonstrated how changes in blowing affect soot formation and thus, flame structure in the absence of natural convection. Numerical simulations have been performed [7] to determine two regimes (attached and detached flows) of a low Reynolds number flow in the absence of natural convection, function of blowing and thermal expansion, but without radiation due to lack of adequate soot formation model.

Up to now, the exact elementary reactions and species, and chemical processes of soot formation are not fully understood. Currently, for predicting local soot concentration, extensive efforts are devoted in 3D CFD codes to the semi-empirical soot models [8-11], involving the inception, coagulation, surface growth and oxidation processes. Non-buoyant ethylene laminar jet diffusion flames have been modelled by assuming the nucleation and growth of soot are first-order functions of acetylene concentrations [8, 9]. Recent research suggests that the growth of aromatics is the rate limiting step rather than the formation of the first ring [10]. Vovelle et al. [11] showed that in decane flames, aromatic hydrocarbons are formed from acetylene. Anderson et al. [12] has shown that even a small quantity of aromatic hydrocarbons in the fuel can significantly increase soot emissions. Hall et al. [13] in their study of laminar methane flame, developed a soot inception rate based on the formation of two and three-ringed aromatic species from single ring aromatic species using a simplified reaction mechanism. While no guarantees are given that validation of the soot models [8-13] applies to a co-current, boundary type laminar flame under microgravity. Thus this paper tends to evaluate three approaches for soot formation, referred respectively to as the fuel, acetylene and PAH inception models, in such flame. For this, a detailed kinetic reaction mechanism [14] is required for providing the local concentration of soot precursory species, such as acetylene (C_2H_2), benzene (C_6H_6) and phenyl (C_6H_5). The measured soot volume fraction under microgravity conditions, attained by means of parabolic flights [5,6], serves to validate the numerical models.

2. Theoretical model

In order to conform as much as possible to the experimental configuration [5, 6], the starting point of the analysis is the set of three-dimensional elliptic, reacting flow equations that governs the phenomena of interest here. This set consists of the conservation equations of mass, momentum, energy and species. In a heavily sooting flame such as fire, as the radiation spectrum of soot is continuous, it is possible to assume that the gas behaves as a gray medium, and the radiative transfer equation without scattering is solved by using a discrete expression adapted to a Finite Volume Method. The calculation of the gray absorption coefficients, κ, of a sooting diffusion flame (mixture of soot, CO_2 and H_2O) is based on the work of Grosshandler (RadCal) [15]. The simulations were performed using the Fire Dynamics Simulator (FDS) software [16]. The precise formulation of the differential equations describing the model and the numerical technique can be found elsewhere [16], and will not be repeated here. For such low Reynolds number reacting flow, comparison between the numerical results and experimental standoff distances to validate the accuracy of FDS code has been presented in detail by Rouvreau et al. [7]. In this study, for investigating quenching of a laminar flame due to radiation losses, soot models were implemented within the FDS code by the authors.

2.1. Combustion and soot formation

In gaseous combustion, the chemical compositions are found by using an appropriate reaction mechanism. For providing the local concentration of soot precursory species, such as acetylene (C_2H_2), benzene (C_6H_6) and phenyl (C_6H_5), a detailed kinetic reaction [14], including 736 reversible

Journal of Physics: Conference Series **327** (2011) 012038 | doi:10.1088/1742-6596/327/1/012038

reactions and 99 species has to be used to model the combustion for ethylene/air. This reaction mechanism consists of a set of elementary reactions with corresponding rate parameters. The rate coefficient for reaction is calculated from an Arrhenius expression. These rate parameters, together with the elementary reactions and their stoichiometric coefficients, constitute the reaction mechanism. The rate laws for the species can be calculated by using a time consuming stiff ODE solvers. However, any attempt to use Arrhenius expression for a detailed kinetic [14] coupled with a three dimensional elliptic, reacting flow is discouraged due to a problem of computational cost. In this study, the reaction mechanism of NO_x, included in the detailed kinetic [14], is not taken into account. This treatment does not affect the soot precursory species (C_2H_2, C_6H_6, C_6H_5, etc) due to lack of the interaction between the NO_x formation and the reaction of other species.

In combustion, the mixture fraction is a conserved quantity traditionally defined as the (mass) fraction of the gas mixture that originates in the fuel stream. The mixture fraction is a function of space and time, commonly denoted Z(x,t). Therefore, many combustion models employ a detailed, cost effective kinetic reaction by tabulating the kinetics in terms of mixture fraction and only the mixture fraction is tracked in the calculation [17]. This implies that all species of interest of combustion products can be described in terms of the mixture fraction alone. The correspondence between the mass fraction of an individual species and the mixture fraction is called its "state relation". This state relation can be obtained by using the OPPDIF (Opposed Diffusion Flames) computer code [18] which is performed by using subroutines from the chemical program library Chemkin.

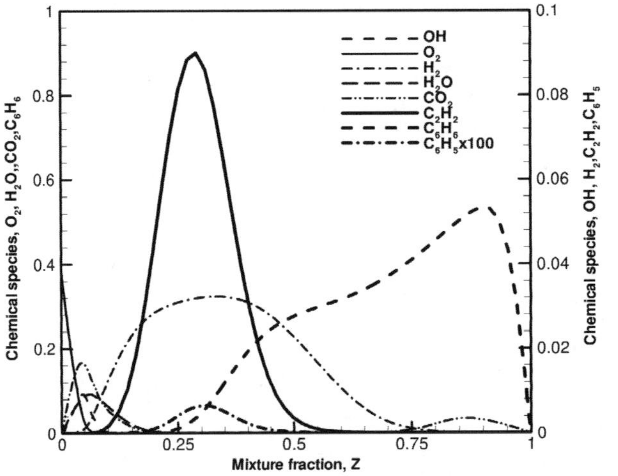

Figure 1. State relation between chemical species and mixture fraction

For obtaining such state relation for combustion of fuel/air, OPPDIF [18] requires the chemical and physical input parameters, such as the fuel type, the initial temperature value of fuel and oxygen, the initial value of fuel and oxygen mass fraction, pressure and strain rate. While this state relation is independent of reactant and oxidizer flow rates. As an illustration, dependence of the main chemical species on mixture fraction is presented in Figure 1. These chemical compositions profiles for combustion of ethylene /air as a function of the mixture fraction (cf. Fig.1) correspond to the temperature (T_F=T_{OX}=300 K), pressure (P=1 atm) and composition (X_{O2}= 0.35) found in the experiment [6]. A small value of the strain rate (<0.0003) which is much lower than the critical extinction value, is assigned, corresponding to the chemical equilibrium conditions. By looking at the relative locations of the chemical species as a function of the mixture fraction, high amounts of the phenyl (C_6H_5), acetylene (C_2H_2) and hydrogen (H_2) take place in fuel-rich region around a mixture fraction value of 0.2. It is found that an abundant benzene is formed around a mixture fraction of 0.9, and maximum of the benzene mass fraction attains up to 0.6. At the start of a simulation, the detailed chemical species involved are tabulated as a function of mixture fraction. During the simulation, the local value of all the species is found by table-lookup according to the local mixture fraction value [16].

The two time-dependent conservation equations for soot particles number, N (particles/m^3) and soot mass density, M (kg/m^3) are incorporated in FDS [16].

$$\frac{\partial \rho N}{\partial t} + \frac{\partial [\rho N(u_i + u_i^{th})]}{\partial x_j} - \frac{\partial}{\partial x_j}\left(\frac{\mu}{Sc}\frac{\partial N}{\partial x_j}\right) = \dot{\omega}_{N,inc} + \dot{\omega}_{N,coa} \quad (1)$$

$$\frac{\partial \rho M}{\partial t} + \frac{\partial [\rho M(u_i + u_i^{th})]}{\partial x_j} - \frac{\partial}{\partial x_j}\left(\frac{\mu}{Sc}\frac{\partial M}{\partial x_j}\right) = \dot{\omega}_{M,inc} + \dot{\omega}_{M,gro} + \dot{\omega}_{M,oxi} \quad (2)$$

where u_i^{th} is the thermophoretic velocity components :

$$u_i^{th} = -0.54\frac{\mu}{\rho}\frac{\partial \ln T}{\partial x_i} \quad (3)$$

The first term in Eq.(1) represents the rate of soot particle inception and the second the soot coagulation. The second term in Eq.(2) represents the surface growth of soot, and the last the soot oxidation rate. Three approaches, referred respectively to as the fuel, acetylene and PAH inception models, are tested.

2.1.1. *Fuel inception model.*

The first approach is developed by Moss et al. [19], which incorporates the essential physical process of soot inception, coagulation and surface growth in terms of the local temperature, T, and the mole fraction of fuel (ethylene) gas, X_F.

$$\dot{\omega}_{N,inc} = C_\alpha N_A \rho^2 T^{1/2} X_F e^{-T_\alpha/T} \quad (4)$$

$$\dot{\omega}_{N,coa} = -\frac{C_\beta}{N_A} T^{1/2} N^2 \quad (5)$$

$$\dot{\omega}_{M,inc} = \frac{C_\delta}{N_A}\dot{\omega}_{N,inc} \quad (6)$$

$$\dot{\omega}_{M,gro} = C_\gamma \rho T^{1/2} X_F e^{-T_\gamma/T} N^{1/3} M^{2/3} \quad (7)$$

2.1.2. *Acetylene inception model*

In the second approach, a conventional soot inception model, for which acetylene (C_2H_2) is assumed the only soot precursor, is proposed by Leung et al. [9] via a simple two-step soot chemistry scheme.

$$C_2H_2 \rightarrow 2C_s + H_2 \quad (8)$$

$$C_2H_2 + nC_s \rightarrow (n+2)C_s + H_2 \quad (9)$$

The soot inception rate is expressed as,

$$\dot{\omega}_{N,inc} = c_1 N_A \left[\rho \frac{Y_{C_2H_2}}{W_{C_2H_2}}\right] e^{-\frac{21100}{T}} \quad (10)$$

Journal of Physics: Conference Series **327** (2011) 012038 doi:10.1088/1742-6596/327/1/012038

where c_1 is a scaling factor, given as c_1=54 s^{-1}, and T the gas temperature. The soot coagulation rate is determined by Lindstedt [20] based on the collision frequency defined by Puri et al.[21].

$$\dot{\omega}_{N,coa} = -\left(\frac{24R}{\rho_{soot} N_A}\right)^{1/2} \left(\frac{6}{\pi\rho_{soot}}\right)^{1/6} T^{1/2} M^{1/6} N^{11/6} \tag{11}$$

$$\dot{\omega}_{M,inc} = \frac{M_P}{N_A}\dot{\omega}_{N,inc} \tag{12}$$

The surface growth of soot is proposed by Frenklach et al. [22].

$$\dot{\omega}_{M,gro} = c_4 \left(\rho \frac{Y_{C_2H_2}}{W_{C_2H_2}}\right) e^{-\frac{12100}{T}} \left[(\pi N)^{1/3} \left(\frac{6M}{\rho_{soot}}\right)^{2/3}\right] \tag{13}$$

where c_4 is a scaling factor, given as c_4=9000.6 kg.m.kmol^{-1}.s^{-1}, M_p $(=12011\,\text{kg/kmol})$ molar mass of a soot nucleus, ρ_{soot} density of a soot particle, R the universal gas constant and W_i molecular weight of species, i.

2.1.3. *PAH inception model*

Several theoretical and experimental works [10,11,13], highlight the importance of poly-aromatics species (PAH) as result of intermediary reactions in soot inception processes. Hall et al. [13] developed a soot inception model, which is based on the formation rates of two and three ringed aromatics ($C_{10}H_7$ and $C_{14}H_{10}$) from acetylene (C_2H_2), benzene (C_6H_6) and the phenyl radical (C_6H_5). The growth mechanism of PAH is composed of consecutive repeated reactions that can be classified in two steps mechanism, called HACA (H-Abstraction-C_2H_2-Addition). The first one is an abstraction mechanism of hydrogen.

$$2C_2H_2 + C_6H_5 \rightarrow C_{10}H_7 + H_2 \tag{14}$$

The second one is the addition reaction of acetylene.

$$C_2H_2 + C_6H_6 + C_6H_5 \rightarrow C_{14}H_{10} + H_2 + H \tag{15}$$

Based on the laminar flame data, the resulting inception rate is expressed as,

$$\dot{\omega}_{inc} = \frac{8N_A}{M_p}\left\{c_2\left[\rho^2\left(\frac{Y_{C_2H_2}}{W_{C_2H_2}}\right)^2 \frac{Y_{C_6H_6}W_{H_2}}{W_{C_6H_6}Y_{H_2}}\right]e^{-\frac{4378}{T}} + c_3\left[\rho^2 \frac{Y_{C_2H_2}Y_{C_6H_6}Y_{C_6H_5}W_{H_2}}{W_{C_2H_2}W_{C_6H_6}W_{C_6H_5}Y_{H_2}}\right]e^{-\frac{6390}{T}}\right\} \tag{16}$$

where $c_2 = 127\times10^{8.88}$ and $c_3 = 178\times10^{9.50}$. The rates for coagulation and surface growth are identical to these in the acetylene inception model (section 2.1.2).

2.1.4. Soot Particles Oxidation

The soot oxidation in Eq.(2) is assumed to proceed through two reaction steps,

$$\frac{1}{2}O_2 + C_s \rightarrow CO \tag{17}$$

$$C_s + OH \rightarrow CO + H \qquad (18)$$

Lee et al. [23] measured and modelled the O_2 and OH dependence of the soot oxidation in a laminar diffusion flame. By assuming that the kinetics of surface reactions are the limiting mechanism, the specific rate of soot oxidation is expressed by,

$$\dot{\omega}_{M,oxi} = -c_5 \tau \rho \frac{Y_{OH}}{W_{OH}} T^{1/2} (\pi N)^{1/3} \left(\frac{6M}{\rho_{soot}} \right)^{2/3} - c_6 \rho \frac{Y_{O_2}}{W_{O_2}} T^{1/2} (\pi N)^{1/3} \left(\frac{6M}{\rho_{soot}} \right)^{2/3} e^{-\frac{19778}{T}} \qquad (19)$$

where $c_5 = 105.81 \text{kg.m.kmol}^{-1}.K^{-1/2}.s^{-1}$, $c_6 = 18903.51 \text{kg.m.kmol}^{-1}.K^{-1/2}.s^{-1}$ and a value of 0.13 is assigned to the collision efficiency, τ.

2.2 Numerical technique

Low speed solvers used in FDS [16] explicitly eliminate compressibility effects that give rise to acoustic (sound) waves for low-Mach number flows (<0.3). The Boussinesq approximation invoked – the fluid is still considered thermally-expandable. Due to the use of the low Mach number approximation, the divergence of the flow plays a very important role in the overall solution scheme. The temperature is extracted from the density via the equation of state. The fidelity of the numerical solution of the entire system of equations is tied to the pressure/velocity coupling thanks to a direct Poisson solver [16]. All spatial derivatives are approximated by second-order finite differences and the flow variables are updated in time using an explicit second-order predictor-corrector scheme. The convective terms are written as upwind-biased differences in the predictor step and downwind-biased differences in the corrector step. The thermal and material diffusion terms are pure central differences in both the predictor and corrector steps [16]. In this study, the fuel temperature of 300 K and the atmospheric pressure (P=1 atm) are imposed at the porous burner boundaries, conforming to the experimental conditions [5, 6]. At the free boundaries, zero gradient conditions are used for the farfield boundary values of the mass fractions.

3. Results and discussions

The sketch of the experimental setup [5, 6] with the relevant burner dimensions is displayed in Fig.2(a,b). In the experiment, microgravity conditions are attained by means of parabolic flights that provide a gravity level of 10^{-3}g during 22 seconds. The experimental set-up corresponds to a square porous burner (L_p=W_p=5 cm, cf. Fig.2b) which is embedded in the plane at z=0, centred in the y direction and at 5 cm from the inlet of a large enclosed combustion chamber. Porous burners have been used regularly in an attempt to simplify experiments by avoiding the coupling between heat feedback from the flame and fuel supply so that longer experimentation time can be achieved. The pyrolysis products of a condensed fuel are simulated by injecting ethylene with a constant velocity, V_F, through a porous burner. The dimensions of the computational domain, as shown in Fig.2(a,b), are 8Lp in the windward direction, and 4Lp in both the lateral and normal directions so that boundary conditions can be set to conform as much as possible to the experimental configuration. The grid system contains 240 x 80 x 60 cells, and cells of 1 mm are concentrated around the reacting zone. The resolution tests indicated that the computations are grid independent when the cell dimensions are lower than 2 mm. Thus this grid system is considered as the best trade-off between accuracy and cost for a three-dimensional reacting flow simulation. A further reduction in the grid size results in a significant reduction in the time step (Δt <0.0005 s) for satisfying the CFL stability condition.

As the experiment, a uniform oxidizer flow at the inlet with a composition of 35% oxygen and 65% nitrogen flows parallel to the surface with a constant velocity, V_{ox}. This type flame can be described by the ratio between buoyant forces and inertia, defined as follows, $\xi = \frac{Gr_x}{Re_x^{5/2}}$, where Re_x is the Reynolds number and Gr_x the Grashof number [24]. The experimental work [24] shows that if the magnitude of the forced flow is small, an almost vertical plume is formed (cf. Fig.3a), indicating a predominant natural convection ($\xi > 1$). While microgravity reduces buoyancy enough to enable a much lower ratio ($\xi \leq 1$) even for a low Reynolds number flow (Re<100), inducing a boundary layer diffusion flame (cf. Fig.3b). From Fig.4 for the velocity vectors in microgravity, it appears that energy release from an exothermic chemical reaction enhances the flow perturbations at the leading edge enough to enable an establishment of a parabolic nature boundary layer behind of that region. A low Reynolds number flow (Re<100) is maintained all along the flame and parallel to the surface. The velocity overshoots in the line where fuel and oxidizer are delivered in stoichiometric proportions as a consequence of thermal expansion and interaction of the fuel and oxidizer flows.

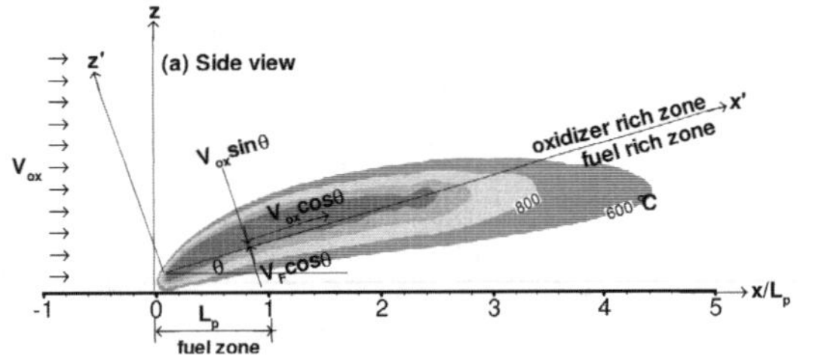

a) calculated iso-contours of the temperature (side view)

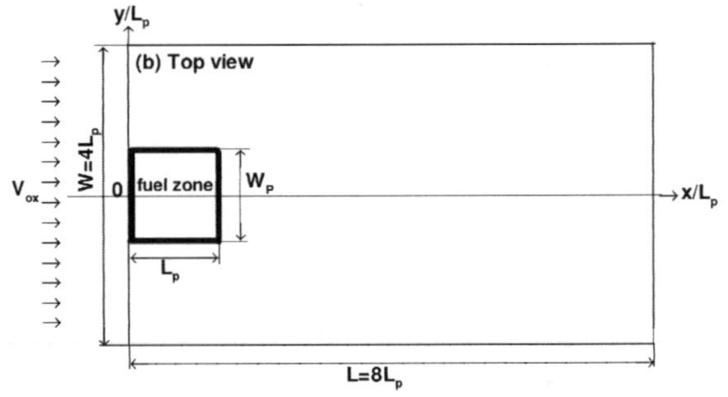

b) disposition of the fuel zone (top view)

Figure 2. Computational domain and the coordinate system

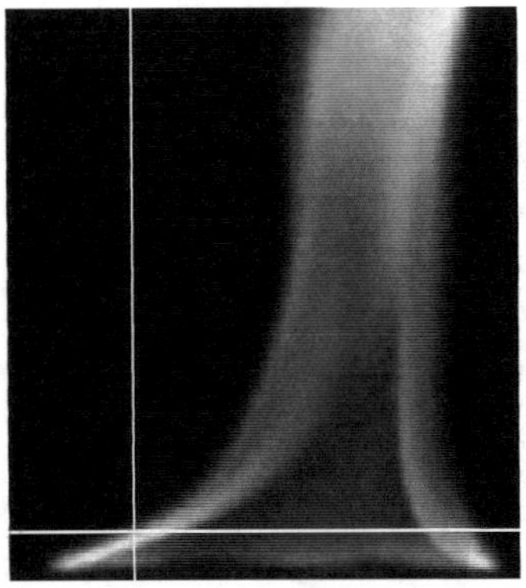

a) normal gravity

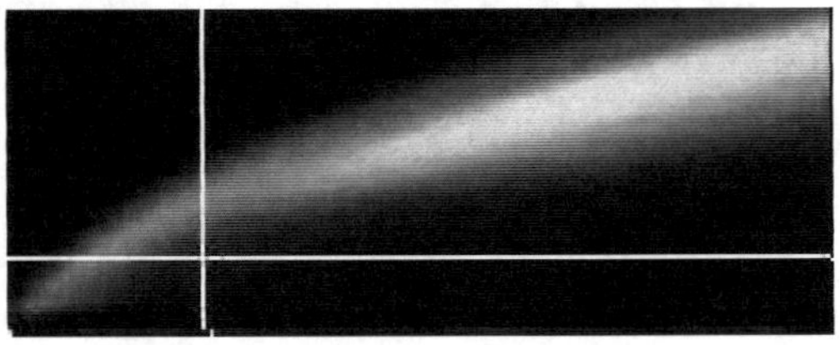

b) micro-gravity

Figure 3. Structure of non premixed reacting flow from experiment

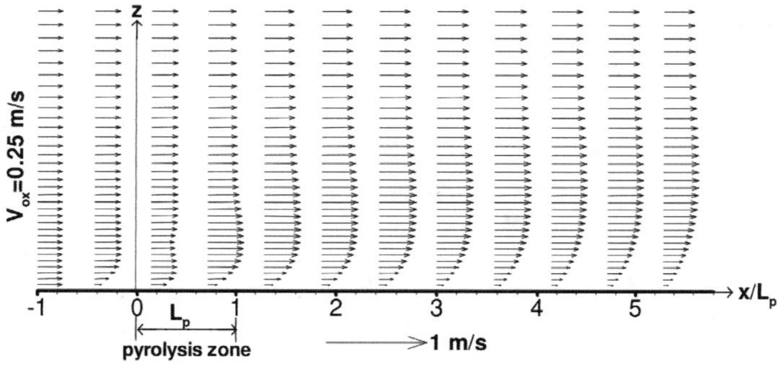

Figure 4. Predicted velocity vectors of a boundary layer type laminar diffusion flame in microgravity

The sooting behaviour of such laminar diffusion flame is characterized by using the laser induced incandescence technique (LII). The basic principle of LII relies on the detection of the thermal radiation from the soot particles that have been heated up to vaporization temperature by means of high energy laser pulse. Given in Fig.5(a-d) is the comparison between the predicted and measured soot volume fractions across the height, z, at various axial locations x/L_p=0.5, 1, 1.5 and 2 for V_{ox}=0.25 m/s and V_F=0.005 m/s. Based on the Moss model [19], both the soot inception and surface growth rates are determined solely from the mole fraction of fuel (ethylene) gas. This simplification gives a relatively good agreement with the measurement only in the fuel injection zone ($x/L_p \leq 1$) where the fuel concentration is important. However, far away from the fuel injection zone (x/L_p>1.5), the Moss model [19] underpredicts by a factor of 5-6 times the measured value. Based on the same soot surface growth model (Eq.13), the predicted soot by using the PAH inception model is an order of magnitude larger than that by using the acetylene inception model. It seems that the PAH inception model reproduces correctly both the general shape and the peak value (about 5 ppm) of experimentally-determined soot volume fraction. By looking at their locations, the measured peak is placed just above the plate surface at z=3 mm, while the calculated one by using the PAH inception model occurs at z=5 mm, meaning an over-prediction of the vertical stratification of the soot layer.

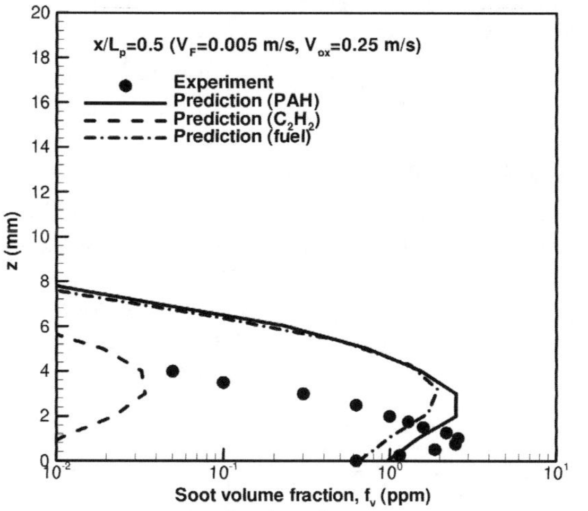

a) x/L_p=0.5

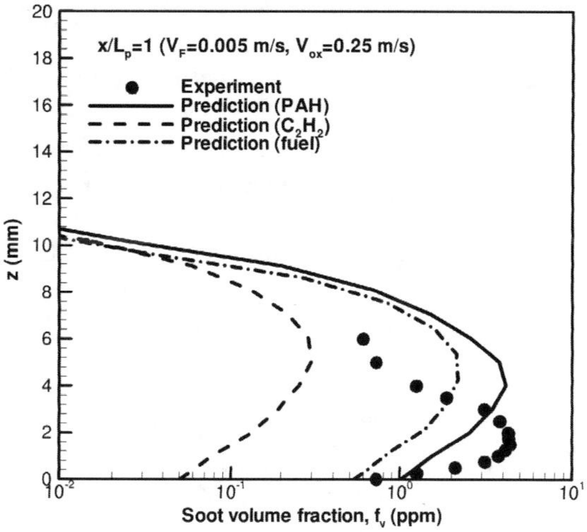

b) x/L_p=1

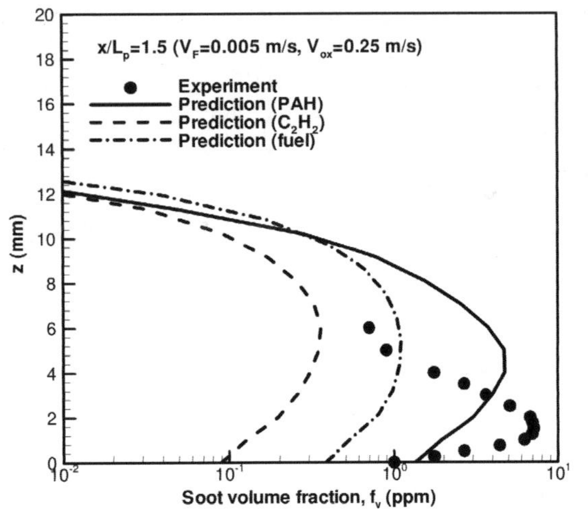

c) x/L_p=1.5

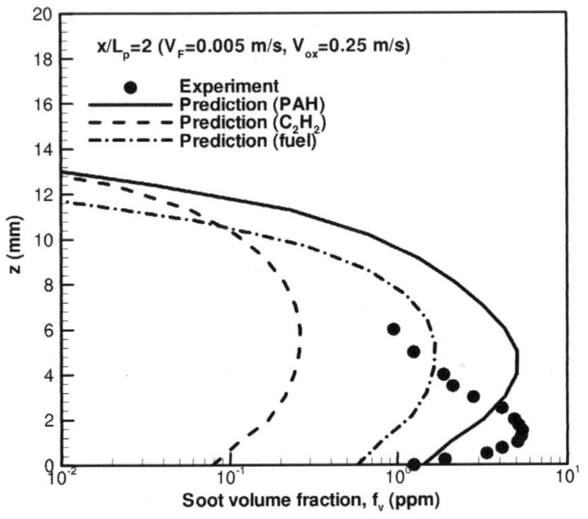

d) x/L_p=2

Figure 5. Comparison between computation and experiment for the soot volume fraction at different axial locations

Soot volume fraction is integrated over the soot layer in the z direction, as follows : $F_{soot}(x) = \int_{z=0}^{z\,max(x)} f_{soot}(x,z)dz$, where Z_{max} (x) is the height so that f_{soot} (x,z)=0 for z>Z_{max}(x) at a given x. A global description of the integrated soot production, F_{soot}, along x-axis as a function of both V_{ox} and V_F is presented in Figs.6(a,b). It seems that the PAH inception model reproduces the aforementioned sensitivity of the experimental soot formation to both V_{ox} and V_F. A decreasing of F_{soot} from 60 to 30 (ppm)mm (cf. Fig.6a) is induced with an increase of the oxidizer flow velocity, V_{ox}, from 0.06 to 0.25 m/s at a fuel injection velocity of 0.005 m/s. An inverse dependence of soot formation on the mainstream flow is predicted, and these tendencies are consistent with the experimental observation of Konsur et al.[2] based on the global residence time, defined as the ratio of the characteristic flame length, L_f, to the oxidizer velocity. While an enhancement of F_{soot} (cf. Fig.6b) from 20 to 38 (ppm)mm occurs with an increase of the fuel injection velocity, V_F, from 0.003 to 0.005 m/s at an oxidizer velocity of 0.25 m/s. The peak values of the integrated soot volume fraction are located far away from the trailing edge (x/L_p>1). It is found that the extent of the measured soot is always within the zone of x/L_p<3. It should be noted that the measured soot volume fraction distribution is converted from LII intensity image, and its accuracy depends on the accuracy of the proportionality constant required for correction due to attenuations of both the laser beam and the collected signal. Besides, for the low fuel injection velocity of V_F=0.003 m/s, the flames considered have the particularity of being mostly two-dimensional. While for high V_F, the flames considered develop strongly three-dimensional features far away from the trailing edge, affecting the collected signal for LII. This analyse is supported by the fact that the predicted soot profile follows closely the experimental trend for the low fuel injection velocity of 0.003 m/s, because the soot formation occurs within the extent of the collected signal for LII. While for the fuel injection velocity beyond 0.004 m/s, general shape of the predicted soot profiles has qualitative agreement with the experimental data. The error in numerical simulation may be linked to the constants used in the semi-empirical soot model, and accuracy of this type soot model seems to be strongly linked to the configuration studied. Generally, only a qualitative agreement with the measured soot volume fraction is achieved by using these standard constants [10]. It is felt that the difference is due to a combination of experimental uncertainties and the error in the numerical simulation.

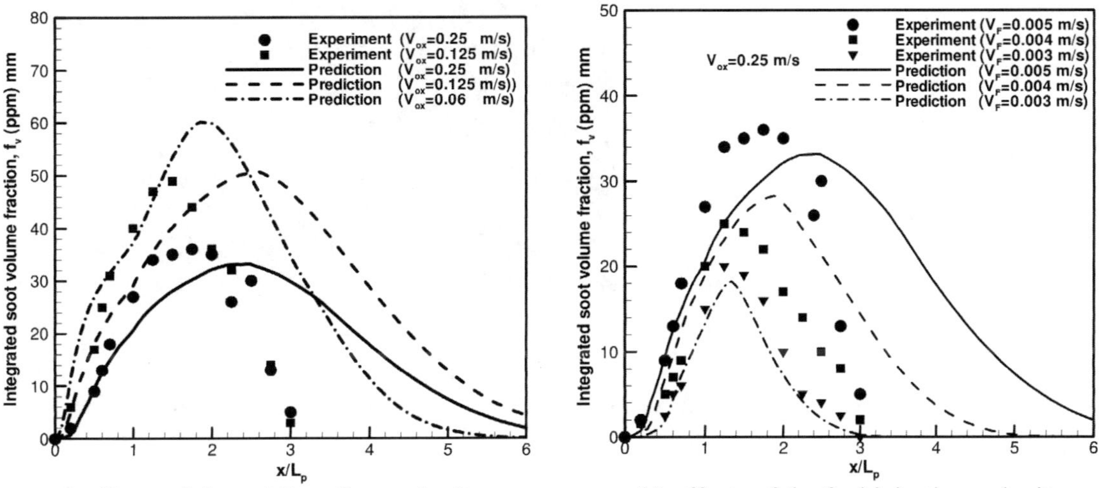

a) effects of the oxidizer flow velocity b) effects of the fuel injection velocity

Figure 6. Comparison between prediction and experiment for the integrated soot volume fraction

Effect of the soot inception model on the geometric characteristics of a co-current flame is evidenced in Fig.7(a-d). As compared to the experiment (cf. Fig.7d), the disagreement with or without (cf. Fig.7a) radiation is carried through the entire flame, becoming most evident in the quenching or extinction region behind the trailing edge. Absence of radiation induces a large flame stand-off distance (cf. Fig.7a), implying the lack of oxygen inside the boundary layer so that excessive fuel is burned into an area of freestream where oxygen is available. Particularly for the flame considered here, presence of soot particles in the combustion gas mixture greatly enhances radiation from the flame. Although, a zero gravity numerical simulation is performed, the fluid and combustion are strongly coupled through the contributions of the heat release rate and the thermal radiation associated with soot formation, inducing a major influence on the flame structure through radiation loss. Therefore, in the previous work [7], 40% of energy had to be subtracted numerically from the heat release rate for providing a correct flame structure due to neglecting thermal radiation. Besides, if the soot is underpredicted (cf. Fig.7b) by using the acetylene inception model, the flame significantly exceeds the measured flame zone at the trailing edge. The experimental flame shape is correctly hold with the PAH inception model (cf. Fig.7c), and thus it is used in the following study.

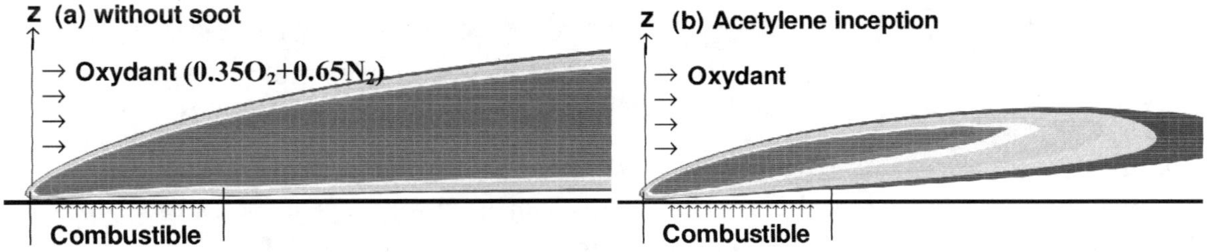

a) without radiation b) C_2H_2 inception model

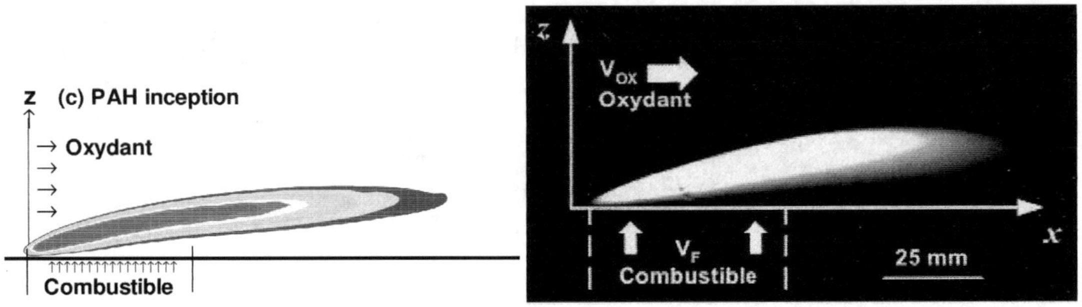

c) PAH inception model d) experiment

Figure 7. Effect of soot formation on the geometric characteristics of a co-current flame

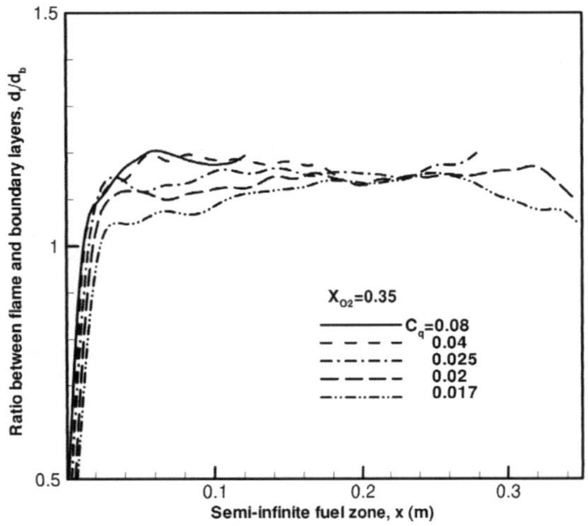

Figure 8. Ratio between flame standoff distance and boundary layer thickness for different C_q along a semi-infinite fuel zone

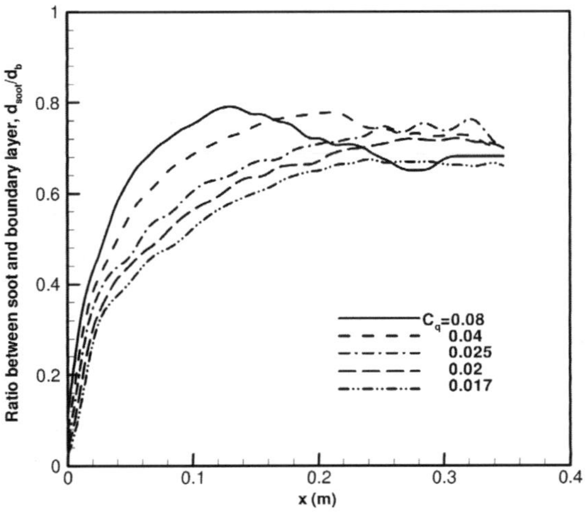

Figure 9. Ratio between soot and boundary layer thicknesses over a semi-infinite fuel zone for different C_q

Most of the work [26, 27, 28] in reactive boundary layer with blowing concentrated on the description of flame standoff over a semi-infinite fuel surface ($L_P \rightarrow \infty$). This case is also selected in this demonstration to clearly illustrate the effect of radiation on the flame characteristics over a large fuel size. For describing the effect of injection in boundary layer with blowing, a dimensionless volume coefficient, $C_q=V_F/V_{ox}$, defined by Schlichting [29], is currently employed. The work of Rouvreau et al. [7] clearly indicated that for the low values of C_q (≤ 0.08) considered here, the low Reynolds number flow remains attached. Therefore, the different assumptions used within the boundary layer formulation should be validated for conditions in which there is no separation. Given in Figure 8 is the ratio, d_f/d_b, between the flame standoff distance, d_f, and the boundary layer thickness, d_b. The boundary layer thickness, d_b, is defined as the distance above the surface where the parallel component of velocity reaches 99% of the main flow velocity. The flame stand-off distance, d_f, is the distance between the plate and the flame sheet where oxygen and fuel is delivered in stoichiometric proportions corresponding to the maximum HRR. Near the leading edge, the flame is positioned very close to the porous plate edge, and the flame establishes inside the viscous boundary layer, denoting the importance of diffusion there. As x increases along an infinite fuel zone (cf. Fig.8), both d_f and d_b move away from the plate into an area of higher velocity, and the ratio, d_f/d_b, converges to one single curve and toward a constant value of 1.2, implying that the flame sheet crosses the momentum boundary layer. While an asymptotical solution without the consideration of radiative loss from the reactive boundary layer model [26] suggests that the flame resides always within the boundary layer, and seems still incapable of predicting this type flame. Moreover, the flame stand-off distance, d_f, is an order of magnitude larger than that predicted by a Shvab-Zeldovich analysis, as described by Emmons [27] for a reacting boundary layer flow with blowing. The works of Andreussi [28] concluded that the Shvab-Zeldovich analysis is only adequate for freestream velocities higher than 1.2 m/s. Consequently, the natural constraints imposed by buoyancy in the absence of radiation heat transfer prevent complete validation of this model.

Contrary to the flame standoff distance along a semi-infinite fuel zone (cf. Fig.8), the maximum soot formation occurs always inside the boundary layer ($d_{soot}/d_b<1$) far away from the stoichiometric line, as shown in Figure 9. Fujita et al. [30] observed that soot particles trajectories inside a laminar boundary layer diffusion flame in microgravity do not follow exactly the gas movement due to thermophoretical effects. This is evidenced by a significant deviation of soot particles from the stoichiometric line due to the dominant thermal expansion which prevents soot particles approaching the flame. The peak soot production is placed just over the plate surface in the fuel rich zone (cf. Fig.5), and its location does not follow the response of flame standoff to changes in mainstream parameters.

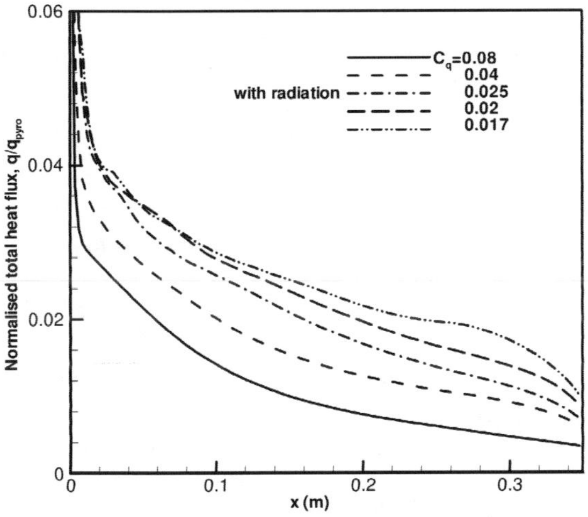

Figure 10. Normalized total heat flux over a semi-infinite fuel surface

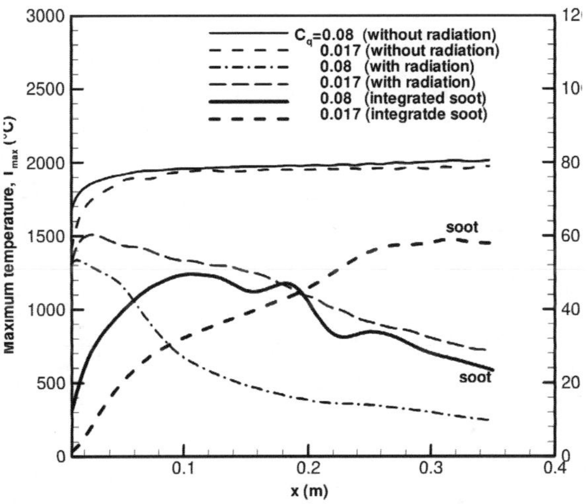

Figure 11. Effect of soot formation on the flame temperature along a semi-infinite fuel zone

Flame standoff is of great importance on the heat flux over a material surface. After a threshold pyrolysis temperature is attained, all heat flux to the material surface is used entirely for pyrolysis. Given in Fig.10 is the normalized total heat flux over a semi-infinite fuel surface, where q_{pyro} is the heat release rate per unit area (kW/m^2) from a flame. Microgravity flames are typically smooth (cf. Fig.3b) with a large heat feedback close to the leading edge where the fraction of the total energy released by the flame to the surface may be higher than 10%. Because heat flux is proportional to the flame temperature level and its location (view factor), with an increase of the flame standoff distance downstream (cf. Fig.8), the maximum temperature moves away from the surface, implying a reduction in the view factor from flame to the wall. This leads to a decrease of the fraction of the total energy to the surface, approaching an asymptotic solution of about 2% for C_q below 0.02.

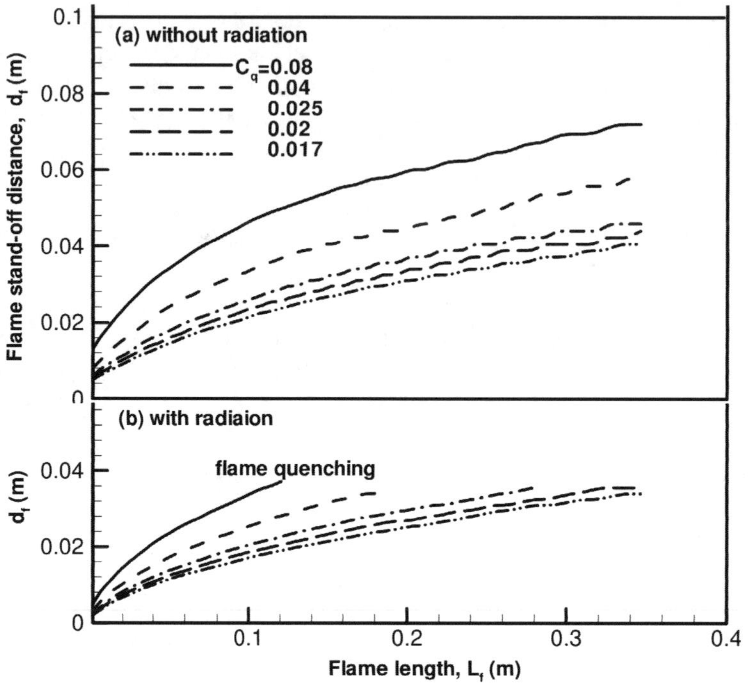

Figure 12. Flame standoff distance and its length along x-axis for different C_q. (a) without radiation; (b) with radiation

Diffusion flame quenching is of both fundamental and practical interest to applications such as fire spread over a condensed fuel surface. Gas phase quenching or extinction phenomena are often described using the Damköhler number, $\overline{D}_a = \frac{\tau_{re}}{\tau_{ch}}$, which is the ratio of a characteristic residence time, τ_{re}, over a characteristic chemical time, τ_{ch}, defined respectively as follows :

$$\tau_{re} = (V_{ox}^2/\alpha_\infty)^{-1} \qquad \tau_{ch} = \left[\rho_\infty A T_f \exp(-E/R T_f)\right]^{-1} \tag{20}$$

Where α_∞ is the free-stream thermal diffusivity, A the pre-exponential factor, and E the activation energy. From equation (20), flame extinction can be characterized as either kinetic (i.e. diffusion) or radiative loss. Kinetic extinction is unlikely in the absence of natural convection where buoyancy is not available to accelerate (strain) the flow field and consequently, to reduce residence times, τ_{re}, by enhancing mixing. Whereas radiative extinction is expected at long residence times, and thus occurs at very low strain rates. Microgravity experiments [5, 6] allowed the first observation of flame quenching, this being for boundary layer diffusion flame. The effect of soot formation on the flame temperature, T_f, which is the dominant factor for the characteristic chemical time τ_{ch}, along a semi-infinite fuel zone ($L_P \to \infty$) is presented in Fig.11. The horizontal flame length, L_f, is defined as the distance of the 600°C contour from the leading edge (cf. Fig.2a), and this definition appears to be very close to the experimental visible flame [5, 6]. As given in Figure 12 is the flame length and its stand-off distance along a semi-infinite fuel zone as a function of C_q. It can be seen that the lack of adequate radiative heat transfer dominated by soot formation, leads to a flame temperature of 2000 °C downstream the leading edge, and to an infinite flame length (cf. Fig.12a). In fact, because of radiative heat loss from this heavily sooting flame, the flame temperature does not surpass 1300ºC and decreases along the x-axis. This implies longer chemical times and thus, lower Damkhöler numbers, up to a flame quenching due to radiative heat loss around a soot concentration of $F_{soot} \approx 50(\text{ppm})\text{mm}$, consistent to the Smoke-Point concept [2, 4]. It is found that a decrease of C_q results in an increase in the flame length (cf. Fig.12b) due to reduction in soot formation (cf. Fig.6a).

4. Conclusions

This numerical study highlights the importance of soot formation on the quenching of a microgravity diffusion flame established in a shear boundary layer. The acetylene or fuel inception model underestimates the soot amount behind the trailing edge, and thus entails an overestimation of flame length. The behaviour of the soot production is reasonably well predicted by using PAH inception model despite the fact that a qualitative agreement with the measured soot volume fraction is achieved for a high value of C_q. It is felt that the difference between prediction and experiment is due to a combination of experimental uncertainties and the error in the numerical simulation linked to the semi-empirical soot model. Both prediction and experiment suggest that soot production is enhanced with decreasing the oxidizer flow velocity or increasing the fuel injection rate, further emphasizing the role of radiation on the geometric characteristics of a co-current, boundary layer laminar flame under microgravity. It can be concluded that a large C_q favours soot formation, and consequently, flame quenching due to a significant radiative heat loss. For a low Reynolds number reacting flow in microgravity, the flame establishes outside the viscous boundary layer, while the soot resides within the boundary layer far away from the stoichiometric line, implying the less importance of diffusion as compared to expansion and convection. A microgravity diffusion flame seems "open-tip" because quenching occurs before the fuel is fully consumed, so that fuel accumulates at the stoichiometric line. As a consequence, the flames significantly exceed at the trailing edge for C_q below 0.02. A maximum in the flame length for oxidizer velocities below those encountered in natural convection, represents a greater hazard for fire safety in spacecraft. These results are valid only for gaseous fuel and oxidizer, and should not be directly extrapolated to condensed fuel burning. Therefore, ongoing work is accounting for the PAH inception model for predicting flame spread over a condensed fuel surface in microgravity.

References

[1] Olson, S.L., T'ien, J.S., 2000, Combustion and Flame, 121(3), p.439.
[2] Konsur, B., Megaridis, C.M., Griffin, D.W., 1999, Combustion and Flame 118 (4), p.509.
[3] Markstein, G.H., De Ris, J., 1985, 20th Symposium (international) on Combustion (Pittsburgh: The Combustion Institute), p.1637.
[4] T'ien, J.S., 1990, Combustion and Flame, 80, p.335.
[5] Legros, G., Joulain, P., Vantelon, J.P., Fuentes, A., Torero, J.L., 2006, Combustion Science and Technology, p.178.
[6] Fuentes, A., Legros, G., Claverie, A., Joulain, P., Vantelon, J.P., Torero, J.L., 2007, 31th Symposium (international) on Combustion (Pittsburgh: The Combustion Institute), p.2685.
[7] Rouvreau, S., Torero, J.L, Joulain, P., 2005, Combustion Theory and Modelling, 9 (2), p.137.
[8] Lui, F., Guo, H.S., Gregory, J., Smallwood, J., Ömer, L., 2002, Journal of Quantitative Spectroscopy & Radiative Transfer, 73, p.409.
[9] Leung, K.M., Lindstedt, R.P., Jones, W.P., 1991, Combustion and Flame, p.289.
[10] Wen, Z., Yun, S., Thomson, M.J., Lightstone, M.F., 2003, Combustion and Flame, 135, p.323.
[11] Vovelle, C., Delfan, J.L., Reuillon, M., 1994, Springer-Verlag, Berlin, p.50.
[12] Anderson, H., McEnally, C.S., Pfefferle, L.D., 2000, 28th Symposium (international) on Combustion (Pittsburgh: The Combustion Institute), p.2577.
[13] Hall, R.J., Smooke, M.D., Colket, M.D., 1997, Gordon & Breach, p.189.
[14] Dagaut, P., Nicolle, A., 2005, Combustion and Flame, 140, p.161.
[15] W. Grosshandler, 1993, NIST technical Note TN 1402, National Institute of Standards and Thchnology, Maryland.
[16] Mcgrattan, K.B., Glenn, P.F., Jason, E.F., 2008, Fire Dynamics Simulator.

[17] Poinsot T. and Veynante, D., 2005, Second Edition, ISBN:1-930217-10-2, R.T. Edwards, Inc., P.O. Box 27388, Philadelphia, PA, 19118 USA.
[18] Lutz, A.E., Kee, R.J., Grcar, J.F., Rupley, F.M., 1997, Sandia National Laboratories Technical Rep. SAND96-8243.
[19] Moss, J.B., Stewart, C.D., Young, K.J., 1995, Combustion and Flame, 101, p. 491.
[20] Lindstedt, P.R., 1994, Springer-Verlag, p.417.
[21] Puri, R., Richardson, T.F., Santoro, R.J., 1993, Combustion and Flame, 92 (3), p.320.
[22] Frenklach, M., Clary, D.W., Gardier, J., Stein, S.E., 1984, 20th Symposium (international) on Combustion (Pittsburgh: The Combustion Institute), p.887.
[23] Lee, K.B., Thring, M.W., Beer, J.M., 1962, Combustion and Flame, 6, p.137.
[24] Torero, J.L., Bonneau, L., Most, J.M., Joulain, P., 1994, 25th Symposium (international) on Combustion (Pittsburgh: The Combustion Institute), p.1701.
[25] Brahmi, L., Vietoris, T., Rouvreau, S., Joulain, P., David, L., Torero, J.L., 2005, AIAA Journal, Vol.43, No.8, p.1725.
[26] Liu, S., Catalin, G., Fotache, 2008, Combustion and Flame, 154, p.378.
[27] Emmons, H., 1956, Z. Angew. Math. Mech., 36, p.60.
[28] Andreussi, P., 1982, Combustion and Flame, 45, p.1.
[29] Schlichting, H., 1979, Seventh Edition, McGraw-Hill.
[30] Fujita, O., Ito, K., Ito, H., Takeshita, Y., 1997, 4th NASA International Microgravity Combustion Workshop, p.217.

Journal of Physics: Conference Series **327** (2011) 012039 doi:10.1088/1742-6596/327/1/012039

Boiling in variable gravity under the action of an electric field: results of parabolic flight experiments

P Di Marco [1], R Raj [2], J Kim [2]

[1]Department of Energy and Systems Engineering, University of Pisa, Largo Lazzarino 1, 56122 Pisa, Italy

[2]Department of Mechanical Engineering, University of Maryland,College Park, MD 20742, USA

E-mail: p.dimarco@ing.unipi.it

Abstract. Results from the variable gravity pool boiling experiments performed during the 52nd ESA parabolic flight campaign are reported in this paper. During a typical parabola, the gravity acceleration changes from $1.8g_E$ (high gravity) to $\sim 0g_E$ (low gravity) and finally back to $1.8g_E$. The two high gravity periods and the microgravity period are each roughly maintained for 20 seconds while the transition from high gravity to low gravity and vice versa occurs over a period of 3–5 seconds. Use of the high feedback frequency microheater array allowed quasi-steady boiling data over the continuous range of gravity levels ($0g_E$-$1.8g_E$). The experimental apparatus consisted of a boiling chamber with a 7x7 mm^2 microheater array in a 10x10 configuration. Each heater in the array was individually controlled to maintain a constant temperature. The array could be operated in a full configuration or a selectively powered reduced set of 3x3 heaters. Experiments were performed with FC-72 as the test fluid, the pressure was maintained at a constant value between 1 and 1.13 atm and the subcooling ranged from 27 to 11 K. An external electric field was imposed over the boiling surface by means of a grid consisting of 4 rods, laid parallel to the surface; voltages up to 10 kV were applied. The electric field was effective in reducing the size of the detaching bubbles, and increasing the heat transfer compared to the values in low-g, although its effectiveness decayed as the heat flux/superheat increased. The current results compared well with previous results obtained in the ARIEL apparatus that was operated in orbital flight.

1. Introduction

Investigating and assessing boiling performance in reduced gravity conditions is of paramount importance for the design of efficient heat exchange devices for future space and planetary missions. The problem has been addressed since the early 60s [1] and recently reviewed by Di Marco, Kim and Ohta [2] and Di Marco [3]. In particular, significant contributions were made by Straub and coworkers [4] and Merte [5] who performed numerous experiments on different platforms. It has been shown that – as could be easily expected –bubble detachment, and mostly vapor removal away from the surface are impaired by the reduction of buoyancy forces; however different forces, including inertia, surface tension and Marangoni convection come into play and partially replace the absent buoyancy force.

Recently, Kim and coworkers [6] took advantage of the injection phase of parabolic flight to operate a heater in quasi-steady condition in variable gravity. As the gravity level changed, a sharp transition in the heat transfer mechanism was observed at a threshold gravity level. Below this threshold (low-g regime), a nondeparting coalesced bubble governed the heat transfer and the effect of residual gravity was small. Above this threshold (high-g regime), bubble growth and departure dominated the heat transfer and gravity effects became more important. Heat flux was found to be heater size dependent only in the low-g regime. Significantly, the slope of the heat flux versus acceleration curve was different in the two regimes, ruling out the possibility of a unified power law dependence across all gravity levels.

Di Marco and Grassi [7,8] widely explored the possibility to apply an electric field to the boiling systems in order to have the buoyancy forces replaced by the electrostatic one. Several experiments on different platforms demonstrated that the boiling heat transfer could be improved in this way. In some cases, the boiling performance with electric field in low-g was similar to what could be obtained in normal boiling in terrestrial gravity.

This paper describes the results obtained in a pool boiling experiment carried out during the ESA 52^{nd}parabolic flight campaign in May, 2010. The experiment apparatus consisted of a boiling chamber filled with FC-72, with a 7x7 mm^2 microheater array in a 10x10 configuration (four corners missing, 96 heaters). Each heater in the array was individually controlled to keep its temperature at a constant value. The response of the heater and electronic control circuitry was as fast as 15 kHz, much faster than the boiling phenomena being studied. The heater could be operated in full configuration (all 96 heaters were powered) or a reduced set of 3x3 heaters could be powered. An external electric field was imposed over the boiling surface by means of a grid consisting of 4 rods, laid parallel to the surface at a distance of 7 mm from it; voltages up to 10 kV were applied.

2. Experimental Apparatus

A schematic of the experiment is shown in Figures 1 and 2. The experiment package used components from a sounding rocket experiment payload that flew previously on a Terrier-Orion sounding rocket. The primary rack (Figure 1) housed a sealed chamber containing the test fluid at nearly atmospheric pressure, a microheater array to measure the heat transfer distribution during boiling, a rack of electronic feedback circuits, two 30 Hz video cameras, a computer, and a LCD display. All components are contained within a Vertical Equipment Rack (VER) provided by NASA and designed for use on their parabolic aircraft. The secondary rack (Figure 2) contained a high-voltage power supply for the electric field measurements, a transformer to convert from 220 VAC to 110 VAC, a triple output power supply, and a 28 VDC power supply. The components were contained in an Amco 50" Equipment Rack (AER) provided by NASA. A third small rack was used to hold a compressed air bottle.

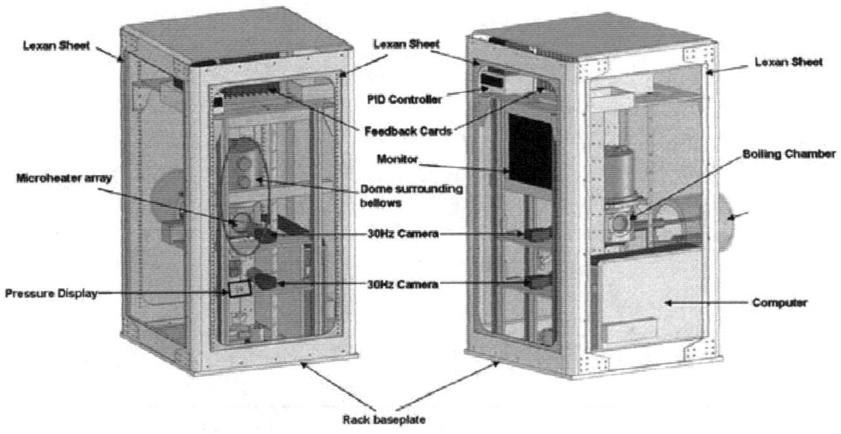

Figure 1. Solid model of the primary rack and its components.

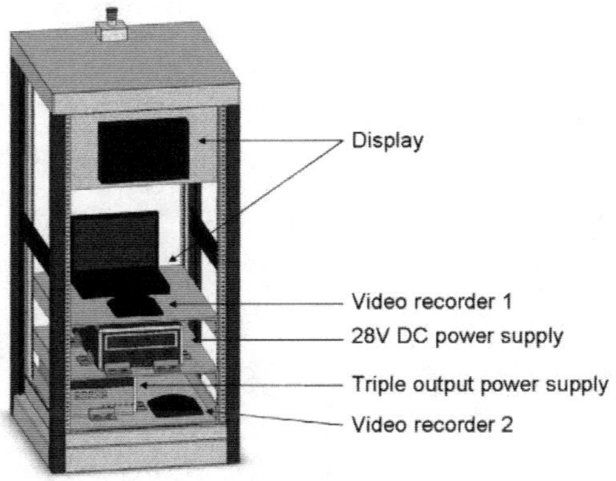

Figure 2. Solid model of secondary rack.

2.1. Description of the subsystems

2.1.1. Test Chamber (primary rack). The test chamber is shown on Figures 3 and 4. The bellows and the surrounding dome allowed the test section pressure to be controlled by varying the air pressure around the bellows. A PID temperature controller, a RTD probe, and Kapton heaters attached to the boiling chamber were used to control the bulk fluid temperature. The chamber was filled with nominally 3 liters of distilled FC-72.

A thorough degassing procedure described in[9] was used to degas the fluid. When degassing the liquid, the test rig was turned upside down. Valves # 1 and #2 were open while valve #3 was closed during the process of filling/emptying the boiling chamber with test liquid (ground procedure). To degas the fluid, valves # 1 and # 2 were open and a vacuum was periodically pulled on the chamber to remove the gas and vapor above the liquid. The final dissolved gas concentration in the liquid was determined using the chamber temperature and pressure, the thermophysical properties of FC-72, and Henry's law. Once completed degassing, Valve # 2 was closed and the rig was brought back in the upright position. Then, valves #1 and #3 were opened momentarily so that the bellows adjusted and the chamber pressure was brought back to 1 atm.

The test chamber included electrodes so electric field effects on boiling could be studied. The cylindrical electrodes (described below) were energized through an insulated conductor fed through a hole drilled into one side of the mating plate between the dome and the lower chamber. The test rig as well as the microheater array was grounded. The electric field was applied when desired between the horizontal microheater array and the electrodes (Figure 5 and Figure 6). The potential difference between the electrode and the heater array could be varied in the range0-10 kVDC.

2.1.2. Microheater Array and Electronic Feedback Circuits (primary rack). A microheater array consisting of 96 platinum resistance heaters deposited in a 10x10 configuration onto a quartz substrate was used to measure the heat transfer distribution during boiling. Each heater in the array was nominally 0.7x0.7 mm^2 in size. Gold power leads 1 μm thick allowed connections to be made to the heaters. Each array element was square with a nominal area of 0.49mm^2. Each heater had a nominal resistance of 200 kΩ and a temperature coefficient of resistance (TCR) of 0.003 K^{-1}. Because the heater lines covered only half of the quartz substrate, images of the bubble could be obtained by looking through the substrate. A photograph of a microheater array is shown on Figure 5.

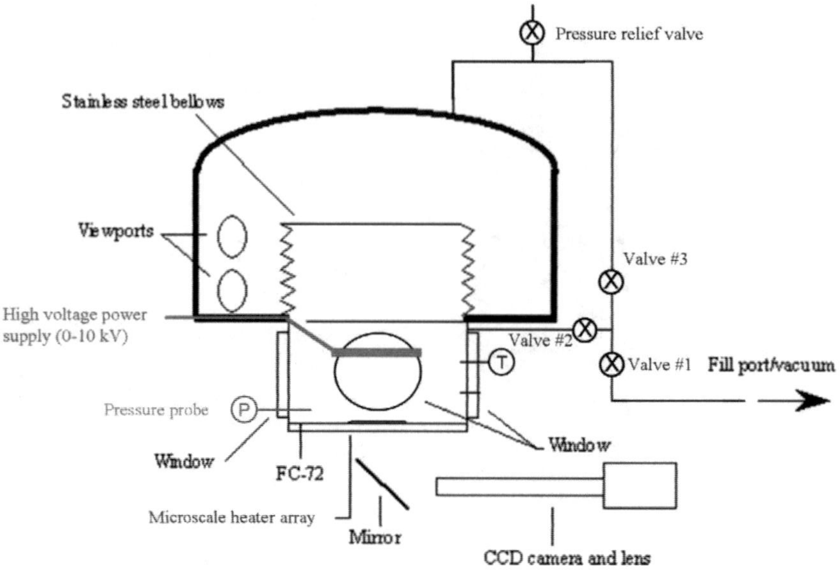

Figure 3. Schematic of the test chamber.

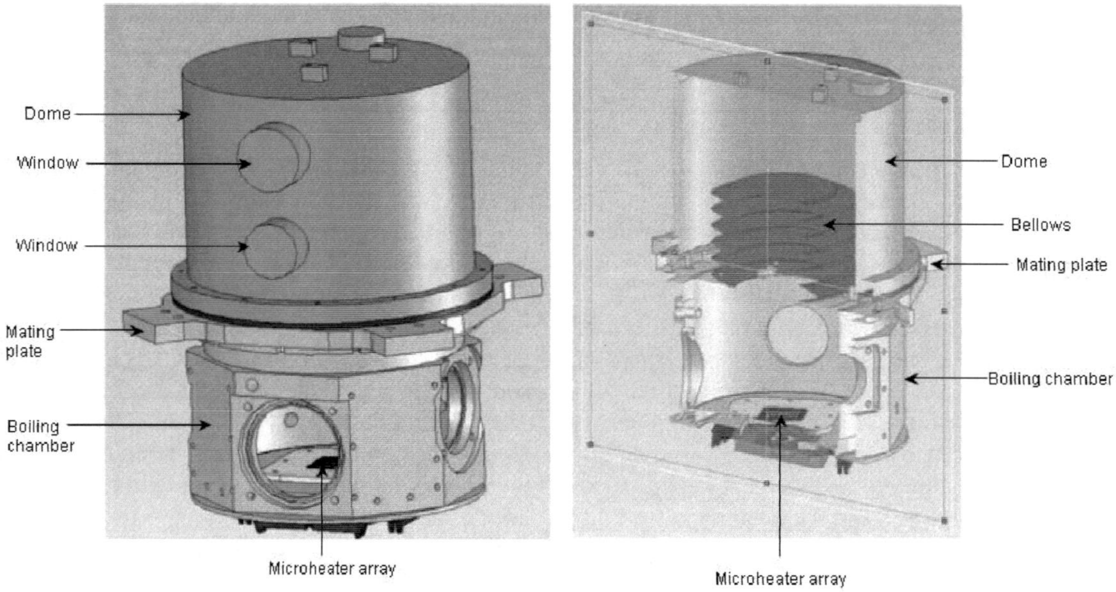

Figure 4. Solid model test chamber.

The power provided to each heater could be individually controlled, therefore it was possible to operate the array in a variety of configurations, and in particular at uniform heat flux or uniform temperature. The constant-temperature mode was chosen for the present series of tests. The temperature of each heater in the array was kept at a specified value by a bank of feedback circuits similar to those used in constant temperature hot-wire anemometry. The reader is referred to Rule and Kim [10] for the details regarding the heater construction and operation. The high feedback frequency (~15 kHz) of the microheater array and a constant temperature boundary condition at the heater surface allowed quasi-steady state data to be obtained during the transition phases of the flight [6].

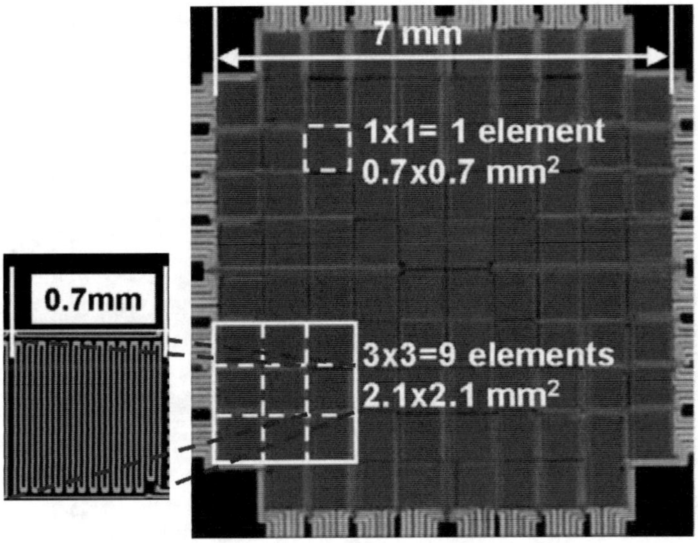

Figure 5. Photograph of microheater array.

2.1.3. High voltage electrode and power supply. The high voltage electrode consisted of a grid of 4 stainless steel rods, 0.9 mm of diameter, 2.1 mm pitch, laid parallel to the heater plane at a distance of 7 mm (from rod axis to heater surface). The grid was fed by a high voltage DC power supply (Brandenburg 590). The high voltage was connected to the rod (positive pole) while the heater acted as ground. The HV power supply was powered at 24 Vdc, and the output high voltage was regulated by an analog input signal 0-10 Vdc. A monitor output indicated the actual value of the high voltage. Figure 6 shows a schematic of the electrode and a plot of the resulting electric field. The maximum field intensity (for an imposed voltage of 10 kV) was 1.2 kV/mm at the surface of electrode rods, which was much lower than the breakdown voltage of 14 kV/mm for FC-72. The average value at the heater surface (in the absence of bubbles) was about 1.22 kV/mm.

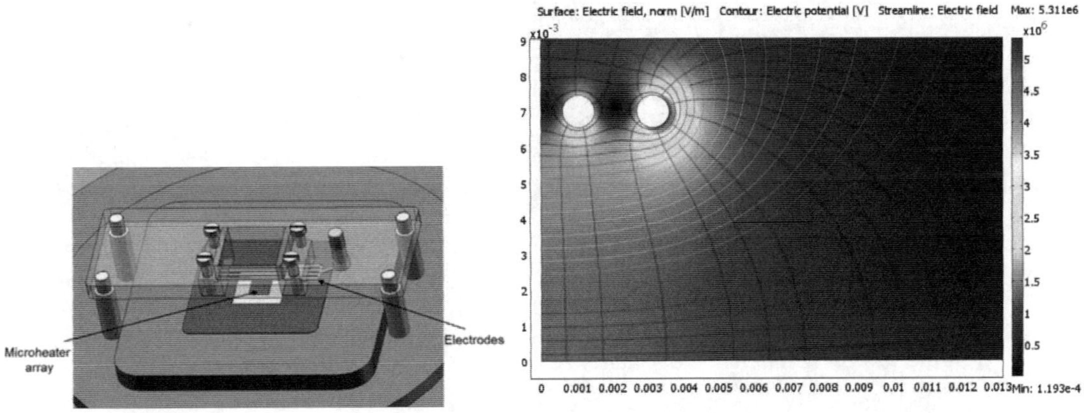

Figure 6. Schematic of the electrode grid and heater (left) and the resulting electrostatic field (right).

2.1.4. Imaging. Two video cameras were used to obtain side and bottom images of the boiling process. One camera used a long focal length lens that imaged the boiling process through the bottom of the heater. The other camera was used to image the bubble life cycle from the side. The frame rate of both cameras was 30 Hz. The output of the video cameras were recorded onto two video recorders housed

Journal of Physics: Conference Series **327** (2011) 012039 doi:10.1088/1742-6596/327/1/012039

in the secondary rack. The video recorder digitized the analog videos and stored them on a USB drive. Two remote controlled DVD players were used as the display for these two cameras.

3. Test Operation

During the present series of test, the apparatus was operated with subcooled fluid conditions (11, 18 and 27 K of subcooling).Due to the thermal dilatation of the fluid, the pressure and hence the saturation temperature increased slightly with decreasing subcooling, as reported in Table 1. The heater temperature was set at an assigned value (from 70 to 95 °C, in steps of 2.5 K) during level flight, and data were acquired through the entire gravitational transient, including macrogravity and transition phases. The heater was used in full configuration (96 heaters) or with a reduced array of 3x3 heater activated; the reduced array was located in between the verticals of two HV electrode rods, i.e. in a zone where the electrical force was directed upwards.

Table 1. Pressure and saturation temperature for the three values of subcooling tested.

Subcooling (K)	Pressure (bar)	Saturation temperature (°C)
27	1.02	57.0
18	1.07	58.5
11	1.12	68.5

An Airbus A300 aircraft flew successive parabolic maneuvers providing ~20 s periods of low-g preceded and followed by ~ 20 s periods of ~1.8g_E. A 1g period lasting a few minutes between successive parabolas was used to set the test parameters. A typical parabola along with the acceleration levels are shown on Figure 7 (left). Thirty parabolas were flown per flight, and three flights were made. The parabolic flight campaign has been primarily used to study phenomena under low-g and hypergravity (>1.5g_E) conditions. However, as can be seen in Figure 7, there was a transition period of approximately 3–5 s when the acceleration varied continuously from hypergravity to low-g and vice versa. This period is generally considered too short for processes to reach equilibrium, and hence unsuitable for making measurements. Nonetheless, data were acquired for this period throughout the transition from hypergravity to low-g and vice versa and was confirmed to be under quasi-steady state conditions as explained below.

The data in Figure7 (right) illustrates the values of gravitational acceleration and heat flux recorded during the pull-up (decreasing gravity) and recovery (increasing gravity) phases along with three steady state experiment results at 1, 1.3, and 1.6 g_E. The data acquired during transition from hypergravity to microgravity (blue gradient symbols) and that from microgravity to hypergravity (red delta symbols) agree, indicating the heat flux depends only on the gravity level and not the direction of change in gravity. Results at steady-state hypergravity conditions (1.3g and 1.6g) along with the reference earth gravity data confirm the quasi-steady boiling during the relatively quick transition period. It was thus possible to obtain quasi-steady boiling data across a continuous range of gravity levels between ~0 to 1.8 g_E.

The electric field was activated before the beginning of a parabola (to assess its effect in variable gravity) or during the micro-g phase, to investigate the flow pattern and heat transfer changes due to its application in reduced gravity.

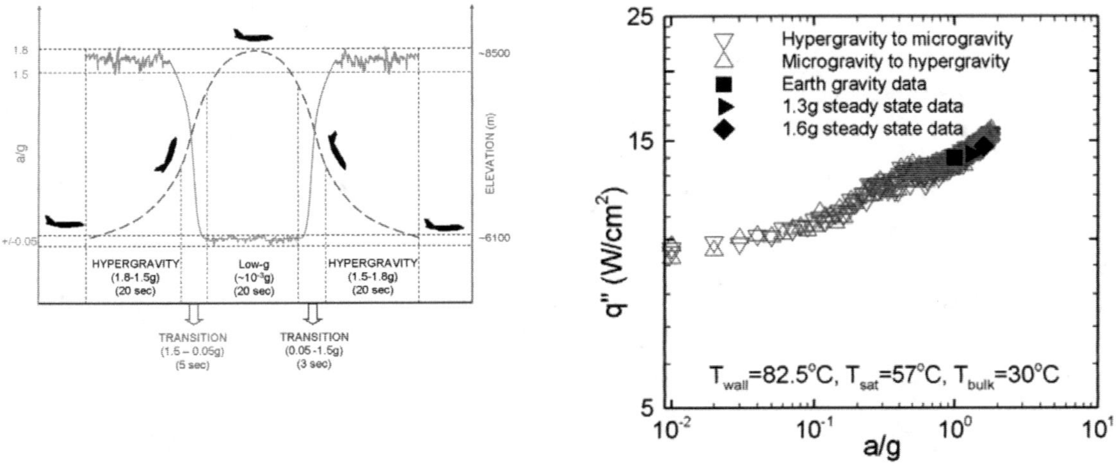

Figure 7. Gravitational transient during parabolic flight in A300 "Zero-G and boiling data (heat flux) during gravitational transient.

4. Results and Discussion

4.1. Large heater- 7x7 mm²

4.1.1. High subcooling data. Figures 8-9 show the effect of the application of electric field during the gravitational transient: heat flux data are reported versus gravity acceleration, with and without electric field. The figures show clearly that the electric field is ineffective above a threshold value of gravity acceleration. While it improves the boiling performance below a value of gravity acceleration of 0.1-0.2 g_E, at least for moderate wall superheat (Figure 8), this effect vanishes at a larger wall superheat (Figure 9). The boiling process is dominated by buoyancy for sufficiently high values of gravity acceleration [6], while it is dominated by surface tension at lower values of g; these two regimes are characterized by a different boiling flow pattern; in the gravity-dominated regime several individual bubble are present and detaching over the heater, while in the surface-tension dominated regime a single coalesced bubble whose size is of the order of the heated surfaces forms over it.

Boiling curves in microgravity are shown in Figure 10. Here, the low-g data (with and without electric field) are compared with the boiling curve in normal gravity, where the application of electric field has little effect. Images of the boiling flow patterns, as observed from the bottom of the semi-transparent heater, are also shown in correspondence with the related data points. It can be noted that the boiling performance in microgravity is worse than in normal gravity whether or not an electric field is applied; however the application of electric field in low-g improves the heat flux of about 13% at intermediate wall superheats, while its effect is lower or negligible at higher and lower wall superheat. The maximum improvement can be associated to a change in flow pattern: the single coalesced bubble residing over the heater is split in smaller ones by the electrical forces.

From the analysis of these data, it may be concluded that the application of electric field, at least in the present geometrical configuration, is effective only in the surface-tension-dominated region, and its effect is strongly reduced or vanishes at all at high values of wall superheat. This last effect may be due to the increased bubble size; when the bubble is larger, electrical forces may be too weak to break up the large coalesced bubble that forms on the heater.

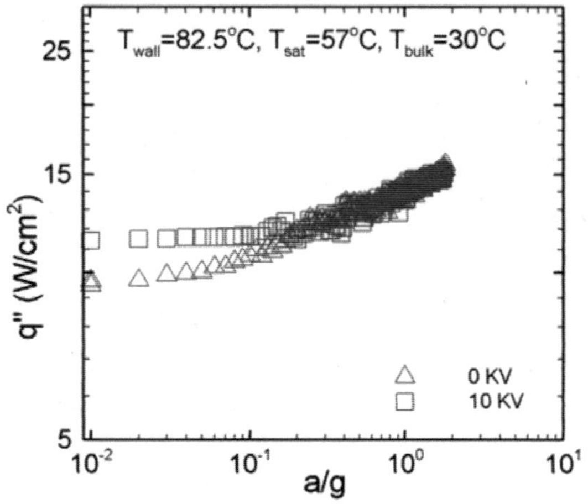

Figure 8. Values of heat flux vs. gravitational acceleration, at intermediate wall superheat and high subcooling.

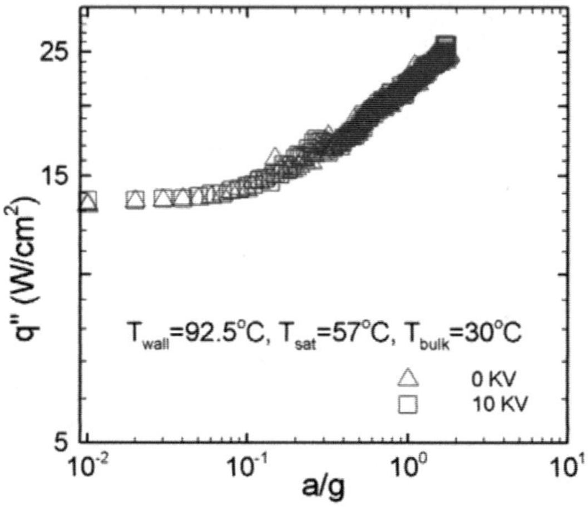

Figure 9. Values of heat flux vs. gravitational acceleration, at high wall superheat and high subcooling.

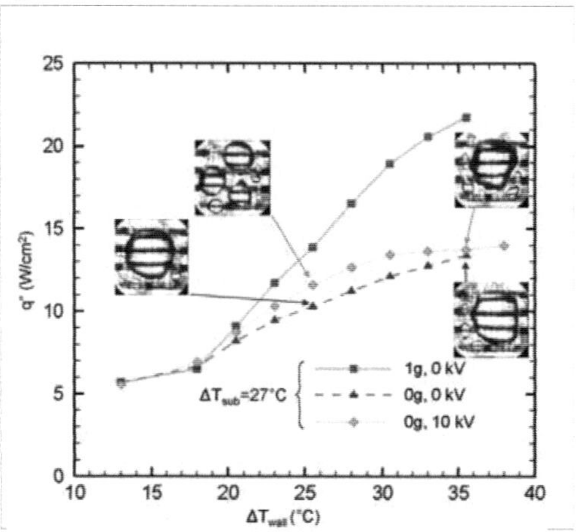

Figure 10. Boiling curves in normal and reduced gravity, high subcooling, full heater.

4.1.2. Intermediate and small subcooling data. Figure 11-13and 14-16 report the same data as in former section for the intermediate (18 K) and small (11 K) subcooling, respectively. Similar qualitative observations as in former section may be made. In particular it may be noted that an improvement due to the application of EF takes place also at high wall superheat. By comparing Figs. 8, 11 and 14, it may be noted that in presence of EF a plateau is attained when gravity drops below about 0.1 g_E. Below 0.1g_E, boiling is surface tension dominated and the effect of gravity acceleration is very small [6]. As a result, the heat transfer is unaffected, however, is larger compared to the case without electric field. In effect, it can be concluded that under low-g conditions, the "equivalent gravity field" due to the applied electric field (10 kV, 7 mm) at intermediate superheats is approximately 0.1 g_E. Furthermore, the improvement due to EF application is more marked with decreasing subcooling, and for intermediate subcooling at low wall superheat the same boiling performance as in normal gravity is restored.

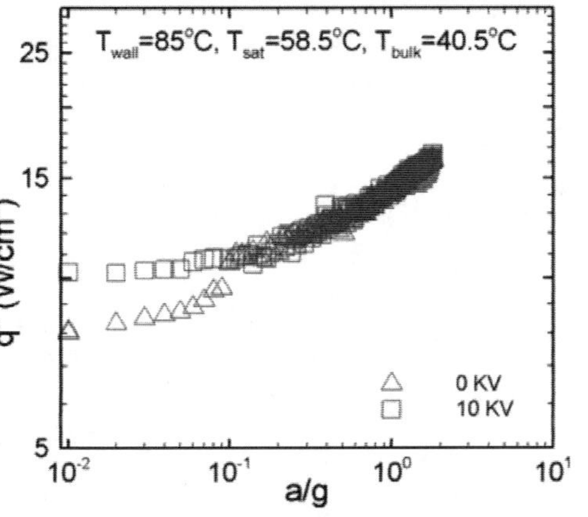

Figure 11. Values of heat flux vs. gravitational acceleration, at intermediate wall superheat and intermediate subcooling.

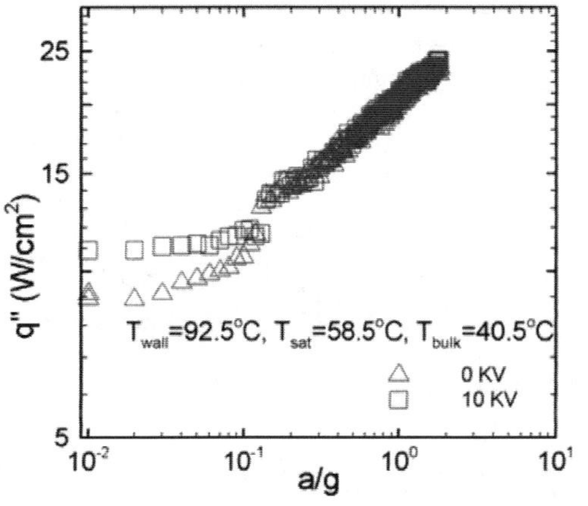

Figure 12. Values of heat flux vs. gravitational acceleration, at high wall superheat and intermediate subcooling.

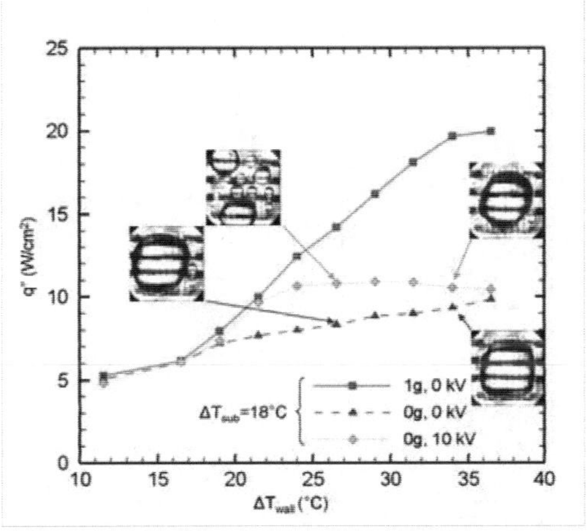

Figure 13. Boiling curves in normal and reduced gravity, intermediate subcooling, full heater.

4.2 Small heater- 2.1x2.1 mm^2

The boiling curves for the small heater (3x3) configuration are shown on Figures 17-19. At high subcooling the application of electric field restores the same heat transfer performance as in normal gravity for the lower wall superheats: the electric field becomes less effective at higher wall superheats. The heat transfer degradation in microgravity is less marked in this case. For intermediate subcooling the enhancement takes place mostly at intermediate wall superheat. Finally, at low subcooling, the application of electric field is less effective. Generally, on the small heater, the bubble does not split into smaller ones under the action of electric field, but its size is reduced due to the contribution of electrical overpressure in the surrounding liquid. The observation of videos shows that convective patterns around the bubble are noticeably enhanced when the electric field is applied: this is likely due to electroconvection.

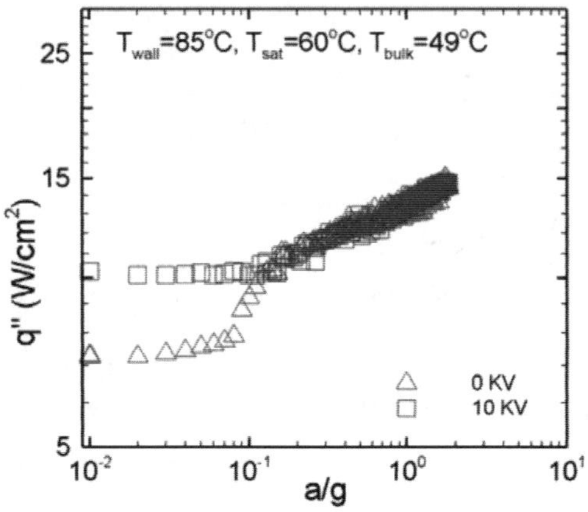

Figure 14. Values of heat flux vs. gravitational acceleration, at intermediate wall superheat and low subcooling.

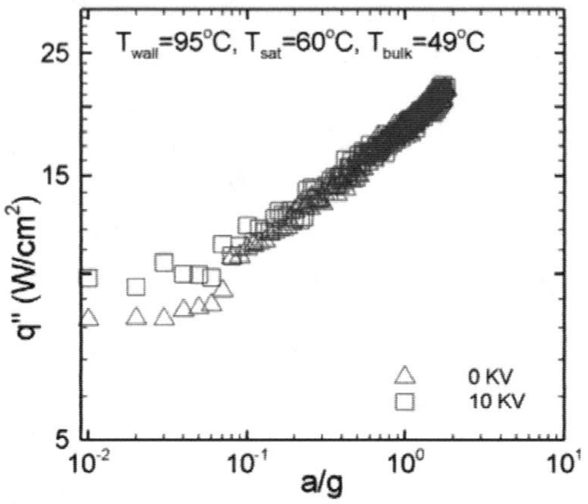

Figure 15. Values of heat flux vs. gravitational acceleration, at high wall superheat and low subcooling.

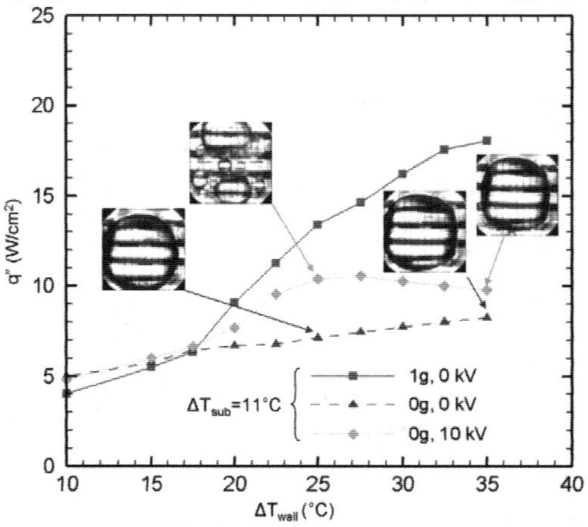

Figure 16. Boiling curves in normal and reduced gravity, low subcooling, full heater.

4.3 Resume of heat transfer improvement

Table 2 summarizes the heat transfer improvements generated by the application of electric field in low-g in different conditions. It can be noted that the increase in performance is more marked for the small heater at intermediate wall superheat.

Table 2. Summary of heat transfer enhancements.

	7x7 mm^2 heater		2.1x2.1 mm^2 heater	
SUBCOOLING	Increase at intermediate superheat	Increase at high superheat	Increase at intermediate superheat	Increase at high superheat
HIGH	13 %	1 – 5 %	15 %	3 – 7 %
MEDIUM	33 %	5 – 10 %	55 %	20 – 30 %
LOW	46 %	15 – 20 %	35 %	20 – 30 %

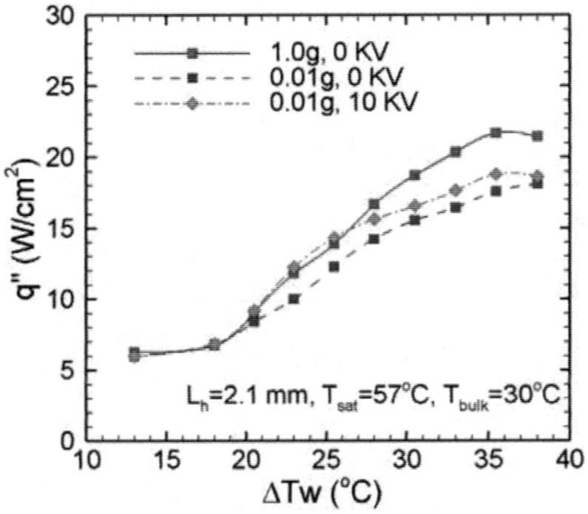

Figure 17. Boiling curves in normal and reduced gravity, high subcooling, 3x3 heater.

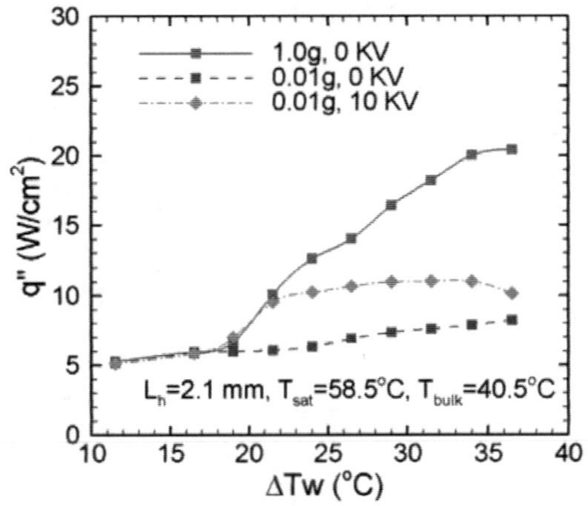

Figure 18. Boiling curves in normal and reduced gravity, intermediate subcooling, 3x3 heater.

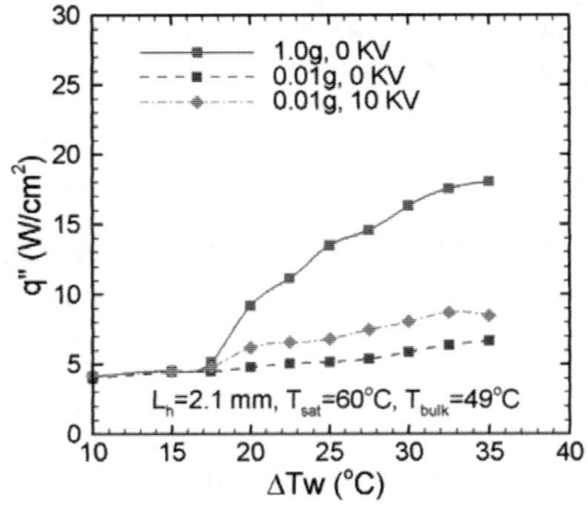

Figure 19. Boiling curves in normal and reduced gravity, low subcooling, 3x3 heater.

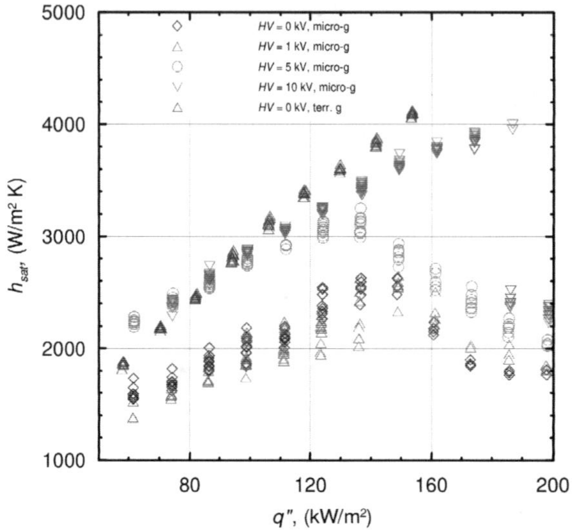

Figure 20. Heat transfer coefficient h_{sat} in normal and reduced gravity in ARIEL apparatus, for a subcooling of 16.6 K, from [8].

4.4. Comparison with former results (ARIEL apparatus)

The ARIEL apparatus [8], operated in orbital flight (gravity acceleration lower than 10^{-4} g_E) during Foton-M2 mission, consisted mainly in an aluminum vessel of about 1 liter volume, containing the test fluid, FC-72. The active heating element consisted of a square uniform layer of 20x20 mm, with a 40-nm thick gold layer deposited via sputtering over a 10-mm-thick ZnS substrate, and was operated at constant power, achieving heat fluxes up to 200 kW/m^2. The pressure was kept constant at 100±3 kPa, with subcooling ranging from 6.6 to 16.6 K. An external electrostatic field with a voltage up to 10 kV dc was imposed by means of a grid of five parallel stainless steel rods, 1 mm in diameter, spaced 5 mm and laid parallel to heater, at a distance 5 mm above it (from rod axis to plate).

Data in microgravity, with no electric field applied, exhibited a degradation of boiling performance: the heat transfer coefficient showed to be closely linked to the boiling pattern, and in particular to the presence and evolution of the big mass of vapor over the heater, originated by bubble coalescence. The heat transfer degradation was reduced or even cancelled with the application of the electrostatic field, that, especially for high applied voltages, contrasted the formation of the large vapor mass, improved the bubble removal from the heated surface and restored a performance more similar to terrestrial one. This improvement, however, was no more effective for increasing heat flux, even for the highest value of the voltage applied during the tests. An example is shown in Fig.20. Although the heater morphology and its mode of operation were different, it may be concluded that there is a good qualitative agreement between the results of the two tests.

5 Conclusions

Results of variable gravity pool boiling with FC-72 have been reported in this paper; in particular, the effects of an external electric field on the bubble dynamics and associated heat transfer. The main experimental results can be summarized as follows.

- Boiling heat transfer is generally degraded in low-g; the degradation is discontinuous when transition from buoyancy dominated to surface-tension-dominated boiling occurs. This discontinuity is associated to a transition in flow patterns (from multiple departing bubbles to a large coalesced non-departing bubble covering the entire heater).
- The effect of electric field is most pronounced at lowest subcooling.

- The effect of electric field at a given subcooling is more pronounced at intermediate wall superheats.
- The baseline heat transfer (0 kV) is already high at higher subcoolings (higher condensation / increased thermocapillary convection) resulting in smaller percentage increase with electric field.
- Both increased subcooling and electric field inhibit bubble coalescence. At higher wall superheats, coalescence is less inhibited by application of the electric field.
- Lower subcooling results in larger coalesced bubbles.
- The coalesced bubble is split by the application of EF into smaller bubbles which slide to the side at lower subcoolings. The coalesced bubble is split into smaller bubbles which reside on the heater (with increased condensation at the bubble top) at larger subcoolings.
- Small (2.1 mm) and large (7 mm) heaters show similar trends; the relative increase in heat transfer is generally higher for the small heater.

In summary, the electric field is effective in reducing the size of the detaching bubbles, and increasing the heat transfer compared to the values in low-g, although its effectiveness decays as the heat flux/superheat increases. From the analysis of flow patterns, it may be concluded that contrasting bubble coalescence leads to an enhancement of boiling in reduced gravity.

The present results compare well with former ones obtained in the ARIEL apparatus [8] operated in orbital flight, and extend the explored range of parameters in terms of heater size, subcooling, gravity and electric field intensity. Further analysis is ongoing to better explain the physical mechanisms governing the boiling phenomena in the presence of electric field or less.

Acknowledgements
The authors would like to thank the European Space Agency for accommodating the experiment on the 52ndESA Parabolic Flight Campaign organized May 2010, and Novespace for the assistance given in setting up the apparatus on board of the aircraft.

References

[1] Siegel R Keshock E G, Effects of reduced gravity on nucleate boiling bubble dynamics in saturated water, *AICHE Journal*, **10**,4, 1964, 509.
[2] Di Marco P, Kim J, Ohta H, Boiling Heat Transfer in Reduced Gravity Environments, *Adv. in Multiphase Flow and Heat Transfer*, L. Cheng and D Mewes Eds,. Bentham Publishing Co. **1**, Ch.2, 2009, 53.
[3] Di Marco P, Influence of force fields and flow patterns on boiling heat transfer performance, keynote lecture, *Proc. of the International Heat Transfer Conference, IHTC14, August 8-13, 2010*, Washington, DC, USA, (CD-ROM), paper IHTC14-23409, 1-18.
[4] Straub J, Boiling Heat Transfer and Bubble Dynamics in Microgravity, *Adv. Heat Transfer*, **35**, 2001, 57.
[5] Merte H, Some parameter boundary governing microgravity pool boiling modes, *Ann. N.Y. Acad. Sci.*, **1077**, 2006, 629-649.
[6] Raj R, Kim J, and McQuillen J,Subcooled Pool Boiling in Variable Gravity Environments, *ASME Journal of Heat Transfer*, **131**, 2009, 091502-1.
[7] Di Marco P, Grassi W, Pool boiling in microgravity: old and recent results, *Multiph. Sci. and Technology*, **19**, 2, 2007, 141.
[8] Di Marco P, Grassi W, Effects of external electric field on pool boiling: comparison of terrestrial and microgravity data in the ARIEL experiment, *Experimental Thermal and Fluid Science*, **35**,5, 2011, 780.
[9] Henry C D, Kim J, and McQuillen J, Dissolved Gas Effects onThermocapillary Convection During Subcooled Boiling in Reduced GravityEnvironments, *Int. J. Heat Mass Transfer*, 2006, **42**, 919.
[10] Rule T D, and Kim J, Heat Transfer Behavior on Small HeatersDuring Pool Boiling of FC-72, *ASME J. Heat Transfer*, **121**(2), 1999, 386.

Journal of Physics: Conference Series **327** (2011) 012040 doi:10.1088/1742-6596/327/1/012040

Experiment on nucleate pool boiling in microgravity by using transparent heating surface – Analysis of surface heat transfer coefficients

C Kubota[1], O Kawanami[2], Y Asada[1], Y Wada[1], T Nagayasu[1], Y Shinmoto[1], H Ohta[1], O Kabov[3], P Queeckers[3], S Chikov[3] and J Straub[4]

[1]Department of Aeronautics and Astronautics, Kyushu University, 744 Motooka, Nishiku, Fukuoka 814-0395, Japan
[2]Department of Mechanical and System Engineering, University of Hyogo, 2167 Shosha, Himeji, Hyogo 671-2201, Japan
[3]Physical-Chemistry Department of the Faculty of Applied Sciences, Université Libre de Bruxelles, Av. F.D. Roosevelt 50c, B-1050 Bruxelles, Belgium
[4]Technische Universität München, Lehrstuhl für Thermodynamik, Bolzmannstrasse 15, D-85748 Garching, Germany

E-mail: kubota@aero.kyushu-u.ac.jp

Abstract. Investigation of mechanisms in nucleate boiling under microgravity conditions is essential for the development of the cooling systems handling a large amount of waste heat. A transparent heating surface with multiple arrays of 88 thin film temperature sensors and mini-heaters was developed for the clarification of boiling heat transfer mechanisms in microgravity. To investigate gravity effects on the microlayer behaviors and corresponding local heat transfer coefficients, images of liquid-vapor behaviors underneath attached bubbles and local heat transfer data were simultaneously obtained in microgravity pool boiling. The present paper reports the analysis of the data measured during the ESA parabolic flight campaign. It was found that the liquid-vapor behaviors were strongly affected by the direction and the level of residual gravity. Various patterns of liquid-vapor behaviours and corresponding enhancement or deterioration of the heat transfer are observed.

1. Introduction

Nucleate boiling is one of the most effective modes of heat transfer due to the latent heat transport during liquid evaporation. Investigation of mechanisms in nucleate boiling is essential for the design of the thermal management systems handling a large amount of waste heat. Although many researches in the relevant subject are exited, the heat transfer mechanism was not yet enough clarified for various bubble behaviors observed under different combination of dominant parameters [1],[2].

Because bubbles are growing up in a large size under microgravity conditions due to the reduction in buoyant force, the employment of observation techniques and measurement methods become easier keeping the essence of elementary processes in nucleate boiling unchanged. In such a situation, behaviours of liquid microlayer underneath vapor bubbles dominate the heat transfer. The model of

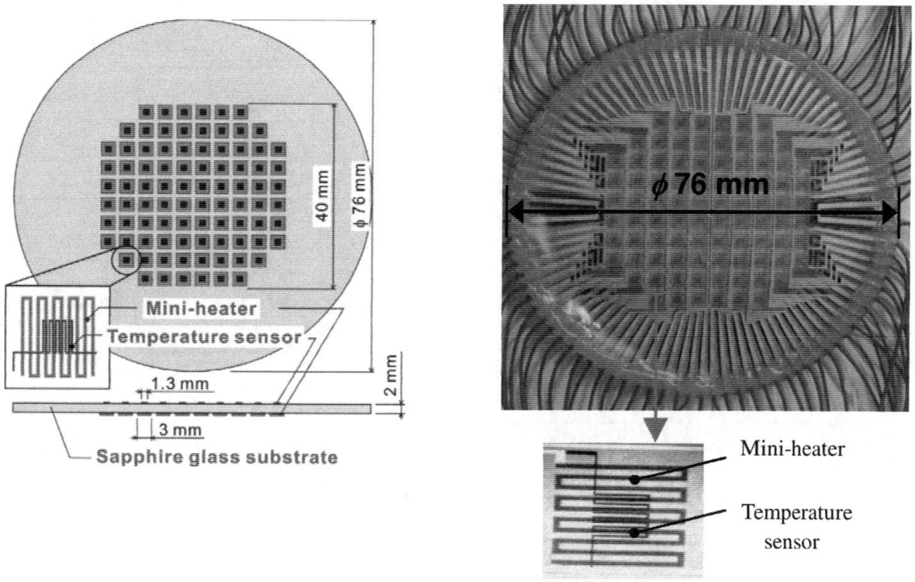

Figure 1. Schematic of transparent heating surface.

Table 1. Specifications of temperature sensors and mini-heaters.

	Temperature sensor	Mini-heater
Materials	Ti/Pt	Ti/Au
Size	1.3 × 1.3 mm	3 × 3 mm
Thickness	0.1 μm	0.04 μm
Nominal resistance	800 Ω	350 Ω

heat transfer around a single bubble in nucleate boiling was analyzed by Stephan and Hammer [3] using the concept of Wayner et al. [4]. Distribution of liquid microlayers underneath vapour bubbles is unsteady and their thickness varies along local positions on the heating surface. Hence, to investigate the relation between the microlayer behaviors and local heat transfer characteristics becomes a key to solve problems on the mechanisms of nucleate boiling also in microgravity.

In the present study, to investigate gravity effects on the relation between local heat transfer and liquid-vapor behaviors, pool boiling experiments in microgravity are conducted by using a transparent heating surface with multiple sensors and heaters array during the 49th parabolic flight campaign by ESA. In this paper, based on the analysis of detailed distribution of local surface heat transfer coefficients, the change of local heat transfer is discussed relating to the observed liquid-vapor behaviors.

2. Transparent heating surface

A transparent heating surface as shown in Figure 1 was developed for the measurement of local heat transfer characteristics. The heating surface, 76 mm in diameter and 2 mm in thickness, was made of sapphire glass with very high thermal conductivity (λ = 42 W/mK) to realize quick control of heating conditions from backside and the observation of liquid-vapor behaviors simultaneously. Artificial cavities were located at the center of heating surface for the enhancement of nucleation. The present heating surface with effective heating area of 40 mm in diameter located on the center is quite enlarged from the surface developed by Kim et al. [5] who have already developed the structure of the sensor/heater array for the clarification of local heat transfer characteristics using the heater with a size

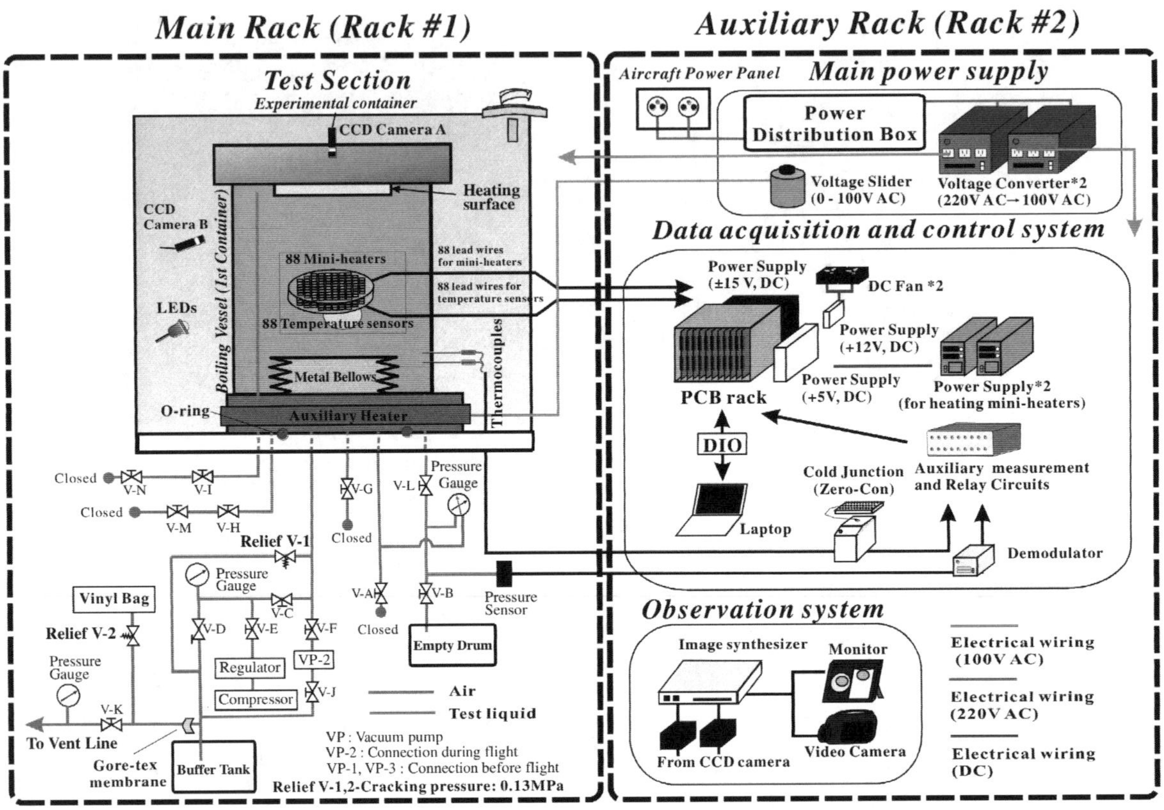

Figure 2. Outline of experimental setup.

of 2.7 mm × 2.7 mm. The present heating surface has 88 pairs of temperature sensors and mini-heaters coated directly on the both side of the substrate. The sensor and the heater are 1.3 mm × 1.3 mm and 3 mm × 3 mm in size, respectively. Temperature sensors are operated as resistance thermometers and they are made by platinum thin films. The temperature coefficient is 0.0012 - 0.0013K^{-1}. Mini-heaters are made of gold. They are heated individually by the electric circuits controlled by a computer. The specifications of temperature sensors and mini-heaters are summarized in Table 1. Each pair is connected to the feedback circuits controlling the relation between local surface temperature and mini-heater power. The substrate of heating surface is transparent, which allows to relate the observed liquid-vapor behaviors underneath bubbles directly to the measured local heat transfer. More detailed description about the heating surface was presented in our previous papers [6],[7].

3. Experimental setup and conditions

The outline of experimental setup for aircraft experiments is shown in Figure 2. The setup consists of 2 racks. A main rack consist of a boiling vessel as a core part of the experiment, the pressure regulation system of the boiling vessel and the experimental container operated as an air bath. An auxiliary rack consists of the measurement and control system for the heating surface, i.e. the measurement system of liquid temperature and liquid pressure, and the observation system. The boiling vessel with 120 mm in diameter and 260 mm in length is filled with 3 L of degassed FC-72 as a test fluid. Two CCD cameras are installed above the heating surface and at the side of the boiling vessel to observe liquid-vapor behaviors. The orientation of the transparent heating surface, installed at the top of the chamber, is facing downward. Detailed illustration about the experimental apparatus was reported by a previous paper [8].

Table 2. Experimental conditions.

Test liquid	FC-72 (deaerated)
Pressure range in boiling vessel	$P = 0.06 - 0.1$ MPa
Target temperature of heating surface	$T_{target} = 50.0 - 85.0$ °C
Liquid subcooling	$\Delta T_{sub} = 3.0 - 17.0$ K
Heating mode	Uniform surface temperature

Experiments described in this paper were conducted under the heating mode of uniform surface temperature. The target temperature T_{target} and liquid subcooling ΔT_{sub} are selected as experimental parameters. These experimental conditions are shows in Table 2.

A reduced gravity environment is obtained by the flight of Airbus A300 Zero-G through a series of parabolic maneuvers, which result in approximately 20 second periods of microgravity condition. Each parabola is initialed with a 1.8g pull up and terminated with a 1.8g pull out. During 1g period, power is supplied to mini-heaters to keep the surface temperature near a desired value. The data measurement and the video recording were synchronized to relate the heat transfer data to the observed liquid-vapor behaviors directly. The sampling period of the data measurement system is 108 ms.

4. Analytical method

To obtain local heat transfer coefficient, the heat flux distribution on the liquid side of heating surface is calculated from heat conduction across the substrate by using the three-dimensional analysis tool, MSC.Nastran. The three-dimensional grids for heat conduction analysis are shown in Figure 3. The grids for this analysis are generated by using the function of finite element modeling of MSC.Patran. The structured mesh was used in the central heated area and the unstructured mesh was used around it. The division number in z-direction is ten. The details of the finite element model are shown in Table 3. The temperature distribution on the entire surface area is obtained by the cubic spline interpolation of measured surface temperatures at the prescribed locations on the central heated area and at the periphery of the sapphire disk. The boundary conditions are shown in Table 4. The time-series temperature distribution data on the entire surface area is used to give one boundary conditions for the calculation of temperature distribution in the heated disk which, in turn, gives the local surface heat flux distribution at each instances of data acquisition. The time-series heat flux data from mini-heaters located on the central heated area, which is equated to the power input to each mini-heaters element taking account of heat loss to the surrounding air, is also used as a boundary conditions. The boundary condition of the rear surface around the central heated area is given by the convection to surrounding air ($\lambda = 10$ W/mK). The condition of disk periphery is assumed to be thermal insulation, and a uniform temperature along the peripheral direction is given by the averaged value of measured temperatures. The error of temperatures measured by the resistance temperature sensors is less than 0.7 K. The evaluated error of the local heat flux to z-direction on the heating surface is less than 1%.

5. Results and discussion

Various patterns of liquid-vapor behaviors were observed in this experiment [9],[10]. The typical patterns during microgravity period are shown in Figure 4. In the case of low liquid subcooling condition, during positive microgravity period, where bubbles are pushed toward the downward-facing heating surface, generated bubbles became more larger on the surface. The heat transfer seemed to be deteriorated because of the emergence of the dry patches underneath the bubbles (Figure 4-Pattern 1). On the other hand, during negative microgravity period, bubbles were detached without distinct condensation, where bubbles were generated from heating surface with high frequency (Figure 4-Pattern 2). In the case of high liquid subcooling condition, fine bubbles detached from the heating surface during the negative microgravity period, where no marked enhancement or deterioration of heat transfer was expected. (Figure 4-Pattern 3).

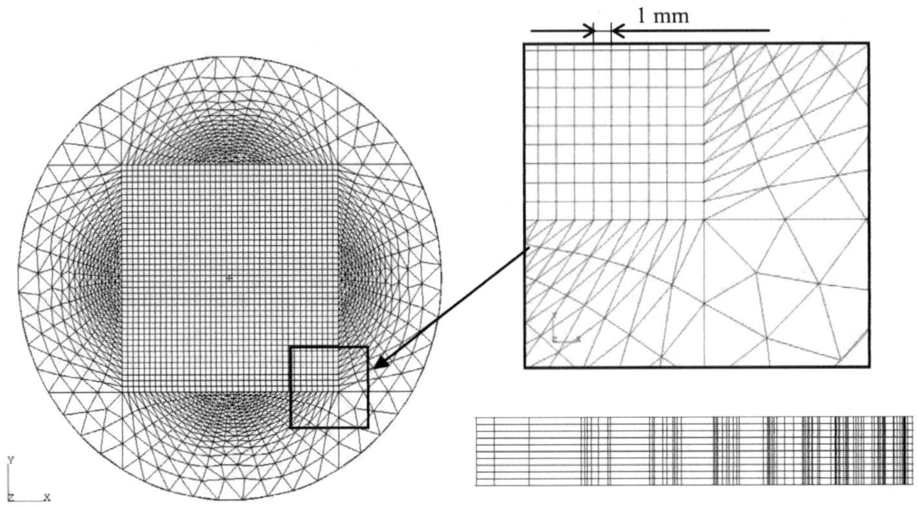

Figure 3. Three-dimensional grids for heat conduction analysis.

Table 3. Details of finite element model.

	Heated area	Periphery of the sapphire disk
Number of elements	15210	20720
Number of nodes	17600	13984
Number of layers	10	10
Material	Sapphire glass	Sapphire glass

Table 4. Boundary conditions.

Location	Boundary condition	Value of parameter
Surface of entire disk	Surface temperatures	variable
Rear heated area	Heat flux from mini-heaters	variable
Rear surface of disk around central heated area	Air convection	$\lambda = 10$ W/mK
Periphery of disk	Thermal insulation	$\lambda = 0$ W/mK
Periphery of disk	Averaged temperature	variable

One of the analytical results for low liquid subcooling condition is described here. The images of liquid-vapor behaviors and corresponding heat transfer coefficient distributions are shown in Figure 5. The transitions of gravity, surface temperatures, local heat fluxes and local heat transfer coefficients at the points A and B on Figure5 are shown in Figure 6 for the same experimental conditions. The nucleation site density depending on the surface texture of heating surface is high around point B, while it is low around point A, because artificial cavities is located at the center of heating surface. Generally, the nucleation site density increase with increasing the surface temperature and the heat transfer is enhanced with the nucleation site density.

In positive microgravity (Figures 5-(a) and 6, Point B, Pattern 1 in Figure 4), boiling bubbles were pressed against the heating surface and grew. Underneath attached flattened bubbles in a large size,

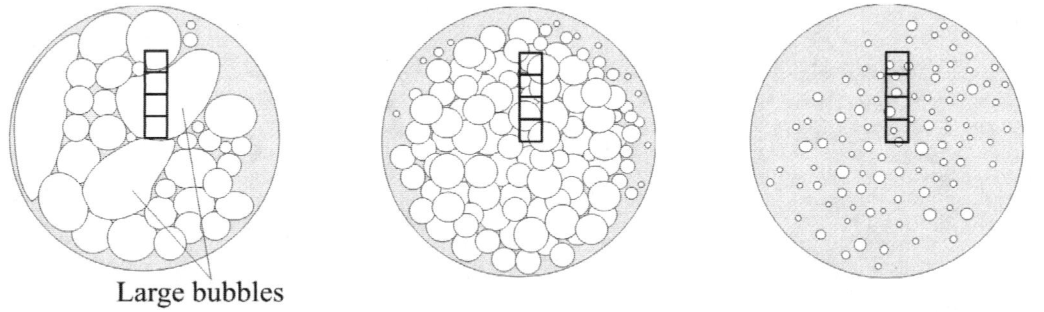

Pattern 1 (T_{target} = 75.0 °C, ΔT_{sub} = 3.0 K, +μg) Pattern 2 (T_{target} = 75.0 °C, ΔT_{sub} = 3.0 K, −μg) Pattern 3 (T_{target} = 75.0 °C, ΔT_{sub} = 11.0 K, −μg)

Figure 4. Schematics of liquid-vapor behaviors on the heating surface during microgravity period.

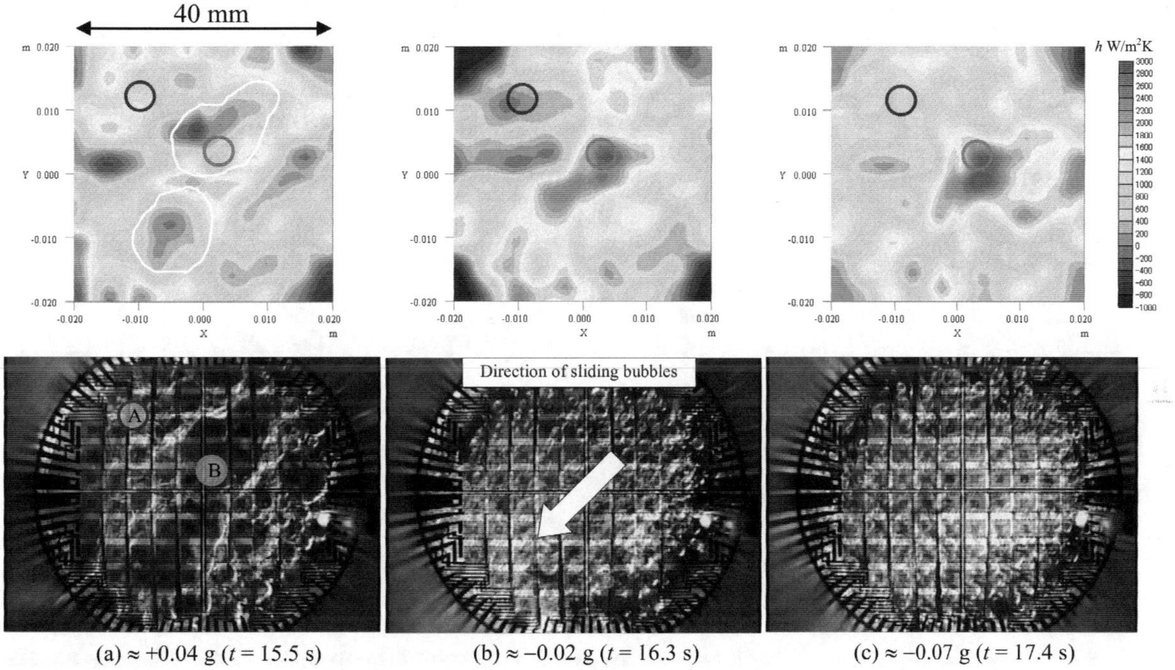

(a) ≈ +0.04 g (t = 15.5 s) (b) ≈ −0.02 g (t = 16.3 s) (c) ≈ −0.07 g (t = 17.4 s)

Figure 5. Liquid-vapor behavior and heat transfer coefficient distribution. (T_{target} = 75.0 °C, ΔT_{sub} = 3.0 K, T_{bulk} = 38.0 °C, P = 0.0613 MPa)

dry patches were extended. As a result, the local surface temperature increased reducing surface heat flux, and marked heat transfer deterioration was observed. On the other hand, under negative microgravity conditions (Figures 5-(b), (c) and 6, Pattern 2 in Figure 4), small bubbles were detached from the surface. At -0.07g (Figures 5-(c) and 6), the periods of detachment were short and the bubble size was very small. As a result, heat transfer characteristics were similar to those on ground and they were dependent on the surface nucleation characteristics. The heat transfer coefficients were relatively lower at the upper left part of the heating surface around Point A because of the shortage of active nucleation sites, while the heat transfer coefficients were relatively enhanced at the center of the surface around Point B with a lot of active cavities. At -0.02g (Figures 5-(b) and 6), the periods of

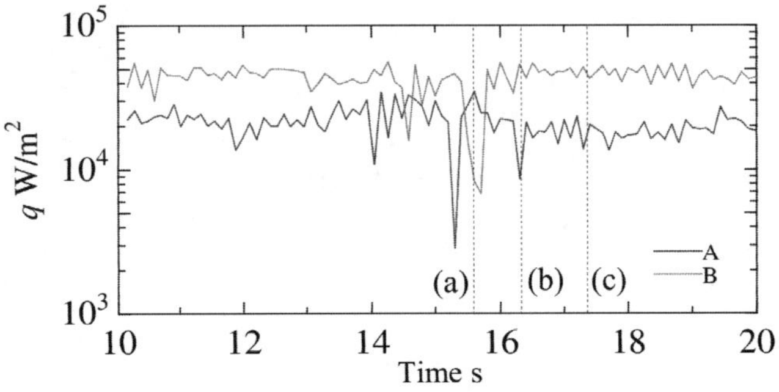

(i) Heat flux

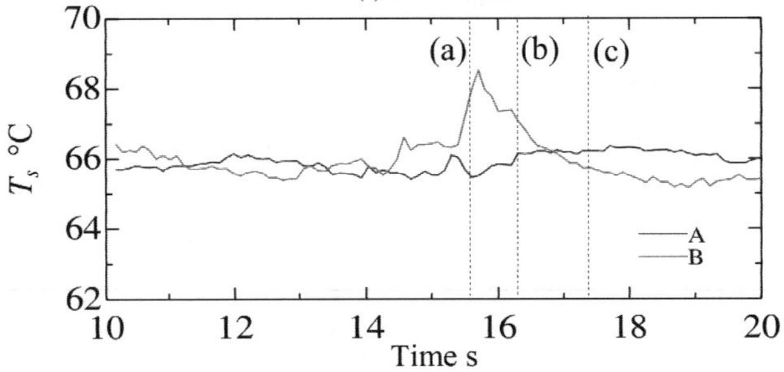

(ii) Surface temperature

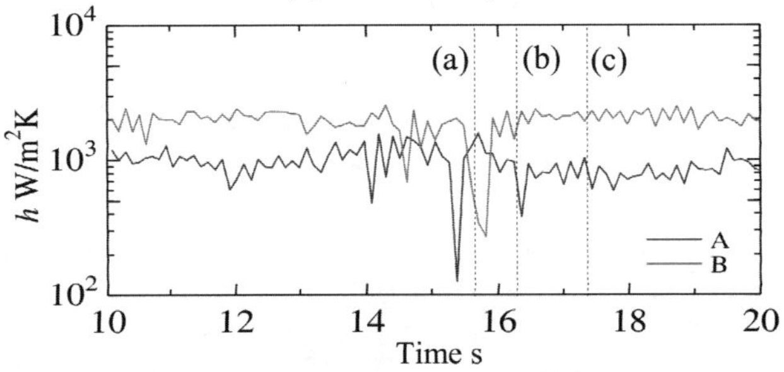

(iii) Heat transfer coefficient

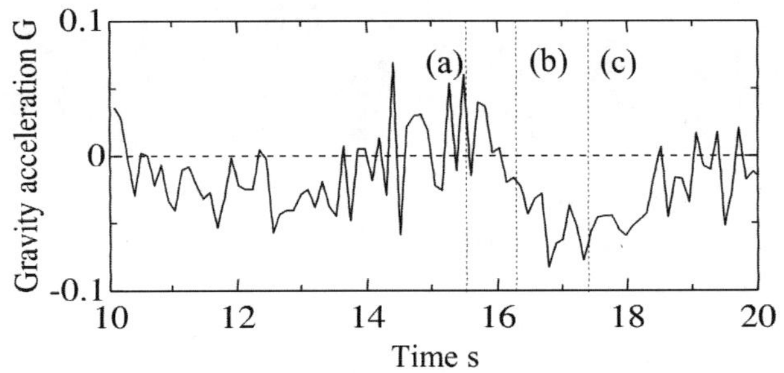

(iv) Gravity acceleration

Figure 6. Transitions of local heat fluxes, surface temperatures, local heat transfer coefficients at points A and B on Figure 5 and gravity acceleration.

detachment became longer compared with those at –0.07g. Generated bubbles grew up slowly sliding from the upper right to the lower left on the heating surface, where the heat transfer was enhanced to same extent along the flow direction of sliding bubbles.

In these figures, negative values of heat transfer coefficient was observed due to caluclated negative surface heat fluxes. Such a situation occured when the heat flow was directed to the backside of the heating surface substrate because of the deteriorated heat transfer or of the heat loss at the corner of the heated area.

6. Conclusions

A transparent heating surface with multiple arrays of 88 thin film temperature sensors and mini-heaters was developed for the clarification of boiling heat transfer mechanisms in microgravity. The images of liquid-vapor behaviors and corresponding data of local heat transfer were simultaneously obtained in microgravity pool boiling experiments by ESA parabolic flight campaign. To obtain local heat transfer coefficient, the heat flux distribution on the liquid side of heating surface was calculated from heat conduction across the substrate by using three-dimensional analysis tool.

(1) It was found that the liquid-vapor behaviors were strongly affected by the direction and the level of residual gravity, which in turn resulted in the variation and the distribution of local heat transfer coefficients.
(2) During a positive microgravity period, bubbles grew being pushed towards the heating surface. The heat transfer was deteriorated because dry patches were extended underneath attached flattened bubbles in a large size.
(3) During a negative microgravity period, bubbles are detached from the heating surface. The heat transfer was dependent on the surface nucleation characteristics. The heat transfer coefficients were relatively enhanced at the center of the surface with a lot of active cavities.
(4) When bubbles grow up slowly sliding on the heating surface, the heat transfer was enhanced along the flow direction of sliding bubbles.

Acknowledgments

The authors express appreciation for the support of the experiment by ESA, NOVESPACE and JAXA in the 49th ESA parabolic flight campaign. This study was carried out as a part of the activity of the "Boiling and two-phase flow in microgravity" research group that is supported by JAXA (HAJK202905).

References

[1] Merte H. Jr 1990 *Progress in Astronautics and Aeronautics* **130** 15
[2] Straub J. 2001 Boiling heat transfer and bubble dynamics in microgravity *Advances in Heat Transfer* **35** 57-172
[3] Stephan P. and Hammer J. 1994 *Warme und Stoffubertragung-thermo and Fluid Dynamics* **30** 119
[4] Wayner P. C., Kao Y. K. and LaCroix L. V. 1976 *J. Heat Mass Transfer* **19** 487
[5] Kim J., Benton J. F. and Wisniewski D. 2002 *Int. J. Heat Mass Transfer* **45** 3919
[6] Sato Y., Inoue K. and Ohta H. 2007 *J. Japan Society of Microgravity Application* **24(1)** 71
[7] Akagi S., Sakata Y. and Ohta H. 2008 *J. Japan Society of Microgravity Application* **25** 327
[8] Sakata Y., Akagi S., Yoshioka S., Shinmoto Y. and Ohta H. 2008 *Proc. 26th International Symposium on Space Technology Science* 2008-h-27 in CD-ROM
[9] Sakata Y., Kawanami O., Kotani Y., Asada Y., Nagayasu T., Shinmoto Y., Ohta H., Kabov O., Queeckers P., Chikov S. and Straub J. 2009 (in Japanese) *J. Japan Society of Microgravity Application* **26(3)** 204
[10] Kawanami O., Ohta H., Kabov O., Sakata Y., Kotani Y., Asada Y., Nagayasu T., Shinmoto Y., Chikov S., Queekers P. and Straub J. 2009 *Microgravity Science and Technology* **21(Suppl 1)** S3

Mass transfer of organic substances in supercritical carbon dioxide

M Hu[1], R Benning[1], Ö Ertunc[1], J Neukam[1], T Bielke[1], A Delgado[1], V Nercissian[2], A Berger[2]

[1] Institute of Fluid Mechanics, Friedrich-Alexander University Erlangen-Nuremberg, Cauerstrasse 4,

[2] Institute of Optics, Information and Photonics, Friedrich-Alexander University Erlangen-Nuremberg, Staudtstr.7 / B2, 91058 Erlangen, Germany

E-mail: miao.hu@lstm.uni-erlangen.de

Abstract. In this work special attention is paid on the direct visualization of the diffusion process of oil droplets in supercritical carbon dioxide as well as better characterization of the process by quantitatively evaluating the important parameter - the diffusion coefficients obtained with a shearing interferometer. Experiments are also to be carried out under microgravity to improve the experiment condition where the influence of gravity-driven convection that usually dominates the transport process is minimized.

1. Introduction

Supercritical fluids have been widely utilized as solvents in the chemical industries, as well known in the material processing procedures, e.g. tea and coffee decaffeination, hops and flavor extraction, pharmaceuticals isolation and fractionation, etc. Supercritical fluids, especially carbon dioxide, water and alcohols have been recognized as excellent alternatives to the conventional chemical solvents in many areas, including separation, chemical reaction, surface processing and waste treatment etc. They can provide clean and sustainable chemical processes with less energy consumption and high efficiency, which keeps up with the increasing awareness of health, environment and safety issues in nutrition and medicine production processes. To achieve a better understanding of the process as well as a more precise system control of the parameters, research into supercritical fluids concerning thermodynamic fundamentals, i.e. fluid dynamics, heat and mass transfer, interfacial phenomena, etc. is very helpful and will contribute to the further development of their large – scale industrial applications [1].

Besides the well known environmental aspects of supercritical carbon dioxide ($SCCO_2$), it is also advantageous in the extraction processes as solvent due to its preferable solubility properties. It has been proved to be selective in the separation of desired compounds, with the relatively modified critical parameters ($T_c = 304.1K$, $P_c = 7.38MPa$) it does not tend to leave any toxic residuals in the extracts or to cause potential thermal degradation of the products being processed. The application of $SCCO_2$ to extract and to isolate compounds from natural materials has been documented for nearly three decades [2].

Journal of Physics: Conference Series **327** (2011) 012041 doi:10.1088/1742-6596/327/1/012041

Although possessing many obvious advantages, the commercial applications are still restricted to very few products due to high investment costs for this novel but unfamiliar treatment. Necessary advances in process, equipment and product design as well as the realization of the potentially economic opportunities in the market of high value products manufacturing could lead to higher interest in the investment and further development in this branch. [3]

2. Experiment implementation

2.1 Concept and experiment design

To exploit the behavior of supercritical fluids near their critical points and their transport properties, which are essential in designing reactors and optimization of the processing procedures, e.g., many efforts have been made to quantify the dissolving process by precisely measuring the transport coefficients. Well known are Taylor-Aris Dispersion (also known as Capillary Peak Broadening and Supercritical Chromatography Technique), Solid Dissolution, Photon Correlation Spectroscopy, Nuclear Magnetic Resonance (NMR), etc. [4] In this work, a new optical approach – shearing interferometer – was applied to visualize directly as well as to analyze quantitatively the phenomenon. It provides a noninvasive observation method with the possibility of being automated and integrated into the industrial scale – setups serving as the inspection or control link.

The concept was to visualize the diffusion of a sample droplet (solute) with a diameter of ca. 2mm in supercritical CO_2 (solvent). In this dimension, it requires great care to minimize the interference of convection, which is largely enhanced by the gravity and also by the sensitivity of supercritical fluids to small changes in temperature and pressure. In order to better control the experiment conditions, an experimental setup was specially built to provide a stable environment. Besides, experiments under microgravity (two parabolic flight campaigns) were also carried out. The proposed setup to implement the experiments is shown in Figure 2.1 below.

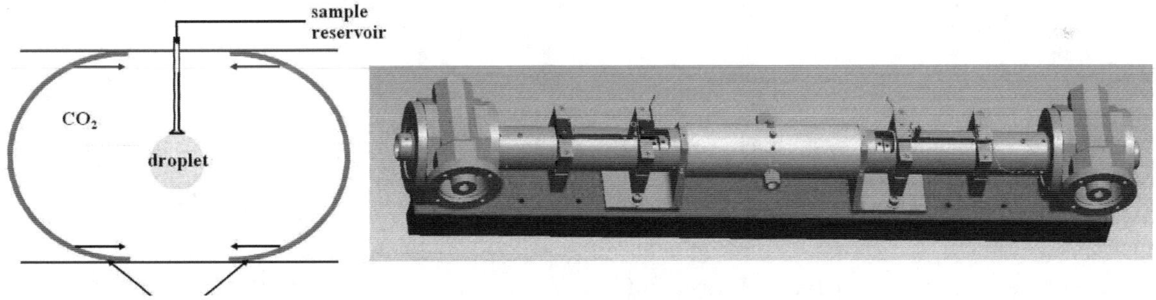

Figure 2.1 Concept for the diffusion experiment (left, side view) and the 3D experiment construction (right) with experiment cell in the middle (the yellow part).

A cylindrical high pressure chamber, as shown in Figure 2.1, with two movable half spherical walls is first filled with sub-critical carbon dioxide. In the middle of the chamber a droplet of sample is produced and fixed (see Figure 3.1, left). The temperature inside the chamber is kept at a constant level above the critical temperature of CO_2 (31°C). By reducing the volume of the chamber (increasing the pressure while compressing), CO_2 reaches its supercritical state and the dissolution process of the sample is optically observed. As the phase transition is very sensitive to the small changes in pressure and in order to obtain measurements with high accuracy, the concept aims to eliminate the density – driven convection, that would affect the pressure distribution inside the chamber. As the sample is a droplet, the chamber walls are specially designed in half spherical shape, to minimize the induced flow during a compression. This way, the direction of the change in density is constrained to a direction perpendicular to the droplet surface and the variations in concentration fields near the droplet are restricted to a minimum.

In the experimental setup the volume of the cell was controlled with two pistons on the sides, which were driven by two servo motors (see Figure 2.1, right). The experiment system was later assembled in two different configurations to meet the safety requirements of a parabolic flight campaign as well as to achieve a better performance of the setup in earth laboratory, respectively.

2.2 System description

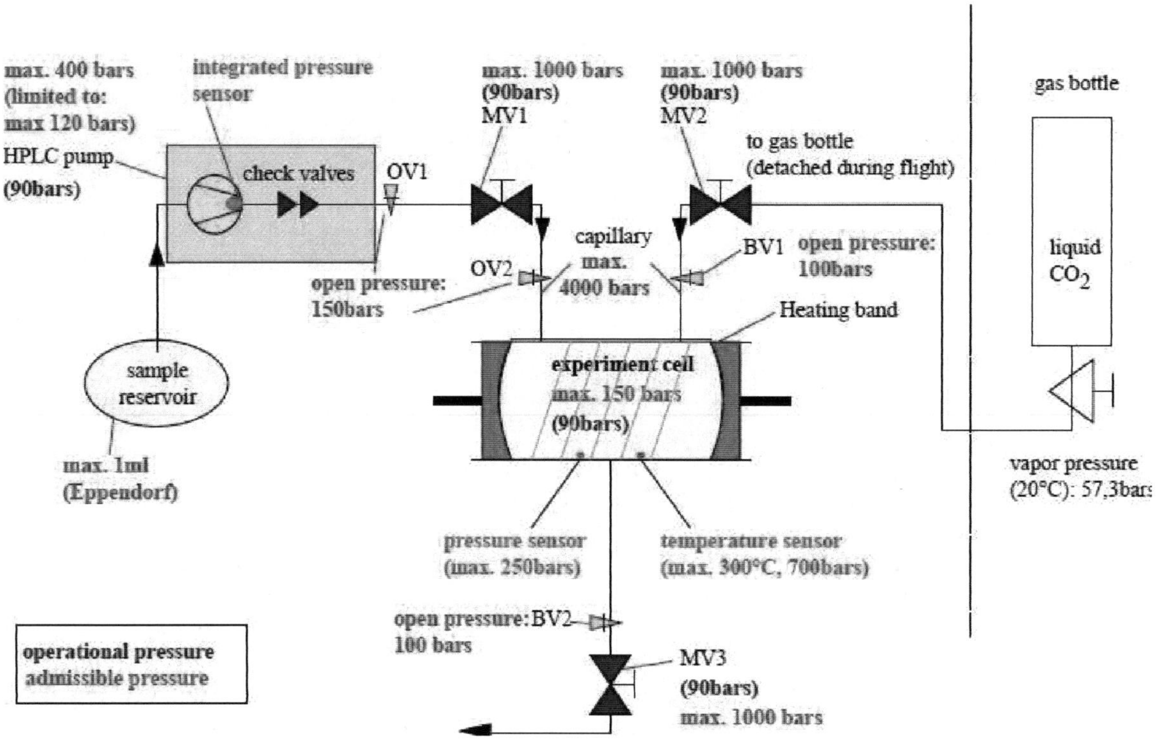

Figure 2.2 Experiment system plan, with "OV" indicating overpressure valves, "BV" burst plate valves and "MV" manual valves.

The experiments took place in the experiment cell – the high pressure compressor which enabled an optical access during experiment. Besides the three accesses to the experiment cell (for medium injection (CO_2), sample injection and waste outlet), on the cell wall there were also mounting holes for the temperature as well as pressure sensors, which served to control and monitor the experimental conditions.

2.2.1 High pressure system

The experiment cell can hold a pressure up to 150bars, which can be achieved by isothermal compression of the cell. The sample droplet was injected by a HPLC pump with an accuracy of flow control down to 0.001ml/min. The high pressure system consisted of three manual valves (see Figure 2.3 below) with the purpose to reduce the dead volume inside the high pressure capillaries. To fulfill the safety requirements for such high pressure systems in parabolic flight campaign, two overpressure valves (OVs, with an opening pressure of 150bars) and two burst plate valves (BVs, with a burst pressure of 100bars) were also installed into the system according to the "two-fault-tolerance" rule.

In the configuration of the experimental setup in earth laboratory the safety valves (OVs and BVs) were dissembled to further minimize the dead volume during experiments to improve the experiment conditions.

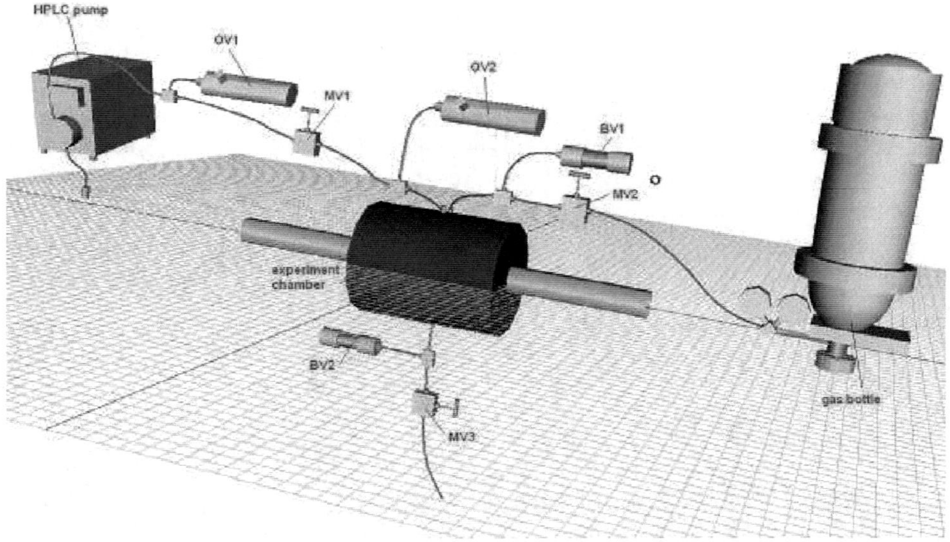

Figure 2.3 The high pressure system (3D demonstration), with the "OV" indicating overpressure valves, "BV" for burst plate valves und "MV" for manual valves.

2.2.2 Optical system

During an experiment run the pressure and the temperature inside the experiment chamber were monitored and registered, and the observation of the diffusion process was accomplished by means of a shearing interferometer with a 12mW He-Ne-Laser (wavelength of 632.8nm), two Wollaston prisms as the shear-unit and a high speed camera as detector (with 1024x1280 pixels and 12μm x 12μm/pixel, with a max. frame rate of 636fps), see Figure 2.4 below. The "infinite fringe alignment" – arrangement was applied for the measurements.

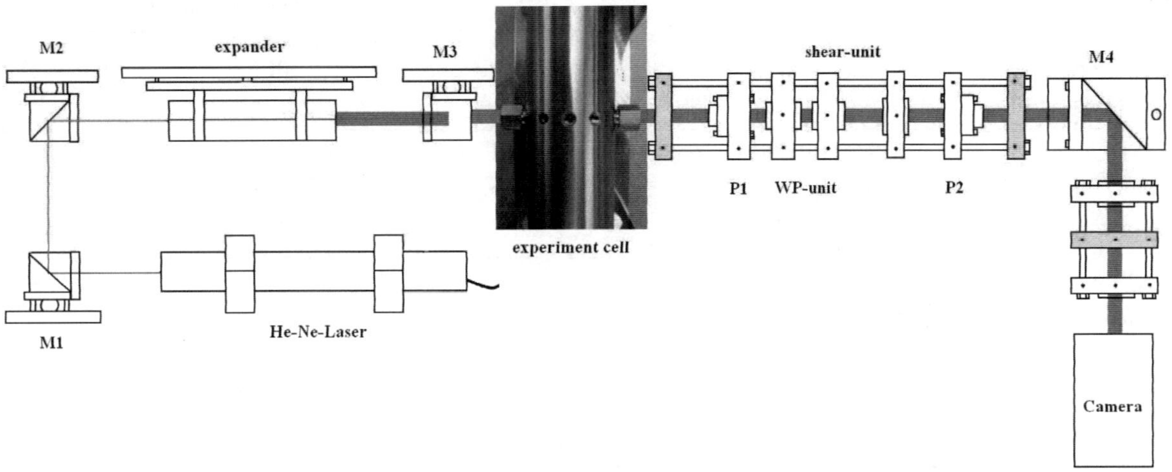

Figure 2.4 Shearing interferometer setup. "M" indicates mirrors, "P" for polarizers and "WP" for Wollaston prisms.

This construction was under the consideration of experiment conditions in flight as well as the actual space in the experiment rack, in the configuration in earth laboratory after flight experiments unnecessary components (e.g. multiple mirrors, etc.) were dissembled, which would bring in additional errors to the system.

Journal of Physics: Conference Series **327** (2011) 012041 doi:10.1088/1742-6596/327/1/012041

2.2.3 Mechanical and electrical system

Optical methods especially those concerning lasers are well known for their accuracy and at the same time their sensitivity to their mechanical constructions, i.e. their own fixations, the vibrations during experiments, etc. Considering the experiment conditions in an airplane, the shearing interferometer setup was chosen because of its satisfying stability agsainst strong mechanical vibrations which were not avoidable. According to the different requirements in flights as well as in earth laboratory two mechanical configurations were constructed (see Figure 2.5 below) to guarantee the proper experiment functionality of the whole system.

Figure 2.5 The experiment configurations for parabolic flight campaign (left) and in earth laboratory (right), respectively.

For experiments in flight the configuration was safety-orientated with double-containment-rule that all laser components were covered with protection plates and laser adsorbing material was attached on the inside. In the configuration in earth laboratory the laser components were assembled on a standard optical bank to better assure the maximum measurement capacity of the whole setup.

Electrical systems are mainly data acquisition systems, which consist of temperature-/pressure acquisition and the image acquisition from the interferometer (also: interferogram), as well as the motor control system. For the flight configuration a fuse box was built in to meet the safety requirement.

The technical data of the equipment used in the experiments are summarized in Table 2.1 below.

	Temperature sensor	Pressure sensor	He-Ne-Laser	High speed camera
Sample rate	40Hz	40Hz	Wavelength: 632.8nm Power: 12mW	Resolution: 1280 x 1024 pixels (12μm x 12μm) with 1 x 1 binning, max. 636fps
Measurement range	0 – 200°C	0 – 250bars		

Table 2.1 Technical data of the equipment used in the experiment.

3. Experiment plan and setup calibration

In the overall experiments three test-oil samples were used – rose oil (an aromatic oil for household as air fresher, commercial use), a lubricant oil (Croda, a gear oil in automobiles, commercial use) and DL-alpha-tocopherol (Merck, 99.8% purity, research use only). The three test-oils exhibited increasing density and viscosity. And after the pretests, they had been proved to be able to demonstrate three different scales of transport phenomena characterized by different diffusion velocities (diffusion coefficient, i.e.) into the solvent, which were directly observable from the fringe patterns of the interferograms.

The experimental conditions were first set to assimilate the conditions that usually encountered in the industrial procedures of supercritical fluids processing, i.e. the temperature was near room temperature (35-45°C) and the pressure was adjustable in a relatively wide range (60-100bars, limited by the capacity of the experiment cell). The relatively small dimensions of the experiment cell defined the usable amount of solvent, and therefore limited the dissolubility of the solute in this certain amount of solvent. In order to obtain the most advantageous observation a wider range of temperature was set in the later experiments. The overall experiment region is illustrated in Figure 3.1 below.

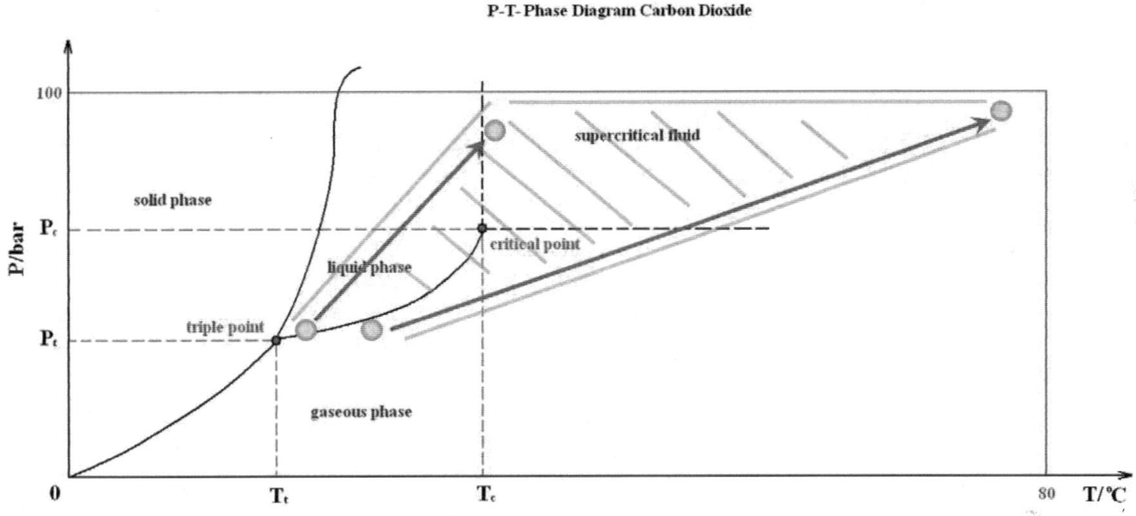

Figure 3.1 Experiment region illustrated in a P-T-phase diagram.

The experimental setup requires a proper validation and scale up so that with direct observations of the interferograms in real-time, one can conveniently categorize the sample oil into the correspondent property class. This can be realized by testing the reference-oil samples with the experiment setup, whose transport properties were well studied and reported in the literatures. As the first tests, in order to obtain obvious observations under relatively low temperature condition, some substances with higher diffusion coefficients were chosen – acetone, benzene and toluene, with increasing molecular weight as well as density. In addition, samples as naphthalene and oleic acid methyl ester were taken as alternatives (see Table 3.1 below). Depending on the results, further reference-oil samples with smaller diffusion coefficients would be applied.

The diffusion coefficients demonstrated in the table above were taken from the data provided by different authors for the same substance with different measurement methods ([5] [6] [7]). Some of them show a nonlinear behavior with increasing pressure at a constant temperature.

Besides diffusion phenomena it is also of interest to observe the dynamic of the fluid behavior in its supercritical state as well as during the phase transition. For this purpose isothermal compression and expansion were also carried out.

	Molar weight [g/mol]	Density [kg/m^3]	Temperature [K]	Pressure [bar]	Diffusion coefficient [10^{-8} m^2/s]
Acetone	58.08	792.5	303-333	110-160	1.4-2.7
Benzene	78.11	876.5	303-333	110-160	1.3-2.5
Toluene	92.14	866.9	313-373	100	1.5-4.5
Naphthalene	128.17	1140	303-333	72-90	3.8-1.22-1.1
Oleic acid methyl ester	296.5	870	313	80-110	1.7-1.1-0.8

Table 3.1 Properties of reference-oil samples.

4. Results and discussions

4.1 Calibration

Calibration experiments with the reference-oil samples (acetone, benzene, toluene, etc.) were carried out in earth laboratory, the experiment temperature was first set to a lower value (35-40°C) and the pressure was varied between ca. 65bar and 90bar. Example images of fringe patterns in the interferograms are shown in Figure 4.1 below.

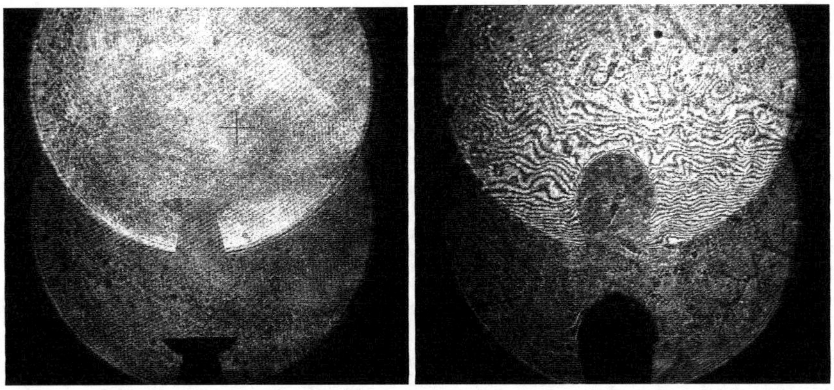

Figure 4.1 Example images of interferometer, after laser adjustment – infinite fringe alignment (left) and experiment with acetone droplet in earth laboratory (right, 308K and 78bar).

Experiments with the reference-oil samples (acetone, benzene, toluene, etc.) were carried out in earth laboratory and processed. The characteristics of the interferograms, e.g. number and width of the fringes, the distances between the fringes around the droplet, additionally also the changes of the droplet size in time were registered as references. Also for the purpose of setting the references of the results, the outlet process of the medium from the experiment cell through the sample injection line were also captured and processed, where the changes of the velocity field (i.e. the density changes) observed were easily predictable by aerodynamic relations for a pressure balancing process. Example images for an outlet process are shown in Figure 4.2 below.

4.2 Results from the earth laboratory

The diffusion phenomena were observed in different time scales of ca. 1.5s for the rose oil and ca. 15s for the lubricant oil, respectively. In the case of tocopherol which has a typical observable diffusion time of about 30min to one hour, no simultaneous density changes were observed besides the changes of the sample droplet's size. All experiments were carried out under isothermal condition under different pressures, i.e. the medium was heated to a defined temperature, which was kept constant while the pressure was adjusted to new values – isothermal compression and expansion.

Figure 4.3 below shows two example image sequences of rose oil and lubricant oil with the most obvious observations of density changes occurred at ca. 333K and 75bar.

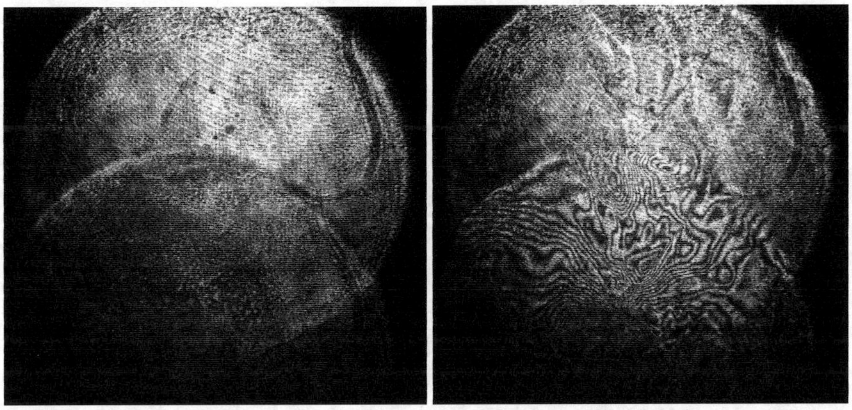

Figure 4.2 Example images of an outlet process (through the injection capillary in the middle) as reference for density changes in the medium, with a start pressure at ca. 65bar.

Comparing the two samples, rose oil had the smaller viscosity and diffused faster (density changes in ca. 1.5s compared to lubricant oil in ca. 15s), which could be directly evaluated from the interferograms.

4.3 Observation under microgravity

One of the goals of the project is the observation of diffusion under microgravity to avoid the influence of the gravity-driven convection that overlaps the diffusion process on earth. For the flight experiments in the most recent campaign (54th ESA parabolic flight campaign) only tocopherol as sample was certified, so no significant changes of the droplet as well as in the medium concerning a diffusion process was observed. Example images of an arbitrary parabola during the flight campaign are shown in Figure 4.4 below.

From the recordings of the flight experiments it can be seen that the microgravity phase stabilized after the first 3-5s. Image 4 is the most optimal condition for the observation of a convection free diffusion process.

Other goals of the flight campaign were to test the recording of the behavior of $SCCO_2$ in microgravity phase and to successful carry out the experiment procedures (droplet injection in both 0g and 2g phases, experiment cell volume control by driving the pistons, etc.). The functionality of the experiment setup was proved and confirmed to be able to provide a stable environment for observation of diffusion also under microgravity conditions.

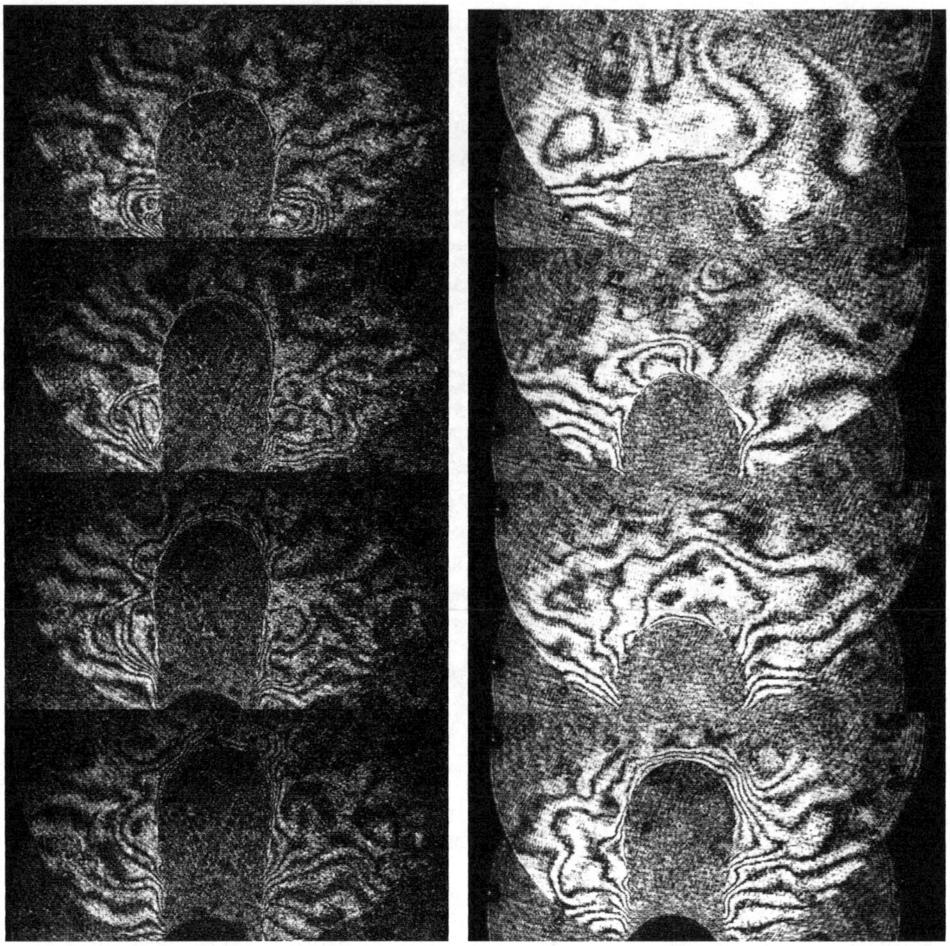

Figure 4.3 Example image sequences from experiments with rose oil (left, in ca. 1.5s) and lubricant oil (right, in ca. 15s), respectively.

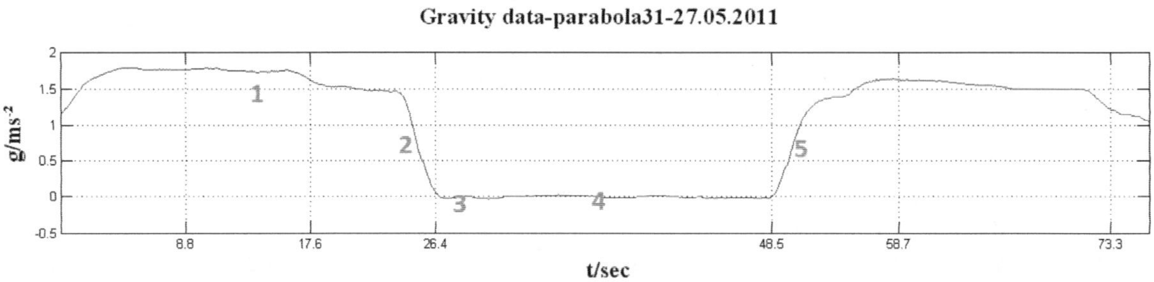

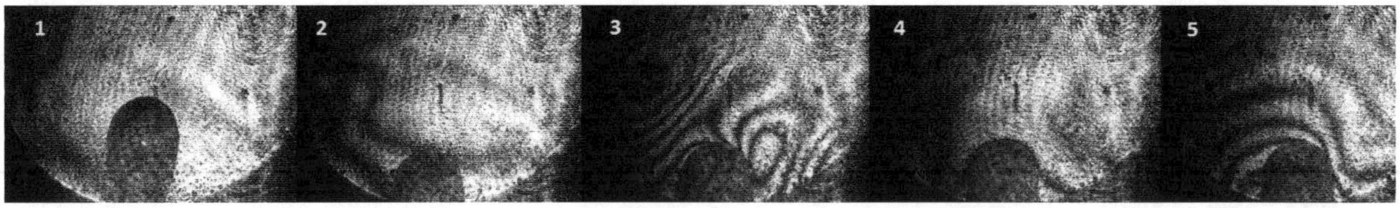

Figure 4.4 One image sequence for a parabola, images 1 to 5 were from ca. 2g to 0g.

5. Data processing and model fitting

5.1 Interferogram analysis

A shearing interferometer links the variations in refractive index gradients observed as fringe shifting in an interferogram to the variations in density gradients, with one fringe shift corresponding to one wavelength λ of the laser used by

$$\frac{\Delta l}{\lambda} = \frac{\Delta S}{S} . \qquad (1)$$

with Δl being the optical length change, ΔS the fringe shifting and S the fringe distance before shifting.

Under the assumption that the test object (the experiment cell) is a 2-D object (considering the 12mm distance between the windows, and the ca. 2mm sample droplet), refractive index, i.e. the density, does not vary in the laser propagation direction, the optical length change Δl of the laser beam after passing the test object is obtained by

$$\Delta l(x, y) = d \cdot \int_{z_1}^{z_2} \frac{\partial n(x, y)}{\partial \xi} dz = d \cdot \frac{\partial n(x, y)}{\partial \xi} \cdot L \qquad (2)$$

with

d – the shear-length of the interferometer setup,

n – the refractive index,

L – the length of the test object in the laser propagation direction,

ξ – the integration direction.

The Clausius-Mosotti relation [8], which links the refractive index to density, yields

$$\frac{n^2 - 1}{n^2 + 2} = \frac{2}{3} K \cdot \rho \qquad (3)$$

where,

ρ – the density,

K – the Gladstone-Dale constant.

Combination of Eq. (1) and (2) leads to

$$\frac{\partial n}{\partial \xi} = \frac{\lambda}{d \cdot L} \cdot \frac{\Delta S}{S} . \qquad (4)$$

With the following form the density gradients are obtained

$$\frac{\partial \rho}{\partial \xi} = \frac{\partial \rho}{\partial n} \cdot \frac{\partial n}{\partial \xi} \qquad (5)$$

To calculate the diffusion coefficient Fick's 1st Law [9]

$$J = -D \frac{\partial c}{\partial \xi} \qquad (6)$$

is applied, where,

Journal of Physics: Conference Series **327** (2011) 012041 doi:10.1088/1742-6596/327/1/012041

J – the flux,

D – the mutual diffusion coefficient,

c – the mole concentration of the sample.

It is assumed that the diffusion coefficient is constant for each image in a sequence and that there are only small concentration changes between images (adjustment by setting the delay between the images) compared to the whole process. As in the experiments the droplet injection was carried out in a continuous mode, the surface concentration of the droplet remains unchanged. According to mentioned criteria the following relation is obtained,

$$\frac{\partial \rho}{\partial t} \cdot \Delta s_i = D_i \cdot \frac{\partial \rho}{\partial \xi} \qquad (7)$$

with, Δs_i – the diffusion length in each image,

ρ – the density of the medium,

i – the image number in one sequence.

For all images in a sequence a diffusion coefficient is calculated by analyzing the fringe shiftings (see Figure 5.1).

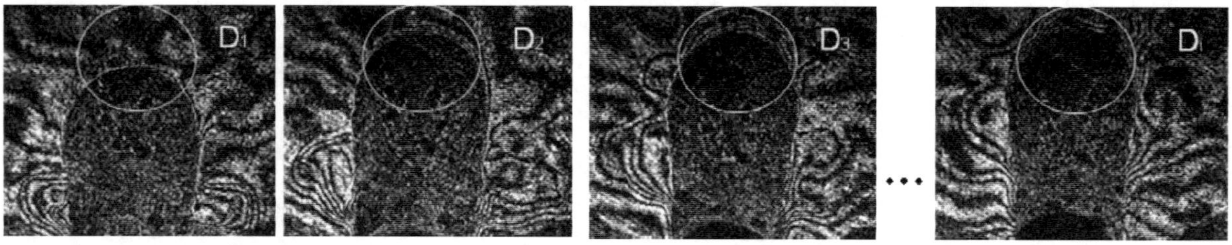

Figure 5.1 The diffusion coefficient was calculated for each image in one certain area. Due to the vertical shear (highest sensitivity) the top of the droplet was chosen.

5.2 Evaluation of the droplet size

When the test-oil samples had a relatively small diffusivity (by ca. 10^{-9} m^2/s) at the moderate temperature, or after many injections of the oil samples the solvent had almost reached its saturation limit, no obvious fringe patterns in the interferograms could be observed. In these situations the data processing could be accomplished by evaluating the changes of the droplet size during an isothermal compression. An example is shown in Figure 5.2 below.

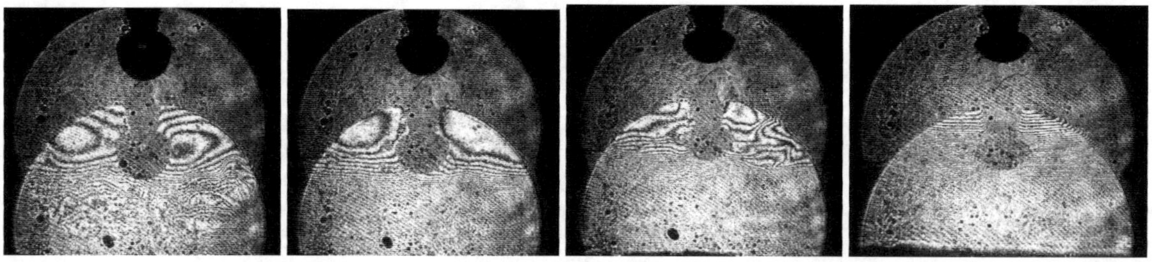

Figure 5.2 Example images of solute dissolution during an isothermal compression, which enabled the evaluation of the transport property by analyzing the droplet size changes in time.

The requirement of this approach was to properly choose the compression in a relatively small pressure range (within 5bars) to make sure that the diffusivity could be assumed constant and at the same time an optimal size change could be captured.

By applying the Fick's first law for diffusion in a radial symmetrical geometry:

$$\frac{\partial c}{\partial t} = D \cdot \frac{1}{r^2} \cdot \frac{\partial}{\partial r}\left(r^2 \cdot \frac{\partial c}{\partial r}\right) = D \cdot \frac{1}{r^2} \cdot \left(2r\frac{\partial c}{\partial r} + r^2 \frac{\partial^2 c}{\partial r^2}\right) \tag{8}$$

with

c – the mole concentration of the solute,

t – the diffusion time (compression time),

D – the diffusion coefficient of the solute,

r – the integration direction.

Considering the dimension and condition of the experiment, that the droplet disappears completely in the end, the following relation is valid,

$$\frac{\partial c}{\partial r} \approx \frac{c_0}{R} \tag{9}$$ and

$$\frac{\partial c}{\partial t} \approx \frac{c_0}{T}, \tag{10}$$

while

c_o – the initial mole concentration,

R – radius of the droplet,

T – the experiment time.

By means of equations (9) and (10) together with (8) the following relation can be obtained,

$$\frac{c_0}{T} \approx D \cdot \frac{1}{R^2} \cdot \left(2 \cdot R \cdot \frac{c_0}{R} + R^2 \cdot \frac{c_0}{R^2}\right) = D \cdot \frac{1}{R^2} \cdot 3 \cdot c_0. \tag{11}$$

So the diffusion coefficient could be calculated by

$$D \approx \frac{1}{3}\frac{R^2}{T} \tag{12}$$

5.3 Semi-empirical diffusion model based on experiment data

A nonlinear diffusion model was applied and all the parameters were chosen in order to approximate the experiments. According to this fitted model the diffusion process was simulated.

Nonlinear diffusion is a further approach to the real diffusion situation, in which the diffusion coefficient varies with the density gradient. According to the Fick's 2nd Law,

Journal of Physics: Conference Series **327** (2011) 012041 doi:10.1088/1742-6596/327/1/012041

$$\frac{\partial c}{\partial t} = \frac{\partial}{\partial \xi}(D_{nl} \cdot \frac{\partial c}{\partial \xi}), \qquad (13)$$

where $D_{nl} = D_{nl}(\frac{\partial c}{\partial t})$ is the function of the concentration gradient,

c – the mole concentration of the medium,

D_{nl} – nonlinear diffusion coefficient, which can be describe by the following form,

$$D_{nl} = 1 - \exp(-\frac{C_m}{(g/\lambda)^m}) \text{ for } g > 0 \text{ and } D_{nl} = 1 \text{ for } g <= 0. \qquad (14)$$

The parameters C_m, λ and g are chosen according to the experimental observations and conditions as well as considering the dependence between the three.

The diffusion calculated with the model is displayed in Figure 5.3 below.

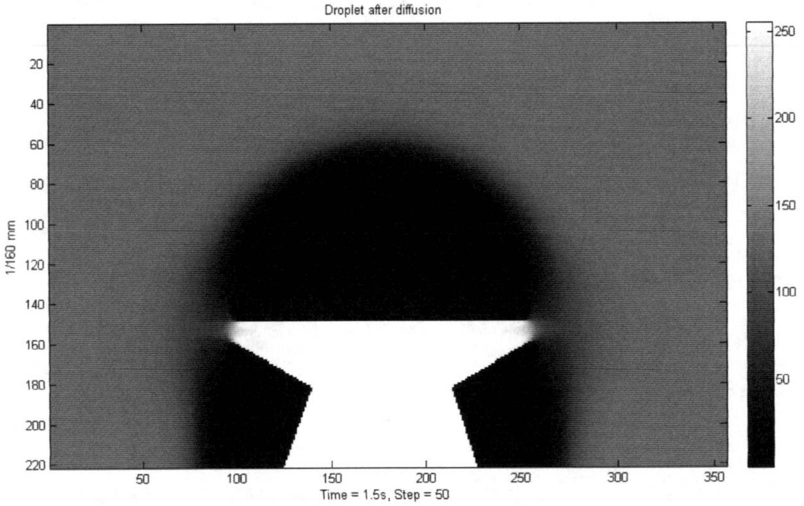

Figure 5.3 The diffusion layers were calculated according to the experiment conditions (rose oil) as well as considering the form of the injection capillary (in the middle) and the real density difference between the droplet and the medium. [10]

After a nonlinear diffusion process a variation of the diffusion velocity with time was observed and the diffusion fronts were mainly the propagation of the contour lines of the droplet (which were the same as the fringes in an interferogram). The deviations from the droplet contour were observed both in the model and in the experiment at the sharp edges of the injection capillary, where the density gradient varies faster than that in the other regions. This phenomenon in the experiment is shown in Figure 5.4 below.

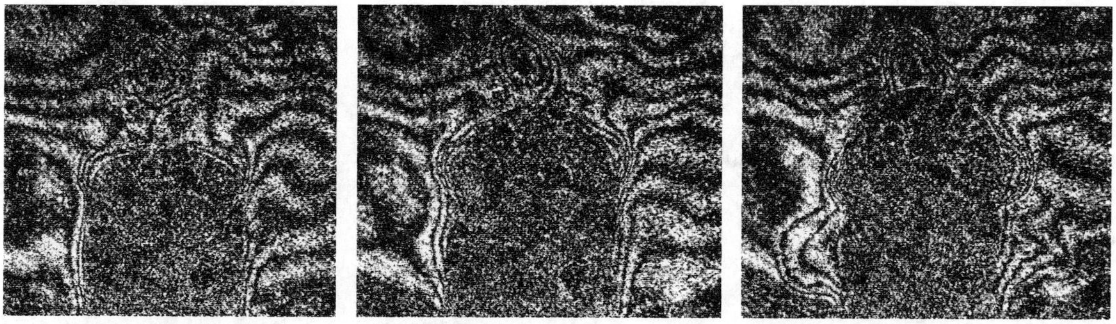

Figure 5.4 The deviation of the diffusion fronts from the droplet contour at the sharp edges of the injection capillary in an image sequence (rose oil).

For experiment results from earth laboratory with the influence of convection the transport process is to be characterized with transport coefficient, instead of diffusion coefficients. The obtained values from one image sequence of rose oil as well as the calculated results from the model were plotted in Figure 5.5 below.

From the plot in Figure 5.5 the aberration between the data set and the model is obvious. The measured diffusivities are generally faster than those from the calculations, as in earth laboratory the dissolving process had been the combination of diffusion and convection. And the fluctuations of the data set indicated that small variations of the density in the bulk medium were not avoidable.

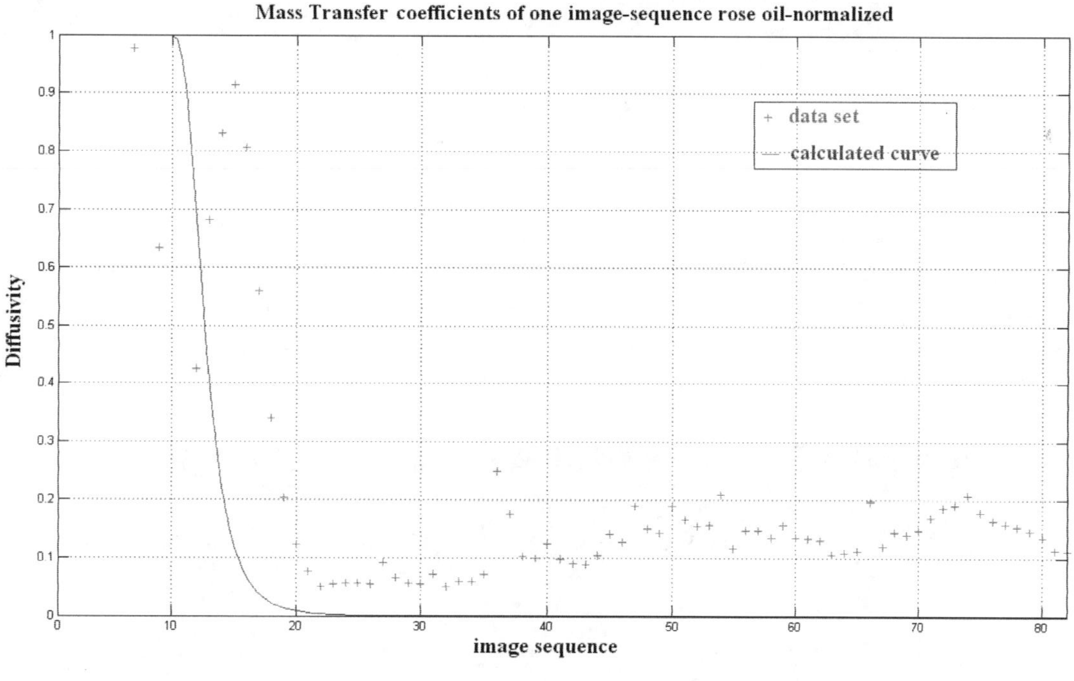

Figure 5.5 Plot of the diffusion coefficients from experiment and from calculation (for better comparison the diffusivity was normalized).

6. Conclusions and further aspects

In order to observe a relatively fast diffusion process (usually in seconds) in compressible fluids interferometry was proved to be a suitable method considering the accuracy and also the simple technical realisation. The only requirement on the experiment setup is the optical access. With the reference tests of the oil samples that have the identified transport properties, the experiment setup can be proper scaled and therefore be well validated. By in situ observation of the interferograms the transport properties if the test-oil samples are to be directly evaluated, the correspondent coefficients can be approximately estimated according to the established catalogue of references.

Besides, the saturation limit of a certain solute under a certain pressure and temperature can also be directly observed from the interferometer, e.g. in an isothermal compression or expansion procedure. With precisely controlled amount of mediums it is also possible to quantitatively estimate the solubility data with the setup, which shows the new applicability of this optical measurement method. An example is shown in Figure 6.1 below.

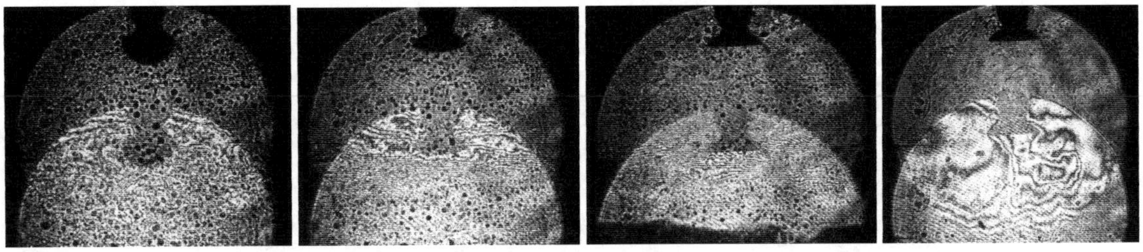

Figure 6.1 Observation of the solute dissolution during an isothermal compression. With the pressure increasing, the solute condensed on the windows gradually dissolved into the medium (lubricant oil, start pressure: 70bar, end pressure: 90bar).

With the interferometer setup besides the dynamic of fluid under various experiment conditions many other phenomena could also be observed, e.g. the transport behaviour of a mixture of oil samples, or the injection of a gas bubble into the liquid medium, which proves the applicability of the experiment setup in the research of numerous different kinds of fluids interactions, as shown in Figure 6.2 below. Considering its easy-to-use advantages and high accuracy comparing to the other measurement methods to obtain material properties, it opens a new aspect of both visualizing and analyzing the thermodynamic processes.

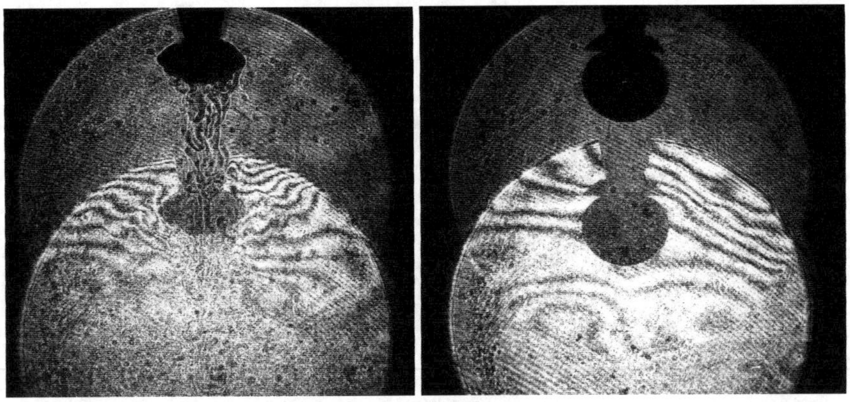

Figure 6.2 Example images of a mixture droplet of rose oil and benzene (left) and the injection of a gas bubble into a liquid reservoir.

The nonlinear diffusion model can be applied to describe the diffusion process in supercritical fluids. The necessary parameters can be obtained from comparison with the experimental results and later an improvement by generalization for different samples was also possible.

As further aspects, more experiments under compensated gravity are desired with possibly more test-oil samples in order to obtain enough data to be compared with the results from the earth laboratory. The data analysis was accomplished analytically until now. A further approach is to use numerical methods, which also can be applied to model fitting and further process simulations. From the observation and evaluation of the interferograms (i.e. acquisition of transport properties) obtained from experiments it also opens the opportunity to characterize the interactions of solute and solvent particles, which will provide necessary feed backs for simulations at molecular level.

References

[1] Arai Y, Sako T and Takabayashi Y (Eds.) 2002 *Supercritical Fluids, Molecular Interactions, Physical Properties and New Applications* (Springer-Verlag Berlin und Heidelberg)

[2] Mohamed R S and Mansoori G A June 2002 *Featured Article – Food Technology Magazine The World Market Research Centre (London UK)*

[3] Stahl E and Quirin K. W 1983 *Fluid Phase Equilibria* 8 p 93

[4] Liong K K, Wells P A and Forster P R 1991 *Journal of Supercritical Fluids* 4 p 91

[5] Sassiat R, Mourier P, Caude M H and Rosset R H 1987 *Analytical Chemistry* 59 p 1164

[6] Dahmen N, Kordikowski A and Schneider G M 1990 *Journal of Chromatography* 505 p 169

[7] Higashi H, Iwai Y, Takahashi Y, Uchida H and Arai Y 1998 *Fluid Phase Equilibria* 144 p 269

[8] Merzkirch W 1987 *Flow Visualization* (2. Ed., Academic Press, New York and London)

[9] Crank J 1985 *The mathematics of diffusion* (2.Ed. Oxford Science Publications, Clarendon)

[10] *Nonlinear Diffusion Toolbox* (Mathworks Inc.)

Acknowledgements

The project is funded by DLR (German Aerospace Agency). The authors would like to thank both DLR and ESA for the opportunity of conducting experiments in the two parabolic flight campaigns (16th DLR and 54th ESA parabolic flight campaign).

Microgravity experiments on ISS in order to examine a new atomization theory discovered through normalgravity and microgravity environments

J Osaka[1], S Suzuki[1], Y Suzuki[1] and A Umemura[1]

[1]Department of Aerospace Engineering, Nagoya University,
Furo-cho, Chikusa-ku, Nagoya, 464-8603, Japan

E-mail: osaka@nuae.nagoya-u.ac.jp

Abstract. In order to elucidate turbulent atomization processes, many studies by the use of a liquid jet issuing from a circular nozzle have been conducted for a long time. Although Rayleigh's instability has been regarded as the only determinant for the breakup of the liquid jet, the source of the disturbances has been unclear and thus the physical explanation of experimental results was impossible. From our experimental and numerical approaches under normalgravity and microgravity environments, it was found that the breakup by the short-wave mode occurs around the tip of the liquid jet without any disturbances. The long-wave mode is caused by the existence of a nozzle exit through a self-closed breakup cycle sustained inherently by the capillary waves emanated from the tip of the liquid jet after every breakup. Our further experiments revealed the existence of the relaxation region which gives a reasonable explanation of the extremely large breakup length. In addition, the two-valueness of the breakup length was found through a lot of experimental results. Establishment of a new breakup theory enable to explain all of experimental results requires long-period microgravity environments and the currently-projected experiments on ISS are introduced in the present paper.

1. Introduction

Liquid atomization processes are very important in engineering applications using liquid fuel such as rocket engines, gas turbines including jet engines, automobile engines and so on. In addition, this technology is essential to ink-jet printers. Against such practical backgrounds, many studies on liquid atomization processes have been conducted. A snapshot representing typical atomization processes is depicted in figure. 1. A high-speed liquid jet is injected from a circular nozzle. Liquid atomization processes are categorized into three phases. The nondimensional number characterizing each phase is gas Weber number defined as

$$We_G = \frac{\rho_G U^2 a}{\sigma} \tag{1}$$

where ρ_G, U, a and σ correspond to gas density, velocity difference between gas and liquid, a radius of liquid jet and surface tension coefficient, respectively. Gas Weber number corresponds to the ratio of gas dynamic pressure to surface tension in two-phase flow fields. Just downstream of the nozzle exit, denoted as the region A in figure. 1, the small-scale disturbances are induced on the liquid surface.

Since gas Weber number in the region A is sufficiently larger than **O**(1) due to the large velocity difference, the effects of surface tension is small and the formation of such disturbances are dominated by the characteristics of turbulence. Such disturbances cause the formation of the fine liquid ligaments in the region B, where gas Weber number is **O**(1) due to the small scale of the liquid ligaments and the small local velocity difference. Therefore, the effects of surface tension dominate the behaviour of the liquid ligaments and their breakup. Gas Weber number of the flow field occupied by many small droplets formed from liquid ligaments is smaller than **O**(1). In order to predict the characteristics of droplets (e.g. drop-size distribution), the flow field in the region B must be correctly understood. Since such flow field is too high-speed and small to be observed directly, a certain approach has been adopted for a long time. A fine liquid ligament is replaced by a low-speed and large-scale liquid jet in accordance with Weber's analogy. In conventional experiments, a liquid jet was injected from a circular nozzle along the direction of gravity as depicted in figure 2. Although the effects of gravitational acceleration must be taken into account, they have been unnoticed for a long time. The corresponding nondimensional number is Bond number, which is expressed as follow.

$$Bo = \frac{\rho a^2 g}{\sigma} \tag{2}$$

ρ is liquid density. The scale of the original liquid ligaments is so small that the corresponding Bond number is essentially small. However, such traditional replacement can make Bond number larger than that of the original liquid ligaments, which has led to the misunderstanding of the physics underlying

Figure 1. A schematic of typical liquid atomization processes. The region surrounded by a dotted circle corresponds to the target of the present study.

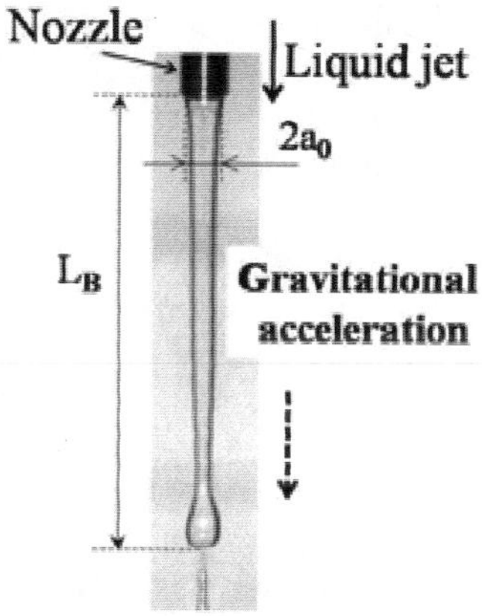

Figure 2. A typical snapshot obtained through conventional experiments.

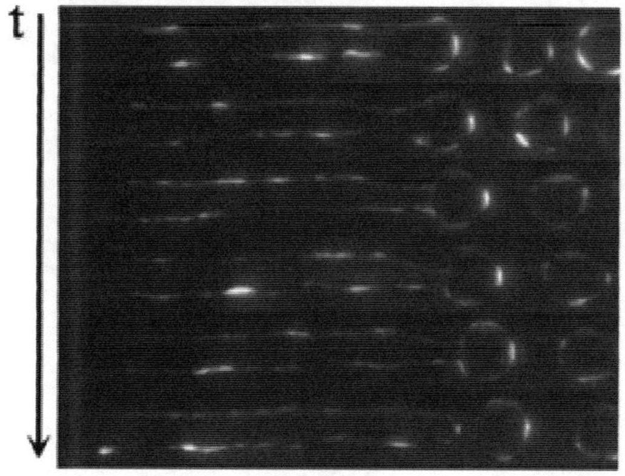

Figure 3. Successive images obtained from a microgravity experiment under high-pressure condition.

the liquid atomization processes for a long time. The most simple and effective way is to remove the effect of gravity. From this viewpoint, Umemura et al [1] carried out the experiments as illustrated in figure 2 under microgravity environments. In order to realize the flow field with small gas Weber number and composed of the liquid jet large enough to get snapshots with sufficient temporal and spatial resolution, liquid SF_6 was injected from a circular nozzle into high-pressure N_2 atmosphere. The typical snapshots obtained from the microgravity experiments are shown figure 3. In figure 3, the radius of the nozzle and the ambient pressure were 0.2 mm and 6.0 MPa, respectively. Gas Weber number is 0.133 and thus the flow field around liquid ligaments in turbulent atomization was well simulated. The mechanism conventionally used for the explanation of liquid jet disintegration is that proposed by Rayleigh [2]. According to the theory of Rayleigh, the long-wave surface deformation is the only determinant for the breakup of the liquid jet in a laminar case. However, the wavelength observed in the microgravity experiments is smaller even if the effect of gas dynamic pressure and viscosity are considered [3-5] and the breakup configuration is so different from that derived from the theories of Rayleigh. This breakup configuration was named as 'short-wave mode' by Umemura [6]. What is important is the fact that the breakup in the short-wave mode occurs without any disturbances as long as there exist the tip of the liquid jet. On the other hand, the breakup due to the long-wave disturbances, namely 'long-wave mode', inevitably requires some kind of disturbances. The breakup in the long-wave mode is usually observed even in laminar flow fields. In such cases, the sources of the disturbances were unclear in the theories of Rayleigh and have been usually attributed to the ones existing inside of a nozzle or in nature according to the previous researches. Depending on the experimental conditions, such approaches lead to the incorrectly small initial amplitude of the disturbances at the nozzle exit. This value is unable to give reasonable answers to the repeatability of the experimental results. In order to investigate the breakup processes of a liquid jet in detail, authors have been carried out the experiments as illustrated in figure 2 by the use of pure water not only on normalgravity but also on microgravity environments. From our experimental results and numerical approaches [6], the periodic breakup of a liquid jet is sustained through a self-closed breakup cycle requiring inherently no external disturbances. Keys to the self-closed breakup cycle are the capillary waves emanated from the tip of liquid jet and the behavior of the waves at nozzle exit. In addition, the further experiments elucidated the determinants other than the waves existing on the liquid jet. In order to investigate the further physics underlying the breakup of the liquid jet, the long-term microgravity experiments on International Space Station (ISS) are currently projected with the aid of JAXA. In the present paper, an outline of our breakup theory and the currently-projected space experiments are provided along with the several experimental results.

2. An outline of a new breakup theory

2.1. Capillary and unstable waves

First of all, let us consider waves on a liquid jet issuing from a circular nozzle by the use of a flow field of a liquid column with an infinite length. Here the assumptions are adopted that the effects of gas pressure, viscosity and velocity distribution of a liquid jet are negligible for simplicity. In addition, the coordinate moving at the same speed as that of the liquid jet are used. These assumptions give the similar analyses proposed by Rayleigh [2]. This analysis corresponds to the pressure imbalance between neck and bulge parts of the small-amplitude surface deformation on an infinite liquid column. The surface deformation δ is expressed in a nondimensional form as follow,

$$\frac{\delta}{a} = \frac{\varepsilon}{a} exp\{i(kx - \omega t)\} \tag{3}$$

where a, ε, k and ω are an initial radius of a liquid column, an initial amplitude, wavenumber and angular frequency, respectively. The following dispersion equation is derived by the linearization of the equations of continuity and motion around the basic flow field and by the appropriate boundary conditions.

$$\left(\omega\frac{a}{V}\right)^2 = -(ka)\left\{1-(ka)^2\right\}\frac{I_1(ka)}{I_0(ka)} \quad (4)$$

$$V = \sqrt{\frac{\sigma}{\rho a}} \quad (5)$$

$I_n(x)$ is the n-th modified Bessel function. ω is a pure imaginary number and corresponds to the temporal growth rate of the disturbances. Due to the local curvature of the generating line and the circumferential directions, the pressure differences between the neck and the bulge parts are generated. When the nondimensional wavelength $\lambda/a = 2\pi/(ka)$ exceeds 2π, the flow field is unstable. Physically speaking, the pressure at the neck parts exceeds that at the bulge parts. This pressure difference generates the liquid flows from the neck parts into the bulge parts and finally leads to the breakup of the liquid column. Such waves are called as unstable waves. According to the linear stability analysis, the disturbance with $\lambda/a = 9$ gives maximum growth rate. Growth of the unstable wave by numerical simulation is illustrated in figure. 5. The governing equation adopted is the axially-symmetric incompressible Navier-Stokes equation for nonevaporating liquid and gas. The effect of the surface tension was expressed by the continuous surface force method [7]. The flow field was assumed to be periodic in the axial direction. The details of the numerical methods were described in [8]. A sinusoidal surface deformation with $\varepsilon/a = 0.01$ was added to a stationary liquid column. The axially-elongated surface configuration of the neck part and the small satellite formed after the breakup between the main droplets are characteristics of the breakup in the long-wave mode. The breakup configuration illustrated in figure 4 is quite different from that in figure 3 as noted above.

In the case where λ/a is less than 2π, the solutions corresponding to the neutral propagative capillary wave can be obtained. Their phase speed c_R can be obtained from the equation (4) and expressed as follow.

$$\frac{c_R}{V} = \sqrt{\frac{I_1(ka)}{I_0(ka)}\left(ka - \frac{1}{ka}\right)} \quad (6)$$

From (6) it is clear that the capillary waves with the shorter wavelength give the larger phase speed and this property is important for the realization of the self-closed breakup cycle noted later. In the linear stability analysis, only the solutions corresponding to a single capillary wave component is obtained. Although the properties of the unstable and the capillary waves are explained here, it is still unclear how such waves, in other words, 'disturbances' are generated. In the next subsection, the sources of disturbances and the breakup in the short-wave mode are investigated through a problem related to contraction of a liquid column.

2.2. Short-wave mode and a source of capillary waves

In order to investigate the short-wave mode and the role of the capillary waves driven by a liquid tip, a

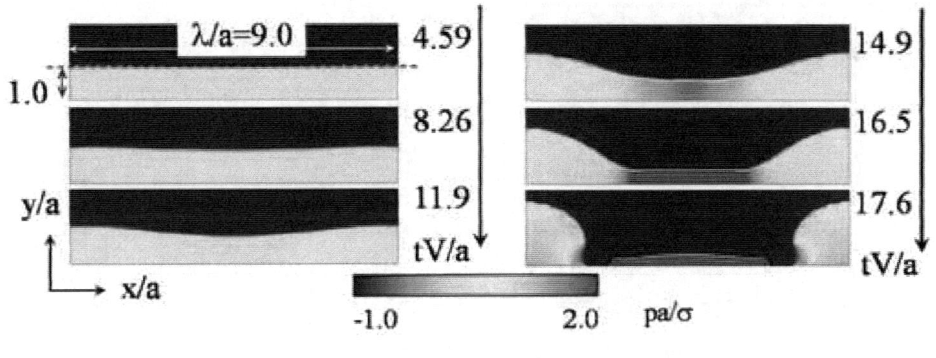

Figure 4. Growth of the most unstable wave on a liquid column. Gray-scale corresponds to pressure inside of the liquid column.

numerical simulation of an initially static liquid column were carried out as shown in figure 5 (a). The same governing equations in the previous subsection were adopted. The conditions of the flow field are similar to those of our microgravity experiments. The definitions of the characteristics parts are illustrated in figure 5 (b). The breakup of the liquid column occurs at approximately $tV/a = 20$ and the surface configuration around the breakup point is quite similar to the experimental results in figure 3. Since the tip bulb is attracted due to surface tension of the liquid column, it moves upstream with 'swallowing' its upstream liquid and enlarging its size. The resulting pressure decrease in the tip bulb observed in figure 5 (a) is necessary for the breakup in the short-wave mode. As a necessary condition for the breakup in the short-wave mode, the radius of the tip bulb must be larger than 2a. In addition, the linear surface configuration just upstream in the 1st neck observed in figures 3 and 5 (a) also plays an important role in sustaining the liquid flow into the tip bulb through its sufficient pressure rise. The breakup in the short-wave mode has no direct relationship to liquid issue speed and has a specific frequency.

As the tip contraction occurs, the capillary waves are emanated and propagate upstream as illustrated in figure 5. Although inherently a group of waves propagates due to its dispersive nature, the propagation speed of each capillary wave component follows the equation (6) [9]. By focusing on the locations of the neck parts, the movement of the capillary wave components is tracked as illustrated in figure 5 (c). Due to their dispersive nature, the capillary wave components gradually increase its wavelength with decreasing their propagation speed. Finally their propagation speed gradually reach an asymptotic value [6], in other words a steady-capillary wave is formed. V in the equation in (5) corresponds to the analytically-derived asymptotic speed and nearly equal to the contraction speed of the liquid column. The gray line in figure 5 (c) corresponds to the contraction speed and its agreement with the analytical one is satisfactory. During the contraction process, the total surface area of a liquid column decreases in accordance with the minimum surface area principle. The decrease in the surface energy is transferred upstream through the capillary wave [6]. By adding the jet issue speed U to the flow field noted above, we can discuss a liquid jet issuing from a nozzle. However, unless the effects of nozzle remain clear, the further discussion is impossible. The conversion of the capillary waves reaching at a nozzle exit is discussed in the following subsection.

2.3. A self-closed breakup cycle

It was clarified that capillary waves propagate upstream due to contraction of a liquid tip in the previous subsection. The Capillary waves with large propagation speed and small wavelength can

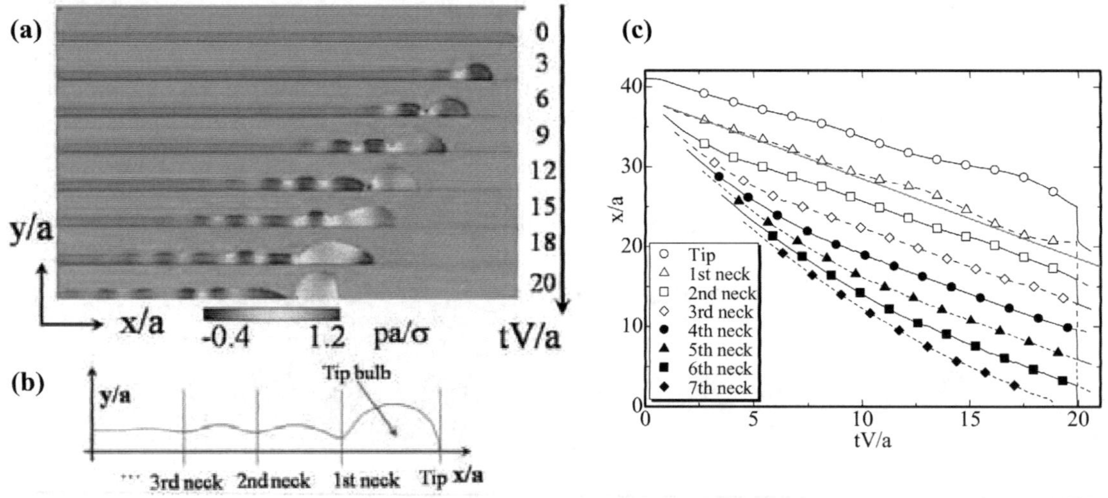

Figure 5. (a) Numerical simulation of contraction of a liquid column. (b) Definitions of the characteristic parts on the contracting liquid column. (c) Upstream propagation of capillary waves emanated from the liquid tip.

reach a nozzle exit in the case of a liquid jet. In addition, although the situation including only one breakup is considered in the previous subsection, capillary waves would be emanated after each breakup not depending on breakup modes. Through Doppler-shift at the nozzle exit the wavelength of the capillary waves becomes larger [6, 9]. The properties of the resulting wave can be described through the following equations,

$$2\pi\left(\frac{fa}{U}\frac{U}{V}\right)=\left\{\frac{c_R(ka)}{V}-\frac{U}{V}\right\}(ka)=\left\{\frac{c_R(k'a)}{V}+\frac{U}{V}\right\}(k'a) \qquad (k'a>1) \qquad (7\text{-}1)$$

$$2\pi\left(\frac{fa}{U}\frac{U}{V}\right)=\left\{\frac{c_R(ka)}{V}-\frac{U}{V}\right\}(ka)=\left(\frac{U}{V}\right)(k'a) \qquad (k'a<1) \qquad (7\text{-}2)$$

where $k'a$ and f correspond to nondimensional wavenumber of a wave component subjected to Doppler-shift and its frequency, respectively. In addition, fa/U corresponds to Strouhal number. For example, U/V = 2.0 case gives a solution of λ/a = 1.10 under the assumption that the most unstable waves noted above dominate the breakup characteristics. At least in laminar injection cases, there always exist the capillary waves converted into the unstable waves through Doppler-shirt. In other words, the unstable waves are spontaneously generated at the nozzle exit without any external disturbances. In addition, 'reproduction' of the waves at the nozzle exit also has an important role for sustaining the formation of the unstable waves [6]. Once the waves convecting (or propagating) downstream are formed just at the nozzle exit, the same waves continues to be formed. The amplitude of the waves formed owning to the reproduction process becomes smaller as time passes in the case of $U/V > 1$. However, the continuous generation of the unstable waves is possible by the capillary waves emanated after each breakup. The concept explained above corresponds to a self-closed breakup cycle of a liquid jet proposed by Umemura [6]. By looking for answers not to unidentified disturbances but to the capillary waves generated by the liquid jet itself, the repeatability of experimental results can be explained well. However, not all of experimental results can be explained only by the use of the self-closed breakup cycle. The additional characteristics of a liquid jet are described in the section 4 along with our experimental results.

3. Experimental setup and image processing

3.1. Experimental setup

Almost all of the following experimental apparatuses were commonly used in both our normalgravity and microgravity experiments. A schematic of the experimental setup is illustrated in figure 6. A back-light method was used for obtaining clear successive images of a liquid jet. As a light source, an array of high-intensity LEDs made by Opto Supply was arranged in the test section. Luminosity and diffusion angle of each LED are 6000 mcd and 60 deg, respectively. In order to obtain brightness distribution as uniform as possible, a holographic diffuser made by Edmund Optics was adopted. A variable resistance incorporated in the LED circuit can finely modulate the brightness. The light source is driven by direct current supplied from the experiment facilities. Images of a liquid jet were taken by a high-speed camera. A planar aluminum mirror was used to obtain an appropriate optical path. A high-speed camera made by Photron was adopted owning to its performance and impact resistance necessary for the experiments in a drop tower. Exposure time was set as 1/15000 s and frame rate was set as 4000 fps. The length able to capture in the liquid issue direction can be changed approximately from 130 mm to 200 mm by modulating the optical system. Images stored in the high-speed camera were downloaded into a PC through an optical fiber cable. Frame rate 4000 fps corresponds to approximately 4.1 s. After flow field reaches a quasi-steady state, a start trigger was transferred to the high-speed camera. Pure water was stored in a syringe and issued through a nozzle part. The syringe and the nozzle part were connected by the use of a tube made by Teflon. Water installed in the syringe was pushed by a rack-and-pinion system made by Oriental motor. There is a possibility that controlling flow rate by the use of a flow meter cannot follow pressure fluctuations at a nozzle exit. Since the radii of the syringe and the nozzle exit can be measured in advance, water flow

Journal of Physics: Conference Series **327** (2011) 012042 doi:10.1088/1742-6596/327/1/012042

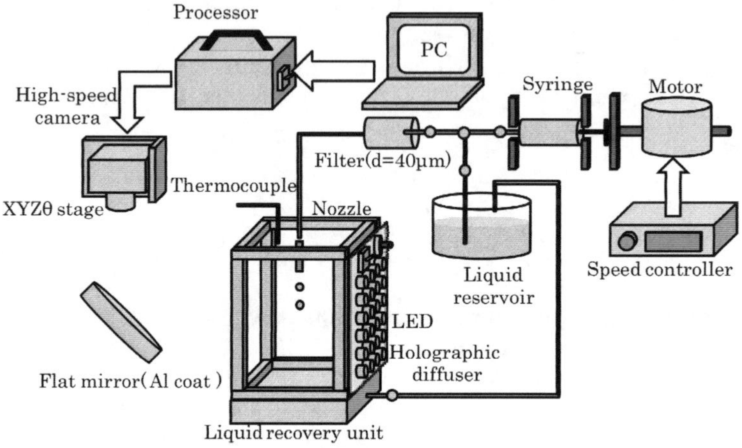

Figure 6. A schematic of experimental setup.

rate itself can be controlled by controlling the moving speed of the rack. It takes at most 2 s to achieve a constant-speed movement of rack. A liquid recovery unit was used only in the normalgravity experiments. The nozzle parts made of stainless were composed of an upstream entrance region and a downstream nozzle section. The length and the radius of the entrance region are 20 and 4.35 mm, respectively. The thickness of the nozzle section is 0.5mm. The radii of the nozzle exit are 0.39, 0.56, 0.80 and 1.03mm, respectively. Although the experimental space on ISS is a little smaller than that on the drop tower and the parabolic flight facilities the authors have used so far, the fundamental arrangement can be applied directly to the space experiments.

3.2. Image processing

Many images with resolution $j_{max} \times i_{max}$ = 1024×128 pixel can be obtained after each experiment as depicted in figure 7. In order to extract information related to the breakup process of a liquid jet, its surface configuration must be correctly captured. Since the number of successive images obtained from one experiment exceeds 10,000, virtual judgement is impractical. Appropriate methods tracking a liquid tip are necessary. The tip-tracking method previously used in our studies is described in [10] in detail. By the use of the minimum and the maximum (strictly speaking, the mode value of the background) brightness of raw images, new ones with the larger contrast were used for detecting the location of the liquid tip in this method. However, in the less resolved case, unphysical oscillations of the tip location were caused by the small non-uniformity of backlight and by the existence of small satellites formed between main droplets. In addition, there is a difficulty in determining the threshold value of the brightness determining the liquid tip. Instead of this method, another method was adopted in which the new images are constructed by subtracting the background brightness from the original images as illustrated in figure 7 (b). This procedure reverses the brightness of the original images and thus the liquid surface is expressed by the white region. The minimum brightness at the j th location, $B_{min}(j)$, can be obtained by comparing the brightness from i = 1 to i = i_{max}. Detecting a sudden decrease in $B_{min}(j)$ from $j = j_{nozzle}$ to j_{max}, the liquid tip can be specified as j_{tip}. It was found that the images obtained through this method are less susceptible to the non-uniformity of the backlight and to the effects of the satellites. A typical time history of the tip location obtained through the procedures denoted above is shown in figure 7 (c). During one experiment, breakup occurs many times. A sudden decrease in the tip location corresponds to one breakup. Breakup length is defined as the length from the nozzle exit to the liquid tip just after a certain breakup as illustrated in figure 7 (b). One breakup length plotted as one point in graphs in the section 4 is the mode value evaluated by the use of all breakup information obtained from one experiment. Together with the mode value, the standard deviation is also evaluated and expressed as error bars.

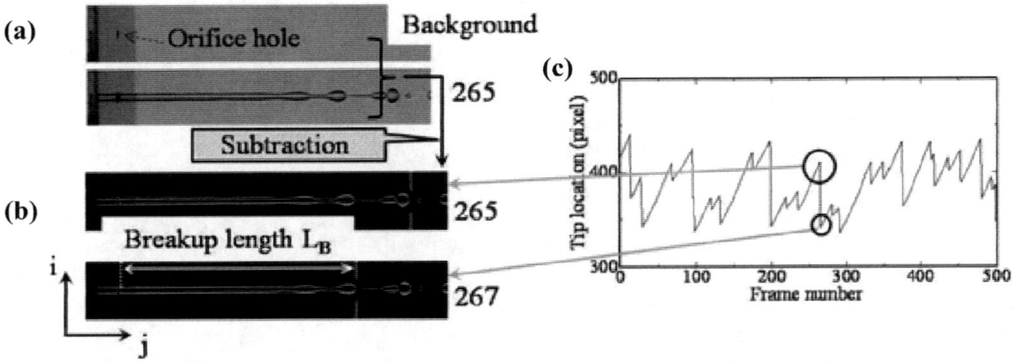

Figure 7. Schematics of image processing. (a) Typical images obtained from an experiment. (b) Images obtained through background elimination. The white vertical lines indicate the tip locations judged by our program. The figures on the right upper side correspond to frame number. (c) Typical time history of the tip locations.

4. Experimental results

4.1. Difference in breakup configuration due to gravity

Typical snapshots obtained from the experiments under normalgravity and microgravity environments are listed in figure 8. The nozzle of a = 1.03 mm was used in each experiment. In all images, the brightness is modulated for improvements in the visualization. Hereinafter, U/V is defined as the ratio of cross-sectional average liquid issue speed at the nozzle exit to the analytical tip contraction speed expressed in the equation (5) unless otherwise noted. Under microgravity environments, the critical U/V determining the dominant breakup mode is nearly 1.58. This value can be derived by considering the displacement of the 1st neck part and the liquid flow rate entering the tip liquid bulb during one breakup [6]. In the U/V = 1.98 and 2.49 cases, the long-wave mode fixes the breakup length. The significant difference in the breakup configuration is observed in the U/V = 1.23 and 1.50 cases depending on the existence of gravitational acceleration. Local U/V gradually increases in the axial direction due to gravitational acceleration (Bond number is approximately 0.146 in figure 8) and finally exceeds 1.58. In this case, the unstable waves convecting downstream increase their amplitude and finally the breakup length is fixed by the long-wave mode even in the case of U/V < 1.58. In the case where the nozzle of a = 0.39 mm was used, the short-wave mode can fix the breakup length even under normalgravity environment due to the small Bond number [6].

The Larger liquid issue speed causes the larger breakup length intuitively. However, this intuition

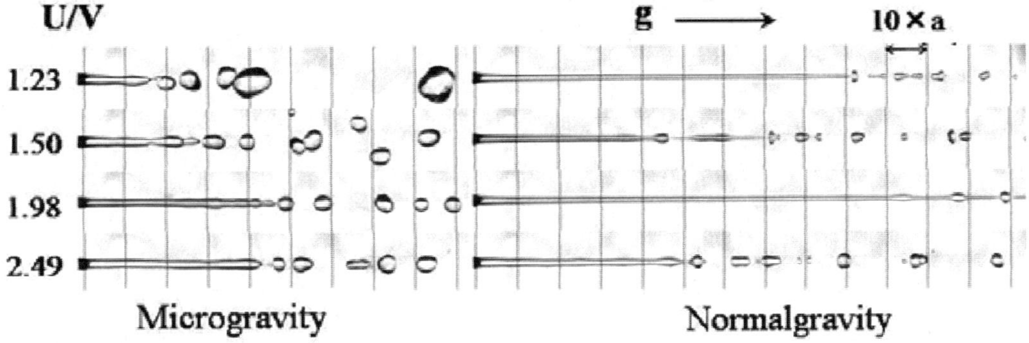

Figure 8. Difference in breakup configurations caused by gravitational acceleration. A nozzle of a = 1.03 mm was used in each experiment.

does not always hold judging from the images shown in figure 8. Either the effects of the gas dynamic pressure or viscosity [3-5] are unable to explain this phenomenon correctly. The flow field shown in figure 8 is inherently laminar according to the turbulence standard of Reynolds, $Re_D < 2300$. This is true judging from their smooth surface configuration. Turbulence inside of the nozzle alone is not the correct answer to such phenomenon. In order to investigate such phenomenon, a lot of experiments were carried out by the use of the several nozzles and the the results are shown in the next subsection.

4.2. Existence of relaxation region

The breakup length obtained from normalgravity experiments by the use of the nozzles of $L_n = 30$ mm is shown in figure 9. Experiments by the use of a certain nozzle were repeatedly conducted at least twice in order to examine their repeatability. The experimental results in figure 9 show a sufficient agreement with the ones conducted by the other researchers [11, 12], which suggests the existence of the breakup mechanism independent of the disturbances inside of nozzles at least in laminar cases. The jumps in the breakup length around small U/V are caused due to gravitational acceleration as noted in the previous subsection. Figure 9 shows that a lot of experimental results can be classified into two breakup modes, namely long distance breakup (LDB) and short distance breakup (SDB) modes. The breakup length for the nozzles of the small radius has a tendency to belong to the LDB mode and vice versa for the nozzles of the large radius. This tendency is deeply related to the flow field inside of the nozzle. Flow entering a circular tube goes downstream with developing its velocity distribution. If length of a tube is sufficiently large, the velocity distribution approaches that of well-known Hagen-Poiseuille flow perfectly in a laminar case. Since this tendency is characterised by Reynolds number $Re_D = 2Oh/(U/V)$, the velocity distribution of the liquid issuing from the nozzles of smaller radius is likely to approach such velocity distribution. In the previous researches the effects of velocity distribution was suggested [5, 13]. However, the instabilities induced by such flow field have not been investigated in detail. Umemura [6] performed the linear stability analysis as to the liquid flow with such velocity distribution. It was found that the increase in $\Delta(U/V)$ gives the smaller growth rate of the unstable waves and finally the unstable waves disappear in the case of $\Delta(U/V) > 2.0$. Here $\Delta(U/V)$ corresponds to the nondimensional maximum difference of the parabolic velocity distribution. $\Delta(U/V) = 0$ corresponds to the uniform velocity distribution. This effect is understood as a situation where a rigid body (corresponding to a limit of flow with larger momentum) is inserted into liquid. In addition, even if the velocity distribution is perfectly that of Hagen-Poiseuille flow, viscosity of liquid relaxes the velocity distribution as goes downstream and finally nearly uniform velocity distribution is achieved. This suggests that the breakup length shown in figure 9 includes a 'relaxation region' which is inherently independent of the growth of the unstable wave. In figure 9, a Boussinesq line is also shown on the assumption that the velocity relaxation is the opposite phenomenon to the development of the velocity distribution inside of the nozzle. Although this is a crude assumption, it is found that the relaxation region occupies the large part of the breakup length. Since the liquid 'surface' velocity increases gradually within the relaxation region, the capillary waves propagating upstream are subject to Doppler-shift and their wavelength gradually become larger during their upstream propagation. The larger wavelength corresponds to the smaller propagation speed. The decrease in the propagation speed of the capillary waves, which is caused by the increase in their wavelength, changes their propagation direction into the downstream direction. The increase in the wavelength continues during their downstream propagation and finally their nondimensional wavelength exceeds 1 [6]. Such waves coming back to a uniform velocity region obtains the nature of the unstable wave. This suggests that the relaxation region is identical to the nozzle exit. In fact, if we evaluate an initial amplitude without considering the relaxation region, an invalid value smaller than an intermolecular distance is obtained for the experimental results by the use of the nozzle of $a = 0.39$ mm. Such unphysically small value cannot explain the sufficient repeatability and the independency of experimental apparatuses.

As U/V increases, Re_D approaches 2300 which is the criterion classifying laminar and turbulent pipe flows investigated by Reynolds. The intermittent change in the breakup length, which corresponds to

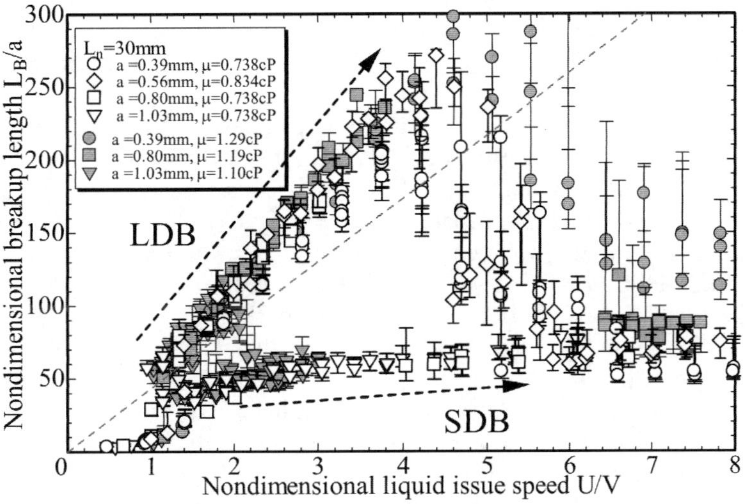

Figure 9. Difference in breakup process caused by gravitational acceleration. A nozzle of a = 1.03 mm was used in each experiment. LDB and SDB correspond to long breakup distance and short breakup, respectively.

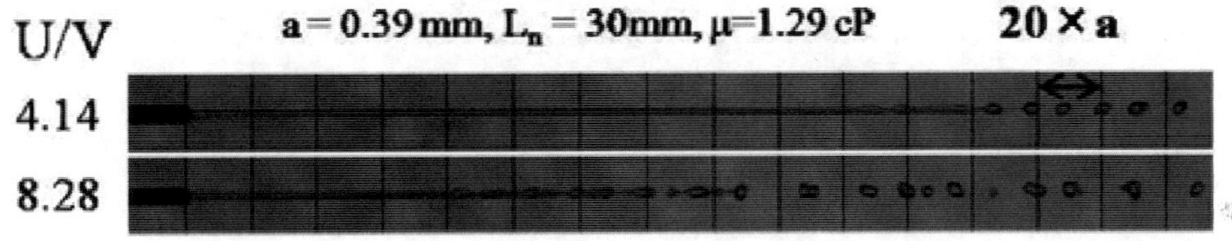

Figure 10. Typical images obtained from experiments by the use of a nozzle of a = 0.39mm.

the large error bars in figure 9, is thought to be caused by the vortices shed from the separation bubble formed at the nozzle entrance region. In the case where *U/V* exceeds a certain value, the transition into the fully developed turbulent flow occurs within the nozzle. In addition to the more uniform velocity distribution, the disturbances shed from the nozzle dominate the breakup of the liquid jet and the sudden decrease in the breakup length accompanied by so irregular and large-amplitude surface deformation is observed as illustrated in figure 10. The value of the critical Reynolds number 2300 may not be directly applicable to our experimental results since this value was determined from the experimental results by the use of sufficiently long pipes. In our experiments, the length of the nozzle is comparatively small and the flow field inside of the nozzle is subject to the disturbances coming from downstream. On the contrary, the SDB mode is not always dominated by the turbulence inside of the nozzles. The surface configuration of the liquid jet classified into the SDB mode seems different depending on the radii of nozzles. In order to examine the breakup characteristics in the SDB mode, let us focus on the breakup length obtained from the experiments by the nozzle of a = 1.03mm.

4.3. Two-valueness of breakup length

It is observed that the breakup length for the nozzle of a = 1.03 mm takes two values for a certain liquid issue speed as illustrated in figure 11. Each value is categorized into either the LDB or the SDB mode. The typical images for U/V =1.78 are shown in figure 12. The critical U/V corresponding to Re_D = 2300 in the case of μ = 1.10 cP (Oh = 4.00×10^{-3}) is 4.60. Since U/V is smaller than the half of the critical value and both of liquid jet seems laminar judging from the smoothness of the surface configurations and the regular breakup pattern, it is unlikely that the SDB mode is not always dominated by turbulence. Such laminar SDB case is observed in the experimental results of the nozzles of comparatively small L_n/a. If the velocity distribution inside of the nozzle does not fully develop at the nozzle exit, the length of the relaxation region would be small and the SDB mode is more realizable. In fact, the smaller viscosity (μ = 0.738 cP case) gives the narrow U/V range where the two-valueness is observed as illustrated in figure 11. Although the feasibility of the breakup length in the laminar SDB mode is described as above, the two-valueness is not fully explained yet. One of the possibilities is that a slight difference in the start conditions of the liquid injection can affect the velocity distribution inside and eventually outside of the nozzle. The energy transferred from the liquid tip to the nozzle exit through the capillary waves propagating upstream is comparatively large [6]. Since capillary waves with larger wavelength have larger amplitude due to their dispersive nature, the energy disturbing the liquid flow inside of the nozzle is larger at the beginning of liquid injection than that at the quasi-steady breakup state. There is a possibility that such large energy affects separation bubbles at the entrance of the nozzle parts and their change can lead to more uniform velocity distribution at the nozzle exit than that of the LDB mode. Put another way, the 'coupling' of the internal and the external flow fields can cause the laminar SDB mode. If the energy sufficient enough to affect the flow inside of the nozzle is not transferred to the nozzle exit at the beginning of injection, the velocity distribution would be dominated only by viscosity inside of the nozzle and the LDB mode appears. However, the laminar SDB mode case was not observed. The detailed condition for the realization of the laminar SDB mode has not been clear yet. In order to validate and control the realization of the two-valueness, a lot of experiments in the same condition are necessary.

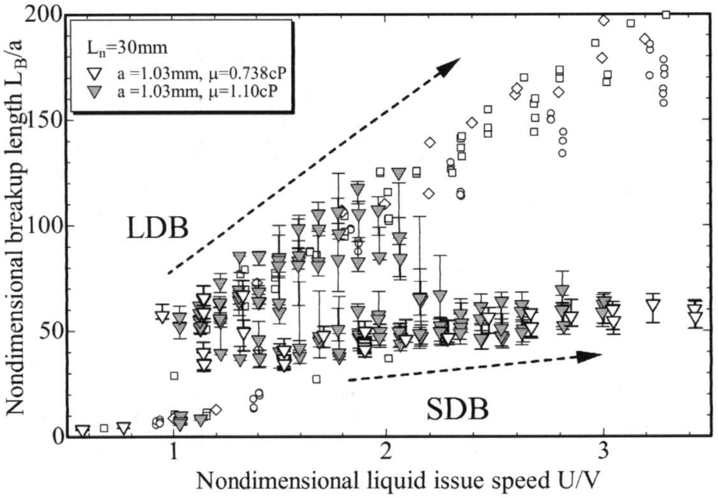

Figure 11. Two-valueness of breakup length on a certain liquid issue speed. The results for a nozzle of a = 1.03mm is emphasized in this figure. Symbols corresponding to other experimental conditions are the same as ones in figure 11. LDB and SDB correspond to long distance breakup and short distance breakup, respectively.

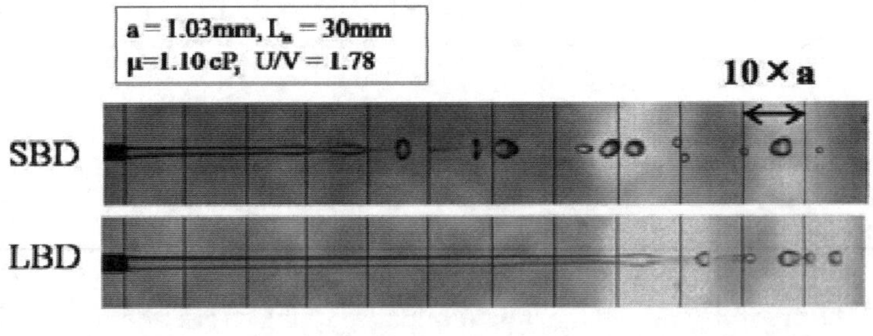

Figure 12. Typical images indicating two-valueness of breakup length.

5. Experiments on ISS

As noted above, gravity affects the generation of the unstable waves through acceleration of liquid issuing from a nozzle. In addition to the destabilization of the capillary wave, the unstable wave cannot keep its wavelength during its convection. These effects appear dominantly in the small U/V case. In addition, it is predicted that the length of the relaxation region becomes smaller due to the gravity in the small U/V and large Bo cases. Since our new breakup theory must be examined through experimental results, it is desirable to remove what affects the breakup configuration, gravity in this case. The best solution is the microgravity experiments. However, the limitation to the number of experiments has prevented from obtaining the sufficient number of the microgravity experimental results and then the repeatability has not fully confirmed yet. In particular, the existence of two-valueness has only been confirmed only on the normalgravity condition and must be also examined on the microgravity environment through multiple experiments.

Several experiments are projected where the liquid issue speed is changed through several stages as illustrated in figure 13. Initial U/V is approximately 1.10 for all of the experiments. Before obtaining the maximum liquid issue speed $(U/V)_{max}$, the liquid issue speed is increased through 20 phases by an interval of $\Delta(U/V)= (U/V)_{max} /20$. Since then the liquid issue speed is decreased by $\Delta(U/V) = (U/V)_{max}/10$ through 10 phases. In order to distinguish several types of experiments clearly, such experiments and the ones described in section 4 are named as the continuous and the single injection experiments, respectively. On the basis of the previous experiments, the time interval of a certain constant liquid issue speed is determined as approximately 4 s. The total experimental time amounts to approximately 120 s. Since all of the images from the beginning to the end of the liquid injection are unable to be obtained due to the limitation of memories installed on a high-speed camera, the images corresponding to approximately 50 breakups after reaching the steady liquid issue speed are realized for every U/V. Such long-period experiments are realizable only on ISS. High-speed ($(U/V)_{max} = (U/V)_{cr}$, in section 4) and low-speed ($(U/V)_{max} = 2$) experiments are projected. The low-speed case includes the region where the breakup length is dominated by the short-wave mode which is peculiar to microgravity environment ($U/V < 1.58$). The breakup in the short-wave mode is affected by the large-amplitude capillary waves reflected at the nozzle exit. Therefore, around the nozzle exit the strong interference of the capillary waves can be observed in detail and the waves formed on the liquid surface can differ from those for the single injection experiments. It follows that the low-speed continuous injection experiments are suitable to investigate the effects of the time histories of capillary waves. In the high-speed continuous injection case, along with effects of the time histories, the two-valueness of the breakup length can be examined. There is a possibility that even on the experimental conditions where both SDB and LDB are observed in the single injection experiments, SDB only realizes in the continuous ones. In addition, by the use of the nozzles of the several radii and lengths, the effects of nozzle will be examined.

Furthermore, the experiments aimed at obtaining the correlation functions for the time histories of the liquid tip location are planned. As noted above, the capillary waves emanated from the liquid tip

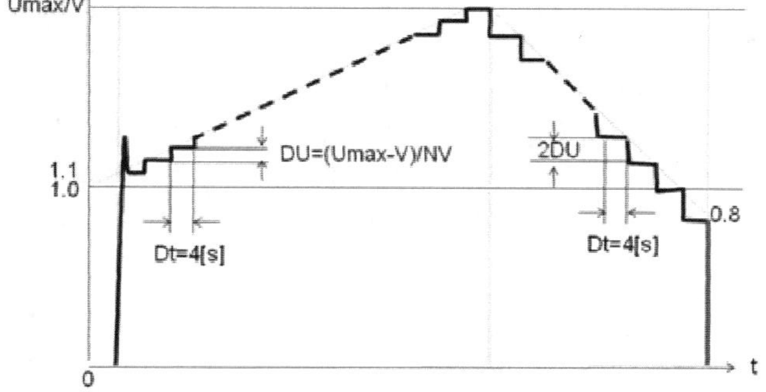

Figure 13. A schematic of representative microgravity experiments on ISS.

Journal of Physics: Conference Series **327** (2011) 012042 doi:10.1088/1742-6596/327/1/012042

have the roles of upstream energy transfer and generation of the resulting unstable waves at the nozzle exit. The correlation of the time histories of the liquid tip can give the information related to the wave structures. The small *U/V* cases (< 2.0) are targets of this kind of experiments. Since the waves at nozzle exit can be comparatively visible in the small *U/V* case, the statistic characteristics obtained through the time histories of the breakup length can be examined by the use of the information of waves around the nozzle exit. Concretely, the images including several times breakups are taken at the specified time intervals during a long-period and single injection experiment.

As noted above, almost all of our experimental setup can be directly applied to the experiments on ISS. There is a possibility that the droplets scattering from the base part of the test section prevents from obtaining images of liquid jet during such long-period microgravity experiments. One of solutions for such a problem is the use of sponge blocks with high water-absorption performance. Considering the necessity to reduce experimental instruments and to avoid excess procedures, automatic water-absorption by sponge blocks can be one candidate. The behavior of the scattering droplets and water-absorption performance of sponge blocks were investigated through microgravity experiments by the use of parabolic flights. PVA sponge blocks made by AION with approximately 80 μm pores were adopted and installed at the bottom of the test section. Typical figures around the sponge block are shown in figure 14. Large droplets scattering from the blocks were not observed. The images obtained approximately 16 s after the beginning of liquid injection are shown in figure 15. Vitiation due to the fine droplets near the base part is observed only in the high-speed injection case. Such contamination of the test section can be avoided by making a hole on the sponge block. Water-absorption capacity of PVA sponge block during the long-period experiments was confirmed through other normalgravity experiments. Of course, other methods can be used for collection of water and

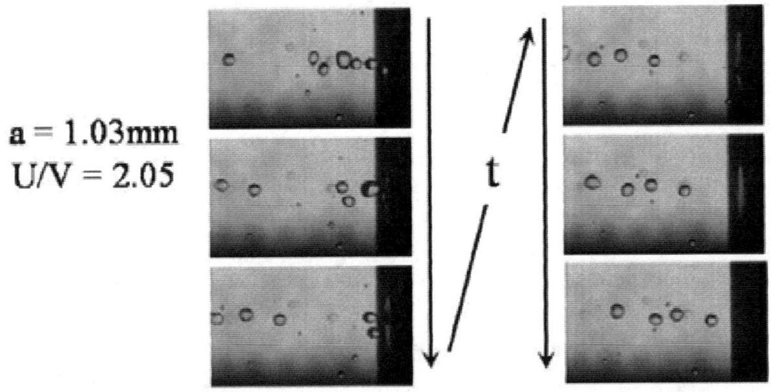

Figure 14. Typical successive images around surface of a sponge block located at the bottom of the test section.

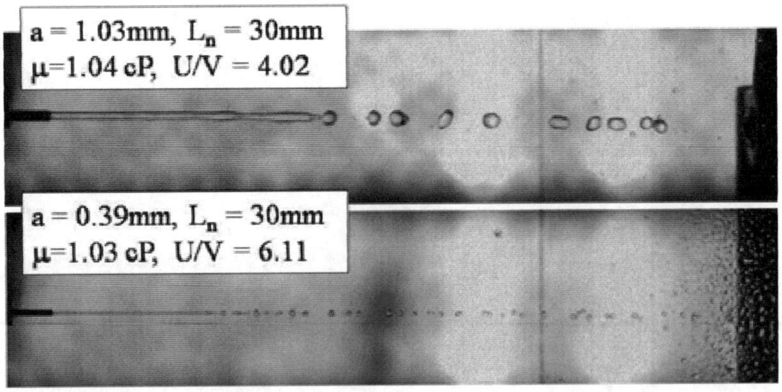

Figure 15. Images indicating scattering of fine droplets from a sponge block.

final solution is currently under consideration.

6. Conclusion

Breakup process of a liquid jet issued from a circular nozzle, which has been attributed to 'some kind of disturbance existing in nature and inside of a nozzle', has been investigated through a new breakup theory. The spontaneous breakup occurs by the short-wave mode as long as the tip of the liquid jet exists. The steady breakup length is determined only by the short-wave mode if the case of $U/V < 1.58$. If U/V exceeds 1.58, the breakup by the short-wave mode cannot fix the breakup length and the unstable waves have important roles for the realization of the steady breakup length. Even in such a case, the main determinant for the liquid breakup is the capillary waves emanated from the liquid tip after each breakup. The reproduction and the reflection of the capillary waves at the nozzle exit can generate the unstable waves with the same amplitude as that of the capillary waves. A self-closed breakup cycle is established by the wave structures. From normalgravity experiments, it was found that the breakup length is classified into the LDB and the SDB modes. The breakup length in the LDB mode is occupied by the long relaxation region. While the separation bubbles or turbulence formed inside of the nozzle causes the turbulent SDB mode, the uniform velocity distribution due to the coupling of the flow fields inside and outside the nozzle is thought to be necessary for the realization of the laminar SDB mode. Although gravitational acceleration is inherently negligible for turbulent atomization, it can generate the unstable waves from the capillary waves and change the length of the relaxation region in normalgravity experiments. This is why microgravity experiments are desirable for establishing a comprehensive breakup theory. In addition, the effect of time histories of the waves on the breakup of liquid jet and the appearance of the two-valueness of the breakup length is suggested from our breakup theory. In order to investigate such unclear points, the microgravity experiments on ISS which can provide long-period microgravity environments are currently projected. The experimental setup of our previous experiments is directly applicable to such experiments and thus the high feasibility of such experiments is expected.

Acknowledgements
This study was supported by the Grant-in Aid for Scientific Research (A) from the Ministry of Education, Culture, Sports, Science and Technology-Japan, and carried out as a part of 'preparation and implementation of researches utilizing Kibo' carried out by Japan Space Forum under the contract with Japan Aerospace Exploration Agency.

References

[1] Umemura A and Wakashima Y 2002 *Proc. Combust. Inst.* **29** 633.
[2] Raylengh L 1878 *Proc. Lond. Math.***4** 4.
[3] Arthur M S and SLEICHER C A 1975 **68** 477.
[4] Fenn R W and Middleman S 1969 *AIChE J.***15** 379.
[5] Leroux S, Dumouchel C. and Ledoux M. 1997 *Int. J. Fluid Mech. Res.* **24** 428.
[6] Umemura A 2011 *Phys. Rev. E* **83** 04637.
[7] Brackbill J U, Kothe D B and Zemach C 1992 *J. Comput. Phys.* **100** 335.
[8] Shinjo J and Umemura A 2010 *Int. J. Multiphase Flow* **36** 513.
[9] Umemura A 2008 *J.Jpn. Soc. Microgravity Appl.* **25** 50.
[10] Kawanabe S 2010 Master's thesis, Department of Aeroengineering, Nagoya Univeristy.
[11] Amagai K 2000 Atomization **9** 261.
[12] Grant R P and Middleman S 1966 *AIChE J.***15** 669.
[13] Phinney R E 1972 *AIChE J.***18** 432.

Particle temperature measurement using pair distribution function in complex plasmas

Satoshi Adachi[1], Hiroo Totsuji[2], Kazuo Takahashi[3], Yasuaki Hayashi[3], Masahiro Takayanagi[1]

[1] Institute of Space and Astronautical Science, Japan Aerospace Exploration Agency, 2-1-1 Sengen, Tsukuba-shi, Ibaraki-ken 305-8505 Japan
[2] Okayama University, 1-2-1 Tokuyoshicho, Naka-ku, Okayama 703-8291 Japan
[3] Department of Electronics, Kyoto Institute of Technology, Matsugasaki, Sakyo-ku, Kyoto 606-8585 Japan

E-mail: `adachi.satoshi@jaxa.jp`

Abstract. In order to obtain more reliable particle temperature of complex plasmas, a new method of obtaining it in a solid phase is investigated. Since a conventional method uses a velocity distribution function (VDF), both high spacial resolution and sufficiently large observation area are required for sufficient measurement accuracy. It is often difficult, however, to satisfy them simultaneously due to necessities of highly magnified observation for the former requirement and low magnified one for the latter one. Therefore, we investigate another method of estimating particle temperature in the solid phase by use of a pair distribution function (PDF). This method has a feature that only one observation magnification is required. From comparison between the experimentally obtained PDF's and calculated ones, mean square displacement (MSD) is determined. The temperature is estimated by using the MSD if the particle motion is regarded as harmonic oscillation. It is found that the obtained temperature from the PDF is consistent with that from the VDF.

1. Introduction

Solid phase formation in complex plasmas was predicted in 1986 [1]. After that, many laboratories begun researches of observation of the solid phase. In 1994, the solid phase was successfully observed [2, 3, 4, 5]. Today this solid phase is called the Coulomb crystal or Plasma crystal.

There are many scientific issues in the complex plasmas researches, for example, Coulomb crystal formation mechanisms, the critical point in charged systems, particle and power balances, morphology and equilibrium form of the Coulomb crystals, solid-solid phase transition and so on. Since the particles usually have equally negative charges, it is often considered that the Coulomb Crystal is similar to the Winger crystal [6]. In addition to the Coulomb repulsive force, however, other forces working as attractive forces have been reported [7, 8, 9, 10]. It is considered that some attractive forces such as the ion drag force [10] are large enough to affect the crystal morphology. If particle kinetic energy, or temperature, is large enough to flee from the attractive forces, the morphology may be changed.

The critical point in charged systems has been predicted [11] though is may be difficult to observe it in usual plasmas. The complex plasmas may be suitable for the observation and thus theoretical investigations have been carried out [12, 13, 14, 15, 16]. Since the particle temperature is one of the most important parameters for the critical point, the precise temperature measurement is necessary. Investigations on the particle and power balances also require it.

The temperature is usually estimated by using a velocity distribution function (VDF). The VDF requires sufficiently large number of particles in an observation area to get statistical reliability. On the other hand, velocity resolution depends on spacial resolution. This means that the high spacial resolution is required, especially for a small velocity region. It is often difficult to satisfy these requirements simultaneously.

Therefore, we investigate another method to obtain the particle temperature by using a pair distribution function (PDF). The PDF includes the temperature effect. This effect is called the temperature factor or the Debye-Waller factor, $\exp(-M)$. In this paper, the experimentally obtained PDF, PDF_{exp}, is compared with the calculated PDF, PDF_{cal}, to determine the temperature factor. The PDF_{exp} is obtained from the two-dimensional particle coordinates based on the ground-based experiments in this paper. The PDF_{cal} is also calculated from the two-dimensional coordinates with a certain deviation of a normal distribution. From the determined temperature factor, the mean square displacement (MSD) is obtained. Thus the particle temperature is estimated from the MSD by assuming the harmonic oscillation of the particles. Although the new method is applicable to the temperature measurement in the solid phase at present since we assume the harmonic oscillation, this is still useful because the crystalline is easily formed on the ground in many experimental conditions. We also discuss the temperature estimated from the PDF by comparison with that from the VDF.

2. Theory for temperature measurement

General expression of the pair distribution function (PDF) of a Coulomb crystal in three-dimension is described as

$$g(r) = \frac{1}{V\rho}\sum_h\sum_k\sum_\ell F_{hk\ell}\exp\left\{2\pi i\left(hx+ky+\ell z\right)\right\}\ , \tag{1}$$

where where V is volume of a unit cell, ρ average number density of particles and $F_{kh\ell}$ the structure factor. The variables h, k and ℓ are the Miller indices, x, y and z the positions in the unit cell and $r=\sqrt{x^2+y^2+z^2}$. The structure factor is described as

$$F_{hk\ell} = \sum_m \exp\left(-M\right)\exp\left\{-2\pi i\left(hu_m+kv_m+\ell w_m\right)\right\}\ , \tag{2}$$

where where u_m, v_m and w_m the coordinates of the m-th particle in the unit cell. The term of $\exp(-M)$ is called the Debye-Waller factor or the temperature factor. This is expressed as

$$\exp\left(-M\right) = \exp\left[-2\pi^2\left\{{\sigma_{as}}^2+\left\langle(\Delta x)^2\right\rangle\right\}\left(h^2+k^2+\ell^2\right)\right]\ , \tag{3}$$

σ_{as} is the standard deviation of the atomic scattering factor, $\left\langle(\Delta x)^2\right\rangle$ the MSD. Please pay attention to the following two points in Eqs. (2) and (3). One is that the atomic scattering factor should not appear in this equation since the particles are completely classical. The other one is that the extinction rule does not work due to direct observation of the particles. This means that the summation of $\exp\left\{-2\pi i\left(hu_m+kv_m+\ell w_m\right)\right\}$ should be replaced to ρV. Therefore, the PDF for the complex plasmas should be expressed as

$$g(r) = \sum_h\sum_k\sum_\ell \exp\left\{-2\pi^2\left\langle(\Delta x)^2\right\rangle\left(h^2+k^2+\ell^2\right)\right\} \times\exp\left\{2\pi i\left(hx+ky+\ell z\right)\right\}\ . \tag{4}$$

Although the PDF can be obtained from Eq. (4) in principle, it is not easy to calculate it due to slow convergence of the summation on the Miller indices. The better calculation method for the complex plasmas exists, that is, direct calculation from the coordinations of test particles. This method is not applicable to an actual crystal but to the Coulomb crystal due to visual sizes of the particles. The PDF is easily calculated from the coordinates though sufficiently large number of particles is required for statistical reliability. The three-dimensional PDF is described as

$$g(r) = \frac{\frac{4}{3}\pi\, {r_{max}}^3 \sum_{i=1}^{N_j} N_i(r-\Delta r/2,\, r+\Delta r/2)}{N_j\, N_i^{total}\, \frac{4}{3}\pi \left\{(r+\Delta r/2)^3 - (r-\Delta r/2)^3\right\}} = \frac{{r_{max}}^3 \sum_{i=1}^{N_j} N_i(r-\Delta r/2,\, r+\Delta r/2)}{3\, N_j\, N_i^{total}\, r^2\, \Delta r}, \tag{5}$$

where r_{max}, N_i, N_j, and N_i^{total} are the radius of the PDF calculation area, the number of particles in a spherical shell from $r - \Delta r/2$ to $r + \Delta r/2$, the number of calculation areas and the total number of particles in one calculation area, respectively. The two-dimensional PDF is similar formula to the three-dimensional one like

$$g(r) = \frac{{r_{max}}^2 \sum_{i=1}^{N_j} N_i(r-\Delta r/2,\, r+\Delta r/2)}{2\, N_j\, N_i^{total}\, r\, \Delta r}. \tag{6}$$

In this case, the shell is not spherical but concentrically circular. In the ground-based experiments, the Coulomb crystal are usually observed as two-dimensional crystals. Actually, the Coulomb crystal observed in our apparatus typically has a few layers. Therefore, we use Eq. (6) to calculate the PDF in this paper.

To calculate Eq. (5) or (6), the particle coordinates are needed. In this paper, to set the particle at a certain position, we assume that probability of the displacement Δx from a lattice point is described by a normal distribution like

$$f(\Delta x) = \frac{1}{\sqrt{2\pi}\,\sigma} \exp\left\{-\frac{(\Delta x)^2}{2\,\sigma^2}\right\}, \tag{7}$$

where f and σ are the probability and the standard deviation, respectively. In this case, the MSD is expressed as

$$\left\langle (\Delta x)^2 \right\rangle = \int_{-\infty}^{\infty} x^2 f(x)\, dx = \sigma^2 . \tag{8}$$

Therefore, the square root of MSD is equal to the standard deviation.

The Box-Muller transform [17] is a well-known method of generating random numbers satisfying the normal distribution. Thus the particles are set at the positions determined from this method. The obtained two-dimensional particles are shown in Fig. 1 (a). The dashed triangular lines and closed circles in this figure represent perfect lattices and set particle with σ of 0.1, that is, $\left\langle (\Delta x)^2 \right\rangle = 0.01$. The elementary lattice in this case is an equilateral triangle. It is confirmed that some particles are close to the lattice point, while some particles are far away from it.

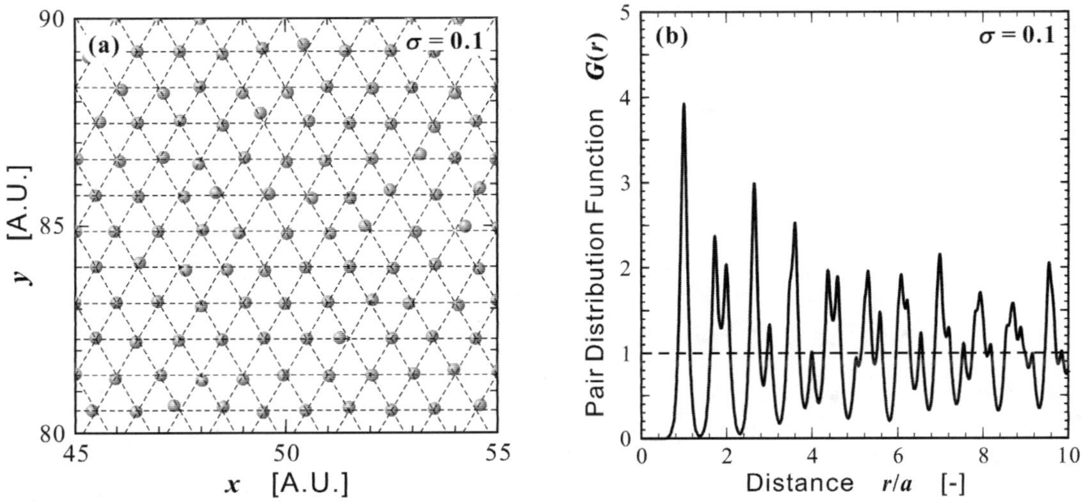

Figure 1. Particles with displacement obeyed by the normal distribution and calculated pair distribution function; (a) example and (b) PDF. The dashed lines and closed circles in (a) represent ideal lattices and particles, respectively. In this case, the standard deviation of the normal distribution is 0.1.

The MSD is closely connected to the Lindemann melting law, that is,

$$\chi = \frac{\sqrt{\left\langle (\Delta x)^2 \right\rangle}}{a}, \tag{9}$$

where $\left\langle (\Delta x)^2 \right\rangle$ is the MSD and a is the lattice constant. It is known that the value of $\chi = 0.3 \sim 0.35$ is the upper limit for the solid phase. In the Fig. 1 (a) case, χ is equal to 0.1. The value of $\left\langle (\Delta x)^2 \right\rangle \simeq 0.15$ may be the limit of the solid phase of the complex plasmas if it has the similar property to the usual crystals but it is not clear at present.

By using Eq. (6), the PDF can be calculated and is shown in Fig. 1 (b). The vertical and horizontal axes are the PDF value and the distance normalized by the nearest neighbor inter-particle distance, respectively. From Fig. 1 (b), it is found that the PDF clearly shows the solid phase since the peaks being far from the origin exist.

The particle displacement is expressed like Eq. (7) but is the two-dimensional function. Here, we assume that the force working between the particles is expressed as kr. This is the same as a spring or a harmonic oscillator. Average energy can be calculated in this case, that is,

$$\overline{E} = \int_0^\infty \frac{1}{2} k\, r^2 \frac{1}{2\pi\, \sigma^2} \exp\left(-\frac{r^2}{2\, \sigma^2}\right) dx = k\, \sigma^2 . \tag{10}$$

On the other hand, when the velocity distribution function (VDF) is expressed by the two-dimensional Maxwellian, the average energy is known as $\overline{E} = k_B T$. In the case of the three dimensions, the average energy is $\overline{E} = \frac{3}{2} k_B T$. By comparing these energies, we obtain the relationship of

$$k\, \sigma^2 = k_B T . \tag{11}$$

Journal of Physics: Conference Series **327** (2011) 012043 doi:10.1088/1742-6596/327/1/012043

where k_B is the Boltzmann constant. Therefore, the temperature is obtained from this equation if the spring constant is known. If the displacement is regarded as the simple oscillation like $x = A\exp(i\omega t)$, the oscillation frequency ω is described as $\sqrt{\frac{k}{m}}$. Thus the spring constant is expressed as

$$k = m\,\omega^2\,. \tag{12}$$

By substituting Eq. (12) to Eq. (11), we obtain the relationship of

$$k_B\,T = m\,\omega^2\,\sigma^2 \equiv m\,\omega^2\,\left\langle(\Delta x)^2\right\rangle\,. \tag{13}$$

Therefore, the temperature is obtained from the MSD and the oscillation frequency. This is the same expression of the harmonic oscillation. This means that this method is applicable to only the particle temperature measurements in the solid phase. To obtain the liquid phase temperature, it will be necessary to describe Eq. (13) in another expression. To rewrite Eq. (13) for the liquid phase, further investigation on the particle oscillation in the liquid phase is needed. Please pay attention that our model is basically applicable to any Coulomb crystals. Its accucary, however, may be affected by the crystal quality, especially dislocation density.

It is also difficult to obtain the temperature by applying the conventional VDF method to the liquid phase since drift influence must be taken into account. The VDF should be expressed as the shifted Maxwellian at least like,

$$f(v_x) = n_0\sqrt{\frac{m}{2\,\pi\,k_B\,T}}\,\exp\left\{-\frac{\frac{1}{2}m\,(v_x - v_0)^2}{k_B\,T}\right\}\,, \tag{14}$$

in the one dimension. In this case, the averaged kinetic energy is equal to

$$\bar{\epsilon} = \frac{1}{2}\,k\,T + \frac{1}{2}\,m\,{v_0}^2\,. \tag{15}$$

This indicates that drift influence must be eliminated to obtain the particle temperature though it is not easy. The particle temperature measurement in the liquid phase needs further investigations in both methods. Thus we focus on the temperature in the solid phase in this paper.

3. Experimental apparatus

The inside of the apparatus is shown in Fig. 2 (a). There are two RF electrodes of 10 cm in diameter. The distance between the electrodes is 10 cm. Therefore, the coupling between the electrodes is not strong. Each electrode has two RF feeders so that the RF current can flows along the electrode surface. The electrode is flat to achieve flatter potential profiles near the electrode. The nominal diameter of the vacuum chamber is 16 inches. This also helps to keep the flatter potential profile near the edge of the electrode. Figure 2 (b) shows the generated helium plasma without particles. It is shown that the bright area exist near the electrodes. In the region between the electrode and the bright are, there are darker regions. The particles are levitated near the boundary of the bright and dark regions.

One Langmuir single probe is set though it is not shown in this figure. Two CCD cameras are also set. One is set for the horizontal view and the other for top view. The view line of the top view, however, is slanted since the top electrode is an obstacle for the observation.

Intensity of the scattered light from the particles strongly depends on the viewing angle due to the Mie scattering. Therefore, we use macro-lenses for 35 mm cameras to obtain brighter

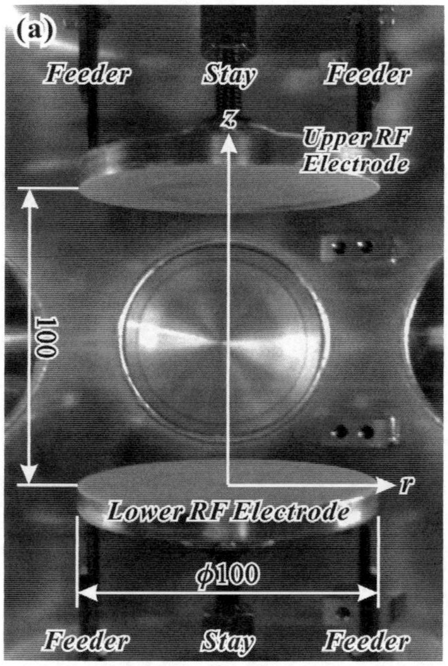

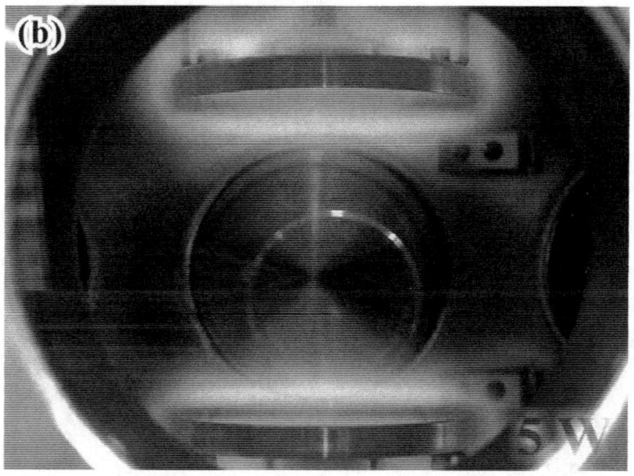

Figure 2. Inside of experimental apparatus; (a) main parts for plasma generation, dimensions and coordinates and (b) snapshot of RF plasmas. There are two RF electrodes. Each electrode is flat and has two RF feeders. The nominal diameter of the vacuum chamber is 16 inches.

Figure 3. Typical observation result of Coulomb crystal from top view camera on ground. The bright points are the particles of 5 μm in diameter. The shadow at the upper-left region is the upper electrode. The grids are the shallow ditches made on the back surface, which are used for a scale.

field of views. The magnification in such the lenses is sensitive to the distance between the focal plane and the particles. In addition, images must be corrected to the orthogonal coordinates since those obtained from the top view is slanted. Therefore, We use a specially made glass as a scale. This glass is optically flat ($\lambda/4$) on both surfaces and shallow grids at 5 mm intervals are made on the downward surface touching the lower electrode surface.

The typical observation result of the Coulomb crystal by using this apparatus is shown in Fig. 3. The helium gas pressure and RF oscillator power is set to 0.136 Torr (18 Pa) and 70 W, respectively. The bright points are the particles of 5 μm in diameter. The shadow at the upper-left region is the upper electrode. The shallow grids are also shown in this picture. It is confirmed that the image is slanted due to the slanted viewing angle of the top view camera. The corrected image is shown in the next section.

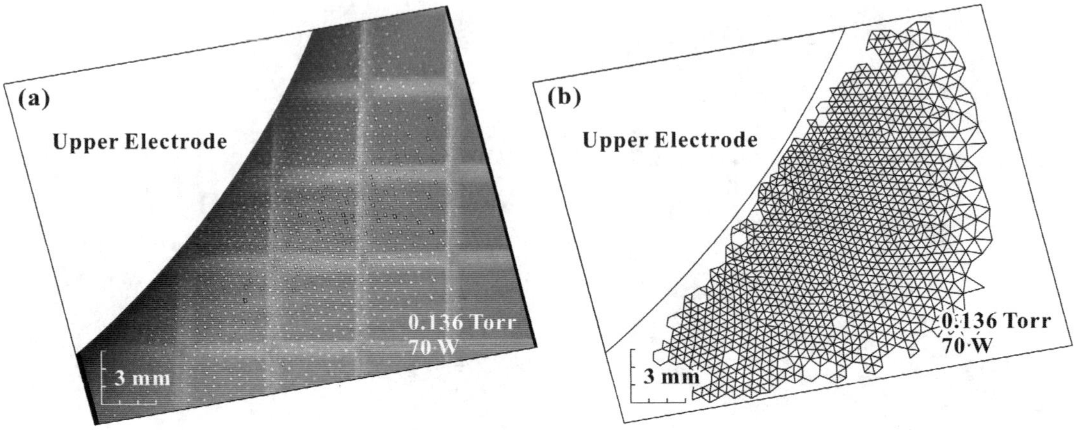

Figure 4. Orthogonal image and Delaunay diagram; (a) corrected to orthogonal coordinates and (b) the Delaunay diagram from (a). The image (a) is obtained by rotating the original image, skewing it and magnifying it vertically. Figure (b) shows that several dislocations and point defects exist.

4. Results and Discussions

As mentioned previously, the top view data is slanted as shown in Fig. 3. Hence, the data is corrected to the orthogonal coordinates as shown in Fig. 4 (a). This image is obtained by rotating Fig. 3, skewing it and magnifying it vertically. It is confirmed that the grids on the opposite side of the glass is corrected to the rectangular. The shadow of upper electrode is covered by a white sector, which is the same as Fig. 4 (b). Figure 4 (b) is the Delaunay diagram of Fig. 3 (a). It is confirmed that there are some area with highly regular lattices, while there are some dislocations and point defects. Since the highly regular area is a single grain, such the area is suitable for calculating the PDF. From these figures, it is found that the elementary lattice is an equilateral triangle. This is the same structure as that shown in Fig. 1.

The two-dimensional PDF using Eq. (6) is shown in Fig. 5. The solid, dashed and dot-dashed lines represent the experimentally obtained PDF and calculated ones in cases of $\sigma = 0.1$ and 0.15, respectively. In this case, the data in the case of RF power of 100 W and helium gas pressure of 0.136 Torr (18 Pa) are used since more power makes an inter-particle distance shorter. From comparison between the experimental result and the calculations, it is found that the first peak of the experimental PDF (PDF_{exp}) agrees well with the calculated PDF (PDF_{cal}) in the case of the deviation σ of 0.1. The PDF_{exp} apart from the origin, however, does not agree with the PDF_{cal} in the case of $\sigma = 0.1$ but with the PDF_{cal} in the case of 0.15, though the peak position of the PDF_{exp} is well reproduced by these calculations. The good reproducibility of the peak positions can be understood from the coordination number shown in Fig. 6. This figure clearly shows the coordination number of 6.2. This number indicates hexagonal structure.

Although the σ of 0.15 may be almost the limit of the solid phase from the Lindemann's law, the Coulomb crystal seems to still keep the solid phase since the particles do not freely move around but stay and oscillate. This suggests that the limit value of χ in Eq. (9) may differ from that of actual solids consisting of usual atoms.

If the defects exist, the displacement from the lattice point occurs due to the distortion. Such the displacement is accumulated as the distance increases. Hence, the deviation of $\sigma =$ 0.1 at the first peak is considered to indicate the temperature effect in this paper. To calculate the temperature, the oscillation frequency is required. Therefore, the particle trajectories are traced. The typical trajectory of the particle is shown in Fig. 7. In Fig. 7 (a), the horizontal

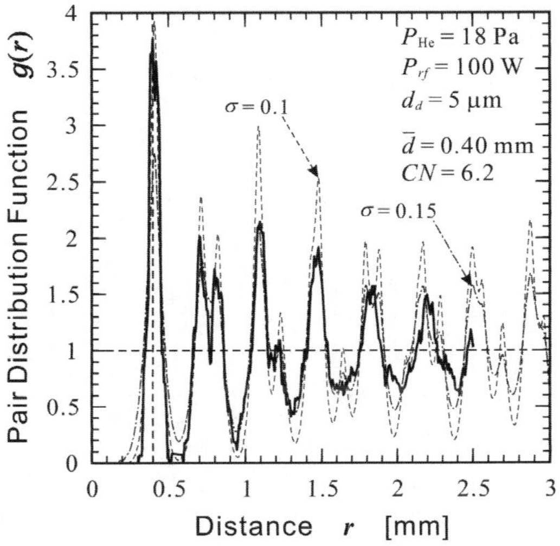

Figure 5. Comparison of experimentally obtained PDF with calculated ones. Solid, dashed and dot-dashed lines represent an experimental result, calculated results in cases of $\sigma = 0.1$ and 0.15 in Eq. 7, respectively.

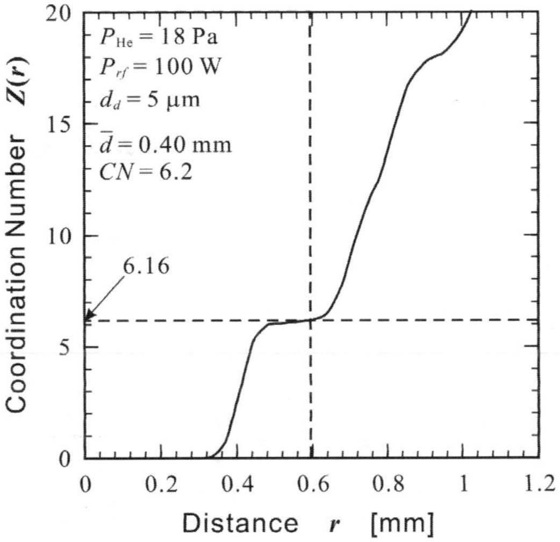

Figure 6. Coordination number obtained by integrating PDF. The coordination number is 6.2. This result indicates hexagonal lattice.

and vertical lines are the relative x and y coordinates in the observation area, respectively. The closed circles represent the particle positions, which vary over time. The positions are discrete due to the spacial resolution of the CCD camera. The time evolution of the position is shown in Fig. 7 (b). The offset and the trend of the position is already eliminated. It seems that very low and high frequencies exist. The former one is caused by the particle drift, which includes circular motion. The latter one should relate to the particle oscillation. Since it is difficult to obtain the oscillation frequency directly from Fig. 7 (b), we carry out the Fourier transform to find out the oscillation frequency. The amplitude spectrum is shown in Fig. 8. There are three Fourier transform results in this figure. Thin solid, dashed and dot-dashed lines correspond to these results. By considering the harmonic oscillation, the peak frequency being common throughout these data should exist. The bold dashed lines are located at 0.64, 0.88, 1.29 and 1.64 Hz. The frequencies of 1.29 and 1.64 Hz may be the second harmonics of 0.64 and 0.88 Hz. Since several common peaks exist, Eq. (13) is rewritten as

$$k_B T = m \left\langle (\Delta x)^2 \right\rangle \sum_n {\omega_n}^2 . \tag{16}$$

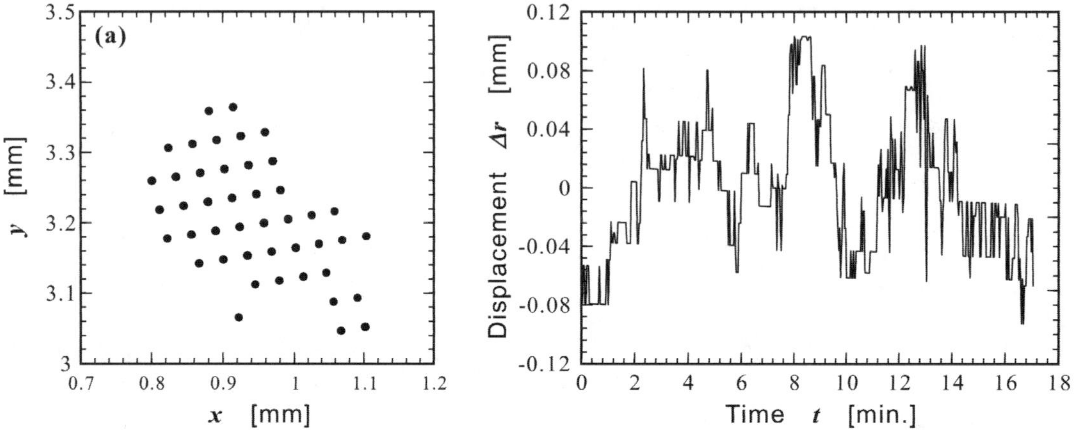

Figure 7. Particle trajectory and time evolution of particle displacement; (a) single particle trajectory and (b) time evolution of the displacement of the particle (a). The closed circles in (a) represent the particle position. The position is obtained as discrete data due to the spacial resolution of the CCD camera. The offset and trend in (b) are corrected.

From $\sigma = 0.1$ and the nearest neighbor distance is 0.4 mm, the actual value of σ is equal to 0.04 mm. Thus the MSD $\left\langle(\Delta x)^2\right\rangle$ is equal to 1.6×10^{-9} m^2. The mass of the particle is about 1.31×10^{-13} kg from the diameter of 5 μm and density of 2000 kg/m^3. The particle temperature is obtained by substituting these parameters to Eq. (16) as

$$k_B T = 0.29 \quad [\text{eV}]\,. \tag{17}$$

To compare this result with the temperature obtained from the VDF, the VDF is obtained and shown in Fig. 9. The open circles are the experimental data. The solid and dashed lines are the Maxwellian. The solid one represents an intermediate line, while the dashed ones the possible minimum and maximum ones. From Fig. 9, it is found that the average velocity is about $\overline{v} = 1.0 \pm 0.15$ mm/s. This velocity corresponds to the temperature of $\overline{E} = 0.42 \pm 0.12$ eV. This is about 40 % higher than the temperature estimated from the PDF. It is found, however, that non-zero VDF data is located at a few velocity points, namely, the low resolution in velocity space. This is mainly caused by the spacial resolution of the CCD camera as mentioned in section 1. As a result, the temperature from the VDF has an estimation error of 30 %. The lowest possible temperature from the VDF is 0.30 eV. This is consistent with the temperature from the PDF.

A weak point of the VDF method is to satisfy both the spacial (velocity) resolution and statistical reliability simultaneously. On the other hand, a weak point of the PDF method is to require the Coulomb crystal with low defect density and an observation duration being long enough to get high frequency resolution of the Fourier transform. Fortunately, these weak points are different from each other. Therefore, both methods can compensate each other and thus the more precise particle temperature measurement will be achieved.

5. Conclusions

The particle temperature in complex plasmas is sometimes regarded to be around the room temperature. The data from our apparatus, however, indicate that the temperature is always higher than that from the VDF. The particle temperature affects some important scientific issues in the complex plasmas such as the critical point and the power and particle balances.

Journal of Physics: Conference Series **327** (2011) 012043 | doi:10.1088/1742-6596/327/1/012043

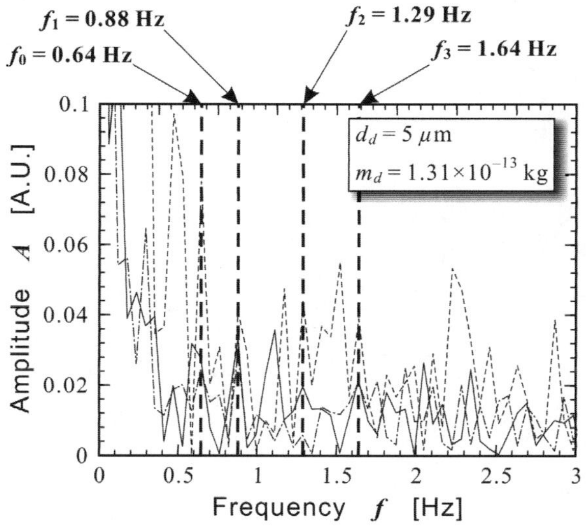

Figure 8. Fourier spectra of time evolution of particle positions. Three data are plotted in the figure. Several common peaks seem to exist.

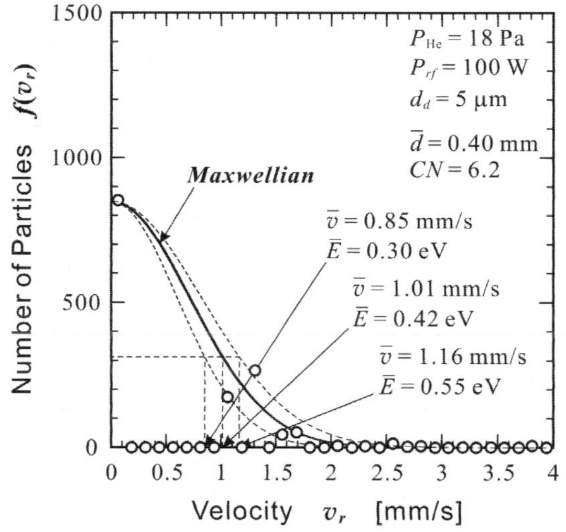

Figure 9. Experimentally obtained velocity distribution function. Open circles are the experimental data. The solid and dashed lines represent intermediate and possible minimum and maximum lines, respectively.

Therefore, we establish another method of the temperature measurement by using the PDF. First, the MSD is determined by comparing the PDF_{exp} with the PDF_{cal}. Second, the particle oscillation frequency is determined from the time evolution of the particle coordinates. Then the temperature is estimated by using the MSD and the frequency. It is found that the estimated temperature of 0.29 eV is consistent with that by the VDF method. Since each method has different features and weak points, combination of both methods will lead to compensation of each other's weak points and the more precise temperature measurement.

Although two or more cameras are needed in the VDF method to get high resolution and large area observations, only one camera is basically needed in the PDF method. Therefore, the PDF method may be more suitable for an apparatus for the microgravity experiment because allowable space and weight are always severe.

References

[1] Ikezi H 1986 *Phys. Fluids* **29** 1764
[2] Thomas H, Morfill G E, Demmel V and Goree J 1994 *Phys. Rev. Lett.* **73** 652

[3] Chu J H and I Lin 1994 *Physica* A **205** 183
[4] Melzer A, Trottenberg T and Piel A 1994 *Phys. Lett.* A **191** 301
[5] Hayashi Y and Tachibana K 1994 *Japan. J. Appl. Phys.* **33** L804
[6] Wigner E 1938 *Trans. Faraday Soc.* **34** 678
[7] Vladimirov S V and Nambu M 1995 *Phys. Rev.* E **52** R2172
[8] Ignatov A M 1996 *Plasma Phys. Rep.* **22** 585
[9] Tsytovich V N, Khodataev Y and Bingham R 1996 *Comments Plasma Phys. Control Fusion* **17** 249
[10] Khrapak S A, Ivlev A V, Morfill G E and Thomas H M 2002 *Phys. Rev.* E **66** 046414-1
[11] Totsuji H and Ichimaru S 1974 *Prog. Theor. Phys.* **52** 42
[12] Totsuji H 2006 *J. Phys.* A: Math. Gen. **39** 4565
[13] Khrapak S A, Morfill G E, Ivlev A V, Thomas H M, Beysens D A, Zappoli B, Fortov V E, Lipaev A M and Molotkov V I 2006 *Phys. Rev. Lett.* **96** 015001-1
[14] Totsuji H 2008 *Plasma Fusion Res.* **3** 046-1
[15] Totsuji H 2009 *J. Plasma Fusion Res. SERIES* **8** 257
[16] Totsuji H 2009 *J. Phys.* A: Math. Theor. **42** 214022
[17] Box G E P and Muller M E 1958 *Ann. Math. Stat.* **29** 610

International Symposium on Physical Sciences in Space IOP Publishing
Journal of Physics: Conference Series **327** (2011) 012044 doi:10.1088/1742-6596/327/1/012044

Photophoretic forces on chondrules in drop tower experiments

Alexander Hesse, Jens Teiser, Gerhard Wurm

Faculty of Physics, University Duisburg-Essen, Lotharstr. 1, D-47048 Duisburg, Germany

E-mail: alexander.hesse@stud.uni-due.de

Abstract. Studies of meteoritic and cometary materials give evidence for radial mixing of solid material within the early solar system. Among other transport mechanisms, photophoresis is a proposed mechanism for radial material transport in protoplanetary disks. Within this study, the photophoretic motion of chondrules in microgravity is investigated in drop tower studies. Chondrules form the meteorite Bjurböle were exposed to an intense laser beam. The acceleration due to photophoresis was observed and tracked with two cameras. The experimental results give evidence that photophoresis indeed is an efficient transport mechanism for chondrules and other small particles in the solar nebula. The measured values for the photophoretic force correspond well to the results of earlier studies and show clearly, that the models for particle transport by photophoresis are in agreement with the experimental data.

1. Introduction

In the early solar system the young sun was surrounded by a protoplanetary disk, the Solar Nebula [1]. It consisted of gas and embedded dust particles. This medium initially was optically thick for radiation of the visible wavelength range. With the dust being subject to different processes of accretion, e.g. the formation of planetesimals, protoplanets, terrestrial planets etc., the dust reservoir in the disk diminished and the disk eventually became optically thin. The gas molecules, too, were subject to processes that decreased the gas content of the disk, for example photoevaporation or solar wind [2]. However, in the standard model of giant planet formation, gas is accreted onto a solid nucleus of several earth masses, which implies that the gas still has to be dense while the dust is already further evolved into large aggregates. This implication is equivalent to the conclusion that the protoplanetary disk has a state, where it is at least partly optically thin and yet still rich in gas. Combinations of optically thin inner disks and optically thick outer disks are well observed these days [3]. In these transitional disks with inner holes extending up to several AU there is no doubt that the particles at the inner edge of the outer disk are illuminated by starlight but are embedded in a gaseous environment. The existence of such a disk-state gives way to a particle sorting and concentration process that is based on photophoresis, a light induced force besides radiation pressure, Poynting-Robertson drag and the Yarkovski effect. While the latter are based on momentum transfer by photons, for photophoresis gas molecules in the vicinity of the observed particle act as mediators. The principle of photophoresis is shown in figure 1 for the simple case of a spherical particle:

The particle-side facing towards the radiation source is heated to the higher temperature T_2 and a temperature gradient in direction of the incident radiation builds up depending on

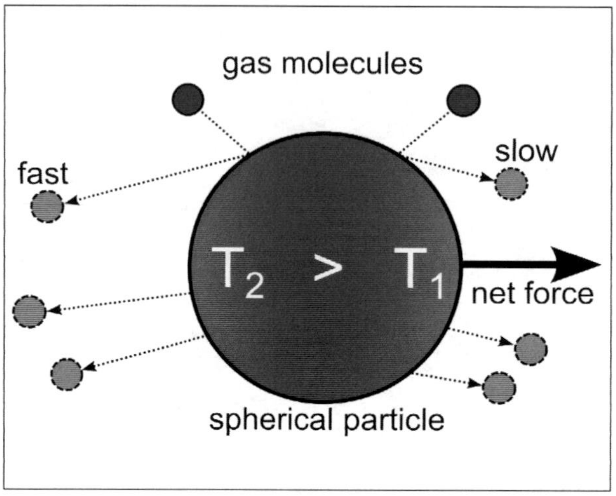

Figure 1. Principle of photophoresis: the ideal spherical particle is heated by radiation from the left side resulting in a temperature gradient in direction of the incident light. Gas molecules sticking to the surface leave the particle at different thermal velocities (faster on the warmer side). To conserve momentum the particle has to balance this mismatch and is accelerated away from the light source.

the thermal conductivity k_{th} of the particle. Gas molecules then hit and stick to the surface, taking on the local temperature, and leave the body with the associated thermal velocities, which are lower on the colder side of the particle than on the warmer side. In accordance with the conservation of momentum, the particle has to balance this mismatch and is accelerated. Due to the assumed isotropy of the ambient gas atmosphere and the rotational symmetry of the temperature distribution around the incident direction of the light, only the momentum components along this direction of light account for the change in momentum of the particle. Hence the acceleration of the particle is directed away from the light source or outward in protoplanetary disks. Quantitative descriptions for this process exist for simple cases as, for example, the aforementioned perfect sphere with an absorption coefficient of 100%. The photophoretic force is [4,5]

$$F_{Ph} = \frac{\pi a^2 I p J_1}{\frac{3k_{th}T}{a} + 12\sigma T^4 \epsilon + p\sqrt{\frac{18kT}{\pi m_g}}} \approx \frac{\pi a^3 I p}{6k_{th}T} . \quad (1)$$

Here the particle parameters are the radius a and the thermal conductivity k_{th} and the properties of the ambient gas are the temperature T, the molecule mass m_g and the pressure p. k and σ are the Boltzmann- and Stefan-Boltzmann-constant and I is the radiative flux (or intensity). J_1 is the asymmetry parameter, taking into account the light-absorption properties of the particle, $J_1 = 0.5$ for perfect absorption at the front side, respectively. At a distance of 1 AU to the star the pressure is $p = 1$ Pa and the temperature $T = 280$ K in the minimum mass solar nebula [6]. Assuming a thermal conductivity of $k_{th} = 0.1$ W/(m K) one finds for the radiation pressure $F_{rp} \approx F_{ph} * 10^{-3}$ assuming $F_{rp} = I/c\pi a^2$ [7]. As Poynting-Roberston drag and the Yarkovski effect are based on radiation pressure they are still smaller.

In comparison photophoresis turns out to be the dominant force and the other forces can be neglected if the ambient conditions (namely an optically thin medium with still high enough gas density) allow for photophoresis to take effect.

The aim of this paper is to show our recent experimental developments to determine the absolute strength of photophoresis on certain particle classes. Based on these, eventually, one goal is to study the sorting and concentration efficiency of photophoresis concerning a special type of particles: chondrules. Chondrules are sub-mm- to mm-sized, almost spherical particles, consisting mainly of Mg-Fe silicates. They occur at concentrations of up to 80% in specimens of a class of stony meteorites called chondrites, embedded in a fine-grained matrix [8]. There is a consensus that chondrules are formed from precursor materials by flash heating and adjacent fast cooling in the early solar nebula, whereas the source of heat for this process is still subject to discussion. Studies have shown that the overall formation of the chondrule reservoir in the protoplanetary disk takes about two to three million years, while individual chondrites can contain chondrules with up to one million years of age difference [9]. An interesting fact is that chondrules partially seem to be sorted by size [10]. This shows the necessity of a locally effective transport and concentration mechanism.

To characterize the motion of particles in the protoplanetary disk under the influence of photophoresis, let, without loss of generality, the gas density profile be

$$\rho = 1.4 \times 10^{-6} \left(\frac{R}{1\mathrm{AU}}\right)^{-11/4} \frac{\mathrm{kg}}{\mathrm{m}^3} \tag{2}$$

according to the minimum mass Solar Nebula model of [6] (with R being the radial distance to the star). The radial density gradient causes a pressure gradient which imposes a force upon the gas molecules. This results in an altered (in this case slowed down) rotational speed than predicted by Keplers laws, to establish a stable orbit. The subsequent gas drag decelerates the solid particles to the velocity of the gas. However, as they are too dense to be supported by the pressure gradient, according to [11] they are subject to a residual gravity

$$F_D = \frac{n R_{Gas} T m_P}{\mu R}\,, \tag{3}$$

with $n = 11/4$ the exponent of the density profiles power law, R_{Gas} the gas constant, μ the molar mass and m_P the mass of the particle. Opposing this inward drift is the outward directed photophoretic force. The ratio of the photophoretic force and the residual gravity is given as [7,12]

$$F_{ph}/F_D = \frac{\mu I p R}{k_{th} \rho_p 8 n R_{Gas} T^2}\,. \tag{4}$$

If this value is larger than 1, particles move outwards. This is the case for particles close to the star. If the value is smaller than 1 a particle moves inwards which is the case for particles far away from the star. For $F_{ph}/F_D = 1$ equilibrium exists where particles are concentrating and no further radial transport takes place. If the strength of the photophoretic force varies for different kinds of particles, so does the position of the equilibrium. This way a simple sorting is possible. With k_{th} and ρ_p, the thermal conductivity and the mass density sorting can depend on particle properties. Also, assuming that the chondrules might already be covered in a dust mantle of the fine-grained matrix-material during the transport, the review by [13] shows that the thermal conductivity of a mixture is dominated by the size of the larger constituent. So essentially the sorting by photophoresis is a possible mechanism to explain the meteoritical evidence.

Unfortunatly, little data exists on the photophoretic forces on chondrules as the details depend on a large number of parameters not easily accessible. Direct measurement for free particles allow to quantify the strength of photophoresis but they need microgravity. In an earlier drop tower campaign we tested the feasibility of determining the photophoretic force [14]. Indeed we could show that rough estimates were correct and that the photophoretic force on dust mantled chondrules is stronger than on bare chondrules and that pure dust aggregates experience an even stronger force.

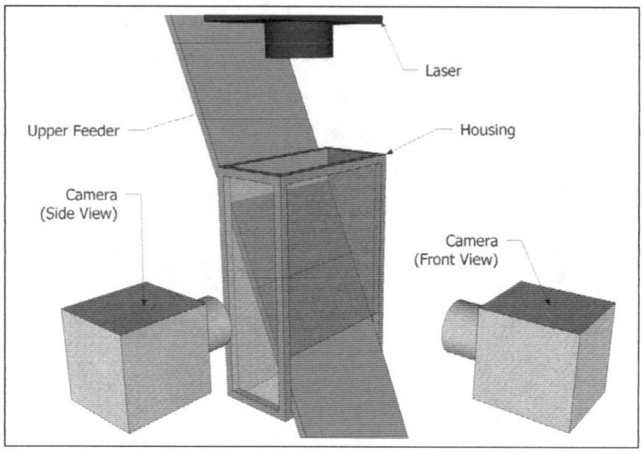

Figure 2. Overview of the experimental setup: The chondrules are loaded into the lower feeder and the feeders are closed.

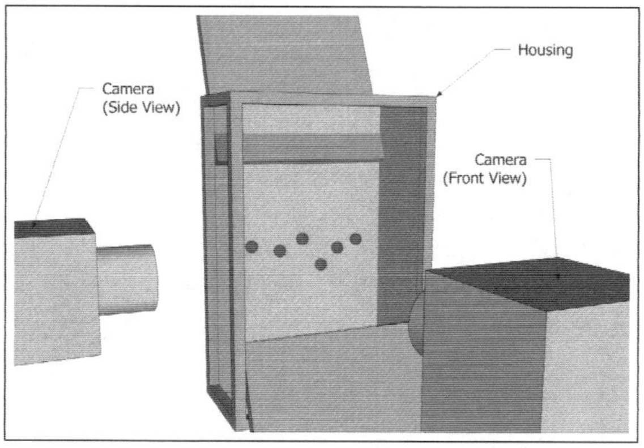

Figure 3. After microgravity is established in the drop capsule, the chondrules are released and exposed to the laser beam.

2. Microgravity experiments

To quantify the effect of photophoresis on chondrules, a set of 16 microgravity-experiments was carried out at the ZARM drop tower in Bremen, each providing - in catapult mode - up to 9.3 seconds of weightlessness with a residual acceleration of $10^{-6}g$ or less. This is necessary, because at typical and feasible experimental parameters, Earth gravity is about three orders of magnitude stronger than the expected photophoretic acceleration.

In figures 2 and 3 a sketch of the experimental setup is shown, displaying, with the exception of the enclosing vacuum chamber, the basic components: the housing for the chondrules, the laser light source and the video observation. The housing confines the chondrules movement to an observable space and facilitates their recovery after the experiment. It has four glass windows, two for video observation and one, at the top, as inlet for the laser beam. The fourth window is placed at the bottom of the housing to prevent heating of the housing by laser radiation. As the laser light is then only absorbed by the particles, no thermophoretic forces occur. Inside there are two retractable feeders, the lower one with nine cavities for the chondrules, to keep them in position at the center of the housing until microgravity is established in the drop capsule. They are released automatically by command of the gravity sensor, that is part of the capsule electronics. The transition from normal gravity to microgravity leads to an initial velocity of the particles, which is damped by collisions with the cavity walls. However, the movement of the feeders can itself induce particle rotation and linear motion. The feeders are retracted within 4 ms, which leads to a variety of initial movement of the chondrules in the range form 0 cm/s to about 1 cm/s. A red laser (1W, 655 nm ± 10 nm) is used as light source, but, deviating from the sketch, is placed outside the vacuum chamber. Furthermore the laser beam is deflected four times to maximize the optical path and the profile is shaped by two cylindrical lenses to broaden and homogenize the laser profile. Video observation is realized with two synchronized high-speed cameras aligned perpendicular to each other, allowing for the motion of the chondrules to be tracked in all three spatial directions.

To characterize the intensity profile of the laser beam, the light output was measured with a bolometer for surface elements of $4\,\mathrm{mm}^2$ at the position of the experiment chamber's inlet window. This measurement showed a central peak with intensities of up to $27000\,\mathrm{W/m}^2$.

For the experiment, nine selected chondrules are loaded into the cavities of the lower feeder and the feeders are clamped. In some experiments transparent 1 mm glass spheres are added to rule out the existence of possible disturbing effects (charging, ...). The pressure in the experiment chamber is set to a predefined value. Experiments were carried out at pressures between 0.01

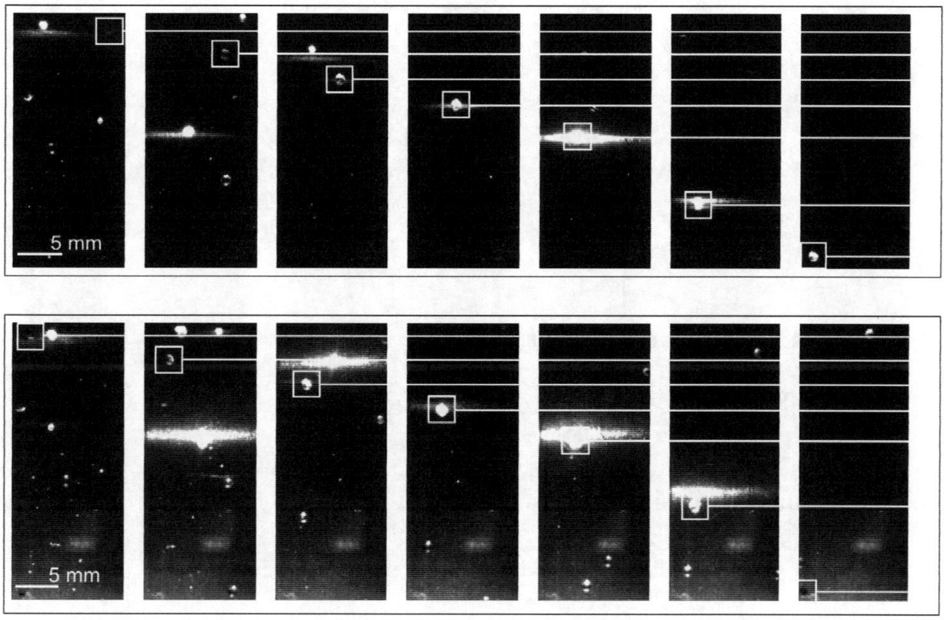

Figure 4. Tracked particle in front (above) and side view (below) in time steps of 0.6 s: Overexposure marks the exposition to the intense central peak of the laser beam.

mbar and 0.1 mbar. In this pressure range the coupling time of the chondrules are large enough to prevent any influence of gas movement, which might be induced be the retracting feeders. After retrieval of the experiment the chondrules are recovered.

The chondrules used in the experiments originate from the meteorite "Bjurböle", a chondrite of the L/LL4-class. After separation they were processed in a grinding mill to remove all remains of matrix material on the surface and then photographed and weighed.

3. Results

The acceleration acting on the chondrules when they are exposed to the intense central peak of the Laser profile is shown in figure 4 using the example of one selected trajectory.

To quantify the particle motion the chondrules were allocated to their respective positions in the feeder at the beginning of the recording and then tracked in the image files of both cameras. In figure 5 the trajectory is plotted component-wise (distance from the reference point against time) and the area of acceleration is marked with a box. The trajectories in the non-illuminated parts of the experiment volume clearly show that no other forces act on the particles, as particles move linearly. Electrical charges and magnetic forces can be ruled out, as the particle motion is not influenced by neighbor particles, despite of particle collisions.

To determine the average acceleration in every direction, the constant velocity-components before and after the acceleration area are derived by applying linear regressions to the data set. This leads to the average acceleration

$$a_{avg} = \frac{v_2 - v_1}{t_2 - t_1} \tag{5}$$

With the given particle masses this can be translated to a photophoretic force F, plotted in figure 6. We get values between 7.9×10^{-12} N and 3.7×10^{-8} N which is consistent with earlier data by [14]. The systematic deviation of the measured forces from the calculated values of one

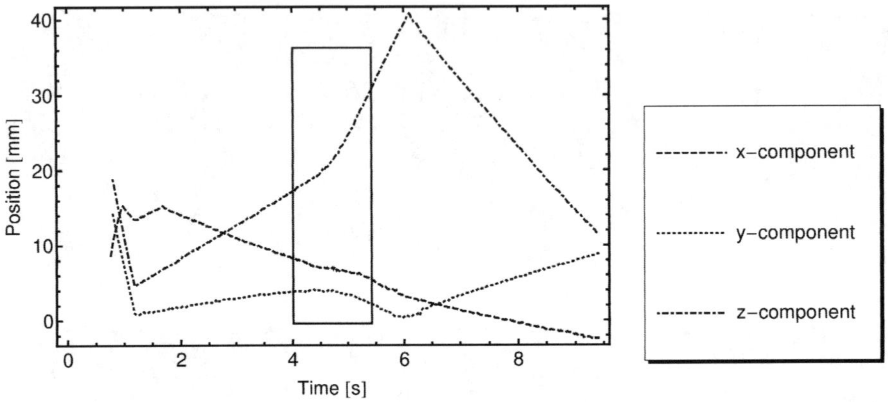

Figure 5. Plot of a chondrule's trajectory (distance to the reference point in x-, y- and z-direction): The box marks the time span shown in figure 4.

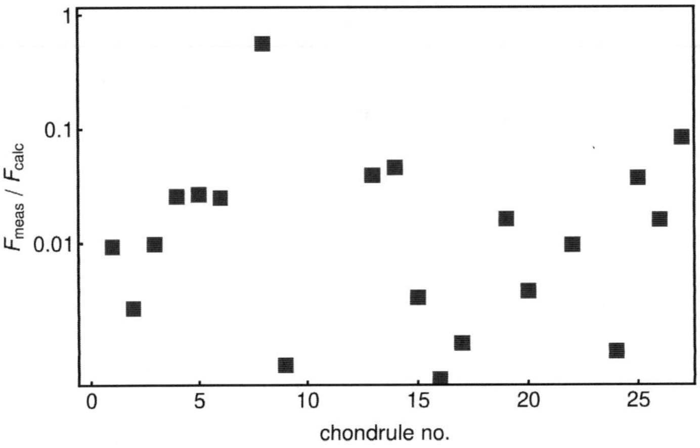

Figure 6. Ratio of the measured photophoretic force and the photophoretic force calculated using equation 1 with $k_{th} = 1$ W/mK, $J = 0.5$ and $\alpha = 1$

to two orders of magnitude can be attributed to the simplified equation used in the calculation and of course to the non-idealistic particle parameters of the real chondrules (e.g. non-spherical shape, variable thermal conductivity, albedo, ...).

4. Conclusions and Outlook

In [14] a set of photophoretic measurements on chondrules, dust mantled chondrules and dust was tested for the first time and allowed us to verify simple estimates of absolute forces on chondrules as they have been used previously [12]. The idea behind the experiments reported here was to achieve data at a qualitatively much better level by using a longer duration of microgravity by using the catapult (9 s instead of 4.7 s), by using new particle cavities to keep chondrules slow and to use a laser as light source to better quantify the light flux. Indeed the particles could be tracked much better and longer and therefore with better accuracy than in the original experiment. In a following campaign, just carried out, the light source was improved providing an almost homogeneous illumination over the entire chamber. These data are not

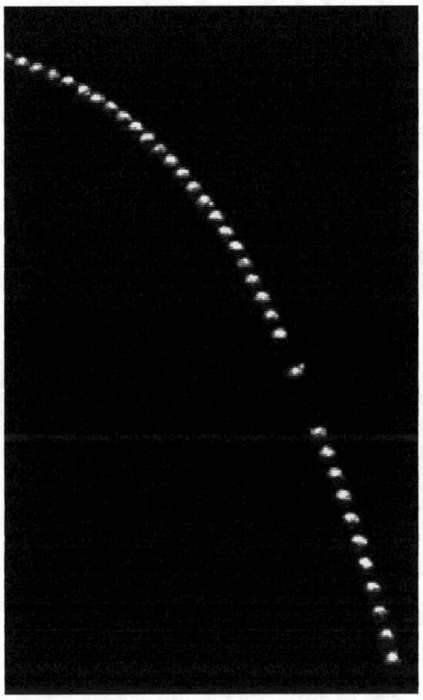

Figure 7. High accuracy trajectory of a chondrule in time steps of 0.1 s with the most recent setup.

analyzed yet but we currently estimate the measurement errors to be in the ten % range. This is much less then the spread in data in the earlier experiments by [14] or reported above where individual values even for a single chondrule could vary by a factor of a few. A first example of a particle trajectory is shown in figure 7 which is a perfect parabola.

We now have a setup to measure the photophoretic strength and variations for different particles in detail and we do have data but this has just been started to be analyzed at the time of these proceedings. Specific applications to protoplanetary disks therefore have to await this further data analysis but should allow us to specify the variations in photophoretic strength between chondrules in much more detail. As we also used dark material from the Allende meteorite in the recent campaign (low albedo) this will enable us to draw first conclusions from experiments on the capability of photophoresis to transport and sort particles. Last not least we also observed and resolved the rotation of the particles. From the data it can be specified how rotation changes the photophoretic strength or how it is changed by photophoresis.

Acknowledgments
Access to the drop tower was granted by the European Space Agency. The meteoritic material was provided by Addi Bischoff and the Institut für Planetologie, Münster. Also many thanks to the kind support of the drop tower team in Bremen.

5. References

[1] Dullemond, C. P., Hollenbach, D., Kamp, I., and DAlessio, P., 2007. In Reipurth, B., Jewitt, D., and Keil, K., editors, *Protostars and Planets V*, 555.
[2] Alexander, R. D. and Armitage, P. J., 2007. *Monthly Notices Roy. Astr. Soc.*, **375**, 500.
[3] Sicilia-Aguilar, A., Hartmann, L., Calvet, N., Megeath, S. T., Muzerolle, J., Allen, L., DAlessio, P., Mern, B., Stauffer, J., Young, E., and Lada, C., 2006. *ApJ*, **638**, 897.
[4] Beresnev, S., Chernyak, V., and Fomyagin, G., 1993). *Phys. Fluids A: Fluid*, **5**, 2043.
[5] Rohatschek, H., 1995. *Journal of Aerosol Science*, **26(5)**, 717.

[6] Hayashi, C., Nakazawa, K., and Nakagawa, Y., 1985. In: Black, D. C. and Matthews, M. S., editors, *Protostars and Planets II*, 1100.
[7] Krauss, O. and Wurm, G., 2005. *ApJ*, **630**, 1088.
[8] Wadhwa, M., Amelin, Y., Davis, A. M., Lugmair, G. W., Meyer, B., Gounelle, M., and Desch, S. J., 2007. In Reipurth, B., Jewitt, D., and Keil, K., editors, *Protostars and Planets V*, 835.
[9] Scott, E. R. D., 2007. *Annual Review of Earth and Planetary Sciences*, **35**, 577.
[10] Grossman, J.N., Rubin, A.E., Nagahara, H., King, E.A., 1988. *Meteorit. Early Solar Syst.*, 619.
[11] Weidenschilling, S. J., 1977. *Monthly Notices Roy. Astr. Soc.*, **180**, 57.
[12] Wurm, G. and Krauss, O., 2006. *Icarus*, **180**, 487.
[13] Presley, M. A. and Christensen, P. R., 1997. *J. Geophys. Res.*, **102**, 6551.
[14] Wurm, G., Teiser, J., Bischoff, A., Haack, H., Roszjar, J., 2010., *Icarus* , **208**, 482.

Journal of Physics: Conference Series **327** (2011) 012045 doi:10.1088/1742-6596/327/1/012045

Electrodeposition experiments in microgravity conditions

K Nishikawa[1], **Y Fukunaka**[2], **E Chassaing**[3] **and M Rosso**[1]
[1] Laboratoire de Physique de la Matière Condensée, UMR CNRS 7643, Ecole Polytechnique, 91128 Palaiseau Cedex, France
[2] Graduate School of Energy Science, Kyoto University, Sakyo Ku, Kyoto 6068501, Japan
[3] IRDEP, UMR CNRS 7174, 6 quai Watier, 78401 Chatou Cedex, France

E-mail: michel.rosso@polytechnique.fr

Abstract. Metal electrodeposition may produce irregular deposits with various morphologies. For liquid electrolytes a precise study of these deposits may be complicated by convective motion due to buoyancy. Zero-gravity (0g) conditions provided by parabolic flights give a straightforward mean to avoid this effect: we present here 0g electrodeposition experiments, that we compare to ground experiments (1g).
Two electrochemical systems were studied by laser interferometry, allowing to measure concentration variations in the electrolyte: copper deposition from copper sulfate aqueous solution and lithium deposition from an ionic liquid containing LiTFSI. For copper, concentration variations were in good agreement with theory. For lithium, an induction time was observed for the concentration evolution at 1g: due to this induction time and to the low diffusion coefficient in the ionic liquid, concentration variations were hardly measurable in the parabolic flight 0g periods of 20 seconds.

1. Introduction

It has been known for decades that metal electrodeposition in the absence of a supporting electrolyte may produce irregular deposits with various morphologies. This phenomenon has attracted much interest in the last 50 years. From the viewpoint of fundamental electrochemistry, dendritic electrodeposits may provide simple model systems to study Diffusion Limited Aggregation (DLA) phenomena [1]. On the other hand, dendritic growth was also intensively studied for practical reasons, in particular because of its detrimental role in battery technology [2]. This phenomenon has limited the use of all-solid-state lithium batteries with metallic lithium negative electrodes: although one of the most promising technologies to meet the requirements of electric power sources in novel applications (portable electronic devices or electric vehicles), this technology is not yet available because dendritic growth may appear during battery charging. Hence, controlling the morphology of the metal deposited on a lithium metal electrode is still an important issue in order to avoid dendritic growth in lithium batteries.

However, for liquid electrolytes, and whatever the electrochemical system under concern, a precise study of dendritic growth mechanisms may be complicated by convective motion due to buoyancy [3]: even in thin, quasi-two-dimensional horizontal cells with vertical electrodes, electrodeposition is accompanied by a gravity-induced fluid flow at the electrodes (figure 1). This effect is due to the electrolyte density decrease at the cathode (due to the concentration decrease) or increase at the anode (concentration increase): it has been extensively studied in the recent literature, both theoretically [4-6] and experimentally [7-12]. Convective motion mixes the electrolyte and tends to homogenise the concentration. The effect depends on cell configuration, salt concentration, and current density [4]. This convection driven by buoyancy was shown to increase the instability of the system [12].

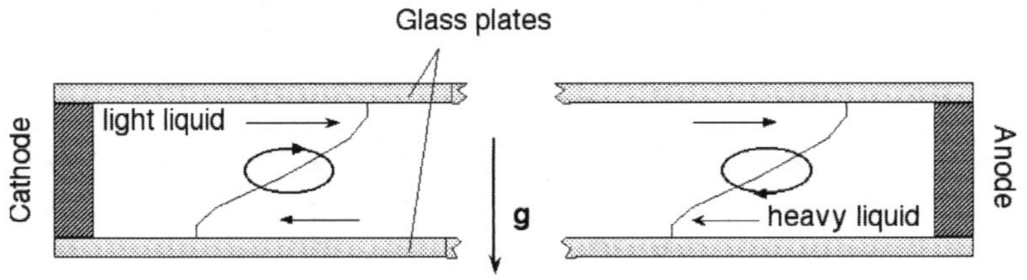

Figure 1. Schematic view of an electrochemical cell in the vicinity of vertical electrodes: a convective motion is induced in these regions, due to the electrolyte density decrease at the cathode (concentration decrease) or increase at the anode (concentration increase).

Obviously, zero-gravity experiments [13] should give a straightforward mean to avoid this effect: apart from gravity, no other parameter is altered. However, these experiments are relatively difficult to work out and of short duration (around 5 - 25 s in the most affordable facilities: drop tower and parabolic flight).

Hence, several other methods have been proposed to minimise the convection driven by buoyancy:

- very thin cells (cell thickness less than 30 μm) and/or electrolytes with very low concentration, below 0.01 mol L^{-1} [9]. However in practical cases, electrolyte concentrations are much higher than this limit: for example, electrolytes with concentration over 1.0 mol L^{-1} are used in Li batteries. The limited concentration range mentioned above thus restricts the interest of this method.
- gel [7], or electrolytes with increased viscosity [14] allowed for measurements without the above limitation. However, the chemistry involved in these electrolytes may be markedly different from that involved in the absence of additives.
- several authors [7, 9, 12, 15-18] used thin cells in a vertical position, with a plane cathode set horizontally on top of the cell, and facing downward. It is called "cathode over anode" (C/A) type cell. For example, Marshall et al. [12] showed that this configuration enables to obtain a stable regime. Using this method, Morisue et al. could evidence the gravitational effect on nucleation phenomena in copper electrodeposition onto a TiN substrate [19]. González et al. could measure impedance diagrams of the electrolytic cell at the onset of dendritic growth [18].

These experiments, with a cathode set horizontally on top of the cell, are easy to handle, and they have no time limitation. However, obviously gravity is still present, and convective motion may appear, in particular when the current density distribution (hence the concentration distribution) is not uniform along the electrodes, because the electrodeposits increase the surface roughness.

In conclusion, zero-gravity experiments may appear as a unique way for avoiding the convective motion driven by buoyancy in electrodeposition in conditions well controlled and free from any of the above mentioned parasitic effects.

Previous studies have shown that even short zero-gravity experiments may provide interesting information on the preliminary stages of dendritic growth in electrodeposition. In particular, Fukunaka et al. performed copper electrodeposition experiments from $CuSO_4$ aqueous solution [13]. The experiments were carried out in a drop shaft allowing free falls for more than 8 seconds under zero-gravity environment. Ionic concentration profile in the electrolyte was measured by interferometry: the diffusion layer thickness was shown to increase linearly with the square root of time, in good agreement with theoretical calculation. Due to natural convection, the diffusion layer thickness was shown to increase much faster at 1g than at 0g. Also, at 0g, the deposit consisted in fewer, larger grains than at 1g.

We present in this paper a study of electrodeposition in the zero-gravity environment provided by parabolic flights. This project was supported by the Centre National d'Etudes Spatiales (CNES): it planed copper electrodeposition experiments on copper electrode and lithium electrodeposition on Ni electrodes in the Airbus A300 from Novespace. During these flights, almost zero-gravity conditions were available over 20 to 25 seconds. The project included concentration measurements by laser interferometry, and similar but longer experiments performed at 1g in the cathode over anode configuration. Both series of experimental results were to be compared with numerical calculations.

2. Experimental

The project presented here consisted in carrying out electrochemical experiments during zero-g periods provided by parabolic flights. Although limited, the duration of these experiments indeed allowed to explore ranges of parameters hitherto inaccessible. They were supplemented by experiments in cells whose geometry (C/A cell) allowed longer experiments, while minimizing the effect of concentration gradients.

The experiments were carried out in a pseudo 2 D-cell (figure 2). Care was taken to obtain a well defined cell geometry: see Ref. [20] for details. Copper electrodes were used for the copper electrodeposition and nickel electrodes were used for lithium electrodeposition: they were held between two glass plates enabling to observe *in situ* the electrolyte (figure 2).

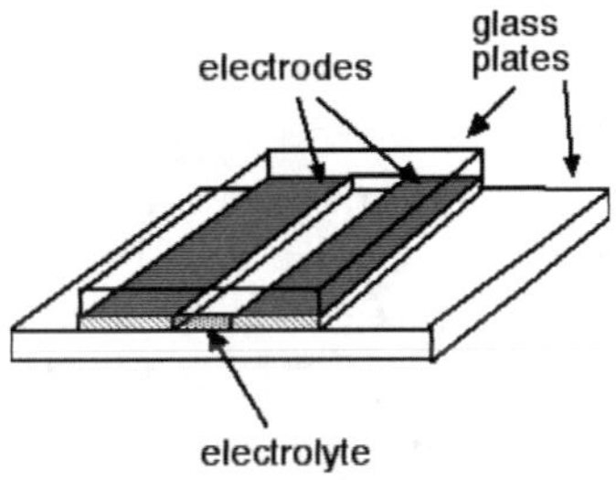

Figure 2. Schematic view of the cell

The active part of the cell had a parallelepipedic shape, with dimensions L, the active length of the electrodes, *l,* the inter-electrode distance, and d, the thickness of the electrodes, respectively. Practical values were: L = 0.8-1.5 cm, $l \approx 0.09 - 0.3$ cm, d $\approx 0.012 - 0.1$ cm. The lateral faces of the electrodes were coated with a hydrophobic film which limited the invasion of the electrolyte in between the electrodes and the glass plates. After filling the cell with the electrolyte, we sealed it with a two-component resin before taking off: this allowed to keep it under almost constant conditions for several hours [20]. During the flight, the temperature in the plane was 18°C. Two electrochemical systems were studied: copper deposition from 0.1 and 0.2 mol L^{-1} aqueous $CuSO_4$ solutions and lithium deposition from an ionic liquid. The ionic liquid was N-methoxymethyl-N-methylpyrrolidinium bis (trifluoro-methane-sulfonyl)imide containing 1.0 mol L^{-1} LiTFSI.

In these microgravity experiments, a concentration variation of the electrolyte induced a variation of the refractive index which was measured by a common path laser interferometer.

Laser interferometry is a well known technique for measuring the concentration changes in electrochemical processes [21-25]. In this study, we used the interferometer described in Ref. [13] to measure *in situ* the variation of the ionic concentration in the electrolyte between cathode and anode. This interferometer, provided by Olympus, was designed for 0g experiments. Because we used thin cells (0.1 or 1 mm), we expected light deflection due to refractive index gradients [26] to be negligible. The laser source was a laser diode with 685 nm wavelength. The change in the refractive index was given by

$$d\Delta n = S\lambda \tag{1}$$

where d is the optical path length in the electrochemical cell (here the electrode thickness), Δn is the change of refractive index, S is the shift of interference fringes, and λ is the laser wavelength. The relationship between refractive index and concentration is given by:

$$\Delta n = \left(\frac{\partial n}{\partial C}\right)\Delta C \tag{2}$$

In the case of copper in a $CuSO_4$ aqueous solution, we used (∂n/∂C) = 0.0279. This value was obtained at 23°C [24]. We kept the same value at 18°C, because the temperature coefficient of the refractive index of water is approximately 10^{-4}/°C at this temperature [27]. In the case of lithium in the liquid ionic, (∂n/∂C) = 0.005 [28] was taken. Using equations (1) and (2), the change of concentration could be determined from the measurement of the shift of the interference fringes. The evolution of this shift was recorded by means of a CCD camera. An experiment thus consisted in applying a constant current or a constant potential to the electrochemical cell, while capturing a series of pictures showing the evolution of the interference fringes.

In the ground experiments a Mach Zehnder interferometer was used [29], with a laser diode with 638 nm wavelength.

3. Results

3.1. Copper deposits

We have reported elsewhere [7, 11, 29] concentration measurements obtained in 1g conditions, both for vertical electrodes, and for horizontal electrodes in the C/A configuration. In the latter case, concentration profiles observed before the onset of dendritic growth were in good agreement with calculations neglecting the presence of gravity [7, 29]. On the other hand, in experiments performed in cells with vertical electrodes concentration profiles were markedly different, due to the presence of buoyancy induced convection [7-11].

We present here concentration measurements in the electrolyte at 0g. Figure 3 shows an example of interference fringes recorded during one parabola, before and 15 seconds after the beginning of cell polarization. The fringes are bent in the vicinity of the cathode, at the bottom of the figure, because of the variation of concentration.

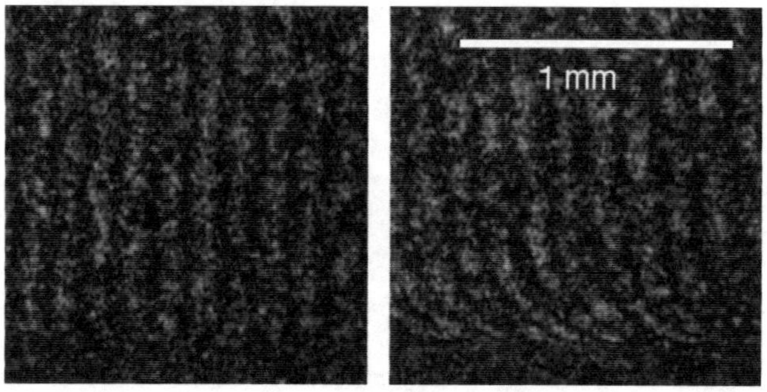

Figure 3. Interference fringes recorded before and 15 s after the onset of Cu deposition. The fringes are bent in the vicinity of the cathode (bottom of the figure) because of the concentration variation. The shift to the right hand side corresponds to a decrease of the refractive index, hence of the concentration. Electrolyte concentration is 0.1 mol L^{-1}; current density is 8 mA cm^{-2}. Cell thickness is 0.1 cm.

Journal of Physics: Conference Series **327** (2011) 012045 doi:10.1088/1742-6596/327/1/012045

This can also be seen in figure 4, showing the position of the interference fringes as a function of position at three different times.

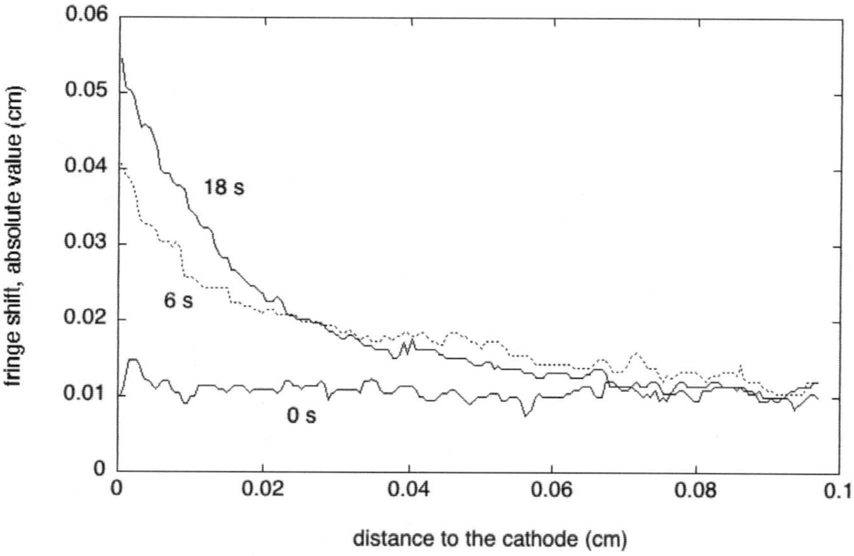

Figure 4. Shift of the interference fringes as a function of the distance to the cathode at three different times after the onset of Cu deposition. Same conditions as in figure 3.

These concentration variations were also numerically calculated, using general equations describing transport phenomena in electrochemical systems [31], and taking into account the variation of the diffusion coefficients with concentration in $CuSO_4$ aqueous solution [32]. A detailed presentation of the application of these equations to the Cu/$CuSO_4$ system is given in Ref. [7]. Variation of the surface concentration at the cathode as a function of time during and after one parabola is shown in figure 5, and compared with theory. As shown in figure 5, we found a good agreement between experimental and theoretical results, thus confirming 1g results for the C/A cell. After the end of the parabola, the concentration at the cathode rapidly returned to its initial value (figure 5): this was due to the large gravitational level (~ 1.5g) imposed at this time in order to drive back the plane to its normal flying conditions.

Our experiments did not enable us to observe dendritic growth of copper in 0g conditions, because the dendrites only appear after an induction time t_s the so-called Sand time [30]. Sand time varies as

$$t_s = \frac{\pi e^2 D(1 + \mu_c/\mu_a)C_0^2}{J^2} \tag{3}$$

where e is the electronic charge, D the diffusion constant, μ_c and μ_a the cationic and anionic mobilities respectively, C_0 is the initial ionic concentration and J the current density.

To obtain a Sand time shorter than 10 seconds for example, would require a high current density (larger than 20 mA cm^{-2}) and still, the time available for observing dendritic growth would be very short. Experimental environment did not provide such conditions in this parabolic flight campaign. Further experiments are thus necessary to explore dendritic growth itself.

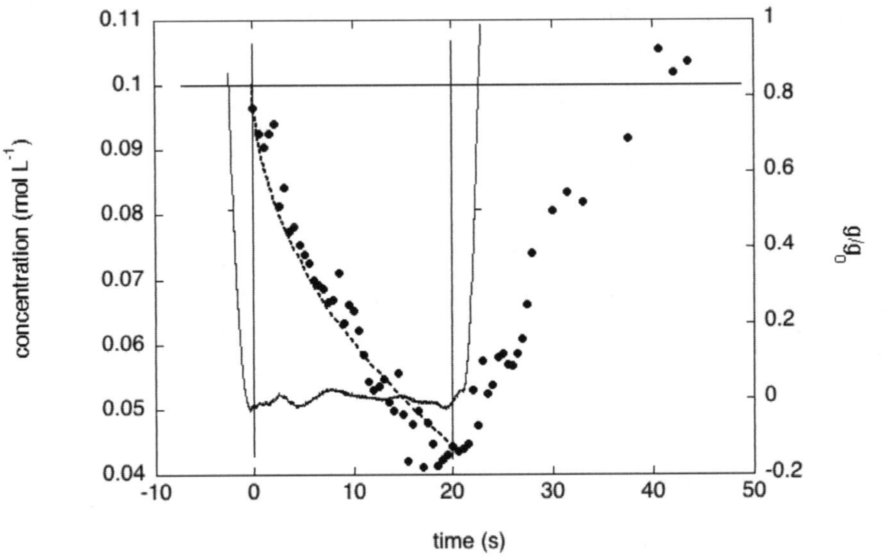

Figure 5. Variation of ionic concentration in copper sulfate at cathode during one parabola: experimental results (closed circles), calculated results (dashed line). The variation of g/g_0 is shown as a continuous line. The horizontal line shows the initial concentration, the two vertical lines show the beginning and the end of the cell polarization. Same conditions as in figure 3.

3.2. Lithium deposits

As mentioned earlier, a series of experiments were performed at 1g. A typical result is given in figure 6, showing the variation of the cathode surface concentration as a function of the square root of time. In this experiment, the applied current density was 1 mA cm^{-2} and cell thickness was d = 0.1 cm. One clearly sees a first stage, where the concentration variation is very small. Then, after 90 s, the variation is much more rapid. Such an induction time was already reported [17]. In the present case, the observed induction time might be due to side reactions of the electrodeposited lithium metal.

The diffusion coefficients in ionic liquids are very low. From the variation of the diffusion layer thickness in the experiment illustrated in figure 6, we estimated the following value for the diffusion coefficient: $D \sim 10^{-7}$ cm^2 s^{-1}. This value is in good agreement with that measured by other methods [33]. The concentration variations are compared in figure 6 with theoretical values [31], which are calculated without taking into account the concentration dependence of the diffusion coefficient. The result obtained taking into account the above value of D and a transport number of 0.35 for Li^+ ion is shown as a straight continuous line.

At 0g, we could not observe any variation of the concentration profiles in the electrolyte. Due to the very low diffusion coefficient and to the induction time mentioned above, the concentration variations due to Li deposition during the first 20 seconds were very small, and the concentration boundary layer thickness was at most 10 μm: thus concentration variations were hardly detectable by our interferometer. The Li deposition experiment would thus require longer zero-g times.

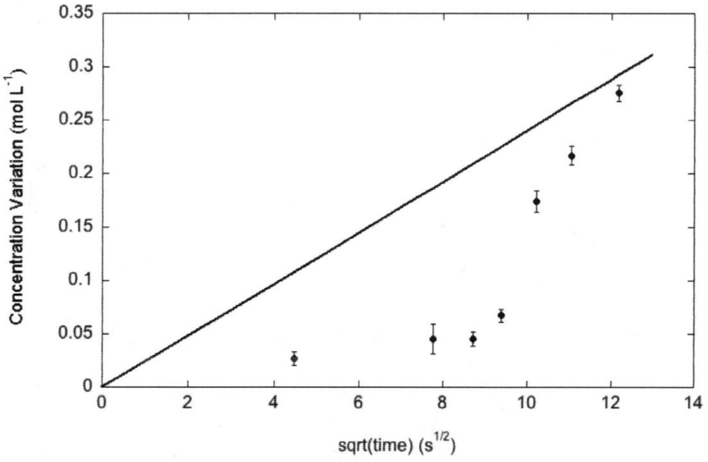

Figure 6. Deposition of lithium from ionic liquid: variation of cathode surface concentration as a function of the square root of time after the onset of polarization: experimental results (filled circles), calculated results (straight continuous line). Current density is 1 mA cm^{-2}. Cell thickness is 0.1 cm.

4. Conclusion

We could perform electrodeposition experiments in the microgravity conditions provided by the parabolic flight facilitity operated by Novespace. The evolution of concentration profiles measured in the electrolyte by laser interferometry was consistent with theory and with previous experiments performed at 1g for cells in the C/A configuration. However, because of their limited duration, parabolic flight experiments could only provide results on the early stage of electrodeposition. A more complete study would require further zero-g experiments: for studying dendritic growth of copper, or for deposition of lithium from ionic liquids, longer times would be necessary (such as those available in sound rockets or in the International Space Station). Such study should give access to experimental conditions different from those available on ground, even in the C/A configuration, in particular in the case where the current density distribution (hence the concentration distribution) is not uniform along the electrodes. These improvements are considered in the framework of a project developed by an "Electrochemical Nucleation & Growth" ESA Topical Team.

Acknowledgements

The authors thank Drs. H. Yasuda (Osaka University), Y. Inatomi (Japan Aerospace Exploration Agency - JAXA), K. Kinoshita (JAXA), M. Takayanagi (JAXA), S. Yoda (JAXA) and Mr. T. Maki (Olympus Corporation), for providing the interferometer especially designed for zero-g experiments that we used in this study. The authors also wish to thank Dominique Clément for the mechanical realization of the experimental setup and for the many technical solutions which he brought to the problems involved in the operation of this experiment in parabolic flight. Centre National d'Etudes Spatiales is acknowledged for its financial support and Novespace for help in preparing the experiments.

References

[1] Witten T A and Sander L M 1981 *Phys. Rev. Lett.* **47** 1400
[2] Tarascon J M and Armand M 2008 *Nature* **451** 652
[3] Laitinen H A and Kolthoff I M 1939 *J. Am. Chem. Soc.* **61** 3344

[4] Chazalviel J-N, Rosso M, Chassaing E and Fleury V 1996 *J. Electroanal. Chem.* **407** 61
[5] Marshall G, Mocskos P, Swinney H L and Huth J M 1999 *Phys. Rev. E* **59** 2157
[6] Kawai S, Nishikawa K, Fukunaka Y and Kida S 2007 *Electrochim. Acta* **53** 257
[7] Rosso M, Chazalviel J-N, Fleury V and Chassaing E 1994 *Electrochim. Acta* **39** 507
[8] Barkey D P, Watt D, Liu Z and Raber S 1994 *J. Electrochem. Soc.* **141** 1206
[9] Huth J M, Swinney H L, Mc Cornmick W D, Kuhn A and Argoul F 1995 *Phys. Rev. E* **51** 3444
[10] Argoul F, Freysz E, Kuhn A, Léger C and Potin L 1996 *Phys. Rev. E* **53** 1777
[11] Rosso M, Chassaing E and Chazalviel J-N 1999 *Phys. Rev. E* **59** 3135
[12] Marshall G, Mocskos E, González G, Dengra S, Molina F V and Iemmi C 2006 *Electrochim. Acta* **51** 3058
[13] Fukunaka Y, Okano K, Tomii Y, Asaki Z and Kuribayashi K 1998 *J. Electrochem. Soc.* **145** 1876
[14] González G, Marshall G, Molina F and Dengra S 2002 *Phys. Rev. E* **65** 051607
[15] Sagués F, López-Salvans M Q and Claret J 2000 *Phys. Reports* **337** 97
[16] González G, Soba A, Marshall G, Molina F V and Rosso M 2007 *Electrochim. Acta,* **53** 133
[17] Nishikawa K, Fukunaka Y, Sakka T, Ogata Y H and Selman J R 2007 *Electrochim. Acta,* **53** 218
[18] González G, Rosso M, Chassaing E and Chazalviel J –N 2007 *Electrochim. Acta* **53** 141
[19] Morisue M, Fukunaka Y, Kusaka E, Ishii R and Kuribayashi K 2003 *J. Electroanal. Chem.* **559** 155
[20] Rosso M, Chassaing E, Chazalviel J –N and Gobron T 2002 *Electrochim. Acta* **47** 1267
[21] McLarnon F R, Muller R H and Tobias C W 1976 *Electrochim. Acta* **21** 101
[22] Denpo K, Okumura T, Fukunaka Y and Kondo Y 1985 *J. Electrochem. Soc* **132**, 1145
[23] Eriksson R 1995 *Electrochim. Acta* **40** 725
[24] Konishi Y, Tanaka Y, Kondo Y and Fukunaka Y 2000 *Electrochim. Acta* **46** 681
[25] Fukunaka Y, Nakamura Y and Konishi Y 1998 *J. Electrochem. Soc.* **145** 3814
[26] Beach KW, Müller R H and Tobias C W 1973 *J. Opt. Soc. Am.* A **63** 559
[27] Hawkes J B, Astheimer R W 1948 *J. Opt. Soc. Am.* **38** 804
[28] Nishikawa K, private communication
[29] Nishikawa K, Chassaing E and Rosso M 2011 *Electrochim. Acta* **56** 5464
[30] Sand H J S 1901 *Phil Mag* **1** 45
[31] Marciano S L and Arvia A J 1983 Diffusion in the absence of convection: steady state and nonsteady state *Comprehensive Treatise of Electrochemistry* Ed Yeager E, Bockris J O'M, Conway B E, Sarangapani S (New York: Plenum Press) vol.6, pp. 65-132,
[32] Miller D G, Rard J A, Eppstein L B and Robinson R A 1980 *J. Solution Chem.* **9** 467
[33] Nishikawa K, Naito H, Kawase M and Nishida T 2011 *Lithium Batteries Discussion* Arcachon (France)

Journal of Physics: Conference Series **327** (2011) 012046 doi:10.1088/1742-6596/327/1/012046

Microgravity experiments in the field of physical chemistry in Japan

M Natsuisaka[1], K Tsujii[2], M Shimomura[3], H Yabu[3], Y Hirai[3], T Mashiko[4], S Deguchi[5], S Mukai[6], Y Inoue[7], Y Nishiyama[8], M Sawada[9], K Okumura[10] and K Sakamoto[11]

[1] Institute of Space and Astronautical Sciences (ISAS), Japan Aerospace Exploration Agency (JAXA), 2-1-1 Sengen, Tsukuba, Ibaraki 305-8505, JAPAN
[2] Research Institute for Electronic Science, Hokkaido University, Kita-ku N21 W10, Sapporo, Hokkaido 001-0021, JAPAN (Retired)
[3] Institute of Multidisciplinary Research for Advanced Materials (IMRAM), Tohoku University, 2-1-1 Katahira, Aoba-ku, Sendai, Miyagi 980-8577, JAPAN
[4] Departm ent of Mechanical Engineering, Faculty of Engi neering, Shi zuoka Univertisty, 3-5-1 Jouhoku, Naka-ku, Hamamatsu, Shizuoka 432-8561, JAPAN
[5] Institute of Biogeosciences, Japan Agency for Marine-Earth Science and Technology (JAMSTEC), 2-15 Natsushima-cho, Yokosuka, Kanagawa 237-0061, JAPAN
[6] Faculty of Sciences, Kyushu University, 6-10-1 Hakozaki, Higashi-ku, Fukuoka, Fukuoka 812-8581, JAPAN
[7] Department of Applied Chemistry, Osaka University, 2-1 Yamada-oka, Suita, Osaka 565-0871, JAPAN
[8] Graduate School of Materials Science, Nara Institute of Science and Technology (NAIST), 8916-5 Takayama, Ikoma, Nara 630-0192, JAPAN
[9] COSMO OIL Co., Ltd, JAPAN
[10] Faculty of Sciences, Ochanomizu U niversity, 2-1-1 Ohtsuka, Bunkyo-ku, Tokyo 112-8610, JAPAN
[11] Facult y of Pharm acy, Chiba Institute of Science, 15-8 Shiomi-cho, Choshi-city, Chiba 288-0025, JAPAN

E-mail: natsuisaka.makoto@jaxa.jp

Abstract. Ja pan has been o perating ' KIBO' ('hope' in Japanese) as J apanese ex perimental module on ISS (International Space Station) to perform researche s on physical, life, m edical, space sciences in space. Several research racks and facilities have already been accommodated in th e pressurized m odule of 'KIBO' and so me researches ha ve alrea dy been ca rried out. Japan c urrently has 17 I SS flight projects (i ncluding 4 candidates) i n t he field of physical sciences a nd also i ncubates ne w projects t hrough a re search s upport p rogram na med as 'research WG (Working Group)', wh ere 25 r esearch WGs ar e activ e in t he field of physical sciences. T hose i nclude 1 fl ight ca ndidate and 2 res earch WGs i n t he fi eld of physical chemistry. The article introduces those to promote international collaborations.

1. Current status on microgravity experiments in Japan

Japan has been operating 'KIBO' as Japanese experimental module on ISS to perform researches on physical, life, medical, space sciences in space. Several research racks and facilities have been accommodated in the pressurized module of 'KIBO' (see figure 1) and researches on Marangoni convection, ice crystals have already been carried out.

Japan currently has 17 ISS flight projects (including 4 candidates) and 25 research WGs in the field of physical sciences. A research aiming at ISS/KIBO experiment has to follow a process step by step (see figure 2). Activities in research WGs annually solicited are the first step of the process. A research WG discusses new ideas and develops a flight proposal to submit it to a future flight AO (Announcement of Opportunity) to be released on biannual basis. Some research WGs might try BBM (Bread Board Model) tests including parabolic flight experiments to assess feasibilities of their observations and operations and verify the validities of their science requirements, if necessary . A research, after passing a flight AO, will try science definitions as a flight candidate for a couple of years and has to pass the project definition review to step up to a flight project.

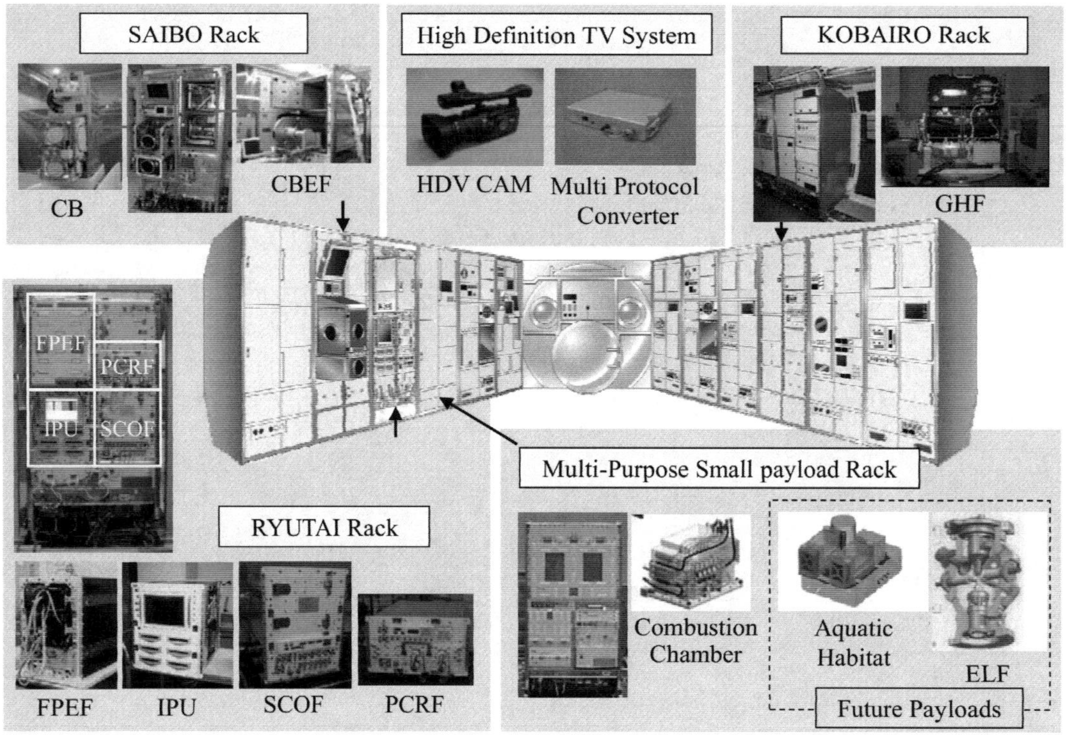

Figure 1. Racks and facilities accommodated in KIBO pressurized module. The acronyms of the facilities are as follows; CB Clean Bench, CBEF Cell Biology Experiment Facility, GHF Gradient Heating Furnace, FPEF Fluid Physics Experiment Facility, IPU Image Processing Unit, SCOF Solution Crystallization Observation Facility, PCRF Protein Crystallization Research Facility, ELF Electrostatic Levitation Furnace. (c) JAXA.

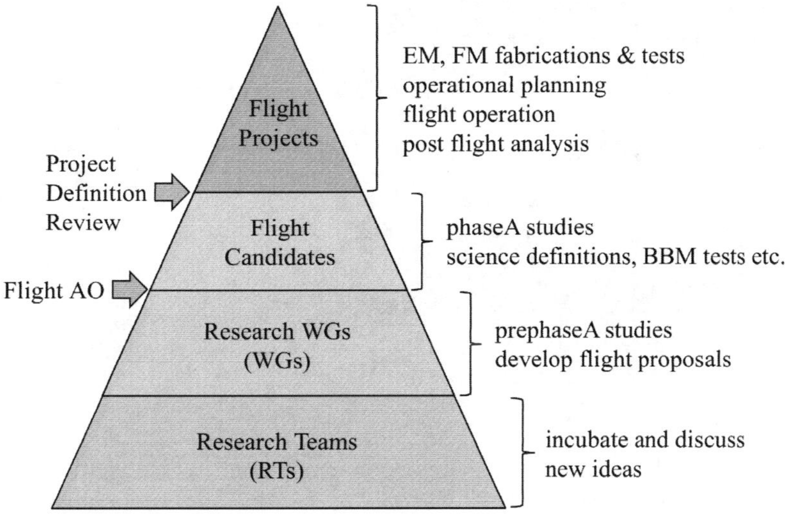

Figure 2. Japanese implementation process of ISS experiments. The acronyms in the figure are as follows; EM engineering m odel, FM flig ht m odel, BBM bread bo ard m odel, AO announcement of opportunity.

2. Current status on researches in physical chemistry in Japan

There are 1 flight candidate and 4 research WGs (2 WGs + 2 RTs) in the field of physical chemistry. Current status on those will be described in this secession.

2.1. Flight projects

A research proposal on colloid physics was selected as a candidate of KIBO experiments in 2010 and has been doing phaseA studies.

2.1.1. Structure analysis on colloid crystals with a laser diffractometer and researches on particle interactions. The research team organized by Ikuo Sogami is interested in interactions in charged colloid dispersions. Crystal formations in those dispersions which occur when the salt concentrations are decreased were normally explained by DLVO (Derjaguin and Landau, Verwey and Overbeek) theory, but some observations like measurements of lattice spacing of the crystals [1][2], observation of gas-liquid coexistence state in colloid dispersions [3][4], and observation of void formations in colloid dispersions[5]-[7] can not be explained by DLVO theory and suggest existence of long range attractive interaction. The research team has been developing ISS research facility to be accommodated in Multi-purpose small payload rack (see figure 1). CCD (Charge Coupled Device) camera of the facility, having several objective lenses, can observe void formation and its dynamics as well as whole image of a cell. The key element of the facility is 'Kikuchi-Kossel' laser diffractometer developed by the team [8][9]. Although this is simpler and more compact than small angle X-ray diffractometer, it can precisely measure crystal structures and lattice spacing (see figure 3). This will be only one facility that enables such precise measurements of crystal structures on colloids on ISS. The members of the project are Ikuo Sogami (Kyoto Sangyo University), Masayuki Tanigawa (Kyoto Sangyo University), Tsuyoshi Yosiyama (Kyoto Sangyo University), Yuji Ikeda (Kyoto Sangyo University), Tadatomi Shinohara (Kyoto Sangyo University), Masamichi Ishikawa (RIKEN), Kensaku Itoh (University of Toyama), Junpei Yamanaka (Nagoya City University), Hideo Oka (Tayca Corporation), Mitsuhiro Matsumoto (Kyoto University), B. V. R. Tata (Indira Gandhi Centre for Atomic Research), and Naokiyo Koshikawa (JAXA).

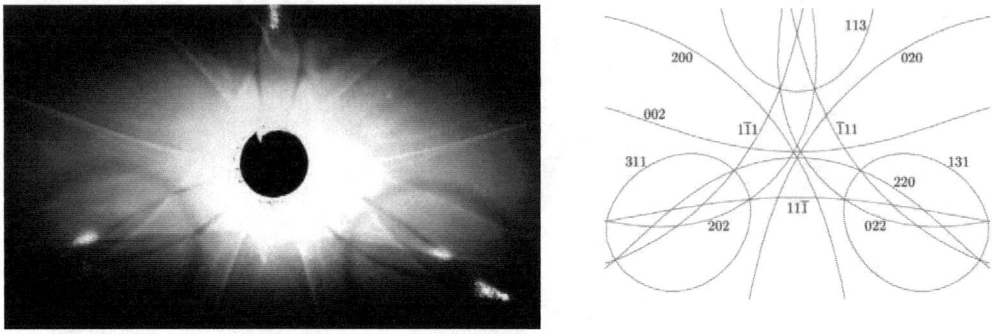

Figure 3. 'Kikuchi-Kossel' diffraction pattern taken by BBM. The left is an actual picture and the right indicates Miller indexes for the individual lines shown in the left picture.

2.2. Research WGs

The research WGs are categorized into RTs (Research Teams) and WGs (see figure 2). The former is one for new comers to discuss their preliminary ideas and the latter is one for more experienced members to define experimental requirements and research plans. In addition to team meetings, WGs carry out BBM tests to clarify the validity of their experiment techniques and requirements. Some WGs might perform parabolic flight experiments, if necessary. 2 WGs and 2 RTs are currently active in the field of physical chemistry.

2.2.1. *Mesoscopic chemistry under microgravity WG.* The WG coordinated by Kaoru Tsujii is aiming at incubating new ideas associated with chemistry and various ideas proposed by various members have been discussed.

(1) *Honey comb film formation under microgravity.* A member of the WG, Masatsugu Shimomura, found that dissipative structures where pores having a uniform diameter are regularly ordered can be obtained if steam is added to casted polymer solutions [10][11]. Although the protocol is very simple, the uniformity and ordering of the pores are quite well (see figure 4). Medical applications have been investigated and larger films are desired for the applications, but a kind of domain formation like a grain boundary for a polycrystalline is observed in some cases. The members speculated that this might arise from natural convections due to gravity and carried out hyper gravity experiments with a centrifuge [12] and has tried low gravity experiments with parabolic flights. The members of the group are Kaoru Tsujii (Hokkaido University (retired)), Masamichi Ishikawa (RIKEN), Masato Sano (Yamagata University), Masatsugu Shimomura (Tohoku University), Hiroshi Yabu (Tohoku University), Yuji Hirai (Tohoku University), Kenichi Yoshikawa (Kyoto University), Nobuyuki Magome (Kyoto University), Takashi Mashiko (Shizuoka University), and Makoto Natsuisaka (JAXA).

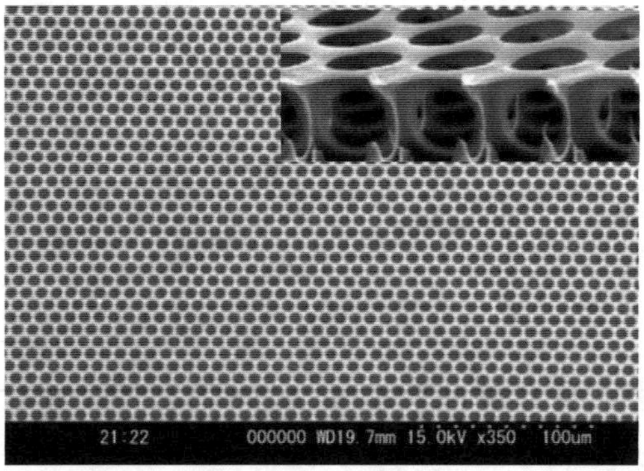

Figure 4. A fine structure of a 'honey comb film'.

2.2.2. *Chemical processes in critical density fluctuation WG.* The WG organized by Shigeru Deguchi has been discussing two major topics described below. The systems and phenomena on the both topics are fully different, but the WG is considering that density fluctuations in vicinity of critical points play an important role on the both phenomena. The members of the group are Shigeru Deguchi (JAMSTEC), Masamichi Ishikawa (RIKEN), Yoshihisa Inoue (Osaka University), Kaoru Tsujii (Hokkaido University (retired)), Makoto Natsuisaka (JAXA), Keiko Nishiyama (Chiba University), Yasuhiro Nishiyama (NAIST), Toru Maekawa (Toyo University), Takeshi Morita (Chiba University), Sadaatsu Mukai (Kyushu University), and Takehiko Wada (Tohoku University).
(1) *Colloid dynamics in the vicinity of critical points.* Shigeru Deguchi etc. found that silica particles dispersed in super critical water take a distance longer than Coulomb's interaction predicts. Since the observation was made on ground, the particles were precipitated onto the base plate and the phenomena were restricted in two dimension. The team has proposed microgravity experiments to be free from such constraint and enable more precise observations.
(2) *Chiral chemical reactions in near- and super- critical CO_2* A member of the WG, Yoshihisa Inoue, revealed that photosensitized enantiodifferentiations (see figure 5) would occur in various systems and under various conditions. In particular, he found that it would be enhanced if the reactions were given in super critical carbon dioxide (see figure 6) [13]. The research team expects that this might be more emphasized under microgravity and has proposed microgravity experiments. The team has tried parabolic flight experiments by developing a very compact and automatic super critical reaction chambers.

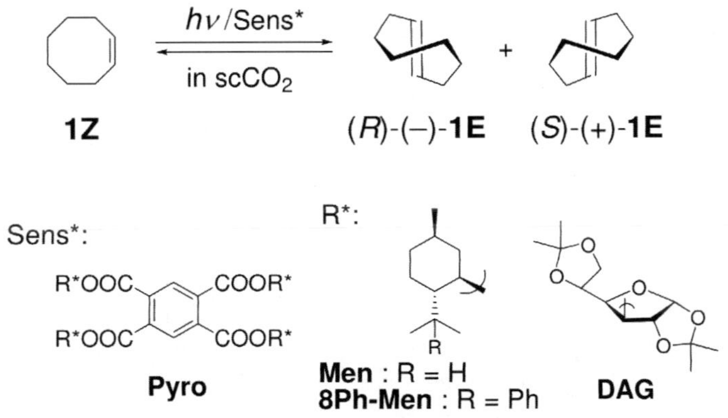

Figure 5. Target reactions - photosensitized enantiodifferentiating isomerization.

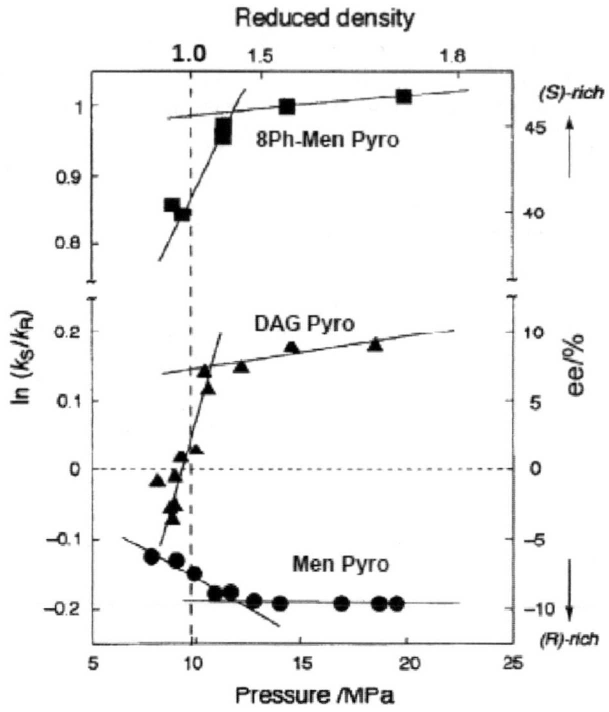

Figure 6. Pressure dependence of enantiomeric excess (%ee).

2.2.3. *Exploration of emulsion stability under microgravity RT.* The RT organized by Kazutami Sakamoto is aiming at observation of emulsion formation and its stability under microgravity. This would promisingly help understand the basic mechanisms of emulsification and stabilization. The outcome of this trial would bring better designing and quality assurance of emulsion products as dispersed colloidal system, which are indispensable for household and institutional applications. The team found a long-term stable emulsion system with new amphiphilic material through a basic research [14] and believes that this could contribute to researches in the field. The team has started to communicate with one of the European topical teams, PASTA (Particle Stabilized Emulsions and Foams). The members of the team are Kazutami Sakamoto (Chiba Institute of Science), Hideki Sakai (Tokyo University of Science), Kenichi Sakai (Tokyo University of Science), and Makoto Natsuisaka (JAXA).

2.2.4. *Impressionistic physics of bubbles, drops and foams under microgravity RT.* The RT organized by Ko Okumura studies the dynamics of bubbles, drops and foams under microgravity, via impressionistic physics which is a powerful tool to find out simple laws from complex phenomena. Ko Okumura has started to attract considerable attention in the field through experiments performed with the Hele-Shaw cell (a quasi two-dimensional cell) [15][16] so that he has been invited to join a number of topical teams of ESA; FOAM-C, PolarDrop and DOLFINE. The members of the team are Ko Okumura (Ochanomizu University), Yumiko Yoshitake (Tokyo Denki University), and Makoto Natsuisaka (JAXA).

3. Conclusions

There are 1 flight candidate and 3 resea rch WGs in th e field of phy sical chemistry in Japan. They all welcome communication with and/or participation of researchers of international communities.

References

[1] Ise N and Okubo T 1980 *Acc. Chem. Res.* **13** 303
[2] Dosho S, Ise N, Ito K, Iwai S, Kitano H, Ma tsuoka H, Nakam ura H, Okumura H, Ono T, and Sogami I S, Ueno Y, Yoshida H and Yoshiyama T 1993 *Langmuir* **9** 394
[3] Arora A K, Tata B V R, Sood A K and Kesavamoorthy R 1988 *Phys. Rev. Lett.* **60** 2438
[4] Tata B V R, Rajalakshmi M and Arora A K 1992 *Phys. Rev. Lett.* **69** 3778
[5] Kose A, Ozaki M, Takano K, Kobayashi Y and Hachisu S 1973 *J. Colloid Interface Sci.* **44** 330
[6] Ito K, Yoshida H and Ise N 1994 *Science* **269** 66
[7] Yoshida H, Ise N and Hashimoto T 1995 *J. Chem. Phys.* **103** 10146
[8] Yoshiyama T and Sogami I S 1996 *Ordering and Phase Transition in Charged Colloids* ed A K Arora and B V R Tata (VCH Publishers, Inc.) chapter 3 p 41
[9] Shinohara T, Yamada H, Sogami I S, Ise N and Yoshiyama T 2004 *Langmuir* **20** 5141
[10] Shimomura M 20 05 *Nanocrystals Forming Mesoscopic Structures* ed M P Pileni (Weinheim: WILEY-VHC Verlag GmbH & Co. KGaA) chapter 6 p 157
[11] Yabu H and Shimomura M 2005 *Chem. Mater.* **17** 5231-5234
[12] Yabu H, Hirai Y, Shim omura M, Natsuisaka M and Tsujii K 2010 *Jap. J. Appl. Phys.* **49** 110210
[13] Saito R, Kaneda M; Wada T; Katoh A and Inoue Y 2002 *Chem. Lett.* 860
[14] Sakai K, Ikeda R, Sharma S C, Shrestha R G, Ohtani N, Yoshioka M, Sakai H, Abe M and Sakamoto K 2010 *Langmuir* **26** 5349-5354
[15] Eri A and Okumura K 2007 *Phys. Rev. E Rapid Communication* **76** 060601
[16] Eri A and Okumura K 2010 *Phys. Rev. E Rapid Communication* **82** 030601

Light Induced Erosion of Dusty Planetesimals and Mars: μg Experiments

C. de Beule, T. Kelling, G. Wurm, J. Teiser and T. Jankowski
Universität Duisburg-Essen, Fakultät für Physik, Lotharstrasse 1, 47057 Duisburg
E-mail: caroline.de-beule@stud.uni-due.de

Abstract. In a number of recent experiments it was found that dust beds continuously eject small, micron sized particles under illumination at low ambient pressure. The ejection is caused by temperature gradients within the illuminated dust bed, which induce lifting forces based on thermal creep or photophoresis. In microgravity experiments we observed that the ejection rate depends inversely on the gravity level. Therefore, the effect is much more effecient in low gravity environments. This mechanism can help to understand processing of dust in protoplanetary disks, where it leads to the ejection of particles from an inward drifting body which would be lost to the star otherwise. The mechanism also supports entrainment of dust into the martian atmosphere. Windspeeds are in general too low to pick up dust from the surface by gas drag alone. In contrast to gas drag light induced drag increases in strength at low ambient pressure and has a maximum in the mbar range prevailing on Mars. The interplay of gas drag and light induced lift might explain the existence of dust devils and dust storms on Mars.

1. Introduction

Dust particles composed of μm sizes are frequently observed in the martian atmosphere and often related to dust devils and (global) dust storms [1]. To date, it is not fully understood how adhering dust particles can be lifted from the martian surface into the atmosphere at the low atmospheric pressures (mbar) on Mars. Particle pick up by gas drag would be the classical mechanism but the mean wind speeds on Mars are in general too low to lift particle by gas drag alone [2–4]. The need for additional lift finds special emphasis by the fact that even at high elevations of several km like e.g. inactive volcanoes active dust devils were found [5]. Here, the ambient pressure is only 1 to 2 mbar. Additional lifting support has to be present, one suggestion, e.g. being a pressure decrease in the center of active dust devils [6]. However, Wurm and Krauss (2006) found a lift of dust particles which are illuminated at low ambient pressure of a few mbar [7]. Applying the experimental findings it was shown that the interplay of atmospheric gas drag and photophoresis might lead to dust particle eruptions from the martian surface [8]. The physical mechanisms are photophoresis and sub-surface pressure increase by thermal creep which obviously can efficiently pull on particles from dusty surfaces [7, 9]. Photophoresis is strongest if the particle size is comparable to the mean free path of the gas molecules [10] – which are just the conditions on Mars: mbar pressure and dust aggregates composed of μm sized particles.

The threshold for dust entrainment depends on gravity which has to be overcome. This is also true for photophoretic particle ejections which was verified in earlier microgravity experiments [8]. While these experiments determined the threshold intensity of the light flux needed to eject particles at all, the reduced martian gravity should also enhance the mass rate (ejected particles per time) of the photophoretic ejections. To study this is subject of the experiments reported here.

Before doing so, it might be noted that Mars is not the only environment where the light induced ejection mechanism is of importance. Dust particles of μm size also play an important role in planet formation processes in protoplanetary disks [11]. Collisions of dust particles in the early phase of planet formation lead to particle gowth [12–14]. At a certain size (decimeter to meter regime) bodies rapidly drift inward by gas drag (1 AU in hundred years) and might be accreted by their host star [15]. However, Kelling et al. (2011) showed that dusty bodies might lose up to kg/s by photophretic and thermal creep ejections at a light flux of several 10^4 W/m^2 [9] . Also here the pressure is in the ideal range for dust ejections with mbar or less, noting that strong variations might be possible [16]. Inward drifting planetesimals are therefore not lost by accreation but are (partially) eroded by the light induced ejetion mechanisms. The ejected particles are transpoted, e.g. by photophoresis upwards or outwards in the disk and are hence reintroduced into the planet formation process [17]. In addition, Kelling and Wurm (2011) showed that small particles are produced by the photophoretic ejections which might be transported by turbulence to the surface of the protoplanetary disk [18]. Hence, the light induced ejection mechanisms might also explain why dust can be observed over the whole liftime of a protoplanetary disk.

In this article we present μg exeriments carried out on a parabolic flight addressing the effeciency of the light induced erosion mechanism to complement the earth bound experiments and to estimate the efficiency of the erosion mechanism on bodies with reduced gravity (0.38g on Mars and μg on planetesimals, g = 9.81 ms^{-2}).

2. Experiment

The setup of the experiment is shown in Fig.1. Basalt powder ($< 100\mu$m grain size) is placed within a vacuum chamber. The dust bed is illuminated by a red laser (light intensity is ~20 kW/m^2) at ambient pressure of $\sim$ 5 mbar. The number of ejected particles over time is observed – Fig.2 shows an expample of the ejected dust particles at the different g-phases and Fig.3 depicts the normlized average values of the released particles.

It is observed that the number of released particles increases with decreasing g-level.

3. Model

Similar to Kelling et al. (2011) we attribute the continuous ejections to a photophoretic force [9]. Photophoresis is a result of an interplay between the gas molecules and an inhomogeneously heated surface of a particle suspended in the gas (Fig.4). Gas molecules accomodating to the cold side of the suspended particle leave the particles surface with a smaller momentum than the gas molecules on the warm side. Hence, a force accelerates the particle in general in the direction from warm to cold. The temperature gradient over the particle's surface is often established by illumination, therefore the name photophoresis. The photophoretic force strongly depends on the pressure and has its maximum if the mean free path of the gas molecules is comparable to the size of the particle [10]. For dust particles of μm size this corresponds to mbar pressures. The photophoretic force is [10]

$$F_{ph} = \frac{2F_{max}}{\frac{p_{max}}{p} + \frac{p}{p_{max}}}, \tag{1}$$

where p is the pressure and F_{max} is the maximal photophoretic force, which appears at the pressure p_{max}. The coefficients F_{max} and p_{max} are discussed in [10] and depend on gas and

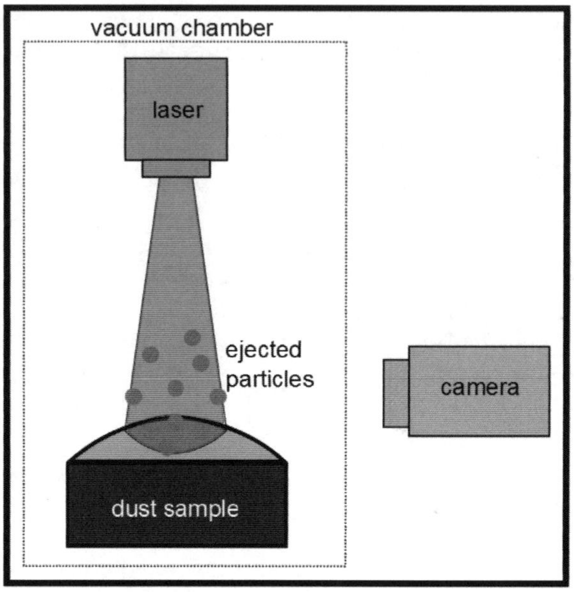

Figure 1. Experimental setup: Basalt powder (< 100 μm) illuminated by a laser (~20 kW/m^2) from the top at 5 mbar pressure. The number of ejected particles over time were observed.

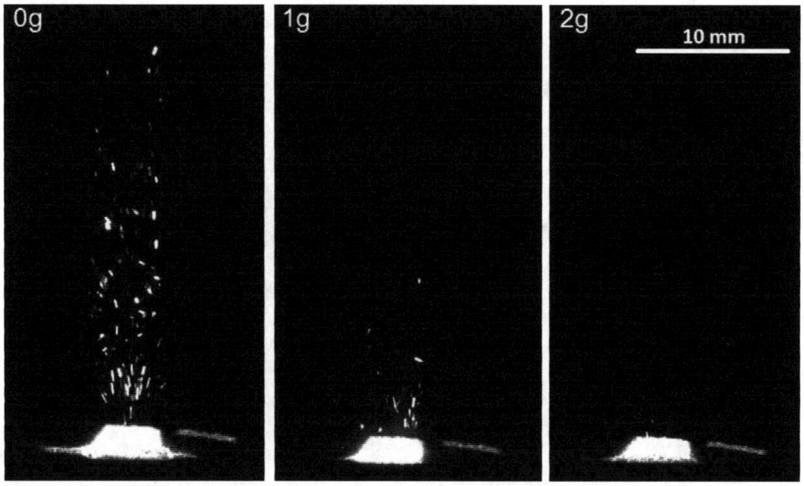

Figure 2. Dust eruptions at different g-phases; 5 mbar ambient pressure; ~20 kW/m^2 light intensity on the dust bed; Basalt powder of < 100 μm size.

particle properties.
Kocifaj et al. (2000) showed that illuminated dust beds develop an inverse temperature gradient pointing from a sub-surface temperature maximum towards the cooler surface [19]. The light heats the upper dust layers due to absorption. Only the upper most particle layers (the surface) of the dust bed cools by thermal radiation while the deeper layers transport the heat mainly through conduction. Consequently a sub-surface temperature maximum of several Kelvin at typical depth's of some 100μm below the surface is established. Fig.5 shows an example of the temperature distribution within an illuminated dust bed [19]. Due to the reversed temperature gradient the surface aggregates in our experiments are effected by a photophoretic force pointing

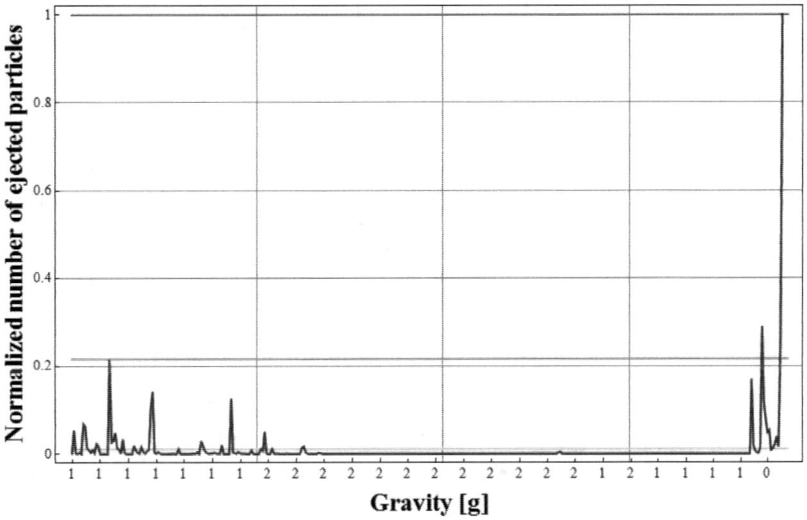

Figure 3. Normalized number of ejected particles over time. Blue line (trend for normalized number of ejected particles), red line (value of the maximum in the 0g phase), orange line (average of the 1g phase), green line (average of the 2g phase). 0g, 1g and 2g mark the different g-phases.

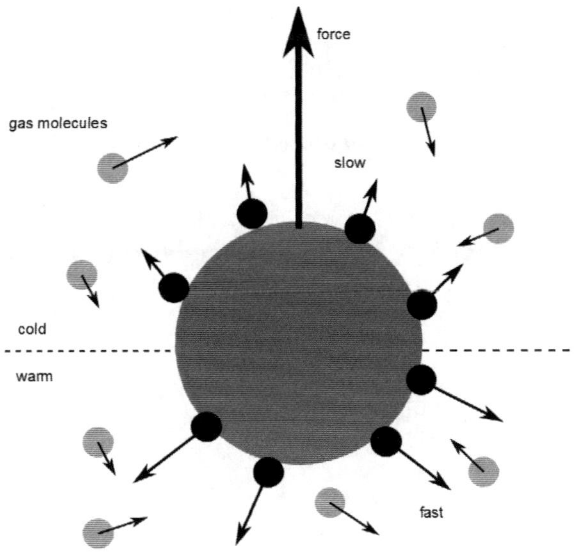

Figure 4. Photophoresis: Gas molecules accomodating to the cold side of the suspended particle with a temperature gradient over its surface leave the surface with a smaller momentum than the gas molecules on the warm side. As a result, a force accelerates the particle in the direction from warm to cold.

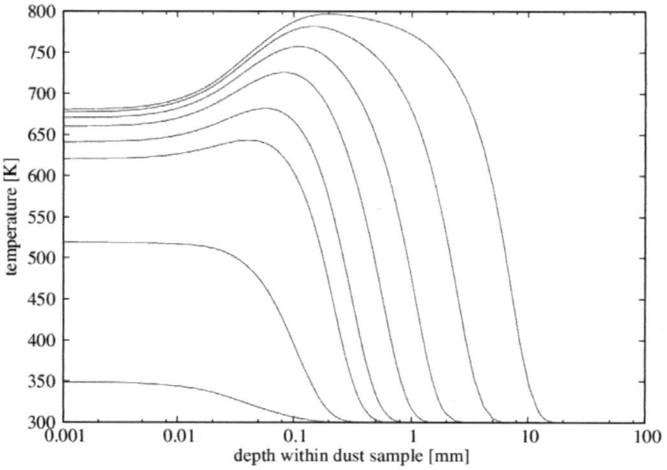

Figure 5. Evolution of the temperature within an illuminated dust bed after a light source of $I = 10$ kW m^{-2} is switched on. Times in seconds are (lines from bottom to top): 0.1, 1.0, 5.0, 10.0, 30.0, 100.0, 500.0 and 3600.0 [19].

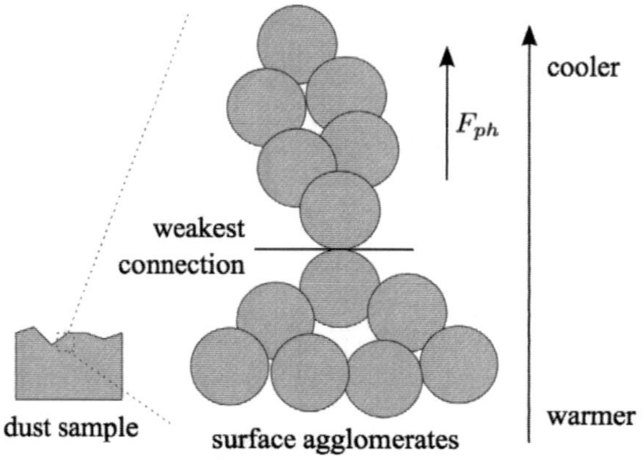

Figure 6. Principle of the photophoretic ejections: The continuous ejections of single particle agglomerates are caused by photophoresis. If the induced photophoretic force overcomes gravity and the cohesion forces at the weakest connection, a surface agglomerate is released into the surroundings [9].

in the direction from warm to cold and hence away from the surface (Fig.6). If the photophoretic force F_{ph} overcomes the cohesion between the aggregates F_c and gravity F_G

$$F_{ph} > F_c + F_G, \tag{2}$$

dust particles are conntinuously ejected from the surface into the surroundings.
In contrast to the linear gravity dependency of the threshold of the particle ejections found by Wurm et al. [8] the analysis of the experimental data of the effeciency (particles per time at the different g-levels, see also Fig.3) of the particle ejections revealed an $1/g$ dependence on the gravity level (Fig.7).

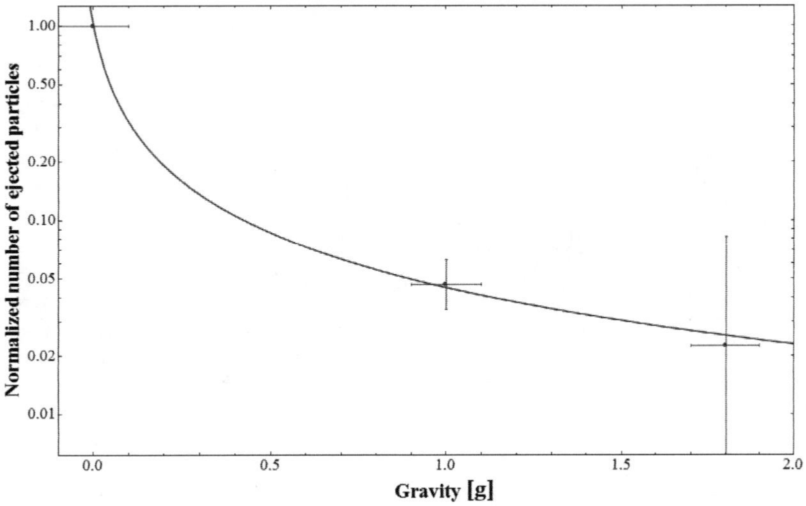

Figure 7. Averaged and normalized number of ejected particles over gravity with standard deviation. The fitted function is of the form $N = \frac{1}{1+\xi \cdot g}$, where g is the gravity and $\xi = 21.2507$ s/m^2 the fitted parameter.

For a given dust sample the dependency of the number of ejected particles (mass loss rate) N can be written as

$$N \propto \frac{1}{1+\xi \cdot g}, \tag{3}$$

where g is the gravitational acceleration and ξ is a factor containing all other parameters like e.g. gas parameters. Due to the complexity of the process which depends on particle size, porosity, optical constants, adhesion, thermal conductivity etc. we do not argue in more detail on the fitting parameter ξ here and leave this to future studies.
The size dependence of the effect is still unclear. While the adhesion of a small particles might dominate over gravity large aggregates only bound by few contacts might be lifted more easily.

4. Applications and Conclusion

In common planet formation scenarios inward drifting bodies are rapidly lost by accretion (meter-size barrier, [15]). As Kelling et al. (2011) and Kelling and Wurm (2011) showed, inward drifting bodies are (partially) eroded by the dicussed erosion mechanism [9, 18]. The produced small particles are reintroduced into the planet formation process as they are transported radially outwards (by e.g. photophoresis) in the disk where they can take part in a re-aggregation process (Fig.8). With the help of the light induced particle ejections the problem of the meter-size barrier is somewhat eased. Small ejectd particles might also be transpoted upwards (by e.g. turbulence) to the disks surface which gives an explanation why dust is observed over the whole lifetime fo a protoplanetary disk. The results of the μg experiments show that the erosion mechanism is stronlgy enhanced towards lower g-levels. Hence, the efficiency of the erosion of inward drifting bodies must be corrected towards a faster, more effective erosion. Wurm et al. (2008) showed that dust particles on Mars can be ejected from the surface by photophoresis – if the cohesion is reduced [8]. With the results of our μg experiments one can draw the conclusion that also here the mass rate of ejection benefits from the reduced gravity and might explain how dust gets entrained into the Martian atmosphere with sufficient rate though details are subject to future analysis and work.

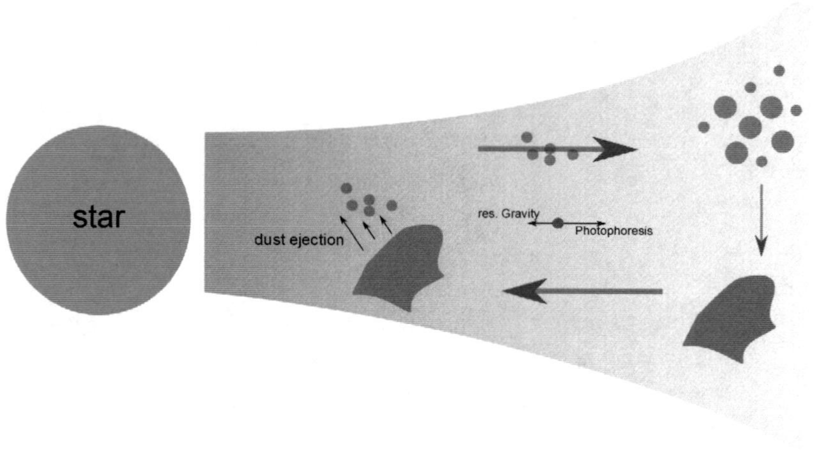

Figure 8. Principle of re-aggregation process in protoplanetary disks. An inward drifting body ejects small particles, which can be transported outwards again by e.g. photophoresis.

5. Acknowledgments

This work was supported by the Deutsches Zentrum für Luft- und Raumfahrt (DLR), the European Space Agency (ESA) and the Deutsche Forschungsgemeinschaft (DFG).

References

[1] Fisher J A, Richardson M I, Newman C E, Szwast M A, Graf C, Basu S, Ewald S P, Toigo A D and Wilson R J 2005 *Journal of Geophysical Research (Planets)* **110** E03004

[2] Greeley R, Leach R, White B, Iversen J and Pollack J B 1980 *Geophysical Research Letters* **7** 121

[3] Stanzel C, Pätzold M, Williams D A, Whelley P L, Greeley R, Neukum G and The HRSC Co-Investigator Team 2008 *Icarus* **197** 39

[4] Hess S L, Henry R M, Leovy C B, Tillman J E and Ryan J A 1977 *JGR* **82** 4559

[5] Reiss D, Lüsebrink D, Hiesinger H, Kelling T, Wurm G and Teiser J 2009 High Altitude Dust Devils on Arsia Mons, Mars: Testing the Greenhouse and Thermophoresis Hypothesis of Dust Lifting *Lunar and Planetary Institute Science Conference Abstracts* (*Lunar and Planetary Institute Science Conference Abstracts* vol 40) p 1961

[6] Greeley R, Balme M R, Iversen J D, Metzger S, Mickelson R, Phoreman J and White B 2003 *Journal of Geophysical Research (Planets)* **108** 5041

[7] Wurm G and Krauss O 2006 *Physical Review Letters* **96** 134301

[8] Wurm G, Teiser J and Reiss D 2008 *Geophysical Research Letters* **1**

[9] Kelling T, Wurm G, Kocifaj M, Klačka J and Reiss D 2011 *Icarus* **212** 935

[10] Rohatschek H 1995 *Journal of Aerosol Science* **26** 717

[11] Blum J and Wurm G 2008 *Annu. Rev. Astron. Astrophys.* **46** 21

[12] Wurm G and Blum J 1998 *Icarus* **132** 125

[13] Dullemond C P and Dominik C 2005 *Astronomy and Astrophysics* **434** 971

[14] Teiser J and Wurm G 2009 *Astronomy and Astrophysics* **505** 351

[15] Weidenschilling S J 1977 *Monthly Notices of the Royal Astronomical Society* **180** 57

[16] Wood J A 2000 *Space Science Reviews* **92** 87

[17] Wurm G 2007 *Monthly Notices of the Royal Astronomical Society* **380** 683

[18] Kelling T and Wurm G 2011 *Astrophysical Journal* **733** 122

[19] Kocifaj M, Klačka J, Wurm G, Kelling T and Kohút I 2010 *Monthly Notices of the Royal Astronomical Society* **404** 1512

Journal of Physics: Conference Series **327** (2011) 012048 doi:10.1088/1742-6596/327/1/012048

Electrostatic precipitation of dust in the Martian atmosphere: Implications for the utilization of resources during future manned exploration missions

C I Calle[1], S M Thompson[2], N D Cox[2], M R Johansen[1], B S Williams[3], M D Hogue[1], and J S Clements[2]

[1] Electrostatics and Surface Physics Laboratory, NASA, Kennedy Space Center, FL 32899, USA

[2] Department of Physics and Astronomy, Appalachian State University, Boone, NC 28608, USA

[3] University of South Alabama, Mobile, AL 36688, USA

E-mail: carlos.i.calle@nasa.gov

Abstract. Future human missions to Mars will require the utilization of local resources for oxygen, fuel, and water. The In Situ Resource Utilization (ISRU) project is an active research endeavor at NASA to develop technologies that can enable cost effective ways to live off the land. The extraction of oxygen from the Martian atmosphere, composed primarily of carbon dioxide, is one of the most important goals of the Mars ISRU project. The main obstacle is the relatively large amount of dust present in the Martian atmosphere. This dust must be efficiently removed from atmospheric gas intakes for ISRU processing chambers. A common technique to achieve this removal on earth is by electrostatic precipitation, where large electrostatic fields are established in a localized region to charge, precipitate and collect dust particles. This technique is difficult to adapt to the Martian environment, with an atmospheric pressure of about one-hundredth of the terrestrial atmosphere. At these low pressures, the corona discharges required to implant an electrostatic charge to the particles to be collected is extremely difficult to sustain and the corona easily transitions to a glow/streamer discharge, which is unsuitable for particle charging. In this paper, we report on our successful efforts to establish a stable corona under Martian simulated conditions. We also present results on dust collecting efficiencies with an electrostatic precipitator prototype that could be effectively used on a future mission to the red planet.

1. Introduction

Plans for the human exploration of Mars include the concept of living off the land, exploiting the planet's natural resources for fuel, water, and oxygen. One of those resources is the Martian atmosphere itself, which is composed mainly of carbon dioxide. Using simple chemical processes, these commodities can be extracted in chemical reactors. But before that can be accomplished, dust, which is always present in the atmosphere, must be removed from the atmospheric intakes.

An electrostatic precipitator (ESP) is the obvious choice for this task [1,2]. However, the low pressure of the Martian atmosphere, at 7 to 10 mbar, prevents the large electrostatic potentials

[1] To whom any correspondence should be addressed.

normally used in terrestrial electrostatic precipitators. Although the Martian atmospheric dust is expected to be electrostatically charged, effective dust collecting at relatively low potentials would require that the dust particles be charged to saturation and with the same polarity. A corona discharge that could implant this charge could do the trick, but at the low pressures of Mars, the corona easily transitions to a glow/streamer discharge as the applied voltage is increased, and which is unsuitable for this purpose. Therefore almost no studies have been performed on low pressure CO_2 ESPs with the exception of Pang et al [3] who reported results for a non-customary ESP-type system using a discharge wire *external* to a tube or torus to collect dust settling out of the Martian atmosphere onto a solar panel.

In this paper, we report on our efforts to design a traditional wire-cylinder electrostatic precipitator system that could sustain a charging corona for the challenging Martian atmospheric environment. We also present data on particle charging and dust collection with this precipitator design.

2. Living Off the Land: In-Situ Resource Utilization

NASA's In Situ Resource Utilization (ISRU) project seeks to use space resources to reduce the mass, cost, and risk of space exploration. The specific goals of the ISRU project for Mars are to greatly reduce the expense of a human mission to the planet and to help establish the self-sufficiency of long-duration manned bases. The major areas of the Mars ISRU project are: Mission consumable production (propellants, fuel cell reagents, life support consumables, and feedstock for manufacturing and construction), surface construction (radiation shields, landing pads, and habitats), manufacturing and repair, and space utilities and power [4].

A recent evaluation of ISRU technologies recommends the continuation of the Mars atmospheric gas processing methods that are currently at a relatively high level of development [5]. These methods for mission consumable production on Mars will require the collection of dust-free Martian atmospheric gas for processing. The electrostatic precipitator reported here is being designed to efficiently remove dust from the intakes feeding the atmospheric gas processing chambers.

3. The Martian Atmospheric Environment

3.1 The dusty atmospheric environment

The surface of Mars is entirely covered with dust with a similar composition and size throughout the planet due to its constant redistribution during global dust storms and localized dust devils. These dust storms and dust devils uplift dust into the rarefied atmosphere (Fig 1). Calculations based on optical instrumentation on spacecraft in orbit around Mars and on rovers on the surface have yielded estimates for the average diameter of dust particles in the atmosphere. Using this data, Landis *et al* believe that Martian atmospheric dust has a three-component particle size distribution: *Atmospheric dust* suspended for long periods of time, with radii ranging from 1 to 2 μm, with an average value of 1.5 ± 0.2 μm; *settled dust*, raised into the atmosphere by wind and dust devils, with a radii under 5 μm; and *saltating particles*, with radii greater than 40 μm [6].

3.2 Electrical properties of the Martian atmosphere

As will be shown in detail in Section 4, atmospheric dust removal with an electrostatic precipitator requires the establishment of an electric potential difference between two metallic conductors. However, the composition and low pressure of the Martian atmosphere greatly limits the potentials that can be sustained for any length of time. At the 7 to 10 mbar atmospheric pressure, Townsend avalanche breakdown occurs at relatively low potentials [7].

In Townsend breakdown, an electron can appear in the electric field region between two electrodes as a result of a cosmic ray particle knocking it from the cathode or ionizing a gas molecule. This electron is accelerated in the electric field and is able to ionize another molecule in the gas, releasing another electron. These two electrons, in turn, are accelerated by the field, further ionizing additional molecules, and creating an avalanche of electrons accelerating toward the anode while the heavier ions just generated accelerate toward the cathode. When the ions reach the cathode, additional *secondary electrons* are ejected in the collision. These new electrons create their own avalanches, creating a new cycle.

Figure 1. A dust devil photographed by the Spirit rover [Courtesy NASA]

If the electric field is weak, the number of electrons in each cycle decreases, and the charged particles that periodically appear recombine in the electrodes. In a strong field, however, the number of electrons in each cycle increases and, as a result, the avalanches increase, filling the space between electrodes with electric charges, and expanding the avalanche region. When the entire space between the electrodes becomes ionized, gap breakdown takes place.

The breakdown voltage in Townsend breakdown for a uniform electric field depends on the product of gap length d and gas density. This relationship is known as a Paschen's law [8]. Historically, the gas pressure p at standard temperature is plotted instead of gas density, but these curves must be adjusted for the low temperature regions on Mars. Breakdown experiments with uniform fields have been performed for most elemental and molecular gases, including CO_2. Note that voltage polarity is usually not reported because it does not matter when using identical (symmetric) electrodes that are far from a ground plane. In 2004, we performed experiments with a premixed gas to emulate the Martian atmosphere, composed of a mixture of 95% carbon dioxide, 2.7% nitrogen, 1.6% argon, 0.13% oxygen, and 0.07% carbon monoxide [9]. Data was taken at pressures ranging from 270 mbar to 466 mbar (200 mtorr to 350 torr) with gaps of 5 mm, 7.5 mm, and 10 mm. Fig. 2 shows that breakdown potentials in CO_2 and in the Martian gas mixture are extremely similar. It appears as if CO_2 dictates the Paschen breakdown and that the other gases, present in relatively low concentrations, do not affect it substantially. The Martian gas breakdown voltages are equal to or slightly smaller than the CO_2 voltages by an average of 15 volts, perhaps due to the known lower breakdown potentials of nitrogen and argon, present in smaller concentrations. Based on these results we used either a 95% CO_2/5% air mixture or pure CO_2 to approximate the Martian atmosphere in the present study.

The Mars atmospheric gas has a minimum breakdown potential of 500 V at 5 torr·mm [9]. We also obtained the same value for CO_2. Earlier experimental values for CO_2 give a potential of 420 V at 5.1 torr mm [10]. For the typical Mars atmospheric pressure of 7 mbar (5.2 torr), the breakdown potentials that we obtained in a uniform field are 725 V for a 5 mm gap, 800 V for a 7.5 mm gap, and 895 for a 10 mm gap. From our Paschen plot, we expect potentials of the order of 2.8 kV for a 10 cm gap. At the 10 mbar upper end of the Martian atmospheric pressure, these voltages range from 800 V for a 5 mm gap to 3.2 kV for a 10 cm gap in a uniform electric field.

In addition to gas breakdown, dust particle charging and migration is also affected by the low gas pressures on Mars. The roughly 100× lower gas pressure on Mars results in molecular mean free paths λ that are ~ 100× longer than those at atmospheric pressure, e.g., for CO_2, λ = 44 nm at one atmosphere [11] and ~4 μm at 5 mbar. The mean free path can be adjusted for temperature and pressure using the equations in Schmid [11] with the Sutherland constant for CO_2 = 220. There are three particle charging regimes that are defined based on the Knudsen number Kn, which is simply the ratio of λ to particle radius. The regimes are continuum (Kn << 1), free molecular (Kn >> 1), and transitional (Kn ~ 1). The low pressure on Mars shifts dust particle charging from the continuum regime to the transitional. Drag on the particle as it moves through the gas is also affected. Stoke's drag is proportional to the particle radius and the gas viscosity. Although viscosity only depends on temperature and not pressure, the Stoke's formula still has to be adjusted using the Cunningham slip correction factor to account for the particle "slipping" between the gas molecules due to their large mean free paths.

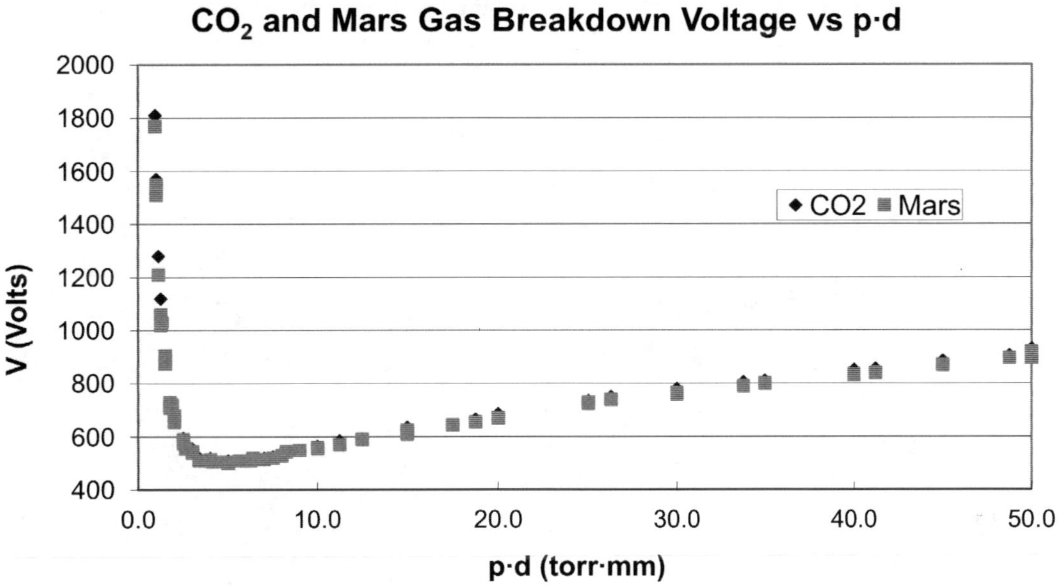

Figure 2. Paschen breakdown potentials versus pressure-distance for a Martian gas mixture (red squares) and for CO_2 (blue triangles) [9].

4. Precipitator Theory

An electrostatic precipitator (ESP) consists of two electrodes set at an electrostatic potential difference that can drive charged particles to one of the electrodes for collection. There are two general types of electrostatic precipitators: Single-stage or Cottrell precipitators, where particle charging and particle collection take place in a single state; and two-stage or Penney precipitators, with a pre-charging stage and a collecting stage. In both precipitator types, dust particles are charged using corona generation around the high voltage discharge electrode, which ionizes gas molecules. These ions are accelerated by the electric field E in the region between the electrodes, but numerous collisions with gas molecules results in a constant drift velocity characterized by an ion mobility $b = E/\text{velocity}$. The ions transfer charge to dust particles encountered in their path as they drift to the grounded collecting electrode. The ions form space charge which modifies the applied E between the electrodes. The electric field for a wire-cylinder geometry without space charge is given by:

$$E(r) = \frac{V}{r ln(\frac{R}{a})} \tag{1}$$

where V is the applied voltage, r is the distance from the wire, a is the wire radius, and R is the cylinder radius. [12] contains an equation for the case with space charge:

$$E(\mathrm{r}) = \sqrt{\frac{I}{2\pi\varepsilon_0 Lb} + \left(\frac{a}{r}\right)^2 \left\{\left(\frac{V}{a \ln\frac{R}{a}}\right)^2 - \frac{I}{2\pi\varepsilon_0 Lb}\right\}} \tag{2}$$

where V is the applied voltage, I is the ion current, L is the length of the wire-cylinder, R is the cylinder radius, a is the wire radius and b is the ion mobility. Larger ion currents yield more space charge with increases the applied electric field substantially.

Since the atmospheric gas intakes for the ISRU processing chambers will likely be cylindrical, we chose a cylindrical precipitator geometry (Fig. 3). When the voltage between the electrodes is raised gradually, the glow of the corona appears in the inner electrode, where the electric field is stronger and is able to ionize the gas. The ionic current is weak and thus not able to reduce the voltage between the electrodes. When the voltage is further increased, the corona current increases with the square of the applied voltage. At higher potentials, the glowing corona envelope begins to break up into filaments that appear and disappear within a few microseconds forming a streamer corona. For our single-stage design, we are interested in obtaining the highest possible charge on the dust particles without reaching the unstable streamer corona region.

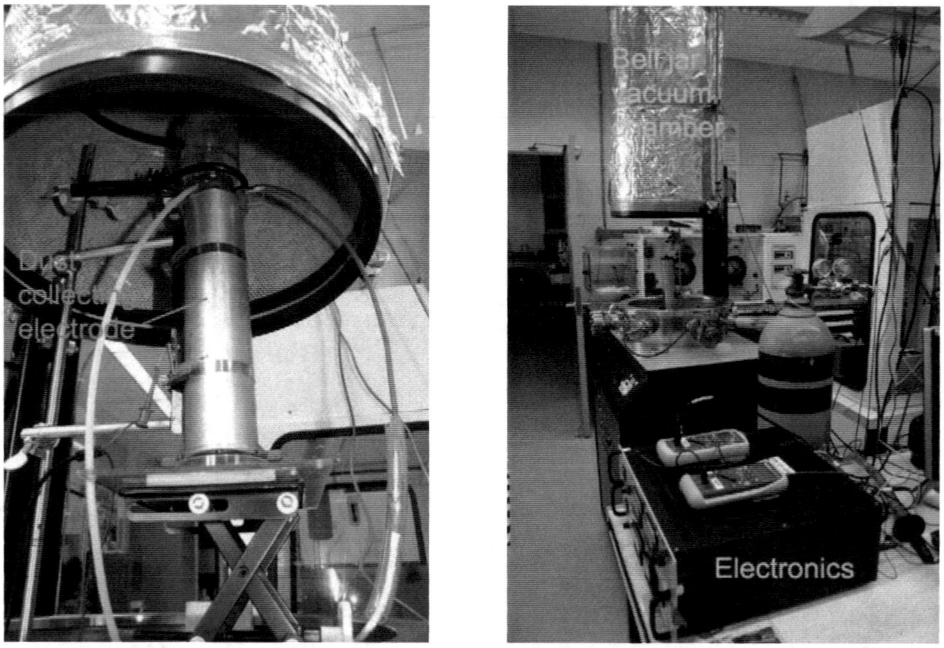

Figure 3. Electrostatic Precipitator assembly (left) and supporting equipment (right).

Gas ions formed by the strong electric field in a precipitator charge dust particles in two ways: *field charging* and *diffusion charging*. In field charging, the drifting ions follow the electric field lines to the dust particles. The number of possible ions attaching to the surface of a particle is clearly proportional to the square of the particle radius. The charging process continues until charge saturation q_s is achieved, which takes place with a characteristic time τ given by

$$\tau = \frac{4\varepsilon_0}{N_0 e b} \tag{3}$$

where ε_0 is the permittivity of free space, N_0 is the ion number density in m^{-3}, e is the fundamental charge, and b is the ion mobility. The particle charge as a function of time $q(t)$ is given by:

$$q(t) = q_s \frac{t}{t+\tau} \tag{4}$$

The saturation charge can be calculated from the external field and the distortion to the field by the charged particle. This calculation was first performed by Pauthenier and Moreau-Hanot [13]. Field charging is known as Pauthenier charging and is valid in the continuum regime.

The Pauthenier saturation charge for a spherical particle of radius a is reached when the force due to the field E and the force or repulsion from the charged particle are balanced [14]. This saturation charge is

$$Q_s = 12(k/k+2)\pi\varepsilon_0 E a^2 \tag{5}$$

where k is the relative permittivity of the particle. The term $(k/k+2) = 1$ for conductive particles.

In diffusion charging, ions diffuse through the gas due to thermal motion, colliding with the dust particles at random. Technically, diffusion charging does not have a saturation limit because the charging does not stop but instead the rate slows as the particle acquires charge which repels the incoming ions. The ions have a distribution of velocities and energies, and ions with too little energy are repelled. Eventually only a very few ions in the high energy tail of the distribution can reach the particle, and the amount of additional charging becomes insignificant. The diffusion charge acquired by dust particles can be calculated using standard kinetic theory of gases. This charge depends on the ion number density, N_0, their mean velocity, c_i, and the cross sectional area of the particle, πa^2. Cross [9] reports the expression developed by Arent [15] in the continuum regime for the diffusion charge q as a function of time t

$$q(t) = \frac{4\pi\epsilon_0 kT}{\mathrm{e}} \ln\left(\frac{aN_0\,\mathrm{e}^2 c_i t}{4\epsilon_0 kT} + 1\right) \tag{6}$$

where k is Boltzmann's constant, and T is the temperature of the gas.

The mean ion velocity c_i for CO_2 at 9 mbar (7 torr) and 273 Kcan be calculated from kinetic theory using the following expression [16]

$$c_i = \left(\frac{8RT}{\pi M}\right)^{\frac{1}{2}} = 362\ \mathrm{m/s} \tag{7}$$

where R is the universal gas constant and M the molecular mass of CO_2. The ion number density N_0 can be calculated using the relationships

$$N_0 = \frac{J}{ebE} \tag{8}$$

$$J = I/A \tag{9}$$

where J is the current density, I is the measured current flowing from the corona wire to the ground electrode, and A is the area the current flows through. The ion number density N_0, half way between the electrodes for the cylindrical precipitator geometry with a typical potential difference of 1.7 kV, is 4.4×10^{13} ions/m^3.

At a pressure of 1 atm, field charging dominates for particle sizes above 10 μm and diffusion charging dominates below 1 μm (Fig. 4). The region between requires a theory which incorporates both mechanisms at the same time [17,18], but simply adding the two is a fair approximation. As stated in Section 2.1, the diameters of particles in the Martian atmosphere range from 2 to 10 μm. In this case, both field and diffusion charging contribute to the charging mechanism, with field charging contributing more to the charge of the larger particles uplifted into the atmosphere during dust storm and dust devil activity, and both mechanisms contributing to the 2 to 4 μm normally found in the Martian atmosphere. Note that continuum regime theory only provides an order of magnitude estimate because transitional regime theory is more appropriate for this particle size range at Mars gas pressures. In fact, the longer mean free paths at the low Mars pressures may shift the dominant diffusion charging size range to larger dust particles. Unfortunately, transitional regime theory is quite complicated and several different ones, all requiring numerical solution, have been proposed in the literature [19-21].

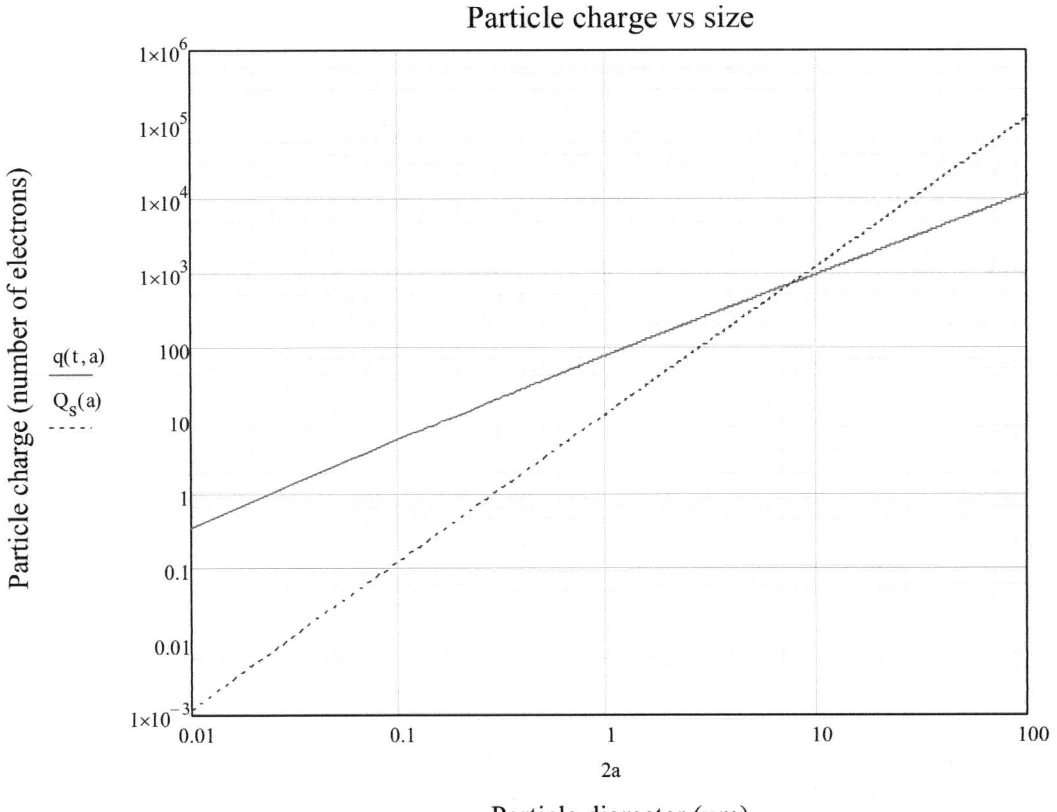

Figure 4. Continuum regime field (Pauthenier) saturation charge (dotted line) and diffusion charge (red line) for particles in CO_2 at 9 mbars with $E = 0.23$ kV/cm and an exposure time of 10s, as given by equations (3) and (4).

Once the dust particles are charged to a level q, they experience an $F = qE$ force which drives them to the collecting electrode. They drift towards the electrode with a terminal velocity (migration velocity) determined by the gas drag. In the Stoke's regime (Reynold's number less than one), the drag force is given by Stoke's law with the Cunningham slip correction factor C_c required for small Knudsen numbers, ie., when the particle size is not much greater than the gas molecule mean free path [16]. The migration velocity w is given by [13,16]

$$w = C_c \frac{qE}{6\pi\eta a} \tag{10}$$

with

$$C_c = 1 + K_n[\alpha + \beta e^{\frac{-\gamma}{K_n}}] \tag{11}$$

Journal of Physics: Conference Series **327** (2011) 012048 doi:10.1088/1742-6596/327/1/012048

where η is the gas viscosity at that gas temperature, a is the particle radius, and α,β,γ are parameters that depend on the gas composition and particle shape, etc. For CO_2 at STP η=1.47*10^{-5}, and α=1.15, β=0.479 and γ=0.92 for solid spherical particles in CO_2 [11]. The Knudsen number $K_n = \lambda/a = 8.5$ for 1 µm diameter particles in the Mars atmosphere at 273 K. The Cunningham correction factor for K_n = 8.5 in CO_2 is approximately 15. The low gas pressure on Mars increases w because of the Cunningham slip factor, but also decreases w because E and q are limited to low values due to gas breakdown. For example, a 1 µm diameter particle in an ESP on Mars with E = 0.23 kV/cm with a roughly estimated charge of about 100 electrons (see Fig. 4) would have a migration velocity w = 0.4 m/s, which is higher than the typical value on Earth of 0.1 m/s for 1 µm particles [1]. For 10 um particles with 2000 electrons, K_n = 0.85, and C_c = 2, and w = 0.01 m/s. which is much lower than on Earth where w = 1 m/s for 10 µm particles typically. Therefore an ESP should work better than on Earth for 1 µm particles due to the very low drag for small particles only, but much worse for 10 µm particles because of the low E.

The precipitator collection efficiency can be estimated using the Deutsch-Anderson equation which applies to turbulent (high Reynold's number) flow in the ESP. The collection efficiency, CE, is given by White [1], Oglesby et al [2], and Cross [9], as

$$CE = 1 - \mathrm{e}^{-\left(\frac{A}{V}\right)w} \tag{12}$$

where A is the total collection electrode area, V is the gas flow rate, and w is the particle migration velocity.

The cylindrical ESP employed in our tests was oriented vertically so that in addition to the gas flow through the ESP, the dust particles settled through it due to the force of gravity. The terminal velocity of the particles occurs when the force of gravity is balanced by the Stoke' drag corrected for Cunningham slip. The equation for terminal velocity, V_t, is given in Hinds [16]:

$$V_t = C_c \frac{\rho d g}{18\eta} \tag{13}$$

where ρ is the particle density and g is the gravitational strength. Mars dust has a density of approximately 2300 kG/m^3 and the strength of gravity is 0.38 of Earth's g. Using the previous values for a dust particle with 1 µm diameter, C_c = 15 and V_t = 0.0005 m/s. In the tests conducted in our lab, V_t would be approximately 2.5× faster due to the stronger gravity on Earth. But these velocities are much slower than the migration velocity w calculated above. For a 10 um particle on Mars, V_t = 0.007 m/s, which is almost as large as w. Therefore the smaller dust particles should be easier to collect in our vertical ESP with downward gas flow, whereas the opposite is usually the case in atmospheric pressure ESPs on Earth [1,2].

5. Current-Voltage Characteristics

Three 30 cm-long aluminium cylinders with inner diameters of 5.2 cm, 7.1 cm, and 9.5 cm, a 0.3-cm rod, and two steel wires with 70 µm and 100 µm in diameter were used for the outer and inner electrodes of the precipitator. All nine combinations were tested in a vacuum chamber evacuated to 5 mbars of carbon dioxide pressure.

Current-voltage (*I-V*) curves were obtained for seven geometries at 5 mbars in CO_2 and in 95% CO_2-5% air mixture, to approximate more closely the Martian atmospheric composition (Figs. 5 and 6). Negative corona, which is normally more stable than positive in atmospheric pressure air, yielded very steep *I-V* curves (not shown) with a narrow region of stability. More importantly, the negative corona only occurred on one location on the discharge wire, and turned into a streamer with a slight increase in voltage. In contrast, positive polarity resulted in a very uniform glow on the entire wire and had a much wider stable voltage range. Because a stable, uniform corona discharge is required for effective particle charging, only positive polarity was used. In addition, a 100 MΩ high voltage 10 W resistor was added in series with the high voltage power supply to help stabilize t he discharge. With the *I-V* curves, voltages for corona onset and streamer corona were determined. *I-V* curves for the seven configurations are shown in Figs. 5 and 6. Fig. 5 shows that *I-V* curves for plain CO_2 and for the carbon dioxide-air mixture are very similar. Thus, we decided to perform our experiments in CO_2 rather than in the mixture. The data for the 0.3 cm rod emphasizes how different the corona discharge is at low pressures compared to one atmosphere. In contrast to the atmospheric pressure case, corona current was easily obtained at low voltages using this large diameter rod and also for a very large 1.26 cm rod (not shown).

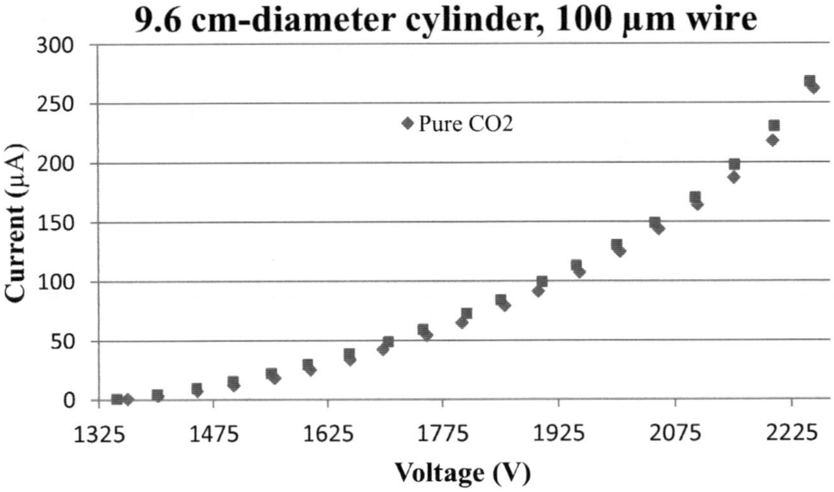

Figure 5. *I-V* curves for one configuration of the precipitator. Data taken at 5 mbar in pure CO_2 and in a 95% CO_2-5% air mixture, show that there is little difference in the *I-V* characteristics between the two environments at this pressure.

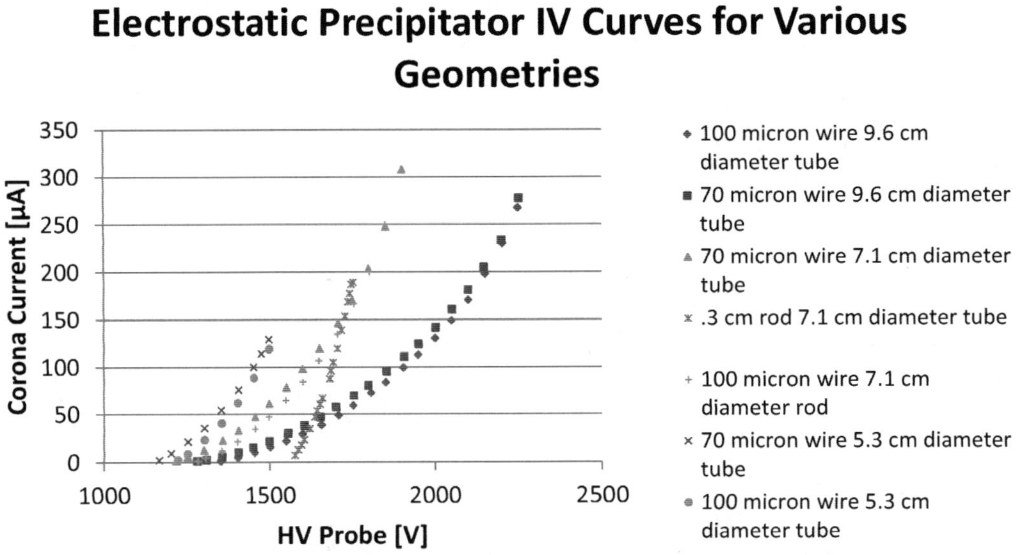

Figure 6. *I-V* curves for seven configurations of the precipitator. Data taken with clean electrodes and positive polarity of the wire at 5 mbar in CO_2.

6. Experiments

6.1 Corona Charging Experiments

Experiments with three different brass spheres were performed to test the charging performance of the precipitator geometry in the continuum regime where Pauthenier field charging theory clearly applies. Spheres with diameters of 0.95 cm, 1.27 cm, and 2.42 cm were lowered half way into the precipitator—away from distorted edge fields—with a thin insulating Teflon thread. The thread was kept equidistant from the two electrodes and care was taken to prevent the oscillations of the thread that would alter the position of the ball. The field was turned on for 10 seconds using a DC high voltage supply. The supply was then turned off and the charged sphere was lowered into a Faraday cup connected through a coaxial vacuum feed-through to a Keithley 6514 Electrometer to obtain the value of the charge acquired by the sphere. The zero check on the electrometer was not deactivated until several seconds after the high voltage was turned off. This prevented stray ions from the corona from influencing the reading. In addition, the electrometer reading was not recorded until after the sphere was raised out of the electrometer so that the charge on the thread was not recorded. The conductive sphere was discharged when it touched the Faraday cup, which enabled the test to be repeated by simply raising the sphere back to the charging position inside the precipitator.

As indicated in Fig. 4, diffusion charging should have little effect on the charging of the brass spheres. Thus, theoretical values for field charging of the spheres using equation (3) with the electric field calculated halfway between the wire and cylinder to compare to the experimental values obtained. Table 1 shows the results for spheres of three different sizes in the 7 cm diameter cylinder with a 100 μm wire. The average of ten tests (charge measurements) and the resulting standard deviation is shown for each condition. Most of the measured values were larger than the theoretical ones, most likely due to upward drift of the electrometer and also the presence of the metal ball shortening the inter-electrode gap and thereby increasing the electric field which increases the Pauthenier saturation charge value. The ball charge data for the 7 cm cylinder with 100 um wire and 0.95 cm ball is plotted vs E field in Figure 7 left side. The right side of Figure 7 has a plot of charge vs the square of the ball radius at E = 0.28 kV/cm. A best fit line is shown for each plot. The charge varies linearly with E and approximately with the square of the radius, as predicted by Pauthenier field charging theory.

Table 1. Brass Ball Charging Experiments in the 7 cm ESP with 100 μm wire at 5 mBar CO_2.

Ball dia [cm]	Average Actual Voltage [V]	Current [uA]	R/2 Ion density N [m^{-3}]	Space Q R/2 E-field [kV/cm]	Field Saturation Charge [pC]	Average Measured Charge [pC]	Standard Deviation Charge [pC]	Percent Difference [%]
0.47	1696	126	6.3E+13	0.28	51	66	6	29
0.95	1318	5	5.2E+12	0.12	93	110	7	19
0.95	1509	51	3.5E+13	0.20	150	166	5	11
0.95	1690	127	6.3E+13	0.28	210	225	7	7
1.27	1433	32	2.6E+13	0.17	231	261	7	13
1.27	1577	79	4.7E+13	0.23	311	323	9	4
1.27	1674	124	6.3E+13	0.28	371	362	9	-2

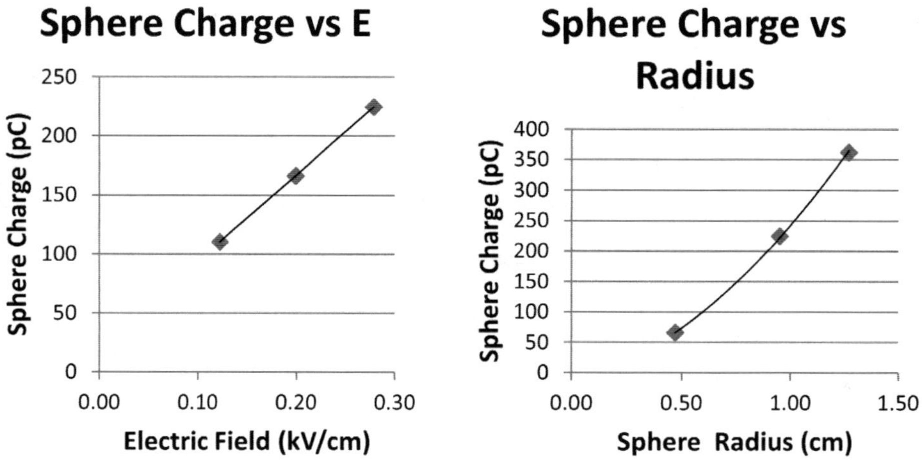

Figure 7. Left: Experimental values of the charge on 0.95 cm brass sphere vs *E*. **(Right):** Experimental charge at *E* = 0.28 kV/cm vs sphere radius. Data taken in the 7 cm diameter cylindrical ESP with 100 μm wire at 5 mbar in CO_2.

6.2 Precipitator Dust Experiments

The *I-V* curves for the seven configurations shown in Fig. 6 show that both the 100 μm and the 70 μ wires with the 9.6 cm diameter tube provided the largest stable voltage range between corona onset and streamer formation, as well as a relatively wide range in the corona current available for particle charging. Since the 100 μm wire is easier to handle, we decided to start our precipitator experiments with that configuration. However, the 9.6 cm width of the tube is perhaps too large for the comparatively small potentials allowed at the low pressure in the chamber. Dust particles falling through the precipitator have a larger distance to travel to the collecting electrode, resulting in little collection. Since the 100 μm wire with the 7.0-cm diameter tube did charge the simulant dust particles, all experiments were performed with that configuration.

About 2 mg of JSC Mars-1 simulant dust particles under 35 μm in diameter were introduced into the chamber through a Tygon tube and gas feed-through. The dust was aerosolized into a 1 liter polyethylene container inside the vacuum chamber located 1 cm above the ESP with a puff of CO_2 gas that raised the pressure in the chamber from 8.5 mbar to 9.3 mbar. A small net gas flow entered the ESP which carried dust through the precipitator, and the dust also fell with a terminal velocity dependent on the strength of gravity and the drag force. The dust collected on a silicon wafer mirror placed 5 cm under the exit of the ESP. Tribocharging of the dust during injection with no field on was measured with a Faraday cup located below the ESP. Appropriate tube and container material selection minimized this charging process to obtain a more controlled charging process in the precipitator.

Images of the dust collected on the silicon wafer with the fields on and off were analyzed to obtain particle size distributions (PSD). Fig. 8 shows PSD data for five experiments with the field turned off. Three, five, and ten CO_2 gas puffs, each carrying about 2 mg of the Mars simulant that sent dust through the precipitator and collected on the wafer were examined and photographed with an optical microscope at 100× magnification. PSD data for these runs show that the largest fraction of particles lies in the under 10 μm size fraction, matching the particle size distribution expected in the Martian atmosphere.

Journal of Physics: Conference Series **327** (2011) 012048
doi:10.1088/1742-6596/327/1/012048

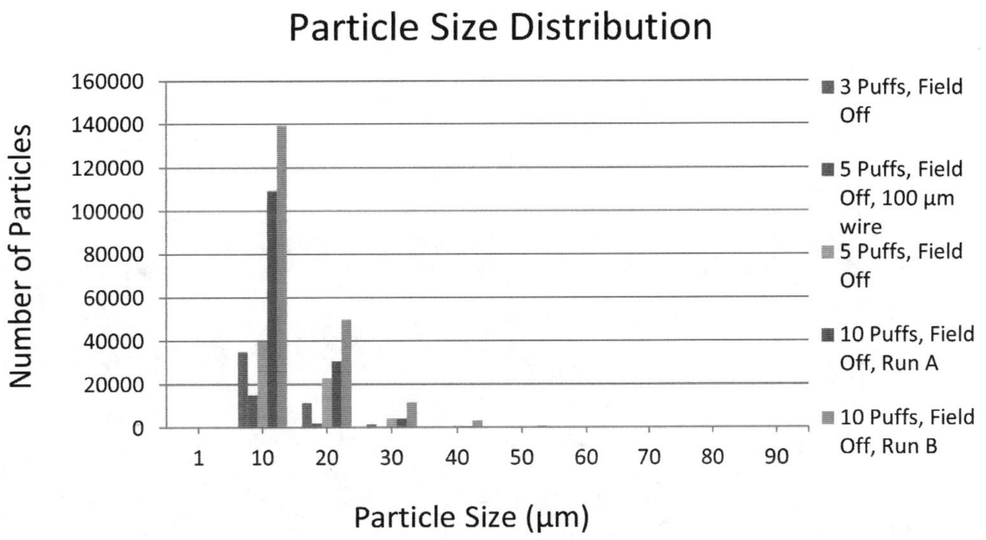

Figure 8. Particle size distribution of JSC Mars-1 simulant dust particles introduced into the chamber with short puffs of CO_2 gas and aerosolized before falling through the precipitator with the field off. Three, five, and ten puffs, each carrying about 2 mg of simulant dust, were supplied. Dust was collected on silicon wafers 7 cm in diameter. Four runs were performed with the 7.0 cm-0.3 cm rod outer-inner electrode configuration and one with the 7.0 cm-100 μm configuration. Smaller numbers of particles with diameters in the 60 to 90 μm size fractions, ranging from 1 to 309, are too small to show up on the graph.

Efforts to standardize the amount of dust introduced in the chamber are underway so that PSD data with and without the field on can be compared for quantification of ESP collecting performance. Microscope images of dust collected on the silicon wafer with and without the field on show promising performance results (Fig. 9). The applied voltage was 1600 V. No determination of possible nanoparticle production in the precipitator was made.

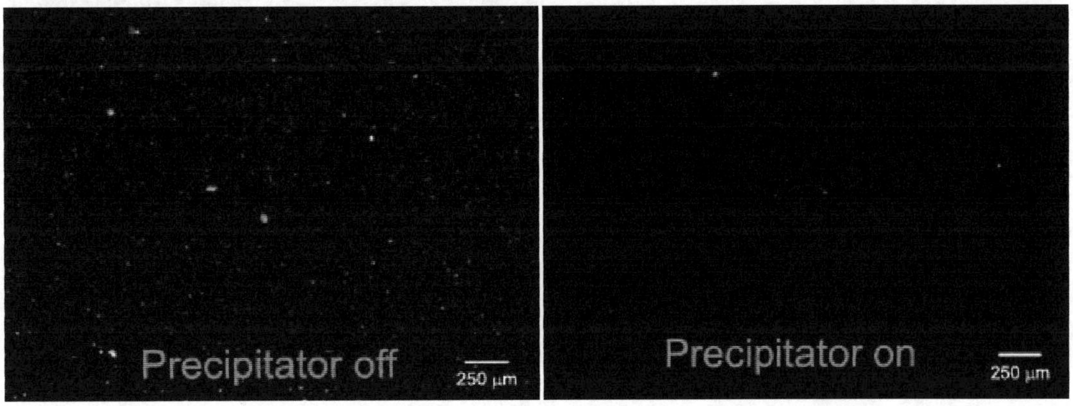

Figure 9. Microscope images at 100× of JSC Mars-1 dust simulant particles aerosolized in the vacuum chamber and sent through the precipitator with the field off (left) and with the potential set at 1600 V (right). The largest particles seen on the image with the field on are outside the range of particles expected in the Martian atmosphere.

It is estimated that the amount of dust in the Martian atmosphere can be as high as 2,000 particles per cubic centimeter. The Materials Adherence Experiment on the Mars Pathfinder mission provided data showing that dust deposition on surfaces averaged about 0.29% per Martian day [22]. Measurements of the visible/near infrared reflectance spectra of the radiometric calibration targets on the Mars Pathfinder rover show dust fall rates of about 20-45 μm per Earth year [23]. Figure 10 shows the Mars Exploration Rover Spirit and Opportunity Color Calibration Targets before landing and after 23 and 246 Martian days. The amount of dust deposited into the ESP will be determined by comparison to these dust deposition rates.

Figure 10. (Left): Clean color calibration target on Mars Exploration Rover Spirit. The target's mirror and the shadows cast on it by the Sun help scientists determine the degree to which dusty Martian skies alter the panoramic camera's perception of color. **(Center)**: Calibration target on the missions' twin rover Opportunity after 23 Martian days (sol). **Right):** Target after 346 sols. [Courtesy NASA/JPL/Cornell University]

7. Conclusions

Human exploration of the planet Mars will require the utilization of local resources to reduce the mass, cost, and risk of those future missions. Among the most important resources are water and oxygen, which can be extracted from the Martian atmosphere, composed mostly of carbon dioxide. Before the Martian atmospheric gas can be processed for the production of these commodities, dust particles with diameters ranging from 2 to 10 μm must be first removed.

The electrostatic precipitator design presented here removes simulated Martian dust particles in the required range in a simulated Martian atmospheric environment. The current-voltage, *I-V*, characteristic curves taken for the nine precipitator configurations at 9 mbars of pressure showed that cylindrical collecting electrode 7.0 cm in diameter with a concentric positive high voltage electrode 100 μm thick provides the best range of voltage and charging corona current.

The charge acquired by brass spheres placed in several different geometries of our precipitator design agreed fairly well with Pauthenier charging theory and showed that the precipitator produces the desired charge. This charging mechanism also contributes to the 2 to 10 μm dust particles expected in the Martian atmosphere. Diffusion charging, the second charging mechanism applicable to the dust particles, has little effect on the large brass spheres.

Mass determination as well as microscopic images and particle size distributions of dust collected on a silicon wafer placed directly below the precipitator with the field on and off showed excellent initial results. We are currently attempting to standardize the amount of Martian simulant dust introduced into the chamber to obtain quantification of precipitator performance.

Acknowledgements

This project was supported by NASA's ISRU project. We thank Dr. M. L. Lowder for her support of our experiments during the last phase of this project and Clara Wright for performing PSD analyses.

References

[1] White H J 1963 *Industrial Electrostatic Precipitation*, (Reading, MA Addison-Wesley)
[2] Olgesby S and Nichols G B 1978 *Electrostatic Precipitation* (New York,: Marcel Dekker)
[3] Pang H L, Atten P and Reboud J L 2009 *IEEE Trans. on Industry Applications* **45** No.1 50
[4] Sanders G B et al (2005) *International Lunar Conference* Toronto (Canada)
[5] Muscatello A C and Santiago-Maldonado E 2011 *Planetary and Terrestrial Mining Sciences Symposium and Space Resources Roundtable Second Joint Meeting* Ottawa (Canada)
[6] Landis G A, Herkenhoff K, Greeley R, Thompon S, Whelley P, and the MER Athena Science Team 2006 *LPS* **37** 1937
[7] Townsend J S 1914 *Electricity in Gases* (New York: Oxford University Press)
[8] Paschen F 1889 *Wied. Ann,* **37** 69
[9] Calle C I, Clements J S, Wills J and Buhler C R 2004 *NASA Technical Memorandum 2004-211535* 86
[10] Thomson J J and Thomson G P 1933 *Conduction of electricity through gases* (London: Cambridge University Press)
[11] Schmid O et al 2002 *Aerosol and Science Technology* **36** 351
[12] Kaiser K L 2005 *Electromagnetic Compatibility Handbook* (Boca Raton: CRC Press)
[13] Pauthenier M M and Moreau-Hanot M 1932 *J. Phys. Radium Ser.* 7 3 500
[14] Cross J 1987 *Electrostatics: Principles and Applications* (London: IOP Publishing)
[15] Arendt P and Kallman H 1926 *Z. Physics* **35** 421
[16] Hinds W C 1999 *Aerosol Technology* (New York: John Wiley and Sons)
[17] Smith W B and McDonald J R 1976 *J. Aerosol Sci.* **7** 151
[18] Liu B Y H and Kapadia A 1978 *J. Aerosol Science* **9** 227
[19] Romai F J and Piu D Y H 1993 *J. Aerosol Science* **23** 679
[20] Filippov A V 1993 J. Aerosol Science 24 424
[21] Marlow W H and Brock J R 1975 *J. Colloid Interface Science* **50** 32
[22] Landis G G and Jenkins P P 1997 *Proc. 26th IEEE Photovoltaic Spec. Conf.* Chicago (USA)
[23] Johnson J R et al 2003 *Icarus* **1163** 330

International Symposium on Physical Sciences in Space IOP Publishing
Journal of Physics: Conference Series **327** (2011) 012049 doi:10.1088/1742-6596/327/1/012049

Atomic Clock Ensemble in Space

Luigi Cacciapuoti
European Space Agency, Research and Scientific Support Department,
Keplerlaan 1, P.O.Box 299, 2200 AG Noordwijk ZH The Netherlands
E-mail: Luigi.Cacciapuoti@esa.int

Christophe Salomon
Laboratoire Kastler Brossel, CNRS, UPMC, ENS
24, rue Lhomond, 75005 Paris, France

Abstract. Atomic Clock Ensemble in Space (ACES) is a mission using high-performance clocks and links to test fundamental laws of physics in space.
Operated in the microgravity environment of the International Space Station, the ACES clocks, PHARAO and SHM, will generate a frequency reference reaching instability and inaccuracy at the $1 \cdot 10^{-16}$ level. A link in the microwave domain (MWL) and an optical link (ELT) will make the ACES clock signal available to ground laboratories equipped with atomic clocks. Space-to-ground and ground-to-ground comparisons of atomic frequency standards will be used to test Einstein's theory of general relativity including a precision measurement of the gravitational red-shift, a search for time variations of fundamental constants, and Lorentz Invariance tests. Applications in geodesy, optical time transfer, and ranging will also be supported.
ACES has now reached an advanced technology maturity, with engineering models completed and successfully tested and flight hardware under development. This paper presents the ACES mission concept and the status of its main instruments.

1. Introduction

Time intervals and frequencies are physical quantities that can be measured with outstanding accuracy. As a consequence, applications of atomic clocks are numerous and very diverse. Most precision measurements, fundamental constants and SI units can be traced back to frequencies including distances, electrical units, and perhaps in the future the mass unit. With the redefinition of the meter 25 years ago and the choice of a conventional value for the speed of light in vacuum, distance measurements have been simply translated into time interval measurements. The most visible application of atomic clocks is the satellite Global Positioning System (GPS). GPS receivers are vastly used in geodesy and Earth monitoring as well as for the comparison of distant clocks and for the generation of international atomic time scales. Precise clocks are also used for the datation of millisecond binary pulsars, which emit almost-periodic electromagnetic signals associated with the rotation of the binary star system. Finally clocks are used to perform precision tests of fundamental physical laws and to challenge our knowledge of the Universe.
Atomic Clock Ensemble in Space (ACES) is a fundamental physics mission of the European Space Agency based on a new generation of clocks to be operated in the microgravity environment of the International Space Station. ACES is a distributed system designed to disseminate a high stability and accuracy clock signal [1, 2]. It consists of a payload generating

Journal of Physics: Conference Series **327** (2011) 012049 doi:10.1088/1742-6596/327/1/012049

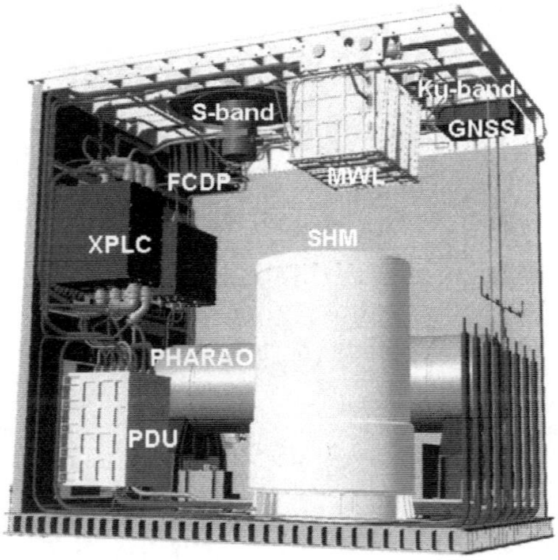

Figure 1. Left: The external payload facility of the Columbus module. Right: Detail of the ACES payload; the ACES payload has a volume of $1.340\times1.117\times1.320$ m^3, a mass of 240 kg, for a total power consumption of 450 W.

an atomic frequency reference in space and a network of ground terminals connected to high-performance atomic clocks on ground. Transported on the International Space Station (ISS) by the Japanese H-II Transfer Vehicle HTV, the ACES payload will be installed at the external payload facility of the Columbus module (see Fig. 1). The ACES payload accommodates two atomic clocks: PHARAO (acronym of "Projet d'Horloge Atomique par Refroidissement d'Atomes en Orbit"), a primary frequency standard based on samples of laser cooled cesium atoms and SHM (acronym of "Space H-Maser"), an active hydrogen maser for space applications. The performances of the two clocks are combined to generate an on-board timescale with the short-term stability of SHM and the long-term stability and accuracy of the PHARAO clock. The on-board comparison of PHARAO and SHM and the distribution of the ACES clock signal are ensured by the Frequency Comparison and Distribution Package (FCDP), while all data handling processes are controlled by the eXternal PayLoad Computer (XPLC). A GNSS receiver connected to the on-board time scale will provide orbit determination of the ACES clocks. One of the main objectives of the ACES mission consists in maintaining a stable and accurate on-board timescale that will be used to perform space-to-ground as well as ground-to-ground comparisons of atomic frequency standards. Two high-performance links will be used for the comparison of distant clocks, a link in the microwave domain (MWL, acronym of "MicroWave Link") and an optical link (ELT, acronym of "European Laser Timing"). The comparison of distant clocks performed by ACES will be the primary data product of the mission. These measurements will be used to test fundamental law of physics and to develop applications in different areas of research.

ACES will be ready for launch in 2014, for a planned mission duration is 18 months. During the first two weeks, the functionality of the clocks and of MWL will be tested. Then, a period of 6 months will be devoted to the characterization and performance evaluation of the clocks. During this phase, a clock signal with frequency inaccuracy in the 10^{-15} range will be available to ground users. Under microgravity conditions, it will be possible to tune the linewidth of the atomic resonance of PHARAO by two orders of magnitude, down to sub-Hz values (from 11 Hz to 110 mHz). After the clocks are optimized, performance in the 10^{-16} range both for frequency

Journal of Physics: Conference Series **327** (2011) 012049 doi:10.1088/1742-6596/327/1/012049

instability and inaccuracy are expected. In the second part of the mission (12 months, possibly extended up to 30 months), the on-board clocks will be compared to a number of atomic clocks on ground operating both in the microwave and optical domain. ACES will perform worldwide comparisons of advanced clocks operating on different atoms or molecules reaching a frequency resolution in the 10^{-17} regime. These measurements will test general relativity and seek for new interactions beyond the Standard Model.

2. ACES Science Objectives

ACES will conduct experiments with cold atoms in a freely falling laboratory, it will perform fundamental physics tests to high resolution, and develop applications in different areas of research.

2.1. High-performance Microwave Clocks for Space

A new generation of space clocks reaching frequency instability and inaccuracy of a few parts in 10^{16} will be validated by ACES. PHARAO will combine laser cooling techniques and microgravity conditions to significantly increase the interaction time and consequently reduce the linewidth of the clock transition. Improved stability and better control of systematic effects will be demonstrated in the space environment. PHARAO will reach a fractional frequency instability of $1 \cdot 10^{-13}/\sqrt{\tau}$, where τ is the integration time expressed in seconds, and an inaccuracy of a few parts in 10^{16}. The reliability offered by active H-masers will be made available for space applications by SHM. SHM will demonstrate a fractional frequency instability of $1.5 \cdot 10^{-15}$ after 10000 s of integration time. Two servo-loops will lock together the clock signals of PHARAO and SHM generating an on-board time scale combining the short-term stability of the H-maser with the long-term stability and accuracy of the cesium clock (Fig. 2).

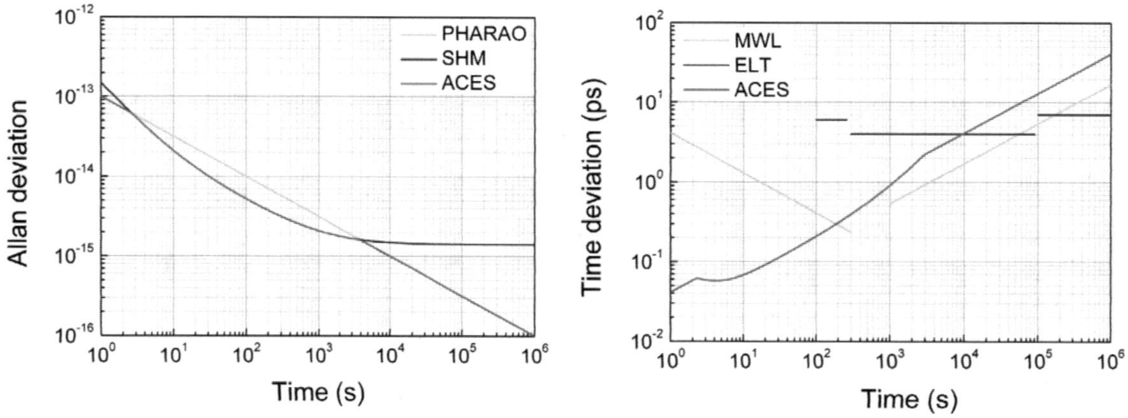

Figure 2. Left: Specified Allan deviation of PHARAO, SHM, and of the ACES clock signal. Right: Specified time deviation of MWL and ELT, compared with the time stability of the ACES clock signal.

2.2. ACES Time and Frequency Transfer Systems

The ACES clock signal will be distributed to a network of ground terminals via a dedicated MicroWave Link (MWL, see Sec. 3.5). Frequency comparisons with time deviation better than 0.4 ps at 300 s, 8 ps at 1 day, and 25 ps at 10 days of integration time will be demonstrated (see Fig. 2). These performances, surpassing existing techniques (Two-Way Satellite Time and

Frequency Transfer and GPS) by one to two orders of magnitude, will enable common view and non-common view comparisons of ground clocks with 10^{-17} frequency resolution after a few days of integration time. Thanks to the development of optical frequency combs [3, 4], the reference signal generated by optical clocks on ground can be easily downconverted in the microwave domain and compared with ACES. In this way, ACES will take full advantage of the progress of optical clocks [5, 6, 7], today reaching instability and inaccuracy levels below $1 \cdot 10^{-17}$. ACES will also deliver a global atomic time scale with 10^{-16} accuracy, it will allow clock synchronization at the 100 ps uncertainty level, and it will contribute to international atomic time scales (TAI, UTC).
ELT, acronym of "European Laser Timing", will provide ACES with an alternative time and frequency transfer system reaching a time stability of 6 ps after 100 s of integration time, down to 4 ps between 300 s and 10000 s and with a long-term stability of 7 ps (see Fig. 2). The system can be calibrated to deliver the ACES time scale with an accuracy better than 50 ps, finding important applications in the distribution of the ACES time reference and in the synchronization of geodetic observatories.
The ACES clocks and links will allow to establish a global network for the comparison and the synchronization of distant clocks on ground.
Time and frequency transfer experiments at the 10^{-16} level require the modeling of relativistic effects in the space-to-ground propagation of electromagnetic signals up to $1/c^3$ terms [8], beyond the $1/c^2$ expansion needed for existing links. Future experiments aiming at the 10^{-18} level require an expansion up to $1/c^4$ [9].

2.3. Fundamental Physics Tests with ACES

According to Einstein's theory of general relativity, identical clocks placed in different gravitational fields experience a frequency shift that depends on the difference between the Newtonian potentials at the clocks positions. The comparison between the ACES on-board clocks and ground-based atomic clocks will measure the frequency variation due to the gravitational red-shift with a 70-fold improvement on the GP-A experiment [10], testing Einstein's prediction at the 2 ppm uncertainty level. In [11] a claim was made that atom interferometry experiments combined with a local measurement of the gravitational acceleration realizes a test of the Einstein's gravitational redshift to $7 \cdot 10^{-9}$. However, as shown in [12], this claim is incorrect so that the GPA-experiment still remains the most precise test to date.
Time variations of fundamental constants can be measured by comparing clocks based on different transitions or different atomic species [13]. Indeed, the energy of an atomic transition can be expressed in terms of the fine structure constant α and the two dimensionless constants m_q/Λ_{QCD} and m_e/Λ_{QCD}, which depend on the quark mass m_q, the electron mass m_e, and the QCD mass scale Λ_{QCD} [14, 15]. ACES will perform crossed comparisons of ground clocks both in the microwave and in the optical domain with a frequency resolution of $1 \cdot 10^{-17}$ in a few days of integration time. These comparisons will impose strong and unambiguous constraints on time variations of the three fundamental constants reaching an uncertainty as low as $1 \cdot 10^{-17}$/year in case of a 1-year mission duration, down to $3 \cdot 10^{-18}$/year after three years.

The foundations of special relativity lie on the hypothesis of Local Lorentz Invariance (LLI). According to this principle, the outcome of any local test experiment is independent of the velocity of the freely falling apparatus. In 1997, LLI tests based on the measurement of the round-trip speed of light have been performed by comparing clocks on-board GPS satellites to ground hydrogen masers [16]. In such experiments, LLI violations would appear as variations of the speed of light c with the direction and the relative velocity of the clocks. ACES will perform a similar experiment by measuring relative variations of the speed of light at the 10^{-10} uncertainty level.

2.4. Applications
ACES will also demonstrate a new geodesy technique to map the Earth gravitational potential. This technique uses a precision measurement of the Einstein's gravitational red-shift between two clocks to determine the corresponding difference in the local gravitational potentials. The possibility of performing comparisons of ground clocks at the 10^{-17} frequency uncertainty level will allow ACES to resolve geopotential differences to 10 cm on the geoid height.
A dedicated GNSS receiver on-board the ACES payload will ensure orbit determination, important for comparing clocks and for performing fundamental physics tests. In addition, the GNSS subsystem will be connected to the ACES clock signal, opening the possibility to use the GNSS network for clock comparisons or remote sensing applications (radio-occultation and reflectometry experiments).
The simultaneous operation of the optical (ELT) and microwave (MWL) links will provide a test bench for the characterization of the two ACES links. Optical versus dual-frequency microwave measurements will also provide useful data for the study of atmospheric propagation delays and for the construction of atmosphere mapping functions at three different wavelengths. In addition, the ACES links will provide absolute range measurements both in the microwave and in the optical domain.

3. ACES Status
All ACES instruments and subsystems have now reached a high degree of technology maturity, as demonstrated by the tests performed on the the engineering models and by the ongoing activities on the flight hardware. The system tests on the ACES Engineering Model (EM) workbench conducted at CNES facilities in Toulouse (see Fig. 3) have represented a major achievement in the ACES development cycle. Their successful completion has indeed closed the EM phase with the performance verification of the ACES clock signal and it has confirmed the adequacy of the ACES design, releasing the manufacturing for the flight hardware.

Figure 3. ACES EM system level tests at the CNES facilities in Toulouse. On the left, the vacuum chamber with the ACES EM workbench before being closed; the PHARAO tube is vertically installed in the chamber. On the right, the RF support equipment and part of the measurement instrumentation.

3.1. The PHARAO Clock

PHARAO is a cesium clock based on laser cooled atoms developed by LNE-SYRTE, LKB, and CNES. Its concept is very similar to ground based atomic fountains, but with a major difference: PHARAO will be operated under microgravity conditions. Atoms, launched in free flight along the PHARAO tube, cross a resonant cavity where they interact twice with a microwave field tuned on the transition between the two hyperfine levels of the cesium ground state (9.192631770 GHz, from the SI definition of the second). In a microgravity environment, the velocity of the atoms along the ballistic trajectories is constant and can be continuously changed over almost two orders of magnitude (5 to 500 cm/s), allowing the detection of atomic signals with sub-Hz linewidth.

The cesium clock PHARAO is composed of four main subsystems: the cesium tube, the optical bench, the microwave source, and the computer control. The engineering model of the PHARAO clock has been completed and tested. On ground, because of gravity, PHARAO can only be operated like a fountain clock, with the atoms launched vertically along the tube. Cesium atoms have been loaded in the optical molasses, cooled down to a few μK, interrogated on the clock transition by the resonant microwave field, and detected by laser-induced fluorescence emission. Microwave resonance signals (Ramsey fringes) with a signal-to-noise ratio of ~ 700 have been recorded. For a typical launch velocity of about 3.5 m/s, the duration of the free flight between the two Ramsey interaction regions is about 100 ms, corresponding to a width of the central fringe of about 5 Hz. Once operated in microgravity, the longer interaction times will allow PHARAO to measure linewidths 10 to 50 times narrower.

The tuning and optimization of the instrument on ground has been performed with the atomic clouds launched vertically against gravity at a speed of 3.56 m/s. In these conditions, an Allan deviation of $3.5 \cdot 10^{-13}/\sqrt{\tau}$ has been measured, where τ is the integration time expressed in seconds. Frequency instabilities of $1 \cdot 10^{-15}$ have been reached in less than 2 days of integration time. PHARAO performance on ground is mainly set by the phase noise of the local oscillator which is sampled by the atoms in the microwave cavity (Dick effect). In space, this effect will be reduced by one order of magnitude because of the longer interrogation time and the narrower resonance width.

PHARAO accuracy evaluation has been completed. Several maps of the magnetic field experienced by the atoms along their ballistic trajectories in the PHARAO tube have been measured by operating the clock on the $m_f = 1 \to m_f = 1$ transition and detecting the Zeeman shift of the atomic resonance with respect to the clock transition. The contribution of the second order Zeeman effect to the clock accuracy budget is at the level of $6.6 \cdot 10^{-16}$ for a bias field along the PHARAO tube of 35 nT. The cold collisions shift has been evaluated to an accuracy of $9.5 \cdot 10^{-16}$. Second order Zeeman effect and collisional shift are indeed the two major contributors to the clock accuracy, presently evaluated to $1.3 \cdot 10^{-15}$. PHARAO accuracy evaluation has been verified by measuring the PHARAO frequency output with respect to the LNE-SYRTE mobile fountain clock FOM (FOntaine Mobile). The result is in agreement with a zero frequency difference within 1 part in 10^{15}.

PHARAO flight model is under manufacturing in SODERN. Figure 4 shows the flight model of the PHARAO tube, recently completed and successfully tested against vibrations.

3.2. The SHM Clock

Because of their simplicity and reliability, hydrogen masers are used in a large variety of applications. Passive and active masers are expected to be key instruments in future space missions, satellite positioning systems, and high-resolution VLBI (Very Long Baseline Interferometry) experiments.

The clock operates on the hyperfine transition of atomic hydrogen at 1.420405751 GHz. H_2 molecules are dissociated in a plasma discharge and the resulting beam of H atoms is state-

Journal of Physics: Conference Series **327** (2011) 012049 doi:10.1088/1742-6596/327/1/012049

Figure 4. Flight model of the PHARAO tube. Fully assembled (including thermal shields, μ-metal shields, fiber-coupled collimators, and harness), the tube has a volume of $990 \times 336 \times 444$ mm^3 and a mass of 44 kg.

selected and sent to a storage bulb. The bulb is surrounded by a microwave cavity that, tuned on the atomic resonance, induces the maser action. Developed by SpectraTime under ESA contract, SHM provides ACES with a stable fly-wheel oscillator. The main challenge of SHM is represented by the low mass and volume figures (42 kg, $390 \times 390 \times 590$ mm^3) required by the space clock with respect to ground H-masers. For this purpose, the number of thermal shields of the clock has been reduced and a dedicated Automatic Cavity Tuning (ACT) system has been implemented to steer the resonance frequency of the maser cavity against thermal drifts. SHM ACT injects two tones, symmetrically placed around the H-maser signal. The two tones are coherently detected and the unbalance between their power levels is used to close a feedback loop acting on the cavity varactor and stabilizing the resonance frequency of the microwave cavity against temperature variations. This method allows SHM to reach fractional frequency instabilities down to $1.5 \cdot 10^{-15}$ at 10^4 s of integration time. Figure 5 shows the H-maser physics package in the different assembly phases. Active oscillation on the hydrogen clock transition at

Figure 5. Physics package of the SHM engineering model during the different assembly phases.

a power level of -101 dBm has been measured. The cavity quality factor is about 44000 and the atomic quality factor is $1 \cdot 10^9$. The H-maser electronics, including the receiver and the ACT system, have been completed and tested showing an Allan deviation in agreement with SHM performance requirements. SHM EM is presently under test. The test campaign will close the EM phase and release the manufacturing of the SHM flight model.

3.3. The Frequency Comparison and Distribution Package

FCDP is the central node of the ACES payload. Developed by ASTRIUM and TIMETECH under ESA contract, FCDP is the on-board hardware that compares the signals delivered by the two space clocks, it measures and optimizes the performances of the ACES frequency reference, and finally distributes it to MWL electronics.
The engineering model of FCDP has been completed and tested. The noise introduced by FCDP on the clock signal distributed at the MWL output has an Allan deviation that decreases as the inverse of the integration time, entering the 10^{-18} regime after 10^4 s of integration time. The noise floor of the FCDP phase comparator has also been characterized. Its Allan deviation decreases as the inverse of the integration time, dropping below $1 \cdot 10^{-17}$ after 10^4 s.

3.4. Ground tests of the ACES Clock Signal

At completion of the ACES engineering model phase, the ACES EM workbench has been integrated at CNES premises in Toulouse with the objective of testing interfaces, functions, and performance. The ACES EM workbench includes: PHARAO EM, FCDP EM, SHM ground model (EM0), an XPLC test crate, and a PDU (Power Distribution Unit) simulator. Both PHARAO EM and FCDP EM were mounted in the CNES chamber and operated under vacuum (see Fig. 3).
As first test step, the ACES clocks were individually powered. The continuous monitoring of the PHARAO and SHM EM0 performance (both Allan deviation and phase noise) has demonstrated that the two clocks do not interfere with each other. Phase noise measurements performed against a cryogenic sapphire oscillator have shown no degradation with respect to stand-alone characterization of the ACES clocks. Then, FCDP was also powered to deliver the ACES clock signal at the MWL output. This sequence completed the so called mutual compatibility tests between the ACES clocks and FCDP.
As second step, the short-term servo-loop, steering the local oscillator of PHARAO on the clock signal of SHM EM0, was closed. After optimization of the servo-loop parameters, the performance of PHARAO and SHM EM0 were measured in different operational modes exercising the system both during standard operational conditions and during calibration measurements.
As last step, the long-term servo-loop was also closed and the ACES clock signal, now reproducing SHM EM0 for short-to-medium integration times and PHARAO on the long-term, was generated. The performance of the ACES clock signal was measured against FOM, the mobile fountain clock of LNE-SYRTE. A long duration measurement was conducted both to characterize the Allan deviation of the ACES signal and to perform a frequency measurement with respect to FOM.
Figure 6 shows the stability of the ACES clock signal (red) measured with respect to FOM. For integration times shorter than the long-term servo-loop time constant (1000 s), the ACES clock signal closely follows SHM EM0, therefore the Allan deviation measurement is limited by the FOM performance. For longer integration times, the long-term servo loop forces ACES on the PHARAO clock signal providing it with the long-term stability and accuracy of the Cs clock.

3.5. The Microwave Link

The ACES microwave link is developed by ASTRIUM, TIMETECH, TZR, and EREMS under ESA contract. The proposed MWL concept is an upgraded version of the Vessot two-way technique used for the GP-A experiment in 1976 [10] and the PRARE (Precise Range and Range-Rate Equipment) geodesy instrument. Its purpose is to compare the 100 MHz signal generated by the ground clock to the signal received from the space clock via the microwave transfer link. The system operates continuously with a carrier frequency in the Ku-band. The high carrier frequencies of the up and down links (13.5 GHz and 14.7 GHz respectively) allow for

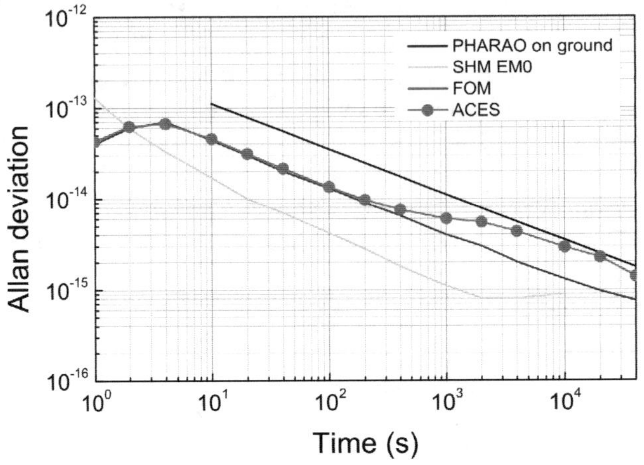

Figure 6. Performance of the ACES clock signal measured on ground with respect to FOM (red) and compared to the performance of FOM (blue), SHM EM0 (green), and PHARAO on ground (black).

a noticeable reduction of the ionospheric delay. A third frequency in the S-band (2.2 GHz) is used to determine the Total Electron Content (TEC) and correct for the ionosphere time delay. A PN-code modulation (100 Mchip/s) on the carrier removes the phase ambiguity between successive comparison sessions separated by large dead times. The system is designed for multiple access capability, allowing up to 4 simultaneous ground users distinguished by the different PN-codes and Doppler shifts.

3.5.1. MWL Flight Segment The engineering model of the flight segment electronic unit has been completed and tested (see Fig. 7). MWL long-term stability is ensured by the continuous calibration of the receiver channels provided by a built-in test-loop translator. For shorter durations (<300 s), time stability is driven by the noise performance of the Ku transmitter and receiver and of the DLL (Delay-Locked Loop) boards. The 100 MHz chip rate allows to reach a time stability better than 0.2 ps already with code measurements. However, ultimate performance is reached with the carrier phase measurements, which can ensure a time stability as low as 70 fs at about 100 s of integration time (see Fig. 7). For longer durations, the time stability remains well below the 1 ps level even in the worst conditions of signal-to-noise density ratio, corresponding to very low elevation angles of the ISS over a ground terminal. The thermal sensitivity of the system has been measured and used to calibrate MWL phase comparison data against temperature variations. The sensitivity to a series of key parameters such as clock input power, received signal-to-noise density ratios, supply voltage, Doppler, Doppler rate, etc. has been measured. The susceptibility of the system to narrowband and broadband interference, as well as to multipath effects has been characterized.

3.5.2. MWL Ground Terminal The MWL ground terminal electronics is similar to the MWL flight hardware, symmetry being important in a two-way system to reduce instrumental errors. The ACES MWL Ground Terminal (GT) is a microwave station interfacing the local clock on ground to the ACES payload for space-to-ground clock comparisons. To reduce phase instabilities due to the tracking motion, the electronic unit of the MWL GT has been rigidly attached

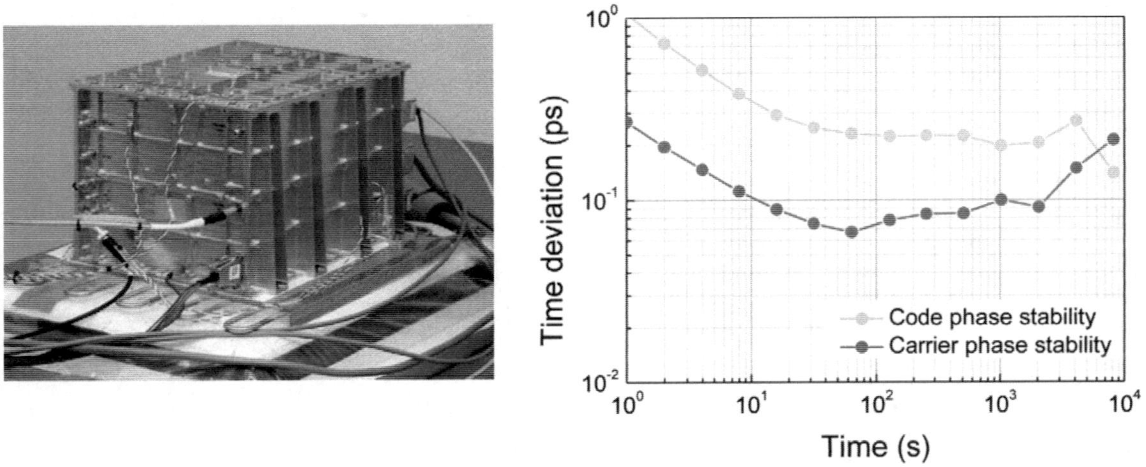

Figure 7. Time stability of the code (green) and carrier (red) phase measurement measured on the engineering model of the MWL flight segment electronic unit.

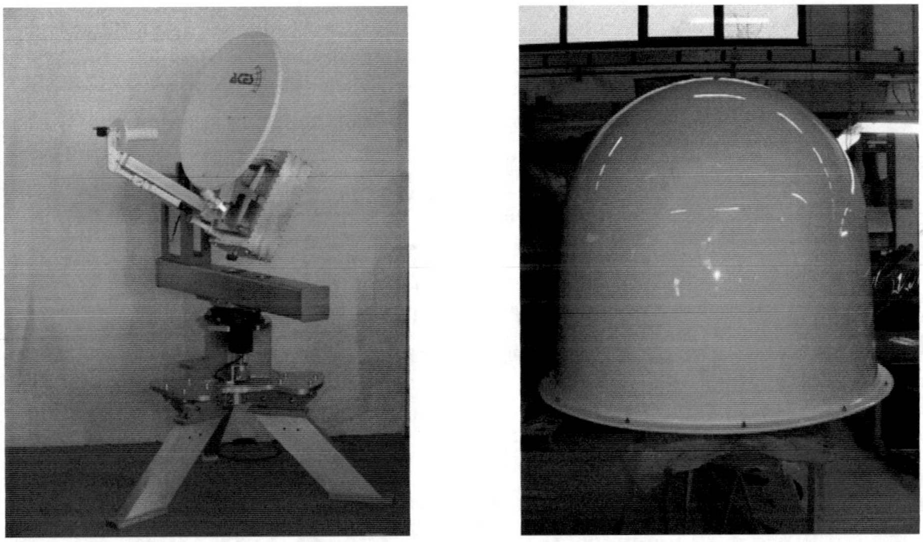

Figure 8. First prototype of the MWL ground terminal in its assembly phase (left) and the radome cupola (right) enclosing the steering unit and the electronics.

to the antenna unit. The Ku-band signal is delivered to the antenna feeder via a waveguide, a high stability RF cable is used for the S-band. The antenna is a 60 cm offset reflector with a dual-band feed system automatically pointed in azimuth and elevation by a steering mechanism. A computer controls the steering unit based on ISS orbit prediction files, collects telemetry and science data both from the local clock and the MWL GT electronics, and interfaces directly with the ACES Users Support and Operation Center (USOC). The system is housed below a protective radome cover, which also allows to stabilize the temperature of the enclosed volume by an air conditioning system, part of a separate service pallet (see Fig. 8). The total weight is 650 kg: 270 kg for the radome pallet and 380 kg for the service pallet and the air-conditioning system. About 10 kW are needed to power the MWL GT electronics and its temperature control system, the steering unit, and the air conditioning below the radome cupola. The thermal design will allow to operate the MWL ground terminals for an external temperature ranging between −30°C and +45°C.

MWL will be validated in an end-to-end test campaign in which the FS and GT electronic units are connected by cables through a signal simulator. The signal simulator will mimic frequency and amplitude variations of the MWL signal (Ku and S-band) according to the orbit dynamics. The link delays will also be measured preparing for the calibration campaign which will take place on the flight hardware and on the ground terminals after on-site installation.

3.6. The ELT Optical Link

ELT is an optical link using laser pulses exchanged between the space segment and Satellite Laser Ranging (SLR) stations on the ground to compare distant clocks. The on-board hardware consists of a Corner Cube Reflector (CCR), a Single-Photon Avalanche Diode (SPAD), and an event timer board connected to the ACES time scale. Laser pulses fired towards ACES by a SLR station are detected by the SPAD diode. The fire and the detection events are tagged in the local clock time scales both in space and on ground. At the same time, the ELT CCR re-directs the laser pulse towards the ground station where it is detected and stamped in the time scale maintained at the SLR station. The measurement of the start and return time on ground and of the detection time in space allows to calculate the desynchronization between space and ground clocks, additionally providing precise ranging information.

A study conducted at the Geodetic Observatory of Wettzell has confirmed the feasibility of the ELT experiment. The full detection chain proposed for ELT has been installed in a second independent detection port of the station and tested against the Wettzell channel both by ranging a ground target and satellites equipped with corner cube reflectors. Detector properties such as the operation at the single photon light level regime, reliable operation under conditions of high background radiation, and detector jitter have been characterized. The results have been found in good agreement with the expected performance levels [17].

In addition, the SPAD diode has been tested at the SLR stations of Wettzell and Graz . A time deviation of 5 ps after 500 s of integration time has been measured with the SPAD diode integrated in the Wettzell detection channel. This measurement was limited by the timing resolution of the Wettzell time tagging system. Time deviations down to about 1 ps after 10 s of integration time have been measured at the Graz station thanks to the higher firing rate and the better timing resolution available there [17, 18].

The SPAD diode is presently being characterized in terms of optical to electrical delay. Recent measurements have shown that the absolute delay between the time of arrival of the single photon onto the detector input aperture and the time when the electrical output signal exceeds the pre-defined level can be calibrated to a precision of about 10 ps [19].

3.7. The On-board GNSS System

A GPS/GALILEO receiver will be part of the ACES payload, directly connected to the ACES clock signal. The system is designed to provide orbit determination and payload positioning for evaluating relativistic corrections in the space-to-ground clock comparison measurements. Additionally, it supports remote sensing applications in the field of radio-occultation and reflectometry, also exploring the use of the new GNSS signals [20]. The GNSS receiver will be accommodated on the ACES payload while the antenna will be installed externally to the ACES box. In the final configuration, the GNSS antenna will be looking along the flight direction of the ISS, with a tilt angle of about 50° with respect to the ISS velocity vector and 30° in the zenith direction. The GNSS hardware will provide ACES with a completely autonomous

system for orbit determination. Even if not optimal in terms of visibility towards the GNSS constellation, the selected accommodation of the GNSS antenna fulfills ACES requirements for orbit determination [21, 22]. In addition, this geometry turns out to be particularly favorable for supplementary GNSS science such as radio-occultation and reflectometry.
The current ACES GNSS subsystem baseline consists of a redundant set of commercial-of-the-shelf (COTS) JAVAD GNSS Triumph TRE-G3TH receiver boards and a power/data interface board. The interface board protects the GNSS receiver against latch-up events, it formats the native receiver data flow into the ACES time-slotted protocol, and converts the 100 MHz ACES clock signal down to 10 MHz generating a reference that can be used for the receiver operation. A series of simulations and tests have been recently performed on the GNSS receiver to assess its performance and evaluate its robustness against radiation. Signal simulation studies show that the selected GNSS receiver, based on the JAVAD Triumph technology, provides measurements with excellent quality at the expected high signal dynamics in low Earth orbit. The performed radiation tests illustrate that the JAVAD GNSS TRE-G3TH receiver boards can withstand the radiation expected during the ACES mission life time.

3.8. The ACES Ground Segment

The ACES ground segment will be integrated in CADMOS-Toulouse within the overall ISS ground architecture providing the communication links between ground and space through the Columbus Control Center (Col-CC) and the NASA ground segment. The main components of the ACES ground segment are the ACES Users Support and Operations Center and the network of MWL ground terminals connected to the ground clocks participating to the ACES mission. The ground terminals are remotely controlled by the ACES USOC, responsible for defining the planning of the space-to-ground clock comparison sessions along the ISS orbit. Link authorization is regulated by the ACES USOC via Code Division Multiple Access (CDMA) to the MWL flight segment hardware. The ACES USOC provides all functions related to online monitoring and control of both the ACES payload and the MWL ground terminals network, it defines the ACES utilization plan, schedules ACES operations, collects the ACES raw data, generates the ACES data products, and archives them. The USOC also manages external interfaces through which access to the ACES data and data products is provided to the users' community. A detailed description of the ACES ground segment and its functionalities is provided in [23].

Acknowledgments

The authors express their warm thanks for their valuable contribution to the ACES science team, to the ACES project team at ESA, ASTRIUM, and TIMETECH, to the PHARAO team at LNE-SYRTE and CNES, to the SHM team at SpectraTime. This paper has been written on behalf of the many scientists and engineers contributing to the mission development.

References

[1] Salomon C *et al* 2002 *C. R. Acad. Sci. Paris* **t.2 Séries IV** 1313
[2] Cacciapuoti L and Salomon C 2009 *Eur. Phys. J. Special Topics* **172** 57
[3] Holzwarth R *et al* 2000 *Phys. Rev. Lett.* **85** 2264
[4] Diddams S A *et al* 2000 *Phys. Rev. Lett.* **84** 5102
[5] Rosenband T *et al* 2008 *Science* **319** 1808
[6] Ludlow A D *et al* 2008 *Science* **319** 1805
[7] Chou C W *et al* 2010 *Phys. Rev. Lett.* **104** 070802
[8] Blanchet L *et al* 2001 *A&A* **370** 320
[9] Linet B and Teyssandier P 2002 *Phys. Rev. D* **66** 024045
[10] Vessot R F C *et al* 1980 *Phys. Rev. Lett.* **45** 2081
[11] Müller H *et al* 2010 *Nature* **463** 08776

[12] Wolf P *et al* 2011 *Class. Quantum Grav.* **28** 145017
[13] Fortier T *et al* 2007 *Phys. Rev. Lett.* **98** 070801
[14] Flambaum V V *et al* 2004 *Phys. Rev.* D **69** 115006
[15] V.V. Flambaum V V *et al* 2006 *Phys. Rev.* C **73** 055501
[16] Wolf P and Petit P 1997 *Phys. Rev.* A **56** 4405
[17] Schreiber U *et al* 2009 *IEEE Trans* **57** 728
[18] Prochazka I *et al* 2010 *Proc. of 2010 European Time and Frequency Forum*
[19] Prochazka I *et al* 2011 *Metrologia* **48** L13
[20] Beyerle G *et al* 2010 *Atmos. Chem. Phys. Discuss.* **10** 28821
[21] Duchayne L *et al* 2009 *A&A* **504** 653
[22] Montenbruck O *et al* 2011 *Journal of Spacecraft and Rockets* DOI 10.2514/1.52657
[23] Daganzo E *et al* 2009 *Proc. of EFTF-FCS 2009* DOI 978-1-4244-3510-4/09

Journal of Physics: Conference Series **327** (2011) 012050 doi:10.1088/1742-6596/327/1/012050

The Space Atom Interferometer project: status and prospects

F Sorrentino[1] K Bongs[2] P Bouyer[3] L Cacciapuoti[4] M de Angelis[1,5] H Dittus[6] W Ertmer[7] J Hartwig[7] M Hauth[8] S Herrmann[9] K Huang [10] M Inguscio[1] E Kajari[11,14] T Könemann[9] C Lämmerzahl[9] A Landragin[12] G Modugno[1] F Pereira dos Santos[12] A Peters[8] M Prevedelli[13] E M Rasel[7] W P Schleich[14] M Schmidt[7] A Senger[7] K Sengstock[15] G Stern[3] G M Tino[1] T Valenzuela [2] R Walser[16] P Windpassinger[15]

[1] Dipartimento di Fisica e Astronomia and LENS, Università di Firenze, Polo Scientifico, via Sansone 1, I-50019 Sesto Fiorentino, Italy
[2] Midlands Ultracold Atom Research Centre School of Physics & Astronomy University of Birmingham, Edgbaston, Birmingham B15 2TT, UK
[3] Laboratoire Charles Fabry de LInstitut dOptique, Centre National de la Recherche Scientifique, Campus Polytechnique Rd 128, 91127 Palaiseau, France
[4] European Space Agency, Research and Scientific Support Department, Keplerlaan 1, 2201 AZ Noordwijk, The Netherlands
[5] Istituto di Fisica Applicata CNR, via Madonna del Piano 10, I-50019 Sesto Fiorentino
[6] Institute of Space Systems, German Aerospace Center (DLR), Robert-Hooke-Strasse 7, D 28359 Bremen, Germany
[7] Institute of Quantum Optics, Leibniz Universität Hannover, Welfengarten 1, D 30167 Hannover, Germany
[8] Humboldt-Universität zu Berlin, Unter den Linden 6, D-10099 Berlin, Germany
[9] University of Bremen, Centre of Applied Space Technology and Microgravity (ZARM), Am Fallturm, D - 29359 Bremen, Germany
[10] Institute of Optics, Physics Department, Zhejiang University, China
[11] Theoretische Physik, Universität des Saarlandes, Postfach 151150, D-66041 Saarbrücken, Germany
[12] Observatoire de Paris - SYRTE 61 avenue de l'Obseravtoire, 75014 Paris, France
[13] Dipartimento di Fisica dell'Università di Bologna, Via Irnerio 46, I-40126, Bologna, Italy
[14] Institut für Quantenphysik, Universität Ulm, Albert-Einstein-Allee 11, D-89081 Ulm, Germany
[15] Institut für Laserphysik, Universität Hamburg, Luruper Chaussee 149, 22 167 Hamburg, Germany
[16] Institut für Angewandte Physik, Technische Universität Darmstadt, Hochschulstr. 4a, D-64289 Darmstadt, Germany

E-mail: `sorrentino@fi.infn.it`

Abstract. This paper presents the current status and future prospects of the Space Atom Interferometer project (SAI), funded by the European Space Agency. Atom interferometry provides extremely sensitive and accurate tools for the measurement of inertial forces. Operation of atom interferometers in microgravity is expected to enhance the performance of such sensors. Main goal of SAI is to demonstrate the possibility of placing atom interferometers in space. The resulting drop-tower compatible atom interferometry acceleration sensor prototype is described. Expected performance limits and potential scientific applications in a micro-gravity environment are also discussed.

1. Introduction

Matter-wave interferometry has recently led to the development of new techniques for the measurement of inertial forces, with important applications both in fundamental physics and applied research. The remarkable stability and accuracy that atom interferometers have reached for acceleration measurements can play a crucial role for science and technology. Quantum sensors based on atom interferometry had a rapid development during the last decade and different measurement schemes were demonstrated and implemented. Atom interferometry is used for precise measurements of the gravitational acceleration [1, 2, 3], Earths gravity gradient [4, 5], and rotations [6, 7]. Experiments on the validity of the equivalence principle [8] and on the measurement of the gravitational constant G [9, 10, 11] have been performed, while tests of general relativity [12] and of Newton's $1/r^2$ law [13, 14] as well as the detection of gravitational waves [15, 16] have been proposed. Accelerometers based on atom interferometry have been developed for many practical applications including metrology, geodesy, geophysics, engineering prospecting and inertial navigation [5, 17, 18]. Ongoing studies show that the space environment will allow us to take full advantage of the potential sensitivity of atom interferometers [19, 20].

Space Atom Interferometer (SAI) is a project of the European Space Agengy (ESA contract n. 20578/07/NL/VJ, AO-2004-064/082), with the main goal to demonstrate the possibility of applying such technology to future space missions.

In analogy to optical interferometers, atomic matter-waves can be split and recombined giving rise to an interference signal. Different schemes have been used for splitting, reflecting and recombining atomic matter waves. In the SAI project, the Raman interferometer is chosen as the basic configuration [1]. This approach was proven to be the most succesful so far in atom interferometry, at least for classical atom sources with respect to applications in precision experiments. Raman transitions change the internal hyperfine state and at the same time the external momentum state of the atom by 2 photon recoils. A sequence of three atom-light interaction processes with time separation T can be applied to coherently split, redirect and recombine the atomic de Broglie wave. After the first beam splitter two atomic de Broglie waves emerge. Their relative momenta differ by the photon recoil transferred to the atom. Choosing the duration of the atom-light interaction properly one may (as in the first and the third zone) equally split the incident waves (so called $\pi/2$ pulse), or (as in the second zone) deflect entirely the incident waves (π pulse). The two exit ports of the interferometer could be addressed and detected separately by means of spectroscopic techniques, like the excitation by laser and detection of the fluorescence.

The sensitivity of such atomic sensors depends on the free evolution time during which the matter wave accumulates phase along the interferometer trajectories. The output phase of a Raman pulse matter-wave accelerometer is

$$\phi = kaT^2 \tag{1}$$

where k is the wavevector associated to the Raman transitions ($1.6 \times 10^7\,\mathrm{m}^{-1}$ in Rb), a is the acceleration and T is the time interval between Raman pulses [2]. The phase resolution $\Delta\phi$ depends on the signal-to-noise ratio (SNR) at detection; for high enough number of atoms, detection is limited by quantum projection noise (QPN) and the phase resolution per shot is $\delta\phi \approx 1/\sqrt{N}$, thus the acceleration resolution per shot is

$$\delta a = \frac{1}{\sqrt{N}kT^2}. \tag{2}$$

Laser cooling is required to reach lower temperatures for a maximum extension of the free evolution time T as well as a high efficiency to achieve large numbers, N, of cold atoms. In absence of gravity the duration of the interaction is limited by the drift time of the atoms through the experimental apparatus due to thermal diffusion, which could be in principle tens of seconds for the lowest atomic temperatures (nK) presently reached. On Earth, gravity is the major problem for cold atom optics: the free fall of about 5 m in one second restricts the measurement time well below 1 s. Sensitivities of $\sim 10^{-8}\,\mathrm{g}/\sqrt{\mathrm{Hz}}$ to acceleration and of $\sim 0.5\,\mathrm{nrad\,s^{-1}}/\sqrt{\mathrm{Hz}}$ to rotations have been demonstrated, which are already at the level of the best optical interferometry sensors. A huge sensitivity improvement is expected in microgravity, where T can be increased by more than one order of magnitude.

2. The SAI project

The main goal of the SAI project is to demonstrate the technology readiness of atom interferometry inertial sensors for space applications. For this purpose, a compact prototype of a single-axis accelerometer based on ultracold ^{87}Rb has been realized (see section 2.1). At the same time, the project investigated new schemes, based for example on quantum degenerate gases as source for the interferometer (see section 2.2).

2.1. Transportable atom interferometry sensor prototype

The design of the SAI sensor prototype has been described in detail in [22]. The prototype is a vertical accelerometer, and allows both operation in microgravity and in ground tests. Design target was to keep the prototype as compact as possible without degradation of sensitivity as compared to existing laboratory instruments. In ground tests, a typical measurement cycle with interferometer duration $2T \sim 100$ ms, repetition rate ~ 5 Hz and QPN-limited detection of $\sim 10^6$ atoms per cycle (SNR ~ 1000) would result in virtual acceleration sensitivity of $10^{-8}\,\mathrm{m/s^2}$ at 1 s. In microgravity experiments, a typical measurement cycle with interferometer duration $2T \sim 2$ s, repetition rate ~ 0.3 Hz and QPN-limited detection of $\sim 10^6$ atoms per cycle would result in virtual acceleration sensitivity of $5 \cdot 10^{-11}\,\mathrm{m/s^2}$ at 1 s.

A modular laser system is connected through optical fibers to the sensor head consisting of an ultra-high vacuum system with attached optics and fiber connectors, surrounded by μ-metal shields to control stray magnetic fields (see fig. 1). In order to demonstrate the possibility of application in microgravity, the sensor was designed to fit into a standard capsule of the Bremen drop tower. Most of the technological solutions have already proved to be compatible with the acceleration loads of the Bremen drop tower [23]. The mechanical structure holding the vacuum system and all necessary optics is contained in a cylindrical volume with diameter ~ 60 cm and height ~ 150 cm. Fig. 2 shows the SAI sensor assembly.

The same vacuum chamber is used for both trapping and cooling the atoms and to detect them at the end of the interferometric sequence. This scheme is mandatory to demonstrate possible future operation in microgravity, where the relative velocity and the rotation between the atoms and the chamber will be ideally null. On the other hand, the possibility of launching the atomic sample is important for ground testing of the instrument. The implementation of a 2D magneto-optical trap (MOT) [21] as source of cold atoms for a standard 3D-MOT ensures ultra-high vacuum conditions in the loading and interferometer regions, important for operating the sensor with coherent atomic sources such as a Bose-Einstein Condensate (BEC) and for testing its performance at long interrogation times.

A short vertical tube above the main chamber allows to tune the interaction time and to operate the sensor as a gravity gradiometer for characterization and performance measurements. Since the interferometer sequence has to happen close or within the main chamber, magnetic shielding of the tube would not be sufficient. A double layer of μ-metal shield encloses the whole vacuum system. The mechanical structure, including magnetic shield, is qualified for drop tower experiments. The mounting of μ-metal layers is rigid enough to withstand 50 g accelerations. In order to find the optimal shape for the magnetic shield, we implemented a numerical model based on a commercial software (ANSYS). According to our calculations, the magnetic field would be

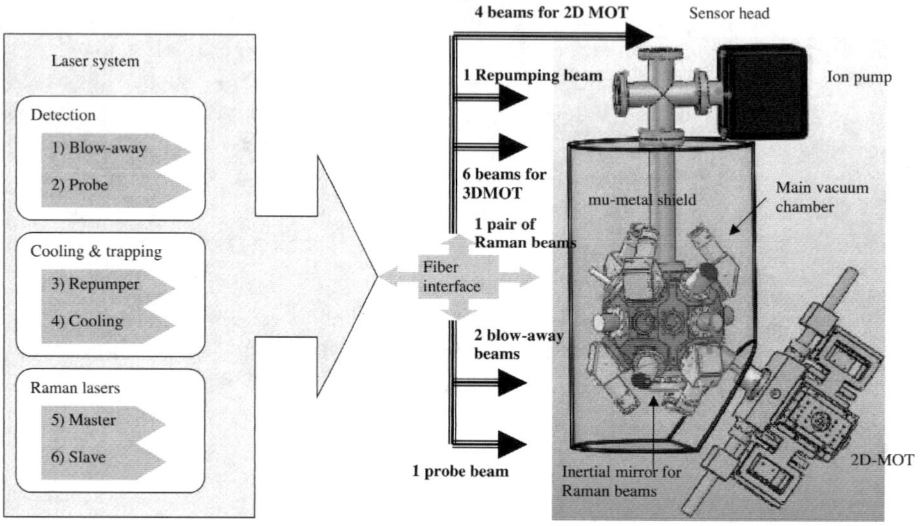

Figure 1. Scheme of the SAI sensor; the main vacuum chamber is employed for 3D-MOT, interferometer sequence and detection; it is separated from the 2D-MOT chamber by a differential pumping tube: background vacuum pressure is of the order of 10^{-9} torr in the main chamber and $10^{-6} \div 10^{-7}$ torr in the 2D-MOT chamber.

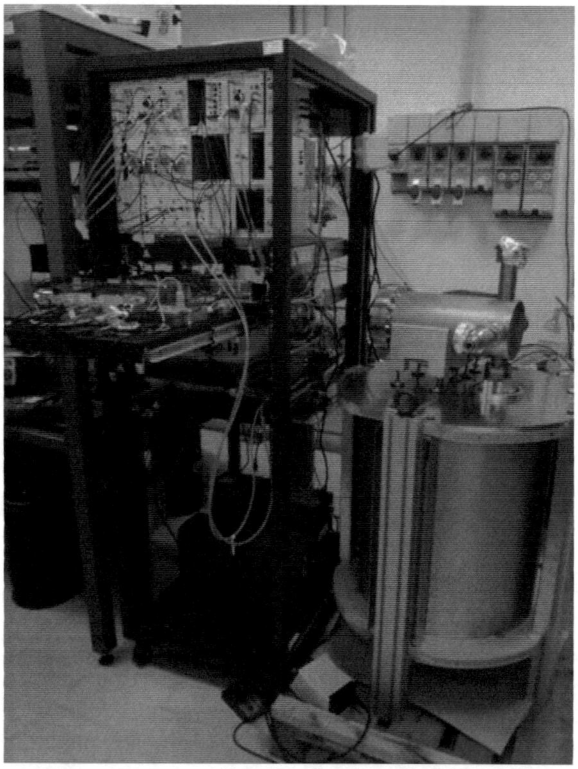

Figure 2. A picture of the SAI sensor assembly; the rack on the left contains the electronics, power supplies and laser modules; the sensor head on the right contains the vacuum system with attached optics enclosed in the double-layer magnetic shield; the mechanical structure of the sensor head is the central cut of a standard drop tower capsule.

Journal of Physics: Conference Series **327** (2011) 012050 doi:10.1088/1742-6596/327/1/012050

Figure 3. A picture of the 2D-MOT system and main vacuum chamber separated by a dual μ-metal layer.

uniform in time and in space within 30 nT along the atom interferometry region during a free fall in the drop tower. The same analysis has been used to find a configuration of coils to produce a uniform bias field in the region where the atom interferometry sequence takes place. The overall mass of the magnetic shield will be about 85 kg.

The 2D-MOT for the Space Atom Interferometer (see fig. 3) is a mixture of the design implemented in the CASI experiment [35] for cold-atom Sagnac interferometer and the ATLAS experiment [36] for an all optical produced Bose-Einstein condensate.

The main vacuum chamber is shown in fig. 3. It has 24 view-ports giving access for cooling laser beams, 2D-MOT atomic flux, laser beams for optical BEC, and detection optics.

The vacuum system is contained in the central part of the drop-tower capsule, between two horizontal platforms that are connected by four vertical stringers. All kind of implemented components are rigidly mounted on the capsule platforms to withstand decelerations up to 50 g during the impact of the drop capsule at the end of the free fall.

The laser system for cooling and detecting Rb atoms, and to generate the laser beams for Raman transitions, is split into six compact modules as shown in fig. 4. In the master module, a laser is stabilised on the Rb FM spectroscopy signal (see fig. 5) and sent to a second module, where it is amplified. This light is used as a reference for an offset-lock in the repumper module and in the Raman laser system; in addition, it is also forwarded to the 2D-MOT and 3D-MOT laser modules. The two MOT modules shift the master frequency to the cooling frequency and amplify it, in addition allowing a control of the laser detuning from resonance and of laser power. Finally, they distribute the cooling and repumping light onto several outputs. Each module was designed and optimised for compactness and robustness. The mounts had already been developed and tested under drop-tower conditions in the framework of the QUANTUS project [23].

The Raman laser optical setup contains two Extended Cavity Lasers (ECLs). The Raman master laser is overlapped on a photodiode with the reference laser. The resulting beat note signal is locked on a 100 MHz reference provided by an auxiliary output of the frequency synthesis

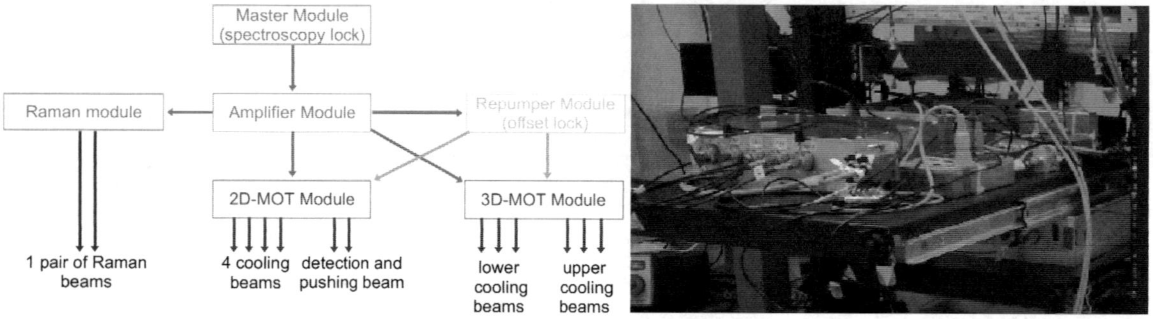

Figure 4. Left: scheme of the modular laser system; right: laser modules mounted on a 19″ rack.

chain described below. The second ECL (slave) is overlapped with the master laser and their beat note is phase locked to the low phase noise signal at 6.8 GHz (hyperfine splitting of ^{87}Rb) provided by the frequency synthesis chain. Since the phase difference between Raman master and slave lasers is imprinted directly on the atoms, phase noise performance of both the frequency reference and the optical phase locked loop are of key importance for the sensor performance. The phase noise spectral density for the optical phase-locked loop (OPLL) stays below a level of -120 dBrad^2/Hz (1 μrad/$\sqrt{\text{Hz}}$) between 100 Hz and 60 kHz (see fig. 5). To our knowledge, the measured OPLL phase noise is among the lowest levels ever reached with diode lasers [25]. If the OPLL phase noise were the main contributing factor to overall accelerometer sensitivity, a single-shot sensitivity $\delta a/a$ of 2×10^{-10} would be achieved.

A microwave frequency source is used to generate the reference for the OPLL of the Raman lasers. The measured phase noise of the 6.8 GHz signal is shown in fig. 5. The contribution of the microwave source onto the phase noise of the interferometer is reduced down to about 1 mrad/shot for $2T = 100$ ms, using commercial ultra low noise quartz oscillators. When extrapolating the performances of an atom interferometer to larger interrogation times ($2T \leq 1$ s), the phase noise at low frequency will be the dominant contribution. Assuming a phase noise scaling as $1/f^2$ for frequencies below 10 Hz, the expected sensitivity per shot of the interferometer scales linearly with T. We calculate an interferometer phase noise of 9 mrad/shot for $T = 1$ s. As for the microwave frequency synthesis, the contribution of the quartz can be reduced to less than 1 mrad/shot by a careful design, independent of T.

The SAI system employs a versatile source of signals able to control the different phases of a cold atom experiment. It is based upon a series of parallel modules, generating parallel signals with only one clock. The output of all modules is generated in a Field Programmable Gate Array (FPGA). The control system is integrated in a stand-alone 19" rack (see fig. 2).

Baseline design of the SAI sensor includes integration of a coherent atomic source. Using a BEC will enable a much longer interaction time when operating the atom interferometer in microgravity. All-optical BEC has been implemented in a separate experiment using a fiber laser at 1565 nm [26], and the SAI system has been designed to be compatible with the production of a coherent source using such technique. The additional subsystems for the optical BEC consist of a commercial 1565 nm high-power, continuous-wave (CW) fiber laser and of two telescopes, which are attached to the main chamber and produce a pair of focused beams crossing at the position of the MOT. We also studied the possibility to obtain a BEC with an hybrid trap, i.e. by employing a single-beam optical trap in combination with a magnetic quadrupole to increase the axial confinement. Run-away evaporation is obtained by lowering the power of the optical trap at constant magnetic gradient [37].

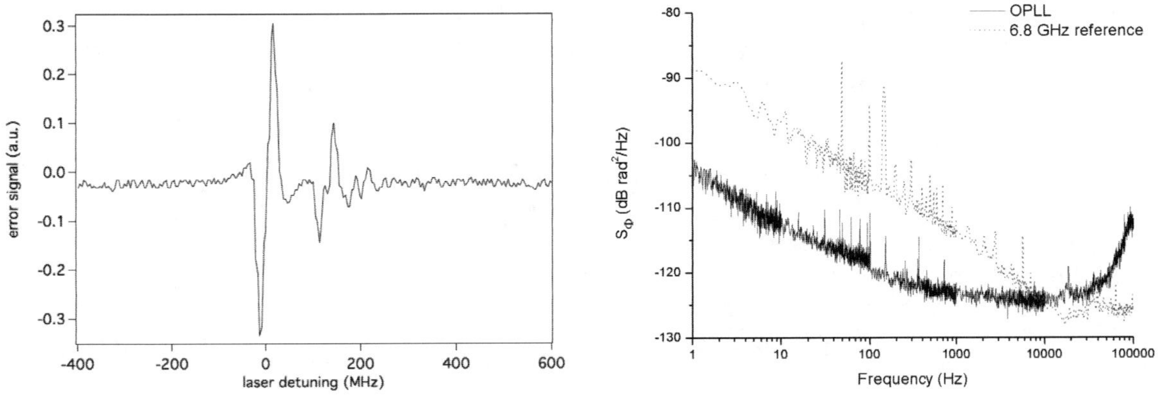

Figure 5. Left: saturated spectroscopy signal from the master laser module on 1 MHz bandwidth; the main line corresponds to the $F = 2 \rightarrow F' = 3$ component of the D2 transition; linewidth is 30 MHz; SNR=1 MHz; right: phase noise spectral density of the reference frequency chain and of the OPLL (from [22]).

2.2. Theoretical studies and alternative schemes

2.2.1. Sensor modeling We studied the different sources of noise and systematic shifts that may limit the sensitivity and the accuracy of the atomic sensors. Such effects may be split into two categories, depending on whether their impact is related to the external degrees of freedom of the atoms.

Effects which do not depend on the external degrees of freedom are the time fluctuations of the phase difference between the Raman lasers, of their intensity, of the magnetic field, of acceleration and rotation. The study is based on the use of the sensitivity function first developed for atomic clocks and that has been modified for atomic interferometers [27]. The calculated sensitivity function in time and frequency domain was experimentally validated on the SYRTE atom gravimeter, showing excellent agreement with measured data [28, 27, 29]. This method has been used to determine the limit due to residual phase noise in the phase locked loop of the Raman lasers. Such calculations have also been important in the design of the reference frequency chain. The results show that a sensitivity of 1 mrad per shot in the interferometer phase is achievable. We applied the same method to study the sensitivity to the frequency fluctuations of the master Raman laser. Calculating the impact of this effect with the sensitivity function yields a limit due to the frequency noise of the Raman laser to 2.4 mrad per shot for an optical path difference of 93 cm between the counterpropagating Raman beams [28]; such result, which is again in good agreement with experimental data, shows the need of a master Raman laser with a narrow linewidth or a small distance between the center of mass of the atomic cloud and the retro-reflecting Raman mirror. The sensitivity function method was also employed to show that the effect of fluctuations of the power of the Raman lasers and time fluctuation of the magnetic field is negligible [29]. The same method can be used to estimate noise sources from vibration and rotation noise [27]. The vibration noise is clearly the limit of the sensitivity on ground and can be overcome by using the signal from a mass accelerometer [28]. Moreover, the use of the auxiliary accelerometer enables us to go beyond the linear range (with acceleration phase smaller than 1 rad), thereby providing the possibility to work in a noisy environment, such as for instance on a platform in a low Earth orbit without a drag free system.

Effects depending on the external degrees of freedom of the atoms, cannot be treated in this formalism. Four effects have been identified: wave front curvature of the Raman lasers, light shift gradients (if not perfectly cancelled), magnetic field gradients and cold atom collisions.

Journal of Physics: Conference Series **327** (2011) 012050 doi:10.1088/1742-6596/327/1/012050

The effect of magnetic field gradients can be reduced by more than two orders of magnitude by exchanging the internal state at the input of the interferometer. By averaging the measurements using the two internal states successively at the input of the interferometer the effect can be reduced to below 10^{-9} g [30]. A non-magnetic vacuum chamber and careful control of external magnetic fields are therefore important to keep this effect below the instrument accuracy. In the geometry of retro-reflected Raman beams, the influence of off-resonant Raman transitions must be taken into account in order to achieve best accuracy and stability of interferometers [31]. The two photon light shift can be drastically reduced by increasing the Doppler effect and/or using colder atoms, allowing to reduce the Rabi frequency during the Raman pulses. In contrast, this effect becomes extremely large for a set-up with intrinsic small Doppler effect, as expected for space applications, and has to be taken into account in the design of the experiment

2.2.2. Advanced schemes The interest in applying atomic quantum gases to atom interferometry resides primarily in their coherence properties, which makes them more appealing than the thermal gases used so far in state-of-the-art interferometers. First, the naturally narrow momentum distribution of degenerate quantum gases represents an advantage with respect to a thermal gas produced in a standard MOT, where an additional selection stage is necessary to produce the required sub-recoil samples Second, the coherence of a quantum gas is the natural starting point to implement quantum techniques to improve the sensitivity of the interferometer, by the use of squeezing and/or entanglement, beyond the standard quantum limit. In this respect Bose gases have a strong advantage over Fermi gases. The kinetic energy of an ideal, i.e. non interacting, Bose gas at zero temperature is indeed of the order of $h\nu$, where h is the Planck constant and ν is the mean oscillation frequency of the trapping potential used to produce and initially store the gas before feeding it to the interferometer. For a Fermi gas in 3D, this quantity grows as $h\nu N^{1/6}$ due to the Pauli exclusion principle, where N is the number of particles in the sample. On the other hand, ultracold Fermi gases of identical atoms are intrinsically ideal because of the Pauli principle, while Bose gases are typically interacting. The presence of the interaction between atoms results in a strong broadening of the momentum distribution and in detrimental phase-shift and phase-diffusion effects. A first milestone in the very promising direction of employing Bose and Fermi gases to advanced atom interferometry is assessing which of the two systems is more appropriate for various kinds of applications. In a first investigation of the potentialities of Fermi gases for atom-interferometry applications, a Bloch-oscillation interferometer was tested on a gas of ^{40}K atoms trapped in an optical lattice [32]. The standard Fermi gas employed had a momentum width of about 1/3 of the recoil momentum, and no evidence of phase diffusion due to the atom-atom interaction was visible. We have then studied the same kind of interferometer with a weakly interacting Bose gas of ^{39}K atoms by tuning the s-wave scattering length, or in other words the contact interaction, almost to zero by means of a magnetic Feshbach resonance [33]. The interaction-induced decoherence of the interferometer is reduced when reducing the scattering length from its natural value by as much as a factor of 1000. In a situation of trapped high-density samples, we have observed a reduction of the decoherence rate from about 100 Hz to below 1 Hz. This is accompanied by an effective momentum width that gets as small as approximately 1/20 of the recoil momentum. Unfortunately, this great improvement of the performances cannot be obtained for free, and the price to pay is to work in presence of a rather large magnetic field. Indeed, the Feshbach resonance of ^{39}K in the $F = 1$, $m_F = 1$ state, is between 350 G and 400 G, and the atoms have a magnetic moment of about $0.95\,\mu_B$. At least two solutions can be envisaged to attack this problem. With a trapped interferometer, there is in principle the possibility to work at a different Feshbach resonance in the ground state, for which the magnetic field is of the order of 10 G. Conversely, for free-fall interferometers, where the interaction-induced shift and diffusion are not the main limit to the sensitivity, one could work at a third resonance around 80 G

[33], where the magnetic moment is zero at a scattering length of about 10 Bohr radii. This would strongly reduce the requirements on the magnetic field gradient, at least in a Bragg-type interferometer.

Finally, when using a high density BEC as source for matter-wave interferometry, the interaction between the atoms can lead to phase diffusion and frequency shifts that arise from an inhomogeneous atomic density distribution. A systematic incorporation of these effects in the phase evolution of the BEC requires a detailed knowledge of the total particle number and the density distribution of the atomic cloud. However, the experimental realization of BECs is often accompanied with significant uncertainties in the particle number. For this reason, we explored the effect of asymmetric beam splitters within the matter-wave interferometer aiming towards a reduction of the uncertainty in the particle number $n \pm \Delta n$ of the prepared atomic ensemble. In particular, we have assumed that a coherent matter wave source in the form of a homogeneous ^{87}Rb BEC is fed into the input arm of the interferometer. Given that binary collisions in the degenerate quantum gas are the source of the collision phase shift, which is proportional to the particle density, we calculated the optimal working point of the interferometer.

We assume that all the atoms are initially in the ground state and denote with p the population of the excited state after the first beam splitter pulse. A detailed analysis [34] of the $p - \pi - \bar{p}$ interferometer pulse sequence shows that the particle number in the output channel is then given by

$$n_e^{out} = n\{\xi - \gamma \cos[2Tn\delta_2(p - 1/2)]\} \quad (3)$$

with the mixing parameters ξ, γ defined in terms of two beam splitting parameters p and $\bar{p}$ according to

$$\xi = p\bar{p} + (1-p)(1-\bar{p}) \text{ and } \gamma = 2\sqrt{p\bar{p}(1-p)(1-\bar{p})} \quad (4)$$

and with n as initial population of the ground state. The splitting parameters p and $\bar{p}$ are realized with Raman laser pulses of specific duration. The central difference $\delta_2 = g_{ee} - 2g_{eg} + g_{gg}$ just follows from the coupling constants g_{ee}, g_{eg} and g_{gg} which appear in the Gross-Pitaevskii equation and are directly connected to the individual scattering lengths of the two particle s-wave collisions. Moreover, we have numerically simulated the experimentally more interesting case of a BEC trapped within a harmonic potential (see fig. 6). In both cases we encounter a promising behavior of this matter-wave interferometer as a number filter. We successfully examined several situations in which the uncertainty in the particle number is found to be reduced due to binary collisions in the interferometer setup.

3. Atom interferometry sensors in microgravity: present and future

3.1. State of the art

At present, worldwide and especially in Europe, several atom interferometry inertial sensors are operating. Gravity measurements with sensitivity of $8 \cdot 10^{-8}$ m/s^2 at 1 s and accuracy better than $3 \cdot 10^{-8}$ m/s^2, gravity gradient measurements with sensitivity of $4 \cdot 10^{-8}$ m/s^2 at 1 s, and rotation rate measurements with sensitivity of $6 \cdot 10^{-10}$ rad/s at 1 s and bias stability $< 70\,\mu$deg/h have been already demonstrated in laboratory systems. Transportable systems are being developed with similar performance. Future advances in atom optics (large momentum transfer beam splitters, high-flux atomic sources, advanced detection schemes based for example on quantum degenerate gases) are expected to largely improve the performance of such devices. However, one major limitation to the sensitivity of light-pulse atom interferometry sensors on Earth is the short interaction time achieved, due to the free fall of atoms under gravity. An improvement of several orders of magnitude is expected in microgravity environment, since the sensitivity scales as the square of the interaction time.

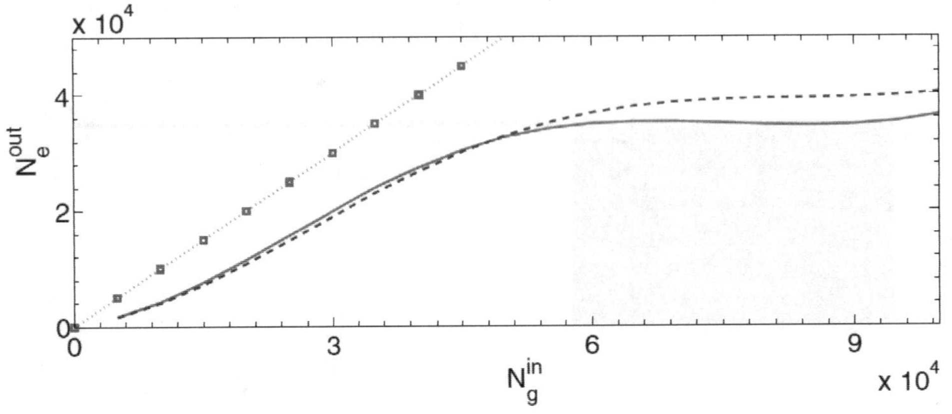

Figure 6. Output particle number N_e out vs the input particle number N_g in a harmonically trapped, two-component BEC after passing through an highly asymmetric interferometer with $p = 0.055$ and $\bar{p} = 0.5$. The (red) solid line displays the full numerical solution of the GP equation, while the (blue) dashed line represents the local density approximation of the homogeneous result. The gray shaded area illustrates the compression of number fluctuations. The (green) dotted line marked with squares is the trivial response of a $0 - \pi - 0$ interferometer (from [34]).

3.1.1. Technological developments for compact atom interferometers. Current research in the field of atom interferometry inertial sensing is proceeding along two main directions: on one side, to improve the instruments accuracy by pushing the sensitivity limits of laboratory systems with advanced atom optics techniques; at the same time, several groups are working to reduce the size, weight and power consumption without degradation of performances, in order to make transportable atomic sensors for both terrestrial and space applications. The developments in the field of commercial optics have produced impressive improvements in size, mass, stability and reliability of lasers and other optoelectronic components. Today, a complete laser system for a rubidium atom interferometer can nearly be assembled with off-the-shelf commercial components of compact size and high reliability. The industrial basis for further developing these commercial laser systems into engineering models is thus well established. Compact and trasportable atom interferometry absolute gravimeters have been developed for geophysical and metrological applications. The group of Kasevich at Stanford has developed a transportable instrument for a full determination of the gravity gradient tensor for inertial navigation [41]. At JPL in Pasadena, a compact gravity gradiometer has been realized with interesting new solutions for compactness and robustness of the apparatus [42]. The SYRTE gravimeter in Paris, intended to provide an absolute measurement of g in the context of a Watt balance experiment, is a compact laboratory apparatus designed to be transportable for comparison campaigns [43]. The atom gravimeter in Berlin, developed within the FINAQS collaboration, is a transportable apparatus for geophysical applications [44]. A second generation of compact atomic sensors is expected to yield a drastic reduction in size by replacing the current technology of free-air optoelectronics with either integrated micro-optics or fiber-based sources. Along this line, the EC STREP program iSense was started in 2010 with the main goal to develop terrestrial small-scale atomic quantum sensors.

3.1.2. Current activities for atom interferometers in microgravity. Many technology developments are presently on-going, both at ESA and at National Space Agencies, to bring atom interferometry instruments to maturity for space applications. Within the European project Space-BEC: Quantum Gases in Microgravity, intensive activities are carried on under national funding. The parabolic flight experiments presently ongoing in the ICE project (Interferometry with Coherent Sources for Applications in Space) of the French space agency CNES, and the drop tower tests conducted by the German space agency DLR in the QUANTUS project, have demonstrated the robustness of miniaturised ultra-cold atom technology. Indeed, atom interference fringes have been detected by ICE in the noisy environment of the Zero-G parabolic flight; at the same time, the QUANTUS set-up is able to routinely produce Bose-Einstein Condensates (BEC) during the about 100 m of free fall in the Bremen drop tower and survives decelerations as high as 50 g. It is expected that the use of ultra-cold gases as sources for atom interferometry, combining the progress made with SAI, ICE and QUANTUS, will provide European scientists with unique sensor capabilities for investigations in fundamental physics.

3.2. Evolution of the SAI project

QUANTUS and ICE have been the first experiments demonstrating atom interferometry and BEC in microgravity; the space environment will allow to fully exploit the potentials of quantum sensors by overcoming the limitations of parabolic flights (large acceleration noise) and drop towers (low repetition rate). Evolution of SAI towards a space mission, either on the International Space Station (ISS) or on a dedicated satellite, would have the main goal to demonstrate the operation of a space atom interferometer with interaction time $\sim 1\,\mathrm{s}$ and acceleration sensitivity $10^{-10} \div 10^{-11}\,\mathrm{m/s^2}$ at $1\,\mathrm{s}$. The experiment would implement a dual-species differential measurement using either the ^{87}Rb-^{85}Rb, or the K-Rb pair, and would demonstrate differential acceleration measurement to better than $10^{-14}\,\mathrm{m/s^2}$.

Continuation of the SAI activity between 2011 and 2014 would have the primary goal to develop and test an advanced breadboard prototype, and to develop and test engineering models of the components. In a future evolution of this project, the ground-based prototype could be operated in the Bremen drop-tower laboratories for first tests under microgravity conditions, and/or in underground, seismically quiet laboratories for ground tests and comparison with other acceleration sensors. The SAI ground-based prototype will be important for the definition of the space instrument design and for studying the ultimate limits of an atom interferometer for acceleration measurements.

3.2.1. Scientific objectives: tests of fundamentel physics A first outcome of the space mission would be a test of the Equivalence Principle at the 10^{-15} level, thus improving the current experimental limits by two orders of magnitude [38, 39]. In this case the measurement will be implemented on two different atomic clouds (e.g. ^{87}Rb and ^{85}Rb) on the same spatial trajectory. A Quantum test of the Weak Equivalence Principle (Q-WEP) on the ISS can be considered as the natural joined evolution of SAI and Space BEC. A WEP test is one of the priorities identified in the ESA Roadmap for Fundamental Physics in Space. Following the results of the study on "Atom Interferometry Test of The Weak Equivalence Principle in Space", recently approved by ESA in the frame of the General Study Programme (GSP) and to be started in the second quarter of 2011, Q-WEP will eventually be proposed for a phase A study to be conducted in the frame of the ELIPS programme.

Additionally, the experiment would be able to provide demonstrative tests of fundamental physics in microgravity [40]:

- test of quantum gravity effects: measurement of coherence time in atom interferometers with a BEC might probe models suggesting fundamental decoherence mechanisms arising from Planck scale physics; observation of wave-packet spreading with a BEC in microgravity would represent a test of scenarios with fluctuating space-time; precision measurements of the photon recoil might be used to perform tests of modified energy-momentum relations, which may arise as a possible consequence of a quantum gravity theory;

- test of quantum mechanical effects: a search for possible additional phase shift terms in the atom interferometer response would test non-linearities in the Schrödinger equation, i.e. violations of the superposition principle;
- test of the neutrality of matter: by applying electrical potentials to the atoms in either arm of the interferometer would allow to set upper limits to the neutron charge and to test the equality of the proton and electron charge.

3.2.2. Scientific objectives: other applications On the other hand, operation with two clouds having a spatial separation of $\sim 1\,\mathrm{m}$ can provide a measurement of the local gravity gradient with a sensitivity of $0.1\,\mathrm{E}/\sqrt{\mathrm{Hz}}$. Such result would demonstrate the possible use of atom interferometry based inertial quantum sensors for the exploration of the gravitational field and the mapping of gravity gradients of celestial bodies. The target SAI sensitivity would be better than the expected GOCE performance below 10 mHz by one order of magnitude. Moreover, the differential acceleration measurement at $\sim 10^{-11}\,\mathrm{m\,s^{-2}}/\sqrt{\mathrm{Hz}}$ on spatially separated atomic clouds could represent a first demonstrative experiment for the possible detection of low frequency gravitational waves.

4. Conclusions

The SAI sensor, a compact and transportable prototype of atom interferometry accelerometer, was buildt to demonstrate the possibility to apply such technology in future space missions. The Space Atom Interferometer project will provide a valuable insight into the potential of atom interferometry for space applications.

4.1. Acknowledgments

This work was supported by ESA through the SAI project (contract no. 20578/07/NL/VJ).

References

[1] M. Kasevich and S. Chu. *Appl. Phys. B* **54**, 321 (1992).
[2] A. Peters, K. Y. Chung, and S. Chu. *Nature* **400**, 849 (1999).
[3] H. Müller, S. w. Chiow, S. Herrmann, S. Chu, and K.-Y. Chung. *Phys. Rev. Lett.* **100**, 031101 (2008).
[4] M. J. Snadden, J. M. McGuirk, P. Bouyer, K. G. Haritos, and M. A. Kasevich. *Phys. Rev. Lett.* **81**, 971 (1998).
[5] J. M. McGuirk, G. T. Foster, J. B. Fixler, M. J. Snadden, and M. A. Kasevich. *Phys. Rev. A* **65**, 033608 (2002).
[6] T. L. Gustavson, P. Bouyer, and M. Kasevich. *Phys. Rev. Lett.* **78**, 2046 (1997).
[7] T. L. Gustavson, A. Landragin, and M. Kasevich. *Class. Quantum Grav.* **17**, 2385 (2000).
[8] S. Fray, C. Alvarez Diez, T. W. Hänsch, and M. Weitz. *Phys. Rev. Lett.* **93**, 240404 (2004).
[9] M. Fattori, G. Lamporesi, T. Petelski, J. Stuhler, and G. M. Tino. *Phys. Lett. A* **318**, 184 (2003).
[10] J. B. Fixler, G. T. Foster, J. M. McGuirk, and M. Kasevich. *Science*, **315**, 74 (2007).
[11] G. Lamporesi, A. Bertoldi, L. Cacciapuoti, M. Prevedelli, and G. M. Tino. *Phys. Rev. Lett.* **100**, 050801 (2007). F. Sorrentino, L. Cacciapuoti, Y.-H. Lien, M. Prevedelli, G. Rosi, and G. M. Tino. *New J. Phys.* **12**, 095009 (2010).
[12] S. Dimopoulos, P. Graham, J. Hogan, and M. Kasevich. *Phys. Rev. Lett.* **98**, 111102 (2007).
[13] G. M. Tino. 2001: A relativistic spacetime odyssey. In I. Ciufolini, D. Dominici, and L. Lusanna, editors, *High precision gravity measurements by atom interferometry - Proceedings of JH Workshop, Firenze 2001.* World Scientific, 2003.
[14] G. Ferrari, N. Poli, F. Sorrentino, and G. M. Tino. *Phys. Rev. Lett.* **97**, 060402 (2006).
[15] G. M. Tino and F. Vetrano. *Class. Quantum Grav.* **24**, 2167 (2007).
[16] S. Dimopoulos, P.W. Graham, J.M. Hogan, M.A. Kasevich, and S. Rajendran. *Phys. Lett. B* **678**, 37 (2009).
[17] A. Peters, K. Y. Chung, and S. Chu. *Metrologia* **38**, 25 (2001).
[18] A. Bresson, Y. Bidel, P. Bouyer, B. Leone, E. Murphy, and P. Silvestrin. *Appl. Phys. B* **84**, 545 (2006).

[19] G. M. Tino *et al.* *Nuclear Physics B (Proc. Suppl.)* **166**, 159, (2007).
[20] Atom Optics and Space Physics, Proceedings of the International School of Physics "Enrico Fermi", Course CLXVIII, edited by E. Arimondo, W. Ertmer, E.M. Rasel, and W.P. Schleich, IOS Press, Amsterdam, 2009
[21] K. Dieckmann, R. J. C. Spreeuw, M. Weidemüller and J. T. M. Walraven *Phys. Rev. A* **58**, 3891 (1998).
[22] F. Sorrentino, K. Bongs, P. Bouyer, L. Cacciapuoti, M. de Angelis, H. Dittus, W. Ertmer, M. Hauth, S. Herrmann, M. Inguscio, E. Kajari, T. Könemann, C. Lämmerzahl, A. Landragin, G. Modugno, F. Pereira dos Santos, A. Peters, M. Prevedelli, E. M. Rasel, W. P. Schleich, M. Schmidt, A. Senger, K. Sengstok, G. Stern, G. M. Tino, R. Walser. *Microgravity Sci. Technol.* **22**, 551 (2010).
[23] T. van Zoest, N. Gaaloul, Y. Singh, H. Ahlers, W. Herr, S. T. Seidel, W. Ertmer, E. Rasel, M. Eckart, E. Kajari, S. Arnold, G. Nandi, W. P. Schleich, R. Walser, A. Vogel, K. Sengstock, K. Bongs, W. Lewoczko-Adamczyk, M. Schiemangk, T. Schuldt, A. Peters, T. Könemann, H. Muntinga, C. Lämmerzahl, H. Dittus, T. Steinmetz, T. W. Hänsch, and J. Reichel. *Science* **328**, 1540 (2010).
[24] X. Baillard, A. Gauguet, S. Bize, P. Lemonde, P. Laurent, A. Clairon and P. Rosenbusch, *Opt. Coom.* **266**, 609 (2006).
[25] M. Schmidt, M. Prevedelli, A. Giorgini, G.M. Tino and A. Peters *Appl. Phys. B* **102**, 11 (2011).
[26] J. F. Clément, J. P. Brantut, M. Robert de Saint Vincent, G. Varoquaux, R. A. Nyman, A. Aspect, T. Bourdel and P. Bouyer. *Phys. Rev. A* **79**, 61406 (2009).
[27] P. Cheinet, B. Canuel, F. Pereira Dos Santos, A. Gauguet, F. Leduc and A. Landragin. *IEEE Trans. on Instrum. Meas.* **57** ,1141 (2008).
[28] J. Le Gouët, P. Cheinet, J. Kim, D. Holleville, A. Clairon, A. Landragin, and F. Pereira Dos Santos. *Eur. Phys. J. D* **44**, 419 (2007).
[29] J. Le Gouët, T.E. Mehlstäubler, J. Kim, S. Melet, A. Clairon, A. Landragin, and F. Pereira Dos Santos. *Appl. Phys. B* **92**, 133 (2008).
[30] A. Landragin, and F. Pereira Dos Santos. Accelerometer using atomic waves for space applications. *Atom Optics and Space Physics, Proceedings of the Enrico Fermi International School of Physics "Enrico Fermi"*, Course CLXVIII, Varenna, 2007.
[31] A. Gauguet, T.E. Mehlstäubler, T. Lvque, J. Le Gouët, W. Chaibi, B. Canuel, A. Clairon, F. Pereira Dos Santos, and A. Landragin. *Phys. Rev. A* **78**, 043615 (2008).
[32] G. Roati, E. de Mirandes, F. Ferlaino, H. Ott, G. Modugno, and M. Inguscio. *Phys. Rev. Lett.* **92**, 230402 (2004).
[33] C. D'Errico, M. Zaccanti, M. Fattori, G. Roati, M. Inguscio, G. Modugno, and A. Simoni. *New J. Phys.* **9**, 223 (2007).
[34] G. Nandi, A. Sizmann, J. Fortágh, C. Weiss, and R. Walser. *Phys. Rev. A* **78**, 013605 (2008).
[35] http://www.iqo.uni-hannover.de/ertmer/casiindex/
[36] http://www.iqo.uni-hannover.de/finaqs/index.php?n=WP1.IPulsedBosonicSource
[37] M. Zaiser, J. Hartwig, D. Schlippert, U. Velte, N. Winter, V. Lebedev, W. Ertmer, and E. M. Rasel. *Phys. Rev. A* **83**, 035601 (2011).
[38] S. Schlamminger, K.-Y. Choi, T. A. Wagner, J. H. Gundlach, and E. G. Adelberger. *Phys. Rev. Lett.* **100**, 041101 (2008).
[39] J. G. Williams, S. G. Turyshev, and D. H. Boggs. *Phys. Rev. Lett.* **93**, 261101 (2004).
[40] S. Herrmann, E. Göklü, H. Müntinga, A. Resch, T. van Zoest, H. Dittus, C. Lämmerzahl *Microgravity Sci. Technol.* **22**, 529 (2010).
[41] B. Young, D. S. Bonomi, T. Patterson, F. Roller, T. Tran, A. Vitouchkine, T. Gustavson, and M. Kasevich Atom optic inertial and gravitational sensors. *Laser Science, OSA Technical Digest (USA: Optical Society of America)*, LTuH1, 2007.
[42] N. Yu, J. M. Kohel, J. R. Keollogg, and L. Maleki. *Appl. Phys. B* **84**, 647 (2006).
[43] S. Merlet, Q. Bodart, N. Malossi, A. Landragin, F. Pereira Dos Santos, O. Gitlein, and L. Timmen *Metrologia* **47**, L9 (2010).
[44] M. Schmidt, A. Senger, M. Hauth, C. Freier, V. Schkolnik, and A. Peters *Gyroscopy and Navigation* **2**, 170 (2011).

ATLAS-M and Batt-M: Development of Flight Hardware for MAPHEUS Sounding Rocket

G Blochberger, J Drescher, C Neumann, P Penkert, A Griesche, F Kargl, A Meyer

Institut für Materialphysik im Weltraum, Deutsches Zentrum für Luft und Raumfahrt (DLR, 51170 Köln, Germany

E-mail: `guido.blochberger@dlr.de`

Abstract. Two modules ATLAS-M and Batt-M were especially developed for application in the MAPHEUS (MAterialPHysikalische Experimente Unter Schwerelosigkeit) sounding rocket campaign. They were manufactured and built at the Institute of Material Physics in Space at the German Aeropsace Center (DLR). ATLAS-M is a furnace for short time diffusion measurements achieving heating rates of 7 K/s and cooling rates of 6 K/s. Atlas-M is well suited for investigation of interdiffusion coefficients on a time scale of <200 s. Batt-M is a rechargeable battery module, supplying a peak power of 12.6 kW with a nominal capacity of 4.6 Ah for operation.

1. Introduction

The main challenge for building scientific flight hardware on sounding rockets is to find the best compromise between the constructional and environmental constraints. Compromises have to be made concerning installation space, power consumption, overall weight, environmental conditions and available hardware components. The scientific payload has to fulfill these demands required for fully automated and reliable operation aboard. For Cu self diffusion quasi neutron scattering (QNS) delivers reliable and accurate results [4]. For binary and multicomponent alloys also chemical diffusion occurs. For binary alloys, the measurement of interdiffusion coefficients with QNS is typically not possible. Binary alloys require capillary experiments. These capillary experiments are affected by a number of problems. These problems may be overcome by x-ray tomography [5]. Microgravity is used to surpress bouancy driven convective flow. For these μg experiments, the development of a furnace module for the sounding rocket MAPHEUS [1] was required. The MAPHEUS project aims to research material properties under microgravity using a two-staged sounding rocket. Therefore, the ATLAS-M (*A*tomic *T*ransport in *L*iquid *A*lloys and *S*emiconductors on *M*APHEUS) module was developed, processing a large number of samples simultaneously. For this purpose the Batt-M module was built, allowing a high power consumption.

2. Flight hardware

2.1. ATLAS-M

The ATLAS-M experiment is build to perform short-time annealing of metallic alloy diffusion couples in the liquid state during micro gravity.

Journal of Physics: Conference Series 327 (2011) 012051 doi:10.1088/1742-6596/327/1/012051

The main component of ATLAS-M is the furnace housing. It contains 8 graphite crucibles. A single crucible contains 4 diffusion samples. With this setup it is possible to handle 8 crucibles and 32 samples in total. The crucible (Fig. 1) consists of graphite, milled and machined for this purpose.

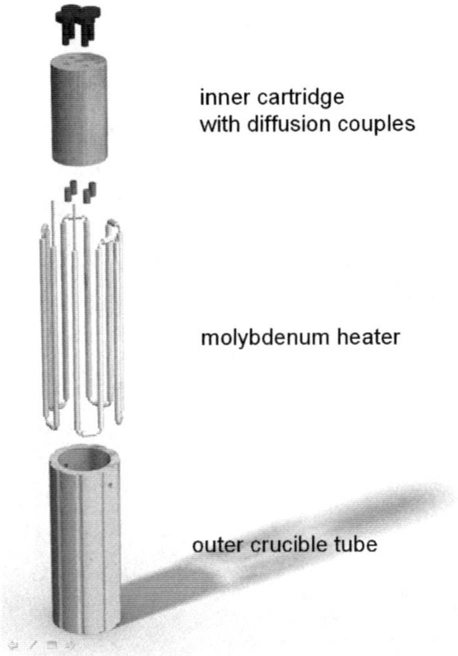

Figure 1. Exploded view of crucible assembly with molybdenum heater.

Figure 2. Exploded view of furnace assembly with 8 crucibles

The used heaters, made of molybdenum wire ⌀0.5 mm and embedded in Al_2O_3 tubes of $\varnothing_o$ 1.3 mm and an inner diameter of $\varnothing_i$ 0.7 mm, are winded in a meandering shape. The whole crucible is wrapped in Al_2O_3 felt for thermal insulation. The heaters are attached to copper feed-troughs, located in the eight top covers, to obtain electric connectivity. A single heater delivers a heating power of 450 W at 30 V. The temperature measurement is accomplished by ⌀ 0.5 mm thermocouples (TC) Type K, two for each crucible. The whole furnace is connected to a vacuum umbilical prior to launch for evacuating the eight (Fig. 2) sample chambers. For redundancy the eight crucibles are divided into two separate measurement units, achieved by discrete construction of the furnace complex. This helps to enhance the reliability in case the pressure increases unexpectedly or electrical problems occur. Before the vacuum umbilical is disconnected during lift-off, vacuum valves close the furnace complex to maintain the vacuum. During flight, the pressure inside the two independent units is measured with piezo pressure transducers. Automated processing of the entire control for the experiments is obtained via National Instruments cRio 9014 FPGA real-time controller using the programmable LabView User-interface (Fig. 3). The cRio-device is capable of handling several different types of data sources and is easily upgradeable through variable expansion cards such as analog and digital I/O devices and for thermocouple measurement. Additionally it records all the measured data and I/O-signals on a internal 2GB solid state flash memory. The operation during flight is stand-alone because μg-time is short and manipulation errors through experimenters would be

Journal of Physics: Conference Series **327** (2011) 012051 doi:10.1088/1742-6596/327/1/012051

fatal. To keep the crucibles in a isothermal state (Fig.5), the temperature regulation is realized by pulse width modulation (PWM). Switching of the heating power, inert gas pressure- and vacuum-valve operation is managed by 12 solid state relays, located at a round shaped circuit board underneath the furnace.
The experiment work flow of ATLAS-M is as follows:

- Preheating prior to launch
- Lift-Off
- Heating close to the target temperature
- Melting of diffusion couples in μg-time
- Isothermal annealing at target temperature
- Rapid cooling to solidify the samples before end of μg-phase

To preserve the samples and ideally freezing in of the liquid diffusion profile before re-entry of the payload, cooling is supported by flooding the furnace with helium gas, provided by small gas pressure reservoirs. This enhances heat transfer into the aluminum furnace which serves as a heat sink. After recovery and disassembly of the payload, the crucibles are removed from the furnace for transportation and laboratory analysis.

Figure 3. ATLAS-M module cutout illustration. On the lower side of the assembly is the furnace housing, surrounded by two helium pressure reservoirs and gas valves. On the upper platform is the National Instruments cRio 9014 controller with up to eight measurement and/or digital/analog I/O controller boards.

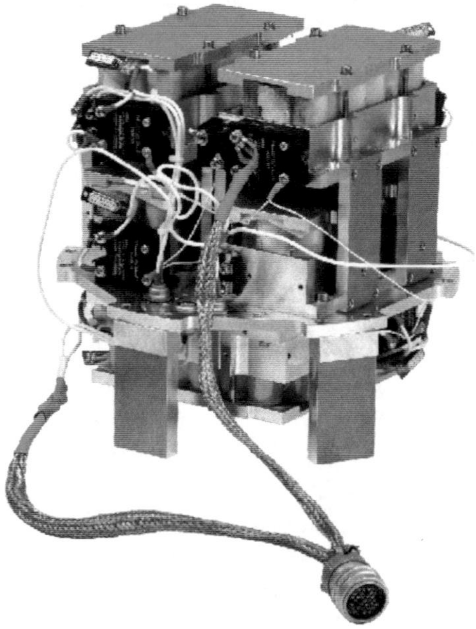

Figure 4. Interior assembly (without outer hull) of Batt-M with six rechargeable battery packs and umbilical connector cable

The ATLAS-M module was first launched on MAPHEUS-01 (lift-off 22.05.2009) with 8 graphite crucibles containing 16 diffusion couples. For MAPHEUS-02 (lift-off 27.10.2010) the crucible design was further enhanced. It enabled to process 8 crucibles with 32 diffusion samples in total. Additionally the existing circuit board was altered to achieve better heating performance at the expense of energy reserves in the batteries.

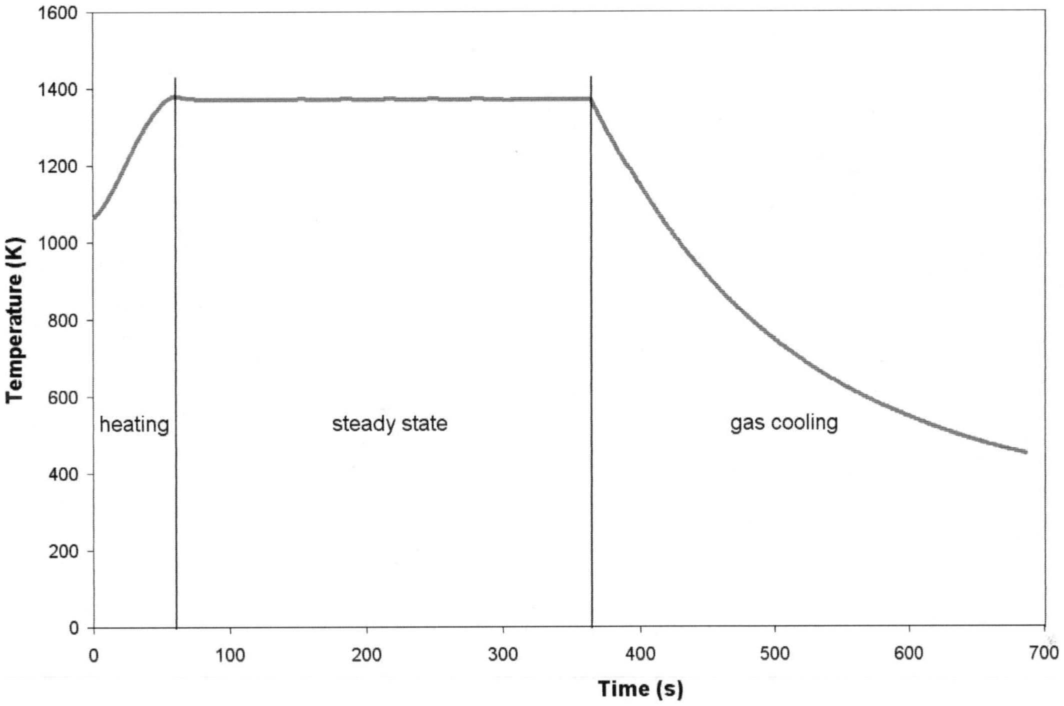

Figure 5. Typical heating curve for ATLAS-M experiment. Target temperature 1373K.

2.2. Batt-M

The rechargeable battery module Batt-M (Fig. 4) provided electrical power for the experiments accomplished on the sounding rocket campaigns MAPHEUS-01 and -02. Especially built as a power supply for high temperature furnace experiments, the Batt-M module provides sufficient capacity and ample power reserves for the accomplished experiments.

The main battery system of the so called Service System of MAPHEUS is capable of providing about 30W (30 V, 1 A permanent, depending on battery charge status) for each experiment, and therefore is not suitable for experiments with high power consumptions [2].

Therefore a rechargeable battery was designed, housed in a separate module to supply the needed power to the experiments for MAPHEUS. The term battery is normally used for primary cells, these are non rechargeable and for single time use only. For flight hardware and laboratory use, primary cells offer only restricted possibilities due to their high costs and maintenance (exchange and disposal of used/empty cells). Secondary cells on the other hand are rechargeable and offer the same or even better performance as primary cells and require less frequent mechanical module maintenance. Secondary cells can be mainly distinguished by their technical relevance

Table 1. Secondary cell types and classification

Description	Abbreviation	Single cell nominal voltage
Lead acid	Pb	2.0V
Nickel cadmium	NiCd	1.2V
Nickel metal hydride	NiMh	1.2V
Lithium ion	LiIo	3.7V
Lithium polymer	LiPo	3.7V
Lithium iron phosphate	$LiFePO_4$	3.3V

and single cell nominal voltage (see Tab. 1).
To evaluate different representatives of cell types, laboratory tests were conducted. These tests demonstrated relatively poor discharge results and the lowest energy density for the tested lead (Pb) cells. For consumer application nickel metal hydride (NiMh) cells almost completely displaced nickel cadmium (NiCd) cells due to their heavy metal electrode material. However the main problem of NiMh cells is the comparatively low discharge current. Lithium Ion- (LiIon) and Lithium Polymer-cells (LiPo), up to now, are the most advanced secondary cells on the consumer market. They offer a low self discharge rate, very good performance to weight ratio and a high energy density for a reasonable price. This type is widely-used in notebooks, mobile phones and other portable devices. Problematic, especially for LiPo-cells, is their affinity to incinerate during recharging or misuse (short circuit) and their behavior to decrease in performance at low temperature conditions. As a subspecies of the LiPo-technique, lithium iron phosphate ($LiFePO_4$) cells evolved. Strictly, it is a LiIon-cell using lithium iron phosphate as a cathode material. In contrast to the LiPo-technique these cells offer a better thermal stability and uncomplicated handling during charging or in case of a misuse / faulty conditions. The major benefit is the high continuous discharge current.

Moreover for MAPHEUS the vacuum capability of the used cells is vital for safety and operation because the Batt-M module is not pressurized and exposed to ambient pressure during flight. For Batt-M the cells of A123-Systems offered good mechanical and thermal stability [3]. A123 single cells are housed in an electron beam welded aluminum enclosure. Tests in a vacuum chamber ($p_{ambient} \approx 10^{-4}$ mbar) showed no influence on the cells, the thermal stability was tested cooling the cell to a core temperature of $\approx$ -40°C. Afterwards the cell was discharged with a test-current showing the nominal capacity of 2.3 Ah. The mechanical strength of the cells was tested during shaker tests reaching g-levels of 12.7 g_{RMS} and primary succeeded in the MAPHEUS-01 and -02 campaigns.
The Batt-M module consists of 108 single $LiFePO_4$ cells, assembled in six separate cell packs containing 18 cells each. The battery arrangement is called 9S2P, reaching a capacity of 4.6 Ah with a nominal voltage of 29.7 V. The six cell packs are connected to solid state relays with a switching current of 100 A. Prior to lift off, the battery module is connected to a stationary power supply to feed the experiments (without draining the internal power) via a umbilical connector, located at the outer hull of the module. Because charging of the cells is complicated and needs special equipment, it has to be done before assembly of the payload. Correct charging of $LiFePO_4$-cells is crucial. For proper operation, charging is done with a balancer, serving each single cell separately. The critical point is the maximum end-of-charge voltage for a single cell. It is not allowed to exceed this end-of-charge-voltage by any means. To accomplish this, the so called constant current / constant voltage (CC/CV) principle is used for recharging [3]. When the cell is empty, a constant current is applied to charge the cells. If the end-of-charge voltage

is reached (3.6 V for single $LiFePO_4$-cells) the charger continues with a constant voltage and the charging current is reduced (due to reaching the final single cell end-of-charge voltage). The charging process ends when the current reaches nearly zero, this indicates the complete charging process. This charging process results in expanded lifetime, consistent capacity and more charging cycles over lifetime. Six independent 29,7 V-cell packs are housed in aluminum cases which are framed to a baseplate. The weight of the assembled module including the outer hull is 28 kg including all peripheral components.

3. Conclusion

The ATLAS-M module processed 32 short time diffusion samples based on Al and Ge alloys during the MAPHEUS-02 flight. ATLAS-M reached a heating rate of 7 K/s with a maximum temperature of 1373 K and cooled down the furnaces, prior to end of μg-phase, achieving a cooling rate of 6K/s.The analysis and examination is done at the Institue of Material Physics in Space.
The Batt-M module supplied sufficient power to the operated high temperature furnace experiments.

4. Qutlook

ATLAS-M and Batt-M are designated for the MAPHEUS-03 campaign, scheduled for fall 2011. To achieve improved performance, minor technical modifications with respect to temperature control and reprogramming of the LabView interface is done. Both modules are beeing refurbished for summer 2011. Both modules can be used on future MAPHEUS flights but might also be used, with minor modifications, on other sounding rocket platforms.

5. References

[1] Stamminger A *MAPHEUS - The Maiden Flight of a New Vehicle for Microgravity Experiments*, **IAC Proceedings, 2009**
[2] DLR / SSC,*REXUS User manual ver. 7.0.* **Euro Launch, 2009**
[3] Retzbach L *Akkus und Ladetechniken*, **Franzis, 2008**
[4] Meyer A 2010 *Phys. Rev.* B **81** 012102
[5] Zhang B, Griesche A, Meyer A 2010 *Phys. Rev.* Lett. **104** 035902
[6] A123 Systems, www.a123systems.com, *Datasheet High Power Lithium Ion ANR26650M1*, **A123, 2009**

Compact High-Temperature Shear-Cell Furnace for In-Situ Diffusion Measurements

C Neumann[1], E Sondermann[1], F Kargl[1] and A Meyer[1]

[1] Institut für Materialphysik im Weltraum, Deutsches Zentrum für Luft- und Raumfahrt (DLR), 51170 Köln, Germany

E-mail: christian.neumann@dlr.de

Abstract. A compact-sized shear cell for *in-situ* interdiffusion measurements in liquid alloys at high-temperature is presented. It is designed for being used with X-ray radiography in microgravity environment and also in ground-based experiments. Within outer dimensions of 40x40x20 mm, the shear cell contains six samples in long capillaries which remain spatially separated during melting, homogenisation and optional μg-injection. The diffusion is then initiated by shearing both material samples together. Operating temperature is as high as 1500 °C. Stable temperature distribution is found and no sample leakage occurs. The thin layout of the shear cell results in a good X-ray contrast supported by making use of graphite heating foils. First results of ground experiments are presented.

1. Introduction

The measurement of diffusion coefficients in liquid alloys faces a number of experimental challenges of which flow effects during melting of the diffusion couple, buoyancy-driven convection while diffusion annealing, and the formation of shrinkage holes as well as microstructure formation during solidification are the most prominent [1, 2]. For a few sample systems these effects have small influence on the results in standard long capillary experiments, particularly if carried out in μg. An already successfully applied platform for such experiments in μg is the ATLAS-M [3] experiment on the sounding rocket MAPHEUS [4]. For other sample systems the measurement may be noticeably disturbed by the mentioned effects. The challenges to reduce the influence of these can be met by combining a shear-cell technique with X-ray radiography [5, 6] under μg-conditions.

X-ray radiography (XRR) allows time-resolved *in-situ* experiment which are unaffected by microstructure formation and shrinkage holes. It requires though sufficient X-ray contrast in the sample system and a flat and thin furnace design consisting of only weakly X-ray absorbing materials. Within a shear cell the two parts of a diffusion couple are melted separately and homogenized before the diffusion measurement is started by shearing both parts together. Therefore flow effects and segregation of higher melting point phases during heating do not affect the measurement [7]. Furthermore, the diffusion time is well known. To maximize the number of diffusion pairs which can be processed during a flight of the sounding rocket, neighbouring diffusion pairs share one sample material. This allows to investigate a maximum of six diffusion couples simultaneously. Since buoyancy-driven convection poses problems in ground based experiments, the μg-environment on the sounding rocket MAPHEUS is used for

Journal of Physics: Conference Series **327** (2011) 012052 doi:10.1088/1742-6596/327/1/012052

the measurement of reference values. MAPHEUS offers a μg-time of up to 4 minutes and can therefore be used for fast diffusing samples such as Ge-based or Al-based alloys. For the operation on the sounding rocket, the shear cell employs a compact, light weighted and sufficiently rugged design.

2. Shear-Cell Furnace Characteristics

The cuboid shaped shear cells' overall dimension count 40x40x20 mm. The crucible is entirely made out of graphite. Its high thermal conductivity is a key for a constant temperature regime over the samples. Further graphite is fairly inert with a large number of sample systems at moderate temperatures. For other more reactive systems graphite is a good choice for rapid diffusion experiments enabled by XRR when the chemical reaction between graphite and melt is slow. The chosen sample size of thin rods with 1.5 mm in diameter and 15 mm in length corresponds to previous diffusion experiments that have been performed in classic long capillary experiments. As a result, six different diffusion couples can be measured at the same time and temperature, restricted by the condition that a diffusion couple contains one sample from each of its neighbours. This graphite crucible is a set of two components. The upper part

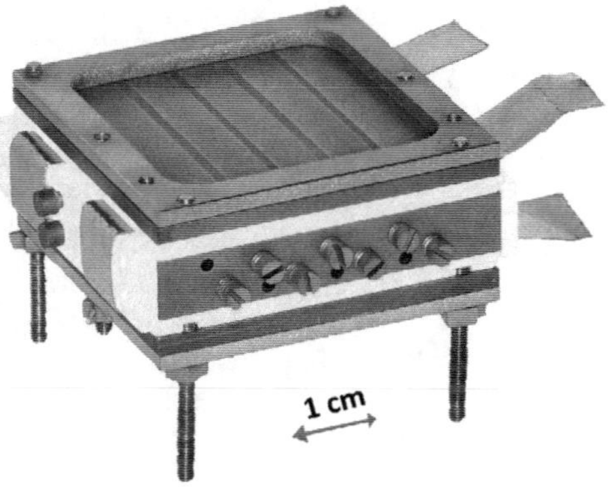

Figure 1. Photography of the assembled shear cell

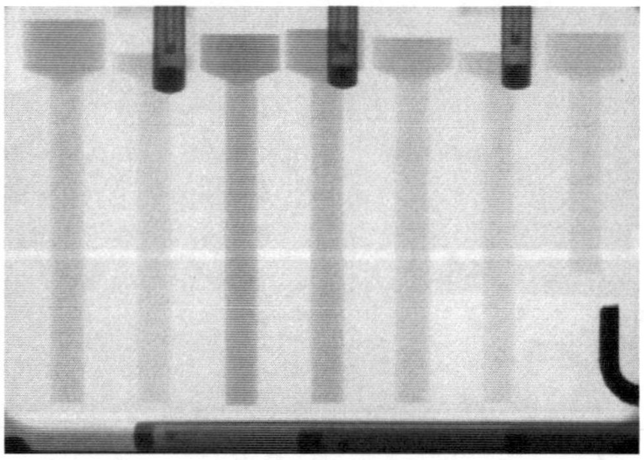

Figure 2. X-ray image showing configuration of the shear cell with liquid $Al_{1-x}Ni_x$ (x=0..12.5 at%) homogenized shortly before shear.

contains half of each capillary as well as the reservoir to compensate volume changes during melting and tolerances in the manufacturing process. The reservoir is sealed by a graphite piston which in turn is driven by some graphite felt compressed with graphite screws. The lower part contains the matching other half of each capillary in the form of a blind hole. Both elements put together deliver seven independent containers, each sealed and with congruent cross section at the separation plane. There is a lateral distance of 4 mm between neighbouring capillaries. The lower part is meant to perform the shear movement and therefore runs in a U-shaped rail preventing any lateral movement. From beneath it is supported by another set of graphite felt to pressurize the separation plane within the divided crucible. By this, leakage is inhibited and at the same time shear movement is allowed. The temperature is measured with three pairs of thermocouples (TCs) being positioned on the left, in the centre and on the right of the shear cell as shown in figure 2. Sheathed TCs of diameter 1.5 mm are used. Each two of them are arranged in a straight line between two samples on upper and lower end to confirm constant temperature. For low temperature operation $\leq$1100 °C Type-K TCs are currently in use whereas for temperatures up to 1500 °C Type-S TCs will be used. To enable compensation of any arising gradients in temperature along the diffusion couples, the shear cell is equipped with two independently controlled heating zones, one covering the top and the other one the bottom part. The material chosen for the heaters is pure expanded graphite foil of 0.15 mm thickness and has the ability to cope with high rates of heat flux. Further, it shows negligible X-ray absorption which is beneficial for the *in-situ* XRR experiments. Each heater provides 750 W of heating power. To electrically insulate the heater from the crucible the heaters are placed between thin layers of alpha boron nitride. The heater covers both sides of the crucible and passes over at the short end. Oppositely the heaters are linked a to power junction by niobium connectors. The shear cell is sandwiched between additional layers of bulk graphite and frames of niobium connected among each other by six 2 mm-rods. By that this shear cell gets rugged and prepared for sounding rocket environment. It is held in place by four of the niobium rods acting as its legs, reducing the local heat discharge to a minimum amount. The entire setup is placed within a thermal insulation of niobium-reflector and aluminiumoxide-fibres in an aluminium-box, each fitted with windows for the X-ray beam. The experiment takes place in high vacuum. The shear movement is regulated by an external servo drive controlled by a microcontroller-based electronics unit. The movements are transmitted by a wire of molybdenum to the crucibles shifting lower part.

3. Experimental Results

Several test experiments have been carried out where the overall functionality of the shear cell has been demonstrated. The maximum set temperature in first tests has been 1100 °C requiring a heating power of about 240 W to hold this temperature. At a setpoint of 900 °C a spatial temperature gradient of approximately 1 K in samples is observed. The heating rate is about 2 K/s at 900 °C, when the heating power is limited to 375 W. First experiments on aluminium-alloys proved *in-situ* X-ray functionality as well. Figure 3 shows a grey value profile of $Al_{95}Ni_5$ vs. $Al_{85}Ni_{15}$ obtained by using the presented shear-cell setup. The difference in Ni-concentration of 10 at% is resolved with about 500 units in grey value. By this method profiles for diffusion couples with only 2,5 at% differences in Ni-concentration can be analysed. To gain diffusion profiles from these data, the measured grey values at each position have to be normalized [6]. Then diffusion profiles following the expected error function as a solution of Fick's law of diffusion are obtained. For the measurements a microfocus X-ray source with 100 kV and up to 10 W

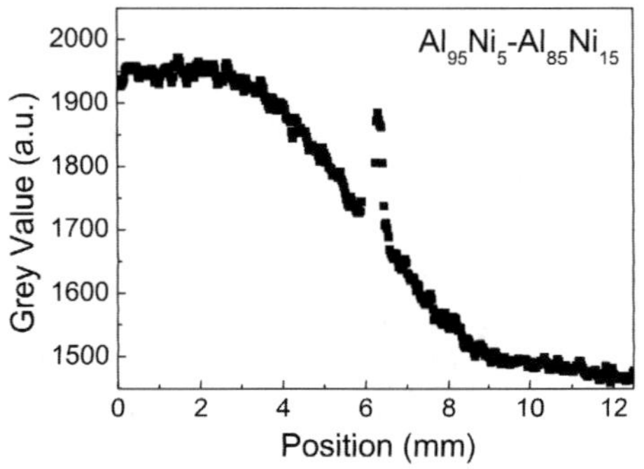

Figure 3. Gray value profile of the diffusion couple $Al_{95}Ni_5$ vs. $Al_{85}Ni_{15}$ as obtained using the presented shear-cell setup without normalization. The peak around 6 mm results from the necessary slit between the U-shaped rail and the upper part of the crucible as can be seen in fig. 2.

in combination with a scintillation detector was used. The detector frame rate was set to 1 Hz, which gives good contrast-properties. To reduce artefacts caused by X-ray scattering, the beam was collimated by an aperture with a diameter of 0.6 mm in front of the X-ray source. Sample-source-detector distances were set such that a total resolution of 29 μm corresponding to sample plane was achieved. Further more extensive tests of the shear movement were carried out. First the cell was tested and calibrated in cold state. In addition it also proved its functionality in a

Journal of Physics: Conference Series **327** (2011) 012052 doi:10.1088/1742-6596/327/1/012052

900 °C Al-Ni experiment. However, the reliability in the precision of the shear movement which is important to obtain correct results [8] turned out to require further development. Incidents of leakage of the sample-material at the shear front did not occur. The whole setup has been already successfully tested for flight readiness in a shaker test performed at the ZARM-Centre in Bremen, Germany.

4. Conclusion

A novel shear cell for the use with X-ray radiography on a sounding rocket has been designed, successfully tested, and space qualified. It allows the simultaneous measurement of six diffusion couples at the same temperature. The shear-cell furnace offers an excellent X-ray contrast. Pressure on reservoirs is adjusted such that neither free sample surfaces nor sample leakages occur. The shear cell is designed for temperatures of up to 1500 °C which will enable to measure higher melting point alloys as Si-rich Si-Ge and Ni-rich Al-Ni in future.

References

[1] Meyer A 2010 *Phys. Rev. B.* **81** 012102
[2] Masaki T, Fukazawa T, Matsumoto S, Itami T and Yoda S 2005 *Meas. Sci. Technol.*. **16** 327
[3] Blochberger G, Drescher J, Neumann C, Penkert P, Griesche A, Kargl F and Meyer A 2011 *J. Phys.: Conf. Series.* (submitted)
[4] Stamminger A *et al Proc. 60th International Astronautical Congress, Daejeon, Korea, 2009.*
[5] Zhang B, Griesche A and Meyer A 2010 *Phys. Rev. Let.*. **104** 035902
[6] Griesche A, Zhang B, Solórzano E and Garcia-Moreno F 2010 *Rev. Sci. Instrum.*. **81** 056104
[7] Griesche A, Kraatz K H and Frohberg G 1998 *Rev. Sci. Instrum.*. **69** 315
[8] Uchida M, Watanabe Y, Matsumoto S, Kaneko M, Fukazawa T, Masaki T and Itami T 2002 *J. Non-Cryst. Solids.* **312-14** 203

MSL compatible isothermal furnace insert for high temperature shear-cell diffusion experiments

D Heuskin[1], F Kargl[1], A Griesche[2], Ch Stenzel[3], D Mitschke[3], D Bräuer[4] and A Meyer[1]

[1] Institut für Materialphysik im Weltraum, Deutsches Zentrum für Luft- und Raumfahrt (DLR), 51170 Köln, Germany
[2] Bundesanstalt für Materialforschung- und -prüfung, 12205 Berlin, Germany
[3] Astrium - Space Transportation, Physical Sciences (TO 5), 88039 Friedrichshafen, Germany
[4] Institut für Automatisierungstechnik, TU Bergakademie Freiberg, 09596 Freiberg, Germany

E-mail: florian.kargl@dlr.de

Abstract. For long-time diffusion experiments shear-cell techniques offer more favourable terms than the traditional long capillary techniques. Here, we present a further developed shear-cell that enables the measurement of diffusion coefficients up to temperatures of 1600 °C. Hence, diffusion experiments can be carried out at temperatures not accessible until now by conventional capillary or shear-cell techniques. The modified shear-cell, which can contain up to six samples of a total length of 90 mm and a diameter of 1.5 mm, is built of 30 shear discs of 3 mm thickness each. It is operated in an isothermal furnace insert which can be accommodated in the Materials Science Laboratory of the International Space Station. This provides the opportunity that the shear-cell can be applied to microgravity and to ground-based experiments, respectively. The heater insert with an overall length of 518 mm and a diameter of 210 mm consists of four heating zones with a total power of 3.5 kW. Temperature homogeneity along the graphite sample compartment is better than 2 K at 1600 °C. Details of the new design are discussed and results of first successfully performed heating and shearing cycles are presented.

1. Introduction

Diffusion is a vital topic in solid-state physics, chemistry, metallurgy and materials science [1]. A thorough understanding of diffusion in the liquid state of multicomponent metallic alloys and semiconducting alloys is crucial for materials development and engineering [2]. Traditionally, diffusion coefficients are measured using long capillary techniques [2]. This technique, whereby the spatial distribution of a tracer or the spatial dependence of composition along a thin rod-like sample is determined after a full heating, annealing, and cooling cycle, proved highly successful for solids [1]. However, for determining diffusion coefficients in liquids a number of problems are encountered that negatively influence the accuracy of the data. Segregation during heating, microstructure and shrinkage hole formation during solidification and density changes between liquid and solid are just a few of the important ones. For binary systems accurate diffusion coefficients can be obtained by in-situ monitoring the long capillary with X-ray radiography [3]. However, for alloys with no X-ray contrast, alloys that feature a complex phase diagram upon heating with e. g. largely differing melting points for the two samples constituting the diffusion couple, or multicomponent alloys, the use of the so-called shear-cell method is indispensable to determine accurate diffusion coefficients. Successful shear-cell experiments on measuring self diffusion in liquid mercury [4] and impurity diffusion in liquid Sn and Cu were reported in the past [5]. Moreover, for multicomponent alloys shear-cells were used to determine chemical diffusion coefficients [6, 7]. For binary samples with X-ray contrast the shear-cell technique can now also be combined with X-ray radiography [8]. In a shear-cell the two different samples of a diffusion couple are separately melted. After some time for homogenization the samples are brought into contact with each other starting the diffusion process. In case of non in-situ imaging with X-ray radiography the diffusion process is terminated by shearing of the diffusion couple into numerous sections after some annealing time. Thereby the diffusion profile of the liquid is frozen in. The individual sections of the solidified sample are then analyzed for their composition. Even so all of the previously mentioned problems encountered for classical long-capillary experiments on liquid samples can be overcome by the shear-cell technique, the measured diffusion coefficients might still be erroneous due to buoyancy-driven convective flow. The latter can only be overcome by performing experiments under reduced-gravity conditions which allows purely diffusion conditions [6]. To this end, a shear-cell has already been successfully used aboard a FOTON satellite up to maximum temperatures of 770 °C [7] and up to 850 °C on ground [9]. The FOTON setup with its small diameter relied on direct heating via a molybdenum-wire resistance heater brought in contact with the cartridge surface. The cartridge and heater were tightly wrapped in Al_2O_3-felt insulation [10]. In order to reach higher temperatures and to be able to process more samples simultaneously the original FOTON design has to be modified. Here, a further developed version of a shear-cell is presented that now enables to simultaneously measure diffusion coefficients for up to six samples at temperatures of up to 1600 °C. In order to achieve these higher temperatures and to enable processing of the required larger-sized shear-cell an isothermal furnace with a large bore diameter has to be used. Details of the complete experimental facility developed by Astrium GmbH and partners and performance parameters from first tests are presented.

2. Isothermal Furnace

The isothermal furnace with an overall length of 518 mm and a diameter of 210 mm offers a bore of 68 mm diameter for the experimental cartridge insertion. It consists of four independently operated heating zones with a total power of 3.5 kW. The heating zones consist of four meandered graphite heaters. Using these heaters a maximal temperature of 1600°C at the sample position can be achieved at a maximal heating rate of $60\,\mathrm{K\,min^{-1}}$. With this maximal temperature determining diffusion coefficients of high melting point alloys like Al-Ni and Si-Ge becomes possible. As verified experimentally the furnace provides a homogeneous temperature zone of 90 mm length at 1600 °C with a gradient of less than 2 K axially. This equals to a gradient of less than $0.02\,\mathrm{K\,mm^{-1}}$ axially. The furnace can be operated under vacuum or under inert gas atmosphere (Ar or He), respectively. Typically, the vacuum is better than $9{\cdot}10^{-5}$ mbar at ambient temperature and $1.2{\cdot}10^{-4}$ mbar at experimental temperature. All measured data are monitored and saved by a closed loop control process computer. The process unit also controls several experiment environment values and allows online telecommanding. The isothermal high temperature furnace insert, shown in figure 1, is fitted in a process chamber fully compatible to the one used in the Materials Science Laboratory (MSL) aboard the International Space Station (ISS). The furnace is further fully compatible with power, cooling, vacuum, and gas supplies available in the MSL. For ground-based diffusion experiments the process chamber is mounted on a decoupled optical table and can be brought into a vertical position. This is required for aligning the capillary axis parallel to the gravitational force. As a result a stable density layering of the liquid column is achieved. This helps to suppress convective flow in ground-based experiments.

Figure 1. Process chamber with integrated isothermal high temperature furnace and open ended experimental cartridge in horizontal position. As described in the text the entire chamber can be tilted vertically for the experiment.

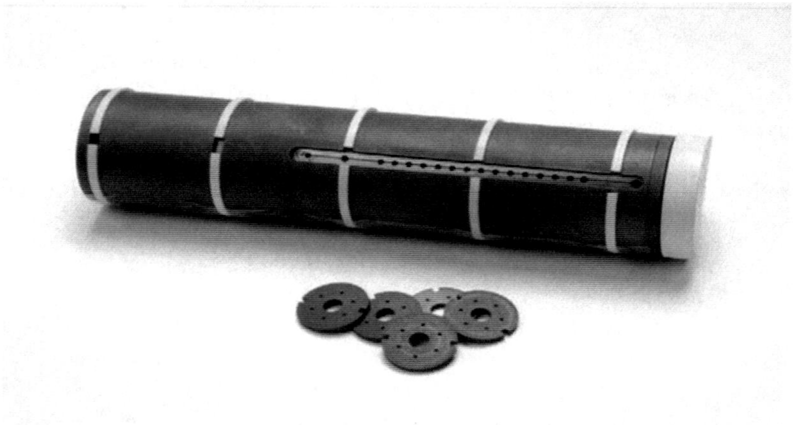

Figure 2. Newly designed high temperature shear-cell. The graphite articulation tube with lead-through for holding bolts is shown. The front part shows individual shear discs as they are integrated in the shear-cell. Cut-outs for disc motion as well as holes for diffusion samples are visible.

3. Shear-Cell

The here presented modified shear-cell is based on the original FOTON design [11, 12]. Its size and operating mechanism were modified according to MSL requirements. The novel shear-cell consists of 30 graphite discs each with a thickness of 3 mm and a diameter of 30 mm. Graphite is chemically inert for a large number of melts of interest. It can be precisely machined and shows moderate friction for shearing. The discs have six equally spaced holes of 1.5 mm diameter at a distance of 8.5 mm from the disc centre. Hence, a total of six different diffusion couples can be simultaneously processed. If required, the diameter of the capillary holes can be modified within certain limits for future application without compromising functionality or stability of the shear-cell.

The shear discs and reservoirs are mounted on a middle axle and integrated into a slotted, articulated tube, shown in figure 2. The middle axle also carries three type-S thermocouples at

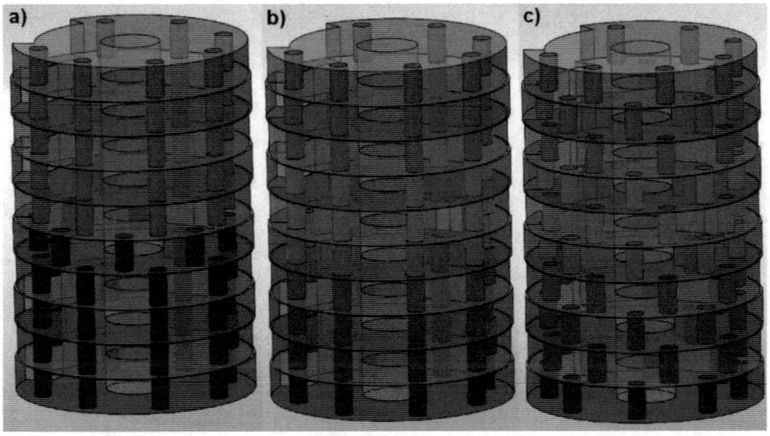

Figure 3. Visualisation of the shear-cell mode of operation: a) Separated diffusion couples with one shear-disc out of alignment for separate melting during heating and followed by a homogenisation phase. b) Diffusion pair in contact during diffusion annealing. c) Termination of the diffusion process by shearing of every other disc, effectively freezing in the liquid diffusion profile. The last step is followed by cooling.

either end and at the center of the diffusion couple. Graphite bolts are gearing though the slots into cut-outs on the shear-discs, limiting the elbow-room of each disc. A mechanical rotation of the articulated tube by a boron nitride driving shaft moves every other disc on the middle axle. In a typical starting position all discs but one are aligned at the end position in clockwise rotation direction. Hence all but one of the capillary holes are aligned for each diffusion couple. The one disc located e. g. in the center of the disc stack is displaced by 30° counterclockwise [cf. figure 3 a)]. With the samples fully melted and homogenised clockwise rotation of this displaced disc brings into contact the two initially separated samples. As a result diffusion is initiated and a diffusion profile develops with time [cf. figure 3 b)]. Counterclockwise rotation of every other disc by 60° terminates the diffusion process [cf. figure 3 c)]. The total length of the capillary rods is 90 mm with reservoirs on each end. The reservoirs are pre-stressed by graphite felt pushing on pistons to balance volume expansion or shrinkage and to minimize free surfaces. Free surfaces can otherwise cause Marangoni convection significantly altering the measured diffusion coefficient [13, 14].

The shear-cell assembly is housed in a rhenium cartridge of an outer diameter of 60 mm and an overall length of 510 mm. The cartridge can be inserted into the previously described isothermal furnace. Boron nitride spacer bars (cf. figure 2) are keeping the graphite cell at distance from the rhenium tube surface to avoid corrosive reaction of graphite with the metal. The driving shaft is connected to a five-stage epicyclic gear which is driven by a direct-current motor. Both are mounted at the far end of the cartridge. To minimize heat flow from the heating zones to the drive system unit the boron nitride shaft is embedded in a solid graphite insulation foam itself separated from the cartridge surface by a boron nitride slotted tube. Three cartridges have been designed which are fitting the furnace. These are shown in figure 4. Cartridges FM-01 and FM-02 enable high temperature operation up to 1650 °C. Cartridge EM01 is available for demonstration purposes. Cartridge FM-01 is fully MSL compatible. It can be vacuum tightly sealed with a leakage rate of less than $10^{-8}\,\mathrm{mbar\,l\,s^{-1}}$. It can be operated both with vacuum or inert gas inside the cartridge. In additional a thermocouple and a signal cable lead-through are available. Cartridge FM-02 has the same size and specifications as FM-01. However, it is designed with open access to the driving unit. Cartridge FM-02 is easier to assemble and to handle than cartridge FM-01. Hence it is mainly used for ground-based experiments. Also

Figure 4. High temperature cartridges. Left to right: FM-01 fully MSL compatible; FM-02 for ground-based experiments; EM-01 engineering model.

modifications during the test phase can be easily carried out. Cartridge EM-01 has a slotted tube to enable visible inspection of the shear-cell in demonstration mode.

4. Performance

A thermal furnace heater characterisation up to 1600°C with the aim to calibrate the three type-S thermocouples inside the shear-cell together with showing proper shear-cell function were carried out (cf. figure 5). Temperature homogeneity along the 4 heating zones was currently better than 2 K over the entire temperature range. This equates to a temperature gradient of better than 0.022 K/mm. However temperature gradient on the shear-cell amounted to 5 K. A homogeneous temperature along the shear-cell is required. Individual tuning of the heating elements generated a homogeneous temperature difference on the shear-cell better than 1.2 K axially. With the initial heating cycles the ceramic insulation foam of the process chamber was baked out. The insulation is highly susceptible to moisture. Hence, even with storage under inert atmosphere, the short opening period for cartridge installation is sufficient for moisture ingression into the ceramic insulation. This is verified by slight pressure surges upon heating (cf. figure 6).

A reliable shear-cell actuation was verified. As verified in on-bench tests a hinge moment of 0.8 to 0.9 Nm is required to actuate the empty and cold shear cell in the cartridge for clockwise rotation of the single initially displaced disc. The counter-clockwise rotation of every other disc requires 2.5 Nm of torque. These moments can be achieved by the driving shaft being connected to a five-stage epicyclic gear (1526:1) which is driven by a 12 V direct-current motor. This driving unit provides a maximal torque of 9 Nm which is more than sufficient for a reliable shear-cell actuation even at the highest temperatures. In order to avoid damage by overtightening the shear-cell, a mechanical drive assembly dead stop made of high grade steel is installed. Moreover, the motor current is limited. Hence, on reaching the dead stop the current increase reaches the set limit and the motor is automatically switched off. Increasing shear-cell temperatures should cause higher actuation moments due to thermal expansion of the graphite components. Higher thorque should simultaneous cause higher motor currents. Against all odds motor current decreases on increasing shear-cell temperatures (cf. table 1).

Table 1. Decreasing actuation motor current on high shear-cell temperatures

shearing temperature (°C)	motor current(mA)
20	95-164
700	72-120
1100	68-112
1300	67-90
1500	55-90

First experiments were performed on Al-Cu and Al-Cu-Ag alloys at a diffusion temperature of 700 °C. These particular sample materials have been chosen to enable direct comparison with long capillary experiments carried out in the laboratory using post-mortem analysis and for Al-Cu also with the in-situ technique of X-ray radiography. The volume compensation compartments worked nominally with minimal sample leakage between the discs. Further, the shear-cell actuation worked reliably. The samples were processed with 2 h pure diffusion time. This was based on earlier experience obtained through long-capillary experiments. Diffusion coefficients of $5{\cdot}10^{-9}\,\mathrm{m^2\,s^{-1}}$ were expected [3, 15]. The processing time could be estimated using the appropriate solution of Fick's second law of diffusion, the estimated diffusion coefficient and the capillary length of 90 mm. The requirement was a sufficient spread of the profile however without reaching the capillary ends. The diffusion experiments consisted of a heating phase, followed by a sample homogenisation phase (cf. figure 6). At the end of the homogenization phase the two separately melted samples were brought into contact and the diffusion process was started. In the experiments on Al-Cu and Al-Cu-Ag the axial temperature homogeneity was better than 1.2 K. After the annealing time the diffusion process was terminated by shearing of the discs and thereby freezing in the diffusion profile of the liquid. This second shearing

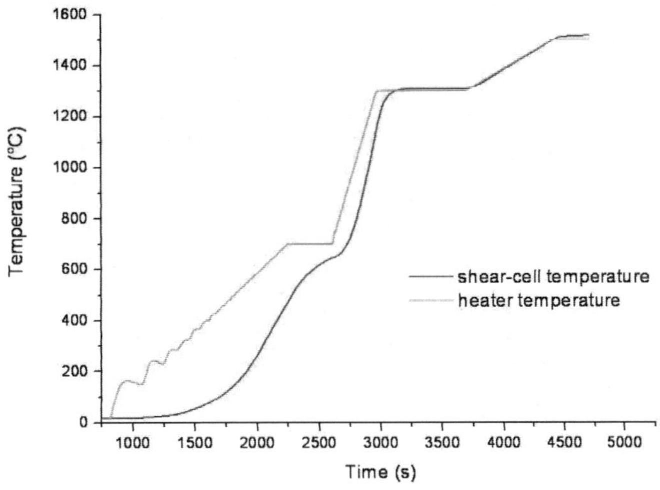

Figure 5. High temperature characterisations and shearing tests at 1500 °C.

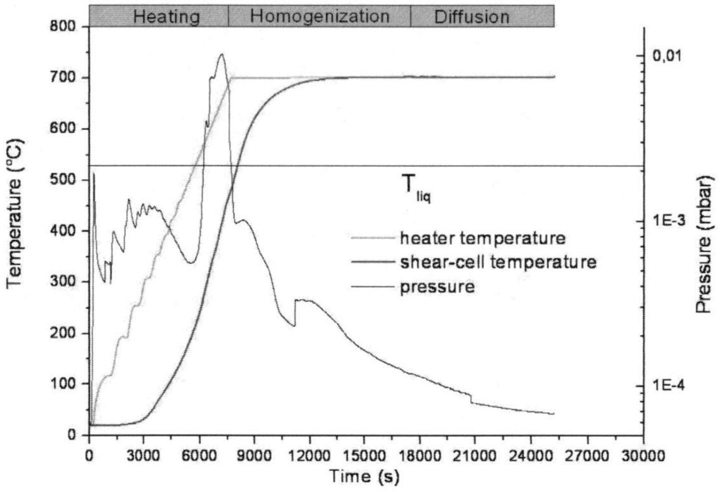

Figure 6. Temperature and pressure processing conditions for a chemical diffusion experiment on a Al-Cu-Ag diffusion pair at 700 °C from start to early stages of diffusion.

processes was followed by shut-down of the heater. As a result the sample was cooled down and solidified. The disassembled small sample cylinders have been weighted for the verification of the compartment filling degree (cf. table 2) and are currently analysed for their composition. Densities calculated for the endmember compositions dividing the measured weight by the compartment volume agree well with densities for Al-Cu melts published in literature [17].

Table 2. Weight of sample cylinders for AlCu7.5-AlCu17.5 after disassembling

Compartment #	1	2	3	4	5	6	7	8
Sample mass (mg)	14.60	14.60	14.64	14.69	14.64	14.74	14.65	14.52
	9	10	11	12	13	14	15	16
	14.99	15.10	15.16	15.60	15.55	16.03	16.77	16.60
	17	18	19	20	21	22	23	24
	16.23	17.08	16.95	16.99	16.84	17.30	16.99	17.09
	25	26	27	28				
	17.07	17.08	17.08	12.61				

For the binary alloy X-ray computed tomography is used as a non-destructive method followed by chemical analysis using atomic absorption spectroscopy. For the ternary sample only chemical analysis is used. Analysis of the samples of these first experiments is currently work in progress.

5. Conclusion

A novel isothermal furnace for the MSL with a large diameter bore capable of steady isothermal conditions for a zone of 90 mm length at temperatures up to 1600°C has been presented. First successful ground-based heating and shearing cycles have been performed using the furnace with a newly designed graphite shear-cell. The experiments have shown the functional capability of the new isothermal furnace. Proper function of the shear-cell device has been demonstrated. In particular, the proper function of the shear-cell driving unit has been shown. Adequate pressure of the pre-stressed sample reservoirs in the shear-cell minimizes free surfaces in the capillaries. Only marginally material leakage between the shearing discs was observed. Temperature homogeneity of better than 1.2 K along the diffusion couples has been achieved at 700 °C. The samples of the first experiments are currently being analysed for composition. In future experiments on Al-Ni and Si-Ge at higher processing temperatures are planned. This contains also tests of melt containory reactions. Again results of these experiments can be directly compared with in-situ results on diffusion obtained by means of X-ray radiography. For online monitoring of the shear process, further developments of the shear-cell actuator will consist of the integration of a feed-forward control, linked to a rotary angle encoder and a multiphase stepper motor.

Acknowledgments

We thank the German Ministry for Economy and Technology (BMWi) for funding of this project through the Economic Stimulus Package I.

References

[1] Mehrer H 2007 Diffusion in Solids (Heidelberg: Springer)
[2] Shimonji M and Itami T 1986 *Def. Diff. Forum* **43**
[3] Zhang B, Griesche A and Meyer A 2010 *Phys. Rev. Let.* **104** 035902
[4] Nachtrieb N H and Petit J 1956 *J. Chem. Phys.* **24** 746
[5] Bruson A and Gehrl M 1980 *J. de Phys.* **41** 533
[6] Frohberg G 1995 in *Diffusion in Liquids, Scientific results of the German Spacelab Mission D-2* edt. Sahm P R (WPF/DLR Cologne) 275
[7] Griesche A, Macht M P, Suzuki S, Kraatz K H and Frohberg G, 2007 *Scripta Mater* **57** 477
[8] Neumann Ch, Sondermann E, Kargl F and Meyer A 2011 *J. Phys.:Conf. Series* (submitted)
[9] Suzuki S, Kraatz K H, Frohberg G, Rosu R, Wendl W and Müller-Vogt G, 2006 *Ann N Y Acad Sci.* **1077** 380
[10] Suzuki S, Kraatz K H, Griesche A and Frohberg G, 2005 *Microgravity Sci. Technol* XVI-1
[11] Griesche A, Kraatz K H and Frohberg G 1998 *Rev. Sci. Instr. Meth.* **69** 315
[12] Griesche A, Kraatz K H, Frohberg G, Müller H, Müller-Vogt G, Garandet J P, Dusserre P, Praizey J P and Mathiak G 2001 *ESA Spec. Pub.* **454** 985
[13] Müller H and Müller-Vogt G 2003 *Cryst. Res. Technol.* **38** 707
[14] Rosu-Pflumm R, Wendl W, Muller-Vogt G, Suzuki S and Frohberg G 2009 *International Journal of Heat and Mass Transfer* **52** 6042
[15] Kargl F, Engelhardt M, Yang F, Weis H, Schmakat P, Schillinger B, Griesche A and Meyer A 2011 *J. Phys.: Condens. Matter* **23** 254201.
[16] Masaki T, Fukazawa T, Watanabe Y,Kaneko M, Yoda S and Itami T 2007 *Journal of Non-Crystalline Solida* **353** 3290
[17] Brillo J, Egry I and Westphal J 2008 *Int. J. Mat. Res.* **99** 162

Journal of Physics: Conference Series **327** (2011) 012054 doi:10.1088/1742-6596/327/1/012054

Entering "A NEW REALM" of KIBO Payload Operations - Continuous efforts for microgravity experiment environment and lessons learned from real time experiment operations in KIBO -

K Sakagami[1], **M Goto**[2], **S Matsumoto**[3] **and H Ohkuma**[4]

[1] Utilization Engineering Department, Japan Manned Space Systems Corporation, 2-1-1, Sengen, Tsukuba-shi, Ibaraki, Japan
[2] Japanese Experiment Module (JEM) Mission Operations and Integration Center, Japan Aerospace Exploration Agency, 2-1-1, Sengen, Tsukuba-shi, Ibaraki, Japan
[3] Institute of Space and Astronautical Science, Japan Aerospace Exploration Agency, 2-1-1, Sengen, Tsukuba-shi, Ibaraki, Japan
[4] Space Station Engineering Department, Advanced Engineering Services Co., Ltd., 2-1-1, Sengen, Tsukuba-shi, Ibaraki, Japan

E-mail:sakagami@jamss.co.jp

Abstract. On January 22nd, 2011(JST), KOUNOTORI2 (H-II Transfer Vehicle: HTV2) was successfully launched from Tanegashima Space Center toward the International Space Station (ISS) and two new JAXA payload racks, Kobairo rack and MSPR (Multi-purpose Small Payload Rack) were transferred to ISS/KIBO (Japanese Experiment Module: JEM). In addition to Saibo rack and Ryutai rack which are already in operation in KIBO, in total 4 Japanese experiment payload racks start operations in KIBO. Then KIBO payload operations embark on a new realm, full utilization phase. While the number and variety of microgravity experiments become increasing, simultaneous operation constraints should be considered to achieve multi-task payload operations in ISS/KIBO and ever more complicated cooperative operations between crewmember and flight control team/science team are required. Especially for g-jitter improvement in ISS/KIBO, we have greatly advanced cooperative operations with crewmember in the recent increment based on the microgravity data analysis results. In this paper, newly operating Japanese experiment payloads characteristics and some methods to improve g-jitter environment are introduced from the front line of KIBO payload operations.

1. Introduction

Since August, 2008, utilizing two JAXA payload racks (SAIBO rack and RYUTAI rack), variety of experiments have been performed in the International Space Station (ISS)/Japanese Experiment Module (JEM), or KIBO. In addition, JAXA exposed payloads and NASA payload racks (MELFI and

[1] To whom any correspondence should be addressed.

Express Racks) are constantly in operating in KIBO. On January 22nd, 2011(JST), KOUNOTORI2 brought two new JAXA payload racks, KOBAIRO rack and MSPR (Multi-purpose Small Payload Rack) to KIBO. Then KIBO payload operations embark on a new realm, full utilization phase.

In order to perform variety of experiments simultaneously in KIBO, the system resources of KIBO such as power, coolant, communication and tracking, command and data handling, and some other resources which experiment support equipments can provide, should be carefully planned and delivered to each experiment payload. Moreover, it is important that microgravity environment should be carefully controlled based on the microgravity data collection and analysis and cooperation between crew and flight control team are essential key to success for good microgravity environment in a manned space system facility like ISS/KIBO. JAXA flight control team, microgravity analysis team and Marangoni experiment science team have been continuing an effort to achieve good microgravity environment in ISS/KIBO with crew and International Partners since Increment 17 and we have recently acquired new lessons learned regarding possible correlation between crew motion and g-jitter during crew sleeping time. These finding are expected to be utilized to reduce g-jitter ISS/KIBO crew sleeping time.

2. New JAXA payload racks

2.1. Kobairo rack overview

KOBAIRO rack consists of Gradient Heating Furnace (GHF) and rack outfitting. The GHF is a vacuum furnace that contains three heating blocks for one sample cartridge. Their positions and temperatures can be independently controlled, and various temperature profiles in the direction of sample cartridge axis can be realized. This facility will be mainly used for high quality crystal growth experiments using unidirectional solidification. GHF has an automatic sample exchange system that can be accommodate up to 15 samples to reduce crew operation. The three independent heating zones can realize various temperature profiles to meet with variety of experiments from researchers. Various experiments can be realized with high temperature (Max.1600 deg. C) and highly stable temperature accuracy. Experiments can be performed continuously by installing 15 sample cartridges (Max.) in Sample Cartridge Automatic Exchange Mechanism. Specification and outlook of KOBAIRO rack is shown in table 2.1-1. and figure 2.1-1. [1].

Table 2.1-1. Kobairo rack specification

		Specifications
GHF Characteristics	Heating Temperature Range	End Heater : 500 to 1600 deg. C (Moving distance : Max. 200 mm) Central Heater : 500 to 1600 deg. C (Moving distance : Max. 250 mm) Auxiliary Heater : 500 to 1150 deg. C (Moving distance : Max. 250 mm)
	Temperature Stability	±0.2 deg. C or better
	Temperature Accuracy	±0.4 % or better
	Temperature Gradient	Up to 150 deg. C/cm (at 1450 deg. C)
	Heating Block Speed	0.1 to 200 mm/hr or 600 mm/hr
	Speed Stability	±1 % or better (10 to 200 mm/Hr) ,

		±10 % or better (0.1 to 10 mm/Hr)
	Insertion Slot Diameter	φ 40 mm
	Sample monitoring	Temp.: Up to 10 points * 2 ranges (Hi/Lo) 5 points (standard) * 2 ranges (Hi/Lo) Pressure : Diaphragm pressure sensor, Pirani gauge transducer, Cold cathode transducer
	Marking Device	Pulse current interface is available (Marking).
	Continuous operating time	300 Hours at maximum temperature
	Max. Power	Up to 5300W
Cartridge	Dimension	Boss : 93 mm Cartridge : 505 mm (φ 34.4 to 36.1 mm)
	Mass	Up to 6 kg
	Max. sample dimensions	φ 31 mm x 370 mm L (common sample cartridge)
	Temperature Monitoring	5 points (Max.10 points)

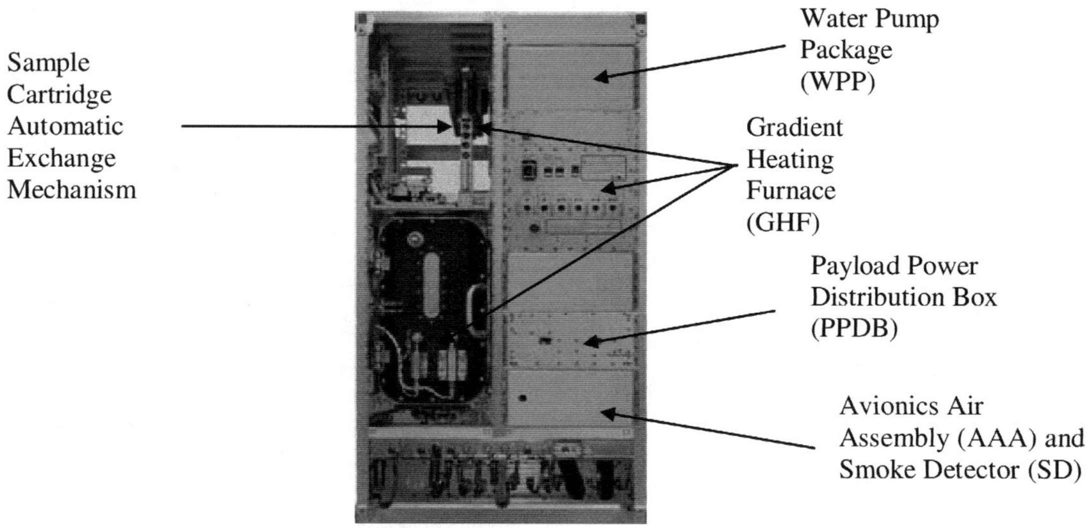

Figure 2.1-1. This figure shows appearance of Kobairo rack.

2.2. MSPR overview

Multi-purpose Small Payload Rack (MSPR) is the rack to accommodate the researcher-made experiment equipment in JPM. MSPR provides Combustion Chamber for Experiment (CCE) which is used to conduct a combustion experiment. MSPR has interfaces with KIBO resources such as JEM Exhaust line, GN2 line and Moderate Temperature Water line. These facilities are used inside of Work Volume (WV) which is the work space to be conducted combustion experiment and other experiments as well.

MSPR consists of DC/DC Converter Unit (DCU), Multi Protocol Converter (MPC), Hub Unit (HBU), Video Compression and Recording Unit (VCRU), and CCE. Laptop PC is used by sending commands to experiment devices. Specification and outlook of MSPR is shown in table 2.2-1. and figure 2.2-1. [1].

Table 2.2-1. MSPR specification

		Specifications
Workspaces Table	Work Volume (WV)	A workspace for the experiments, such as Aquatic Habitat experiments and the combustion experiments. And also accommodate other equipments for science and educational missions within its function. Dimensions : 600mm(H)*900mm(W)*660mm(D)(356[L]) Functions & Properties : N2 gas supply and exhaust line of JEM are available. : Heat rejection, smoke detection from the user's equipments. : Cutting off the power supply at emergency. : Relieving the requirement level of noises and electric emissions from user's equipments.
	Small Experiment Area (SEA)	A volume for small size experiment which do not need much resource such as crystallization experiments. Dimensions : ≧300mm(H)*440mm(W)*516mm(D) (68[L]) Functions &Properties : Heat rejection, smoke detection of the user's equipments. : Cutting off the power supply at emergency. : Relieving the requirement level of noises and emissions from user's experiments.
	Workbench (WB)	A work table for crews maintaining user equipments, exchanging samples and reviewing data etc. Dimensions : 900mm(W)*600mm(D) (0.54[m2]) Functions & Properties : Putting in storage into the rack when it is not in use.
Enclosed Chamber (can be accommodated in WV)	Chamber for Combustion Experiments	A double enclosed vessel for the combustion and other experiments that needs enclosed and/or exhaust systems in JEM. Capacity : :More than 100[L] Functions & Properties : No damage is found in the case of 100[L]gas (at normal temperature and pressure) leakage. : All functions and resources for WV are available.
User's Interfaces	Power	Supplied interfaces : 28[VDC], 16[VDC], 5[VDC] Power Supply : Work Volume and Small Experiment Area･･･Total 500[W] : Workbench 100[W] (16[VDC]; for LAPTOP PC)
	Videos	As MPSR, one of followings is available (recorded,

		downlink). NTSC video*3[ch] or High-definition television(HDTV)*1[ch]
	Communication	Protocol: Ethernet, USB User's equipments ⇔ LAPTOP PC, LAPTOP ⇔ PC Users on the ground

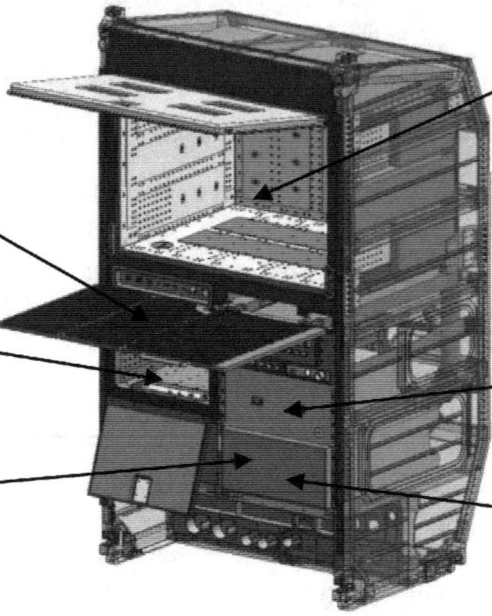

Figure 2.2-1. This figure shows appearance of MSPR.

3. Continuous effort for microgravity environment in KIBO

3.1. Background and past microgravity measurement in KIBO

3.1.1. Marangoni experiment overview. Marangoni convection is a surface-tension-driven flow. A liquid bridge of silicone oil is formed into a pair of disks in Fluid Physics Experiment Facility Experiment Cell (FPEF-EC) [2]. The liquid bridge is very sensitive to g-jitter because it is not contained and is sustained by the only surface tension between supporting disks [3]. Therefore, this experiment is performed during crew sleeping period (21:30 - 06:00) when g-jitter is expected to become slightly calm.

3.1.2. Past microgravity measurement results summary since increment 17

a) *Increment 17*

The first Marangoni Experiment was performed in KIBO. The experiment was going well part of the way, but at the final phase of the experiment series, liquid bridge breakup occurred for the first time. After analyzing the microgravity data, it was found some g-jitter was generated at the timing of liquid bridge breakup during crew sleeping time. And also the analysis showed that the taller liquid bridge tends to vibrate more greatly and that strong g-jitter might have the possibility of causing the liquid bridge breakup [4] [5]. Even though the disturbance is occurred in other modules than in KIBO, it may affect Marangoni Experiment from g-jitter perspective.

b) *Increment 19/20*

The number of crewmembers in the ISS has been increased from 3-person operations to 6-person operations since May, 2009. In order to provide a crew sleeping location pro tempore in preparation for 6-crew person operations start in ISS, the crew quarters equipment needed to be temporarily installed in KIBO. We coordinated this issue with NASA operations team and agreed to measure micro G data for all possible crew sleeping configurations in KIBO. Finally, the analysis result showed that crew quarters configuration did not affect Marangoni experiment and g-jitter with crew sleeping in KIBO was lower than ISS microgravity acceleration performance (SSP41000) [6].

c) *Increment 23/24*
We performed some series of Marangoni experiment continuously and still observed long liquid bridge vibrating largely sometimes. At the expedition crew debriefing, the crewmembers advised us to call-up a heads-up message to all crewmembers at the evening Daily Planning Conference (E-DPC) just before we perform long liquid bridge experiment at night.

d) *Increment 25/26*
Per crew advice, we identified the special nights which we needed the most quiescent micro-g environment for long length liquid bridge experiments with 50 or 60 mm tall of 50 mm diameter. Then we have started to call up a heads-up message to crew at E-DPC just before we perform long liquid bridge experiment. It was very effective to obtain cooperation from all of the crew including Russian cosmonauts. G-jitter level was definitely reduced compared to previous increments because crews might pay attentions when they moved around during Marangoni experiment with taller liquid bridge.

3.2. Microgravity measurement during crew activity time in KIBO

3.2.1. During Increment 25, we measured microgravity data in KIBO during crew working time with KIBO internal camera view and by using MMA (Microgravity Measurement Apparatus) in order to investigate the following correlation between crew motion and g-jitter.
a) What kind of crew motion may generate critical disturbance to our experiment?
b) How much g-jitter will be generated by crew motion?
By analyzing the microgravity data and KIBO internal camera view, we are expecting to reduce g-jitter which is generated by crew motion in future experiment.

3.2.2. Microgravity measurement condition is as follows. MMA consists of TAA (Tri-Axial Accelerometer), RSU (Remote Sensor Unit), NCU (Network Control Unit) and MLT (MMA Laptop Terminal). Measurement parameter is shown in table 3.2.2-1.

Table 3.2.2-1. Microgravity data measurement parameters

Measurement Time	2010/10/07 08:03:32 - 2010/10/08 00:03:33
Sampling Rate	1220.7 Hz
Filter Cutoff	300 Hz
MMA Sensors (RSU and TAA)	RSU ID: 3011 TAA ID: 3005

3.2.3. Analysis results are shown in Table 3.2.3-1. This table shows the summary for the correlation between g-jitter over 50 μG and crew motion. 50 μG is approximately equivalent to ISS standard microgravity requirement. Each axis for micro-G data is related to KIBO direction as follows. X axis (Overhead-Deck), Y axis (Starboard-Port) and Z axis (Forward-Aft). Refer to figure 3.2.3-1.

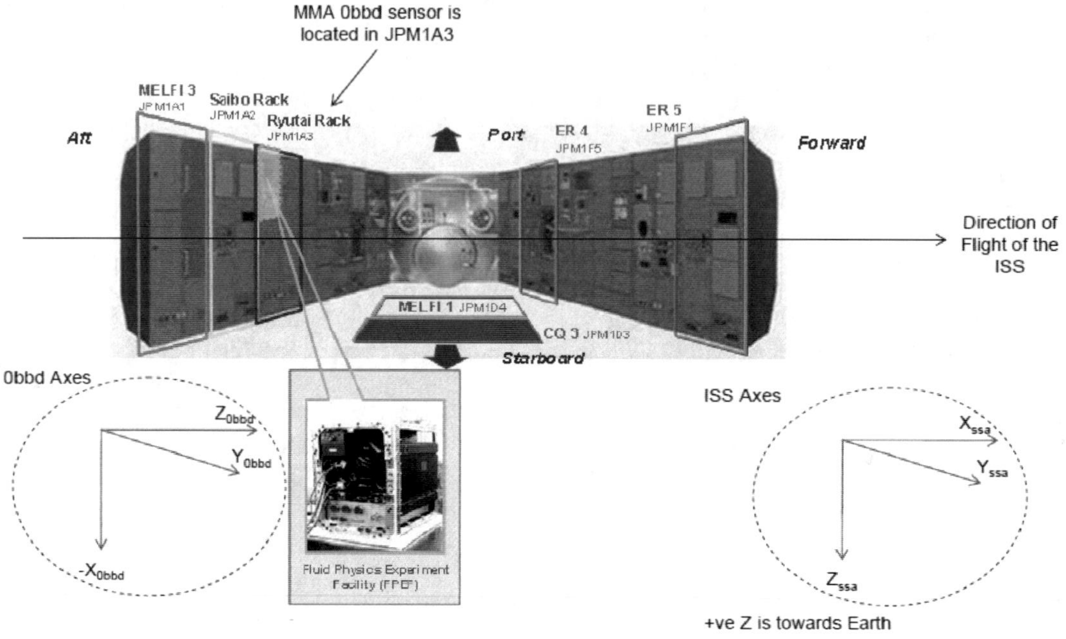

Figure 3.2.3-1. This figure shows MMA sensor and ISS/KIBO axes.

Table 3.2.3-1. This table shows MMA measurement data analysis results (crew motion vs. g-jitter).

KIBO Internal Camera View		**MMA Data**	
Time	**Crew Motion**	**Time**	**G-Jitter**
14:31:50	Moving towards KIBO by pushing Deck side in Node2	14:31:53	70 μG (X axis) 120 μG (Y axis)
14:31:54	Grasping the handrail at O3	14:31:55	120 μG (X axis)
14:33:56	Moving in front of JPM1 Airlock (Aft to Fwd)	14:33:57	150 μG (Y and Z axis)
14:35:40	Kicking the deck side in front of Airlock	14:35:40	150 μG (Y and Z axis)
14:38:16	Grasping the handrail at A5 to stop	14:38:19	70 μG (X axis) 100 μG (Y axis)
14:38:18	Touching the rack at A7		
14:43:36	Touching the fwd side in front of Airlock to stop	14:43:36	100 μG (Z axis)
14:44:24 – 14:44:28	Unstowing something from F8 rack	14:44:25	100 μG (Z axis)
14:49:15	Kicking the deck side in front of Airlock to move	14:49:14	50 μG (X axis) 150 μG (Y axis)
14:50:55	Touching the handrail at O3	14:50:56	250 μG (X axis) 100 μG (Y axis) 50 μG (Z axis)
14:50:56	Grasping the handrail at O6 to stop		
14:54:59	Putting CTB in front of Airlock	14:54:59	100 μG (X axis)
14:57:07	Grasping the handrail at O7	14:57:06	100 μG (X axis)
14:57:22	Grasping the handrail at O7 to move (Fwd to Aft)	14:57:20	50 - 100 μG (Y axis) 50 μG (Z axis) continuing for about 30
14:57:35	Pushing the handrail at O7 to		

	move downward		seconds
14:57:36	Grasping the handrail at D6		
14:57:37	Grasping the handrail at D4		
14:57:52 –14:57:59	Grasping the handrail at deck side to move	14:57:52	200 μG (X axis) 100 μG (Y axis) 150 μG (Z axis)
14:59:48	Kicking the rack at D7 to move upward	14:59:48	100 μG (Y axis) 80 μG (Z axis)
15:02:17	Touching the rack at D8 by foot	15:02:15	120 μG (X axis) 80 μG (Y axis)
15:03:35 –15:03:38	Grasping the handrail at O8 thru O5 to move	15:03:37	100 μG (Y axis)
		15:03:42	150 μG (Z axis)
15:05:02	Touching the rack at A5	15:05:03	100 μG (X axis) 120 μG (Y axis) 150 μG (Z axis)
15:05:06	Pushing the rack at A5	15:05:05	250 μG (Y axis) 350 μG (Z axis)
15:05:07	Touching the rack at D4 by foot	15:05:08	50 μG (X axis) 130 μG (Y axis) 80 μG (Z axis)
15:05:08	Pushing the rack at D4 to move		
15:05:10	Touching the rack at D6 to stop		
15:06:13	Kicking the rack at D8 to move	15:06:12	100 μG (X axis) 150 μG (Y axis)
15:06:14	Touching the rack at D5		
15:10:56	Grasping the handrail at O3 to stop	15:10:57	100 μG (X axis) 130 μG (Y axis) 250 μG (Z axis)
15:13:15	Touching the rack at F7 by foot	15:13:15	80 μG (Y axis) 60 μG (Z axis)
15:14:52	Touching the rack at D7 by foot	15:14:52	90 μG (X axis) 80 μG (Y axis) 50 μG (Z axis)
15:16:16	Touching the rack at D7 by foot	15:16:17	150 μG (X axis) 100 μG (Y axis)
15:16:17	Kicking the rack at D7 to move		
15:16:19	Grasping the handrail at D4	15:16:20	80 μG (Y axis) 120 μG (Z axis)
15:16:24	Pushing the forward side in Node2 and grasping the aft side handrail	15:16:28	80 μG (X axis) 120 μG (Y axis) 90 μG (Z axis)
15:19:22 –15:19:27	Grasping the overhead handrails to move	15:19:27	150 μG (X axis) 100 μG (Y axis) 90 μG (Z axis)
15:20:15	Touching the rack at D8 by foot	15:20:15	150 μG (X axis)
15:22:33 –15:22:45	Working around F7 (Unstowing something and so on.)	15:22:39	50 μG (X axis) 50 μG (Y axis) 110 μG (Z axis)
		15:22:43	120 μG (X axis)

			90 μG (Y axis) 150 μG (Z axis)
15:23:35	Working around F7	15:23:37	80 μG (X axis)
15:23:39	Closing the rack panel at F7	15:23:40	60 μG (X axis) 80 μG (Y axis)
15:23:45 –15:24:44	Handling CTB in front of Airlock	15:24:43	90μG (X axis)
15:48:09	Touching the rack at A8 by foot	15:48:06	70 μG (Y axis) 60 μG (Z axis)
15:48:13	Touching the rack at D8 by foot	15:48:16	90 μG (X axis)
15:58:31 –15:58:41	Shaking body with the foot restraint at D5	15:58:31	100 μG (X axis)
16:00:58	Kicking the rack at D8 and grasping the Airlock upper side to move	16:00:56	60 μG (Y axis) 80 μG (Z axis)
16:04:35	Touching the rack at D8 by foot	16:04:35	100 μG (X axis) 80 μG (Y axis)
16:16:17	Grasping the handrail at O6	16:16:18	100 μG (X axis)
16:16:49	Touching the rack at D8 by foot	16:16:51	100 μG (X axis)
16:19:59	Grasping the handrail at O5	16:19:59	80 μG (X axis)
16:35:08	Opening the panel at F7	16:35:11	100 μG (X axis)
16:41:27 –16:41:38	Stowing CTB at F7 and closing the panel	16:41:42	100 μG (X axis) 130 μG (Y axis) 100 μG (Z axis)
16:42:18	Grasping the handrails in Node2	16:42:18	90 μG (Y axis)
16:50:29	Touching the rack at F7	16:50:27	150 μG (X axis) 70 μG (Y axis) 150 μG (Z axis)
17:26:55 –17:27:35	Working around F7 (Stowing something and so on.)	17:27:09	70 μG (X axis) 60 μG (Y axis) 130 μG (Z axis)
		17:27:22	100 μG (X axis)
		17:27:35	100 μG (X axis)
17:27:59	Touching the rack at D7 by foot	17:28:01	100 μG (X axis) 70 μG (Y axis) 110 μG (Z axis)
17:47:01	Grasping the handrail at O7	17:47:02	100 μG (Y axis)
17:48:51 –17:48:59	Touching the rack at D7	17:48:53	50 μG (X axis) 150 μG (Y axis) 200 μG (Z axis)
		17:48:59	180 μG (X axis) 750 μG (Y axis) 950 μG (Z axis)

3.2.4. Based on the analysis result, the following lessons learned are acquired.

a) When crew moves grasping the overhead handrails, g-jitter over 100 μG often occurs.

b) When crew touches the rack at deck side by hand or by foot, strong g-jitter does not seem occurred. But when crew moves from overhead to deck side and then touches the deck side, or when crew kicks the rack at the deck side to move, g-jitter over 100 μG is generated, especially in X axis.

c) When crew was working in the vicinity of Airlock, strong g-jitter was not observed.
d) When crew used the voice loop to communicate ground team, g-jitter was not observed.
e) Even though crew sometimes used a laptop terminal at A7, g-jitter was not observed at that time.
f) When crew stowed/unstowed something to/from Cargo Transfer Bag (CTB), some g-jitter occurred. But operations in the JLP did not generate g-jitter.

3.3. Waste and Hygiene Compartment (WHC) usage and g-jitter

In addition to micro-G data measurement in crew day time, we also measured micro-G data in crew sleep time during Marangoni experiment. To monitor liquid bridge motion real-timely during experiment, we observed downlinked image of liquid bridge via Image Processing Unit (IPU). Measurement parameter is shown in table 3.3-1.

Table 3.3-1. Microgravity data measurement parameters

Measurement Time	2010/10/27 22:00 - 2010/10/28 05:00
Sampling Rate	398.6 Hz
Filter Cutoff	120 Hz
MMA Sensors (RSU and TAA)	RSU ID: 3011 TAA ID: 3005

We still observed liquid bridge was vibrating so much unexpectedly for several times during experiment, especially when we performed long liquid bridge experiment. The next day after we observed some big disturbances to liquid bridge during crew sleeping, we confirmed with crew and NASA/Houston Flight Control Team (HFCT) whether they had something which rang a bell on midnight g-jitter.

According to crew answer and HFCT's confirmation on the related telemetries, we found that some crew motion related to the Waste Hygiene Compartment (WHC) may generate g-jitter which may impact on Marangoni Experiment. (See figure 3.3-1.)

Figure 3.3-1. This is a figure an appearance of Waster Hygiene Compartment (WHC). WHC is located at Node 3.

By imposing WHC current telemetry on the micro-G data at the time when we observed big disturbances, the following features were inferred. (See figure 3.3-2. and figure 3.3-3.)

1) Big g-jitter occurred just before/after WHC usage. In case of figure 3.3-2 and figure 3.3-3, the vibration level was about ± 20 μG (from 0.1 Hz through 0.5 Hz) in Y and Z axis. This value is almost equivalent to the ISS microgravity acceleration performance (SSP41000), however, it is larger than usual during crew sleeping timeframe. It means WHC itself is not the direct vibration source, but

crew moving or door opening/closing may generate some g-jitter which may cause big vibration on the liquid bridge. (Vibration continuous duration was about 30 seconds for each just before/after WHC usage.)

2) By analyzing crew motion with downlinked KIBO internal camera video, crew movement velocity is thought to be about 8/7 [m/s]. Per this velocity, crew can move 34.3 [m] in 30 seconds, for example, from Service Module to Node 3.

3) The similar features were confirmed in other timings when big g-jitter occurred. The g-jitter level was almost $\pm 20\ \mu G$ in all cases.

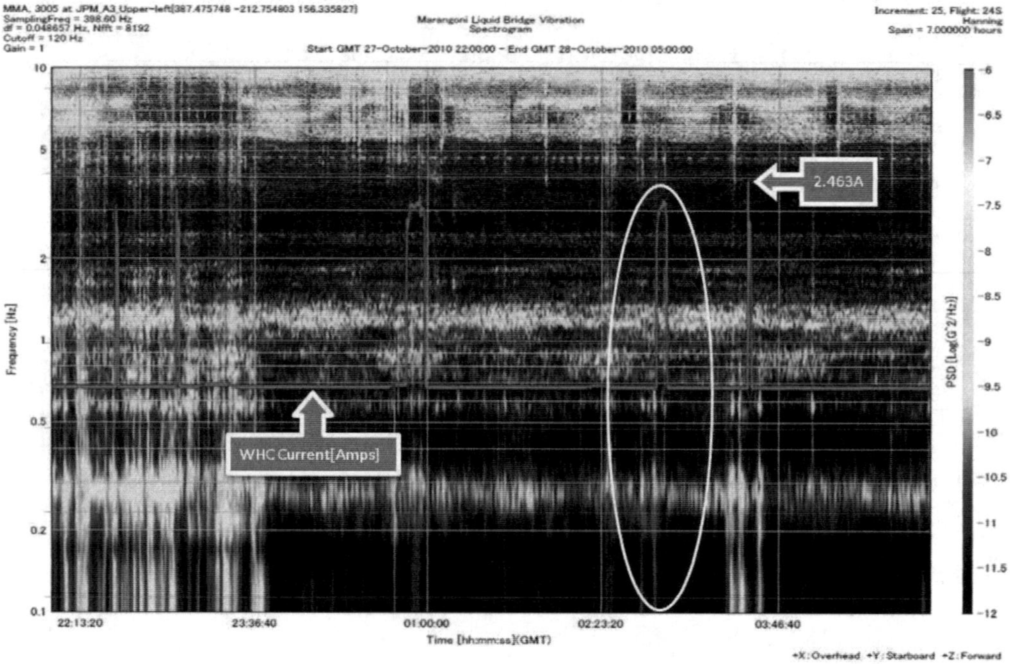

Figure 3.3-2. This figure shows correlation between big disturbances and WHC current telemetry (GMT300-301). When WHC current value increased, low frequency (less than 0.7 Hz) g-jitter was also increased. For example, typical g-jitter related to WHC usage (from GMT 301/0:41 – 1:07) is shown in Figure 3.3-3.

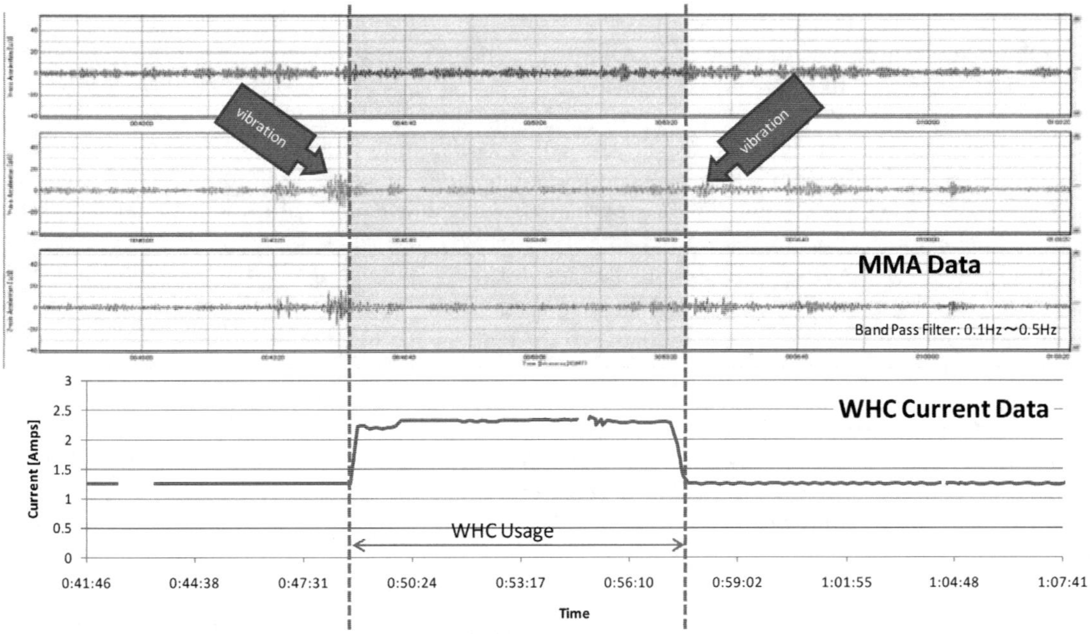

Figure 3.3-3. This figure shows MMA data and WHC current data on GMT301. Just before/after WHC usage timing, g-jitter occurred. It means that WHC is not the direct vibration source but some crew motion related to WHC may generate g-jitter. The vibration level was almost $\pm 20\ \mu G$ (from 0.1 Hz through 0.5 Hz) in Y and Z axis.

Because all crew members have kindly cooperated to be quiet at night when we perform long liquid bridge experiment, we would like to reduce too much crew constraints during the night based on these new findings. From the above results, we will ask crew to open/close the WHC door/shutter softly or to grab the handrails softly, and move quietly and slowly in order to reduce a big g-jitter may be caused by crew motion related to WHC activities.

3.4. *Future work plan*

For future work to improve microgravity environment in KIBO, we are planning to conduct more challenging experiment and microgravity data measurement. For example, in the Increment 29/30, we will try to perform Marangoni experiment during crew working time in order to assess whether we can conduct Marangoni experiment during crew working time without liquid bridge breakup or not. Because we have no plan to use the Experiment Cell in future again which we'll use in this trial, we do not need to request any cleanup activity by crew even though liquid bridge breakup would occur.

In addition, we are thinking about the possibility to measure microgravity data with support of International Partners while the ROBONAUT is operating during crew sleeping time which may simulate crew motion. We are expecting that microgravity data analysis can be more quantitative since ROBONAUT can move predetermined forces.

4. Summary and future KIBO payload operations

Microgravity environment in KIBO has been greatly improved over the past year due to advanced microgravity measurement and analysis, and of course, crew cooperation during sleeping time is highly appreciated. Moreover, these microgravity measurement and analysis required close cooperation by both International Partners and crewmembers, and then this cooperation has resulted in building a good relationship and enhancing the bond among us in order to achieve good science success in ISS/KIBO as a team.

Journal of Physics: Conference Series **327** (2011) 012054 doi:10.1088/1742-6596/327/1/012054

Through comparing the MMA data and other related data such as onboard internal camera video and the WHC telemetry, we are expecting to narrow down the possible causes of g-jitter during crew sleeping time to crew motion related WHC usage. For example, we'll ask crew to reduce g-jitter may be caused by crew motion related to WHC usage such as grabbing the handrails to move or opening/closing the door/shutter of WHC. This request will be applied to reduce g-jitter when we are planning to conduct Marangoni Experiment in crew day time for trial. We would like to share the lessons learned we have got through these experiences with crewmember and International Partners to reduce crew constraints during the special night when we perform long liquid bridge Marangoni experiment. For the future, we will aim to perform further challenging experiments utilizing good microgravity environment in ISS/KIBO.

References

[1] JAXA (Japan Aerospace Exploration Agency) Web page: http://iss.jaxa.jp/en/
[2] Kawamura H., Nishino K., Matsumoto S. and Ueno I. 2010 *ASME Conf. Proc. IHTC14* **8** 343
[3] Wickramasinghe D., Kawaji M. and Das K. 2010 *Proc. 61st International Astronautical Congress* **IAC-10.A2.6.7** 1
[4] http://pims.grc.nasa.gov/pims_iss_index.html
[5] http://idb.exst.jaxa.jp/english/basic-data_e.html
[6] Sakagami K., Goto M. And Matsumoto S. 2010 *Proc. 61st International Astronautical Congress* **IAC-10.A2.5.4** 1

Journal of Physics: Conference Series **327** (2011) 012055 doi:10.1088/1742-6596/327/1/012055

Proposal of Experimental Setup on Boiling Two-phase Flow on-orbit Experiments onboard Japanese Experiment Module "KIBO"

S Baba[1], T Sakai[1], K Sawada[1], C Kubota[1], Y Wada[1], Y Shinmoto[1], H Ohta[1], H Asano[2], O Kawanami[3], K Suzuki[4], R Imai[5], H Kawasaki[6], K Fujii[6], M Takayanagi[6] and S Yoda[6]

[1] Department of Aeronautics and Astronautics, Kyushu University, 744 Motooka, Nishi-ku, Fukuoka, Japan

[2] Department of Mechanical Engineering, Kobe University, 1-1 Rokkoudai-chou, Nada-ku, Kobe, Hyogo, Japan

[3] Department of Mechanical and System Engineering, University of Hyogo, 2167 Shosha, Himeji, Hyogo, Japan

[4] Department of Mechanical Engineering, Tokyo University of Science, Yamaguchi, 1-1-1 Daigakudori, Sanyo-Onoda, Yamaguchi, Japan

[5] IHI Corporation, 1 Shinnakahara-chou, Isogo-ku, Yokohama-shi, Kanagawa, Japan

[6] Japan Aerospace Exploration Agency, 2-1-1 Sengen Tsukuba-shi, Ibaraki, Japan

E-mail: s_baba@aero.kyushu-u.ac.jp

Abstract. Boiling is one of the efficient modes of heat transfer due to phase change, and is regarded as promising means to be applied for the thermal management systems handling a large amount of waste heat under high heat flux. However, gravity effects on the two-phase flow phenomena and corresponding heat transfer characteristics have not been clarified in detail. The experiments onboard Japanese Experiment Module "KIBO" in International Space Station on boiling two-phase flow under microgravity conditions are proposed to clarify both of heat transfer and flow characteristics under microgravity conditions. To verify the feasibility of ISS experiments on boiling two-phase flow, the Bread Board Model is assembled and its performance and the function of components installed in a test loop are examined.

1. Introduction

Recent increase in the size of space platforms requires the management of larger amount of waste heat under high heat flux conditions and the transportation of it along a long distance to the radiator. Flow boiling applied to the thermal management system in space attracts much attention as promising means to realize high-performance heat transfer and transport. In microgravity two-phase flow phenomena are quite different from those under normal gravity conditions because buoyancy effects are significantly reduced and surface tension becomes dominant.

Although a number of experiments on boiling heat transfer in microgravity were conducted worldwide from the 1970s to the present, and reviewed by Di Marco [1], Kim [2] and Ohta [3]. Ohta [4] investigated for flow boiling in a transparent heated tube coated with gold thin film for different gravity levels along the parabolic trajectory of an aircraft. In the annular flow regime, at heat flux not high enough to initiate the nucleate boiling, the heat transfer coefficient in microgravity deteriorates

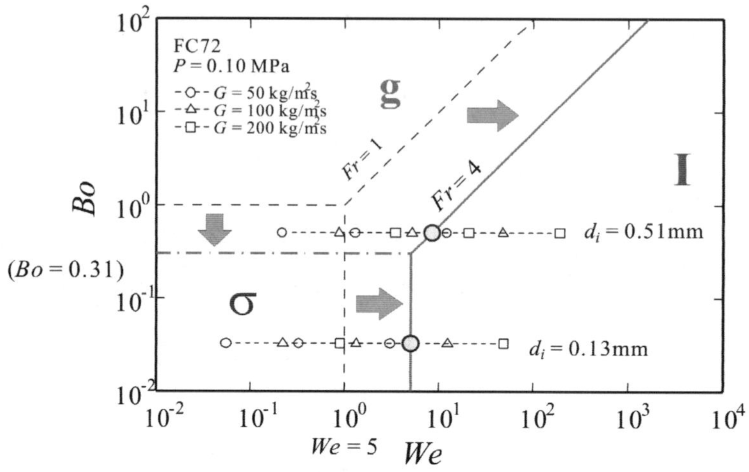

Figure 1. Existing results on dominated force regimes map (Black dashed lines) and latest boundaries (Red solid lines).

due to the increase in the thickness of annular liquid film and the reduction of turbulence in it. No marked gravity effect on the heat transfer is observed when the heat transfer is dominated by nucleate boiling.

On the other hand, the effects of direction of gravity vector on flow boiling were conducted by the researchers. Zhang et al. [5] changed orientation of narrow channel and examined the influence on the critical heat flux. Flow boiling experiments were conducted by using the rectangular narrow channel with heating surfaces of lengths 101.6mm, width 30mm and gap size 2.5mm. At low velocity experiments, especially for saturated conditions, critical heat flux for downward flow was much smaller than for upward flow. Kandlikar and Balasubramanian [6] conducted effects of channel orientation on flow boiling heat transfer by using the rectangular narrow channels with 1.054mm width, 0.197mm depth and a hydraulic diameter of 0.333 mm. They also observed heat transfer coefficients in the vertical downward flow are higher than those in vertical upward flow and horizontal flow. Bower and Klausner [7] proposed the gravity dependent/independent regime map on subcooled flow boiling heat transfer by changing orientations of in rectangular narrow channel. They employed Jakob number *Ja* and dimensionless flow rate parameter ψ as important parameters for defining the gravity dominant regime.

However, gravity effects on the two-phase flow phenomena and corresponding heat transfer characteristics have not been clarified in detail. The experiments onboard the International Space Station (ISS) is proposed to clarify systematically the data and corresponding mechanisms for both of heat transfer and of flow characteristics under microgravity conditions. The establishment of dominated force regime maps by the introduction of appropriate dimensionless groups is also an important objective, because the design of space thermal devices under the conditions of no gravity effect becomes more reliable based mainly on the ground test. In Figure 1, red lines in the figure are the proposed boundaries based on the recent data for flow boiling in mini-channels [8]. Bond number *Bo*, Weber number *We* and Froude number *Fr* are defined here

$$Bo = \frac{(\rho_l - \rho_v) g d_i^2}{\sigma} \tag{1}$$

$$We = \frac{\rho_m u_m^2 d_i}{\sigma} = \frac{G^2 d_i}{\rho_m \sigma} \tag{2}$$

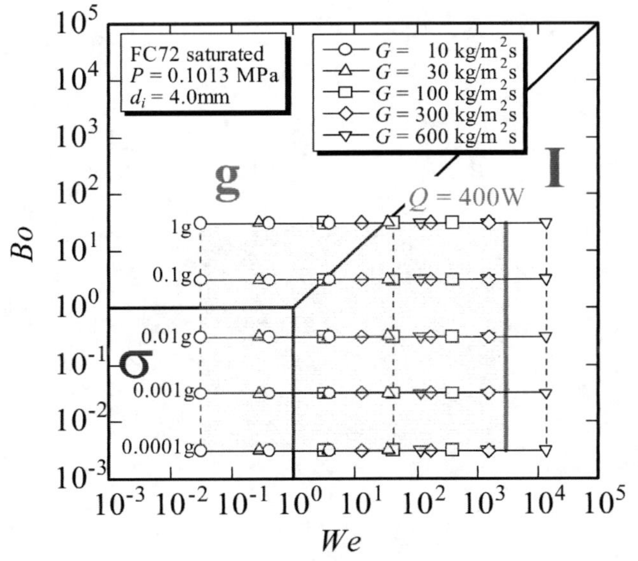

Figure 2. Experimental conditions for the ISS and ground experiments on dominated force regime map.

$$Fr = \sqrt{\frac{We}{Bo}} = \frac{G}{\sqrt{\rho_m(\rho_l - \rho_v)\, g\, d_i}} \tag{3}$$

where, d_i: tube inner diameter, G: mass velocity, u_m: mean velocity, $\rho_m = 1/[x/\rho_v+(1-x)/\rho_l]$: mean density of liquid and vapor, x: vapor quality. Therefore, *Bo*, *We* and *Fr* represent ratios of buoyancy to surface tension, inertia to surface tension and inertia to buoyancy, respectively. In the body force dominated region, effects of gravity level and flow direction become significant. In the Inertia dominated region, effects of mass velocity and vapor quality are observed, and in the surface tension dominated region, effects of gravity, mass velocity and vapor quality become small. There are four points to be clarified concerning the present regime map.

1. Boundaries of regimes are doubtful.
2. Definitions of parameters should be revised.
3. Parameters and boundaries should be changed by different targets such as heat transfer coefficient, flow patterns, pressure drop and critical heat flux.
4. Influence of vapor quality is unknown. Increase in vapor quality increases inertia force due to the increase in liquid-vapor mixture velocity despite of decrease in mixture density.

It is doubtful that the effect of gravity level is equivalent to the effect of the gravity direction on definition of regime boundary between the body force dominated and the surface tension dominated through Bond number. The microgravity experiments are useful for the verification of the effect of gravity level on these regimes. Figure 2 shows points of experimental conditions for the ISS and ground experiments on the dominated force regime map. In further studies, the results obtained under microgravity conditions are to be reflected to confirm the effect of gravity level on the boundaries at very low Bond numbers.

In the present paper, a test loop is proposed for ISS experiments under the limited resources of volume size and power supply, and preliminary experiments on flow boiling are performed on ground.

Figure 3. Bread Board Model.

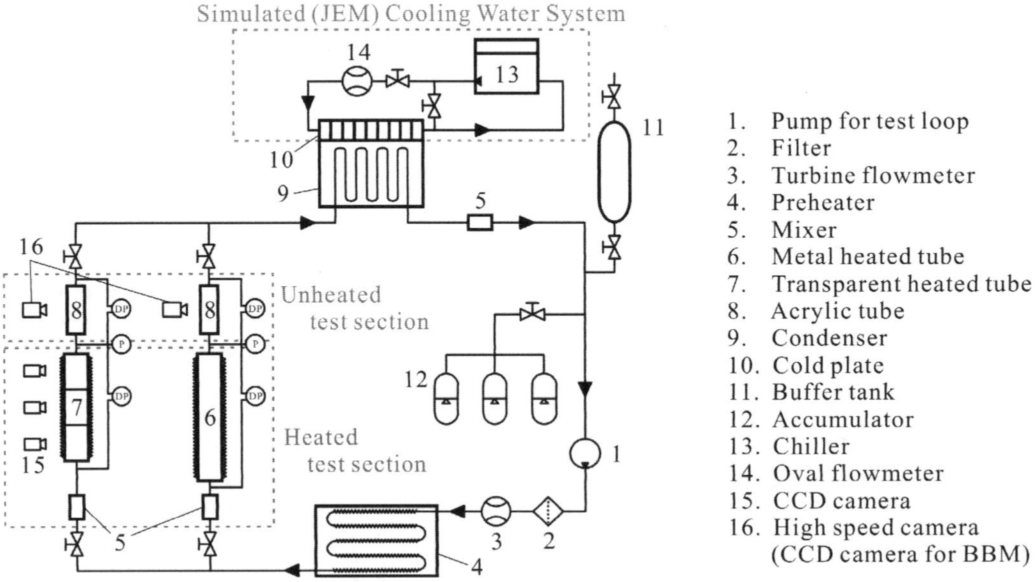

Figure 4. System components of Bread Board Model.

2. Experiment

2.1. ISS experimental overview

The experiments are conducted using the Multi-purpose Small Payload Rack (MSPR) onboard Japanese Experiment Module (JEM) "KIBO" in ISS and the experimental apparatus developed for MSPR. MSPR is resource platform providing user equipment; space, power, communication (Ethernet, IEEE1394), video, cooling and fluid interface (nitrogen, vacuum). Experiment utilizing MSPR expands from life science to fluid science. They are launched in HTV (H-IIA Transfer Vehicle). The

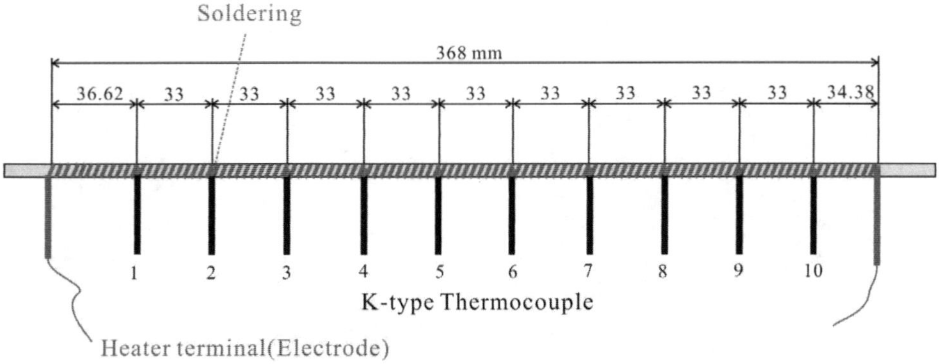

Figure 5. Metal heated tube.

Table 1. Experimental conditions for ISS.

Tube inner diameter	$d_i = 4.0$ mm
Working fluid	FC72 (deaerated)
Mass velocity	$G = 30–600$ kg/m^{2}s
Heat flux	$q = 1–100$ kW/m^2
Inlet condition	$\Delta T_{sub,in} = 0–10$ K, $x_{in} = 0–0.9$

Table 2. Experimental conditions in the present study.

Test section	Metal heated tube
Tube inner diameter	$d_i = 4.0$ mm
Tube outer diameter	$d_o = 10.0$ mm
Heated length	$L = 368$ mm
Working fluid	FC72 (deaerated)
Outlet pressure	$P_{in} = 0.10–0.25$ MPa
Mass velocity	$G = 40–500$ kg/m^{2}s
Heat flux	$q = 0.5–60$ kW/m^2
Inlet condition	$\Delta T_{sub,in} = 0$, 10 K, $x_{in} = 0.1$
Flow direction	$\phi = 0°$ (Horizontal flow)
	$\phi = 90°$ (Vertical upward flow)

flight model is installed in Work Volume (WV) which is a workspace for the experiments; 600mm height × 900mm width × 700mm depth. WV can provide cooling water system through contact with cold plate installed on the bottom panel of WV and avionics air for cooling the equipment.

2.2. Experimental apparatus on ground

Outline of BBM test loop similar to the flight model is shown in Figure 3 and 4. The components are assembled horizontally inside a two-dimensional rectangular frame to replace the components easily and to change the flow direction against the gravity. The experimental apparatus has two different heated test sections of a transparent heated tube and a metal heated tube, and unheated test sections of acrylic tubes placed at downstream of each heated tube. Inner diameter of these tubes is 4 mm. Thermal conditions at the inlet of the test section are adjusted by a preheater. The total power input is

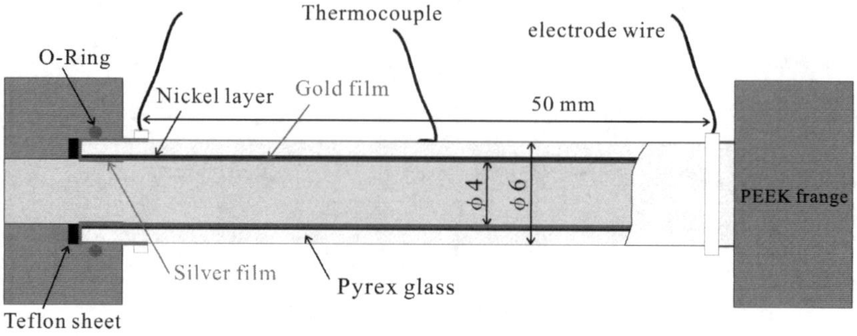

Figure 6. Transparent heated tube, one segment.

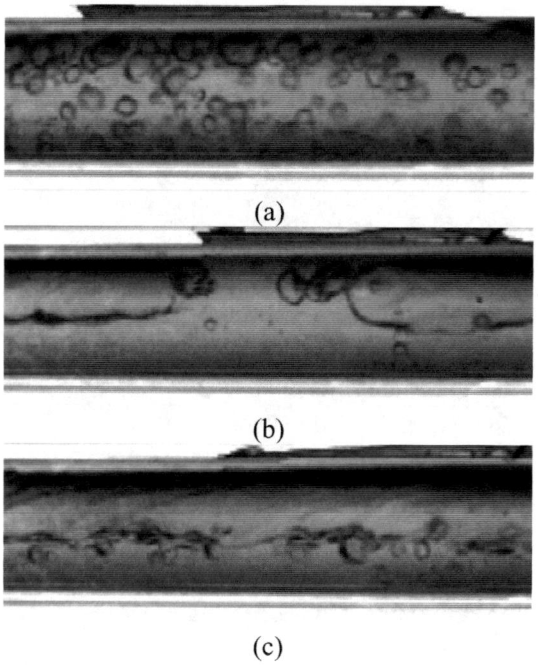

Figure 7. Images of liquid-vapor behaviors thorough a transparent heated tube in horizontal flow for different flow patterns. (a) Bubble flow, (b) Slug flow, (c) Separated flow.

removed in a condenser to obtain the steady state data, where inlet cooling water temperature and flow rate were regulated simulating the system of cooling water circulated in KIBO. The experimental conditions for ISS and these in the present study are shown in Table 1 and 2, respectively. The pressure at the accumulator is kept at near atmospheric pressure, where the saturation temperature is 56 °C for fluorinert FC72 employed as a test fluid. Nevertheless, outlet pressure of the tube is the range from 0.10–0.25 MPa depending on the pressure drop downstream. Flow boiling heat transfer with FC72 in a small diameter tube is influenced by dissolved air, as mentioned in our previous report [9]. Test fluid in the reservoir was initially degassed by preheating of the experiments.

2.2.1. Metal heated tube. The metal heated tube is used for the measurement of critical heat flux and the detailed distribution of local heat transfer coefficients along a tube axis. Test tube is shown in Figure 5. The distribution of inner wall temperature, heat flux, fluid temperature, vapor quality, and heat transfer coefficient along the heated section is calculated as discrete values at each location of 10 thermocouples based on the measured outer wall temperatures, flow rate, inlet liquid temperature, and inlet/outlet pressure. The heat loss from the outer tube wall to the ambient is evaluated by the difference between the outer wall temperature and ambient temperature, which is calibrated in advance. By the introduction of insulation and thin thermocouples, the heat loss is minimized to 45% even at the lowest heat flux tested. The inner wall temperature and the inner wall heat flux at each thermocouple location are evaluated by the radial heat conduction from the measured outer wall temperature and the local heat generation rate, taking into account the heat loss.

The inlet liquid temperature is measured by a K-type sheathed thermocouple inserted. The local heat transfer coefficient α at the location of each thermocouple is defined by

$$\alpha = \frac{q}{T_w - T_f} \tag{4}$$

where, q: inner wall heat flux to fluid, T_w: inner wall temperature, T_f: mean fluid temperature. T_f is evaluated from the heat balance equation. The saturation temperature T_{sat} is used as T_f in the vapor quality region $x > 0$.

A value of two-phase pressure drop in the vapor quality region is obtained by subtracting the value in the unheated section and the subcooled region from the pressure drop across the entire test section. The single-phase pressure drop is evaluated in the subcooled region, ignoring the possibility of partial boiling. In the vapor quality region, linear pressure drop is assumed using the pressures at the end of the subcooled region and at the outlet of the test section. Distribution of saturation temperature in the vapor quality region is obtained by the pressure distribution. All of the instruments used are carefully calibrated. The uncertainty in temperature measurement is ± 0.15 K, pressure measurement ± 0.5%, and flow rate setting ± 0.25%. The accuracy of heat transfer coefficients is evaluated to be ± 6.4%.

2.2.2. Transparent heated tube. Transparent heated tubes, which enable the heating, the observation through the glass tube and the measurement of inner wall temperature simultaneously, are introduced to clarify the mechanisms of heat transfer by the relationship between liquid-vapor interfacial behaviors and heat transfer characteristics. The tube is made of pyrex glass, and it has an ID of 4 mm and a wall thickness of 1 mm. The heated length of the tube is approximately 50 mm, and its inner wall is uniformly coated with a transparent gold film with a thickness of the order of 10 nm. Both ends of the tube are coated with a thick layer of silver film for use as electrodes and are in contact with a brass flange, as shown in Figure 6. The thin gold film for heating is operated as a resistance thermometer for the temperature measurement. As only the averaged heat transfer characteristics along the heated length can be obtained because of its structure, the test section is assembled by three segments in series, i.e., total heated length is 150 mm. At each segment of transparent heated tube, the measurement of the heat transfer coefficient and the observation of liquid-vapor interfacial behaviors by CCD camera are performed. Figure 7 shows an example of images of liquid-vapor behaviors in horizontal flow for different flow patterns recorded by CCD camera through the glass tube wall.

2.2.3. Unheated test section of acrylic tube. The unheated sections enable to observe detailed liquid-vapor interfacial structures, especially of void fraction and the distribution, and behaviors of annular liquid film, additionally to the measurement of pressure drop. The tube is made of polymethyl methacrylate, and the tube length between installed pressure taps is 80 mm.

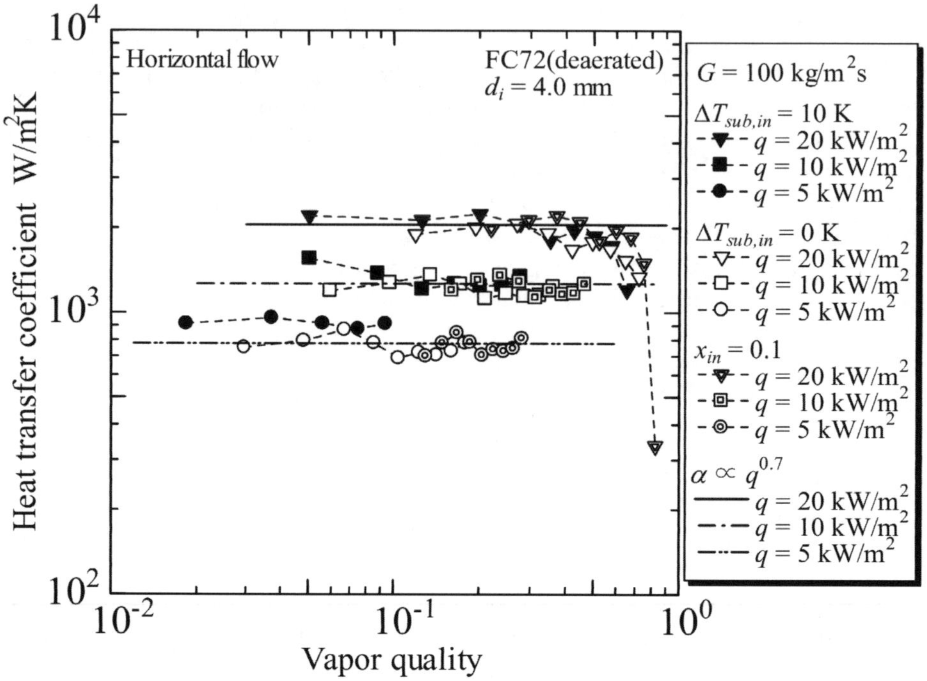

Figure 8. Heat transfer coefficient versus vapor quality at a constant mass velocity G = 100 kg/m^{2}s in horizontal flow.

3. Experimental result and discussion

3.1. Heat transfer coefficient at metal heated tube

Figure 8 shows the relation between the heat transfer coefficient and vapor quality at G = 100 kg/m^{2}s for different values of inlet condition in horizontal flow. Almost no effect of inlet liquid subcooling and inlet vapor quality is observed, and the heat transfer coefficient does not depend on the heating history in the upstream by the use of the metal heated tube installed in the test loop for the ground test.

Figure 9 shows the relation between the heat transfer coefficient and vapor quality for subcooling of inlet liquid $\Delta T_{sub,in}$ = 10 K at the test section inlet under all mass velocity conditions in horizontal flow and vertical upward flow. In the range from low to moderate vapor quality at G = 40–300 kg/m^{2}s and in low vapor quality $x < 0.2$ at G = 500 kg/m^{2}s, heat transfer coefficients is almost independent of mass velocity and vapor quality for both tube orientations. On the other hand, under the moderate vapor quality region, the trends of transition from the nucleate boiling region to two-phase forced convection region where the heat transfer coefficient increases with vapor quality are not clearly observed. This is explained as follows. Under low heat flux conditions, the length of heated tube is too short to realize the moderate or high vapor quality even at the outlet of the tube. On the other hand, under high heat flux conditions, nucleate boiling is dominant even in annular flow regime at moderate and high vapor qualities. It is needed to perform experiments in high vapor quality under lower heat flux conditions by adjusting inlet vapor quality in further studies. In addition, under the highest heat flux condition for G = 40–300 kg/m^{2}s, the deterioration of heat transfer due to dryout is observed. The trend in the nucleate boiling region is confirmed by the existing researches, for example, for Freons by using the tubes with 1.10–4.26 mm inner diameter [10–12]. The relationship between the heat transfer coefficients and heat flux in the nucleate boiling region is shown in Figure 10 for both tube orientations. For the nucleate boiling, the relation of heat transfer coefficient α and heat flux q

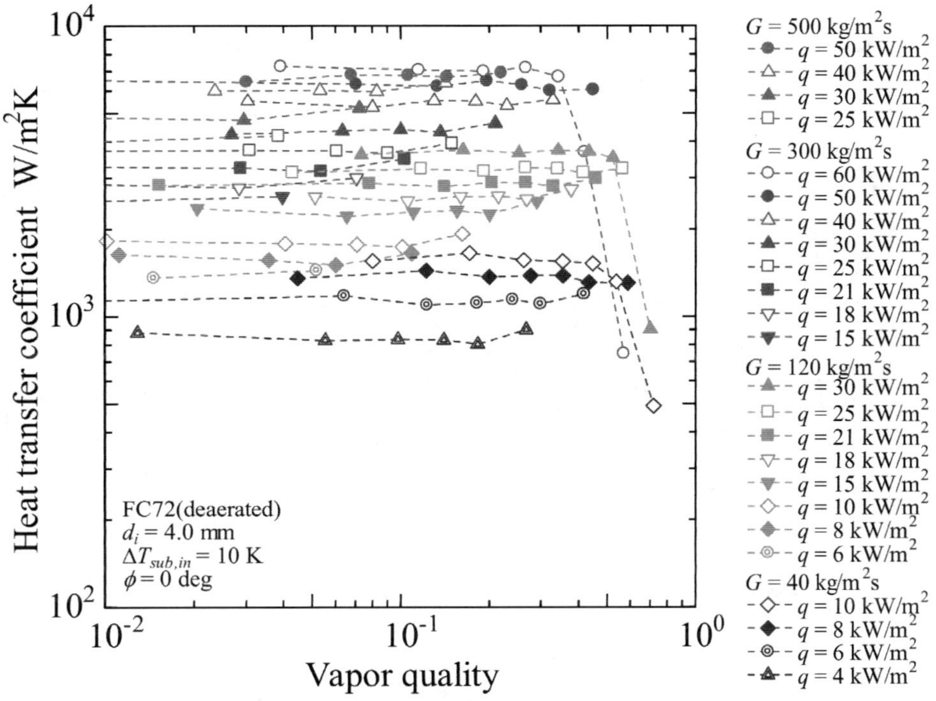

(a)

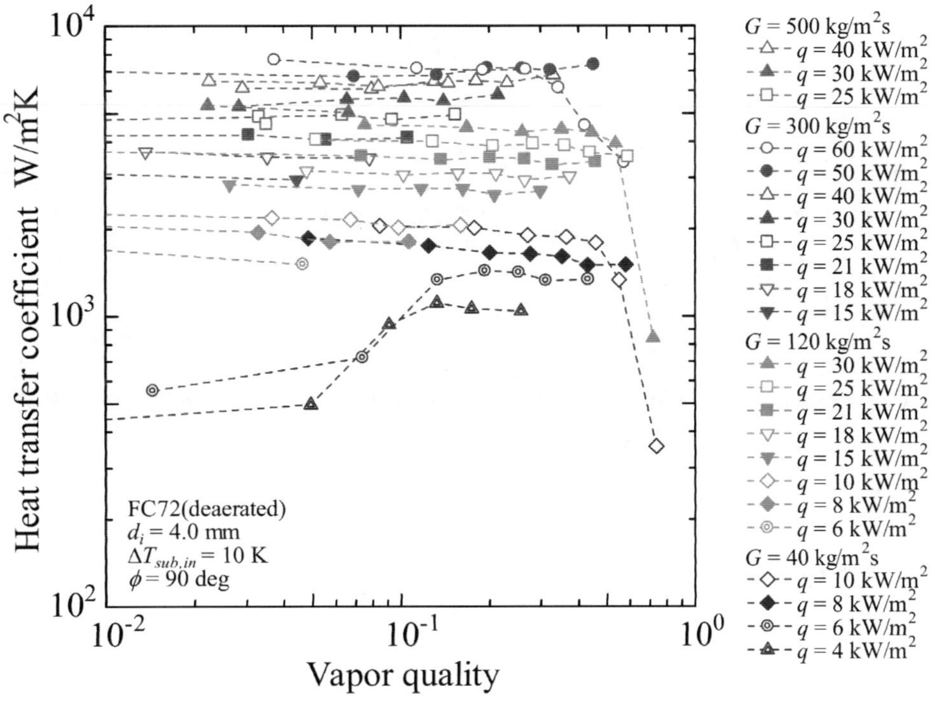

(b)

Figure 9. Heat transfer coefficient versus vapor quality. (a) Horizontal flow, (b) Vertical upward flow.

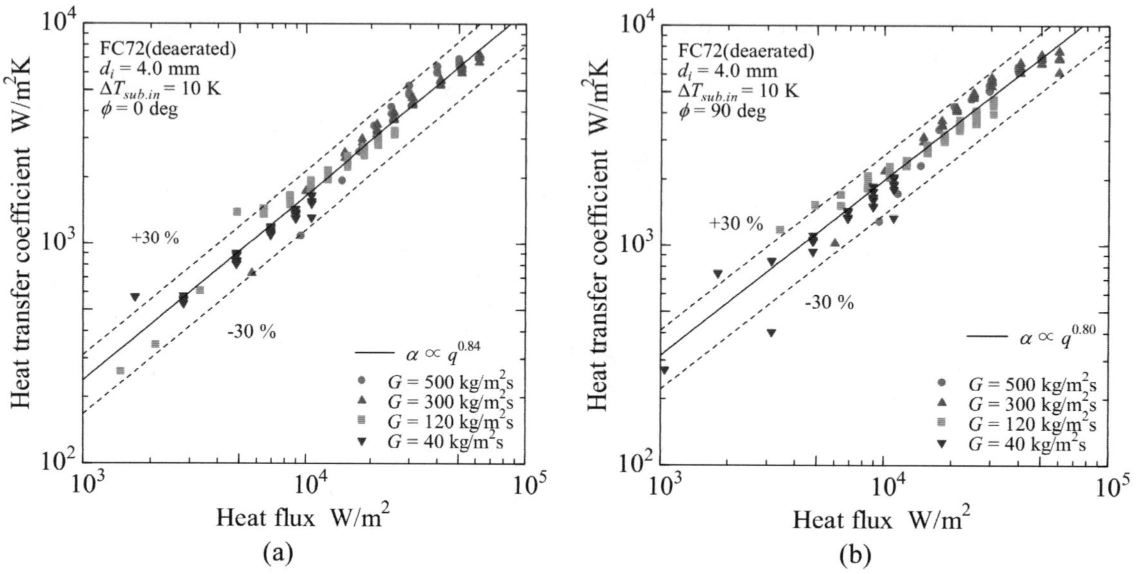

Figure 10. Heat transfer coefficient versus heat flux. (a) Horizontal flow, (b) Vertical upward flow.

can be expressed by the existing relation Eq. (6) within the error ± 30%.

$$\alpha \propto q^n \tag{6}$$

Horizontal flow: $n = 0.84$
Vertical upward flow: $n = 0.80$

Incidentally, at G = 500 kg/m^{2}s, the two-phase forced convection region and the dryout region is not observed because no test are conducted under higher heat flux conditions because of the restriction of the limited power supply 400W from the candidate facility in MSPR.

3.2. Visualization of flow regime at unheated tube

Figure 11 shows images of flow pattern transition recorded by a CCD camera at vertical upward flow, G = 120 kg/m^{2}s for an example. Four principal flow patterns, i.e., slug flow, churn flow, semi-annular flow and annular flow, were observed. Figure 12 shows flow patterns at vertical upward flow in present study and existing flow pattern map proposed by Barnea et al. [13]. Superficial liquid velocity J_l and superficial vapor velocity J_v are calculated from the test results as follows;

$$J_l = \frac{(1-x)G}{\rho_l} \tag{7}$$

$$J_v = \frac{xG}{\rho_v} \tag{8}$$

The results are in good agreement with extrapolating their air-water map to the two-phase mixture of FC72.

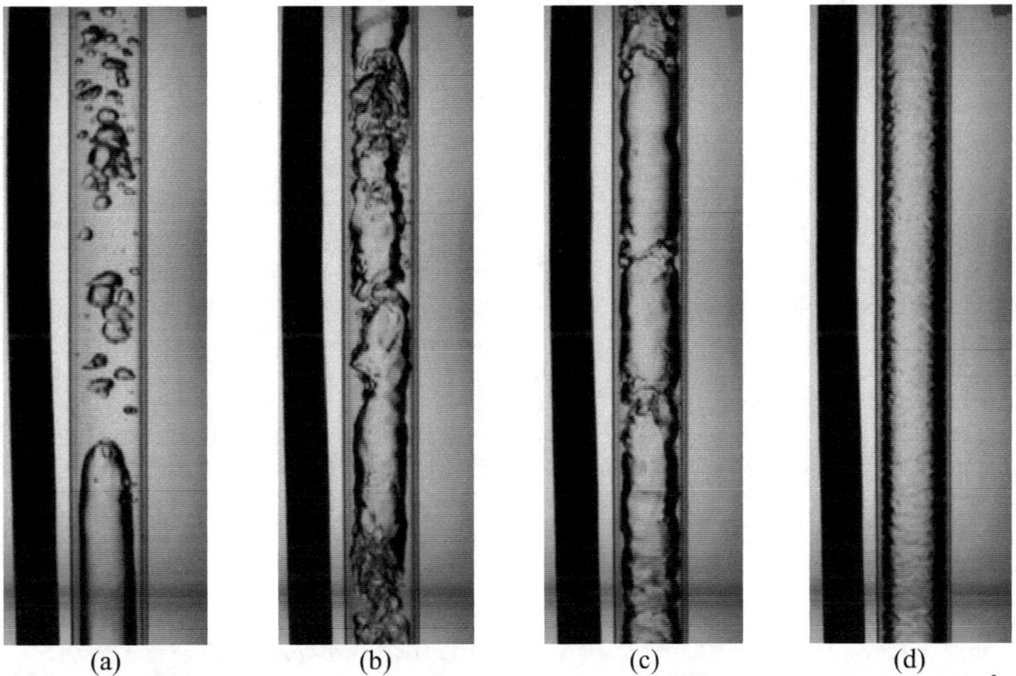

Figure 11. Images of flow pattern transition recorded by a CCD camera at G=120 kg/m^{2}s in vertical upward flow. (a)Slug flow. (b)Churn flow. (c)Semi-Annular flow. (d)Annular flow.

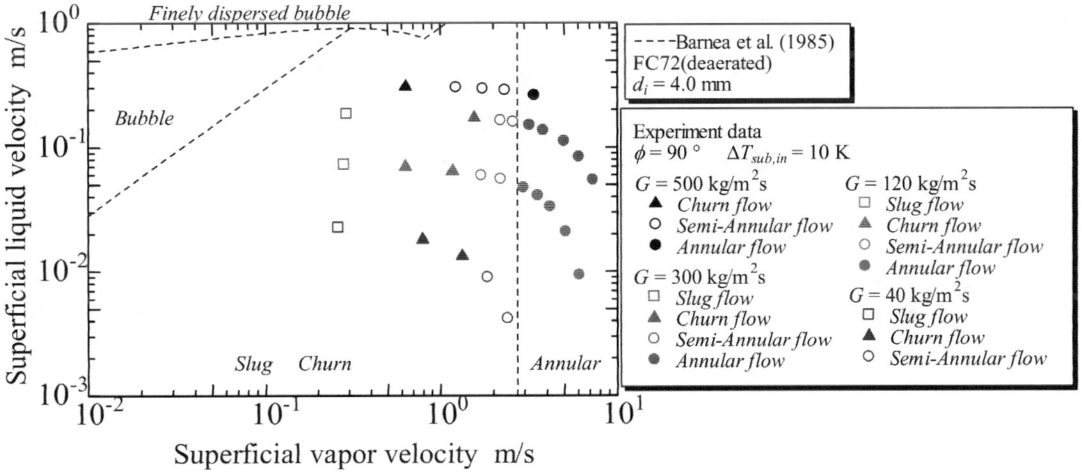

Figure 12. Comparison of the observed flow patterns at vertical upward flow with the transition lines by Barnea et al [13].

4. Conclusions

In order to verify the feasibility of the experiments on boiling and two-phase flow onboard ISS, a test loop similar to the flight model was developed and preliminary ground experiments on flow boiling were performed.

1. The operation and performance of all components installed in the test loop was confirmed.
2. The tendency of heat transfer coefficient obtained from the metal heated tube and images of flow patterns were checked in detail and they coincide with those of existing researches despite of the restrictions in the length of test sections. Then, the flight model with similar specifications will produce the reliable reference data for the normal gravity condition if it is operated on ground before the experiment onboard ISS.

Acknowledgments

The present authors express appreciation for the support by the following project members. Koutaro Tanaka (Shibaura Inst. Tech.), Satoshi Matsumoto, Atsushi Okamoto, Kengoh Ohkubo (JAXA), Shinichi Shinozaki, Kazumi Kogure (JSF), Atsushi Murakami, Yukihiro Ueda, Toshiharu Oka, Yoko Nakagawa (IHI Aerospace), Nobuo Ohtani, Takahiro Hayashida, Kazuaki Ae and Tomoki Hirokawa (Kyushu Univ.).

References

[1] Kim J 2003 *Journal of the Japan Society of Microgravity Application* **20 (4)** pp 264
[2] Di Marco P 2003 *Journal of the Japan Society of Microgravity Application* **20 (4)** pp 252
[3] Ohta H 2003 *Journal of the Japan Society of Microgravity Application* **20 (4)** pp 272
[4] Ohta H 1997 *Nuclear Engineering and Design* **175** pp 167–180
[5] Zhang H, Mudawar I and Hasan M M 2002 *International Journal of Heat and Mass Transfer* **45** pp 4079
[6] Kandlikar S G and Balasubramanian P 2005 *Journal of Heat Transfer* **127 (8)** pp 820
[7] Bower J S and Klausner J F 2006 *Experimental Thermal and Fluid Science* **31** pp 141
[8] Baba S, Ohtani N, Kawanami O, Inoue K, and Ohta H 2011 *Eurotherm Seminar on Gravitational Effects on Liquid-vapor Phase Change* **92**
[9] Ohta H, Inoue K, Ando M and Watanabe K 2009 *Heat Transfer Engineering* **30 (1–2)** pp 19
[10] Bao Z Y, Fletcher D F and Haynes B S 2000 *International Journal of Heat and Mass Transfer* **43** pp 3347
[11] Huo X, Chen L, Tian Y S and Karayiannis T G 2004 *Applied Thermal Engineering* **24** pp 1225
[12] Shiferaw D, Karayiannis T G and Kenning D B R 2009 *International Journal of Thermal Sciences* **48** pp 331
[13] Barnea D, Shoham O, Taitel Y, Dukler A E 1985 *Chemical Engineering Science* **40 (1)** pp 131

Journal of Physics: Conference Series **327** (2011) 012056 doi:10.1088/1742-6596/327/1/012056

Cyclonic Two-Phase Flow Separator Experimentation and Simulation for Use in a Microgravity Environment

N Hoyt, MF Kang, A Kharraz, J Kadambi, Y Kamotani

Department of Mechanical and Aerospace Engineering,
Case Western Reserve University, 10900 Euclid Avenue Cleveland, OH 44106, USA

E-mail: yasuhiro.kamotani@case.edu

Abstract. Devices designed to replace the absent buoyancy separation mechanism within a microgravity environment are of considerable interest to NASA as the functionality of many spacecraft systems are dependent on the proper sequestration of interpenetrating gas and liquid phases. Cyclonic separators provide the gas-liquid separatory action by swirling the multiphase flow – causing the gas to accumulate along the axis of the vortex as the denser liquid is forced to the walls – thereby allowing segregated extraction of the respective phases. Passive cyclonic separators utilize only the inertia of the incoming flow to accomplish this task. In the current work, combined experimental, numerical, and scaling analyses have been performed to quantitatively assess and delimit the operability of these separators. Specifically, steady-state features including velocity profiles have been examined experimentally and compared to computational fluid dynamics results, scaling laws for the gas core size have been created, and the transient behavior of the device with respect to both device and system-level conduct has been modeled.

1. Introduction

Phase separators are devices designed to separate two-phase gas-liquid flows into individual liquid and gas streams. Phase separation is essential to many fluid flow systems in microgravity. While droplet and bubble removal is a spontaneously occurring phenomenon in most terrestrial situations, the absence of the gravity-induced separatory action in a microgravity environment often results in situations where two disparate phases have no distinct inclination to separate from one another. This can present many problems with respect to spacecraft and space station operations as proper control and sequestration of the individual phases is often essential to the successful functionality of many system-critical components.

Multiphase conditions arise in these flow systems for different reasons. Some systems are unavoidably multiphase in character, such as those involving boiling. Multiphase conditions may also impose themselves upon nominally single phase systems. In order to ensure continued functionality, these systems must therefore be necessarily made tolerant of the encroachment of other phases. Other systems may have possible single phase guises, but the addition of multiphase flow can often result in substantial performance enhancement.

Regardless of the origin of the two phases, all of these systems have critical subcomponents that will be intolerant of multiphase flow. Phase separation thus can be an end unto itself, or it can be a necessary function in order to exploit the advantages of multiphase flow – permitting multiphase behavior where it is advantageous within a given flow system while preventing its drawbacks elsewhere in the system. Inasmuch, the number of spacecraft and space-station systems where these factors come into play is vast, ranging from Environmental Control and Life Support Systems (ECLSS) to power conversion.

Many phase separation systems have been designed and employed in the past to provide this needed phase separation. Broadly speaking there are five main types: capillary separators (hydrophilic/hydrophobic screens, porous media, face wicking), membrane separators (hydrophilic and hydrophobic), centrifugal/rotary (active), inertial (elbows and impingement separators), and passive centrifugal (also known as vortex separation). While all approaches have shown merit, the passive centrifugal option represents a promising technology that has been, to date, underutilized.

One such device, NASA Glenn Research Center's Cascade Cyclonic Separation Device (CSD-C) was designed to provide separation amenable for the use in a variety of spacecraft systems. It is a passive approach to separation in that it affords the phase separation without the requirement of moving mechanical components or power input. It is classified as being static in the sense that it has no moving parts (that would otherwise entail rotating seals and bearings). This static passivity combined with long-duration operability is the device's greatest advantage compared to the other approaches – issues limiting the operational lifespan of other devices, such as clogging of membranes, are not present. This approach employs the inertia of the multiphase flow itself to provide the desired phase-parting action. The conceptual basis of these devices involves the creation of a swirling flow by tangential injection of a multiphase fluid stream into the separator device. This eccentric injection thereby creates a buoyancy-like separation action via the pressure gradient that arises to maintain the necessary centripetal acceleration of the fluids as they swirl within the device housing. A gas core forms along the axis of the device, and the separate phases are removed via their respective outlets. The general concept is shown in figure 1.

The CSD-C has received experimental and numerical attention in the past [1,2]. It was demonstrated to function adequately over the duration of a start-up cycle that subsequently proceeds toward steady state within the time allotments provided by microgravity aircraft research. While experimental apparatuses of the CSD-C have, at minimum, shown efficacy for certain specified input conditions, failures with off-design conditions (such as mist development, etc.) and the incomplete delimitation of what constitutes "off-design" dictate that the flow characteristics of this device be more fully investigated.

Even though passive cyclonic separation is a seemingly mature technology, large knowledge-gaps exist with respect to the performance that can be expected from these types of devices — gaps large enough to preclude their use from mission-critical spacecraft systems. This research seeks to provide the complete analysis and understanding of the fluid physics phenomena that is necessary to make this mission-enabling technology available for use within multiphase space systems.

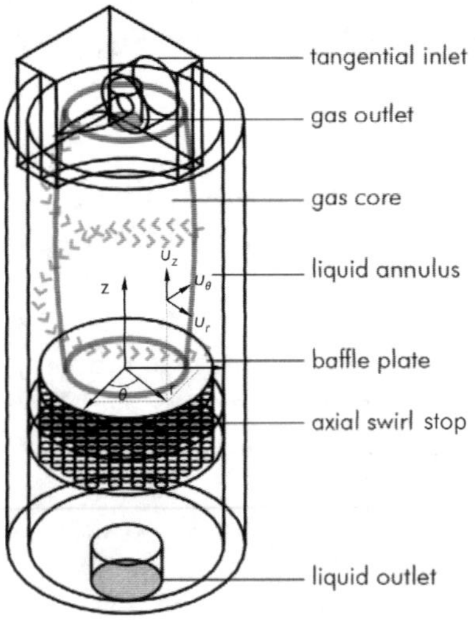

Figure 1. Passive cyclonic separator concept. Two-phase flow is injected tangentially into the separator housing; the baffle plate allows for the stable existence of an annular liquid layer and a cylindrical gas core.

1.1. Experimental Setup

The experimental system has been developed at Case Western Reserve University (CWRU) with instrumentation to allow discernment of typical passive separator device operations. The current work has consisted of ground-based experimentation operating in a high flow rate, gravity-independent mode. The flow loop was constructed to provide metered quantities of gas and liquid to the test separator (see figure 2). A steady flow of liquid is provided from a water tank and circulated by a centrifugal pump. The flow rate of city water is measured by two turbine flow meters (TM series, GPI, Inc. with an accuracy of $\pm 3\%$) installed in parallel. The flow ranges of these two turbine flow meters are 0-4 L/min and 4-16 L/min. Both the gas and liquid flows are introduced to a T-junction mixer after which they are allowed to develop in a representative two-phase flow piping section. The two phase mixture is then injected into the separator.

The pressure and temperature are measured prior to entering the test separator and at a location immediately upstream of two phase flow pipe. The data acquisition and control system software is written in LabView. The control of flow loop functions as well as the data acquisition and storage are done by a National Instruments data acquisition board with a workstation.

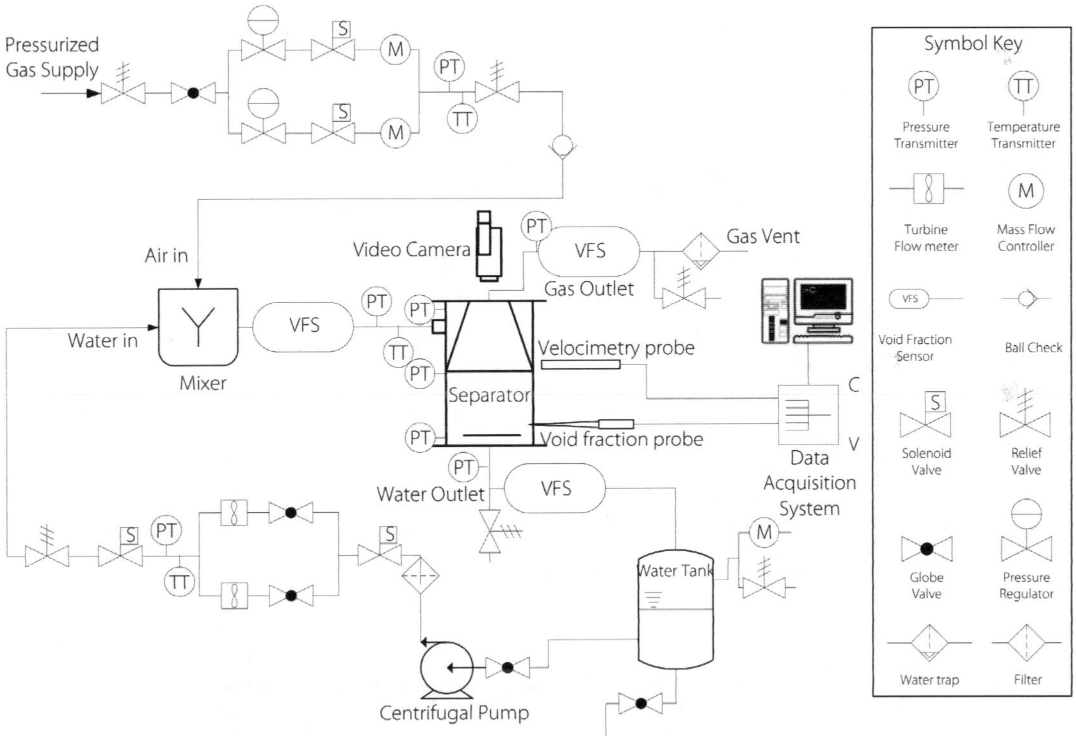

Figure 2. CWRU Experimental Two-phase Flow Loop

Experimentation with simplified straight-cylinder vortex separators has been performed using this loop to understand the fluid physics and to permit validation and verification of CFD approaches under development. Two different designs have been tested.

The schematic of the first separator is given in figure 3. It was manufactured from transparent acrylic. Swirl flow is generated in the test separator due to tangential injection of liquid from an injection nozzle. The characteristics of the swirl flow depended on the flow rate, the inner diameter of the separator and the injection nozzle. The present experiments were carried out under the conditions of flow rate between 4 to 15 L/min, as these proved to be necessary to operate independently of gravity. The inner diameters of the separator and injection nozzle were 57 mm and 5.6 mm, respectively. A honeycomb type of axial swirl-stop is located at 100 mm from the top wall to eliminate vortex flow extending to downstream.

An alternative design including a radial inflow swirl-stop has also been manufactured and undergone testing. In this design, the flow is constrained to move inward under the baffle plate through a series of radially oriented straightening vanes that extend to a fraction of the baffle plate radius. Both swirl-stop concepts are shown in figure 3. It will be shown that the differing pressure balancing properties of these approaches has a significant effect on the equilibrium size of the gas core that develops.

Ground testing is generally performed in a high-flow rate, gravity-independent manner with the separator axis situated perpendicular to the earth's gravitational vector. The separator would still function if operated with the axis vertically aligned, however many features of the flow would be disaffected by gravity.

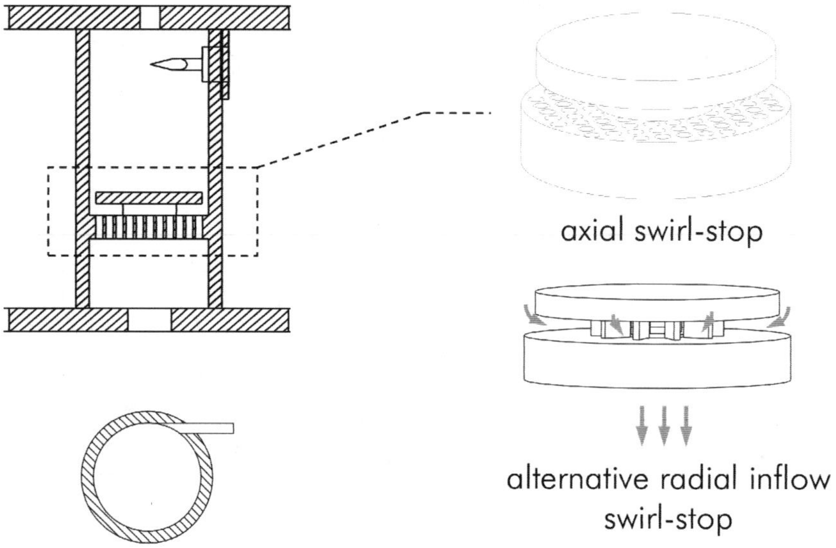

Figure 3. Baseline separator design for fluid physics studies and CFD validation and verification – shown with axial swirl-stop. Details of axial swirl-stop and radial inflow swirl-stop design showing path of liquid under the baffle plate shown to the right (separator walls have been omitted for clarity)

The measurements of axial velocity and the tangential velocity distributions in the liquid layer were carried out using pulsed Doppler ultrasonic velocimetry (PDUV). In ultrasonic Doppler velocimetry a piezoelectric probe emits ultrasound pulse, and the echoes are detected by the same probe. By sampling the incoming echoes at the same time relative to the emission of the pulses, the variation of the positions of scatters are measured. The measurement of the time lapse between the emission and the reception of the pulse gives the position of the particles. Measuring the shift in frequency, the target velocity component along the probe axis is detected. The ultrasonic Doppler velocimeter used in this study is Model DOP 2000 manufactured by Signal Processing S.A. A 4 MHz probe with a diameter of 8 mm and a Doppler angle of 80° were used in the present experiments. Uncertainties in the measurements are mostly in the ± 2.5 % range. A photograph of the experimental set-up is presented in Figure 4.

Figure 4. Photograph of the baseline passive cyclonic separator; the PDUV probe is shown mounted to measure the tangential velocity distribution.

1.2. Computational Modeling

With respect to numerical modeling, the current work entails the extension of past computational fluid dynamics studies by the authors. The multifluid simulations that worked satisfactorily in the past have been improved; in particular, some of the work of Rusche [3] have been included to handle the high volume fractions of gas that are often present in separators. This particular approach is used where the separation process itself is of interest, but it is deficient when capillary forces become important.

For those scenarios, a custom phase-field surface implementation has been developed and has been used to simulate some of the capillary instabilities that affect the functionality of the separator – features that could not be captured with the previous modeling efforts. Phase field is a diffuse interface method that has recently found use in the simulation of free-surface fluid mechanics; the approach has in particular been developed and applied by several authors in the study of fluids in microgravity. The structured/staggered grid approaches used by these previous researchers [4,5] to study microgravity fluid motion using phase-field methodologies have been adopted and made amenable to unstructured, cell-centered grids – the type of mesh characteristics required for complicated separator geometries.

In order to make the models pertinent to the complete separation process, hybridization of this code with a multifluid Eulerian-Eulerian treatment for dispersed phase constituents has been begun. This will allow the simultaneous calculation of the bubbly, unseparated two-phase flow at the inlets along with the surface tension phenomena associated with the stratified fluids – thereby providing modeling capability for all flow conditions (in a similar manner to the continuum surface forcing methods that were investigated previously [1]).

Device level analyses, however, are insufficient to classify the performance that can be expected of a separator. System-level conduct has been found to be crucial to the proper functionality of the separator – as such, efforts have been made to model the upstream and downstream components that are invariably present. To do so, one-dimensional two-phase flow modeling techniques have been coupled to the device-level three-dimensional simulations. Compressible homogeneous equilibrium method (HEM) and multifluid approaches have both been employed. The results from various microgravity experimental efforts [6] have permitted the inclusion of a number of general system components within the simulation framework. For example, microgravity data exists for many of the individual components including pipes, valves, quick-disconnects, etc. [7]. Large knowledge gaps exist, but it is hoped that this approach will create a general predictive framework for the simulation of two-phase flow systems in microgravity – one that should ultimately allow for the prediction and analysis of multiphase instabilities. Our experimental testing has not yet included microgravity testing, and, hence, for the current analyses the subcomponent models are handled in a manner that is generally consistent with terrestrial multiphase thermal-hydraulic codes (e.g. [8, 9 ,10]). The experimental two-phase flow loop is thus modeled as in figure 5. Typically several hundred cells are used for the 1-D flow sections, 50,000 are used for the 2-D axisymmetric representations of the separator geometry, and several hundred thousand are used for the full 3-D mesh.

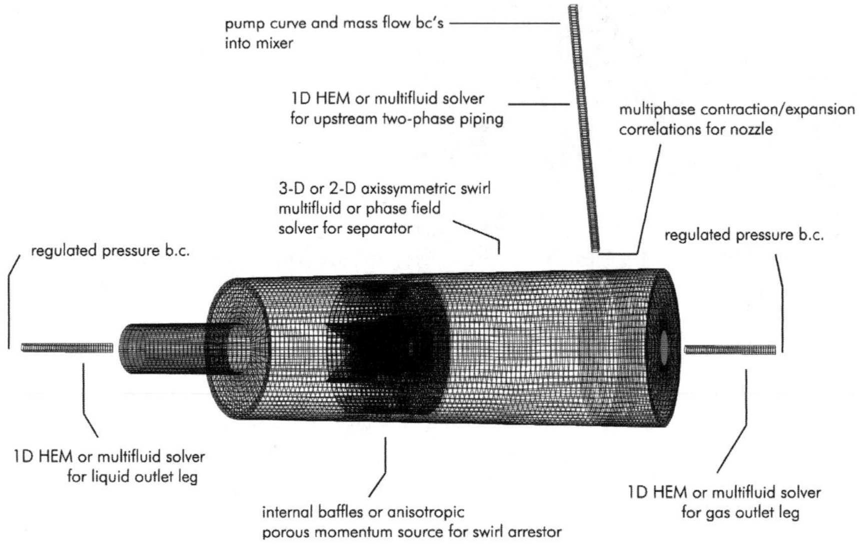

Figure 5. Coupled 1-D and 3-D CFD solvers (not to scale).

2. Steady Device-level Conduct

While the previous CFD work by the authors was found to at least be qualitatively correct, there was no rigorous validation or verification of the modeling approach. This work has sought to address that shortcoming. The first issue investigated was the device-level performance of the separator. While global metrics for separator performance were examined previously, it is important to examine the fundamental aspects of the fluid physics as that is from where the separatory action arises.

2.1. Velocity Profiles

As the velocity distributions are so fundamental to the separatory process, agreement between experimental and numerical velocity profiles within the separator were first sought in order to ensure that the complex effects of streamline curvature were being properly captured by the modeling procedure.

Figure 6 shows the computed tangential velocity profiles along a radius at the midplane of the separator (i.e. as compared to the experimental values for various injection flow rates). The cylindrical coordinate system is indicated in figure 1 wherein the axial coordinate is measured from the baffle plate, and the accompanying subfigures show the location of the ultrasonic transducer and the beam direction used to capture the Doppler shift. Excellent agreement between experiment and simulation is found near the wall, but the experimental data drops off quickly when approaching the free surface of the liquid annulus. This is due to the impedance change from the liquid into the gas core; the reflection effects therein make the error large and the returned values spurious.

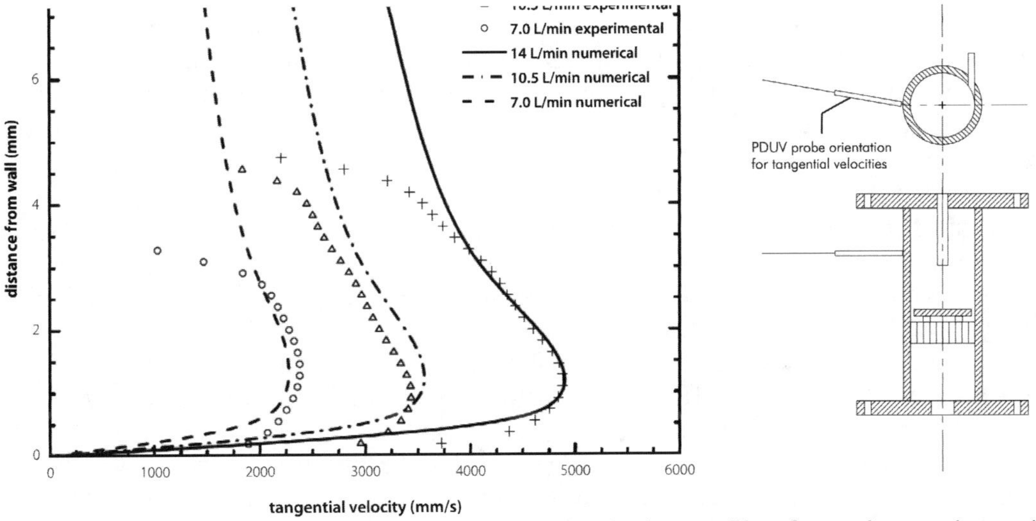

Figure 6. Numerical and experimental tangential velocity profiles for various volumetric flow rates plotted along the indicated radius at the midplane of the separator (i.e. half way between the top wall and the baffle plate — a distance of 50 mm from the top wall).

Figure 7 shows the axial velocity profiles. The agreement is generally good near the wall, but, again, the experimental values fall away rapidly toward smaller radii. Due to the prevailing tangential direction of the flow, the axial velocity experimental results did show large sensitivity with respect to the imposed beam angle; the axial velocities were found to be easily overwhelmed by the incursion of the comparatively larger tangential velocities when there are small experimental errors in placement. The overall error did become unsatisfactorily large when the velocities were low, as can be seen for the lowest flow rate case of 7.0 L/min liquid injection.

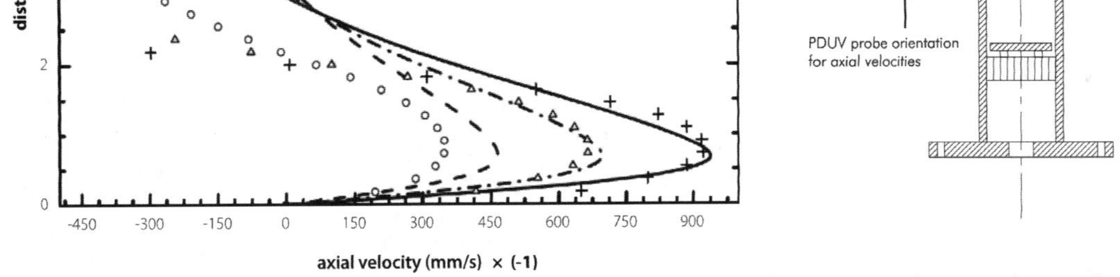

Figure 7. Numerical and experimental axial velocity profiles for various volumetric flow rates plotted along the indicated radius at the midplane of the separator (i.e. half way between the top wall and the baffle plate — a distance of 50 mm from the top wall).

Journal of Physics: Conference Series 327 (2011) 012056 doi:10.1088/1742-6596/327/1/012056

The axial velocity profiles show strong peaks near the wall and areas of reverse flow near to the free surface. The thinness of the liquid annulus in these specific experimental runs, however, makes the reverse flow relatively slow. Any inward flow is accordingly also small. The effects thereof are apparent in the tangential velocity profiles where, as shown, the flow adopts a solid-body rotational character as opposed to that of a Rankine vortex.

Another issue of great importance in separator design [11] is the decay of the swirl along the axis of the device. In order to aid in the design of cyclonic separators, correlations are often used in order to describe the swirl decay pattern along the axis of such devices. These correlations however are generally valid only for axial length to diameter ratios greater than 2; as the separator designs considered in this paper are shorter than minimum, it is important to ensure that CFD is able to predict the swirl decay accurately (a difficult task for turbulence models to calculate). Figure 8 shows the values of the maximum of the tangential velocity plotted versus the axial distance. The swirl is concentrated around the 7 cm location as this is where the injector is situated. Experimental values only exist for the liquid-only injection case, and good agreement is found. Also included in the figure are profiles for different injection volumetric qualities, β, which is the ratio of the volumetric flow rate of the gas compared to the total volumetric flow rate. In general, the tangential velocity (which is providing the separatory action) can, as expected, be seen to be dependent on the total volumetric flux and not simply the superficial velocity of the liquid alone. This will later be shown to affect the behavior of the separator.

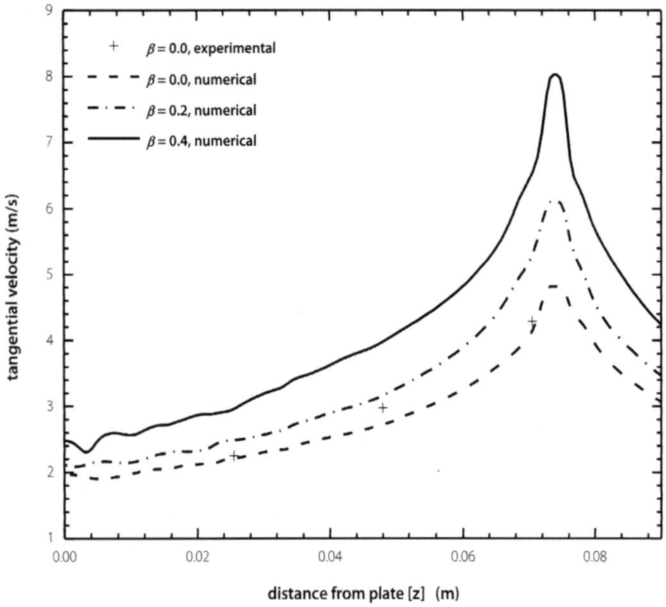

Figure 8. Swirl decay as a function of axial distance for fixed 7.0 L/min volumetric flow rate of water and differing injector volumetric qualities.

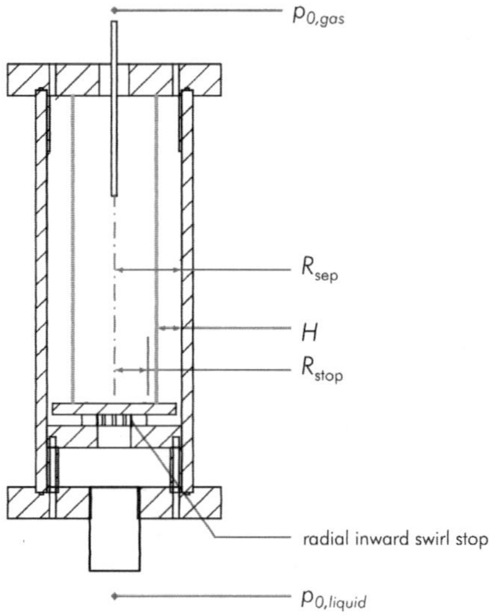

Figure 9. Mechanistic model nomenclature (geometry shown includes the radial swirl-stop)

2.2. Steady Flow Scaling Model

Alongside the velocity profiles, another value critical to separator performance is that of the depth of the liquid annulus. Maintaining a properly sized liquid annulus is clearly essential to maintaining separation and preventing carry-over or carry-under (i.e. allowing liquid to encroach on the gas outlet or vice versa).

The sensitivity of the liquid annulus thickness to the injection rates can be determined in a mechanistic model similar to the one performed by Arpandi, et al [12] In general, the depth is determined by balancing the pressure drops through the liquid and gas outlets with respect to the variable free surface depth. Figure 9 shows the geometry and symbols of interest. The liquid thickness H can be determined simply by ensuring that the pressure drops along the paths to the respective outlets are consistent with the common pressure at the injector.

The path to the gas outlet will be considered first. The overall pressure drop, Δp_1, is the difference between the pressure at the wall of the separator and the pressure at the gas outlet. This quantity is imposed by $\Delta p_{\text{annulus},1}$,which is the centripetal gradient across the annular distance H, and $\Delta p_{\text{gas flow}}$, which is the drop from the gas core pressure to $p_{0,\text{gas}}$ (i.e. the downstream pressure at the gas outlet that is regulated or open to atmosphere). These terms can be expressed as

$$\Delta p_1 = \Delta p_{\text{annulus},1} + \Delta p_{\text{gas flow}} = \int_{R_{\text{sep}}-H}^{R_{\text{sep}}} \frac{\rho_l u_\theta^2}{r} \,\mathrm{d}r + C_g \rho_g Q_g^2 \tag{1}$$

where Q_g is the total volumetric flow rate of the gas, and C_g is a generalized pressure loss coefficient, R_{sep} is the radius of the separator, ρ_l is the liquid density, and ρ_g is the gas density.

The balancing pressure drop across the path to the liquid outlet can be expressed similarly, but the included terms are dependent on the type of swirl-stop. The overall pressure drop Δp_2 is the difference between the pressure at the wall of the separator and the pressure at the liquid outlet. For the radial swirl-stop, this quantity is imposed by the centripetal gradient that is allowed to persist under the baffle plate up until the presence of the vanes, $\Delta p_{\text{annulus},2}$, and $\Delta p_{\text{liquid flow}}$, which consists of the minor losses proceeding to $p_{0,\text{liquid}}$, (i.e. the downstream pressure for the liquid outlet). Hence,

$$\Delta p_2 = \Delta p_{\text{annulus},2} + \Delta p_{\text{liquid flow}} = \int_{R_{\text{stop}}}^{R_{\text{sep}}} \frac{\rho_l u_\theta^2}{r} \mathrm{d}r + C_l \rho_l Q_l^2 \tag{2}$$

where Q_l is the total volumetric flow rate of the liquid, C_l is a generalized pressure loss coefficient for the components of the liquid outlet leg, and R_{stop} is the radial extent of the radial swirl-stop. R_{stop} is the lower limit of the centrifugal integral as at smaller radii the tangential velocity of the liquid is constrained to be zero by the straightening vanes.

The pressure drop for the axial swirl-stop is constructed in the same way, but it lacks the centrifugal pressure drop term as there is no contributing swirl beneath the baffle plate (as the flow is immediately constrained into the axial direction). Thus, the centrifugal term is absent and the pressure drop is instead expressed as

$$\Delta p_2 = C_l \rho_l Q_l^2 \tag{3}$$

Knowledge of the velocity distribution (perhaps through the use of correlations) along with specification of the various parameters allows for the computation of the H that allows this balance to be satisfied. The number of parameters involved can obfuscate the basic physics of what is occurring. A scaling analysis approach that yields much the same information has been done. Several simplifying assumptions must be made.

First, the analysis will be done for the inward radial swirl-stop. It is assumed that the liquid layer is small relative to R_{sep}. Furthermore, the centripetal gradients are assumed to be scaled by u_{injector} which is the injection velocity based off of the total volumetric flux through the injector nozzle. The scaling of the pressure drop supported by the liquid annulus above the plate (i.e. the first term of the right-hand side of equation 1) is thus

$$\Delta p_{\text{annulus},1} \sim \frac{\rho_l u_{\text{injector}}^2 H}{R_{\text{sep}}} = \frac{\rho_l (Q_l + Q_g)^2}{A_{\text{injector}}^2} \frac{H}{R_{\text{sep}}} \tag{4}$$

A_{injector} is the area of the injector and hence dictates the characteristic velocity along with the respective liquid and gas volumetric flow rates.

The scaling of the pressure drop supported by the liquid annulus below the plate can be performed similarly. Assuming that the profiles have a similar character (which requires a properly sized baffle plate gap so as not to effect large peaks in the tangential velocity profile), the term is scaled by the same velocity as before. However, as mentioned, the gradient is only created up until the vane location, and, hence, only acts over the distance $R_{\text{sep}} - R_{\text{stop}} = L$. Specifically,

$$\Delta p_{\text{annulus},2} \sim \frac{\rho_l u_{\text{injector}}^2}{R_{\text{sep}}} \frac{L}{} = \frac{\rho_l (Q_l + Q_g)^2}{A_{\text{injector}}^2} \frac{L}{R_{\text{sep}}} \tag{5}$$

The general philosophy of the inward radial swirl-stop design is to make the two centripetal terms dominant. To do so, the gas outlet and liquid outlet components have been constructed as large as possible in order to remove the frictional wall losses that result from those components. Thus, with the C_l and C_g terms in equations 1 and 2 tending to zero, and with the downstream pressures regulated to be equal (i.e. $p_{0,\text{gas}} = p_{0,\text{liquid}}$), $\Delta p_{\text{annulus},1}$ is exactly balanced by $\Delta p_{\text{annulus},2}$. For this equality to hold, the necessary depth of the liquid annulus can be found by equating equations 4 and 5 to obtain the simple result

$$H \sim L = R_{\text{sep}} - R_{\text{stop}} \tag{6}$$

This result is saying that, for the radial swirl-stop design, the liquid level becomes invariant of the flows that are injected — at least within the context of the various assumptions that have been made. The slower flow rates permissible in microgravity entail numerable problems (capillary effects, insufficient separatory gradients, etc.) which make maintenance of this equality unlikely. Nonetheless, the ability of the design to maintain a fixed gas core size is extremely advantageous for the separation process, helping to prevent carry-over and carry-under dysfunction.

The axial swirl-stop behaves less optimally. There is no underflow centripetal pressure gradient present nor is there any pressure recovery from the honeycomb. In this case, the annular pressure drop is now balanced by the now contributing term of the frictional pressure drop on the liquid leg (the minor losses on the overflow leg again being assumed negligible). Thus, the scaling therefore takes the form

$$\frac{\rho (Q_l + Q_g)^2}{A_{\text{injector}}^2} \frac{H}{R_{\text{sep}}} \sim \rho C_l Q_l^2 \tag{7}$$

with the net result that

$$H \sim \frac{Q_l^2}{(Q_l + Q_g)^2} A_{\text{injector}}^2 \, C_l \, R_{\text{sep}} \tag{8}$$

or, equivalently,

$$H \sim (1-\beta)^2 A_{\text{injector}}^2 C_l \, R_{\text{sep}} \tag{9}$$

The result is not a constant as before. Moreover, the coefficient C_l has a liquid flow rate dependence at low flow rates where viscous forces dominate. Use of the axial swirl stop is therefore predicted to allow greater undesired variation of the gas core size.

For the experimental device, it should be noted that, when run with the axial swirl-stop, the liquid outlet requires a valve to be slightly closed to create a sufficiently large liquid-leg pressure drop term to complete the balance. If the liquid outlet had not been adjusted this way, the C_l term would be too small to counterbalance the annular-induced pressure drop, and the liquid layer would shrink into a small film thereby allowing carry-under.

To examine the utility of this scaling analysis, figure 10 shows numerically predicted liquid annulus depth results for a single fixed liquid flow rate. Data for both the axial and radial swirl-stop designs are shown. As predicted, the liquid layer depth for the radial swirl-stop is very nearly constant and, in fact, is close to the vane clearance L of the separator (which is indicated by the solid line). The data for the axial swirl-stop also follows the expected trend; by changing the injector area or flow quality (and also keeping the rest of the separator geometry identical), the inconsistency in the balancing pressure term for the axial swirl-stop is revealed. Specifically, because the centripetal gradient and the post-swirl-stop losses are not scaled by the same velocity measure, H must adjust to account for the differences, whether they result from a change in the injector area or from the inclusion of gas flow.

Journal of Physics: Conference Series **327** (2011) 012056 doi:10.1088/1742-6596/327/1/012056

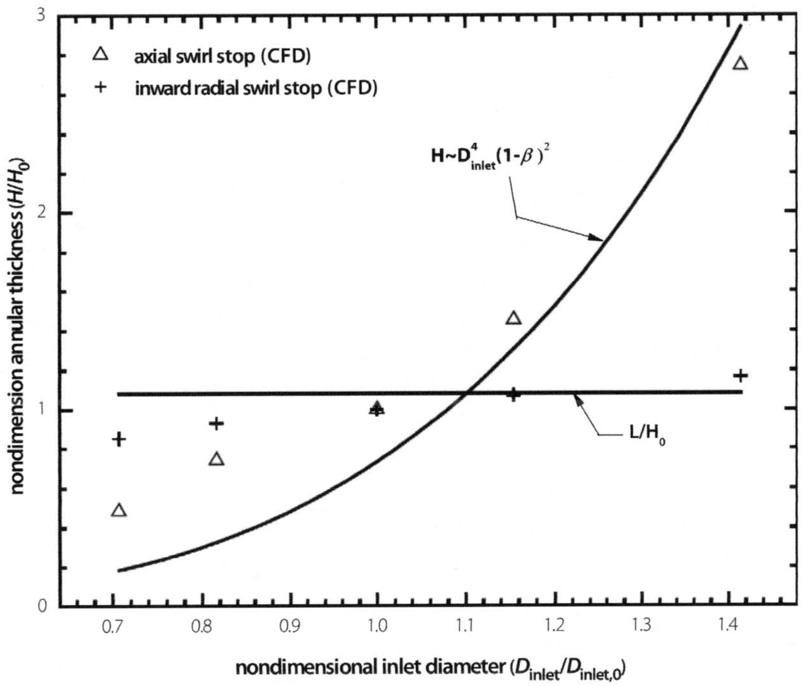

Figure 10. Variation of the liquid annulus thickness for a fixed volumetric flow rate (7.0 L/min) injected through a variable sized diameter nozzle (CFD compared to scaling model). Nondimensionalization has been performed with respect to baseline nozzle diameter $D_{\text{inlet},0}$ (for a circular injector) and the annular thickness thereof H_0

It must be emphasized that this simplified mechanistic model assumes pressures downstream of the gas and liquid outlets are matched through exhausting to atmospheric pressure or through regulation to a common pressure; if this cannot be assured, large system-level pressure differences can easily overwhelm any separator's ability to maintain an adequate liquid level and active control valves must be used to assist the balance.

3. Unsteady Separator Response

The above experimentation, mechanistic modeling, and simulations were performed for nominally steady flow injection into the two-phase separator. Under these conditions, the radial swirl-stop design was predicted to have a very stable gas core size. Steady behavior, however, does not tell the complete story; the dynamic response of the separator is important not only for transient conditions such as set-point changes, but it is critically essential in general multiphase systems where unsteadiness and unstable flow are common.

As was shown in the steady flow results, the separator can handle different flow rates through the interplay of the balance of the pressure drop terms via mechanisms such as thickening or thinning of the liquid layer. Variability in injection flow rates can be handled by a similar balancing process, but in unsteady flow there are now differing time scales that come into play. For example, while during steady operations the wall-frictional pressure drops may be precisely countered by the pressure drop across the swirling annular layer, the spin-up or spin-down of the annular layer will take much longer than the time necessary for the flows in the piping sections to adjust. Thus, rapid set-point changes and unstable injection can be expected to be associated with large variability of the liquid thickness and, perhaps, failure to successfully separate the flows.

As both a demonstration of the ability of the separator to handle transient injection and as a demonstration of the ability of the coupled CFD solvers to simulate unsteady multiphase flow, the experimental two-phase flow system was run with a set of flow rates such that a pulsating dynamic instability was set up in the inlet pipe, which is a 2 m long, 12.7 mm diameter two-phase development pipe upstream of the single injection 2 mm diameter nozzle into the separator. The gas flow rate was fixed by the mass flow controller while the water flow rate was dictated by a centrifugal pump set to a constant rotational velocity. Under pure liquid injection situations, the pump produced 3 L/min, but the flow delivered at the separator inlet was observed to drop to as low as 0 L/min during the course of the instability.

The nature of the instability that arises is that of a heavily pulsating density-wave oscillation (DWO). The instability caused large oscillations of the gas and liquid flow rates at the injector that proved challenging for the separator. While this particular instability is limited to special combinations of the respective gas and liquid superficial velocities in relation to the specific pump curve of the experimental apparatus, the occurrence is representative of the potentially challenging injection conditions that the separator is expected to ingest. Furthermore, while design tools and methodologies exist to create terrestrial systems that do not acquire the pulsating nature of the above flows, the knowledge to predict and avoid of microgravity multiphase instabilities remains limited; as such, it is essential to ensure that the passive cyclonic separator can handle such flows.

The compressible HEM solvers used for the two-phase flow in the upstream piping components proved able to capture the essence of the instability. The CFD in turn predicts substantial volumetric flow rate variability that must be handled by the separator. The large oscillations of the injection velocity at the injector are shown in figure 11.

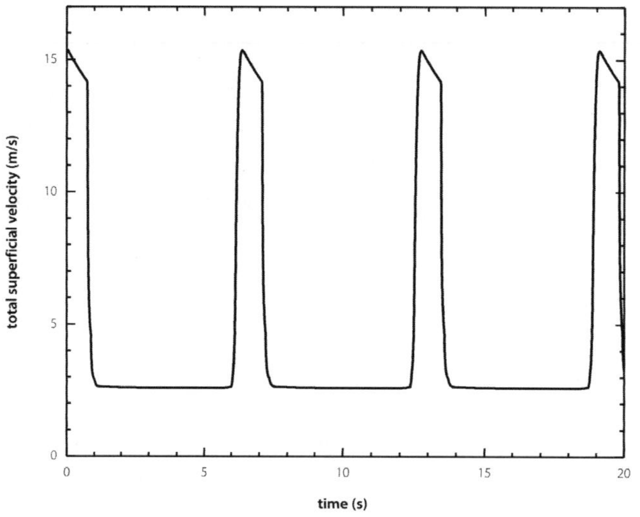

Figure 11. Predicted total injector volumetric flow rate versus time for the representative pulsating dynamic instability

Both experiment and CFD showed that the design featuring the radial swirl-stop works well enough that it is able to successfully handle the highly variable flow rates. Figure 12 shows the behavior therein: even during the very high quality gas injection periods the liquid layer is able to persist. Shown in the figure is the radial extent of the swirl-stop's straightening vanes, and the location of the core's free surface can be seen to deviate little from its expected value. The precise evolution of the gas holdup is shown in figure 13. In this plot, the average gas volume fraction within the separator (a quantity representative of the gas core size) is plotted versus time. As shown, the gas core expands, contracts, and is able to accommodate the imbalances caused by the pulsating flow.

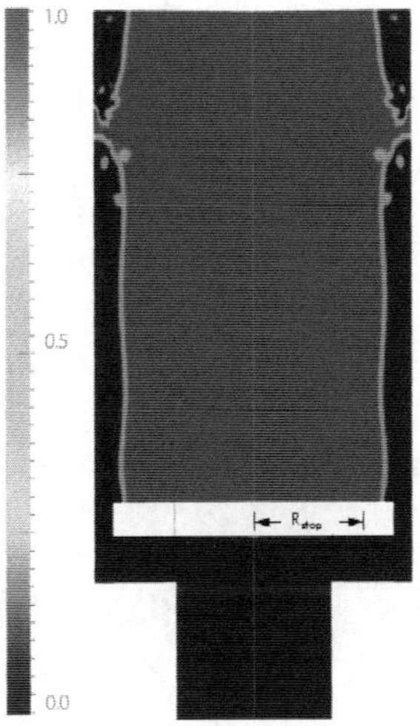

Figure 12. Contours of volume fraction for separator during gas slug injection (multifluid, axisymmetric swirl case)

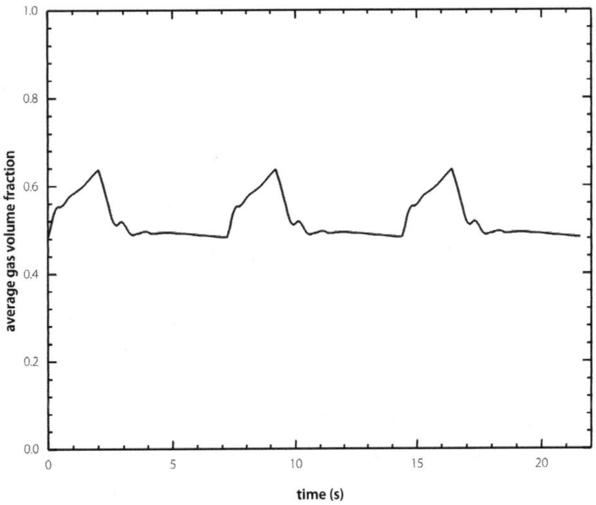

Figure 13. Gas holdup within the separator (as expressed by the average gas volume fraction) versus time during representative unstable operation

4. Conclusion

Gas-liquid separation is an important multiphase flow issue that arises in the absence of gravity. The fluid physics within one such family of devices, cyclonic gas-liquid microgravity separators, is quite complicated and represents a considerable challenge to study. In the present work, combined experimental, numerical, and scaling analyses have been performed to quantitatively assess and delimit the operability of these separators.

First, steady-state features including velocity profiles were examined experimentally and compared to the results from numerical modeling. The results, in general, showed satisfactory agreement and give confidence in both the modeling approach and in the use of PDUV for the study of the flow fields in gas-liquid separators. Next, scaling laws for the gas core size were created. Comparison with CFD results showed that the variation of the liquid annulus thickness did match the general character of the scaling relationships. Finally, the transient behavior of the device with respect to both device and system-level conduct was modeled. The coupled 1-D and 3-D CFD solvers proved able to model transient multiphase flows and demonstrated the ability of the separator to accommodate highly variable liquid and gas flow rates.

It is hoped that the resulting insight into phase separation, distribution, and control offered by this, and future, work will help to afford NASA designers the latitude to take greater advantage of the benefits offered by the use of multiphase systems in spacecraft applications.

5. Acknowledgements

We acknowledge Dr. Michael Hölling of Carl von Ossietzky University of Oldenburg, Oldenburg, Germany for providing us with DOP 2000, the pulsed Doppler ultrasonic velocimeter.

This work is being performed under NASA Grant No. NNX09AI31G with supervising project scientist John McQuillen of Glenn Research Center.

References

[1] Hoyt N, Kamotani Y, Kadambi J, McQuillen J, Sankovic J 2008 Computational Investigation of the NASA Cascade Cyclonic Separation Device *Proceedings of the AIAA 41st Aerospace Sciences Meeting and Exhibition* AIAA-2008-0809

[2] Schrage D, Shoemaker J, McQuillen J, 1998 Passive Two-Phase Fluid Separation *Proceedings of the 36th Aerospace Sciences Meeting and Exhibition (Reno, NV)*

[3] Rusche H 2002 Computaional Fluid Dynamics of Dispersed Two-Phase Flow at High Volume

Journal of Physics: Conference Series **327** (2011) 012056 doi:10.1088/1742-6596/327/1/012056

Fractions *PhD Thesis, Department of Mechanical Engineering,* Imperial College of Science, Technology and Medicine, London
[4] Jaqmin D 1999 Journal of Computational Physics **155** 96-127
[5] Chato D 2004 Penetration of Free Surface by Liquid Jets in Microgravity *PhD Thesis, Department of Mechanical and Aerospace Engineering*, Case Western Reserve University, Cleveland, OH
[6] Colin C, Fabre J, McQuillen J 1996 *Chemical Engineering Comm.* 155-173
[7] Marsden K, Kurwitz C, Best F 2005 *Journal of Thermophysics and Heat Transfer* **19** 486-96
[8] Bayless P, et al 2005 RELAP5-3D Code Manual *INEEL-EXT-98-00834*
[9] Levy S 1999 Two-Phase Flow in Complex Systems (New York, NY: John Wiley and Sons)
[10] Issa R, Kempf M.H.W. 2003 *International Journal of Multiphase Flow* **29** 69-95
[11] Gomez L, Mohan R, Shoham O 2004 *Journal of Fluids Engineering* **126** 935-42
[12] Arpandi I, Joshi A, Shirazi S, Shoham O, Kouba G 1995 Hydrodynamics of Two-Phase Flow in Gas-Liquid Cylindrical Cyclone Separators *SPE Journal: SPE 70th Annual Meeting* 427-436

Journal of Physics: Conference Series **327** (2011) 012057 doi:10.1088/1742-6596/327/1/012057

EML - An Electromagnetic Levitator for the International Space Station

A Seidel[1], **W Soellner and C Stenzel**

Astrium GmbH, Dept. TO54, Claude-Dornier-Str., 88090 Immenstaad, Germany

Email: achim.seidel@astrium.eads.net

Abstract. Based on a long and successful evolution of electromagnetic levitators for microgravity applications, including facilities for parabolic flights, sounding rocket missions and Spacelab missions, the Electromagnetic Levitator EML provides unique experiment opportunities onboard ISS. With the application of the electromagnetic levitation principle under microgravity conditions the undercooled regime of electrically conductive materials becomes accessible for an extended time which allows the performance of unique studies of nucleation phenomena or phase formation as well as the measurement of a range of thermophysical properties both above the melting temperature and in the undercooled regime. The EML payload is presently being developed by Astrium Space Transportation under contracts to ESA and DLR. The design of the payload allows flexible experiment scenarios individually targeted towards specific experimental needs and samples including live video control of the running experiments and automatic or interactive process control.

1. Motivation

Electromagnetic levitation is a powerful technique for containerless processing of electrically conducting samples such as metals and semiconductors. Eddy currents are induced in the sample by the electromagnetic field generated by the levitation coils. The sample is heated by ohmic losses of the induced eddy currents, whereas the interaction of these eddy currents with the electromagnetic field leads to a displacement force on the sample towards the center of the coil system which is used to control the position of the sample inside the coil.

The prospect of processing metallic samples without contact to a container wall represents the striking advantage of electromagnetic levitation. This means that experimental investigations can be performed with a free floating liquid droplet in a wide range of temperatures above and below the melting temperature of the material.

In particular, it is possible to undercool the sample significantly. Undercooling describes a non-equilibrium state where the sample is liquid at temperatures below its equilibrium melting temperature. In order to achieve significant undercooling nucleation has to be reduced as much as possible. This is achieved mainly by the absence of any container walls in contact with the liquid sample, a feature inherent in the electromagnetic levitation technique, in combination with an ultraclean process environment, e.g. ultra-high vacuum or high-purity noble gases.

---

[1] To whom any correspondence should be addressed.

Journal of Physics: Conference Series **327** (2011) 012057 doi:10.1088/1742-6596/327/1/012057

Patented by Muck in 1923 [1], electromagnetic levitation is one of the oldest levitation techniques used for containerless processing experiments and has been applied in material science investigations for decades. Other techniques comprise aerodynamic and acoustic levitation [2], as well as electrostatic levitation [3]. In this technique levitation forces arise from electrodes located above and below the sample that contains a surface charge, whereas heating is performed e.g. via lasers. Electromagnetic levitation is the most mature of all containerless processing techniques and has been applied both in ground-based experiments as well as in microgravity experiments with a wide range of alloys for decades [4]...[9].

In order to levitate a sample in a ground-based experiment strong levitation fields are needed in order to counteract gravity. This leads to a number of drawbacks and limitations for the experiments to be performed. Due to the strong levitation fields required for positioning against gravity positioning and heating are strongly coupled, the achievable minimum temperature is high which excludes samples with low melting points, the samples are deformed by the electromagnetic pressure, and strong convection is induced.

These drawbacks can be eliminated if levitation is performed under microgravity conditions where only small levitation forces are required to compensate for the residual accelerations. With a coil system optimized for microgravity conditions heating and positioning can be performed almost independently. Since the required positioning forces are very low the low temperature regime becomes accessible, convection is reduced and deformation of samples eliminated. As a result, many more sample materials can be processed and their thermophysical properties can be determined with higher accuracy.

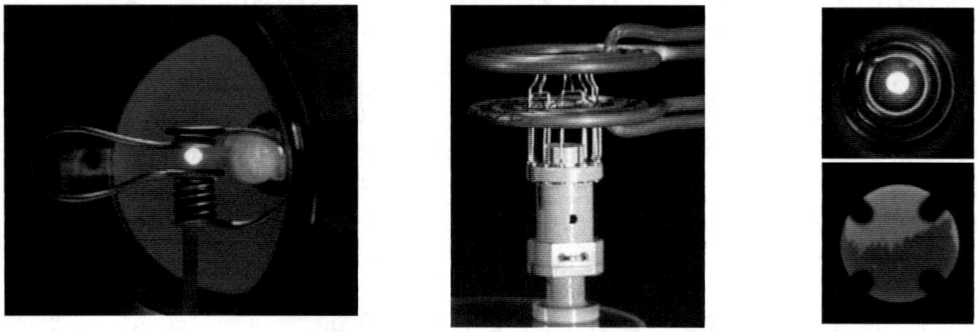

Figure 1. Electromagnetic Coil Systems and Samples.
Left picture: ground levitation coil (courtesy of DLR).
Middle picture: µg levitation coil with sample holder.
Top right picture: heated sample in µg levitation coil (axial view).
Bottom right picture: solidification of sample levitated in µg during parabolic flight.

There is a wide range of opportunities to perform electromagnetic levitation experiments under microgravity each with its unique advantages and disadvantages. One factor to consider is the microgravity quality: Since the coils are optimized for low positioner forces the sample is only weakly mechanically coupled to the electromagnetic field. This mechanical system can be described as an almost ideal harmonic oscillator with resonance frequency in the order of 1-3 Hertz and almost no damping. Hence mechanical disturbances in this frequency range must be minimized to achieve a stable sample positioning.

Other factors are the duration of the microgravity periods, and the possibilities for the experimenter to interactively control the experiment process during the runs. Electromagnetic levitators have been developed for a wide range of carriers and missions, as follows:

- Parabolic flights provide several periods of reduced gravity in the range of several milli-g for about 20 seconds each providing the experimenter direct access to the experiments during the flight. The environment is suitable to perform individual melt cycles of dedicated samples for which such a cycle can be performed in the limited available time [10], [11]

- Sounding rocket flights provide a very quiet microgravity environment (about 10^{-5} g) for about 6 minutes which allows performing multiple melt cycles. Experiments are controlled remotely via telemetry [12].
- Spacelab flights as performed in the 1990s provided a mostly quiet microgravity environment, sometimes interrupted by crew exercise activities (10^{-4}-10^{-5} g). Due to the mission duration which was in the order of 1-2 weeks, several experiments with durations of several hours each could be performed, where each experiment with a given sample included several melt cycles with that sample. Experiments were pre-programmed; interactive access via telemetry was possible. [13].
- ISS missions are essentially an extension of the former Spacelab missions and provide long-term access to a good microgravity environment. Several batches of samples can be uploaded to the facility, processed, and returned to ground. Experiments are pre-programmed; interactive access via telemetry is possible. An exhaustive discussion of experiments that can be performed with EML on ISS is given in [14].

2. Heritage of Electromagnetic Levitators

Electromagnetic levitation under microgravity has been successfully applied on a number of hardware carriers for more than 25 years. [4] … [8]. Starting with a laboratory model developed by Astrium in the mid-1980′s, a long-term development program was established which provided electromagnetic levitation facilities with increasing performance and versatility. These electromagnetic levitation facilities were collectively known under the name TEMPUS which is a German acronym (*Tiegelfreies Elektromagnetisches Prozessieren unter Schwerelosigkeit*). A TEMPUS parabolic flight facility and a TEMPUS version for the TEXUS sounding rocket targeted towards longer microgravity periods and reduced microgravity disturbances offered by sounding rocket missions were developed and had their maiden flights in the late 1980s. The next big step forward was the development of a TEMPUS version for Spacelab which offered mission durations of 1-2 weeks and thus allows processing numerous samples for an extended time during one mission. TEMPUS was operated on three Spacelab missions (IML-2 in 1994, MSL-1 and MSL-1R in 1997) and provided a wealth of scientific data.

Since the first parabolic flight and sounding rocket missions these facilities have been continuously upgraded. Numerous parabolic flights and sounding rocket missions have been performed with a dual focus on both implementing and testing of improvements and new functions of the system and on conducting scientific experiments. Meanwhile these missions are routinely performed, the last TEXUS mission was completed in April 2011, and the parabolic flight facility is currently being upgraded in preparation of future missions.

With the advent of the ISS new opportunities for electromagnetic levitation are now available, in particular much longer mission durations compared to a Spacelab mission (years instead of days). In order to exploit these opportunities ESA and DLR have decided to join forces and to develop an ISS version of TEMPUS, now called EML or *Electromagnetic Levitator*. The EML facility is currently under development and will be described in detail in the following chapters.

Funded and contracted by DLR Space Administration Astrium has been responsible for the industrial implementation of the electromagnetic levitators from the beginning of the development program [5] … [8]. The technology of all major elements such as the RF generator, the oscillating circuit and coil system, the ultrahigh vacuum chamber and associated mechanisms, the scientific diagnostic elements and so forth has been developed by Astrium in a long-term cooperation with a number of key partners such as the DLR Institute of Materials Physics in Space, and Pink GmbH. As Prime Contractor responsible for the overall system Astrium has manufactured, tested, and delivered, upgraded and refurbished all above mentioned electromagnetic levitators and is involved the mission operations of these facilities in cooperation with the DLR Microgravity User Support Center. The

Journal of Physics: Conference Series **327** (2011) 012057 doi:10.1088/1742-6596/327/1/012057

development of the EML facility for ISS is carried out by Astrium and the established industrial consortium under joint funding of DLR and ESA.

During this long-term development program Astrium has acquired unique expertise in the field of electromagnetic levitation facilities which is underlined by the long-term continuity of the project staff.

3. EML on ISS: Payload and Mission

3.1. Overview

In order to accommodate the ISS version of EML inside the *European Drawer Rack* (EDR) of the *International Space Station* ISS Columbus module the facility is composed of four modules designed to fit into drawers and lockers of EDR, as shown in figure 2. The main functions of the 4 modules are given in table 1. The EML is designed as class 2 payload meaning that EML will use the resources provided by EDR, i.e. power (120 & 28 VDC), cooling (water and air), data and video interfaces, pre-vacuum and vent line as well as dry nitrogen. On the other hand EML will also provide dedicated resources as e.g. high-purity noble gases (Ar, He), ultra-high vacuum generated by a turbo-molecular pump, mass memory for data and video as well as its own data management & control system.

Table 1. EML Modules and their purpose.

EML Module	Main Function
Gas Supply Module (GAS)	Storage and distribution of the noble gases
Levitation Power Supply/ Water Pump Module (LPS/WAT)	Provision of RF voltage to the RF coil generating the electromagnetic fields. Secondary cooling loop to subsystems
Experiment Module (EXM)	Core experiment: comprising e.g. vacuum chamber, RF coil, diagnostics, mechanisms, samples
Experiment Controller Module (ECM)	Data management & control, power distribution

The modular design of EML allows not only the exchange of the above mentioned complete modules but also supports exchange of subsystems such as the optical diagnostic systems, video control & compression units, mass memory, and electronic units controlling motors or pumps, and filter cartridges in the gas system. The concept of in-orbit replaceable units (ORU) allows both, maintenance of modules or subsystems by the crew if needed (e.g. replacement of empty Gas Module) as well as future upgrade of optical diagnostic systems (cameras and/or pyrometer) with specific diagnostic tools developed for achieving dedicated experiment goals.

Crew involvement in facility operation is limited and only needed for facility installation into EDR, exchange of sample chambers and transfer of samples from the sample chamber into the Sample Download Container. The experiment operation itself is fully automated and does not require crew involvement. Instead, the experiment control concept is based on pre-defined parameter sets which are developed and validated on ground. In addition the control concept foresees re-programming capabilities for the change of the pre-defined parameters due to lessons learned from previous experiment runs.

According to the experiment scenario the facility will be configured by ground commanding prior to an experiment run. However, intervention from ground is also possible during automated experiment processing. The need for such ground interventions may arise in case of positional instabilities of the sample. The facility design and operational concept allows such ground intervention with short reactions times. Hence the experiment operation can be described as near real-time operation requiring live video transmission for ground monitoring as well as near real-time interactive capabilities for ground commanding to interrupt or alter the pre-defined automated process. This automated process is sub-divided into short time intervals in which facility parameters are set according to the loaded parameter set to define the temperature-time profile of the sample as well as the settings of the diagnostics and stimuli observing and affecting the sample.

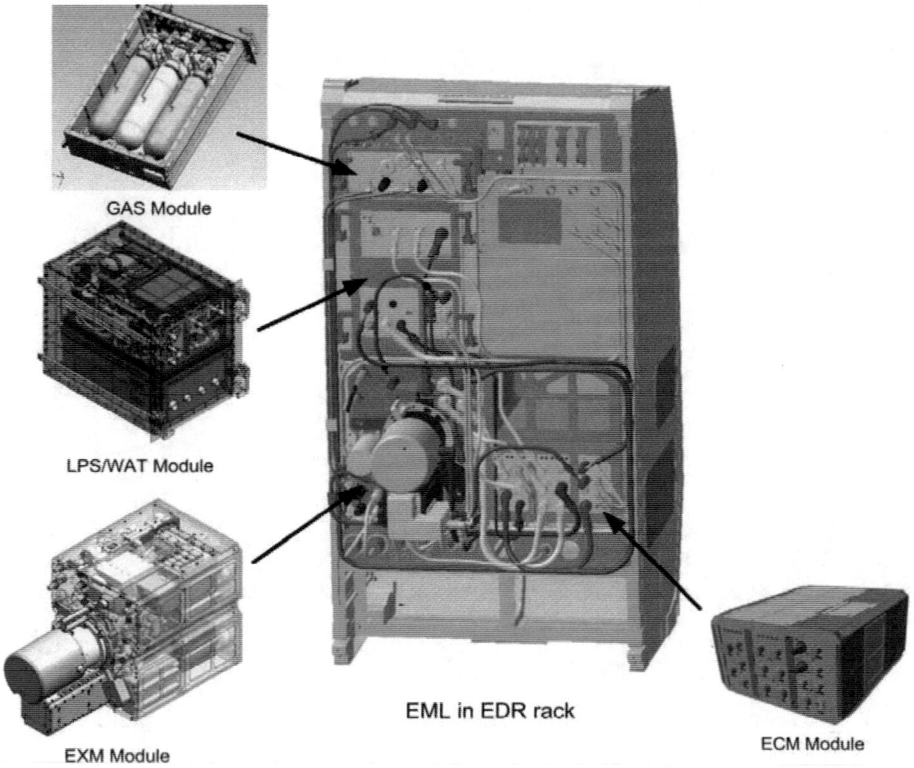

Figure 2. EML Modules located inside the EDR rack, shown with interconnecting harness and lines

3.2. EML Payload Design

The Experiment Module (EXM) houses the core experiment with all diagnostics, vacuum chamber, RF coil system, mechanisms and samples and defines the payloads capabilities with respect to the experiment. Figure 3 gives a functional overview of the various subsystems which are implemented in the EXM as described hereafter.

3.2.1. *Process Environment.* Central part of the EXM is a stainless steel ultra-high vacuum chamber with all-metal gaskets. The exchangeable sample chamber will be flanged to it, whereas the vacuum interface is secured by two vacuum locks. The process atmosphere is either ultra-high vacuum or high-purity (99.9999%) noble gas Ar or He or a mixture of these. The pressure inside the 2 chambers can be varied between 10^{-8} mbar and 400 mbar during experiment runs. To establish the ultra-high vacuum conditions a turbo-molecular pump is used interfacing the ISS vacuum line. The noble gases are provided by the EML GAS module which stores up to 500 Nl of He and 1000 Nl of Ar. The module can be exchanged if necessary. The EML gas distribution system also offers the possibility of a closed loop gas circulation by a specially designed UHV-tight gas pump. This feature can be used to either introduce a smooth gas flow onto the sample for scientific purposes or to flush sample particles or condensates into dedicated filters. The EML also possesses an interface to the ISS nitrogen supply, which might be used for flushing purposes or during storage times when EML is not operated.

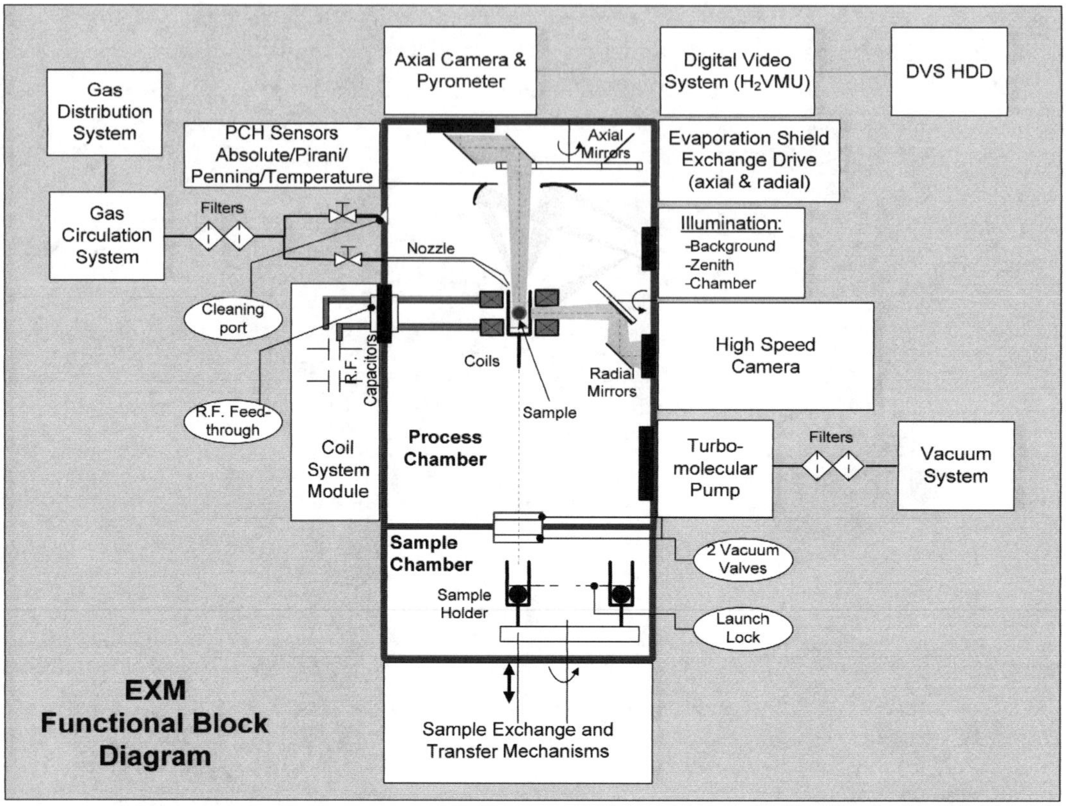

Figure 3. EML Experiment Module Functional Block diagram

The vacuum chamber also houses two double mirror systems which protect the optical viewports from evaporated sample material. Both mirror systems allow exchanging one of the mirrors by its respective sophisticated mechanics, if the reflection due to accumulated material is too low.

3.2.2. *RF Heating and Positioning System.* The RF heating and positioning system consists of the Levitation Power Supply providing the RF power and the Coil System Module (network of coil and capacitors, mounted to the vacuum chamber) generating the electromagnetic fields. Both, positioning and heating are done at the resonance frequencies of the positioning and the heating circuit, respectively. Thus the LPS only needs to compensate for the power losses within the self-oscillatory circuits with easy-to-handle currents in the order of 10 A. The design of the coil is such that the RF currents generating the positioning and heating fields are superimposed within the same copper coil which leads to a very high concentricity of both fields and therefore to an optimized positioning stability. The positioning field is of quadrupole type with low heating efficiency and operates at about 140 kHz whereas the heating field is of dipole type with low forces on the sample and operates at about 380 kHz. Depending on sample material and process environment heating rates of about 100 K/s can be applied such that sample temperatures of up to 2100°C can be reached within less than 1 minute. To meet scientific requirements heater and positioner voltages can be varied as linear functions of time. In addition, the heater circuit allows a harmonic voltage or power modulation as well as short pulses (0.1 s) as stimuli for the sample.

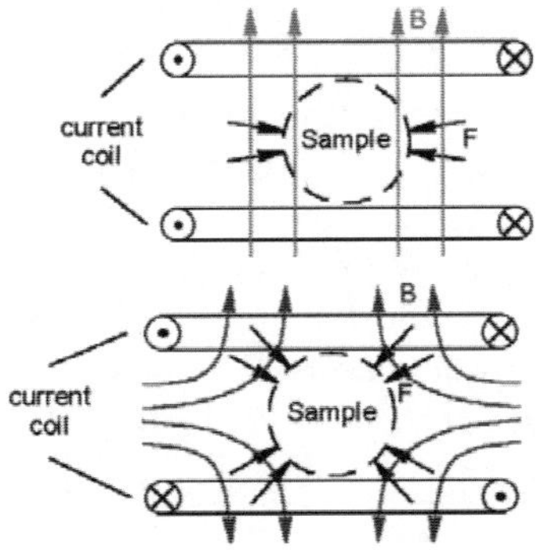

Figure 4. Superimposed electromagnetic fields in EML (note the diagram shows the two fields separately whereas just one coil is physically integrated in EML).

- Top: dipole heating field (red field lines)
- Bottom: quadrupole positioning heating field (blue field lines)

3.2.3. *Sample Holders and handling:* EML can accommodate 18 samples (5-8 mm diameter) per sample chamber. The samples are housed within individual sample holders (inner diameter 14 mm) mounted to a carousel within the sample chamber. By this mechanism, a selected sample can be transferred from the storage position to process position within the r.f. coil. The sample holder design provides a containerless environment for the sample during nominal processing while a contact of the levitated sample with the RF coil is prevented in case of any crew or station induced excessive accelerations, and also sample containment during launch is assured.

Two generic types of holders will be available; cage and cup with slit openings for observation from axial and radial direction by the EML diagnostic instruments. Common to both types is a ceramic foot on top of which the sample confinement is mounted. Depending on the scientific goal and/or the expected sample behavior (e.g. evaporation) one or the other type will be chosen. In order to prevent undesired coupling to the electromagnetic fields the sample holders are made mainly from ceramics. The selected silicon nitride is thermo-shock resistant and compatible with highly reactive melts. The cage design uses tungsten-rhenium wires with a geometry that minimizes coupling to the r.f. fields. The samples will be integrated into the sample holders in a glove box on ground and will then be transferred to the sample chamber under noble gas atmosphere until launch and mounting to the EML facility in orbit.

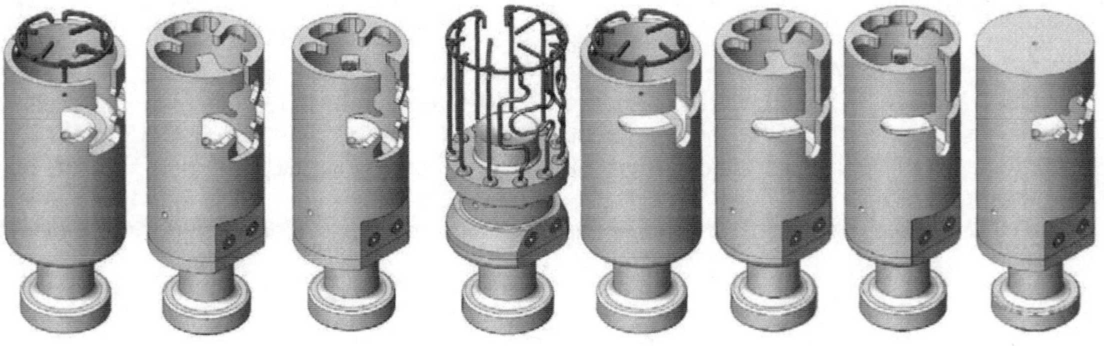

Figure 5. Proposed Sample Holder types under development.

3.2.4. *Sample Stimuli.* EML offers various methods for sample stimulation such that the sample's reaction on the stimulus can be observed by the diagnostics to derive the scientific goal. The various stimuli are:

- Short pulse of r.f. heater field: the force on the sample's surface created by the dipole field squeezes the sample and thus excites shape oscillations from which the surface tension and viscosity can be derived

Journal of Physics: Conference Series **327** (2011) 012057 doi:10.1088/1742-6596/327/1/012057

- Modulation of r.f. heater field: the sinusoidal varied input power leads to a respective sample temperature response from which e.g. the thermal capacity can be derived
- Trigger needle: a coated Rhenium needle integrated into the sample holder lid is used to touch the sample (by moving the sample holder relative to the free floating sample) and acts as nucleation trigger. Hence nucleation can be induced at defined undercooling temperatures and the solidification direction is defined and can be imaged by the orthogonally arranged video systems.
- Chill Cool Plate: ceramic plate integrated as top lid to the cup-type sample holder acting as heat sink for increasing the cooling rate at the sample surface when being touched with the chill cooling plate. The plate offers a small drilling (1.0 mm) to observe the temperature of the interface plate/sample.
- Gas flow: a gas circulation system allows generating an adjustable gas flow around the sample to impose dedicated cooling or convection conditions.
- Sample damping: the cone shape of the sample holder bottom can be used to mechanically dampen translational oscillations of the sample by moving the sample holder relative to the levitated sample

3.2.5. *Diagnostics - Pyrometry.* EML offers a highly sensitive digital pyrometer integrated together with the axial camera in one instrument. Both instruments are using the same optical path where the NIR spectrum is separated from visible light by a beam splitter. The instrument called ACP (axial camera & pyrometer) observes the sample in axial direction of the coil system. Key pyrometer performance data are listed in table 2 below.

Table 2. Key Performance Data of EML Pyrometer

Measurement rate:	100 Hz	Temperature resolution (for ε = 0.05) ▪ ≤0.1 K @ T > 600°C ▪ <0.5 K @ 500°C < T < 600°C ▪ <1.5 K @ 400°C < T < 500°C ▪ <3.0 K @ 300°C < T < 400°C
Measurement range:	300-2100°C	
Emissivity ε :	0.05 ...1.00	
Integration time:	5 ms	
Measurement spot diameter:	0.8 mm	
Wavelength:	1.45 - 1.80 μm	

3.2.6. *Diagnostics - Axial Video System.* For visual observation a digital high resolution video camera is integrated in the above mentioned instrument ACP together with the pyrometer. The camera is able to observe fast sample oscillations for measurement of surface tension and viscosity. The optics is also capable to detect relative sample size changes of $2x10^{-4}$ supporting thermal expansion measurements. By its telecentric property the lens supports the mentioned measurements even for samples slightly moving +/- 1 mm back and forth along the optical axis. Key performance data of the ACP camera are given in table 3 below.

Table 3. Key Performance Data of EML Camera ACP

Sensor: CMOS, resolution 1280x1024 pixel	Various fields of view, resolutions & max. frame rates, e.g.: ▪ 10x10 mm: 704x704 pixel, 50 Hz ▪ 10x10 mm: 352x352 pixel, 150 Hz ▪ 8x8 mm: 280x280 pixel, 200 Hz ▪ 18x14 mm: 1280x1024 pixel, 15 Hz (for inspections)
Digitalization: 10 bit/pixel	
Global shutter, fixed or auto exposure	
Frame rate enhancing by sub-sampling 2x2 (i.e. pixel reduction by factor 4)	
Focal point of optics: 1.7 mm above sample equator towards the pole	

3.2.7. *Diagnostics - Radial Video System.* EML offers a second video camera, called HSC (high-speed camera) observing the sample through a 8 mm wide gap in the RF coil windings from the radial direction with respect to the coil. Hence the camera orientation is perpendicular to the one in the ACP such that EML cameras are observing the sample in two orthogonal views. The HSC provides two modes of observation called "oscillating drop mode" and "recalescence mode" which are achieved by a bifocal lens system allowing imaging the constant field of view on two different chip areas. This feature enables the HSC to acquire up to 30.000 frames per second in the high-speed "recalescence mode" at reduced spatial resolution still imaging the entire sample to visualize the solidification front growth. The "oscillating drop mode" achieves a higher spatial resolution at the expense of a reduced time resolution to support measurements on the surface shape oscillations and thermal expansion as described for the ACP camera. Key performance data of the HSC are given in the table below. The video captured by the high-speed camera is stored first in a camera internal ring memory in order to comply with high-speed requirements and in a later step transferred to dedicated mass memory. In order to capture the recalescence event on video the recording is stopped by an automated recalescence detection algorithm based on the temperature reading.

The HSC provides also the possibility of thermal radiation mapping due to the high digitalization which can be used to display temperature distribution across the sample in false colors by offline ground video processing. Minimum temperatures for which samples can be visibly observed are between 500 -1000°C depending on sample emissivity and exposure time.

Table 4 Key Performance Data of EML Camera HSC

Sensor: CMOS, resolution 800x600 pixel	Various fields of view, resolutions & max. frame rates, e.g.: ▪ 10x10 mm: 600x600 pixel, 8.5 kHz ▪ 10x10 mm: 250x250 pixel, 30 kHz ▪ 13x10 mm: 800x600 pixel, (for inspections)
Digitalization: 14 bit/pixel	
Global shutter, fixed or auto exposure	
Frame rate enhancing by pixel reduction	
Real time analog NTSC video output for process control	Recording duration of 2 GB ring memory partition (examples), for entire 8 GB ring memory a factor 4 applies. ▪ 600x600 pixel, 200 Hz: 16.6 sec ▪ 600x600 pixel, 8500 Hz: 0.39 sec ▪ 256x256 pixel, 10 KHz: 1.854 sec ▪ 256x256 pixel, 30 KHz: 0.615 sec
Internal 8 GB ring memory for high-speed video acquisition, download to EML mass memory after experiment	
Triggering of recording by event through experiment controller, e.g. by re-calescence detection algorithm	

3.2.8. *Diagnostics - Coil System.* The coil system module (CSM) not only heats and positions the sample it also delivers significant housekeeping data such as frequency, voltage and current of the oscillating circuit from which the electrical conductivity and inductivity of the sample can be derived. This is possible because the metallic sample inside the r.f. coil act together as a transformer with the sample as a damping element. Thus with changing electrical sample properties the oscillating circuit changes its electrical properties. EML offers the possibility to measure these small changes on comparable big scale r.f. power signals provided by the Levitation Power Supply.

3.2.9. *Data Processing and Downlink.* Process control and data acquisition is performed by the ECM module which contains a dedicated computer system (ECE) with associated interface and power supply electronics. The ECE acquires standard scientific and housekeeping data at rates between 1 and 100 Hz (i.e. pyrometric data). Data downlink is performed in real time by transferring the data to EDR via Express Rack protocol.

Journal of Physics: Conference Series **327** (2011) 012057 doi:10.1088/1742-6596/327/1/012057

Digital video data are acquired separately by two dedicated systems one for each camera. Both video systems interface the EDR Video Management Unit (VMU) via HSSL link for downlinking of the acquired video data.

- In case of the ACP camera the captured video is acquired and processed in real-time by the Digital Video System (DVS). It allows to split the incoming video stream in two simultaneous streams with different image processing applied for the two purposes below. Control of DVS is possible per experiment parameter set to allow synchronization to events determined by the timeline of the melt cycle or per ground commanding.
    - real-time video: highly compressed video to comply with downlink constraints, used for process control purposes by ground operators
    - scientific video: low or uncompressed video stored on DVS own hard disk for later non-realtime downlink
- In order to clear the HSC internal ring-memory the captured video will be transmitted to the dedicated video management unit called HSC-OS (operating system). The HSC-OS is able to compress, store and downlink the recorded video files on request by ground command. For process control purposes the HSC delivers besides high-speed digital recording also a live video stream as analogue NTSC standard which is transmitted to ground via the analogue video interface of EDR VMU.

3.3. Mission Scenario

The EML facility will be launched and transported to the *International Space Station* ISS, where it will be installed in the *European Drawer Rack* (EDR). During its service lifetime of 5 years EML will be operated in the EDR. In order to be able to process a large number of scientific samples during the service lifetime, samples can be uploaded and downloaded between ground and EML on ISS according to the following scenario (see also figure 6):

- The flight samples which include pure metals, alloys and semiconductors, are spherical in shape with diameters ranging from 5 to 8 millimeters and are manufactured by the participating scientists in their labs.
- These samples are then integrated into dedicated sample holders and subsequently into a Sample Chamber (SCH) under very strictly controlled environmental conditions since the scientific results of the levitation experiments depend critically on the purity of the samples. Each Sample Chamber has a capacity of 18 samples which are stored in individual sample holders mounted on a magazine wheel in order to manipulate and select each sample individually for processing.
- The fully loaded Sample Chamber is uploaded to the ISS where it is attached to the EML facility.
- Sample processing is performed according to a sophisticated procedure which not only involves the settings of the levitator but also of diagnostics elements. The data delivered by the diagnostic elements, mainly high-speed and high-resolution video cameras and pyrometers are the main scientific raw data gathered during the mission. Since the duration of a typical melt cycle of an EML sample is in the order of seconds or minutes and the processing procedure is very complex with hundreds of individual settings, the procedure is developed before the mission in a ground based experiment development program, converted into a parameter set for the facility, and is then automatically executed by the facility. Interactive control of running experiments can be performed e.g. to optimize parameters based on the results of the previous runs by tele-commanding from the user center on ground which is permanently monitoring and controlling the running experiments.
- After all samples in a Sample Chamber have been processed, they are transferred to a Sample Download Container (SDC) by the flight crew for download to ground. The design of SDC has been optimized for the tight constraints of the download in terms of mass and size.

- After landing the processed flight samples are removed from the SDC and returned to the scientists for further analysis.

According to the mission scenario up to two Sample Chambers per year will be uploaded to the EML facility, which translates to a total of 36 samples that can be uploaded, processed, and downloaded per year. Therefore, a maximum of 180 samples can be processed during the 5 year service lifetime of EML.

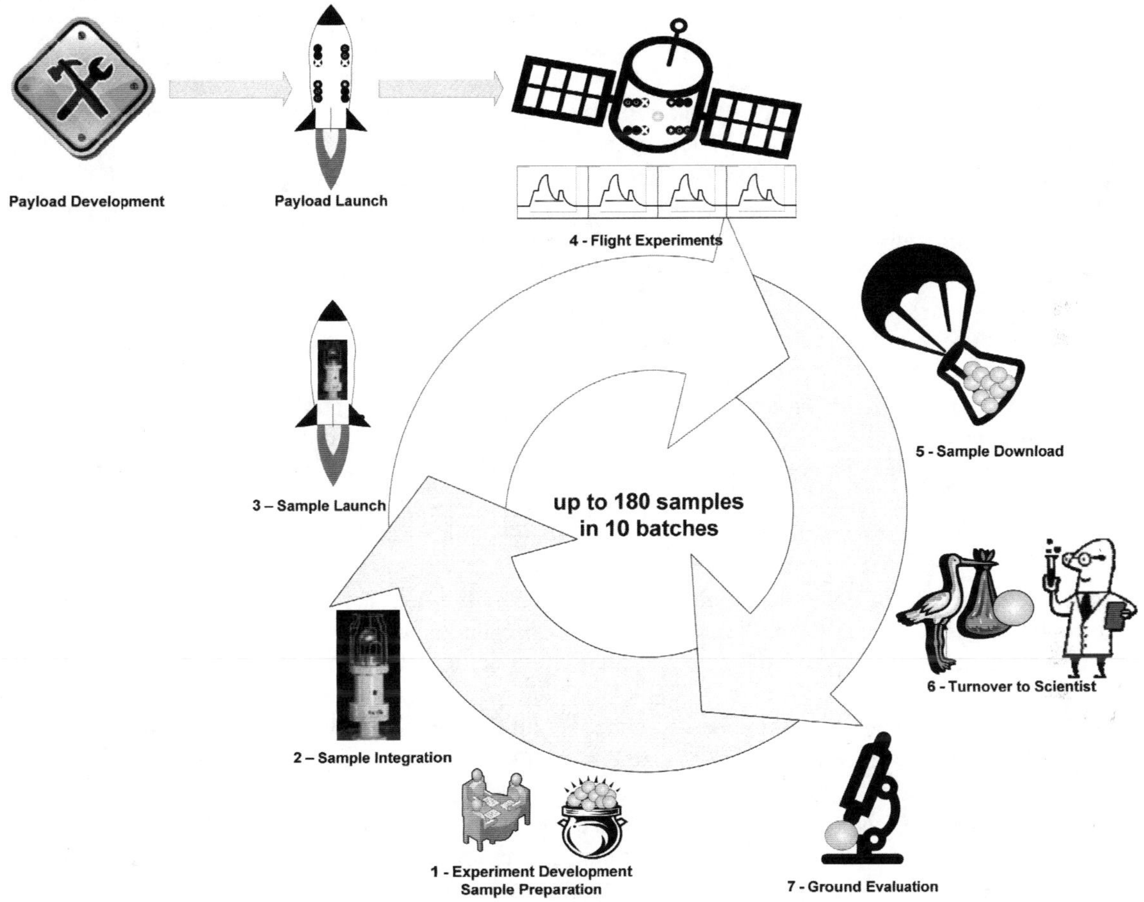

Figure 6. EML Mission Scenario, showing the complete life-cycle of the experiments from experiment preparation to post-flight evaluation.

Limitations of the EML on-orbit lifetime arise mainly from three factors which are all related to the usage of the facility by the experiments: The first limitation comes from the fact that melted samples under ultra-high vacuum conditions will evaporate, which leads to deposition of evaporated sample material on the inner surfaces of the process chamber. While measures have been implemented to protect sensitive elements of the chamber from deposition (e.g. the optical viewports) the deposited layer thickness is limited; if the limit is exceeded the layer may peel off and contaminate the process chamber with flakes which would not be compatible with the ultraclean environment needed by most experiments. The second limitation comes from a similar phenomenon: if samples are processed under a gas atmosphere, the evaporated material forms microscopic dust particles. The amount of the thus produced dust is limited by safety considerations; on the other hand a dust removal system has been implemented in EML which reduces the problem of dust accumulation. The third limitation is related

Journal of Physics: Conference Series **327** (2011) 012057 doi:10.1088/1742-6596/327/1/012057

to the consumption of noble gases stored in EML by those experiments that are running in a gas atmosphere versus the storage capacity of the EML gas supply system.

4. EML Program Overview

EML for ISS is a project jointly funded by ESA and DLR [15]. After going through some pre-development phases the development and manufacturing of the flight model was launched in 2008 as part of a comprehensive program that not only covers the development of the flight facility but also various aspects related to the preparation of the ISS mission, as follows:

- The flight facility is based on the TEMPUS concept but includes several upgrades in particular with respect to diagnostics, and is currently being manufactured by the industrial team under the leadership of Astrium. Delivery is planned for 2012.
- A second model called "Operational Model" OM which is essentially a copy of the flight model is manufactured and will be used for experiment preparation and verification, for sustaining engineering during the mission, and will serve as spare pool for the flight facility.
- A Science Reference Simulator SRS, which is a software based tool designed to support the ground preparation of experiments
- A Training Model TRM, which will be used by the European Astronaut Center to prepare crew operations
- A Ground Support Program GSP with the objective to prepare the flight experiments on ground by performing tests with representative samples in a ground levitation facility and by developing the parameter sets that will control the flight parameter sets
- An Experiment Infrastructure EXI project, which will provide custom-made sample holders for each flight sample and the ground infrastructure needed for integration of the flight samples into the Sample Chambers in a controlled environment
- Development of a Sample Coupling Electronics SCE which is a potential future upgrade of the EML facility which would enhance the diagnostic capabilities of EML to allow a precise determination of the sample's electrical conductivity (see Chapter 5).
- Development of an Oxygen Sensing and Control System OSC which is another potential future upgrade of the EML facility which would allow precisely determining and controlling the oxygen partial pressure in the process atmosphere and would therefore allow improving and characterizing the process atmosphere (see Chapter 5).
- Upgrade of TEMPUS parabolic flight facility PFF in preparation of future parabolic flight missions. Experiments in the PFF are planned as precursors for ISS experiments for samples that have not been previously processed in TEMPUS and for samples which can not be processed on ISS due to safety constraints.

5. Evolution potential

The modular character of the EML facility allows the upgrade of existing instruments and stimuli elements and the implementation of additional systems or the exchange of used ones. Consequently, there are two major classes of possible evolutions:

1. Short-term: Broadening of experimental capabilities by implementation of new diagnostic and stimuli elements
   The modular character of the facility allows for an easy implementation of further diagnostics or stimuli elements like
   - Fast pyrometer (order of MHz)
   - Different video cameras
   - Sample coupling electronics (SCE)
   - Oxygen control system
2. Long-term: Life time extension
   In order to extend the total operation time of the facility a replacement of the experiment

chamber after a certain number of experiments can be envisaged, since the life time is limited by the contamination of the coils. This, however, would indeed impose a large effort on logistics and crew time.

A number of highly peer-ranked scientific proposals require the implementation of two diagnostic/stimuli systems out of the mentioned short-term options. These will be introduced shortly in the following sections.

5.1. Oxygen control system

For the measurement of surface tension and viscosity the control of the partial pressure of oxygen in the ppm region is necessary [16]. This feature is not yet realized in the basic version of the EML facility, but is regarded as an add-on diagnostic system having a high priority.

A space-compatible technology demonstrator of an oxygen sensing and control system has been developed and tested in the frame of an ESTEC technology study [17]. An oxygen measuring and loading system including a control loop to adjust a dedicated oxygen partial pressure in a gas circulation has been developed and tested which consists of a combination of a potentiometric sensor based on Yttrium-stabilized zirconia with an oxygen titration. This system is suited to measure and to adjust the oxygen partial pressure in the sub-ppm range and can be operated in wide range of oxygen partial pressures (10^{-3}-10^{-19} bar). In a laboratory setup a standard deviation of $\log(pO_2/1\text{bar}) < 0.003$ orders of magnitude could be verified for a period of 30 h. Furthermore, field tests of this system performed in the levitation facility of DLR in Cologne have verified the feasibility for operating this system in the harsh environment of a levitation facility.

This concept is also capable to be implemented into the EML facility as an add-on diagnostics. The complete system consisting of sensor, pump, and electronics would be completely accommodated in an additional single locker. The corresponding supplies and control lines are already foreseen by EML.

A further attractive feature which could enhance the technical potential of the EML facility is the capability of the titration pump to remove oxygen from the atmosphere without being limited by any saturation effect. Hence, contamination of the process atmosphere caused by samples or small leaks can be compensated by this pump reducing the risk of a facility shut-down tremendously.

As a next logical step the development of a prototype demonstrator of such an oxygen control system has been proposed to ESA/DLR. This demonstrator shall be compatible with utilization in the TEMPUS parabolic flight facility, terrestrial levitation facilities, and the EML-OM.

Figure 7. Titration pump for adjusting the partial pressure of oxygen in the sub-ppm region as developed under ESA contract.

5.2. Sample Coupling Electronics
In a DLR funded development a laboratory model of a sample coupling electronic (SCE) device has been developed. This unit is connected to the RF oscillating circuit and allows for measuring of thermophysical properties such as electrical conductivity, surface tension and viscosity via the inductive method. Whereas resistivity measurements per induction is relatively straight forward, the instruments high sensitivity and fast response time is such that even sample shape oscillations can be detected from which viscosity and surface tension can be derived. Thus the SCE is a perfect complement to the optical diagnostic tools [18]. The EML facility already foresees the interfaces to accommodate the SCE mechanically, electrically and in terms of data management.

The device is already miniaturized, implements the interfaces to EML and needs a power supply of less than 10 W. During ground tests in the TEMPUS parabolic flight facility a calibration procedure has been developed and first tests with a solid tungsten sample have been successfully conducted showing excellent results.

Acknowledgement
TEMPUS and EML are developed by Astrium GmbH under programs of and funded by the European Space Agency (ESA) and by the Space Agency of the German Aerospace Center (DLR e.V.) with funds of the German Federal Ministry of Economics and Technology.

References
[1] Muck O 1923 *German patent no. 42204*
[2] Nordine P C, Weber R J K and Abadie J G 2000 *Pure Appl. Chem.* 72 No. 11 2127
[3] Mauro N A and Kelton K F 2011 *Rev Sci Instr.* 82 035114
[4] Team TEMPUS 1995 In Materials and Fluids Under Low Gravity: *Proc. IXth Symp. on Gravity-Dependent Phen. in Physical Sciences* ed L Ratke et al. (Berlin: Springer)
[5] Piller J et. al. 1986 Electromagnetic Positioning and Inductive Heating under micro-g: *6th European Symposium on Materials Sciences under Microgravity (Bordeaux)*
[6] Knauf R, Piller J, Seidel A and Dreier W 1992 TEMPUS - An Electromagnetic Levitation Facility for Containerless Processing under Micro-g in Spacelab: *Proc VIIIth European Symposium on Materials and Fluid Sciences in Microgravity (Brussels, Belgium)* ESA SP-333
[7] Geiger R and Thomas J 1993 Resonant High Power Supply for Electromagnetic Levitation and Melting of Metallic Samples *European Space Power Conference (Graz)*
[8] Knauf R et. al. 1994 Containerless Processing Facility TEMPUS for Materials Science Research onboard Spacelab: *6th International Symposium on Experimental Methods for Microgravity Material Science (San Francisco)*
[9] Egry I 1998 *Journal Jpn. Soc. Microgravity Appl.* 15 No. 4 215-24
[10] Motil S M, Harrivel A R and Zimmerli G A 2001 *NASA/TM* 2001-210704
[11] Pletser V 1995 *ESA Bulletin* Nr. 82
[12] http://www.esa.int/SPECIALS/HSF_Research/SEMSPZSTGOF_0.html
[13] Lord, Douglas R, NASA-SP-487. NASA, January 1, 1987.
[14] Seidel A, Soellner W and Stenzel C, "Materials Science Investigations using Electromagnetic Levitation", this proceedings
[15] http://www.dlr.de/rd/desktopdefault.aspx/tabid-2283/3420_read-5126/
[16] Ozawa S, Suzuki S, Hibiya T and Fukuyama H 2011, *J. of Appl. Phys.* 109 014902
[17] Study data package of ESTEC TRP study "Oxygen Sensing and Control" - ESTEC contract no. 19963/07/NL/PM
[18] Lohöfer G 2009 EML Sample Coupling Electronics Test Model Final Report *EML/SCE-TM-Rep 1* (DLR, Cologne)

International Symposium on Physical Sciences in Space IOP Publishing
Journal of Physics: Conference Series **327** (2011) 012058 doi:10.1088/1742-6596/327/1/012058

Magnetic ejection of submilimeter-sized diamagnetic grains observed in a chamber-type drop shaft

K.Hisayoshi[1, 2], C.Uyeda[1], K.Kuwada[1], M.Mamiya[3], H, Nagai[3]
[1]Institute of Earth and Space Science, Graduate School of Science, Osaka University Toyanaka, Osaka, 560-0043, Japan
[2]Kasugaoka High-school of Osaka prefecture, Ibaraki, Osaka, 563-0031, Japan
[3]National Institute of advanced Technology AIST, Tsukuba Central Tsukuba, Ibaragi, 305-8565, Japan

E-mail: hisayoshi@ess.sci.osaka-u.ac.jp

Abstract. Mass dependence of magnetically induced translation is examined for submilimeter-sized diamagnetic grain, which is floated in a diffused gas medium of 70 Pa by the use of microgravity μG condition. Here the grains are ejected from field-center of a pole-piece magnetic circuit by field-gradient force in a direction of decreasing field; maximum field-intensity at initial point is about 0.6T. It was deduced from an energy conservation rule that magnetic potential of the grain at initial position was partially converted to kinetic energy during translation. Therefore, observed velocity was expected to be independent to mass of particle; this is because magnetic potential is proportional to mass of particle. In a given field distribution, the velocity was uniquely determined by intrinsic magnetic susceptibility χ_{DIA} of material. We report here that the magnetic ejection in μG condition is realized for sub-mm sized crystals of bismuth, graphite and magnesia. The χ_{DIA} values obtained from the ejection agreed fairly well with the published values for the three materials. The mass-independent property of translation was examined by observing the translation of two bismuth grains with different sizes. A chamber-type drop shaft having a height of 1.5 m was introduced to produce the μG condition. By the use of this drop shaft, study of field-induced motion of a single diamagnetic particle becomes possible in an ordinary laboratory.

1. Introduction

In terrestrial gravity, magnetic motion of a solid substance is generally considered to occur on materials that bear spontaneous magnetic moments or strong paramagnetic moments. It is believed that dynamic motions of weak magnetic (i.e. diamagnetic and paramagnetic) material require a strong field above B = 10T [1-4]. When a weak magnetic solid was released in μG space with negligible effect of gravity or viscous drag, two kinds of basic motions were expected to occur by low magnetic field. First, translation was caused by field gradient force, $m\chi_{DIA}B(dB/dx)$, where m and χ_{DIA} denoted mass and diamagnetic susceptibility per unit mass of sample, respectively. Secondly, magnetically stable axis of crystal showed rotational oscillation with respect to field direction in a homogeneous field. The two motions were expected to be a common property of a solid material, since all material possessed intrinsic diamagnetic susceptibility χ_{DIA} and anisotropy $\Delta\chi_{DIA}$.

Field gradient force was conventionally used in separating particles that bear strong magnetizations. Magnetic levitation of diamagnetic materials caused by field-gradient force was performed at strong field intensity above 10 Tesla at high magnetic-field facilities [1,2]. Various methods to detect susceptibility of small particles were proposed using a strong field gradient at terrestrial gravity condition in a presence of viscous medium [4]. In these studies based on high magnetic field, magnetization measurements based on free motions of particles in diffused area were not considered.

The above-mentioned properties of magnetic motions expected in μG were recently examined for mm-sized crystals of popular diamagnetic materials, using a conventional μG facility [5,6]. The χ_{DIA} values obtained from the translations were consistent with the published values. By measuring the period of rotational oscillation in μG condition, small $\Delta\chi_{DIA}$ at a level of 1×10^{-9} emu/g was detected [7]. As a step to realize the above-mentioned magnetic ejection at micron-size level, translation was observed in the present study for sub-milimeter sized grain of three diamagnetic materials which had different χ_{DIA} values. A chamber-type drop shaft was adopted to produce μG, for the purpose of studying the above-mentioned magnetic motions in an ordinary laboratory.

2.Experimental

The compact drop shaft used in the present study had a length of 1.5m. The experimental setup to observe the magnetic ejection was set inside a rectangle wooden box which had a size of 30cm×30cm×20cm. The box was attached to the sealing of our laboratory room by an electromagnetic lock. The free fall of the box started shortly after the power supply of the lock was shut down. The duration of μG was below 0.6 seconds. The apparatus installed in the box consisted of a cylindrical glass tube equipped with an electric actuator to operate a sample holder, a signal receptor to control the electric actuator, a magnetic circuit and the high-vision (HV) video camera (Panasonic HDC -SD3-S; spatial resolution 0.004 cm.). The magnetic circuit was composed of a couple of NdFeB magnetic plates having the size of 2.50cm×2.00cm×0.6cm. In order to describe the observed sample positions, an orthogonal co-ordinate as described in Fig.1 was defined in the above-mentioned apparatus. The center of the magnetic circuit was the origin of the co-ordinate. The magnetic line of force, directing from N to S pole at the center of circuit, was parallel to the y-axis; field intensity at the center of circuit was 0.72 T. The x-axis was parallel to the cylindrical axis of the glass tube.

The sample was released in diffused μ G space at a position $x= x_0$ with negligibly small initial-velocity (v_0=0) ; field intensity at $x= x_0$ was defined as $B= B_0$. The B_0 values determined for individual samples are listed in Table 1; these values were determined from initial positions of the sample. The field gradient dB/dx at $x = x_0$ was negative (see Fig.1), so the samples were expected to translate along the x-axis in a positive direction. Translation of the sample was observed from a [+z] direction by the HV camera with a time division of 0.033 s. Before the μG experiment, field distribution along the x-axis was measured with a spatial resolution of 0.1 cm; The result is shown schematically in the lower portion of Fig.1. Field intensity was equal to zero at a position, $x_R = 1.87\pm0.05$ cm, since magnetic line of force turned from [+y] to [-y] direction at this point.

Table I Numerical data of bismuth, graphite and MgO grains that were observed in the present study. Here m and χ_{DIA} denote mass of sample and diamagnetic susceptibility per unit mass, respectively. The samples were cut from a synthetic block with a purity of 99.99%. The B_0 values of individual samples are determined from the observed initial positions. The errors of x_0, B_0 and v_R derive from the ambiguity of sample positions that were determined from the images of the HV camera. The value of χ_{DIA} is obtained from a relationship $v_0^2 = \chi_{DIA}B_0^2$; the errors mainly derive from the ambiguity of the measured B_0 values.

Sample	m [g]	x_0 [cm]	B_0 [T]	v_R [cm/s]	χ_{DIA}[x10^{-7}emu/g] measured	published [8]
Bi-1	(0.50±0.05)x10^{-3}	0.94±0.04	0.658±0.060	9.83±0.1	-16.5±3.0	-13.4
Bi-2	(5.1±0.5)x10^{-3}	0.975±0.043	0.609±0.060	2.74±0.0	- 15.8±3.2	-13.4
G1	(1.6±0.5)x10^{-4}	0.99±0.01	0.629±0.006	13.9±0.1	48.7±1.5	52
M1	(1.2±0.5)x10^{-4}	1.02±0.01	0.609±0.006	2.74±0.02	2.02±0.9	2.6

In a previous study, determination of χ_{DIA} (per unit mass) was based on an energy conservation rule assumed for a sample that was translating in magnetic field, namely $\frac{1}{2}m\chi_{DIA}B_0^2 = \frac{1}{2}m\chi_{DIA}B_1^2 + \frac{1}{2}mv_1^2$; here v_1 denote sample velocity at an arbitrary position $x = x_1$, whereas field intensity at $x = x_1$ is defined as B_1. According to the above equation, χ_{DIA} should be equivalent to the gradient of a linear relationship $v_1^2 = \chi_{DIA}(B_1^2 - B_0^2)$. In case of millimetre-sized crystals measured in the previous study, the χ_{DIA} values obtained from the above-mentioned relationship was consistent with their published χ_{DIA} values [6]. In the present study, χ_{DIA} is calculated from a relationship $v(x_R)^2 = \chi_{DIA}B_0^2$ which is deduced from a energy conservation between position x_0 ($B=B_0$, $v_0=0$) and position x_R ($B=0$, $v=v_R$).

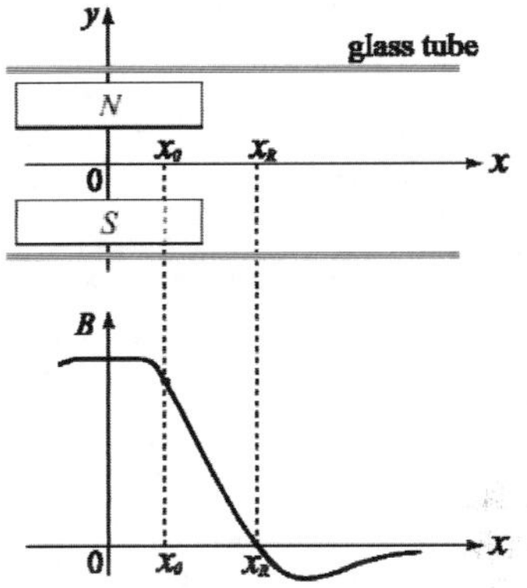

Fig.1 Schematic view of the experimental setup to measure the magnetic ejection of dia-magnetic particle released in μG condition. Details of position x_0, x_R and the x-, y- axes are given in the text. Translation of particle is observe by a HV camera from the +z axis. Field distribution along the x-axis is shown in the lower portion.

3. Results and Discussions

When the effects of gravity and viscous drag of medium was negligible, a particle released in a magnetic field gradient that decreases monotonously along x-axis was expected to show translation that followed a motional equation [5][6] described as,

$$m\,(d^2x/dt^2) = m\chi_{DIA}B(dB/dx). \qquad (1)$$

The field-gradient force is known to be a "volume force" which derived from the localized electrons of individual atoms that composed the material. Accordingly, the two terms in eq.(1) are both proportional to m. It was hence expected that acceleration d^2x/dt^2 of particle was independent to its mass [5]. The equation of energy conservation between magnetic potential and kinetic energy, as described in the previous section, was directly deduced from the above motional equation.

Fig.2 show a time variation of sample position along the x-axis measured for the Bi-1 sample in Table 1. Fig3(a) shows the relationship between sample velocity v and position x obtained from the measured data of Fig.2. From the v-x relationship, the velocity at $x = x_R$ was obtained as $v_R = 0.94$ cm/s. By inserting this v_R and $B_0 = 0.6584$ T (see Table 1) in the above-mentioned equation of energy conservation $v(x_R)^2 = \chi_{DIA}B_0^2$, χ_{DIA} was calculated as $\chi_{DIA} = -15.8 \pm 3.2 \times 10^{-7}$ emu/g.

Fig.2 Time variation of sample position measured along the x-axis of the apparatus; definition of x-axis is given in the text. The relationship was measured for a submilimeter sized sample of bismuth (Bi-1 in Table1). The sample was released at t = 0. The values of x were determined by a scale that was attached along the x-axis. Images of the scale and the sample are recorded by a HV camera. Errors bars of x and t are both within the symbols.

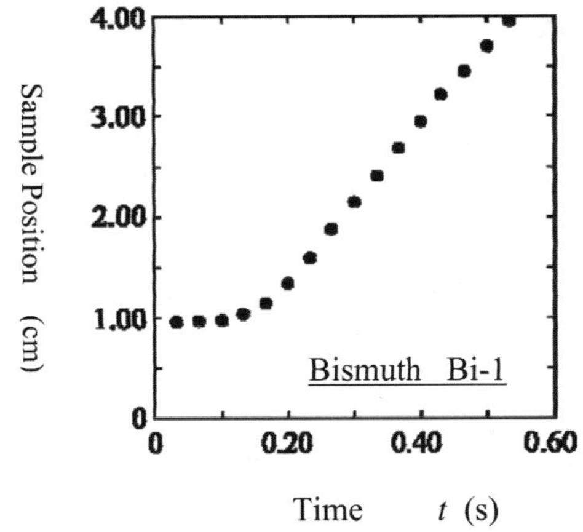

Using the above procedure, the χ_{DIA} values are obtained for two bismuth grains with different m values. The values are compared in Table I with the published value, -13.4x10^{-7} emu/g. For both samples, the deviations of the measured values with respect to the published value are within range of experimental error. It may be concluded that no significant tendency of mass dependence is seen in the field-induced translation at least for bismuth in the range of $m = 0.5 \sim 5.0$ mg.

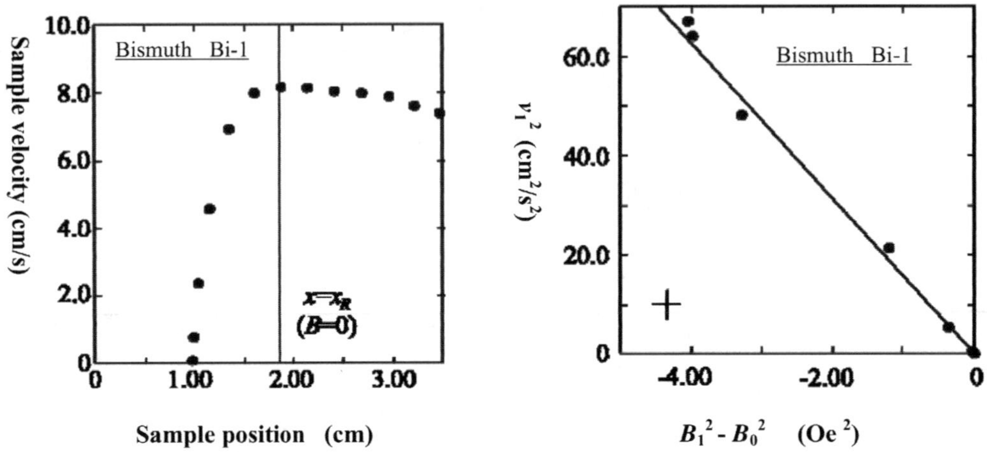

Fig.3 (a) Relationship between sample velocity v and position x. (b) Relationship between v_1^2 and B_1^2 - B_0^2; definitions of v_1, B_1 and B_0 are given in the text. Both relationships are obtained from the measured variation of sample position shown in Fig.1, which were measured for a submillimeter- sized bismuth grain (Bi-1 in Table 1).

The χ_{DIA} values of graphite and magnesia obtained from the above relationship are listed in Table 1. It is seen that measured values agreed fairly well with their published values. The numerical range of χ_{DIA} observed in the present study nearly overlaps with the range of the published χ_{DIA} values compiled in a data book of diamagnetic susceptibility; the smallest published χ_{DIA} value has been reported for indium which is 1.3x10^{-7} emu/g, whereas the largest is reported for graphite which is χ_{DIA}=5.2x10^{-6} emu/g [9] . Hence for sub-millimeter sized particles, efficiency of eq.(1) seems to be confirmed over the χ_{DIA} values of existing materials.

The χ_{DIA} value of Bi-1 was also calculated from the gradient of the v_1^2 v.s. B_1^2 - B_0^2 relationship that is described in Fig.3 (b). The relationship is obtained by determining sets of $v(x)$ and $B(x)$ values at different sample positions in Fig.2. The calculated value is χ_{DIA}= - (1.59±0.32)x10^{-6} emu/g which is consistent with published value; the result indicates that sum of kinetic and magnetic energy is conserved throughout the translation of the submilimeter-sized Bi grain. Whereas the agreement between measured and published χ_{DIA} values compiled in Table 1 show that an accurate χ_{DIA} value is obtained from a calculation based on the equation $v(x_R)^2 = \chi_{DIA} B_0^2$; deduction of χ_{DIA} using the above equation is far more simple compared to the analysis based on the relationship of v_1^2 v.s. B_1^2 - B_0^2, which was performed in previous studies [5][6].

In a conventional method to measure magnetization, measureable sample size is limited by the following two factors; the interference by the background signal emitted from the sample holder, and the difficulty of measuring the mass of sample. The present method is free from the two factors; detection of χ_{DIA} is based on a simple conservation rule between kinetic energy and magnetic potential [6]. Accordingly, magnetization of a single diamagnetic grain smaller than mm-size, which is difficult to perform by conventional methods, becomes detectable by the present method. The present lower-limit of measurable sample size using this system is determined by the spatial resolution of the HV camera, which is about 20 microns. In this sense, the observation of sub-millimetres sized grain achieved in the present study provides a technical step forward to realize the detection of χ_{DIA} in the region of micron size. In order to examine the generality of the above-mentioned properties of translation in the submilimeter region, it is necessary to perform the observation of Fig.3 (a) and (b) on other diamagnetic materials that posses different magnitudes of χ_{DIA}.

The magnetization data on small particle is expected to provide information on electron configurations of solid insulators in the nano-meter region. According to a Pascal rule, which explains the origin of diamagnetic susceptibility of solid insulators, a χ_{DIA} value of a material is equivalent to the sum of the diamagnetic susceptibility assigned to individual orbital that compose the material [9]. It is hence expected that precise χ_{DIA} data of individual particle provide information on the distortion of bulk crystal structure that is considered to increase with the reduction of crystal size [3].

Dynamic motions in nature induced by magnetic field are commonly recognized only for materials that bear spontaneous magnetic moments or strong paramagnetic moments. The effect of field-induced potential energy should be considered in studying the dynamic behaviour of the dust particles in the process of stellar and planetary evolution. When a long duration of μ G is allowed, magnetic ejection may occur on a diamagnetic grain in general even if the field intensity is as small as the of cosmic field. The above conditions are satisfied in a greater part of inter- & intra-stellar regions, since both static field and diamagnetic dust particles are omnipresent in these regions. The mass independent property of the transition in the sub-millimeter region should be examined for dust forming materials such as ice, carbon-dioxide, forsterite and amorphous silica, since upper limit of size distribution in the circum-stellar regions range in the sub-millimetre scale.

The observation of magnetic ejection was obtainable by the compact μG drop shaft because the small magnetic circuits composed of the Nd-Fe-B plates were introduced as the field generator. The compact μG drop shaft is effective to carry out a χ_{DIA} measurement of a single particle as a routine analysis, since the system does not require a long machine time, and the running cost is low enough to repeat a routine process. The studies on field-induced dynamic motion of small particles (i.e. rotation and translation) become possible in an ordinary laboratory.

References

[1] Beaugnon E and Tournier R 1991 *Nature* **349** 470

[2] Ikezoe Y, Hirota N, Nakagawa J and Kitazawa K 1998 *Nature* **393** 749

[3] Kitamura N, Makihara M, Sato T, Hamai M, Mogi I, Awaji S, Watanabe K and Motokawa M 2001 *J. Non-Crystalline Solids* **293-295** 624.

[4] *various papers appearing in* "Magneto-Science", 2006 (eds. Yamaguchi, M., Tanimoto. M., K.Ozeki) Kodansha-Springer, Tokyo

[5] Hisayoshi K, Kanou S and Uyeda C 2006, *J.Phys.:Conf. Ser.,* **156** 012021..

[6] Uyeda C, Hisayoshi K and Kanou S 2010 *J.Phys Soc.Jpn* **79** 064709.

[7] Uyeda C, Mamiya M, Takashima R, Abe T, Nagai H and Okutani T, 2006 *Jpn.J. Appl. Phys.* 45 L124.

[8] Guputa R 1983 "Landort Börnstein" New Series 445

[9] Pauling L 1936 *J. Chem. Phys.* **4** 672

AUTHOR INDEX

Abe, Y. 186, 286
Adachi, S. 416
Addad, S. 216
Adler, M. 216
Amann, B. 224
Angelis, M. 485
Apel, M. 37
Arai, Y. 155
Armour, N. 150
Asada, Y. 378
Asano, H. 529
Baba, S. 529
Balter, M. 102
Beinker, D. 128
Benning, R. 386
Berger, A. 386
Beule, C. 450
Biance, A. 216
Bielke, T. 386
Billia, B. 25, 110
Blochberger, G. 498
Bogno, A. 110
Bojarevics, V. 237
Bongs, K. 485
Bouyer, P. 485
Brauer, D. 508
Browne, D. 25
Cacciapuoti, L. 472, 485
Calle, C. 458
Cannell, D. 203
Caps, H. 216
Carpy, R. 224
Castagnolo, D. 286
Castiglione, L. 224
Celata, G. 286
Cerbino, R. 203
Chassaing, E. 435
Chen, J. 57
Chikov, S. 378
Clements, J. 458
Cox, N. 458
Dagaut, P. 350
Daneke, N. 102
Deguchi, S. 443
Delgado, A. 386
Delon, G. 216
Dettmann, J. 224
Dittus, H. 485
Dost, S. 150
Drescher, J. 498
Easter, S. 237
Egry, L. 196
Ertmer, W. 485
Ertunc, O. 386
Evesque, P. 299, 309
Fecht, H. 83
Feng, S. 51
Fujii, K. 529
Fukunaka, Y. 435
Fukuyama, H. 180
Gandin, C. 25, 57
Gaponenko, Y. 276
Garcia-Escorial, A. 57
Garrabos, Y. 299
Gehrmann, B. 45
Giglio, M. 203
Glicksman, M. 1
Gonzalez-Cinca, R. 230, 250
Goto, M. 516
Griesche, A. 498, 508
Gruhl, T. 102
Gunther, B. 327
Hartwig, J. 485
Hauth, M. 485
Hayashi, Y. 416
Henein, H. 90, 128
Heneine, H. 57
Herlach, D. 57, 90
Hermann, R. 45
Herrmann, S. 485
Hesse, A. 427
Heuskin, D. 508
Hirai, Y. 443
Hisayoshi, K. 569
Hogue, M. 458
Hohler, R. 216, 224
Holland-Mortiz, D. 121
Hou, M. 299, 309
Houltz, Y. 110
Hoyt, N. 541
Hu, M. 386
Huang, K. 485
Hutzler, S. 216
Hyers, R. 83, 121
Ilbagi, A. 57, 90, 128
Iles, G. 327
Imai, R. 529
Inatomi, Y. 155
Inguscio, M. 485
Inoue, Y. 443
Jankowski, T. 450
Johanson, M. 458
Kabov, O. 286, 378
Kadambi, J. 541
Kajari, E. 485
Kamotani, Y. 541
Kaneko, A. 186
Kang, M. 541
Kargi, F. 498, 504
Kargl, F. 102, 508

AUTHOR INDEX

Kawaji, M. 286
Kawamura, H. 259
Kawanami, O. 378, 529
Kawasaki, H. 529
Kelling, T. 450
Kelton, K. 121
Kharraz, A. 541
Khatibi, P. 90, 128
Kihara, T. 155
Kim, J. 365
Kinoshita, K. 155
Konemann, T. 485
Kubota, C. 378, 529
Kubota, Y. 155
Kumar, M. 170
Kuribayashi, K. 161, 170
Kuwada, K. 569
Lammerzahl, C. 485
Landragin, A. 485
Langevin, D. 216, 224
Lecoutre, C. 299
Legros, J. 276
Lengsdorf, R. 57, 90
Li, Y. 309
Lindenkreuz, H. 45
Losch, S. 327
Loser, W. 45
Loth, K. 110
Ludewig, F. 316
Luo, X. 51
Lyubimova, T. 276
Mamiya, M. 569
Mangelinck-Noel, N. 25
Marco, P. 365
Mashiko, T. 443
Mathiesen, R. 110
Matson, D. 83, 121
Matsumoto, S. 186, 259, 516
Mazzoni, S. 203
Melnikov, D. 276
Mendez, N. 250
Merino, J. 350
Meyer, A. 102, 498, 504, 508
Mialdun, A. 276
Minster, O. 224
Mirihanage, W. 25
Mitschke, D. 508
Miyata, H. 155
Modugno, G. 485
Mukai, S. 443
Muroi, R. 76
Nagai, H. 76, 569
Nagayama, K. 161
Nagayasu, T. 378
Natsuisaka, M. 443
Nepomnyashchy, A. 266
Nercissian, V. 386
Neukam, J. 386
Neumann, C. 498, 504
Nguyen-Thi, H. 110
Nishikawas, K. 435
Nishino, K. 259
Nishiyama, Y. 443
Ohkuma, H. 516
Ohnishi, M. 259
Ohta, H. 378, 529
Okumura, K. 443
Okutnal, T. 76
Opsomer, E. 316
Osaka, J. 402
Ozawa, S. 180
Palencia, F. 299
Pascale, F. 110
Penkert, P. 498
Pericleous, K. 237
Peters, A. 485
Picker, G. 224
Pitois, O. 216
Prevedelli, M. 485
Queeckers, P. 378
Raj, R. 365
Ranebo, H. 224
Rasel, E. 485
Reinhart, G. 110
Rio, E. 216
Rosso, M. 435
Saghir, Z. 276
Saint-Jaimes, A. 216
Sakagami, K. 516
Sakai, T. 529
Sakamoto, K. 443
Sakata, K. 155
Sakurai, M. 259
Salomon, C. 472
Santos, F. 485
Sato, M. 286
Saulnier, L. 216
Savino, R. 286
Sawada, K. 529
Sawada, M. 443
Schleich, W. 485
Schmidt, M. 485
Schmitz, B. 327
Schneider, S. 45
Sechenyh, V. 276
Seidel, A. 139, 555
Senger, A. 485
Sengstock, K. 485
Shevtsova, V. 266, 276
Shibayama, H. 155

AUTHOR INDEX

Shimaoka, T. 155
Shimomura, M. 443
Shinmoto, Y. 378, 529
Shuleshova, O. 45
Soellner, W. 139, 555
Sondermann, E. 504
Sone, T. 155
Sorrentino, F. 485
Stenzel, C. 102, 139, 508, 555
Stern, G. 485
Straub, J. 378
Sturz, L. 16, 25, 37
Sunol, F. 230
Suzuki, K. 529
Suzuki, S. 402
Suzuki, Y. 402
Takacs, C. 203
Takahashi, K. 416
Takahashi, S. 180
Takayanagi, M. 155, 416, 529
Tanaka, K. 286
Tanaka, R. 155, 186
Teiser, J. 450
Thi, H. 25
Thome, J. 286
Thompson, S. 458
Tino, G. 485
Totsuji, C. 336
Totsuji, H. 336, 416
Tourret, D. 57
Tsujii, K. 443
Ueno, I. 259
Umemura, A. 402
Uyeda, C. 569
Vaerenbergh, S. 286
Vailati, A. 203
Valenzuela, T. 485
Vandewalle, N. 216, 316
Verga, A. 110
Viardin, A. 37
Vincent-Bonnieu, S. 224
Volkmann, T. 45, 83, 121
Voss, D. 110
Wada, Y. 378, 529
Walser, R. 485
Wang, H. 350
Warashina, Y. 155
Watanabe, K. 161
Watanabe, M. 180
Weaire, D. 216
Williams, B. 458
Windpassinger, P. 485
Winter, J. 224
Woodcock, T. 45
Wurm, G. 450
Wurm, J. Teiser. G. 427
Yabu, H. 443
Yano, T. 259
Yanpei, C. 299
Yoda, S. 155, 529
Yoshikawa, J. 155
Yu, G. 266
Zimmermann, G. 16, 25, 37, 110